Lecture Notes in Networks and Systems

Volume 134

The series "Lecture Notes in Networks and Systems" publishes the latest developments in Networks and Systems—quickly, informally and with high quality. Original research reported in proceedings and post-proceedings represents the core of LNNS.

Volumes published in LNNS embrace all aspects and subfields of, as well as new challenges in, Networks and Systems.

The series contains proceedings and edited volumes in systems and networks, spanning the areas of Cyber-Physical Systems, Autonomous Systems, Sensor Networks, Control Systems, Energy Systems, Automotive Systems, Biological Systems, Vehicular Networking and Connected Vehicles, Aerospace Systems, Automation, Manufacturing, Smart Grids, Nonlinear Systems, Power Systems, Robotics, Social Systems, Economic Systems and other. Of particular value to both the contributors and the readership are the short publication timeframe and the world-wide distribution and exposure which enable both a wide and rapid dissemination of research output.

The series covers the theory, applications, and perspectives on the state of the art and future developments relevant to systems and networks, decision making, control, complex processes and related areas, as embedded in the fields of interdisciplinary and applied sciences, engineering, computer science, physics, economics, social, and life sciences, as well as the paradigms and methodologies behind them.

**** Indexing: The books of this series are submitted to ISI Proceedings, SCOPUS, Google Scholar and Springerlink ****

More information about this series at http://www.springer.com/series/15179

Suresh Chandra Satapathy ·
Vikrant Bhateja · M. Ramakrishna Murty ·
Nguyen Gia Nhu · Jayasri Kotti
Editors

Communication Software and Networks

Proceedings of INDIA 2019

Springer

Editors
Suresh Chandra Satapathy
School of Computer Engineering
Kalinga Institute of Industrial Technology
Deemed to be University
Bhubaneswar, Odisha, India

Vikrant Bhateja
Department of Electronics
and Communication Engineering
Shri Ramswaroop Memorial Group
of Professional Colleges (SRMGPC)
Lucknow, Uttar Pradesh, India

M. Ramakrishna Murty
Department of Computer science
and Engineering
Anil Neerukonda Institute
of Technology and Sciences
Visakhapatnam, Andhra Pradesh, India

Nguyen Gia Nhu
Dean of Graduate School
Duy Tan University
DaNang, Vietnam

Jayasri Kotti
Professor in Computer Science
and Engineering
Lendi Institute of Engineering
and Technology
Visakhapatnam, Andhra Pradesh, India

ISSN 2367-3370 ISSN 2367-3389 (electronic)
Lecture Notes in Networks and Systems
ISBN 978-981-15-5396-7 ISBN 978-981-15-5397-4 (eBook)
https://doi.org/10.1007/978-981-15-5397-4

This Springer imprint is published by the registered company Springer Nature Singapore Pte Ltd.
The registered company address is: 152 Beach Road, #21-01/04 Gateway East, Singapore 189721, Singapore

Editorial Board of INDIA

Preface

This volume contains the papers that were presented at the 6th International Conference on Information System Design and Intelligent Applications (INDIA) organized by the Department of Computer Science and Engineering, LIET, Visakhapatnam, India during November 1–2, 2019. It provided a great platform for researchers from across the world to report, deliberate, and review the latest progress in the cutting-edge research pertaining to smart computing and its applications to various engineering fields.

The response to INDIA was overwhelming with a good number of submissions from different areas relating to Computer Networks, Security, Sensors Technologies, Cloud Computing, etc. After a rigorous peer-review process with the help of program committee members and external reviewers, only quality papers were accepted for publication in this volume of LNNS series of Springer.

Our thanks are due to Dr. Neeraj Gupta, India and Dr. Naeem Hannon, Malaysia for keynote address during the 2 days. We are thankful to Dr. PVGD Prasad Reddy, Hon. VC, Andhra University, Visakhapatnam, for his inaugural address being the Chief Guest. We would like to express our appreciation to the members of the Program Committee for their support and cooperation in this publication. We are also thankful to the team from Springer for providing a meticulous service for the timely production of this volume.

Our heartfelt thanks to the management committee members of LIET for their support for hosting the conference. We place our appreciation to the Principal of LIET and HOD of Department of CSE for continuous support and motivation. Special thanks to all guests who have honored us with their presence in the inaugural day of the conference. Our thanks are due to all Special Session Chairs, track managers, and reviewers for their excellent support. Support of all faculty members and student volunteers of LIET is praiseworthy. Last but certainly not the least, our special thanks go to all the authors who submitted papers and all the attendees for their contributions and fruitful discussions that made this conference a great success.

Bhubaneswar, India	Suresh Chandra Satapathy
Lucknow, India	Vikrant Bhateja
Visakhapatnam, India	M. Ramakrishna Murty
DaNang, Vietnam	Nguyen Gia Nhu
Visakhapatnam, India	Jayasri Kotti

Contents

About the Editors

Suresh Chandra Satapathy is a Professor, School of Computer Engineering, KIIT Deemed to be University, Bhubaneswar, India. His research interest includes machine learning, data mining, swarm intelligence studies and their applications to engineering. He has more than 140 publications to his credit in various reputed international journals and conference proceedings. He has edited many volumes from Springer AISC, LNEE, SIST, etc. He is a senior member of IEEE and life member of Computer society of India.

Vikrant Bhateja is an Associate Professor, Department of ECE in SRMGPC, Lucknow. His areas of research include digital image and video processing, computer vision, medical imaging, machine learning, pattern analysis and recognition. He has around 100 quality publications in various international journals and conference proceedings. He is an Associate Editor IJSE and IJACI. He has been Editor of four published volumes with Springer (IC3T-2015, INDIA-2016, ICIC2-2016, ICDECT-2016) and few other are under press (FICTA-2016, IC3T-2016, ICMEET-2016).

M. Ramakrishna Murty is an Associate Professor, Department of CSE, ANITS, Visakahapatnam. He has varied interests in research in multidisciplinary areas of computing. He has more than 56 publications to his credit in various international journals and conference proceedings. He has been very active in reviewing articles from Springer, IOS press, etc.

Dr. Nguyen Gia Nhu is Dean of Graduate School, Duy Tan University, Viet Nam. He has a total academic teaching experience of 19 years with more than 60 publications in reputed international conferences, journals and online book chapter contributions. His area of research includes healthcare informatics, network performance analysis and simulation, and computational intelligence.

Jayasri Kotti is a Professor, Department of CSE at LIET, Visakhapatnam, India. Her interest lies with software engineering, software safety, software testing and software architecture. She has published quality articles and also acted as reviewer in few international journals and conferences.

IOT-Based Smart Stale Food Detector

B. V. Ramana Murthy, C. Kishor Kumar Reddy, P. R. Anisha, and RajaShekar Sastry

Abstract With the development of innovation and reliance of individuals on advanced mobile phone and expanding requests of snappy and simple method for explaining day by day life assignments, it has turned out to be extremely fundamental to have an innovation that can power over residential and modern applications utilizing IoT. This paper manages the advancements alongside web of things utilizing Aurdino which utilizes the content programming and sensors like DHT sensor, dampness sensor, MQ3 Sensor, Aurdino UNO, and so forth. In this paper, we build up a nourishment quality detecting/distinguishing system. The sensors will be related to Aurdino. Refrigerator is a fundamental nourishment stockpiling strategy that brings down the pace of generation of microorganisms. However, in specific circumstances, we may neglect to see the nourishment things that are not utilized in long haul stockpiling inside it. This paper is created to tackle the issue of nourishment deterioration, with the assistance of sensors which are utilized to identify the stale nourishment by detecting it constantly. The discovery will be appeared through the sign dependent on freshness and nature of nourishment by an alarm message sent to our enrolled cell phone.

1 Introduction

Sustenance prosperity and neatness is an essential stress in order to keep the sustenance wastage. The quality of the sustenance ought to be watched and it must be kept from ruining and decaying by the natural segments like temperature, tenacity, and diminish. Along these lines, it is useful to pass on quality checking contraptions at sustenance stores. These qualities watching devices keep a watch on the regular factor that reason or pace up spoil of the sustenance. A short time later, the normal segments can be controlled by refrigeration, vacuum storing, etc.

B. V. Ramana Murthy · C. Kishor Kumar Reddy (✉) · P. R. Anisha · R. Sastry
Department of Computer Science and Engineering, Stanley College of Engineering & Technology for Women, Hyderabad, India
e-mail: Kishoar23@gmail.com

S. C. Satapathy et al. (eds.), *Communication Software and Networks*, Lecture Notes in Networks and Systems 134, https://doi.org/10.1007/978-981-15-5397-4_1

In this undertaking, a near sustenance quality watching device will be arranged that will keep watch of environmental components like temperature, tenacity, alcohol substance, and introduction to light. The contraption depends on Arduino UNO which is an outstanding prototyping board. The Arduino board is interfaced with various sensors like DHT-11 to screen temperature and tenacity, MQ3 to recognize alcohol substance, and LDR to evaluate introduction to light. This is an IoT contraption and sends the conscious sensor data to an IoT arrangement. The ESP8266 Wi-Fi Modem is interfaced with the Arduino to relate it to the web by methods for Wi-Fi switch. The sensor data similarly appears on a character LCD interfaced with the Arduino UNO. The IoT stage used for logging and checking of sensor data is Freeboard.io. With the force of Internet of Things, the natural components impacting the sustenance amassing can be seen from wherever, at whatever poin,t and from any device.

Various such devices can be presented at a zone for better watching and quality control. The Arduino Sketch running over the contraption executes the various functionalities of the endeavor like examining sensor data, changing over them into strings, demonstrating them on character LCD, and passing them to the IoT organize. The Sketch is created, orchestrated, and stacked using the Arduino IDE.

The IoT is a mechanical vexed that tends to the predetermination of selecting and trades, and its improvement relies on powerful specific movement in various basic fields, from remote sensors to nanotechnology. They are doing imprint everything to see, robotizing, and controlling. Sustenance harming has been the wellspring of endless pollutions and diseases. Utilization of stale sustenance brings about weakened.

2 Relevant Work

In paper [1], "IoT based venture for nourishment quality and checking", the creator proposed a gadget called SmartPlate comprising of an assortment of sensors that are actuated relying upon the nourishment thing. This plate can be set in any utensil and a board can be utilized to choose the sort of nourishment thing. In paper [2], "Nourishment freshness locator utilizing IoT", the creator proposed detecting structures and their explanatory highlights for estimating freshness markers, allergens, pathogens, adulterants, and toxicants are examined with the case of uses. In paper [3],"Smart stockpiling of nourishment dependent on IoT", the creator proposed a kind of capacity which can naturally recharge kitchen things. The utilization of Force Sensitive Resistor (FSR) to identify the nonappearance of fixing in a particular box of the washroom. Arduino UNO is utilized to peruse the sensor esteem.

In paper [4], "Nourishment checking framework dependent on bluetooth low vitality and IoT", the creator proposed the framework utilizes a GSM/GPRS open remote system for remote information move. The blend of web of things development, GSM/GPRS open remote framework advancement, and Internet basically diminishes the cost of the structure, with driving unlimited degree of following affirmation, which overhauls the total execution of the system remarkably. In paper [5],"Food quality and security watching using gas sensor display in sharp packaging", the maker proposed little gas sensors and negligible exertion modified to the kind of sustenance packaging and a particular contraption for transmitting alert respect the customer are enter factors in achieving keen packaging. In paper [6], "Quartz Crystal Microbalance Based Approach for Food Quality" the creator utilized biosensors innovation. There is creating eagerness toward biosensors advancement in light of high distinction, convenience, and smart response.

In paper [7], "Short-lived Food Quality Monitoring—An Internet of Things (IoT) Approach", monitoring of transitory nourishment items and early recognition of corruption will stay away from misfortune because of nourishment wastage and furthermore guarantees the freshness of nourishment. In this situation, remote checking of natural products during transportation from field to retire can guarantee the nature of organic product. In this work, a remote sensor system was intended for observing natural products during transportation and even after capacity. In paper [8], "Plan and usage of nourishment canny observing framework dependent on pH sensor", the creator proposed a pH sensor and an outside examiner that interface to give data on the nature of nourishment items. In paper [9], "Electronic Noses Application to Food Analysis Using Metal Oxide Sensors: A Review" the maker proposed electronic noses which uses different sorts of electronic gas sensors that have midway distinction and a reasonable model affirmation systems fit for seeing direct and complex scents.

In paper [10], "sans battery radio recurrence recognizable proof (RFID) sensors for nourishment quality and security" the creator proposed uninvolved (without battery) radio recurrence ID (RFID) sensors for exceptionally delicate and specific recognition of nourishment freshness and bacterial development. In paper [11], "A Passive Radio-Frequency pH Sensing Tag for Wireless Food-Quality Monitoring" the creator displayed another technique, appropriate for nourishment quality administration by remotely checking pH level changes in nourishment with an adaptable pH sensor inserted in a battery-less radiofrequency (RF) transponder. In paper [12], "Web-based colorimetric detecting for nourishment quality observing" the shading change of a colorimetric sensor was caught with a remote camera and the information was transmitted to a PC, which transferred the data to the web.

The paper [13], "A sensor framework for Automatic recognition of nourishment admission through non-obtrusive observing of biting" shows a straightforward sensor framework and related sign preparing and example acknowledgment strategies to recognize times of nourishment admission dependent on non-intrusive observing of biting. In paper [14], Bakery nourishments were assessed for shading by visual assessment and by improvement of a machine reading framework combined with discriminant examination of the information acquired. An arrangement calculation isolated light from dark-colored bread kitchen food sources.

3 Smart Stale Food Detector Methodology

Sustenance hurting is a significant issue that impacts countless people reliably. Hurting sustenance must be seen in front of calendar to keep a noteworthy issue. The goals of this contraption are to make an electronic device fused with biosensors that can recognize sustenance squander. The use of sensors can check unmistakable parameters of sustenance like pH, clamminess, and ethanol level. The square charts underneath show the model of contraption. The contraption involves a microcontroller Arduino Uno, Bluetooth module, electrical, and biosensor like pH sensor, moistness sensor, and gas sensor. The sustenance to be checked is added to the looking at sensor and the customer can incorporate from Android versatile application, the decision of sustenance thing from application offers course to Arduino Uno with passing on through Bluetooth module. The microcontroller takes readings from the sensor and pick result with a predefined count.

The result is as "Incredible to use" and "Not extraordinary to use" dependent upon the sustenance freshness level. In this endeavor, a sustenance quality watching contraption will be arranged that will keep watch of ordinary/biological components like temperature, clamminess, and alcohol content. The device depends on Arduino UNO which is an unquestionable prototyping board. The Arduino board is interfaced with various sensors like DHT-11 to screen temperature and sogginess and MQ3 to recognize alcohol content. This is an IoT contraption and sends the intentional sensor data to an IoT organize. The ESP8266 Wi-Fi Modem (Node MCU) is interfaced with the Arduino to relate it to the web by methods for Wi-Fi switch. The IoT stage used for logging and checking of sensor data is IOT Watson. With the power of Internet of Things, the biological segments impacting the sustenance storing can be seen from wherever, at whatever point, any place, and from any device.

The nourishment quality observing gadget structured in this venture depends on Arduino UNO. The Arduino has filled in as an IoT board in this task. Different sensors like DHT-11, MQ3 and LDR, and the ESP8266 Wi-Fi Modem are interfaced to the Arduino. Arduino is an open-source gadgets stage dependent on simple to utilize equipment and programming. Arduino sheets can understand inputs—light on a sensor, a finger on a catch, or a Twitter message—and transform it into a yield—initiating motor, turning on a LED, publishing something online. Aurdino having sensors with some basic code, the Arduino can control and collaborate with a wide assortment of sensors—things that can gauge light, temperature, level of flex, weight, nearness, quickening, carbon monoxide, radioactivity, dampness, barometric weight, and so on, you can detect it!

This Arduino-based IoT contraption should be presented in a sustenance store. When it is suitably presented and controlled on, it connects with the web by methods for Wi-Fi modem and starts examining data from the interfaced sensors—DHT-11 temperature and moistness sensor, MQ3 Sensor, and the LDR sensor. DHT11 Temperature and Humidity Sensor is a propelled sensor with inbuilt capacitive tenacity sensor and Thermistor. It moves a ceaseless temperature and soddenness scrutinizing at normal interims. The sensor takes a shot at 3.5–5.5 V supply and can examine temperature between 0 °C and 50 °C and relative soddenness some place in the scope of 20 and 95%. The sensor can't be explicitly interfaced with a propelled stick of the board as it chips away at 1-wire show which must be executed just on the firmware. First the data stick is intended to enter and a start banner is sent to it. The start banner incorporates a LOW for 18 ms sought after by a HIGH for 20–40 microseconds sought after by a LOW again for 80 microseconds and a HIGH for 80 microseconds. In the wake of sending the start banner, the stick is orchestrated to cutting edge yield and 40-piece data including the temperature and clamminess scrutinizing is bolted out. Of the 5-byte data, the underlying two bytes are entire number and decimal bit of examining for relative wetness independently, third and fourth bytes are entire number and decimal bit of scrutinizing for temperature and last one is checksum byte.

For Arduino, standard library for DHT-11 sensor is starting at now open. The data from the sensor can be adequately arranged by calling read11() system for the DHT class. The MQ3 sensor recognizes the transmission of ethanol kind of gases. If the sustenance/common items get demolished, they transmit the ethanol sort of gases. The MQ3 sensor perceives the gathering of such gases and yields a basic voltage comparing to the assembly of the gas. The basic yield is passed to the basic stick of the Arduino which has inbuilt ADC that changes over the easy to cutting edge regard (Fig. 1).

The Arduino assembles data from all of the sensors and convert the characteristics to the strings [15]. The ESP8266 Wi-Fi module related to the Arduino moves the data

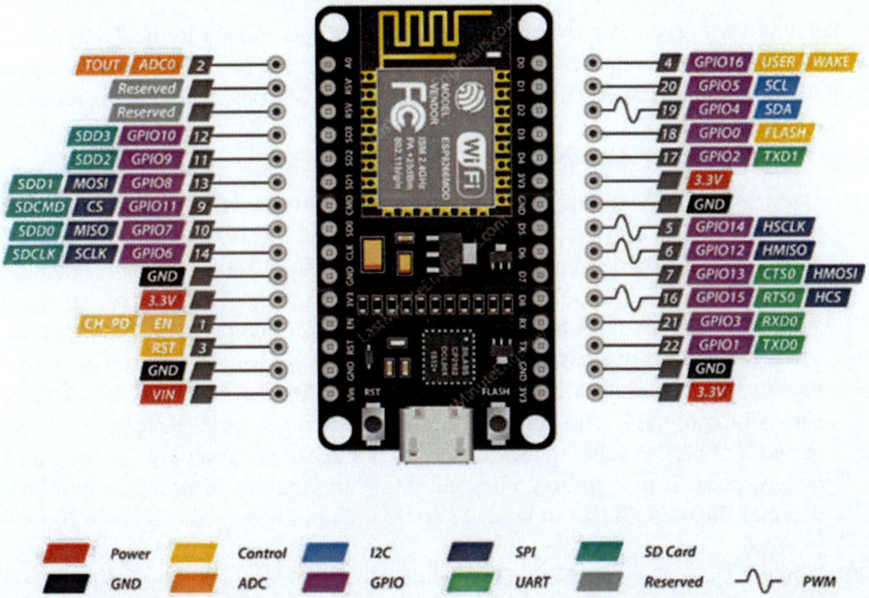

Fig. 1 Aurdino pins

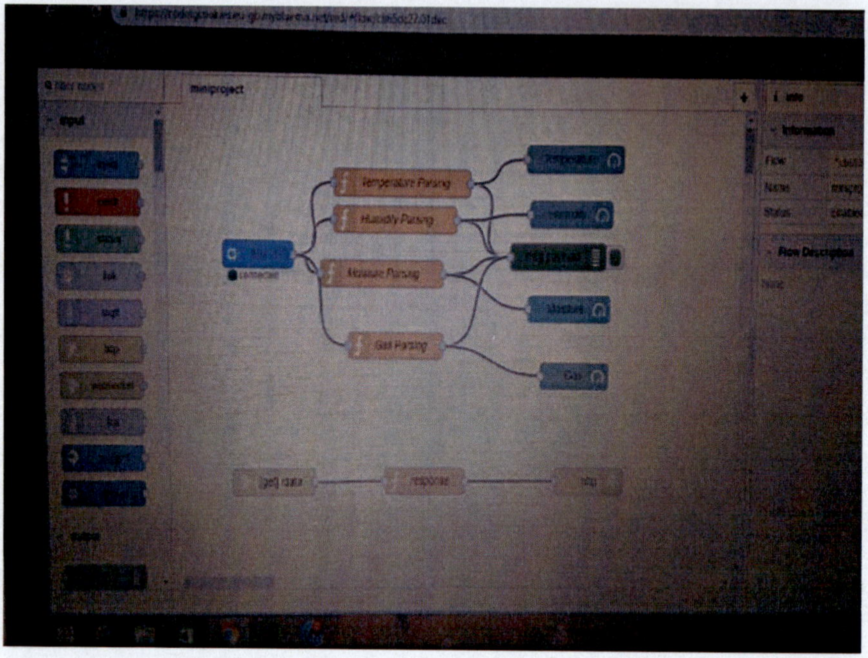

Fig. 2 Smart stale food detector

to ThingSpeak Server. For appearing and checking data moved to the ThingSpeak server, either an automated dashboard or data master is required. In this endeavor, a propelled dashboard called IBM Watson is used to screen the sensor data ostensibly on the web. The IBM Watson uses JSON record to imagine ThingSpeak data. It offers three parts to create a dashboard (Fig. 2).

(1) Data Sources—The data sources get the data from external sources. These external sources can be data authority organizations, JavaScript applications, or JSON reports tolerating substance from an HTTP server. In this endeavor, the data source is a JSON record that gets data from the Thing Speak server.
(2) Widgets—The Widgets help to indicate data in printed or graphical structure. There are various devices open in Freeboard.io like substance, chart, check, etc. version of the volume will be available in LNCS Online. Members of institutes subscribing to the Lecture Notes in Computer Science series have access to all the pdfs of all the online publications. Non-subscribers can only read as far as the abstracts. If they try to go beyond this point, they are automatically asked, whether they would like to order the pdf, and are given instructions as to how to do so.
(3) Panes—These are used to organize widgets. IBM Watson requires sign up and after sign widgets can be created.

4 Discussion and Conclusions

The Aurdino-based gas and dampness sensors had the option to detect along these lines obvious indications of decay, for example, shape or scent was watched. The degree of discharge of gases associated with the level of waste of nourishment. The arduino-based sensors were touchy enough to get low measures of outflows of methane and alkali which is infiltrated by rotting of nourishment. At the point when obvious indications of rot began to appear, the degrees of the gas outflow yery Hign20-fol with respect to control. While I had the option to watch these outcomes in oranges, the level of gases in different nourishments that were tried, for example, rice and milk were a lot lower. Distinguishing normally produced gases, for example, Methane, Ammonia, and Ethylene as nourishments rot can be utilized to identify nourishment waste.

The Aurdino sensors had the option to recognize dampness content from nourishment things once we embed the dampness sensor in it. Using sensors to distinguish the nearness of these unsafe/awful organisms among food sources can help identify nourishment nearness ahead of schedule before anyone expends ruined nourishment. These strategies can be further created to attack the kind of gas sensors and food sources to build the affectability of such location techniques.

The rising occasions of sustenance hurting require an effective system to alert unwary clients of spoilt or stale sustenance things. This finally prompts lesser disorders and low use in regards to investigation and meds. The SmartPlate gives quick and invaluable plans to screen produce/sustenance quality to address sustenance security and waste issues. The SmartPlate is needed to such an extent, that it may be fitted into any standard family unit utensil, along these lines broadening its credibility. It also incorporates an arrangement of sensors, connecting with it to be utilized for various sustenance things. The structure at long last alarms customers concerning the status of their sustenance thing accurately and steadily. The trade-off of an amazingly made versatile application empowers clients to seek after the status of sustenance things at their homes remotely. The course of action is diminished and can be created effectively utilizing effortlessness pieces. A ratty and productive game plan empowers the SmartPlate to be moderate for all parts of the general populace.

References

1. Sumathi MS, Thejaswini S, Pranav Kashyap, ShahinaAnjum, Shashi Shanker, Shreya GK, IoT based project for food quality and monitoring. Int J Recent Innovat. Trends Comput Commun 3(5):3172–3174. ISSN: 2321-8169
2. Mustafa F, Andreescu S, *Department of Chemistry and Biomolecular Science, Clarkson University, Potsdam, NY 13699 USA Chemical and Biological Sensors for Food-Quality Monitoring and Smart Packaging
3. https://create.arduino.cc/projecthub/taifur/drs-enabled-smart-pantry8d2f8b?ref=tag&ref_id=iot&offset=42. Smart storage of food based on IoT

4. Venkatesh A, Saravanakumar T, Vairamsrinivasan S, Vigneshwar A, Santhosh Kumar M (2017) A food monitoring system based on bluetooth low energy and Internet of Things. Int J Eng Res Appl 7(3)(Part-6):30–34.www.ijera.com ISSN: 2248-9622
5. Matindoust S, Baghaei-Nejad M, Abadi MHS, Zou Z, Zheng LR, Food quality and safety monitoring using gas sensor array in intelligent packaging
6. Singh AK, Verma N (2014) Quartz crystal microbalance based approach for food quality. Current Biotechnol 3:000–000
7. Chanthini B, Manivannan D, Umamakeswari A, Perishable food quality monitoring—an Internet of Things (IoT) Approach. http://www.ijpam.eu. ISSN: 1311-8080 (printed version); ISSN: 1314-3395 (on-line version)
8. Xiansheng K, Jing S (2014) Design and implementation of food intelligent monitoring system based on pH sensor, department of computer and information, Xinxiang University, China Xinxiang Radio and Television University, Xinxiang, China. J Chem Pharm Res 6(7):1662–16666
9. Zohora SE, Khan AM, Srivastava AK, Hundewale N (2013) Electronic noses application to food analysis using metal oxide sensors: a review. Int J Soft Comput Eng (IJSCE) 3(5). ISSN: 2231-2307
10. Potyrailo RA, Nagraj N, Tang Z, Mondello FJ, Surman C, Morris W (2012) Battery-free radio frequency identification (RFID) sensors for food quality and safety. J Agric Food Chem 60(35):535–8543
11. Huang WD, Deb S, Seo YS, Rao S, Chiao M, Chiao JC (2010) A passive radio-frequency pHSensing tag for wireless food-quality monitoring. IEEE
12. Hayes J, Dublin City University, Dublin; Pacquit A, Crowley K, Kim Lau (2006) Web-based colorimetric sensing for food quality monitoring. Sensors
13. E.sazonov JM, Fontana A, sensor system for Automatic detection of food intake through non-invasive monitoring of chewing. IEEE Sens J 1340–1348
14. Abdullah MZ, Aziz SA, Dos Mohamed AM (2000) Quality inspection of bakery products using a color-based machine vision system. J Food Quality 23:39–50
15. Wang J, Nakano K, Ohashi S, Kubota Y, Takizawa K, Sasaki Y (2011) Detection of external insect infestations in jujube fruit using hyperspectral reflectance imaging. Biosyst Eng 108:345–351

Detection of Tyre Wear Out Using OpenCV and Convolution Neural Networks–Survey

Allipilli Harshitha and Samyuktha Samala

Abstract Tyre is the only part of vehicle which is in contact with road. The lifetime of tyre depends on tread limit, mileage of car, no. of punctures, and degradation of material. Images of tyres have to be collected and to be trained. Hence, we consider all the above factors, and using OpenCV we find contours and area of image, and we train our model using convolution neural network to detect the wear out of tyre and predict the lifetime of tyre.

1 Introduction

The performance of tyre is very important for maintaining the safety, performance, and efficiency of vehicle. But at some point, tyres will start wearing out and lose their traction and braking ability. Tyre wear causes damage to the vehicle which leads to accidents as sometimes it catches fire. To avoid this scenario, a device is to be developed which tells us when a tyre is going to wear out.

1.1 Parameters to Check When Tyre Is Going to Wear Out

We need to check the condition of tread to know when the tyre is going to wear out. The primary function of tread is to divert water from beneath the tyre to improve traction and avoid hydroplaning on wet roads. If the tread limit is down to 0.16 cm, the tyre is no longer safe. A tyre can run between 40,000 kms and 60,000 kms in ideal conditions. For bad roads and variable driving conditions like congested roads, start-stop traffic, etc., the tyres need to be replaced at a much lesser mileage. A tyre

A. Harshitha (✉) · S. Samala
Department of Computer Science and Engineering, Stanley College of Engineering and Technology for Women, Hyderabad, India
e-mail: allipilliharshita@gmail.com

S. Samala
e-mail: samyuktha.samala@gmail.com

S. C. Satapathy et al. (eds.), *Communication Software and Networks*, Lecture Notes in Networks and Systems 134, https://doi.org/10.1007/978-981-15-5397-4_2

can withhold up to 3 or 4 punctures and maximum of 5, then the tyre needs to be replaced. Also, replacement of tyre depends on material as it is made up of rubber and other parameters and rubber degrades after certain period of time. Check for size of tyre, it should be balanced with car. If the tyre is not fitted properly, then it may cause damage to tyre. Tyre should be properly inflated. If the tread bar is equal to tread, then we say that tyre is weared out. If there is a bulge or bubble in tyre, then we need to replace the tyre. If any objects like stones penetrate into tyre, then we need to replace the tyre. Another parameter for replacing the tyre is the date of manufacture of tyre which is present on tyre as 4 digit no (i.e., month and year of manufacture of tyre). From the date of manufacture the tyre should be replaced after 6 years.

1.2 Dangers of not Changing the Tyre After Wear Out

- Reduce road handling—less control over vehicle.
- Blow out—loss of control over vehicle.
- Aquaplaning—tyres lose grip on certain road conditions.
- Brake will not process at times.

2 Relevant Work

Darekar et al. [19] conducted an experiment on how to detect wear out of tyres using color coding technique with the help of sensors. A color coding is added to tyre with which a human can visualize the color change on the tyre with which a tyre wear can be detected.

Klueppel [20] researched the characteristics that influence the changes in tyre which results in tyre wear out. The author describes some of the parameters like tyre design and model, tyre materials, pressure, temperature, and road surface that results in tyre wear out.

Tyre pressure plays an important role in safety which should be properly monitored. But monitoring with manual gauges is less effective. Inorder to overcome this scenario Sivaraos et al. [21] developed a device called tyre pressure monitoring system (TPMS) which monitors the temperature and pressure of tyres.

2.1 Convolution Neural Network

Convolution Neural Network is a deep learning algorithm which takes an image as input and assign weights and bias to various aspects in the image to be able to classify the images. Convolution neural network was first introduced to recognize handwritten zip code [7], and later it was used to recognize and classify the images such as MNIST data which is a handwritten digits [8]. CNN works only on large datasets, however, it works on small datasets too, but doesn't provide good accuracy. At present, CNN has been successfully applied in every field for object classification [1, 6], object detection [2, 3, 5, 9], and motion detection in video [10].

The CNN layers are the building blocks which are effective in its structuring and connections. The layers are structured much effectively which makes CNN to train the model faster and perform effectively. The layers and their functionality are detailed below.

2.1.1 Convolution Layer

Convolution layer is the first layer in CNN. Convolution layer contains a set of filters such as sobel, and scharr filters. The parameters like height and weight of the filter must be less than the input image. At every position of input, the dot product of height and weight are computed between input and filter. Since the size of filter is less than input image, the size of neuron is small [11].

As the filter is sliding around the input image, it is multiplying the values in the filter with the original pixel values of the image. On summing up the products we get a single number. We need to repeat the same process for every position of the input image by moving the filter to the right by one unit. Every unique position of input produces a number.

2.1.2 Activation Layer

To learn and make sense of something more complicated and nonlinear complex functional mappings between input and response variable, nonlinear properties to our network has been introduced to convert input to output. In neural networks, we apply sum of products of input and weight and apply activation function to get output. Activation function decides whether a neuron should be activated or not in the network.

Several types of activation functions which are broadly used: sigmoid, tanh, relu.

Sigmoid: The sigmoid function is

$$F(x) = \frac{1}{1 + e^{-x}} \tag{1}$$

Sigmoid function is S-shaped curve which returns the output in an interval [0,1].

Tanh: The tanh function is

$$F(x) = \frac{1 - e^{-2x}}{1 + e^{-2x}} \tag{2}$$

Tanh (hyperbolic tangent) function returns the output in an interval [–1,1]. Its output is zero-centered.

Relu (Rectified Linear Unit): The disadvantage of sigmoid and tanh functions is that they easily get saturated to 0, 1 and –1 ranges which slows down the training a model.

Relu's advantage is that it doesn't saturate and gives fast and accurate output. Relu works better with large amount of data like images. Relu function is given as

$$R(x) = \max(0, x) \tag{3}$$

Leaky Relu: The disadvantage of relu is it causes neurons to dead sometimes. To overcome that we use leaky relu. The leaky relu function is given as

$$\mathrm{R(x)} = \begin{cases} x & x > 0 \\ 0.01x & \text{otherwise} \end{cases} \tag{4}$$

Leaky relu allows nonzero values even if the function is not active.

2.1.3 Pooling Layer

Pooling layer is also known as downsampling layer which is used to downsample the features obtained from previous layers. Its function is to progressively reduce the size of image and reduce the amount of parameters and computations in the network retaining the maximum information. It provides robust feature to noise and distortions [4].

There are two types of pooling layers broadly used in CNN. **Average pooling** is to find average value within each pooling window. **Max pooling** is used to find maximum value within each pooling window. A 4×4 matrix is taken and each 2×2 matrix is made as a pooling window and on applying max pooling, it gives the highest value from the window. Max pooling function retains the maximum information of the given image.

2.1.4 Fully Connected Layer

The output from the convolution layer represents high-level features. While that output could be flattened and connected to the output layer. Fully connected layers in neural networks are those that take neurons from previous layers and connect to every activation unit of next layers. The features of convolution layer are vectorized and upon that we apply dense or fully connected layer to connect it to output layer. This layer learns global patterns in the input feature space. After feature extraction, we use activation function such as softmax or sigmoid to classify the data into classes.

2.2 OpenCV

OpenCV (open source computer vision library) is an open source computer vision and machine learning software library. It was built to provide common infrastructure for computer vision applications and to accelerate the use of machine perception in commercial products. OpenCV library has more than 2500 optimized algorithms, which contains computer vision and machine learning-related algorithms. These algorithms can be used to detect and recognize objects, faces and classify human actions in video, extract 3D models of objects, and many more.

Salihbasic et al. [12] conducted experiment on face, age, and gender recognition. A biometric system is used to capture the physical features of humans and comparing them with the images stored in database by using OpenCV library in the form of Android mobile application. The tools used for face detection and face recognition are LBP face features classifier and LBPH model. Trained the model using deep learning algorithm to recognize the face, age, and gender.

Every product in the market will have a unique identity which is nothing but a bar code, by scanning the barcode we get the details of product. Puril et al. [13] researched on detection of bar code using OpenCV library. The first step in detection is scharr gradient which detects the edges and blurs the image in order to reduce noise and frequency disturbances and transforms the image. Then this method is deployed into Android application to scan the bar code.

Due to increase in number of vehicles it is important to detect license plate inorder to avoid traffic rules violation. Palekar et al. [14] experimented on converting image into text format with the help of OpenCV libraries and Tessaract, which is used to recognize character patterns in image.

Now-a-days latest technology is used to identify the injury of skin cells. Li et al. [15] researched on skin cell injuries using OpenCV library. Based on problem the image is divided into segments and each segment is taken and using watershed algorithm of OpenCV, the damage occurred to skin is identified.

Object detection is identification of certain objects in an image and in video capturing. Saxena et al. [16] experimented on object detection using OpenCV and machine learning. MATLAB is a tool used to detect human behavior in videos and further approach is made using machine learning algorithms inorder to implement surveillance systems.

As we could see many accidents are occurring on highways since past years, we must focus on highway traffic accidents. Wafi et al. [17] developed a surveillance system with more attention. Using OpenCV libraries we capture the images and videos of traffic on highway to solve traffic-related issues and detect the factors that cause accidents.

Due to increase in usage of vehicles it is important to count number of vehicles and track them. Inorder to manage traffic—vehicle detection, tracking, and monitoring on highway has to be planned. Uke et al. [18] used Visual C++ with intel OpenCV video stream processing system inorder to detect real-time moving vehicles by capturing video and detecting vehicles movement.

Table 1 Comparitive analysis methodologies proposed by different authors

S.No	Title	Authors	Description
1	Detection of tyre wear using color coding	Praful Darekar, Akash Saste, Shubham Bhandarkar, Nikita Kadam, B. R. Patil	The main aim of this paper is to detect tyre wear out by adding color coding technique with the help of sensors. An end user can detect the tread depth by visualizing the color change on the tyres
2	Wear and abrasion of tyres	Manfred Klueppel	This paper describes the characteristics that influence the changes in tyre which results in tyre wear out. The author describes the parameters like tyre design and model, tyre materials, pressure, temperature, road surface that results in tyre wear out
3	Investigation of tyre pressure drop phenomenon using specially designed real-time data mining and storage system	Sivaraos, M. J. Raguvaran, Aidy Ali, M. F. Abdulla, D. Sivakumar, M. A. M. Ali, A. Hambali, Ms. Salleh, K. Umesh	Since tyre pressure plays an important role. It should be properly monitored. But monitoring it with manual gauges is less effective. The author developed a device tyre pressure monitoring system (TPMS) which monitors the pressure and temperature of tyres

3 Comparitive Analysis

The above are the approaches given by various authors and they haven't concentrated on computer vision techniques. Hence our research is to implement OpenCV and CNN algorithm's on real-time datasets of tyres to predict the wear of tyre (Table 1).

4 Discussion

The main reason behind this proposal is researchers have not yet implemented a device which detects the wear of tyre. As mentioned in Sect. 1, what are the causes of tyre wear and how to identify it and what are the issues if tyre is not replaced on time. Hence, the proposal is that using OpenCv libraries along with convolution neural networks we can classify the images with tyre wear. We explain the process of implementation of proposed methodology (Fig. 1).

The graphical representation of proposed methodology which is appropriate for detecting tyre wear out. First we use OpenCV to segment and threshold the images

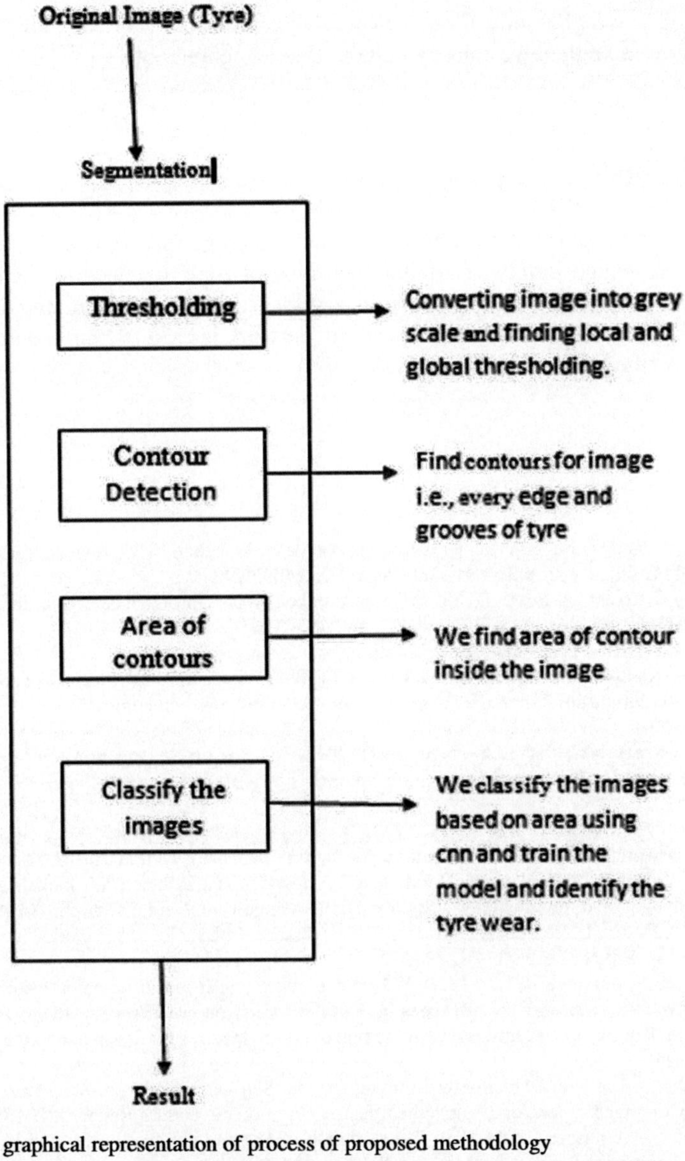

Fig. 1 A graphical representation of process of proposed methodology

and then we find contours and then area of contours, then later we use CNN to train the images and detect the wear of tyre.

In detail about the process: the first step is preprocessing the image for noise removal and smoothening. In the second step, we perform conversion of image into grayscale and apply threshold to convert grayscale image into binary image. In the next step, we find contours for every edge and groove in a tyre and find area of contour and classify the image based on area, i.e., if the area of tyre is much less than the new tyre's area then we can say that the tyre is weared out.

5 Conclusion

Detection of tyre wear out has not been implemented by researchers. As tyre plays a key role in vehicle safety, resolving such type of issue we need some tools and techniques. Open source computer vision provides library for recognizing the object by applying contours and convolution neural network is used to train the model and help identify the tyre wear effectively.

References

1. Szegedy C, Liu W, Jia Y, Sermanet P, Reed S, Anguelov D, Erhan D, Vanhoucke V, Rabinovich A (2014) Going deeper with convolutions. arXiv:1409.4842
2. Szegedy C, Toshev A, Erhan D (2013) Deep neural networks for object detection. In: Advances in neural information processing systems, pp 2553–2561
3. Girshick R (2015) Fast r-cnn. In: International conference on computer vision (ICCV)
4. Jarrett K, Kavukcuoglu K, Ranzato M, LeCun Y (2009) What is the best multi-stage architecture for object recognition? In: IEEE 12th International conference on computer vision. IEEE, pp 2146–2153
5. Erhan D, Szegedy C, Toshev A, Anguelov D (2014) Scalable object detection using deep neural networks. In 2014 IEEE conference on computer vision and pattern recognition (CVPR). IEEE, pp 2155–2162
6. Krizhevsky A, Sutskever I, Hinton GE (2012) Imagenet classification with deep convolutional neural networks. In Advances in neural information processing systems, pp 1097–1105
7. LeCun Y, Boser B, Denker JS, Henderson D, Howard RE, Hubbard W, Jackel LD (1989) Backpropagation applied to handwritten zip code recognition. Neural Comput 1(4):541–551
8. LeCun Y, Bottou L, Bengio Y, Haffner P (1998) Gradient-based learning applied to document recognition. Proc IEEE 86(11):2278–2324
9. Ren S, He K, Girshick R, Sun J (2015) Faster r-cnn: towards real-time object detection with region proposal networks. In Advances in neural information processing systems, pp 91–99
10. Ryoo MS, Rothrock B, Matthies L (2014) Pooled motion features for first-person videos. arXiv:1412.6505
11. Ke Qiuhong, Liu Jun, Bennamoun Mohammed, An Senjian, Sohel Ferdous, Boussaid Farid (2018) Computer vision for assistive healthcare. Comput Vis Pattern Recognit 2018:127–145
12. Salihbasic A, Orehovacki T (2019) Development of android application for gender, age and face recognition using opencv. In: 2019 IEEE, pp 1898–1903
13. Puril R, Jain V (2019) Barcode detection using opencv-python. Int Res J Adv Eng Sci 2455–9024
14. Palekar RR, Parab SU, Parikh DP, Kamble VN (2017) Real-time license plate detection using opencv and tesseract. In: International conference on communication and signal processing (ICCSP). IEEE, pp 2111–2115
15. Li G, Zhang Y, Bo X, Li X (2019) Image analysis and processing of skin cell injury based on opencv. ICPS J Phys: Conf Ser 1237:032003
16. Saxena MR, Pathak A, Singh AP, Shukla I (2019) Real-time object detection using machine learning and opencv. Int J Inform Sci Appl (IJISA) 11(1). ISSN 0974-225

17. Wafi ZNK, Ahmad RB, Paulraj MP (2010) Highways traffic surveillance system (HTSS) using opencv. In: 2010 IEEE control and system graduate research colloquium (ICSGRC 2010). IEEE, pp 44–48
18. Uke NJ, Thool RC (2013) Moving vehicle detection for measuring traffic count using opencv. J Autom Control Eng 1(4)
19. Darekar P, Saste A, Bhandarkar S, Kadam N, Patil BR (2018) Detection of tyre wear using colour coding. Int Res J Eng Technol (IRJET)
20. Klueppel M (2014) Wear and abrasion of tires. In book: Encyclopedia of Polymeric Nanomaterials, pp 1–6
21. Sivaraos MJ, Raguvaran, Ali A, Abdullah MF, Sivakumar D, Ali MAM, Hambali A, Salleh M, Umesh K (2019) Investigation of tyre pressure drop phenomenon using specially designed real-time data mining and storage system. Int J Recent Technol Eng (IJRTE)

Smart Luggage System

L. Lalitha Gayatri and Sai Nidhi Paidimarry

Abstract Inventions are made to reduce human efforts and also to ensure security. Robotic inventions decrease the manual effort and create machine human interaction. This paper discusses one such area where inventions and enablements lead to reducing efforts and increase in security. A baggage that follows the human is designed using one of the magnificent technologies in electronics. It aids auto trailing, lost baggage recovery through GPS and GSM, void battery devices reviving, obstacle detection, alerting and smart unlocking.

1 Introduction

Travelling is exceptionally well known nowadays. As we start travelling more, issues experienced during travelling become unavoidable. Serious issues incorporate lost baggage, the need to revive void battery devices, being worried in carrying heavy luggage, robbery, overlooking significant things and so forth. Lost baggage and robbery disturbs the whole journey and also important documents or things might be lost. Void batteries lead to disconnection from the world. Carrying heavy loads and forgetting significant things leads to physical stress and discomfort. The significant piece of travelling is choosing the correct luggage. The job of luggage can't be underestimated, it is vital. Subsequently, the present study is directed to plan and execute a baggage that could give the necessities of explorers today. This smart luggage is a luggage that guarantee the security of the possessions and provide convenience to the explorer. This study likewise surveyed the explorer's bits of knowledge on the highlights inserted on the baggage.

L. L. Gayatri · S. N. Paidimarry (✉)
Stanley College of Engineering & Technology for Women, Hyderabad, Telangana, India
e-mail: nidhipaidimarry@gmail.com

L. L. Gayatri
e-mail: lalithagayatri11@gmail.com

S. C. Satapathy et al. (eds.), *Communication Software and Networks*, Lecture Notes in Networks and Systems 134, https://doi.org/10.1007/978-981-15-5397-4_3

2 Relavant Work

Sudha Senthilkumar, Brindha. K, Rathi. R, Charanya. R and Mayank Jain through their research proposed a smart luggage system using IoT which can track the luggage bag if it is lost or stolen. Maps are generated to get the location of the misplaced bags. Main components they have used to propose the model are Arduino, Alarm and Map interface [1]. Ms. Sneha Jainwar, Mr. B. HariKishore Rao, Ms. Khyati Varma and Ms. Honey Tamrakar have proposed an intelligent suitcase which is equipped with power unit, ultrasonic sensor to measure the distance of obstacle, GPS module to track the location, Bluetooth module to have communication with trackable device and RFID cards for data. Their model has solved most of the issues faced during travel [2].

Sebin J Olickal, Amal Yohannan, Manu Ajayan and Anjana Alias created a luggage bag which would follow the owner where ever he goes. It includes all the other features in which it can be tracked and through a map, an alarm is set when it moves away from the owner that is hen forgotten or lost. It can be accessed by sending an SMS. The drawback of this bag is it cannot detect the obstacles [3]. Ankush Sutar, Tukaram Kocharekar, Piyush Mestry, Prathamesh Sawantdesai and Suhasini S. Goilkar has solved a major problem of dragging the luggage, provides security and intelligent features which are suitable for this modern era. It has an inbuilt power supply to charge the phone [4].

The luggage bag proposed by Dexter L. Duat, Mario C. Bebelone and Jeffrey M. Gallego has 4 different features which includes biometric lock, tracker, power bank and kick scooter. Biometric lock can be accessed with the fingerprint of the owner. It is equipped with security and good accessibility by the kick scooter feature [5]. Athul P Anand, Deepesh Srivastava, Dushyant Sharma, JyotiRekha Dhal, Arun Kumar Singh and Mahendra Singh Meena integrated this intelligent system to a school bag which would help students in timetable management and security. It stores the timetable and assists students with the books they need to carry to the school. It is also equipped with panic button whenever pressed would alert their parents [6].

An innovative idea by Mrs. Rasika Naik, Sanjana Muppidwar, Pallavi Chavan, Siddhi Medhekar and Pooja Chindarkar on the smart luggage bags was integrating a solar cell. It will be used to charge various devices and can be interfaced to android phones [7].

3 Proposed Model

The Smart luggage is designed to access serial data from the smart device connected with the baggage and output it to controller on the baggage. Bluetooth is used to connect the smart device and the baggage. The baggage follows the human having the smart device. To follow the human either object detection or UV and IR sensors, for finding the lost luggage bag GPS and GSM and to unlock the device fingerprint or face detection techniques are used. An inbuilt power bank is established using which the void battery devices can be charged with ease. Smart alert is also established to keep track of all the significant things taken out of the baggage. The functioning of each is mentioned below:

3.1 Auto Trailing Bag of the Smart Bag

a. **Using Ultrasonic sensors**: The baggage is activated by sending an alert from the smart device or pressing a button on the baggage. Ultrasonic sensors are used to detect the presence of objects in front of the baggage ensuring easy move for the baggage. They are placed in such a way that the transmitter of each sensor is placed in middle and receiver is placed on either side of the baggage. The continuously emitted sound waves travel through air and would come back if it strikes the obstacles on the way. The movement of the baggage is based on the strength of the signal received at the receivers on either side. The baggage turns left if the strength is more at left receiver or right if strength is more towards the right. If the signal strength is equal then the baggage would move forward. Every time it moves the distance between the human and the baggage is also calculated and if it is less than the threshold value the baggage stops moving.

b. **Human following mechanism using two IR sensors**: The two IR sensors are also placed in a similar fashion as mentioned above. Whenever person turns the movement is detected by the IR receivers positioned at the corners of the baggage. This data taken by the IR receiver is sent to the controller. The controller by means of programming rotates the motors in the required direction to follow the person.

c. **Object Detection using Artificial Intelligence**: Obstacles are detected using the object detection algorithms and the baggage moves in the way where there is no obstacle.

3.2 Proximity Detection

Proximity detection is done using Bluetooth. When a theft happens and the baggage moves a certain meter (based on the threshold) away from the smart device, the alarm attached to the baggage starts to sound and sends an immediate notification to the smart device. If the smart device is a wrist band then it is set to vibrate so that the person can easily understand that the baggage has been stolen. Along with the notification to the smart device, the location of the baggage is also sent. This makes it easy for the person to reach the baggage. This prevents theft of baggage and also from situations where the person forgets the baggage.

3.3 Location Tracking

This is achieved using the GPS and GSM modem. The baggage is embedded with the GPS, GSM modem and the controller. GPS coordinate values and the name of the location are stored in a Lookup Table in the controller. When the GSM modem receives a request for the location of the baggage, the microcontroller checks for a closest location that matches inside the LUT with the received GPS coordinate data. The matched location details are sent as a response to the smart device.

3.4 Smart Unlocking

The baggage can be unlocked either using the fingerprint system or using face detection. The smart unlocking system is used to authenticate the person whose fingerprints are pre trained in the controller memory. If the authentication fails for number of times the baggage is locked and an alert is sent to the person having the smart device. This keeps the things inside the baggage safe from being viewed by others.

3.5 Smart Alerting

The smart baggage also acts as an assistant with smart alert system. Humans tend to forget things when taken out from the baggage. Using the smart baggage the problem of missing out important things can be solved. After taking out any important thing, it has to be scanned in front of the baggage camera. Through object detection the object name is stored in the controller memory. List of all the items taken out are stored in the memory and before keeping inside it has to be scanned again to remove from the controller's memory. The remaining things in the list are sent as a text to the smart device before it starts moving from that place. This reminds the person about the things left behind.

3.6 Reviving Void Battery Devices

The smart baggage is provided with an internal power bank which can be used to revive the void battery devices.

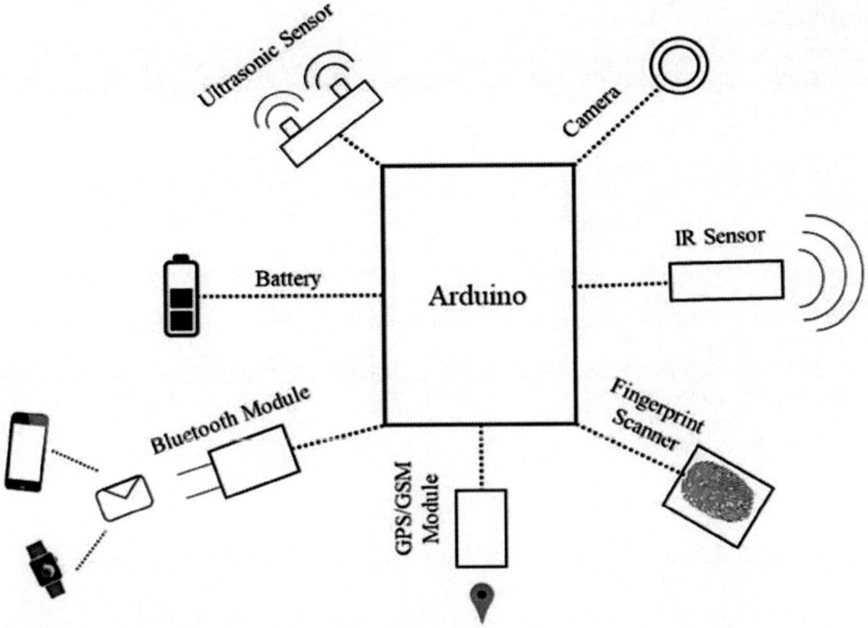

Fig. 1 Block diagram showing all the components connected to the controller

4 Block Diagram

See Fig. 1.

5 Conclusion

The study shows the execution of smart baggage which makes the life of a traveller comfortable and secure. Complications encountered such as carrying heavy loads, robbery of things inside the baggage and recovery of lost baggage are solved using this approach. Apart from this the baggage acts a personal assistant reminding things that are taken out from the baggage. The inbuilt power bank can share power to revive the void battery devices such as mobiles and laptops. This baggage provides security and intelligent features that suits the recent age.

References

1. Sudha S, Brindha K, Rathi R, Charanya R, Mayank J (2017) Luggage tracking system using IoT. Int J Pure Appl Math
2. Jainwar MS, Rao HK, Varma MK, Tamrakar MH (2016) The intelligent suitcase. Int. J Innovat Res Electr Electron Instrum Control Eng
3. Olickal SJ, Yohannan A, Ajayan M, Alias A (2017) Smart bag. Int Res J Eng Technol
4. Sutar A, Kocharekar T, Mestry P, Sawantdesai P, Goilkar SS (2018) Smart bag with theft prevention and real time tracking. Int J Trend Sci Res Dev
5. Duat DL, Bebelone MC, Gallego JM (2017) Design and implementation of 4-In-1 luggage bag. Int J Res Appl Sci Eng Technol
6. Anand AP, Srivastava D, Sharma D, Dhal J, Singh AK, Meena MS (2016) Smart school bag. Int J Eng Sci Comput
7. Naik R, Muppidwar S, Chavan P, Medhekar S, Chindarkar P (2016) Smart bag. J Emerging Technol Innovat Res

Comparative Study of Privacy Preservation of Data in Hybrid Cloud of Combined Clustering and Geometric Data Perturbation Approach and Hybrid Ant Colony Optimization and Gravitational Search Algorithm

Sridhar Reddy Vulapula and Srinivas Malladi

Abstract Cloud computing services provide scalable IT infrastructural facilities to support the processing of various types of big data applications in sectors like health care and other businesses. Data sets like electronic health records contain private and sensitive information resulting in possible privacy issues. A simple method to facilitate data privacy preservation is to anonymize data through generalization to satisfy a given privacy model. However, majority of the existing privacy-preserving methods devised to small-scale data sets turn out to be inadequate in the case big data, due to their limitations or poor scalability. In this paper, two different approaches are discussed and their results are compared with the existing respective methods of Privacy Preservation of Health-care Data in the context of hybrid cloud. These two different approaches are: 1. A Combined Clustering and Geometric Data Perturbation Approach for Enriching Privacy Preservation of Health-care Data in Hybrid Clouds and 2. Privacy Preservation of Health-care Data in Hybrid Cloud Using a Hybrid Meta-heuristics-based Sanitization Approach. This paper focuses on less-known aspects of side effects of the CCGDP approach which was addressed by ACOGSA approach.

1 Introduction

Tele-medicine is an upcoming health-care service where professionals in the field treat a patient with the help of tele-communication technology. Data in health-care system indicates a set of medical health data (X-ray, MRI etc.) which are larger in number. It is not easy to handle the medical-data sets using traditional hardware and

S. R. Vulapula (✉) · S. Malladi
Computer Science Engineering, KLU, Vaddeswaram, Andhra Pradesh, India
e-mail: vsridharreddy19@gmail.com

S. Malladi
e-mail: srinu_cse@kluniversity.in

S. C. Satapathy et al. (eds.), *Communication Software and Networks*, Lecture Notes in Networks and Systems 134, https://doi.org/10.1007/978-981-15-5397-4_4

software. Hence, it is vital to devise a more practical approach to balance the privacy protection and sharing of the health-care data [1].

Cloud computing covers the entire process of the delivery of computing services—including cloud servers, cloud storage, databases, and networking—over the Internet ("the cloud") to enable innovation, flexible resources, and scalable economies. One typically pays only for cloud services one uses, helping to bring down the operating costs, run infrastructure more efficiently, and scale as the needs of the business change from time to time. This is the pay per use model catering to the demands of the users.

Hybrid cloud is a cloud computing environment that uses a mix of private cloud and public cloud services with seamless synchronization between the two platforms. By allowing the data to move between private and public clouds as per the varying computing needs and costs, hybrid cloud provides more flexibility and more data deployment options to the businesses. It combines the benefits of both the private cloud and the public cloud and delivers the private cloud's high-security features along with the easy-to-access features of the public cloud.

In this work, our goal is to give a comparative analysis on enhancing the privacy preservation of health-care data in the hybrid cloud using Combined Clustering and Geometric Perturbation Approach and Privacy Preservation using Hybrid Meta-Heuristics-Based Sanitization Technique. The advantage of the first approach is that it provides privacy without introducing a computation and communication overhead between the private cloud and the public cloud. The K-mean clustering algorithm is used for partitioning the data where each partitioning is treated as a cluster. The GDP algorithm is used to perturb the clustering data, these perturbed values are hard to be decoded. The public cloud stores the perturbed data and the key parameters for randomizing technique and the clustering techniques are stored in a private cloud. Finally, with the help of the data retrieval technique, the original data is recovered from the perturbed data [2].

Various algorithms exist to hide sensitive information. Most of these are applied along with addition and deletion operations to hide sensitive data. Sometimes unauthentic information may be produced or hidden data is to be modified during the time of hiding and securing information.

Data anonymization refers to hiding identity of the user, while in diverse analysis and mining tasks, certain information can be still exposed to data user. The difficulties and the side effects obtained during this process can be overcome by employing a hybrid model for privacy preservation.

In the second model, a hybrid ant colony optimization and gravitational search algorithm (ACOGSA) are used to reduce the side effects in hybrid cloud and also to notify the problem in the hidden sensitive data through transaction deletion.

The rest of the paper is organized as follows. Section 2 provides the Literature Review; in Sect. 3 proposed works are explained; and in Sect. 4 Results and Analysis are explained. Section 5 gives the conclusion.

2 Literature Review

Several techniques are used by researchers in privacy preserving of data cloud systems. Hence, brief evaluations and some important contributions to the existing literature are put forward. Considerable research work has previously existed in literature related to the security-based access control techniques. C. Zhang, L. Zhu, C. Xu, and R. Lu, proposed Privacy-Preserving Disease Prediction technique to improve privacy [3]. X. Liu, R. Lu, J. Ma, L. Chen, and B. Qin, developed Clinical Decision Support System for preserving privacy of the patient data [4]. K. Xing, C. Hu, J. Yu, X. Cheng, and F. Zhang resolved several issues related to Mutual Privacy Protection using K-Means clustering algorithm [5]. Author H. Kaur, N. Kumar, and S. Batra, employed a methodology in order to enhance the health-care recommendation system efficiency, a new algorithm: Privacy-Preserving Collaborative Filtering (PPCF) with Arbitrary Distributed Data (ADD) approach was proposed [6]. Wei Wang et al. have proposed a privacy-preserving framework to transit insensitive data to the commercial public cloud and the rest to trusted private cloud. Nonetheless, the encryption leads to large overhead when answering queries [7]. Abdul Majeed implemented an attribute-centric anonymization to improving the privacy of e-health data. Their approach consists of following steps such as preprocessing the original health records, user ranking, formation and analysis of equivalence classes, attributes classification, and data anonymization. Simulation result shows that the data privacy is improved than existing methods [8]. In 2017 Wu et al. implemented an ant colony system for hiding sensitive item sets. In this process each ant would build a tour for iteration and each tour indicates a deleted transaction in the original database. Their approach includes the following steps: Ant routing map, Termination condition for each tour, Fitness function, Heuristic function, and Delete transactions. Simulation results perform in more complicated computation [9, 10]. Sabin Begum and Sugmar presented a new entropy-based approach for affordable privacy preservation. In their method, adaptive particle swarm optimization was used in optimal entropy value process. Their technique comprises joint entropy model and database difference model. For testing purpose, evaluation metrics were taken from entropy and database difference. In performance their method has minimum access time, better memory usage, and good privacy [11].

3 Proposed Work

3.1 *Privacy Preservation of Health-care Data in Hybrid Clouds Using Combined Clustering and Geometric Data Perturbation (CCGDP) Approach*

In this approach, a combined clustering and geometric data perturbation approach for improving privacy preservation of health-care data in hybrid clouds is planned. An answer that can productively give protection to information sent to the cloud without presenting substantial overhead on both computation and communication is panned in this paper. At first, the data are separated into various parts by adopting the K-mean clustering method where each partition is considered as a cluster. At that point, the mean estimate of each cluster is processed; after that the contrast between each cluster member and the mean of the cluster value is computed. In the next stage, the clustered information is perturbed by utilizing the Geometric Data Perturbation (GDP) algorithm which makes the values difficult to be decoded. These perturbed values are stored in the public cloud and the key parameters for randomizing and clustering are stored in the private cloud. The present approach would contribute to the reduction of large storage on private cloud on the off-chance that the entire sensitive information is stored on private clouds. The experimental results clearly show that the GDP algorithm has better privacy preserving compared to the other methods presently used.

3.2 *Privacy Preservation of Health-care Data in Hybrid Cloud Using a Hybrid Meta-Heuristics-Based Sanitization Approach*

In recent years, the increasing use of cloud services facilitate hospitals and institutions to send their data to the cloud, which provides access-from-anywhere, on-demand, and affordable data services. Even though there are several benefits with the cloud services, certain issues like privacy do exist. Privacy risks rise when outsourcing personal health-care records to cloud due to the sensitive nature of health information and the social and legal implications for its disclosure. Many governments, organizations, and even individuals are concerned about these privacy issues. Because of these issues, privacy-preserving data mining (PPDM) has become a vital issue in the recent times. Our objective is to come up with a privacy-preserving outsourcing framework under the hybrid cloud model. In this work, a Hybrid Ant Colony Optimization and Gravitational Search Algorithm (ACOGSA) is proposed to be employed to address the problem of hiding sensitive data through transaction deletion, while considerably reducing the side effects in hybrid cloud. Substantive experiments will be carried to compare the performance of the designed algorithm with sophisticated methods in terms of the side effects and database integrity. Previously, some heuristic approaches have been proposed to sanitize databases for hiding sensitive information. The method used for the comparison is GA, PSO, ACO, and Firefly framework.

4 Results and Analysis

4.1 CCGDP Approach

Enriching Privacy Preservation of Health-care Data in Hybrid Clouds can lead to finding a solution that can efficiently provide privacy to data stored in the cloud without introducing large overhead on both computation and communication. Here, initially the high-dimensional data set is divided into multiple partitions and each of the sensitive attributes is considered a cluster. In the following stage, an optimal cluster head is selected by using a cluster head selection technique and values are computed considering the relation between each of the cluster members and the cluster head.

In the second stage, the geometric data perturbation technique is applied to the already computed values which make them hard to decode. The perturbed values of each sensitive attribute are then sent to the public cloud while all the key parameters used for clustering and randomizing are kept in the private cloud. The present approach would result in the reduction of large storage in private cloud as the entire sensitive information is stored on private clouds. Other contributions include the reduction of communication overhead between private and public cloud and the delay introduced by communications between private and public cloud.

The database contains large quantities of sensitive information. It also contains critically sensitive information such as medications and immunization. Here, the entire data contain a huge amount of information. In this way, the computation and the communication overhead have taken place. Here, at first, the high-dimensional data set is separated into multiple groups based on their similarity function by utilizing the clustering procedure. The K-mean clustering calculation gives the preferable outcome over other existing techniques.

Data recovery is critical in the information technology. In this system, the health-care database owner gives permission to the three types of users to accept the original database. They are: users, insurance companies, and the physicians. But each of them does not get all the original data about the patients; they can only get data based on the user's specific requirements and all other data are perturbed. When the data is queried, the request is sent simultaneously both to the private cloud and the public cloud. It contains two databases; they are private data and the public data. The private data contains the original data and the perturbed data whereas the public cloud contains only the perturbed data.

To start with, the request query is sent to the system by the user. This query request checks the private cloud data and takes the perturbed value of that. This perturbed value retrieves all the attributes in that row after checking the public cloud data. The GDP algorithm contains both the data perturbation phase as well as the data retrieval phase. Input: Original data P, its size n and delicate characteristic [V].

The evaluation results of the proposed method are described and the performance of the proposed GDP algorithm is compared with the existing ElGamal's encryption algorithm. The GDP technique can be best utilized for perturbing for medical big data by clustering the whole data set making use of K-means clustering and perturbing the clustered data using GDP strategy.

4.2 *ACOGSA Approach*

The experimental section is done by using the proposed method ACOGSA approach. The experiment is performed on a Personal computer with Intel® core (TM) i3-7100, Dual core CPU, Dual core 3.90 GHz processor, 4 GB RAM, 64-bit Windows 7 Operating System and for implementation the JAVA software tool is used. The results of the experiment have been compared both with the Ant colony system-based algorithm (ACS2DT) as well as the Particle swarm optimization-based algorithm (PSO2DT) techniques.

4.2.1 Evaluation Metrics

The performance evaluation is done for the sensitive percentage and minimum support threshold with the runtime, F-T-H (Fail to be hidden), N-T-H (Not to be hidden), and DS (Database similarity).

Fail to Be Hidden (F-T-H)

The F-T-H is defined as the number of sensitive item sets in the original database to the number of sensitive item sets still appearing in the sanitized database. It can be mathematically defined as $F - T - H = \frac{|SI^*|}{|SI|}$, where the sensitive item sets in the original database are denoted as SI^* and the sensitive item sets still appeared in the sanitized database is given as SI.

Not to Be Hidden (N-T-H)

The N-T-H is defined as the number of non-sensitive frequent item sets hidden by the sanitization process. It is represented as $N - T - H = \frac{|FI - SI - FI^*|}{|FI - SI|}$, where FI is the frequent item sets from the original data, SI is the set of sensitive item sets, the term FI-SI is the number of non-sensitive frequent item sets in original database. The term FI* is the item sets that still appeared in the sanitized database, and the non-sensitive frequent item sets that are hidden by the sanitization process is denoted as FI - SI - FI*.

Database Similarity (DS)

For determining the size of the two different databases, the database similarity is used. It evaluates the similarity between the original and sanitized database. It is simply defined as the number of transactions in the original database to the transaction in the sanitized database. It can be expressed as $DS = \frac{|D^*|}{|D|}$, where the original database of transaction is denoted as D and the sanitized database of transactions is given as D*.

4.2.2 Performance Evaluation for Sensitive Percentage

The comparison of the proposed with the existing schemes is analyzed. The runtime of the GS-ACO algorithm is shown in Fig. 1. While the runtime gets increased the sensitive percentage of our proposed gets better value in the runtime of 4900 s (Fig. 2).

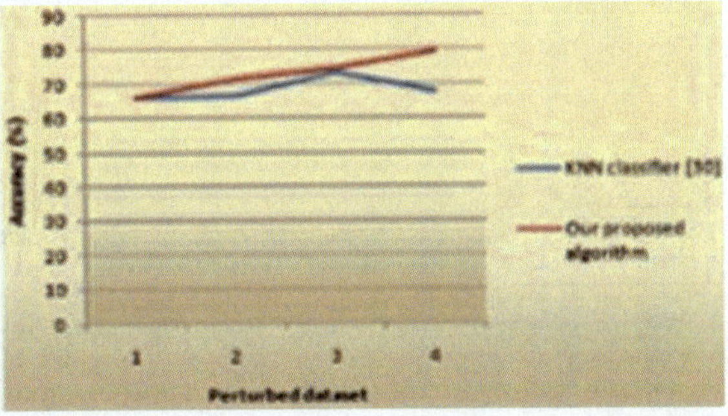

Fig. 1 Accuracy comparisons

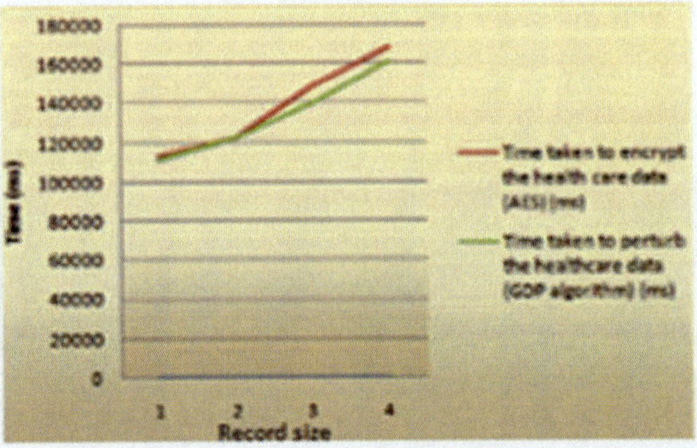

Fig. 2 Time comparison

From the above given Fig. 3, the sensitive percentage is compared for the proposed method with the ACS2DT[19], PSO2DT[20]. The sensitive percentage is maximized for the proposed method and the other two methods also have high sensitive data. For the runtime increases the sensitive percentage is minimized. The overall percentage of the proposed method is 6.28% better for ACSDT and 10.32% better for PSO2DT.

From the above given Fig. 4, the sensitive percentage is compared for the proposed method with the ACS2DT, PSO2DT. The sensitive percentage is maximized for the proposed method and the other two methods also have high sensitive data. For the F-T-H increases the sensitive percentage is minimized. The percentage of the proposed is 8.41% for the 0.1 sensitive percentage, 7.08% for the 0.2 sensitive percentage, 8.17% for the 0.3 sensitive percentage, 7.16% for the 0.4 sensitive percentage, and 9.57% for 0.5 sensitive percentage better of the ACSDT. While comparing with the PSO2DT, the proposed method gets better values for the 0.1, sensitive percentage is 6.88%; for 0.2 the sensitive percentage is 4.8%; for 0.3 the sensitive percentage is 8.82%; for 0.4 the sensitive percentage is 8.16%; and for 0.5 the sensitive percentage is 9.11%.

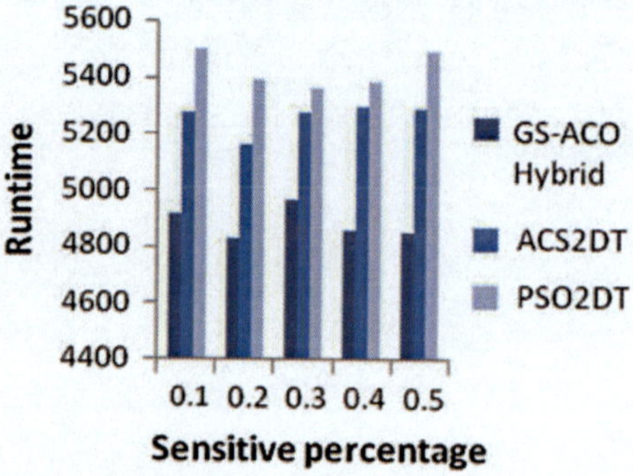

Fig. 3 Comparison of runtime with sensitive percentage

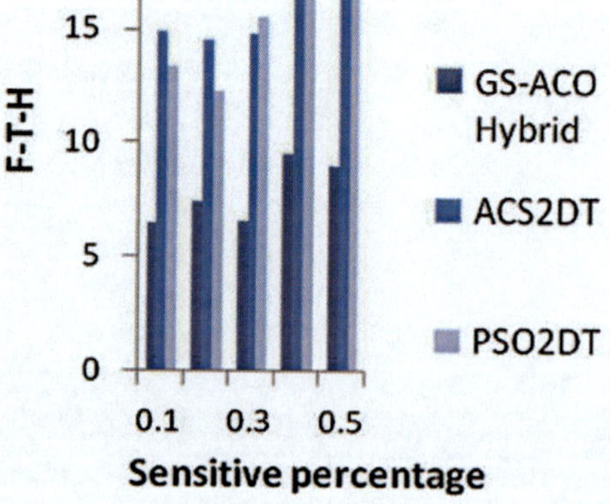

Fig. 4 Comparison of F-T-H with sensitive percentage

From the above given Fig. 5, the sensitive percentage is compared for the proposed method with the ACS2DT, PSO2DT. The sensitive percentage is maximized for the proposed method and the other two methods also have high sensitive data. For the N-T-H increases the sensitive percentage is minimized. The proposed method gets better of ACS2DT in the percentage of 7.29%, 7.67%, 8.17%, 10.71%, and 11.15% for all the sensitive values. While comparing with the PSO2DT the proposed method gets 9.72%, 6.45%, 8.92%, 11.93%, and 12.01% for all the sensitive values.

From the above given Fig. 6, the sensitive percentage is compared for the proposed method with the ACS2DT, PSO2DT. The sensitive percentage is increased for the proposed method and also for the other two methods. As the DS increases the sensitive percentage is also increased. While comparing with the ACS2DT, the proposed method gets better as 1.58%, 1.43%, 1.5%, 1.71%, and 1.25% for all the sensitive percentages and comparing with the PSO2DT the proposed results are improved in percentage of 2.39%, 2.09%, 1.99%, 1.84%, and 1.54% for all the sensitive values.

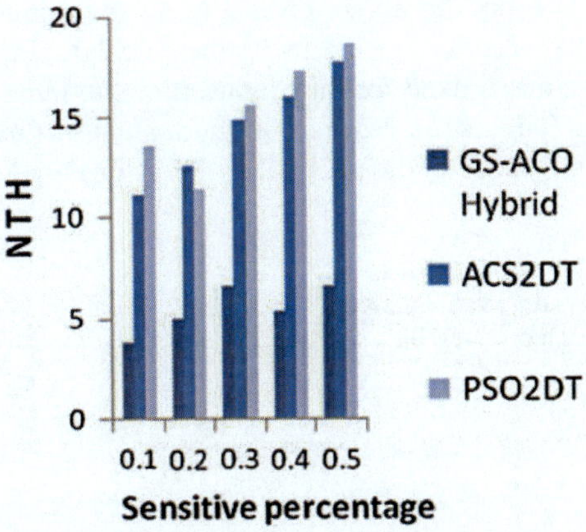

Fig. 5 Comparison of N-T-H with sensitive percentage

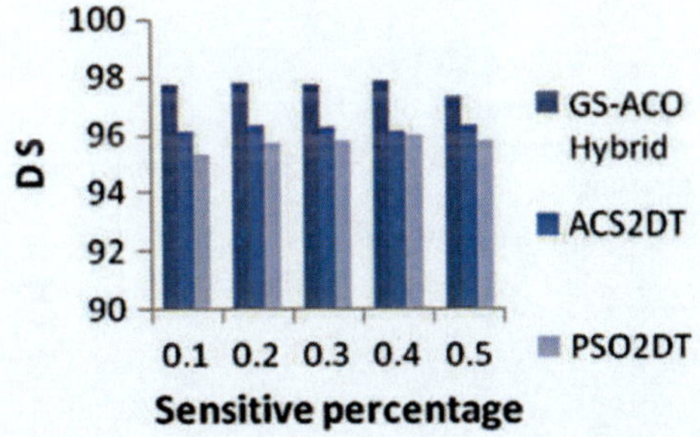

Fig. 6 Comparison of DS with sensitive percentage

4.3 Performance Evaluation for Minimum Support Threshold

The comparison of the proposed methods with the existing schemes is analyzed. The runtime of the GS-ACO algorithm is shown in Fig. 8. While the runtime gets increased the sensitive percentage of our proposed gets better value in the runtime of 4950 s.

The above given Fig. 7 shows that the minimum support threshold is compared for the proposed method with the ACS2DT, PSO2DT. The minimum support threshold is maximized for the proposed method and also for the other two methods. For the runtime increase the minimum support threshold is also minimized. The proposed method gets the better values when compared with the ACS2DT as 7.1%, 3.8%, 5.9%, 6.1%, and 8.8% for all the sensitive values. While comparing with PSO2DT the proposed values get as 10%, 7.1%, 12%, 9.8%, and 9.9% better values in all sensitive percentages.

From the above given Fig. 8, the minimum support threshold is compared for the proposed method with the ACS2DT, PSO2DT. The minimum support threshold is maximized for the proposed method and also for the other two methods. For the F-T-H increases, the minimum support threshold is reduced. The proposed values are better in the ACO2DT: the values are 8.41, 8.84, 7.92, 11.93, 8.87% for all the

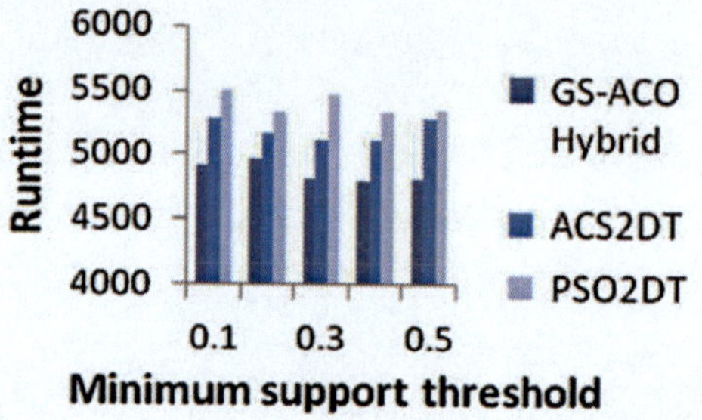

Fig. 7 Comparison of runtime with minimum support threshold

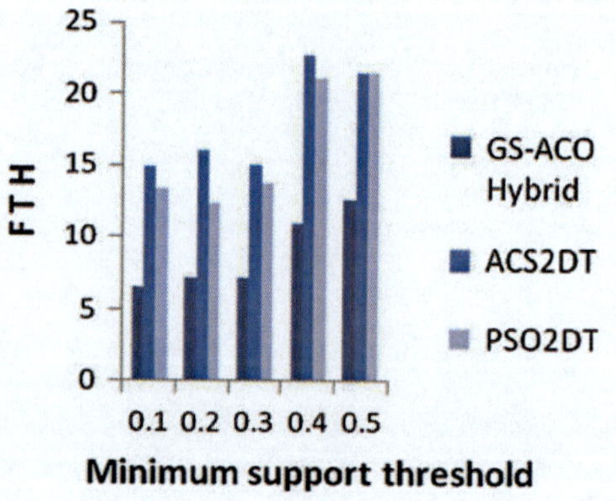

Fig. 8 Comparison of F-T-H with minimum support threshold

minimum threshold values. While comparing with the PSO2DT the proposed values are better off 6.8, 5.25, 6.62, 9.9, and 8.83% for all the minimum threshold values.

From the above given Fig. 9, the minimum support threshold compared with the proposed methods ACS2DT, PSO2DT. The minimum support threshold is maximized for the proposed method and also for the other two methods. For N-T-H increases, the minimum support threshold is reduced. The proposed method gets better than ACS2DT in the percentage of 7.29, 8.69, 8.18, 10.02, and 8.54% for all the minimum support threshold values. While comparing with the PSO2DT the proposed gets 9.72, 9.35, 9.9, 8.72, and 8.91% for all the minimum support threshold values.

From the above given Fig. 10, the minimum support threshold is compared for the proposed method with the ACS2DT, PSO2DT. The minimum support

Fig. 9 Comparison of N-T-H with minimum support threshold

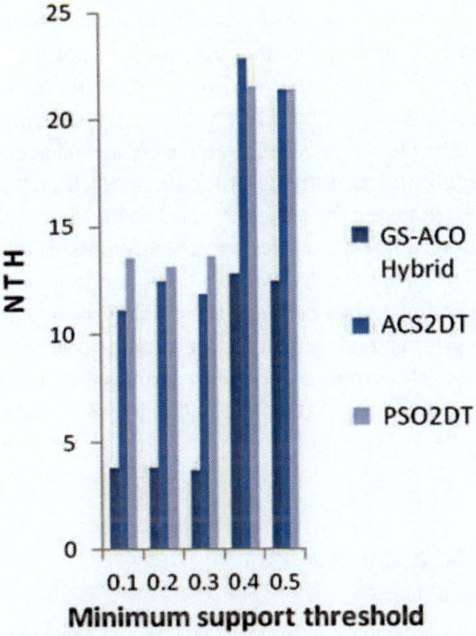

Fig. 10 Comparison of DS with minimum support threshold

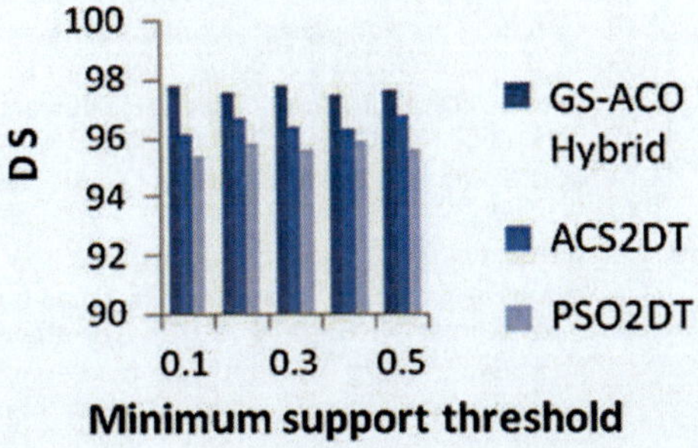

threshold is maximized for the proposed method. For the DS increases, the minimum support threshold is reduced. The proposed method gets better than ACS2DT in the percentage of 1.58, 0.86, 1.4, 1.1, and 0.87% for all the minimum support threshold values. While comparing with the PSO2DT the proposed values get 2.39, 1.71, 2.13, 1.55, and 2.08% for all the minimum support threshold values.

5 Conclusion

In an effort to get over the problems faced while using the existing methods, the proposed method makes use of a combined clustering and random data perturbation approach that enhances the privacy preservation of health-care data in the hybrid cloud. This approach used to totally decrease the computation and communication overhead between the private cloud and the public cloud. Our experimental outcomes demonstrated that this approach is not just safeguarding the accuracy of the framework. Additionally it gives the better privacy guarantee, contrasted with existing perturbation procedures. Here, translation, rotation, and reflection isometrics are used for the proposed approach, so the accuracy is far more reliable than the current methods. But this approach has not focused much on the side effects that followed. To get over the problems and reducing the side effects obtained in privacy preservation, a hybrid model for privacy preservation with Heuristic approach is proposed. The proposed hybrid model is modeled on the lines of ACO and GSA. This work is expected to enhance the privacy policy and reduce the side effects in the existing systems. Using the combination of both the techniques, the proposed system generates the sanitized database via transaction deletion. Compared to existing methods, our proposed work has better privacy and security. The experimental results of the proposed method are compared with the existing methods as ACS2DT and PSO2DT.

References

1. Vulapula SR, Srinivas M (2018) Review on privacy preserving of medical data in cloud computing system. IJPHRD 9(12)
2. Reddy VS, Rao BT (2018) A combined clustering and geometric data perturbation approach for enriching privacy preservation of healthcare data in hybrid clouds. IJIES 11(1)
3. Zhang C, Zhu L, Xu C, Lu R (2018) PPDP: an efficient and privacy-preserving disease prediction scheme in cloud-based e-Healthcare system. Future Gener Comput Syst 79:16–25
4. Liu X, Lu R, Ma J, Chen L, Qin B (2016) Privacy Preserving Patient-Centric Clinical Decision Support System on Naive Bayesian Classification. IEEE J Biomed Health Inform 20(2):655–668
5. Xing K, Hu C, Yu J, Cheng X, Zhang F (2017) Mutual Privacy Preserving k-Means Clustering in Social Participatory Sensing. IEEE Trans Ind Inform 13(4):2066–2076
6. Kaur H, Kumar N, Batra S (2018) An efficient multi-party scheme for privacy preserving collaborative filtering for healthcare recommender system. Future Gener Comput Syst
7. Wang W, Chen L, Zhang Q (2015) Outsourcing high-dimensional healthcare data to cloud with personalized privacy preservation. Comput Netw 88(13):136–148

8. Majeed A (2018) Attribute-centric anonymization scheme for improving user privacy and utility of publishing e-health data. J King Saud Univ Comput Inform Sci
9. Sabin Begum R, Sugumar R (2017) Novel entropy-based approach for cost-effective privacy preservation of intermediate datasets in cloud. Cluster Comput
10. Wu JMT, Zhan J, Lin JCW (2017) Ant colony system sanitization approach to hiding sensitive item sets. IEEE Access 5:10024–10039
11. Lin JC-W et al (2016) A sanitization approach for hiding sensitive item sets based on particle swarm optimization. Eng Appl Artif Intell 53:1–18

Detection of Plant Diseases Using Data Mining Techniques–A Literature Survey

Shivani Munigala, Pravalika Nampelli, and Pallavi Tandra

Abstract Diseases in plants cause major production and economic damages. It also results in drop of both quality and quantity of agricultural products. Disease is the main cause of the death of the plant. It also effects the human health. Farmers experience great difficulties in transferring from one disease control policy to another. Data mining is fairly novel research in agriculture. For detecting the leaf disease, image processing technique is used. This paper presents survey on different data mining techniques for plant diseases and also image processing technique which is used for fast and accurate detection of plant leaf diseases.

1 Introduction

Agriculture is an important source of economic growth in India. India's agriculture sector accounts for 18% of India's gross domestic product and provides employment to 50% of countries workforce. Agriculture helps to gather the basic needs of human's like food, clothing, shelter, and medicine. Many of the studies have shown that quality of agricultural products has been reduced due to plant diseases. These diseases are caused by pathogens, viz., fungi, bacteria, viruses, and adverse environmental conditions. Many facts cause the plants to exhibit poor stamina changes in appearance or even death. Two factors that are abiotic (non-living) and biotic (living) have negative impact on plant health.

Data mining is the process of extracting knowledge from large collection of data or information. The term data mining is named as mining Knowledge from given

S. Munigala · P. Nampelli (✉) · P. Tandra
Department of Computer Science and Engineering, Stanley College of Engineering and Technology for Women, Hyderabad, India
e-mail: pravalikanampelli@gmail.com

S. Munigala
e-mail: Shivanimunigala123@gmail.com

P. Tandra
e-mail: pallavitandra27@gmail.com

S. C. Satapathy et al. (eds.), *Communication Software and Networks*, Lecture Notes in Networks and Systems 134, https://doi.org/10.1007/978-981-15-5397-4_5

data. Data mining is a process of solving problems by analyzing data which is present in database. Data mining techniques are used in agriculture for data characterization, discrimination, and forecasting purposes.

1.1 Plant Diseases

Leaf diseases or plant diseases are categorized into two groups:

- Infectious disease caused by pathogens like Fungi, Bacteria,Viruses, etc.,
- Non-infectious disease caused by mineral toxicities, Soil acidity, Nutrient deficiencies, or environmental factors (like Acid rains)

Bacterial disease symptoms: Pathogenic bacteria cause many serious diseases to vegetables. They do not penetrate directly into plant tissue but enters through wounds or natural plant openings. The disease is identified by tiny pale green spots which soon come into view as water soaked and then they appear as dry dead spot on leaves, blossom, fruits, and stems (Fig. 1).

Viral disease symptoms: Among all the leaf diseases, the diseases that are caused by viruses are difficult to diagnosis. Aphids, leaf hoppers, white flies, and cucumber beetles insects are common carries of this disease. Symptoms are mosaic leaf pattern, wrinkled leaves, yellowed leaves, and plant stunning (Fig. 2).

Fungi disease symptoms: Fungal disease first appears on lower, older leaves like gray-green spots and water soaked. When fungal disease matures these spots darken and then white fungal growth forms on basements. Classification of pathogenic fungi is important for identifying and diagnosing plant disease (Fig. 3).

Fig. 1 Bacterial disease symptom

Fig. 2 Fungal disease symptom

Fig. 3 Viral disease symptoms

2 Relavant Work

The objective of the paper leaf disease classification using artificial neural network written by Syafiqah Ishakais is to capture and analysis data from leaf images for classifying healthy and unhealthy leaves was achieved using image processing method. In the final experiment, the result shows that the RBF network performs better than MLP network [1].

Sushil R. Kamlapukar proposed approach used to identifying plant leaf diseases. In this paper, he used feature extraction for Garbor filter and after performing feature extraction he classified the plant leaf diseases using artificial neural network. The main disadvantage of this paper is computational complexity et al. [2].

Et al. Malvika Ranjan in the paper detection and classification of leaf disease using artificial neural network begins with capturing the images. The present work proposes a methodology for detecting leaf diseases early and accurately, using diverse image processing techniques and artificial neural network [3].

P. Revati, M. Hemalatha proposed an approach which is used to segment the infected part of image using edge detection. They classified the diseases using homogeneous pixel technique for cotton disease detection [4].

In Deep Neural Networks-Based Recognition of Plant Diseases by Leaf Image Classification et al. Srdjan Sladojevic came with a new approach to the development of plant disease detection model, by the use of deep convolutional networks. The experimental results of this model achieved precision between 91% and 98%, for separate class tests, on average 96.3% [5].

Surender Kumar and Rupinder Kaur proposed another approach to identifying the plant diseases; this paper used the segmented infected part based on threshold value, based on the edge, based on similarities, and dissimilarities region segmentation done. After performing segmentation, they extracted feature using the color occurrence method, gray level co-occurrence matrices [5].

Savita N. Ghaiwat, Parul Arora et al. present and K-nearest neighbor (KNN) method for predicting the class of test example. This is time-consuming method but was easy to understand. Also they used Support vector machine (SVM) techniques which have more accuracy in prediction. Another technique they have used is Self Organization Map and Probabilistic Neural Network. It requires large storage space. One more desirable techniques they used is Fuzzy Logic [6].

K. Muthukannan, focused on the neural network algorithm is proposed for diseased plant leaf classification. The neural network techniques such as feed-forward neural network, learning vector quantization, and radial basis function network were tested for two different diseased leaf image classifications such as bean and bitter gourd leaves [7].

3 Data Mining Techniques for Identification of Plant Diseases

There are various data mining techniques but we are mainly using the below two features for identification of plant diseases. They are as follows:

- Classification
- Clustering

3.1 K-Means Clustering

K-means clustering is used in separation of leaf image into four clusters in which one or more clusters contain the disease. The K-means clustering algorithms is used to classify objects based on a set of features into K number of classes. The classification is done by minimizing the sum of squares of distances between the objects and the corresponding cluster or class centroid [7].

Algorithm: *k-means*

Input: Leaf Images

Output: 4 Clusters, with one affected cluster

- Let features of image are denoted as X = {x1, x2,….xn} are data and among these some of them are data centroids.
- Initialize cluster k = 4.
- Randomly select 'c' centers.
- Calculate the center using Euclidean distance formula for each data point x1, x2,… and so on.
- Assign the data points who are nearer to the centers.
- Calculate the new centers.
- New centers (x, y) = (x1–y1)/2, (x2–y2)/2
- Repeat the step until no centroid positions change. Otherwise proceed in calculation on Euclidean distance.

3.2 SVM

Support Vector Machine, is used to arrange samples of data into different clusters. SVM can be utilized for both relegation and regression analysis. Support vector machine (SVM) is an emerging powerful machine learning technique to classify cases. SVM has used in a range of problems and it is successful in pattern recognition in detection of plant diseases [7].

Algorithm: *SVM*

a. Choose the underlying population of individuals randomly.
b. Evaluate the fitness of every individual in that population.
c. Choose another generation of population from the past generation by utilizing selection operator.
d. Take the crossover and mutation operation on the present population, then take the evaluation, selection, crossover, and mutation operation on the incipient breed and proceed.
e. If the fitness capacity estimation of ideal individual is adequately expansive or the calculation has run numerous generations, and the ideal fitness estimation of the individual can't be changed clearly, then we get the ideal estimation of kernel

capacity parameter, punishment component, and coldhearted loss capacity, and we can likewise get the ideal classifier by the preparation datum.

4 Comparative Analysis

S No	Title	Authors	Description
1.	Automatic detection of plant disease using K-means clustering algortihm	Sonal P Patel Prof. Arun Kumar Dewangan	The proposed method can effectively classify the infected plant with high accuracy
2.	Plant leaves disease detection using data mining	Subhadre K Dr. N. Kavitha	Used for detecting color and shape of any type of plants in agriculture fields
3.	Plant disease detection and its solution using image classification	Saradhambal. G Divya R Latha.S R. Rajesh.	This project implements and innovative idea to identify the effected crops and provide remedy measures by the use of K-Means clustering algortihm
4.	Survey on detection and classification of plant leaf disease in agriculture environment	Prof. Patil Ashish Prof. Patil Tanuja	The main approach is to recognize the disease on different plants in agriculture environment where speed and accuracy are the main characteristics of disease detection
5.	An identification of variety of leaf diseases using various data mining techniques.	Dr. N. Sasirekha Dr. N. Swetha	Early information on crop health and disease detection can facilitate the control of diseases through proper management strategies. This technique will improve productivity of crops. This paper surveys the data mining techniques to predict the different leaf diseases of different types of leaf. Various types of leaf diseases also discussed in this paper

(continued)

(continued)

S No	Title	Authors	Description
6.	Leaf disease detection using image processing.	Sujatha R Y Sravan Kumar Garine Uma Akhil	Summarizes major image processing used for identification of leaf diseases are k-means clustering, SVM. This approach can significantly support a precise detection of leaf disease. There are five steps for the leaf disease identification which are verbalized to be image acquisition, image preprocessing, segmentation, feature extraction, relegation
7.	Plant leaf disease detection using K means segmentation	T.Gayathri Devi P. Neelamegam A. Srinivasan	This work focuses on the detection and recognition of leaf diseases using SVM classifier. K means clustering is used for segmentation of images. This method provides a solution to the early recognition of diseases
8.	K-Means segmentation method for automatic leaf disease detection	Jimitha Baghel Prashant Jain	Uses K-means clustering technique for segmentation of image to partition the leaf area, disease area, and background area of the input leaf image in order to calculate the percentage infection of the disease in the leaf and to grade them into different classes

5 Design

See Fig. 4.

6 Conclusion

The accurate disease detection of the plant leaf image is very important for the successful cultivation of cropping and this can be done using image processing. This paper discussed various techniques to extract the features of infected leaf and the classification of plant diseases through various data mining techniques.

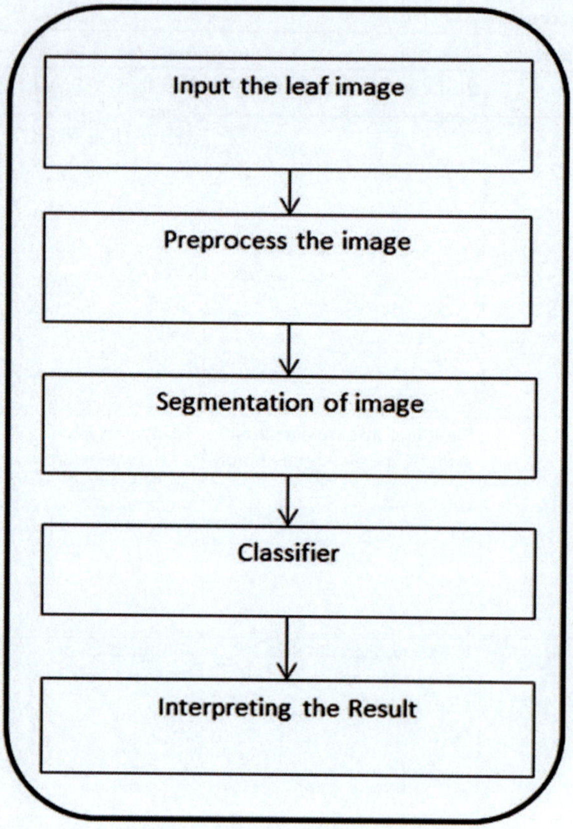

Fig. 4 Architectural diagram

References

1. Ishak S, Rahiman MHF, Aqmariah SN, Kanafiahb M, Saadc H, Leaf disease classification using artificial neural network
2. Barbedo, J.G.A. (2013). Digital image processing technique for detecting, quantifying and classifying plant diseases
3. Naik MR, Sivappagari CMR (2016) Plant leaf and disease detection by using HSV features and SVM classifier
4. Plant leaf diseases detection for datamining—a survey [online] Available: http://www.ijrar.org/papers/IJRAR1944414.pdf
5. Revathi P, Hemalatha M, homogenous segmentation based edge detection techniques for proficient identification of the cotton leaf spot diseases
6. Sladojevic S, Arsenovic M, Anderla A, Culibrk D, Stefanovic D, Deep neural networks based recognition of plant diseases by leaf image classification
7. Muthukannan K, Latha P, PonSelvi R, Nisha P (2015) Classification of diseasedplant leaves using neural network algorithms. ARPN J Eng Appl Sci 10(4):1913–1918
8. Kumar S, Kaur R (2015) Plant disease detection using Image processing-a review. Int J Comput Appl 124(16):6–9
9. Ghawat SN, Arora P (2014) Detection and classification of plant leaf diseases using image processing techniques. Review 3

The New Era of Healthcare: Smart Healthcare Assistant

Zuha Kareem Ansari, Bushra Taher, Nida Arshad, Sadiyya Fayaz, and Fathima Khader

Abstract This paper provides an outline of our work in providing an efficient and accurate Smart Healthcare Assistant at the user's fingertips. The new era gives access to all forms of data, but to what extent can it be considered reliable? One of the biggest issues with such vast availability is that people, today, have substituted everything with such search engines, even professional consultants. The idea for our smart assistant is to provide accurate and definitive information to its users for analyzing symptoms and directing the patients to the right department of doctors for their treatment. Often, even such access to know-the-call requires patients to get recommendations from the informed or many trips to the hospital. But, with a trustworthy source like an authorized smart healthcare assistant such luxury can meet reliability. The implementation of each algorithm ought to be done with consideration to all factors in relation to user and health, while generating data in regular practice, therefore, working out a productive cycle. This model explores a still work in progress, but assures the readers of the potential of technology in providing valuable healthcare assistance.

Z. K. Ansari (✉) · B. Taher · N. Arshad · S. Fayaz · F. Khader
Stanley College of Engineering and Technology for Women, Chapel Road, Abids, Hyderabad, Telangana, India
e-mail: zuhakansari@gmail.com

B. Taher
e-mail: bushrabinttaher18@gmail.com

N. Arshad
e-mail: nida.arshad15@icloud.com

S. Fayaz
e-mail: sadiyyafayaz@gmail.com

F. Khader
e-mail: fathimakhader703@gmail.com

S. C. Satapathy et al. (eds.), *Communication Software and Networks*, Lecture Notes in Networks and Systems 134, https://doi.org/10.1007/978-981-15-5397-4_6

1 Introduction

AI and Machine Learning have opened an entirety to a new phase in the twenty-first century. It's not just robots that have grasped attention from the world; it's much more than that. It is difficult to keep track of all the things that are entering our personal space for our aid. Conventionally, humans have relied on qualified doctors, for curing their ailment, maintaining their good health, and for living a long and healthy life. With current prominence of AI and ML, it has raised bars in fields like Security, Home Assistance, Police, Satellites, and Health care, etc. When the world is moving ahead incorporating technology in everyday life, engineers and doctors have collaborated to move a step further in providing the state of art of health care, and advancing its potential using AI and ML too.

The design focuses on developing a virtual healthcare assistant which can independently be used by any public medical facility. Often, people worry at the slightest cold, concluding it to be a symptom of something awful. With changes to our lifestyles, new emerging health problems have found ways to challenge the health industry and frightening people. There are many sources which patients rely on to give them a possible cause to their sickness and many at times these causes are not verified by the doctors which when seen by them makes them worried and anxious often making them panic in fear. Scientifically, a person can heal better when he his psychologically happy and stable. Considering these factors, this project focuses on getting the most accurate possible cause for their sickness whose data was collected after extensive research along with expert doctors who could best identify the appropriate diagnosis. After which they'll be helped with choosing the right doctor for their illness and, consequently, be directed to book an appointment.

The medical industry is increasing its emphasis on delivering an interface in all possible digital media for person to be acquainted with a provider for their patient access and use [1]. 58% of patients as per reports prefer booking appointments through their devices online. Our healthcare assistant develops with persuasive focus on providing a reliable and trustworthy source for everyone so that the person has this belief that they are at the right place and faith that they'll be led to the right doctor to be cured. We intend to incorporate a great deal of features further, we give in our prime focus on delivering the best, reliable, and accurate healthcare assistant in the industry. The research continues with spark in our motivation.

2 The Working of the Healthcare Assistant

The Smart Healthcare Assistant aspires to provide faithful services to their patients in every step of them discovering their true treatment. It ensures that the patient secures the right department of doctor for their problems. The only task for the patients is to know to review their symptoms with our smart assistant!

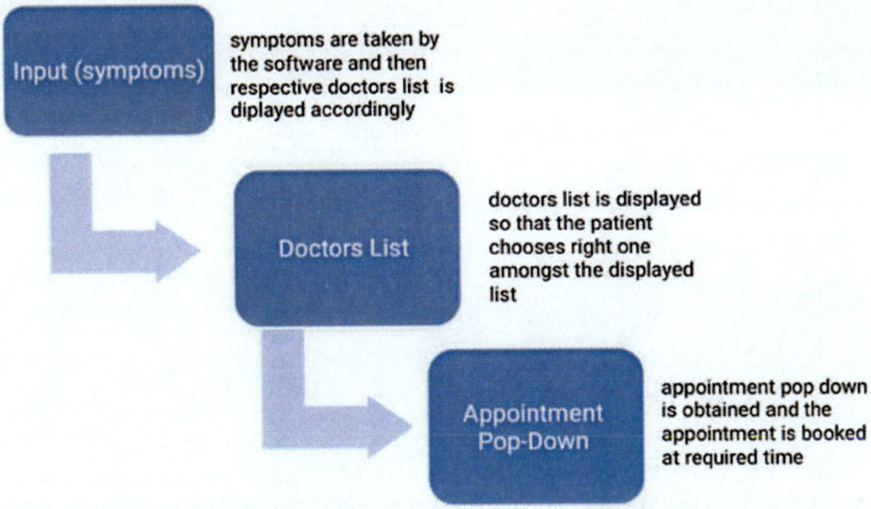

Fig. 1 Basic interface for the assistant model

When the patient doesn't know which doctor he or she wants to consult, then they can open the symptoms pop down, enter the symptoms he or she is facing. The symptoms are then eventually taken down by the web page and respective doctors list is provided to the user according to their symptoms. The user then can click on the preferred doctor where the appointment list pops and then can book the appointment at the well-suited timing according to their convenience. For instance, consider the user is suffering from urinary problem and stomach pain and doesn't know if he or she should go to the general physician or the geriatrician then thus web page suggests for the respective doctors in the urinary disorder field that is the geriatric doctors present in the hospital.

The virtual healthcare assistant comprises Node-RED used for combining the flows and AI services, cloudland, IBM Watson studios, HTML codes for web page creation, and Microsoft Excel sheets for symptom database and their identification process (Fig. 1).

2.1 The Node—RED Flow

The base of the Smart Assistant is made by connecting each component using the Node-RED application from IBM Cloud. Node-RED is a programming tool for wiring together hardware devices, APIs, and online services in new and fascinating ways; it is a browser-based editor where range of nodes in the palette makes the flows to wire together where it can be deployed to get the desired output [4]. Here, we use several nodes to assemble a flow to connect each component of the website (Fig. 2).

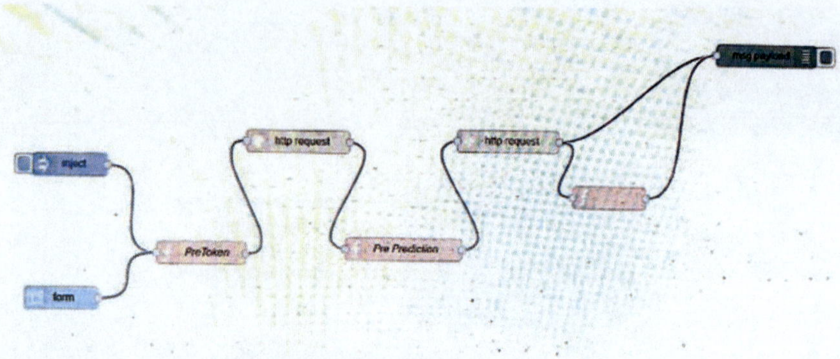

Fig. 2 Flow of Node-RED

The fundamental details of the patient, such as the name and age, are noted through the form node. The program then registers the patient and welcomes them to enter their symptoms. In due course, we have two nodes defined as HTTP request where this node helps to send HTTP requests for the website to be displayed and returns the response required. The Watson Studio service created in IBM cloud permits to create a project within the service by implementing the asset type auto AI experiment. It is a fully automated approach to build a classification or regression model. This sample model runs through analyzing four input symptoms from the patient (Fig. 3).

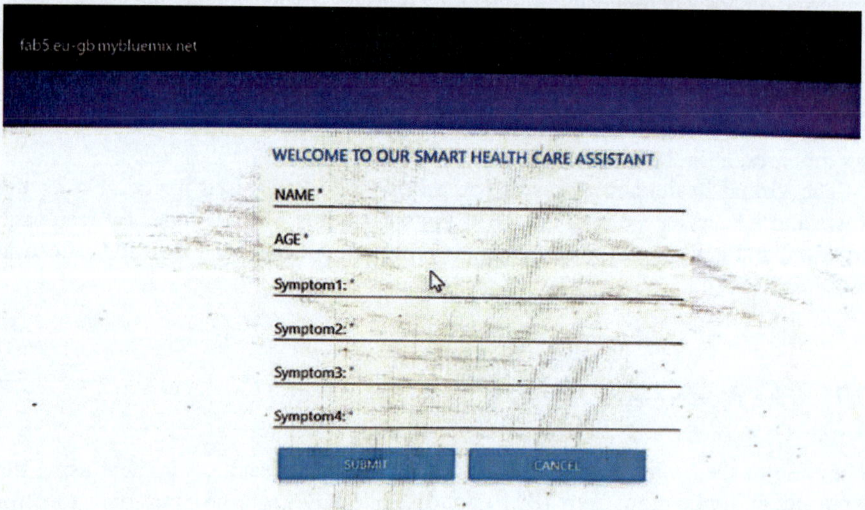

Fig. 3 The figure depicts the page for users to enter their symptoms for the assistant to process

These symptoms are analyzed and differentiated with comparison to the accurately stored data of possible issues. This data is, necessarily, to be made under the guidance of experienced doctors. The data is trained in multi-class classification application in IBM Watson. This model accords a scoring end point which is to be entered in the HTTP request. Subsequently, we take two different function nodes where we pass a code for all the four symptoms to be correlated with variety of diseases and enter the instance id also the API key in the code obtained after training the model.

Thereafter, we again take a function node to enter different types of diseases in a code using control statements so that when the symptoms are to be entered it must be matched with the corresponding disease. Any one of the diseases has to be displayed on the output screen so therefore we use a node known as debug node. This debug node displays the selected message properties in the debug sidebar tab and optionally the runtime log (Fig. 4).

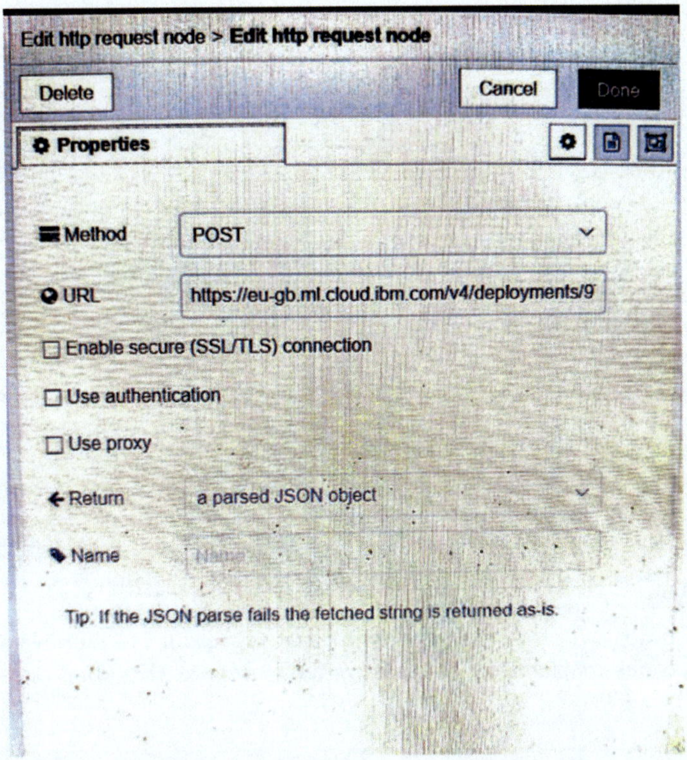

Fig. 4 The figure represents the scoring point request for the analysis-generated result

Fig. 5 Resultant department's doctor list interface in the website

2.2 *Department Database*

A disease to be diagnosed needs to be segregated so that it can be understood to which medical department it belongs. This requires specialists in each department so that the patient can have a lucid idea. For this to proceed, a cloudant is used to create database of doctors where maximum number of doctors under each medical department is mentioned (Fig. 5).

A cloudant is linked with the Node-RED flow where once the presumed patient gives the symptoms and according to that the most accurate disease is diagnosed. On the authority of doctor's suggestion and experience the website can provide a list of specialists of that particular department in descending order where in doctor profile is displayed and after viewing, the patient can click on the ok button.

This idea is similar to the different platforms that exist to assist patients online. The CC-Cruiser is a cloud-based prototype that is connected to an AI system with input data from user along with the clinical actions. Simply, the patients that come to the facility, their demographic data, and medical information is saved into the AI system. These suggestions are pinged to the doctor to assist with the patient's clinical decisions. It also collects feedback and suggestions to improve accuracy. Such necessities are planned to be added in further models (Fig. 6).

2.3 *Appointment Facilities*

Once the patient chooses the required doctor it needs an appointment to be booked. This appointment is again linked to the Node-RED flow. The appointment page is made using the HTML where it gives the options to enter name, age, doctor's name,

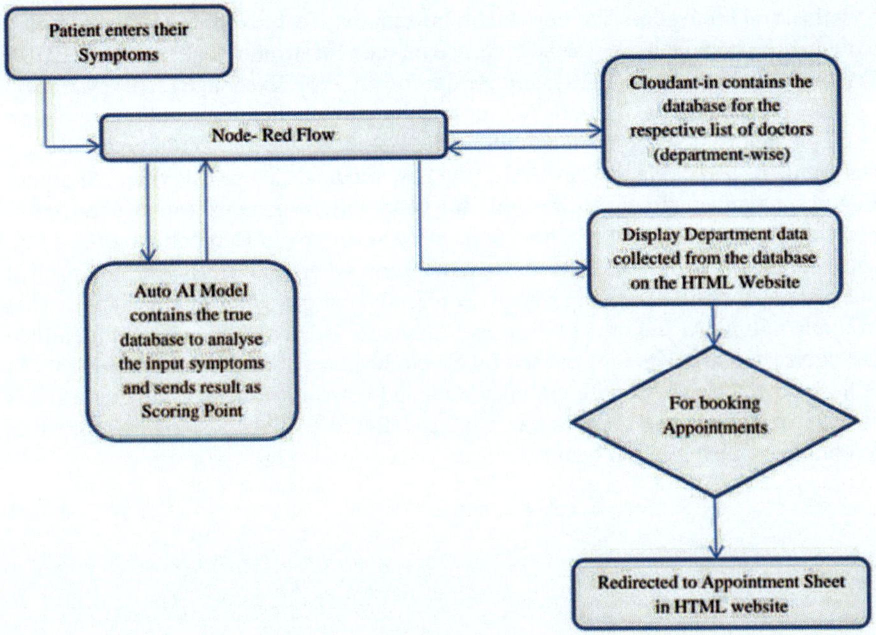

Fig. 6 Working of the smart healthcare assistant

and medical department. Once the patient enters all the details and click on the confirm button and the appointment gets confirm with the respective hospital and a message is been displayed.

3 Enhancements in Healthcare Industry

AI is one of the ever-evolving terms which are hard to define due to its continuous refinement with advancements in technology. Deployment of AI in health care is a big step in effort to advance health care. Although, AI technologies are grasping consideration in healthcare research, implementation in real-life is not as easy and still has a lot of hurdles on its path. One of which is the regulations as the present standard lacks the capability to evaluate its accuracy, safety, and efficacy of these AI systems.

With respect to this the FDA has been continuingly recognizing companies and software as a part of its continued recognition of machine learning and AI in the software as Medical Category that can be used as an aid in predicting and treating diseases. The field of AI is at its peak fame that people believe will find answers better than humans and revolutionizes the field of health care. The important point that healthcare executives and developers should emphasis on as Collaboration, Open

Systems, and Innovation. The key Health Information Technologies (HIT), according to them to be deployed over the next decade include Electronic Health Record (EHR), Personal Health Record (PHR), and Health Information Exchange (HIE) systems.

They projected that by 2020, 80% of health care provider organizations will have implemented EHR systems in the US, and 80% of the general population will have started using PHR systems in the US [2]. A medical device Biovitals Analytics Engine, manufactured by Biofourmis for heart failure patients which uses Artificial Intelligence (AI) and machine learning was approved October 3 by the FDA, according to a new release from the manufacturer where the Biofourmis claims that this new device will have a big role in decreasing emergency hospital visits. Another example of how AI has proved itself is through the IBM Watson, where it identified the rare secondary leukemia caused by Myelodysplastic Syndrome in Japan [3]. AI is thus expected to deal with complex data and close to real-life clinical questions for improved results. Hence, researches have started endeavors along this direction to obtain promising initial results.

4 Conclusion

As this project is developing, we are looking forward to also incorporate Personal Health Record (PHR) into the model, making the website more robust and interactive along with securing this sensitive and confidential data with end to end encryption and an independent authentication method to make sure only authorized personnel and the patient has access to it to make it all in one smart healthcare assistant and make this a remarkable assistive tool for patients and doctors.

As we venture into this, we realize that it requires deep investment of time into the training model, to gather accurate data and to form a realistic approach toward making AI and ML in healthcare efficacious, helping doctors tackle and treat any disease that comes by starting a new era of health care and medicine. Whether the solution provided to the users is truly fit to support its purpose can only be determined by putting it into practice and, simultaneously, continuous training.

References

1. Gervais C (2019) https://www.medicaleconomics.com/news/promise-ai-unlimited-healthcare-what-can-we-actually-do-today
2. Lo'Al AT (Senior Member, IEEE), Mehmood R (Senior Member, IEEE), Benkhlifa E, Song H (Member, IEEE) (2016) Mobile cloud computing model and big data analysis for healthcare applications. https://ieeexplore.ieee.org/stamp/stamp.jsp?tp=&arnumber=7576619
3. Reynolds KA (2019) FDA approves AI—powered monitoring device. https://www.medicaleconomics.com/news/fda-approves-ai-powered-monitoring-device
4. Official Website to access Node Red https://nodered.org/

Usage of Web Mining for Sales and Corporate Marketing

Shivani Yadao and A. Vinaya Babu

Abstract The usage of web has grown to a big extent recently. Web is not only a source of information but is also a great platform to run a business successfully. Some of the major usage of web in E-commerce is retail, sales, and marketing areas. One of the technologies which serve as a tool in these areas is web mining. The process of data mining is to extract the required patterns of information in big database. Web mining is one of the categories of data mining. It uses data mining techniques for extracting interesting information from web. The main goal of this paper is to study the web mining in detail and to learn its usage in corporate sales and marketing.

1 Introduction

There are three different categories of mining the information: data mining, text mining, and web mining. To deal with structured data in the database is data mining. Text mining is to deal with text or unstructured data. Web mining deals between semi-structured data and unstructured data on the web. The automatic extraction of the information from the web documents or services can be done using web mining. It deals with many data types such as contents of web pages, user access information, hyperlinks between pages, and various web resources.

There are three different categories of web mining, which includes.

Web Content Mining (WCM): Finding the valuable information from web has become quicker using WCM. The process of information retrieval from web into a structured form by indexing of information is WCM. It includes web documents which may consist of text, multimedia documents, and html. The information here

S. Yadao (✉)
Computer Science and Engineering, Lincoln University College, Petaling Jaya, Malaysia
e-mail: Shivaniyadao123@gmail.com

A. Vinaya Babu
Computer Science and Engineering, SCETW, Hyderabad, India
e-mail: Avb1222@gmail.com

Lincoln University College, Petaling Jaya, Malaysia

S. C. Satapathy et al. (eds.), *Communication Software and Networks*, Lecture Notes in Networks and Systems 134, https://doi.org/10.1007/978-981-15-5397-4_7

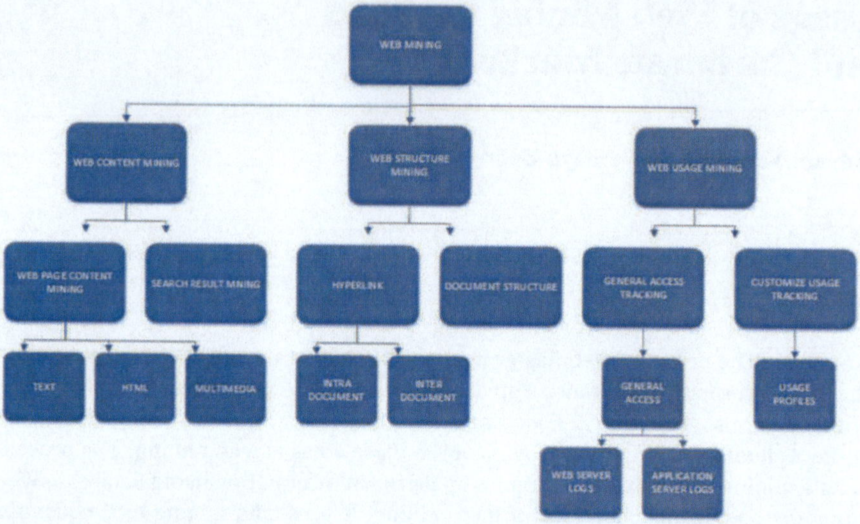

Fig. 1 Classification of Web mining

can be structured or unstructured/semi-structured. WCM uses many algorithms and tools such as Cluster Hierarchy Construction Algorithm (CHCA), Genetic algorithm, and Correlation algorithm. Mozenda, ontology-based tools, screen-scrapper, Web Info Extractor (WIE), web content extractor, and automation anywhere are content mining tools (Fig. 1).

Web Structure Mining (WSM): In the structure of a particular website, the study of interconnected data is the WSM. WSM contains the web graph which consists of the hyperlinks as edges that connects two related pages and web pages or web documents as nodes. WSM is about extracting some of the interesting web graph patterns like complete bipartite graphs, social choice, co-citation, etc. There are two ways it can be performed with: intra-page level or inter-page level. The hyperlink that is used to connect to a different part of the same page is intra-page hyperlink, whereas the hyperlink that is used to connect two different pages is the inter-page hyperlink. Web page is organized in tree structure format depending on HTML tags. The documents here are automatically extracted by the Document Object Model (DOM), Max flow Min cut algorithm, Hypertext Induced Topic Search (HITS) algorithm, Page rank algorithm and ECLAT algorithm are the algorithms of WSM.

Web Usage Mining (WUM): The process of analyzing the behavior of online users is called web usage mining or web log mining. The tracking can be categorized into two: the first is general access tracking and second is customizing usage tracking. Predicting the behavior of customer on web and identifying the users while they interact with web is called the general access tracking. There are three different locations in which web log resides: web proxy server and client browser and web server log. This log contains plain text file (.txt) only. The web log file contains a large amount of irrelevant data. WUM is used in extracting the data which is stored in referrer logs, server access logs, error logs, and agent logs.

Some of the basic data mining algorithms used by WUM are association rule mining, sequential rule mining, clustering, and classification. The tools used are web SIFT, web usage miner, KOINOTITES, speed tracer, INSITE, Archcollect, i-Miner, i-JADE web miner, AWUSA, Web Quilt, SEWeP, webTool, MiDAS, STRATDYN, web mate, DB2Intelligent miner of Data, Poly Analyst version 6.0, WebLog Miner, Clemetine, WEBVIZ, and WEBMINER.

2 Related Work

Sharma and Lodhi [1] worked on a web mining technique in extracting information from web data stream. Their study identified the issue of discovering the relevant information from the abundant information present on the web. The main problem is in identifying how to process the raw data to gather information regarding the website use and filtering the search results in order to present only rules and patterns. To mitigate the abovementioned issues, the data mining method based on the decision tree algorithms is proposed in this paper. The algorithms were developed for mining the web logs temporarily. The proposed method was able to provide useful information for generating the log files and extracting information, rules, and patterns from the web data stream.

Patil and Khandagale [2] reviewed the issue of huge data content on the WWW. A large amount of data present on the web makes it difficult to retrieve some useful information in an easy way. Therefore, the paper presented the standard way for a web developer to identify the user behavior on the server side by accessing the log files. The web log files containing the information about the web navigation are obtained and for proposing the web mining techniques. The predicted behavior of usage was helpful to users to provide efficiency and automated updated links for reducing the time of developer.

Bull et al. [3] provided an open source Software Architecture for Mental Health Self-Management (SAMS) framework. The main purpose of the SAMS is to analyze an individual's usage of computer, for the purpose of storage security, to deliver the non-invasive method. This research was entirely based on the evaluation of the medical professionals that linked both the data and text to shortages in cognitive domains of the dementia features. Along with this, the paper discussed the concern in the previous research and focused on the implementation of text and collection components.

Ristoski and Paulheim [4] defined the Data Mining, and Knowledge Discovery in Database KDD methods with a large amount of data This method performed by applying many approaches that integrated Semantic Web data and Knowledge discovery process This research analyzed an application domain in biomedicine and life science The Linked Open Data (LOD) was used for build content-based recommender systems, but still, the KDD methods were unblocked to be used. An example was also included on how to use the Linked Open data. The paper provides an overview of the approaches used in Knowledge discovery process through different stages.

Raiyani and Pandya [5] provided an overview of the complete preprocessing techniques including the data cleaning, session identification, and user identification activities. The paper introduces a new technique called Distinct User Identification (DUI) which is based on Agent and session time, IP address, and the preferred page on specific time. The Distinct User Identification (DUI) algorithm has been processed which gets the efficient result of log files while web usage mining procedure. The technique helps to mitigate the issues of user security by encountering fraud detection, detection of regular user access behavior, terrorism, detection of unusual access to the confidential data, etc. thus the proposed method enhances the performance and designing of upcoming access of preprocessing step outcome.

Sukumar et al. [6] investigated various data mining techniques and classified them into the effective way and with limitations. The effectiveness and limitations of all the reviewed algorithms are explained in depth that helps to identify an effective technique for the data mining. The web contained a large amount of data, which enhanced the volume and dimension of data day by day. In fact, it was a "Web Structure Mining" method. Several heuristic and preprocessing algorithms were applied by using programming language. The unique users can be then identified using the obtained data. Even session identification is possible when the user identification is successful. However, there was a lack of accuracy for getting a better result in the collected data.

Malviya and Agrawal [7] provided a glance of numerous applications of web usage mining. The paper defines how web usage mining becomes a dynamic region of study in data mining. Defined the Web Usage Mining technique related to the log data files that extracted user's performance, that they used with different applications such as Pre-fetching, Modified services, E-Commerce, etc. Web log data usually has confusing and noisy behavior, but yet preprocessing and pattern analysis technique was needed to be determined. While doing the web usage mining with the web log data files, there were various problems occurred like desirable information not found, personalizeddata was not analyzed, and didn't find the associated information of the gathered data. All the stages in the web usage mining are described in detail along with the problems encountered by various researchers in the same field.

Sunena and Kaur [8] summarized the concept of data mining. For this, comparison of various mining algorithms including the data type generalization was done. Most of the content of the web mining and its categories were based on the Internet. It exposed the comparison between the pattern discovery technique by using its parameters and provided web services to the user. A large amount of data was added to the source in every second, but still, there is a need to discover more the web service related to the user. Along with this, the comparison between the pattern discovery techniques has been provided.

3 Web Mining for Sales and Corporate Marketing

Unlike business having a limited physical space and fixed customers, e-commerce has it all on web. The e-commerce websites have the feature of selling, advertising, and giving different services to the customers online. This is a great advantage of reaching the large customers easily without the concern of time and place. Because of these advantages e-commerce has become a big platform for the business areas like banking, retailing, sales, medical services, transportation, communication, and education. However, a successful business on web can be established only if the website design principles and web engineering techniques are taken care of. To start building the e-commerce website, a big amount of information has to be defined before. For example knowing if the website supposed to attract new customers or increase the sales of current customers, to identify business goals and the way those websites will target those goals, identify the most suitable tools and techniques need to be used/followed in order to target those requirements and to identify if the proposed website will increase the business overall profit. Reflecting a good image of the business and improving it over web is an important part of retail websites.

An important field of study to maintain and improve the functionality of a website is the usage of data mining. Patterns extracted after applying the data mining techniques on data available on web can be used in maintaining websites by improving their usage through simplifying navigation of user and information access and also improving the structure and content of the website in a way to meet the requirements of both user and website owner which will in turn increase the overall profit in the business.

4 Web Usage Mining Techniques

In this paper, we discuss some of the web usage mining techniques that can be used in domains like sales. Some of the data mining techniques in web usage mining are statistical analysis techniques, classification, clustering, association rule mining, and sequential pattern mining. The process of statistical analysis is to apply statistical methodologies on web log files in describing user sessions and navigation, such as in guessing what is the time and length of viewing a navigation path. Statistical prediction is used in predicting when some document or page would be accessed from that particular time. The process of clustering is to partition a given population of items or events into similar element's set. Two main clusters to discover in web usage mining are: pages clusters and usage clusters. Based on some characteristics, to divide an existing set of transactions or events into other existing classes or sets is classification. With respect to the navigation patterns in web usage mining, the classification can be used in grouping users into existing groups in order to develop user profiles that belong to a particular category or class. To discover the attribute values that very often occur in a given set of data is association rule mining. The main usage of web usage mining techniques in association rules mining is either to show within the same user session which pages tend to be visited, or in finding the pages that are usually viewed together. To improve the site search performance, a re-ranking method with the help of website taxonomy is used. It mines for abstract access patterns and generalized association rules of different levels. Next approach mentioned in this section for web log accesses prediction is dependent on association rule mining. This type of mining facilitates the usage of web personalization by identification of navigation patterns or related pages. Determining the sequence of events and actions with respect to time or other sequences is called sequential pattern mining. It can be used in web usage mining in predicting user's visit behaviors in future. There are different analysis tools used for the purpose, such as SpeedTracer and Webminer.

From the above discussion, it can be concluded that the patterns that are extracted by applying different web mining techniques and methodologies on web data can be used in maintaining websites by enhancing their usability through simplification of information accessibility, user navigation, and improving the content and the structure of the website. Thus meeting the requirements of both website users and owners which will in turn increase the profitability of the overall business.

References

1. Sharma S, Lodhi SS (2016) Development of decision tree algorithm for mining web data stream. Int J Comput Appl 0975 8887, 138(2)
2. Patil SS, Khandagale HP (2016) Enhancing web navigation usability using web usage mining techniques. Int Res J Eng Technol IRJET 4:6
3. Bull C, Asfiandy D, Gledson A, Mellor J, Couth S (2016) Combining data mining and text mining for detection of early stage dementia the SAMS framework. LREC Workshop RaPID Portorož Slovenia
4. Ristoski P, Paulheim H (2015) Semantic web in data mining and knowledge discovery a comprehensive survey. Preprint submitted to J Web Semantics
5. Raiyani AG, Pandya SS (2013) Discovering user identification mining technique for preprocessed Web Log Data. J Inform Knowl Res Comput Eng 2(2).ISSN: 0975-760
6. Sukumar P, Robert L, Yuvaraj S (2016) Review on modern data preprocessing techniques in Web Usage Mining (WUM). In: international conference on computational systems and information systems for sustainable solutions. 978-1-50901022-6/16/IEEE
7. Malviya BK, Agrawal J (2015) A study on Web usage mining: theory and applications. In: International conference on communication systems and network technologies. 9781-4799-1797-6/15/IEEE
8. Sunena, Kaur K (2016) web usage mining-current trends and future challenges. In: International conference on electrical, electronics, and optimisation techniques (ICEEOT). 978-14673-9939-5/16/IEEE

A Bird Eye View on the Usage of Artificial Intelligence

P. R. Anisha and Anulekha Polati

Abstract This paper gives a clear cut idea on a question "What actually is AI". In this paper you will come across introduction to AI, different types of AI, comparison between machine learning, deep learning and natural language processing, and applications of AI. In future, AI machines will replace humans in many areas. In last two decades, AI has improved rapidly as it has enhanced human life in many areas and it has become the central core study within computer science. This paper concludes by evaluating the future scope and importance of Artificial Intelligence. The main aim of his paper is to provide a quick idea on AI for beginners.

1 Introduction

"Intelligence", which is exhibited only by the humans, is a boon to the mankind. This intelligence in man leads in finding solutions to many complex problems. One such aspect is providing human intelligence to a computer or making a computer-controlled robot, or software that thinks similar to humans which is known as "Artificial Intelligence". "Artificial Intelligence" can be understood easily from its name, "Artificial"—man-made thing and "Intelligence"—it is considered intelligent if they exhibit behavior similar to human beings. This term was coined by John McCarthy, in 1956 in his conference "Machine Intelligence". He is a computer scientist who is working on Machine Intelligence. The General Purpose problem solver(GPS) program was the first version which was developed by Newell and Simon in 1957. Currently, there are many inventions which are stepping closer to Artificial Intelligence. To the early 1950s, people are not much familiar about Artificial Intelligence. However, with an increasing number of real-world applications over the last few years, interest in AI has been increasing rapidly. In the twenty-first century Artificial

P. R. Anisha (✉) · A. Polati
Stanley College of Engineering and Technology for Women, Hyderabad, India
e-mail: anishanaidu.pushpala@gmail.com

A. Polati
e-mail: anulekhapolati246@gmail.com

S. C. Satapathy et al. (eds.), *Communication Software and Networks*, Lecture Notes in Networks and Systems 134, https://doi.org/10.1007/978-981-15-5397-4_8

Intelligence (AI) has become an important area of research in almost all departments: engineering, science, education, medicine, business, finance, marketing, accounting, stock market, economics, and law. AI plays a vital role in designing machines which mimic human activities such as feelings, decision-making using past experiences, and solving the complexity of problems. AI can also be termed as Machine Intelligence, as its main aim is to create intelligence in machines to fulfill the growing needs of the current scenario. Artificial Intelligence (AI) plays a key role in software development process in the creation, and study of computer systems which outputs some form of intelligence. It is believed that the Artificial Intelligence will become the future. Some of the best programming languages for the Artificial Intelligence field are: Python, Lisp, R, Prolog, Java, etc. Among these Python is considered as the best AI development language [1–3].

AI researches generally use two approaches for creating intelligent machines, namely bottom-up and top-down approach. In bottom-up approach Artificial Intelligence is achieved by creating electronic replicas of human brain. In top-down approach, Intelligence is achieved by computer programs which mimic human brain. It is believed that mind gains its power from its ability to reason symbolically. In brain, knowledge is stored as set of symbols and reasoning is done by manipulating these symbols. As AI is completely based on logical thinking and reasoning, mathematics (algorithms) plays a vital role in development of Artificial Intelligence machines [4–6].

1.1 Foundations of AI

Commonly used Artificial Intelligent techniques and theories are rule-based, statistics, theory of probability, fuzzy logic, decision theory, etc. Since AI has interdisciplinary nature, its foundations are in various fields such as.

Mathematics: AI system uses many logical methods and Boolean logic, probability theory, fuzzy logic, modeling, optimization, linear Algebra, Graph theory, etc.

Neuroscience: This medical science help in studying the functioning of brain. Intelligent machines are developed using complex neural networks. Such systems require parallel computation, remapping, and interconnections to a large extent.

Control theory: Machines can change their behavior according to their surrounding environment. Thermostat regulator, water-flow regulator are some of the examples. This theory helps in building systems that translate from initial state to goal state with minimum energy.

Linguistics: Speech is very important in demonstrating human intelligence. Thought should be taken place to analyze a human language. Languages and thoughts are believed to be interlinked.

1.2 Types of Artificial Intelligent Machines

In the past couple of years, Artificial Intelligence has gained incredible momentum. The present intelligent systems are capable of managing massive amount of data and solving complexity of huge calculations. But these are not sentient machines. Different types of AI have come into existence so as to assist other artificial intelligent machines to work smarter.

Reactive machines:

These are the most basic form of AI and are purely reactive, they do not have ability to create memory and use past experiences to advice current decisions. These are designed for specific jobs. This type of intelligence involves the computer understanding the world directly and acting on what it sees. It doesn't rely on any internal concept of the world. These machines will behave in the same way every time when they come across the same situation [7, 8].

Example:

1. IBM's deepblue which defeated Kasparov at chess.
2. Google's Alpha Go which triumphed over human go championship.

Limited memory AI:

This kind of AI uses past experience and the present data to make decisions. Limited memory AI is mostly used in self-driving cars by detecting the movement of vehicles around them and static data such as traffic lights, lane marks, and curves in the road if any. It takes maximum of 100 s to take decisions thereby avoiding accidents.

Theory of mind AI:

This type of AI is a very advanced technology which can analyze and understand human emotions."Theory of mind" itself represents the understanding of people and things in the world that can have emotions and thoughts which affect their own behavior. This technology is expected to change the world and make our work smarter.

Self-awareness AI:

Self-aware AI is a companion of the theory of mind. This type will be super-intelligent, sentient, and conscious. These types of beings are aware of them, are able to predict feelings of others and know about their internal states.

1.3 Subsets of AI

Artificial Intelligence is divided into three subsets:

1. Machine Learning (ML).

2. Deep Learning (DL).
3. Natural Language Processing (NLP).

2 Machine Learning

With massive amount of exponential growth in AI machine learning is becoming the most trending field of the twenty-first century. Machine learning is considered a subset of AI which provides systems the ability to automatically learn from the environment without being explicitly programmed. Machine learning mainly focuses on development in computer programs which can access the previously stored data and predict the future results. It is given by Arthus Samuel in 1959. Machine learning explores the construction and study of algorithm that can learn from and make predictions on data. These algorithms overcome following strictly static program instructions by making data-driven predictions through building a model from sample outputs. Machine learning is closely related to data modeling and is primarily concerned with development of techniques and algorithms that make computers learn by themselves. The process of learning begins with analyzing data or past experiences and making the predictions on present situations. During this process, system stores the previous experiences in its memory and analyzes the stored data when needed [9].

The two types of machine learning methodologies that are generally used are "Inductive Methodology" and "Deductive Methodology". Inductive Machine Learning methodology's major focus is on extracting information from data automatically by statistical and computational methods. On the other hand, deductive machine learning methodology is mainly concerned with deducing new knowledge from already existing knowledge. The utility of machine learning can be gauged from the wide range of applications such as detection of monetary frauds, bioinformatics, stock market analysis, natural language processing, and game playing.

2.1 Types of Machine Learning

Depending on the nature of outcome, machine learning algorithms are classified into the following categories: Supervised learning, unsupervised learning, and reinforcement learning (Fig. 1).

2.1.1 Supervised Learning

It is a task of learning a function that maps an input to an output based on example input–output pairs. In this type of learning system is trained by providing labeled dataset which contains the details of both input and output. Supervised learning

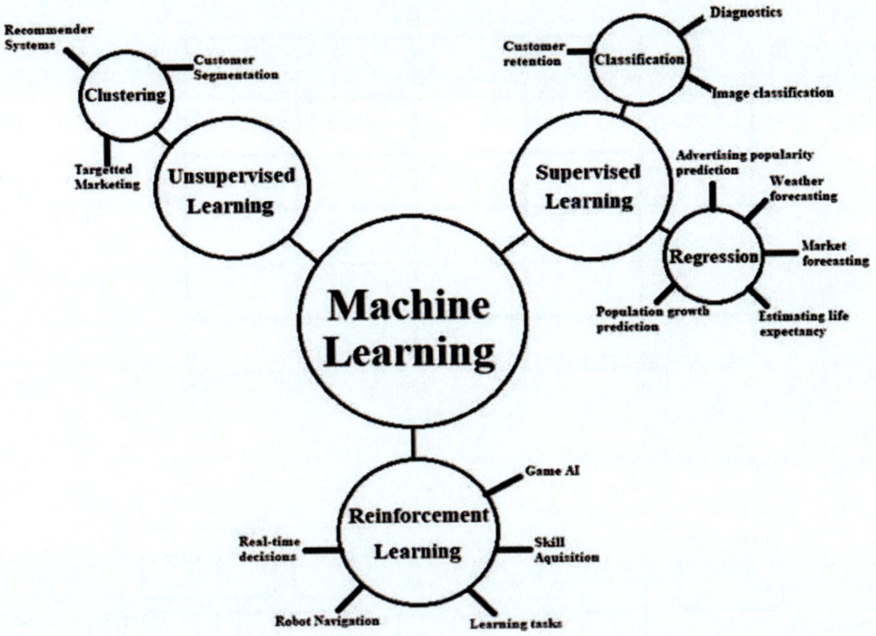

Fig. 1 Types of machine learning

algorithm analyzes the training data and produces an output. It is achieved only with proper guidance and supervision. Example: Athlete taking guidance from his coach.

Supervised learning can solve mainly two categories of problems such as: Regression and Classification. The main difference between them is output for regression is continuous while for classification is discrete.

Regression: Regression algorithms try to estimate the mapping function (f) from input (x) to continuous output variables (y). In this case 'y' is the numerical value.

For example, when system is provided with dataset about houses, and asked to predict their prices then it is a regression task because prices will be continuous output (Fig. 2).

Classification: Classification algorithms try to estimate the mapping function (f) from input (x) to discrete output variables (y). In this case 'y' is categorical in nature.

For example, when the system is provided with the dataset about the climate, and asked to predict whether it will rain or not? In this kind of problem the output can be categorized into: yes or no. These types of problems are considered as classification problems (Fig. 3).

Algorithm choice:

A large number of supervised learning algorithms are available each having its own importance. Supervised learning is not limited to single algorithm, there are different

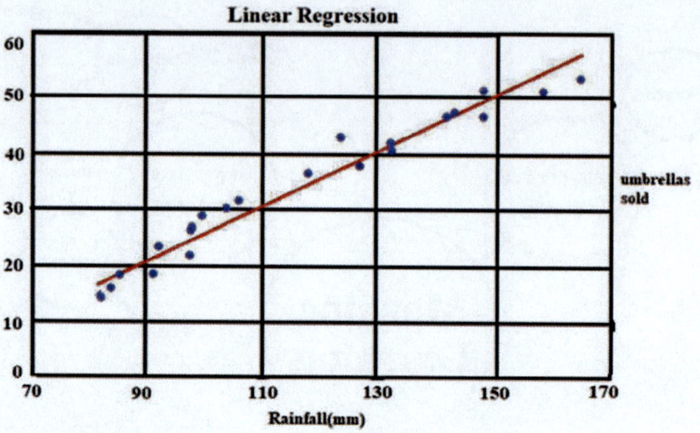

Fig. 2 Regression graph (example)

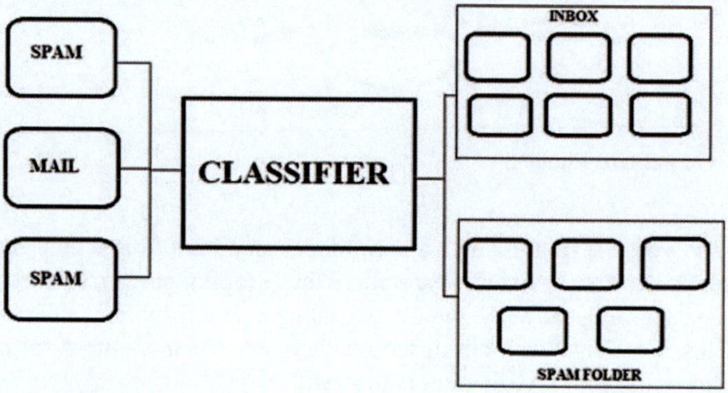

Fig. 3 Classification (example)

algorithms to solve different supervised learning problems. There are four major issues in supervised learning for algorithm choice:

1. Bias-variance tradeoff.
2. Function complexity.
3. Dimensionality of the trading space.
4. Noise in the output values.

1. **Bias-variance tradeoff**: The tradeoff between bias and variance is the first issue. A learning algorithm for a particular input has high variance. A learning algorithm with low bias must be flexible so that it can fit data well. But if it is more flexible each training dataset will fit in different ways and hence have high variance.
2. **Function complexity and amount of training data**: The available training data is relative to the complexity of the true function. If true function is high complex

then the function will only be learnable using a flexible learning algorithm with low bias and variance and from a large amount of training data. If true function is less complex then an inflexible learning algorithm with high bias and low variance will be able to learn it from small amount of data

3. **Dimensionality of input space**: If the input feature vectors have high dimension, then the learning problem can be difficult even if the true function only depends on a small number of those features. High input function requires tuning the classifier to have low variance and high bias.
4. **Noise in the output values**: If the desired output values are often incorrect then the learning algorithm fails to find a function that exactly matches the training examples.

2.1.2 Unsupervised Learning

In machine learning, the problem of unsupervised learning is to find hidden structure in unlabeled data. Unsupervised learning algorithms learn by its own without any guidance or supervision by adopting and discovering, based on the input pattern. In this type of learning algorithms, all input observations are given to the system and no output observations are available. It is the process of training AI algorithm using unclassified or unlabeled information and allowing algorithm to act on that information to make predictions about the output. In this learning the data is partitioned into different clusters and hence it is called as clustering algorithm.

Clustering:

It is considered to be the most important unsupervised inductive problem. Clustering is the process of grouping data or objects based on their similarities and differences. Each cluster contains similar objects, which are dissimilar to objects belonging to other clusters. These type of algorithms deals with more complex models (Fig. 4).

Applications:

1. Market segmentation.
2. DNA classification.
3. Medical records.
4. Cancer diagnosis.

Market segmentation: Companies have huge databases of customer's information. So, unsupervised learning algorithms can look at customer's dataset and automatically group customers and market segments into different market segments. So that the company can efficiently sell or market the different market segments together.

2.1.3 Reinforcement Learning

This learning is based on output with how an agent ought to take actions in an environment so as to maximize some notation of long-term reward. A reward will be

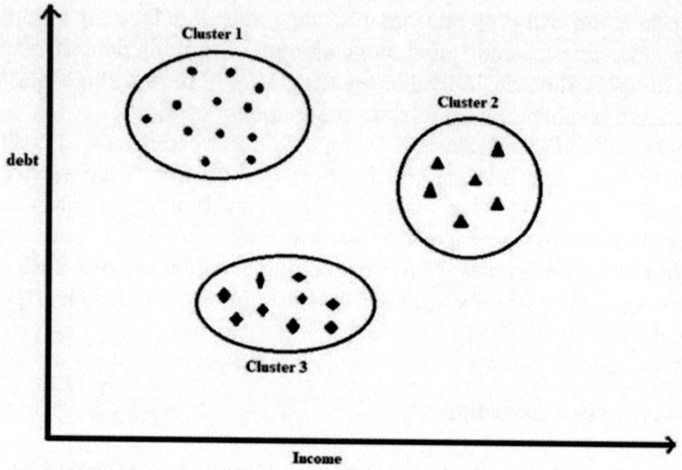

Fig. 4 Clustering (example)

awarded for correct output and a penalty for the wrong output. In this type of learning input depends on the actions we take. In reinforcement learning no predefined data is given to the agent, it starts its work from the scratch. It is inspired by behaviorist psychology (Fig. 5).

Applications:

1. Traffic forecasting service.
2. Computer games.
3. Semantic annotation of learning environments.
4. Stock market analysis.

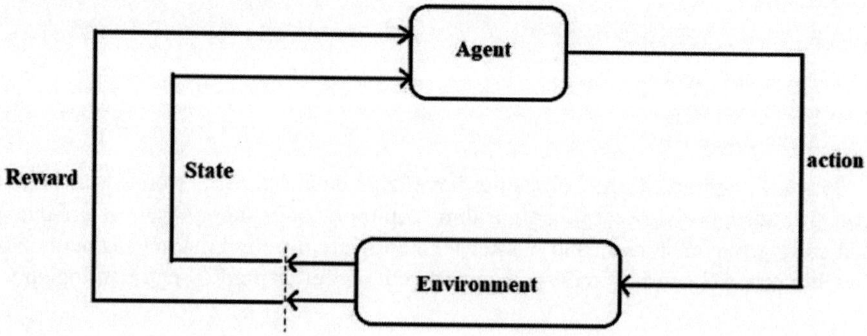

Fig. 5 Reinforcement learning (example)

Traffic forecasting service: The traffic management has become a tough task due to heavy traffic. Machines can be trained and used to solve this problem by predicting about future traffic conditions and provide users with routing options based on that information.

3 Deep Learning

Deep Learning is one of the subset of machine learning. The term Deep Learning was given by Rina Dechter in 1986. It emerged from the limitations of machine learning, as machine learning is not convenient with high-dimensional data consisting of large number of inputs and outputs. This learning enables a computer to analyze things and make decisions from its past experience. Deep learning also includes a very interesting concept of feature extraction. Feature extraction is a concept of learning certain features even if we don't provide features explicitly. The basic motivation behind deep learning is human brain in which "neuron" is the basic functional unit. In order to mimic a human brain we need to have such structures which work like neurons. This is fulfilled by perceptron.

3.1 How Does a Human Brain Work

Human brain has dendrites from which input is given. These input gets summed up in the cell body and axon fires the signals to the next neuron. The distance between these neurons is called synapse. If the signals from the neuron exceed a certain limit, only then the signal fires out. By taking this as guidance artificial neuron called perceptron is created (Fig. 6).

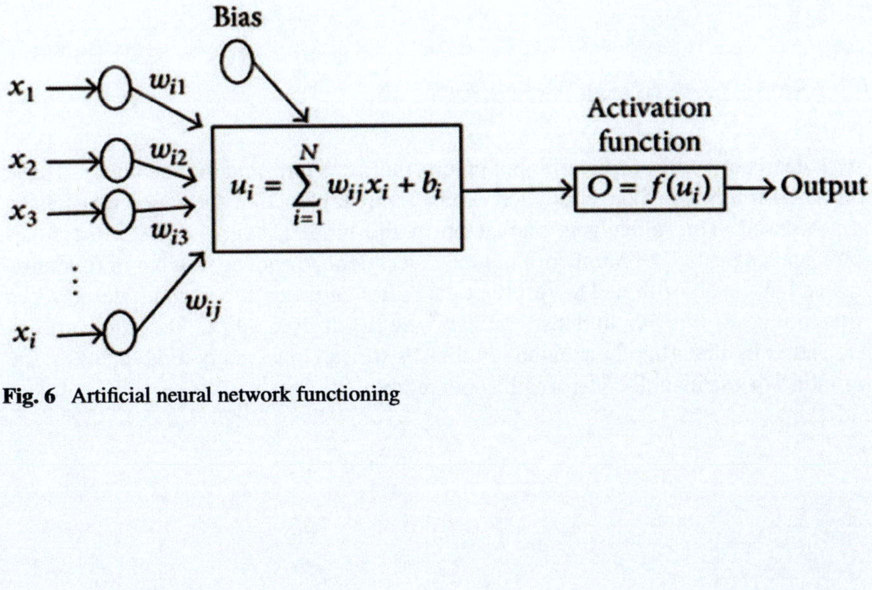

Fig. 6 Artificial neural network functioning

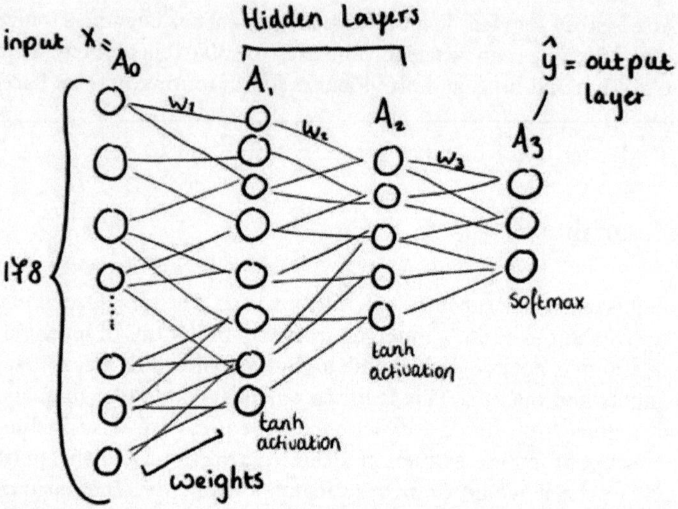

Fig. 7 Deep network layers

Here inputs are $x1, x2, x3, xn$ and $w1, w2, w3, wn$ are weights. These are multiplied and summed together. The result is sent to activating function. Activation function provides a threshold value. So, only about a threshold value a neuron will fire. There can be different artificial functions such as sigmoid function, step function, and sine function. Our human brain works on sigmoid function. If these outputs doesn't match with the real outputs, then the weights get updated based on the difference between outputs and weights. This process repeats and this continues until we get real outputs. This process of updating is called backpropagation method. Deep learning is implemented by deep networks. The combination of multiple perceptrons forms artificial neural network. The neural networks with multiple hidden layers are called deep networks (Fig. 7).

3.2 *Structure of Deep Networks*

These deep networks consist of input layers, output layers, and hidden layers. These layers are interconnected with each other. Input layers take the input that has to be processed. This input gets passed on to the hidden layers where some functioning takes place. The number of hidden layers depends on the number of functions involved in the algorithm. The resultant data after entering into each hidden layer is called output. If this output doesn't match with the desired output, the same process is repeated by updating the weights. In this, as we can have many hidden layers, the limitation of the machine learning has overcome.

Applications:

1. Self-driving cars.
2. Voice control assistant.
3. Automatic language translator.

4 Natural Language Processing (NLP)

Natural Language Processing (NLP) is a term consisting of three words: "Natural"—which is normal and exists in nature, "Language"—that we commonly use to communicate with each other, "processing"—this refers to how that natural language functions. NLP is mainly concerned about the interactions between computers and human. As we know that computers cannot figure out the human language. So, there comes NLP which processes the natural language and ensures that computers understand it.

Example:

1. Google assistant.
2. Sentiment analysis.
3. Machine translation.

4.1 Types of Natural Language Processing (NLP)

Natural Language Understanding (NLU)

Natural Language Generation (NLG) (Fig. 8).

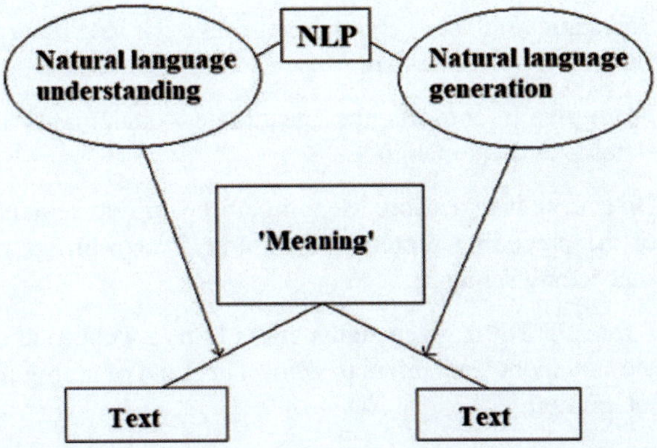

Fig. 8 Classification of natural language processing

Natural language understanding (NLU):

This process attempts to read and understand the meaning behind the given text. This is the most typical part in AI. It splits the given input sentences into words and understands what it actually means. Natural language often develops ambiguity (which means a statement containing more than one meaning), for example, if the given input statement is "call me a cab", this statement has two meanings 1. person addressing someone to call a cab for him, 2. person asking someone to call him as cab. This type of ambiguity occurs in NLU process.

Natural language generation (NLG):

The ambiguity occurred in Natural Language Understanding (NLU) can be solved in Natural Language Generation process (NLG). After understanding the meaning behind the given input statement NLG forms a response to the given statement in comprehensible language that human can understand. NLG systems respond to human with clear and intelligent meaning through deep learning algorithms.

4.2 *Steps for Achieving Natural Language Processing (NLP)*

Lexical analysis: The lexicon itself means collection of words or phrases. It breaks the given input statement into words (Fig. 9).

Example: If the given statement is—"I Eat mango" is divided into three words: I, eat, mango.

Syntactic analysis: It just forms the sentences. It takes words from the lexical analysis step and form sentences which may or may not be meaningful.

Example: In the above-given statement the three words I, eat, mango forms the statements like: mango eat I, eat mango I.

Semantic analysis: From syntactic analysis step it borrows the statements and forms a meaning full statement.

Example: It converts the meaningless statement "mango I eat" to meaningful statement "I eat mango".

Discourse integration: Meaning of the current sentence depends upon the meaning of the preceding sentence. Similarly, it also brings meaning for the immediately succeeding sentence.

Example: If the given statement is "surya went to akash and said he is hungry", in the statement "he" refers to who? This kind of ambiguity can be solved by discourse integration.

Pragmatic analysis: In this step whole process is re-interpreted and the final response is given.

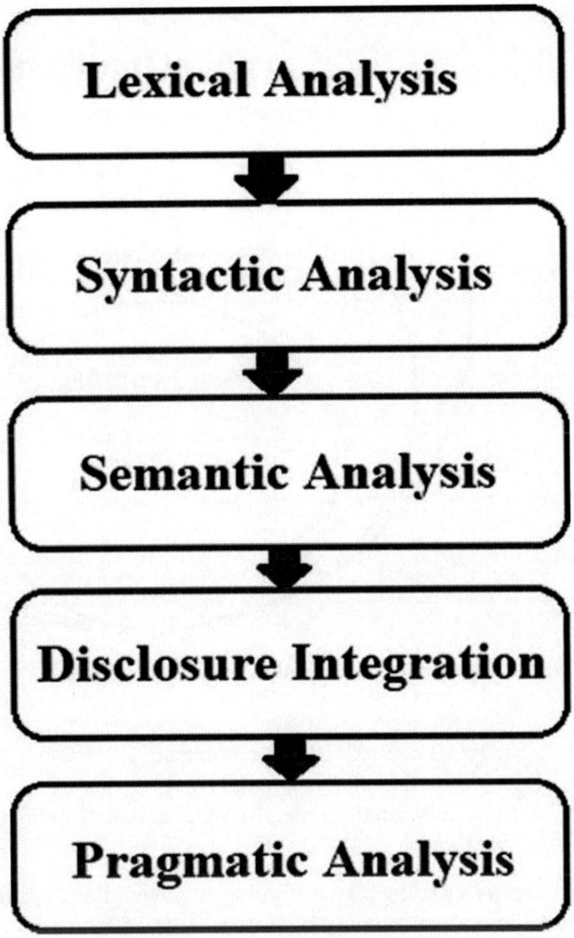

Fig. 9 Steps for achieving natural language processing

5 AI Versus Machine Learning Versus Deep Learning

Artificial Intelligence: It is a broader umbrella under which machine learning and deep learning include. It represents a technique which duplicates the human behavior and takes its decisions from past experiences. Artificial Intelligence can be trained to accomplish specific tasks by providing large amounts of data and recognizing pattern in them (Fig. 10).

Machine Learning: This enables a machine to learn without explicitly coded. Machine learning is found to have many limitations, such as

- It is not capable of working with high-dimensional data which consists of large inputs and outputs.

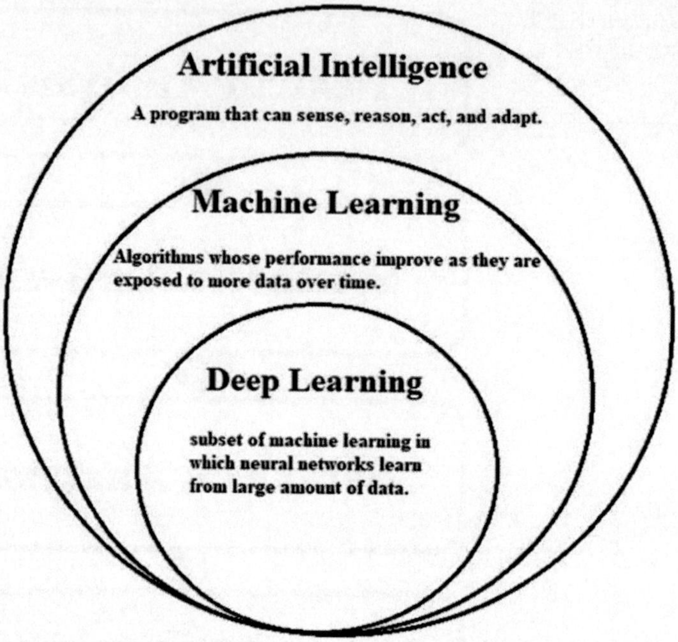

Fig. 10 AI versus machine learning versus deep learning

- Feature Extraction (for the concepts like object recognition and handwriting recognition this plays a crucial role).
- Cannot solve many crucial concepts like Natural Language Processing, Image recognition.

Deep Learning: Deep learning achieves all the limitations of ML. It also provides us with feature extraction. It is capable of focusing on the right features by them providing a little guidance from the programmer. It partially reduces the problem of dimensionality.

6 Applications of AI

Artificial intelligence has applications in almost all fields. The estimated use of AI in different industries has increased rapidly.

6.1 To Improve Hospital Impatient Care

One of the first successful applications of AI is Clinical Decision Support Systems (CDSS), which primarily focus on the diagnosis of patient condition. Mycin an expert system which identifies the bacteria causing infection and recommends the perfect antibiotics to treat these infections. AI is commonly used in detection of tumors, various types of cancer, and heart condition in medical images using MRI scanning.

6.2 College Enquiry Chat Box

This is built using artificial algorithms that analyzes the given queries and understands the input message. This system is an online based web application which provides answer for the queries. It has built-in AI which answers all the queries related to the college. The user does not have to go to the college personally for enquiry. Every student is provided with a chat box in which they can put their queries and no specific format is required for chatting. The system provides answer to the queries as if it is answered by person. Student can ask any question related to college like date of examinations, time of the event etc. This application saves time for both students and faculty.

6.3 Heart Disease Prediction

You might have come across the incidents like person getting stroke and you are in need of doctor immediately, but they are not available due to some reasons. The heart disease prediction system is a web application where user can search for a doctor at any time and get the prescription required for the treatment. This application is fed with various information about the disease and allows user to get the instance guidance. User can share his/her heart disease and get the perfect diagnosis for their health by the intelligent machine. This is very useful in case of emergency.

6.4 Artificial Intelligence Dietician

This application is helpful for people who are very conscious about their diet plan. It is an online artificial dietician which acts similar to the real dietician. Dietician treats people based on their daily activities, height, and weight. This system asks the person all the data, process it, and advice the appropriate diet plan for the user. This application reduces the maximum time for the users.

6.5 Question Paper Generator System

This system is provided with set of questions and their answers for option ticking. It also provides the student with weightage and complexity of the questions which is stored in the database. Once student selects the complexity level of the questions, he/she will be provided with set of questions of the chosen complexity level. In this type of system there is no chance of paper leakage. This system provides unbiased results and saves the time and energy of the administrator.

6.6 Automatic Answer Checker

This application checks and marks written answer similar to the humans. This system is built to check the answers of online examinations and allocate marks to the user as soon as he/she marks the answer. This system is provided with questions and answers, which is stored in the notepad. As soon as the user marks or writes the answer in the given area, it compares with the original answer and allocates marks accordingly. The system analyzes and allocates the marks instantly by saving time and resources.

6.7 Gaming

AI is one of the most commonly used technologies in gaming industry. Games like chess work on this technology, this machine is not much intelligent as humans but they use a brute force algorithm which scans 100s of places and decides for a next move. Artificial intelligence provides the features to the games and solves the complexity of the problem.

7 Conclusion

Artificial intelligence is the theory and development of computer systems to perform tasks requiring human intelligence such as visual perception, speech recognition, decision-making, and translation between languages. This paper is based on artificial intelligence and its applications. There is a tremendous growth in artificial intelligence over last few decades. We have many new inventions based on artificial intelligence like high-tech locking techniques, smarter homes, self-driving cars, humanoid robots like Sofia, speech recognition—Siri, and many more. AI is always coming up with new ideas and innovative products. But, this is not the end; the scope for the artificial intelligence is very wide as it solves the existential questions. Artificial intelligence will continue to play an important role in coming years. The present world is demanding many such solutions as AI is the most promising and profitable field. So, it is likely that AI spreads its roots broader in the coming years.

References

1. Das S, Dey A, Pal A, Roy N (2015) Applications of artificial intelligence in machine. Int J Comput Appl 115(9):0975–8887
2. Oke SA (2008) A literature review on artificial intelligence. Int J Inf Manag Sci 19(4):535–570 (University of Lagos, Nigeria)
3. Anand S, Sinha A, Tiwari U, Research assistance by Sidharth Ray the centre for internet and society. Artif Intell Lit Rev (India)
4. Eugénio Oliveira, Applications of intelligent agent-based systems
5. Shankari KH, Thirumalaiselvi R, A survey on using artificial intelligence techniques in the software development process
6. Shankari KH (2014) Int J Eng Res Appl 4(12):24–33, (Part6). ISSN: 2248-9622
7. Sam Olds: WRTG 3014-April 24, 2014. A literature review of artificial intelligence
8. Borana J, Applications of artificial intelligence & associated technologies
9. Pannu A (2015) Artificial intelligence and its application in different areas. Int J Eng Innov Technol (IJEIT) 4(10)

Early Detection of Diabetes Using Machine Learning Algorithms and Internet of Things: ADPA

P. R. Anisha and C. Kishor Kumar Reddy

Abstract This paper presents another choice tree calculation Advanced Diabetes Prediction Algorithm (ADPA), for the early expectation of diabetes-based absolutely at the datasets The datasets are amassed through the use of IOT Diabetes Sensors, suits of 15000 data, out of which 11250 certainties are utilized for instruction cause and 3750 are utilized for testing intention. The proposed set of guidelines yielded an exactness of 94.23% and slip-ups charge of 5.77%. Thus, the proposed set of standards is as contrasted and existing methods.

1 Introduction

This paper introduces a brand new decision tree algorithm names Advanced Diabetics Prediction Algorithm (ADPA), which uses dataset amassed the usage of IOT Diabetes Sensors [1–11]. Initially, the algorithm is skilled with 75% of the facts and further examined with 25% of the records. In the proposed set of rules, division factors are evaluated based on the C program language period variety. To select the excellent division point, attribute selection degree data gain is followed. The overall performance measure of the proposed set of rules is evaluated using accuracy and error charge. Further, the proposed set of rules is as compared with present published algorithms.

1.1 Our Unique Contributions in This Paper Are Listed Beneath

a. The version is successful to expecting diabetics based totally on the subsequent attributes: Pregnancies, Plasma Glucose, Diastolic Blood Pressure, Triceps

P. R. Anisha · C. K. K. Reddy (✉)
Department of Computer Science and Engineering, Stanley College of Engineering & Technology for Women, Hyderabad, India
e-mail: Kishor23@gmail.com

S. C. Satapathy et al. (eds.), *Communication Software and Networks*, Lecture Notes in Networks and Systems 134, https://doi.org/10.1007/978-981-15-5397-4_10

Thickness, Serum Insulin BMI, Diabetes Pedigree, Age accompanied by way of elegance label: Diabetes, No Diabetes more successfully.

b. A specified evaluation against different prediction selection tree algorithms is carried out that provides a truthful assessment to expose the effectiveness of the brand new model.
c. The new model is computationally green and makes it suitable for small devices including android environment.
d. The proposed model is evaluated with diverse performance measures which includes accuracy and error fee.

The rest of the paper is organized as follows: bankruptcy two introduces Diabetics Prediction Algorithm, Sect. 3 illustrates and focuses on consequences, dialogue, and bankruptcy, and Sect. 4 concludes the paper followed by references.

2 Advanced Diabetes Prediction Algorithm

2.1 Procedure for Evaluating the Split Points and Decision Tree Generation

a. **Read the training dataset T.**
b. **Sort T in ascending order and choose the initial attribute along with the associated class label.**
c. **Evaluate the Division Points, as shown in Eq.** 1.
   i. Evaluate the interval range for choosing the division point.
   ii. If there is a change in the class label for interval range, evaluate the Division Point and the midpoint of changed class labels is the Division Point. For instance, Let V be the initial record and V_i be the second record then division point will be as follows:

$$\text{Division Point} = \text{Mid} - \text{Point}\ (V,\ V_i) \tag{1}$$

d. **Choose the Division Point 1 and apply information gain attribute selection measure and evaluate the information gain value and continue this for all the Division Points obtained for initial attribute and the procedure is as follows:**
   i. Initially, consider attribute and along with its associated class label and evaluate attribute information gain and it is shown in formula (2).

$$\text{Attribute Entropy} = \sum_{i=1}^{N} P_j \left[-\sum_{i=1}^{M} P_i \log_2 P_i \right] \tag{2}$$

where P_i is the probability of class entropy belonging to class i. Logarithm is base 2 because information gain is a measure of the expected encoding length measured in bits.

ii. Further, consider class label and evaluate class information gain and is as follows:
Class information gain is a measure in the information theory, which characterizes the impurity of an arbitrary collection of examples. If the target attribute takes on M different values, then the class entropy relative to this M-wise classification is defined in formula (3).

$$\text{Class Entropy} = -\sum_{i=1}^{M} P_i \log_2 P_i \tag{3}$$

where P_i is the probability of class information gain belonging to class i. Logarithm is base 2 because entropy is a measure of the expected encoding length measured in bits.
Now, compute the information gain: it is the difference of class entropy and attribute entropy and is shown in formula (4).

$$\text{Entropy} = \text{Class Entropy} - \text{Attribute Entropy} \tag{4}$$

e. **The maximum Information Gain is the best Division attribute and becomes the root node, shown in formula 5.**

$$\text{Best Division Point} = \text{Maximum (Entropy)} \tag{5}$$

f. **Finally, if the number of attributes is N, we will get N best division points for individual attributes. As decision tree is a binary tree, there will be only one root node and for this reason, among the N Information Gain values choose one best Information Gain value to form the root node.**

g. **Now, consider the maximum Information Gain value attribute as the root node and take its split point and divide the tree in binary format, i.e., keep the values which are lesser to split point at the left side of the tree and keep the values which are greater and equals to the right side of the tree, and continue the process till it ends with a unique class label.**

3 Results and Discussion

For the experimentation, dataset with 15000 realities and eight qualities are assembled utilizing IOT Diabetes Sensors. At first, the arrangement of standards is taught with 75% of the records and moreover tried with 25% of the realities. In the proposed set of principles, division elements are assessed basically dependent on the interim assortment instead of each time there is an exchange inside the class name. To choose

Table 1 Accuracy comparison with existing approaches

Model name	Accuracy (%)
Random forest (RF)	85.55
Bagging	85.33
Decision tree (DT)	85.09
ANN	84.53
Boosting	84.09
Naïve Bayes (NB)	81.01
SVM	87.6
DPA	93.8
ADPA	94.23

the five-star division point, trademark determination degree actualities increase is pursued. The arrangement of principles is coded the utilization of Net Beans IDE and accomplished in intel i3 processor, four GB RAM.

The precision of the proposed model Advanced Diabetes Prediction Algorithm (ADPA), appeared in Table 1, is in correlation with the current techniques: Random Forest, Bagging, Decision Tree, Artificial Neural Networks, Boosting, credulous Bayes and Support Vector Machines separately. The proposed model yielded a precision of 94.23% higher, while as contrasted and past methodologies. The pictorial portrayal of precision assessment is demonstrated in Fig. 1. The bumbles charge of the proposed model Advanced Diabetes Prediction Algorithm (ADPA), appeared in Table 2, is in examination with the present approaches: Random Forest, Bagging, Decision Tree, Artificial Neural Networks, Boosting, innocent Bayes, and Support Vector Machines individually. The proposed model yielded a missteps pace of 5.77%,

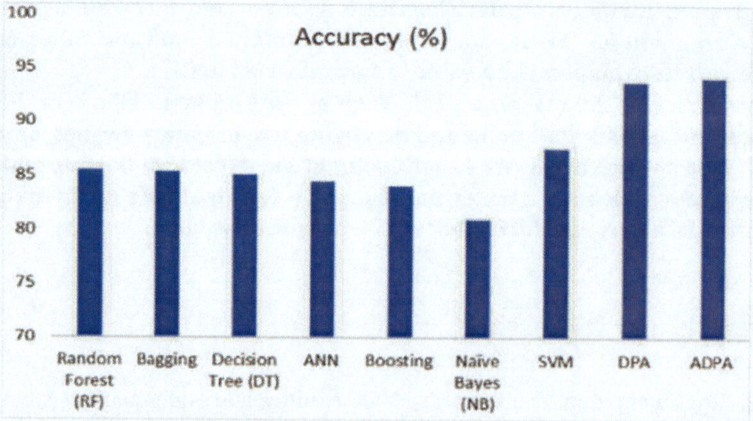

Fig. 1 Accuracy comparison with existing approaches

Table 2 Error rate comparison with existing approaches

Model name	Error rate (%)
Random forest (RF)	14.44
Bagging	14.66
Decision tree (DT)	14.91
ANN	15.46
Boosting	15.90
Naïve Bayes (NB)	18.99
SVM	12.4
DPA	6.2
ADPA	5.77

better when as contrasted and past systems. The pictorial outline of mix-ups charge correlation is demonstrated in Fig. 2.

Further, proposed ADPA is compared with Decision Stump, Hoeffding Tree, Naïve Bayes, and Simple Logistic Algorithms using the data collected from IOT

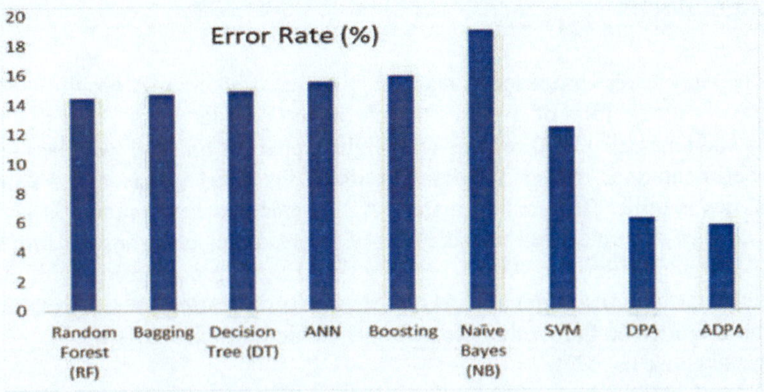

Fig. 2 Error rate comparison with existing approaches

Table 3 Accuracy comparison with other algorithms using WEKA

Model Name	Accuracy (%)
Decision stump	78
Hoeffding tree	87.36
Naïve Bayes	79.36
Simple logistic	79.14
DPA	93.8
ADPA	94.23

Table 4 Error rate comparison with other algorithms using WEKA

Model name	Error rate (%)
Decision stump	22
Hoeffding tree	12.64
Naïve Bayes	20.64
Simple logistic	20.85
DPA	6.2
ADPA	5.77

Diabetes Sensors in terms of accuracy, and the results are shown in Table 3. Here, we used WEKA tool for finding the accuracies of existing algorithms.

Further, proposed ADPA is compared with Decision Stump, Hoeffding Tree, Naïve Bayes, and Simple Logistic Algorithms using the data collected from IOT Diabetes Sensors in terms of error rate, and the results are shown in Table 4. Here, we used WEKA tool for finding the error rate of existing algorithms.

4 Conclusions

In this paper, creators proposed Advanced Diabetes Prediction Algorithm (ADPA) for the early recognition of diabetes the utilization of Internet of Things Diabetes Sensors and amassed 15000 records with eight properties finished two class names. For experimentation, 75% of the data is utilized for tutoring reason and 25p. Cof the realities is utilized for testing reason. ADPA calculation is as thought about in expressions of exactness and mistake cost with present calculations the utilization of WEKA gadget. Further, ADPA set of guidelines is in correlation with current distributed calculations. In general, ADPA calculation is beating when in examination with the majority of the methodologies and yielded an exactness of 94.23% and mistakes expense of 5.77%.

References

1. Akolekar R, Syngelaki A, Sarquis R, Zvanca M, Nicolaides KH (2011) Prediction of early, intermediate and late pre-eclampsia from maternal factors, biophysical and biochemical markers at 11–13 weeks. Prenat Diagn 31(1):66–74
2. Alssema M, Vistisen D, Heymans MW, Nijpels G, Glümer C, Zimmet PZ, Shaw JE et al (2011) The evaluation of screening and early detection strategies for type 2 diabetes and impaired glucose tolerance (DETECT-2) update of the finnish diabetes risk score for prediction of incident type 2 diabetes. Diabetologia 54(5):1004–1012
3. Farran B, Channanath AM, Behbehani K, Thanaraj TA (2013) Predictive models to assess risk of type 2 diabetes, hypertension and comorbidity: machine-learning algorithms and validation using national health data from Kuwait—A cohort study. BMJ Open 3(5):e002457

4. Faust O, Acharya R, Ng EY-K, Ng K-H, Suri JS (2012) Algorithms for the automated detection of diabetic retinopathy using digital fundus images: a review. J Med Syst 36(1):145–157
5. Huang G-B, Zhou H, Ding X, Zhang R (2012) Extreme learning machine for regression and multiclass classification. IEEE Trans Syst Man Cybern Part B: Cybern 42(2):513–529
6. Jensen MH, Mahmoudi Z, Christensen TF, Tarnow L, Seto E, Johansen MD, Hejlesen OK (2014) Evaluation of an algorithm for retrospective hypoglycemia detection using professional continuous glucose monitoring data. J Diabetes Sci Technol 8(1):117–122
7. Kalaiselvi C, Nasira GM, Classification and prediction of heart disease from diabetes patients using hybrid particle swarm optimization and library support vector machine algorithm
8. Karthikeyan T, Vembandasamy K (2015) A novel algorithm to diagnosis type II diabetes mellitus based on association rule mining using MPSO-LSSVM with outlier detection method. Indian J Sci Technol 8(S8):310–320
9. Karthikeyan T, Vembandasamy K (2014) A refined continuous ant colony optimization based FP-growth association rule technique on type 2 diabetes. Int Rev Comput Softw (IRECOS) 9(8):1476–1483
10. Kuo RJ, Lin SY, Shih CW (2007) Mining association rules through integration of clustering analysis and ant colony system for health insurance database in Taiwan. Expert Syst Appl 33(3):794–808
11. Jesmin N, Imam T, Tickle KS, Chen Y-PP (2013) Association rule mining to detect factors which contribute to heart disease in males and females. Expert Syst Appl 40(4):1086–1093

Unique Identification Authority of India BOT: An Intelligent Application Using Artificial Intelligence Approach

V. Prasad, M. Vineeth Kumar, K. Viswanadh, and P. V. N. Mahesh

Abstract In the modern world, the growth of technology is exponential. Artificial Intelligence (AI) and Machine Learning (ML) are evolving rapidly day by day. The evolving AI applications are super smart and definitely usable for layman. In such application, normal bots are famous application for conversation. This project includes an implementation of Unique Identification Authority of India (UDAI) BOT (mentioned as U-bot) of virtual assistant for the conversation in the natural language with the customer, uniquely identified by their Aadhaar number where the details of the Aadhaar card can be detected by using Quick Responsible (QR) code scanner. According to the details of the Aadhaar card, the virtual assistant tries to satisfy the customer by responding for every appropriate query raised in the form of either voice or the text. Thereby, the customer can have a conversation to the maximum possible level to which it is programmed.

1 Introduction to Artificial Intelligence

AI is the simulation of human intelligence processes by machines. AI sharpened both the theory and techniques with advanced implementation. Though there is a fast growth in multidisciplinary fields with the embedded AI, but still AI stood stand alone. The development of recent chatbots made an inspiration to study this scenario. Chatbot deals with message versions and styled in its traditional server mechanisms with prefixed data. Cui et al. [1]. Super-Agent (chatbot Name) deals with leverages large-scale and publicly available e-commerce data. It is unique from other bots. Galitsky and Ilvovsky [2] designed an automated customer support or assistance bot for the user in learning product features and for the tasks which it is usable. Dahiya [3] represented the pattern matching, access information system to provide a predefined acknowledgment for the design and categorical implementation. Bani and Singh [4] the implementation of ALICE chatbot system as an application named as college

V. Prasad (✉) · M. Vineeth Kumar · K. Viswanadh · P. V. N. Mahesh
Department of Computer Science & Engineering, GMR Institute of Technology, Rajam, Andhra Pradesh, India
e-mail: prasad.v@gmrit.edu.in; prasad_v542@yahoo.co.in

S. C. Satapathy et al. (eds.), *Communication Software and Networks*, Lecture Notes in Networks and Systems 134, https://doi.org/10.1007/978-981-15-5397-4_11

enquiry bot, an application which helps students to solve all the problems they face during admissions. Rahman et al. [5] developed a cloud-based chatbots program which acts for the future era of chatbot. Kuhn and De Mori [6] a language model which reflects short-term patterns of word use by means of a cache, a 3 g-gram component was tested on samples drawn from the Lancaster-Oslo/Bergen (LOB) corpus of English text for the speech recognition using natural language processing. Li et al. [7–9] converted a spoken utterance into a feature vector with concurrence accurate statistics in the field of natural language processing, interactive application areas to build a morphological processor for language generation.

2 Literature Review

Khilar [10] discusses on the speech to text appreciation and conversions of the developed application in each stage of classification. Mon and Tun [11] use Hidden Markov Model (HMM) which is applied to train and test the audio files to get the recognized spoken word using MATLAB. Prachi and Bhope [12] developed a human–computer interface system using mother tongue and discusses the various aspects used in speech recognition process and analyzes the same using Raspberry-pi technology. Nafis and Hossain [13] designed a text to speech conversion module using MATLAB, where real-time speech to text, can be defined as accurate conversion of words that represents uttered word instantly after speaking. Kaur and Garcha [14] they developed an algorithm to convert speech to text using Punjabi phonetics. This paper introduces and discusses two popular and different noise reduction techniques by using Java module. Kamble and Kagalkar [15] worked on speech synthesis. This approach is of text to speech where the output is given in Hindi. Shetake et al. [16] conducted survey methods on character recognition as well as text to speech and speech to text and showed the study on Optical Character Recognition with speech synthesis.

3 Problem Statement

Creating a web-based U-bot for adding flavor of speech. This website provides a feasible interface to authorize the user, here the user is an Indian, it is validated using Aadhaar. Every Indian has an Aadhaar that contains the QR code, the website provides an interface to read the QR and validate the user. After validating or authorizing the user, it redirects to the web interface where the U-bot exists. The bot needs to interact with the user through speech, implicates as voice input, and provides the valid response in the form of speech. Bot need to satisfy all the queries related to Aadhaar details and make the user satisfiable. Chatbot should start the conversation and it should continue the conversation until the user want to end the conversation.

4 System Design And Implementation

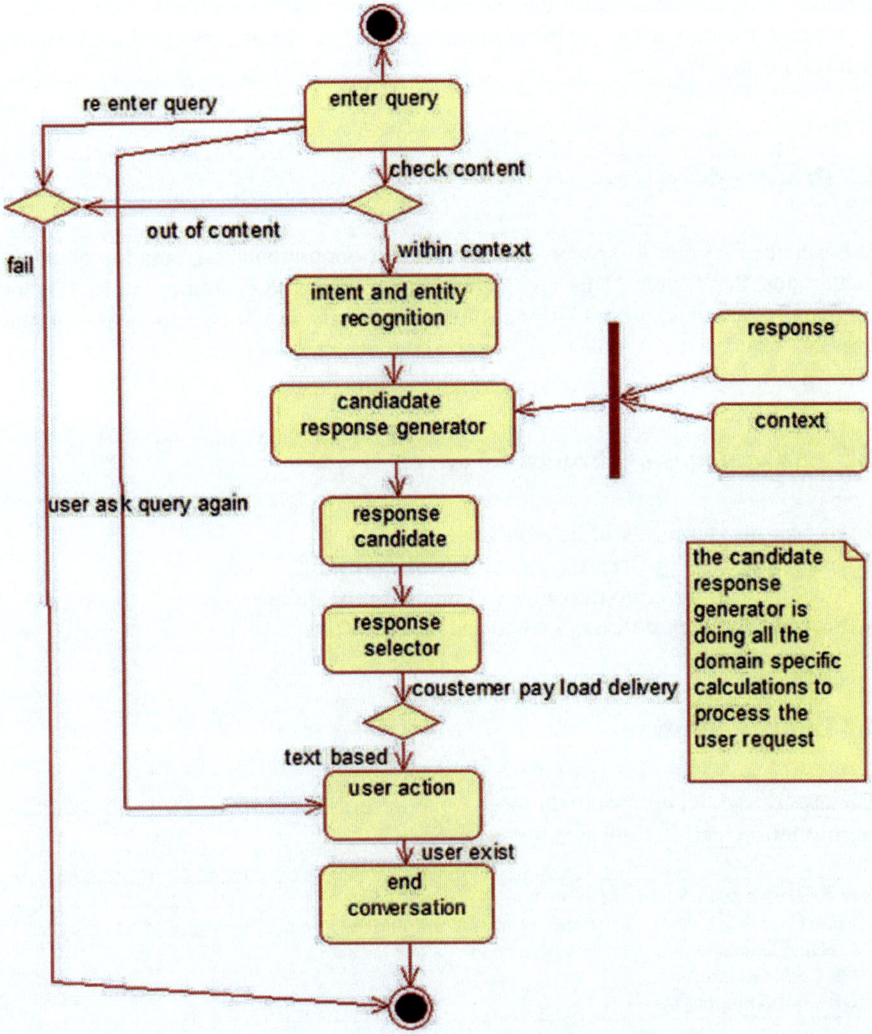

5 Existing System

As per the official website of U Aadhaar services, it doesn't have any chatbot services. The communication between the officials and customer is done through either customer agent service through call or email or offline services which lead to delayed responses.

6 Proposed System

U-bot makes speech to speech and text to text communication possible between human and the system. This application U-bot acts as a customer support agent between Indian citizens and UIDAI Aadhaar Officials. U-bot tries to answer all the queries related to the Aadhaar card details up to its knowledge, i.e., the trained data.

6.1 Advantages of Proposed System

- Human interference will be reduced.
- Available 24 × 7 working hours accessible any time.
- Machine will never go into annoyed mode for the quires.
- Provides the output through voice and cost effective.

7 Implementation

The task is divided into four sub-modules as described below, and the sub-modules are further divided (* Indicates the sub-content)

QR Code Generator and Scanner *Creating Database *QR Code Generation *QR Code Scanning	*Dialog flow Setup* *Creating Intents and Entities *Fulfillment	*Integration with web browser* and *Deploying local server*

QR Generator and Scanner works on the Aadhaar card contains the details of respective citizen, these details are stored in the database server. Accessing government database server is not possible. Hence, manipulation on the database needed at our end considering the local database. Later, the QR Access code accessing the module which needs the QR code for reaching the details of the citizen and thus QR Code Scanning is to authorize the citizen whether he/she is available in Aadhaar database or not. After generating QR code, Scan the QR code, and retrieve the Aadhaar number.

321613251882 321613251883

After execution, a new window with web cam will be popped out displaying the QR code around 30 seconds for all the aadhaar numbers available in the database.

7.1 *Dialogflow Setup And Installation*

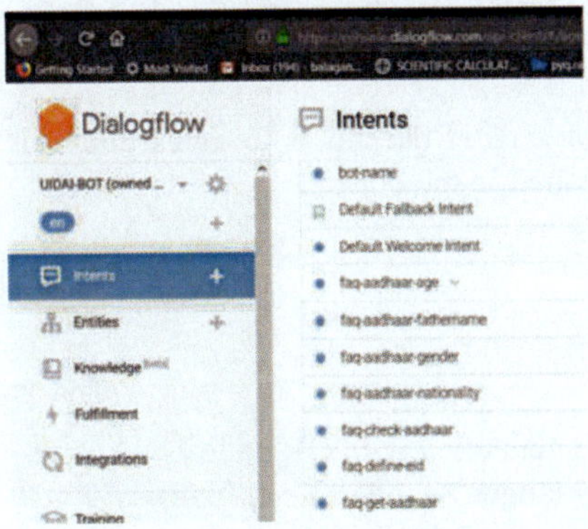

*Access dialog flow

*Extract parameters with entities, which defines the extracted utterances

*Manage the context parameters for the physical display of QR codes

*Integrate with the Google Assistant/ any Virtaul assistant available at your domain, which lets the deployment of Dialog flow agent.

7.2 *Creating Intents and Entities*

In Dialogflow, the basic flow of conversation involves these steps:

- The user input
- Dialogflow agent parsing the input.
- Agent returning a response to the user.

Intent is used to check the responsive progress. Utterances are triggered to extract the responses.

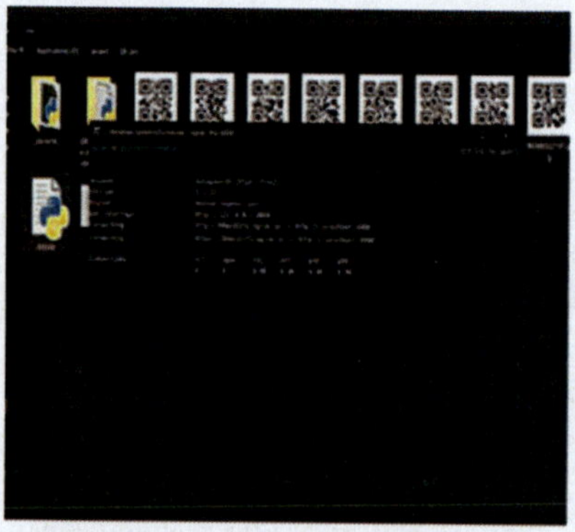

Execute and run the node JS server for generating the index file for seeing this server to listen in port number 5000 as mentioned in the codec.

7.3 *Fulfillment*

All the Aadhaars available in the database are generated with QR code as displayed with the package installed in the programming language codec, i.e., ***pip install Flask*** as NGROK were it will run the code in local server and takes the result into secured server for communication. This flask works at the port 6000 and 4040 exclusively for Text and Speech.

7.4 *Integration with Web Browser*

Execution Impulse to make the RUN state possible as commanded below.

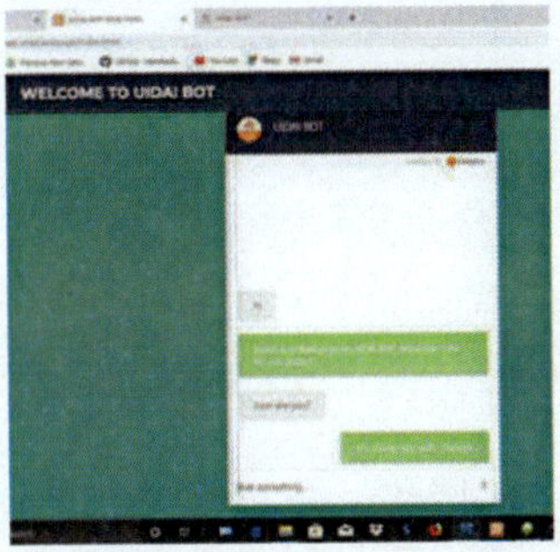

$command = escapeshellcmd ('python D:\project\QR-pics\QR-and-Bar-code-Scanner-and-Generator-master\qrcode3.py');

$output = shell_exec($command);

The figure shows the provision for local host address to the web site and can be used to see that chatbot with speech would be available on the website. This U-bot takes the speech as analog signals and coverts into digital signals using micro-phone and send the information to IO socket, later IO socket converts into text and sends to dialog flow.

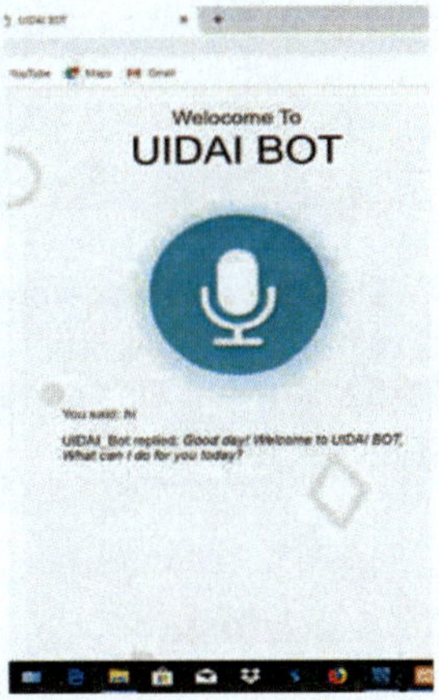

Fig: 5.19

7.5 *Intent Matching*

A typical agent has several intents that represent a range of user observation. Whenever a pop, immediately Dialog flow agent attempts to match the utterance to an intent. Dialogflow matches user utterances to intents using the training phrases you define and the important words, phrases or values you specify within them.

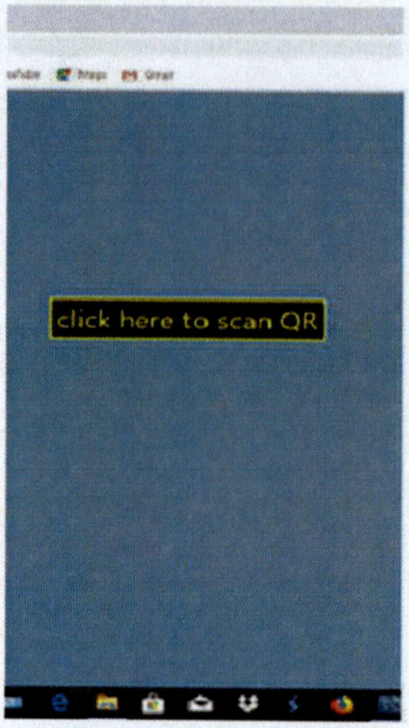

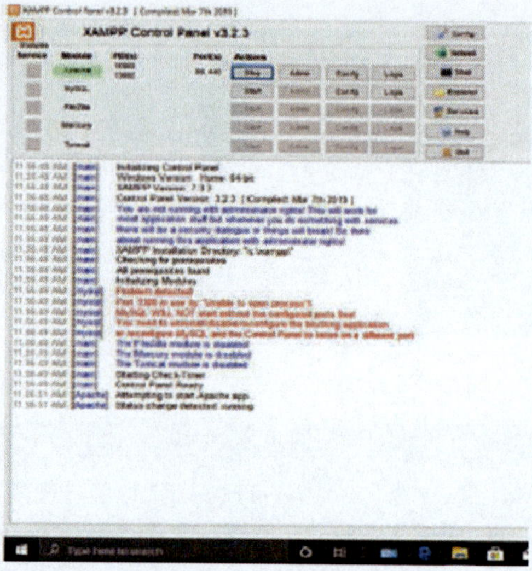

The above figures show the deployment of the server with the dialog flow which is sent as a response int text format. IO socket converts this text format into speech as digital form. This digital signal is converted into analog signal and comes out as output through speakers.

8 Conclusion

Perspective chatbots or smart assistants with artificial intelligence are dramatically changing businesses. There is a wide range of chatbot building platforms that are available for various enterprises, such as ecommerce, retail, banking, leisure, travel, health care, and so on. This work represents to build an UIDAI BOT (U-bot) which helps the Aadhaar clients to know the maximum solutions to the Aadhaar related queries and provides a joyful environment to interact with chatbot through voice as input and provides text and speech as output. It tries to answer the maximum queries up to its knowledge. U-bot can reach out to a large audience on messaging apps and be more effective than humans. They may develop into a capable information-gathering tool soon with various vocal processors.

References

1. Cui L, Huang S, Wei F, Tan C, Duan C, Zhou M (2017) Superagent: a customer service chatbot for E-commerce websites. Association for Computational Linguistics
2. Galitsky B, Ilvovsky D (2017) Chatbot with a discourse structure-driven dialogue management. Association for Computational Linguistics

3. Dahiya M (2017) A tool of conversation: chatbot. Int J Comput Sci Eng (Elsevier)
4. Bani BS, Singh AP (2017) College enquiry chatbot using A.L.I.C.E. Int J New Technol Res
5. Rahman AM, Al Mamun A, Islam A (2017) Programming challenges of chatbot: current and future prospective. In: IEEE region 10 humanitarian technology conference
6. Kuhn R, De Mori R, A cache-based natural language model for speech recognition. IEEE Trans Pattern Anal Mach Intell (IEEE)
7. Li H, Ma B, Lee C-H (2007) A vector space modeling approach to spoken language identification. (IEEE)
8. Bill manaris, Natural language processing: human computer interaction perspective. Elsevier
9. Khilari P, Bhope VP (2015) A review on speech to text conversion methods. IJARCET
10. Mon SM, Tun HM (2015) Speech-to-text conversion (STT) system using hidden markov model (HMM). IJSTR
11. Khilari P, Bhope VP (2015) Implementation of speech to text conversion. IJIRSET
12. Nafis NA, Hossain MS (2015) Speech to text conversion in real-time. Int J Innov Sci Res (IJISR)
13. Kaur M, Garcha L (2017) Speaker independent isolated word speech to text conversion using auto spectral subtraction for punjabi language. Int J Sci Res Publ (IJSR)
14. Kamble K, Kagalkar R (2012) Translation of text to speech conversion for Hindi language. Int J Softw Res (IJSR)
15. Shetake PS, Patel SA, Jadhav PM (2014) Review on text to speech conversion methods. Int J Ind Electron Electr Eng (IJIEEE)
16. Thu CST, Zin T (2014) Implementation of text to speech conversion. Int J Eng Res Technol (IJERT)

CAPTCHA Techniques of Secure Web Authentication: A Survey

Gannavarapu Ananth Kumar and Garg Rishav

Abstract Today, web security is the foremost concern due to the requirement of a validation and verification process that not only authenticates the end user but also prevents the data from various malicious programs. Due to the huge use of internet and electronic devices, the computer bots try to access the authentic data resided at the servers. The CAPTCHA is generally meant for authentication of the websites during login, for securing and getting access of data. A CAPTCHA challenge is kept for doing the conformation whether the process is initiated by an authentic user or by an attacker. If the challenge is successfully chased, it is signified as a human or else a computer bot, and the access will be provided to the right user. Our aim is to make a natural interface between authentic user and web servers so that the communication between the entities will take place in a secured manner. This paper presents a brief summary on CAPTCHA and its necessity in current scenarios. We have discussed a study on various types of CAPTCHAs available, along with their methodologies and the implementation details which help to secure the web data. The paper, moreover, presents some of the important advantages, disadvantages and applications of these techniques and technologies used in different research in addition to the different future scope of CAPTCHAs.

1 Introduction

In the current trend, the usage of the internet became more where the internet can be accessed through many devices. There are many dangerous threats, which are targeting the web data. So, it is time to protect the data from these threats. The main aim of designing the CAPTCHA is to secure the web from the malicious programs

G. Ananth Kumar (✉) · G. Rishav (✉)
G.L.Bajaj Institute of Technology and Management, Greater Noida, India
e-mail: ananth.gannavarapu@gmail.com

G. Rishav
e-mail: rishav0127@gmail.com

S. C. Satapathy et al. (eds.), *Communication Software and Networks*, Lecture Notes in Networks and Systems 134, https://doi.org/10.1007/978-981-15-5397-4_12

like bots by which there is a threat to the data inside it. It also helps to prevent script attacks which are automated, by making more expenditure on breaking it [1].

CAPTCHA acts like a doorway for the user to get access to the web data through the authentication process, and CAPTCHA of any type is challenged to the user, for authentication, to get access to the web data. The CAPTCHA is based on the concept of Turing Test where the human will be judged by posting a question and wait for his response. They help to judge whether the user is human or not [2]. It is like a challenge to the user, to prove himself that he is a human, but not a malicious program like bot. If the challenge is successfully traced by the user, then he is recognized as a human and can access the web data, else considered as a bot. Many schemes related to CAPTCHA have been introduced, and still research is going on, to build more advanced CAPTCHAs, which can defend latest malicious programs which attack the web data.

The recognition of CAPTCHA is generally based on HCI (i.e. Human Computer Interaction) [3]. It must not give any trouble to the user to identify it and easily understand too. The most important thing is that they must be understood to only humans but not the bots. Therefore, the designing of the CAPTCHA is not so simple. It needs to undergo many terms and conditions [4] like easy generation, easy evaluation, must not produce overheads in the network, etc. For the actual implementation of CAPTCHA, it needs such an algorithm, which gives the perfect pattern matching with the input provided in the CAPTCHA and the data stored in database. HCI helps the user to meet all the requirements [3] such that it will be more intractable, high performer, accurate and so on in the recognition of data. There are some properties like automated, open, usable and secure that must be carried out by the CAPTCHA during its development.

The increasing interest in the field of web authentication has made the researchers to do a large number of practicals in which some of the extensive work is given in Table 1. These surveys are directly incorporated into the concept, while some are indirectly related. The primary purpose of the paper is to present the ancient and recent developments in the field of web security, especially authentication through CAPTCHA technology. The paper is organized in various components which cumulate the major work of the field. The major components are classification, hybrid and artificial intelligence-based CAPTCHA followed by pros, cons and various applications of CAPTCHAs.

2 Classification of CAPTCHA

CAPTCHA is an ellipsis for 'Completely Automated Public Turning Test to tell Computers and Humans Apart'. For the first time, the patent of CAPTCHA was registered in 1998 [26]. The work of the CAPTCHA is to authenticate the user and also to distinguish between human and bot. The CAPTCHAs, based on the data used

Table 1 Comparative study of CAPTCHAS

S. No.	References	Year	Type	Challenges	Key points
1	Datta et al. [5]	2005	Hybrid	Interactive interface and automated attack are considered for the end user verification	A user-friendly CAPTCHA is proposed which is solved by a two-step process for authentication. The image is analysed by the human on the clicks of the images and user had to click at the geometric centre of the selected image; a set of words are displayed and solved
2	Elson et al. [6]	2007	Image	The speed with respect to accuracy of image-based CHAPTCHAs	A combination of cat and dog images, named Asirra, is presented for the authentication as CAPTCHA. The methodology was very easy for the human to recognize and performe with the accuracy of 99.6% in less than 30 s
3	Von Ahn et al. [7]	2008	HCI	A human-based recognition using web character	This web character CAPTCHA is known by reCAPTCHA which helps to improve the digitization process. This cannot be recognized by the automated OCR. This uses two words in which one is an unknown word and other is a control word. The user is expected to enter both the words to prove himself as human
4	Srikanth et al. [8]	2009	Hybrid	Some complexity was generated in the CAPTCHA using text and image to make it less vulnerable	The proposed approach displayed an image as the CAPTCHA and user typed the answer related to the question posted in the CAPTCHA for the authentication. If the answer is correct and gets matched, then he is considered as a human, else a bot. A timer is used to enhance the security of the method as a bot cannot solve the question in the limited time

(continued)

Table 1 (continued)

S. No.	References	Year	Type	Challenges	Key points
5	Gupta et al. [9]	2009	Text	The segmentation attack and an issue of authentication in web are carried out	The proposed scheme embedded numbers and text so that segmentation attack could not be possible. The identification of numbers and its logical order created trouble for understating the displays for the bots
6	Vimina and Areekal [10]	2009	HCI	The challenge is relied on activity recognition and thinking of human	In this CAPTCHA, the user must recognize the image and need to predict the activity which is being performed in the image. A user is authenticated by his/her ability of thinking and recognition. The user must select the image with a particular activity to prove as a human
7	Imsamai and Phimoltares [11]	2010	Text	3D fonts in the text as CAPTCHA and pixel count attack against the authentication	The proposed scheme used 3D fonts which were rotated by the angle of 45° so that bots could not interact with server easily. The strength of proposed scheme was measured in terms of resistance to pre-processing, vertical segmentation, colour-filling segmentation and pixel count attack
8	Cui et al. [12]	2010	Video	A motion-based CAPTCHA scheme of the web attacks	The proposed algorithm used an animation-based technology with the help of Zero-Knowledge rule. Random generation of the motion of objects was done at multiple passes and carrier video generation algorithm was used for the same

(continued)

Table 1 (continued)

S. No.	References	Year	Type	Challenges	Key points
9	Gao et al. [13]	2010	Puzzle	A puzzle-based CAPTCHA which was language independent	The displayed image is divided into segments based on row and column size. Two images are missed and other existing images will be in zigzag order. The user must find the missing images and swap the zigzag images in an order to make a complete image
10	Chaudhari et al. [14]	2011	Text	3D-text CAPTCHA, solved by drag-drop operation. The web attack is considered as another issue	The paper proposed a 3D-CAPTCHA with drag and drop operation where the characters are displayed in 3D format. The text in the CAPTCHA contained both characters and numbers that are rotated with an angle ranged from −45° to +45°. This is easy for the human to identify and perform the operation but difficult for the bot to do the same
11	Raj et al. [15]	2011	Image	Introduced issue with text CAPTCHA and proposed remarkable solution	As text-based CAPTCHAs are more frequently attacked, the proposed approach protected the segmentation and secured the web data from the unauthorized access. The proposal is implemented by two phases, namely composite image formation process and composite image distortion process
12	Lee et al. [16]	2012	Text	A text-based CAPTCHA with a high distortion rate	The text of the CAPTCHA contained the combination of cases of the characters with high distortion rate. The technique was easy to interpret by the humans but difficult for the malicious programs

(continued)

Table 1 (continued)

S. No.	References	Year	Type	Challenges	Key points
13	Goswami et al. [17]	2014	Image	Image-based CAPTCHA provides the authentication by the human and non-human faces	On a random background, the CAPTCHA contained the combination of original and fake images of faces, and user must pass through all the tests by clicking on all the original pictures of the faces. The method of adversarial gradient learning is used to solve. The algorithm comprised of estimation of parameters and generation of CAPTCHA using training set
14	Zhu et al. [18]	2014	Image	Web security and protection mechanism was the major issue to be considered and also prevention of replay attacks	A new CAPTCHA, namely CaRP, is proposed which is basically a kind of image-based CAPTCHA. It used alphanumeric characters and images of animals. The method countered the guessing attack online based on the probability and protected the channel at transport layer
15	Ali and Karim [19]	2014	Puzzle	Modern Scenarios of web security insisted a proposal for authentication which is based on the analysis of users	A puzzle-based CAPTCHA was proposed where the displayed images were available for the user as a puzzle in the CAPTCHA. A drag and drop mechanism was adapted to follow the correct order of displayed images. The technique labelled all the images, and randomly allocated images were searched in database to present as well as solve the CAPTCHA

(continued)

Table 1 (continued)

S. No.	References	Year	Type	Challenges	Key points
16	Mehrnezhad et al. [20]	2016	Image	The method considered upright oriented and partially visible image as CAPTCHA, human-made unconscious decision making	The system allowed the users to click distorted and upright-oriented pictures on web. The geometric transformation and random rotations are used for pre-processing of images while unconscious decision making of human behaviour is used for solving the CAPTCHAs
17	Dwivedi et al. [21]	2017	HCI	A human interactive approach is proposed which is based on eye gaze and face expression	The proposed work created the difference between humans and bots using eye gaze and face expression based human interactions. The user followed the sequence of gaze points and generated the sequence of expressions using face. So, it is more dynamic, secure and less vulnerable
18	Kulkarni and Fadewar [22]	2017	HCI	An efficient approach for mobile internet users which can solve the challenge with the foot movement	This is based on the capability of walking or creating acceleration in the mobile. The user must carry mobile phones. They are allowed to walk for minimum five steps in order to log in. All types of disabilities except fully motor-disabled person can use this efficiently
19	Tariq and Khan [23]	2017	HCI	Cognitive abilities of user provided image and audio signals in a CAPTCHA	The audio is listened by the user and then match the respective image with the voice. The selection must match to prove as a human. The results of proposed system are easy for the users and efficient too

(continued)

Table 1 (continued)

S. No.	References	Year	Type	Challenges	Key points
20	Uzun et al. [24]	2018	Audio	Automated attack and fake authentication were handled using a secured audio-based CAPTCHA	The users registered themselves through the device for authentication and the server validated their authentication. A CAPTCHA challenge is managed by the server and time is measured. When CAPTCHA is received at the device in audio format, it checks for the user response, in front of camera of the device. The snapshots are mainly relied on the speech recognition as response to the audio is captured
21	Al-Hammouri et al. [25]	2018	Hybrid	Handling of overloading to the server and DDOS attack at application layer	The proposed approach used CAPTCHA as a firewall in which the requests from the web server are redirected to the CAPTCHA nodes to solve the query and prevent from various attacks. The client will be forwarded to the server to get access only if the CAPTCHA has been solved

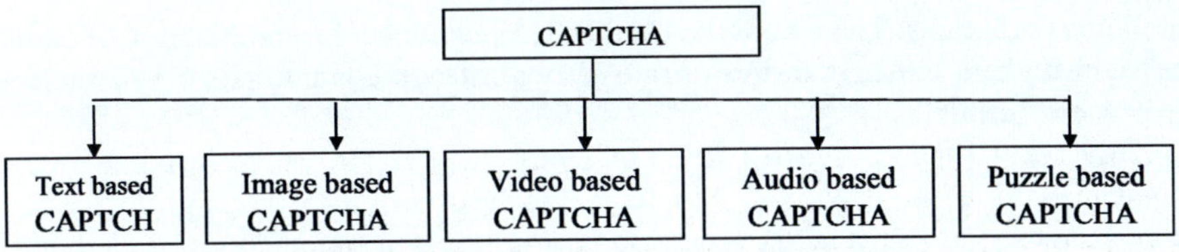

Fig. 1 Various classes of CAPTCHAs

in it, can be categorized as text-based CAPTCHA, image-based CAPTCHA, video-based CAPTCHA, puzzle-based CAPTCHA and audio-based CAPTCHA as shown in Fig. 1.

2.1 Text-Based CAPTCHA

It is one of the easiest CAPTCHAs which can be solved in short time. It contains a text in the form of question as shown in Fig. 2a. The text which is displayed as a CAPTCHA, under the distortion rate, needs to be identified by the user. It is confirmed that the user is a human, only when the given question is solved. This CAPTCHA generally contains text or the numbers or the combination of both, which are needed to be recognized by the humans using the Optical Character Recognition (OCR).

Thakur at el. [27] proposed a Reading-Oriented Overlapping Text (ROOT) based CAPTCHA, which works on the concept of reusability. The security issue of text-based CAPTCHA depends on defending the segmentation attacks and recognition attacks. This CAPTCHA contains the characters containing the alphanumeric, in both computers generated and hand writing. This CAPTCHA can be used many number of times. This CAPTCHA generally focuses on the anti-recognition and anti-segmentation. The CAPTCHA segmentation can be protected by adding noise in the background. The characters of alphanumeric, which are in the range 3–8; are overlapped on the base character, which is a variable that enhances the security. The arcs and dots are added to the CAPTCHA, which act as a noise and become difficult to the bot for identifying them.

Lee et al. [16] proposed a text-based technique for the CAPTCHA with a heavy noise. The user had to recognize the characters in the image that is highly distorted with a tip. This technique was easy to interpret by the humans but difficult for the

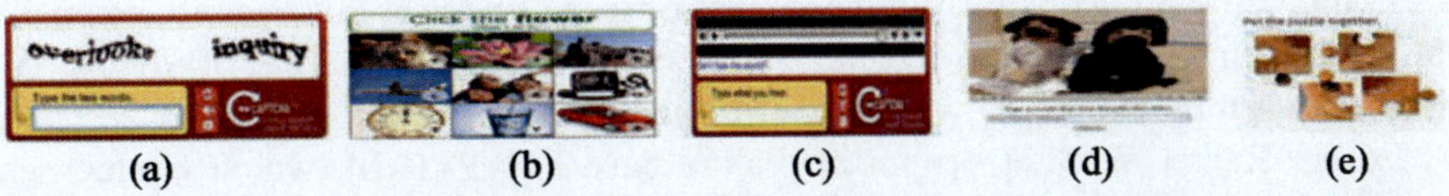

Fig. 2 Prototype of various kinds of CAPTCHA [40] **a** Text CAPTCHA. **b** Image CAPTCHA. **c** Audio CAPTCHA. **d** Video CAPTCHA. **e** Puzzle CAPTCHA

malicious programs. The text of the CAPTCHA contained the combination of cases of the characters with high distortion rate. User must carefully interpret the characters to proceed further.

Gupta et al. [9] proposed a different scheme of CAPTCHA where the numbers are embedded in the text. It leaves two challenges to the users. First is to identify the available characters and next is to trace the logical order on the numbers embedded in it. This paper also generated a new methodology called Sequenced Tagged CAPTCHA (STC) through which the numbers are added to the text-based CAPTCHA. Here, segmentation attack is also not possible because the text and numbers are embedded in one another. The images which are generated are of very good quality and can be used in the websites too for authentication.

2.2 *Image-Based CAPTCHA*

In this type of CAPTCHA, a challenge is kept in front of the user, to recognize a particular image, from the image group, displayed as the CAPTCHA. Image CAPTCHA is based on typical soft computing and intelligence problem. It is difficult to break, as it contains the images, and only can be solved by the human as shown in Fig. 2b.

Kulkarni at el. [22] proposed a very user-friendly design of CAPTCHA for mobile internet users called PEDOMETRIC CAPTCHA. This is based on the walking capability or creating acceleration in the mobile, i.e., the solution of challenge will count the steps taken by the user who is carrying the mobile phone in his/her hand. The count of steps during the walking is incremented by one; on the completion of steps minimally mentioned, the CAPTCHA challenge is solved and user is allowed to login.

Zhu et al. [18] proposed a CAPTCHA with a new security parameter based on unsolved problems of artificial machines. The method, basically, used a graphical password for enabling the CAPTCHA, to enhance its security capability. It countered the guessing attack online based on the probability. The CaRP is a new CAPTCHA where a new image was being generated every time when the user tries to login. It uses alphanumeric characters and images. The CaRP had an advanced protection mechanism which protected the communication channel at transport layer. The CaRP also countered the relay attacks on CAPTCHAs.

Goswami et al. [17] proposed an image-based CAPTCHA where the authentication is done by the human face from the set of database which has the combination of human and non-human faces. On a random background, the CAPTCHA contained the combination of original and fake images of faces, and user must pass through all the tests by clicking on all the original pictures of the faces. The method of adversarial gradient learning is used by which it helps the user to solve it.

Edwin Raj et al. [15] proposed picture-based CAPTCHA which overcomes the problems involved in text-based CAPTCHA. As OCR-based CAPTCHAs are more frequently attacked, the proposed approach adapted a new architecture which protected the segmentation through various parameters like random guessing, edge detection, etc. This proposed scheme consisted of two phases, i.e. Composite Image Formation Process (CIFP) and Composite Image Distortion Process (CIDP). In CIFP, many different pictures will be selected from the database and image transformations will be performed on them like rotation, transparency, etc. by which a composite image is generated

2.3 *Audio-Based CAPTCHA*

In these, the user requires to listen the audio carefully and then needs to answer the given question in the CAPTCHA as shown in Fig. 2c. The response for the question is present in the audio file. The user needs to click on the audio symbol, to listen to the audio clip. This is also a helpful CAPTCHA for the visually disabled people. As they cannot see, they must listen to the audio file and must repeat the same as the CAPTCHA's answer. Uzun et al. [24] proposed a highly secured real-time authentication system by processing the audio signals as the CAPTCHA. The registration is done through the device for authentication, and the connection with server is created. The server will create and send a CAPTCHA challenge to device and measure the time, through the timer.

2.4 *Video Based CAPTCHA*

In this type of CAPTCHA, the user needs to recognize the data present in the video. The video CAPTCHA is a combination of audio, video and sometimes text too. The user needs to solve the question which is asked in the CAPTCHA. The user needs to watch the video completely and answer the question to prove himself as a human as shown in Fig. 2d.

Song chi et al. [12] proposed a CAPTCHA which is based on the identification of the motion of an object. The proposed algorithm used an animation technology where user needs to pass the test of object motion as a CAPTCHA employing Zero-Knowledge rule. Random generation is used in the starting state and continued as a motion in the next state. It can be protected by various attacks. Carrier video generation algorithm is also used for the motion of the objects.

2.5 *Puzzle-Based CAPTCHA*

In this CAPTCHA, the user must combine the parts of the image, in unordered fashion, to form a complete meaningful image, which is displayed as CAPTCHA as shown in Fig. 2e. Ali et al. [19] proposed a CAPTCHA based on the analysis of display pictures where the pictures were available for the user as a puzzle in the

CAPTCHA. The user must drag and drop the available images present in the order in which they are shown as the CAPTCHA. If the user was unable to drag and drop the available images in the given order, then he would be treated as a bot and not human. For the development of this type of CAPTCHA [28], there is no need of any technical requirement like video, audio, etc. Gao et al. [13] proposed a method in puzzle-based CAPTCHA where the image is divided into segments based on row and column size. From the given images as CAPTCHA, two images are missed and other existing images will be in zigzag order. The user must find the missing images and swap the zigzag images in an order to make a complete image.

2.6 *Hybrid CAPTCHA*

The hybrid CAPTCHAs signify the combination of more than one basic kind of CAPTCHA which are discussed earlier. These CAPTCHAs are generated with the amalgamation process of two or more CAPTCHAs. Hybrid CAPTCHAs are not only used to improve the authentication process but also to prevent the various attacks on web.

Al-Hammouri at el. [25] proposed a system for handling the overload of the server and preventing the DDOS attacks. The proposed approach is named as ReCAP in which CAPTCHA is used as a firewall and the requests from the server are redirected to the CAPTCHA nodes to solve the queries passed by the user. Datta et al. [5] proposed an image and text-based CAPTCHA to secure against the automated attack. There was a two-step process of user interface for authentication, which can be solved easily for humans, but for machine, it is expensive. It is based on the imagination of human in which the image is analysed. When the user click at the geometric centre, a distortion is performed, on the image selected, then the set of words are displayed. Shirali-Shahreza et al. [29] proposed a new approach to implement dynamic CAPTCHA. As the users belong to different categories like aged, language, ill, etc., the proposed approach provided the flexibility to choose the type of CAPTCHA from the available which can be solved easily. The best method is used in the final registration process. The multilingual CAPTCHA [30] enabled the users to get the information in their native language after the selection of the language.

2.7 *CAPTCHA Based on HCI*

A CAPTCHA, basically, is used to distinguish a human from the bots. The design of CAPTCHA, i.e. generation and evaluation, must be simple enough to create trouble for bots but easy for human in identification. So, it needs to undergo in such a way that can reduce the network overhead as well as prevent various attacks. Moreover, the actual implementation of CAPTCHA requires an algorithm which gives the perfect pattern matching with the input provided in the CAPTCHA and the data stored in database. In Table 2, some of the comprehensive work has been compared including various parameters like measured attack, technologies used, user comfort level, application complexity level, automation, class along with accuracy and their strength. There are three levels, viz. hard, normal and easy, to be considered for user comfort level and five levels, viz. very slow, slow, average, fast and very fast, for application complexity level.

Human Computer Interaction, abbreviated as HCI [3], helps the user to meet all aforesaid requirements such that it will be more intractable, high performer, accurate and reusable in the recognition of the CAPTCHA. Moreover, some properties like automated, open, usable and secure must be carried out by the CAPTCHA during its development.

Tariq et al. [23] proposed an MSCAPTCHA relied on the cognitive abilities of users. The user needs to match the image displayed with the voice which is listened to prove as a human. This CAPTCHA is based on both the image and the audio. The user must listen to the audio and then match the respective image by which the voice and the image selected must match. Dwivedi at el. [21] proposed an interactive CAPTCHA which used eye gaze and face expression based human interactions so as tohave difference between humans and bots. In this system, commands are generated, and user has to follow the sequence of gaze points and needs to generate the sequence of expressions using face. It enables the dynamic CAPTCHA environment that is more secure and difficult to break. Ahn et al. [7] proposed a reCAPTCHA scheme which helped to improve the digitization process.

3 Pros and Cons

3.1 Pros

CAPTCHA is very much effective to make the web applications secure [28]. It protects the web and cloud data from the unauthorized access by spam and bots. CAPTCHA avoids spam to enter into the website and the authentication prevents them to access the website from malicious programs [31]. CAPTCHA does the differentiation between the humans and other computer programs. It also reduces the possibility of attacks of various dangerous programs by providing the web services with security. It is much useful for the implementation and can be easily embedded in the websites for the protection.

3.2 Cons

Due to more distortion rate, CAPTCHA may become difficult to read and understand [32]. Not all CAPTCHAs are of same type, due to which, there may be some difficulty

Table 2 Comparison of various works considering various parameters

S. No.	Reference	Measured attack	Technologies	User comfort level	App complexity level	Automation	Class	Accuracy	Strength
1	Tariq and Khan [23]	Segmentation and guessing attack	Feature matching, thresholding, image segmentation, crawling, low-level feature extraction	Very easy	Very fast	Fully	Image and audio	92.50%	Cognitive authentication using voice and image
2	Kulkarni and Fadewar [22]	Segmentation and dictionary attack	–	Average	Slow	Fully	Image	92.20%	Use footsteps for authentication
3	Dwivedi et al. [21]	Masquerade and replay attack	–	Easy	Fast	Fully	Image	72.05%	Eye gaze and face expression based authentication
4	Zhu et al. [18]	Guessing, replay and shoulder surfing attack	SHA algorithm, image transformations, CAS and thresholding	Easy	Very slow	Fully	Alphanumeric characters and images of animals	73.60%	Reduce spam emails, password protection, secure communication at transport layer
5	Lee et al. [16]	Segmentation, dictionary and brute force attack	Affine transformation and image warping	Easy	Fast	Partially	Characters, digits and symbols	97%	Malicious text-free channel

(continued)

Table 2 (continued)

S. No.	Reference	Measured attack	Technologies	User comfort level	App complexity level	Automation	Class	Accuracy	Strength
6	Raj et al. [15]	Denial of service and blind attack	Affine transformation, edge detection, thresholding, shape matching and random guessing	Average	Very slow	Fully	Pictures/images	–	Security constraint to web data
7	Gupta et al. [9]	Segmentation and dictionary attack	Median filter, colour-filling segmentation, shearing	Easy	Very Fast	Partially	Digits and characters	–	Segmentation-free authentication
8	Elson et al. [6]	Brute force and denial of service attack	Partial credit algorithm and token buckets	Easy	Slow	Fully	Images of cats and dogs	98.50%	Fast and secure authentication, improve animal welfare
9	Datta et al. [5]	Automated attack	CBIR, FS dithering, integrated, region matching, partitioning and colour remapping	Very easy	Fast	Fully	Image and text	99.54%	Two-Step verification for authentication
10	Vimina and Areekal [10]	–	Random searching, image filtering and image scaling	Easy	Fast	Fully	Image	99.04%	Activity recognition-based authentication

(continued)

Table 2 (continued)

S. No.	Reference	Measured attack	Technologies	User comfort level	App complexity level	Automation	Class	Accuracy	Strength
11	Imsamai and Phimoltares [11]	Pixel count and dictionary attack	Affine transformation, overlapping, distributed noise, colouring and font	Very easy	Very fast	Fully	Alphanumeric text	79%	Authentication free from vertical and colour-filling segmentation
12	Srikanth et al. [8]	–	Fuzzy logic and time-out feature	Easy	Fast	Fully	Image and text	–	Inclusion of timer and analytics in authentication

for some type of users [29], especially for one who is disabled in some manner. Some of the difficulties may arise due to embedding in the web browsers and it takes more time to respond. There will be more effect on the traffic as well as time for decipherment. The background of automation and intelligence are employed more in the CAPTCHA [33]. The CAPTCHA creation is also a risk factor with the parameters in non-distribution environment with respect to the security issue [34].

4 Applications of CAPTCHA

4.1 Online Polling

CAPTCHA prevents the unauthorized access and masquerade of data. During the online polling process, a CAPTCHA is provided to the user so as to make the user authentication such that the voting process is done only by the user but not by a malicious program or bot [35, 36].

4.2 Web Registration

Free registration will be offered by many websites due to which they will be attacked by many malicious programs, and also called as flooding attack. Here, the bots and other malicious programs get registered in such type of websites where there is no security. The CAPTCHA is one type of solution to perform the authentication, during the web registration, by providing security in such a way that only the humans can understand the CAPTCHA and can perform it, for accessing the data, whereas the bots cannot [37–39].

4.3 Dictionary Attacks

Dictionary attacks is a method through which the attacker or hacker tries to access the data by breaking the password. The brute force method will be used with the words present in the dictionary as a password, and try to decrypt the key, to get access. As common words are used as password, a successful access is gotten by the attacker. CAPTCHA prevents this attack as the words in the CAPTCHA are unpredictable, so the dictionary attack is hard to be successful [36].

4.4 MITM Attack

MITM, i.e. man in the middle, attack is an assumption made by attacker to the two end users that they are communicating with each other directly. But, actually, there will not be any suspicion to the end user that there is a man in the middle, who is observing and accessing the data of both the users. By using CAPTCHA as an input parameter in RT MIMT [35] only, the authentication user can answer the CAPTCHA, and the data can be directly communicated, by the users, with high security.

4.5 Phishing Attack

Masquerade attack, where it acts as a trusted party, for getting access to the data, is presented in the network. It is also called as the phishing attack, which is a threat to many companies, whose security is broken and valuable data is accessed. A CAPTCHA with OTP [35] is proposed for eradicating the phishing attack.

4.6 SMS Flooding

Spam is one of the threats that makes the mobile network busy by generating many SMS. This can be avoided by using the CAPTCHA. If the SMS can pass through the CAPTCHA, it will be authenticated and the CAPTCHA will confirm that it is generated by human and not by a bot [37].

4.7 Spam in VoIP

Like spam, the unknown calls will be making the network busy which are generated by the bots. This helps the users to have the calls in low cost as compared with the other calls. If the traffic in the network becomes more, due to these type of spams, then there will be problem in the network. The CAPTCHA will be used to avoid by authenticating the call that is generated from human or any malicious programs like bots [38].

4.8 Online Games

Malicious programs affect the gaming community as many online games are designed day by day; there is an increase in the attacks too. A user plays the online games assuming that there is a user on the other end, and tries to win the game. The CAPTCHA is introduced in the online games to have proper authentication regarding the end user. If the CAPTCHA is solved, then access will be given for playing game by authenticating it as a human [39].

5 Conclusion and Future Scope

This paper presents the journey towards the introduction to the CAPTCHA and the security issues resolved by the challenges. The paper, moreover, includes its enhancements by the usage of different types of methodologies, advantages, weaknesses and various application domains. In summary, our study discussed the different kinds of challenges that can be included in the implemention of CAPTCHAs. Hybrid and HCI-based CAPTCHAs, new age CAPTCHAs, are also discussed in brief.

Many malicious programs are challenging the security aspects in today's scenario, so enhancement in the CAPTCHA is needed to face the attacks from new spams and bots. The complexity of CAPTCHA limited various security concerns with respect to the application, so an improvement in the existing CAPTCHAs is needed. Today, many applications are running on the computer systems in which some are handled by a human and some of them by trend programs, so some betterment is necessary in the technology of CAPTCHA so that level of security can be increased. Visually disabled people, moreover, are unable to handle most of the existing CAPTCHAs, and hence a significant progress is required related to the above-mentioned issue.

References

1. Chellapilla K, Simard PY (2004) Using machine learning to break visual human interaction proofs (HIPs). In: Proceedings of 17th international conference advances in neural information processing systems (NIPS-2004). MIT Press, pp 265–272
2. Von Ahn L, Blum M, Langford J (2004) Telling humans and computers apart automatically. Commun ACM 47(2):56–60
3. Jain S, Sivaselvan B (2012) Usability aspects of HCI in the design of CAPTCHAs. In: Proceedings of international conference on computational intelligence and computing research (ICCIC-2012), Coimbatore IND. IEEE, pp 1–4
4. El Ahmad AS, Yan J, Ng W-Y (2012) CAPTCHA design: color, usability, and security, internet computing. 16(2):44–51 (IEEE)
5. Datta R, Li J, Wang JZ (2005) IMAGINATION: a robust image-based CAPTCHA generation system. In: Proceedings of 13th annual international conference on multimedia (MULTIMEDIA-05), Hilton SGP. ACM, pp 331–334
6. Elson J, Douceur JR, Howell J, Saul J (2007) Asirra: a CAPTCHA that exploits interest-aligned manual image categorization. In: Proceedings of 14th conference on computer and communications security (CCS-2007), Alexandria US-VA. ACM, pp 366–374
7. Von Ahn L, Maurer B, McMillen C, Abraham D, Blum M (2008) recaptcha: human-based character recognition via web security measures. Science 321(5895):1465–1468 (AAAS)

8. V Srikanth, Vishwanathan C, Asati U, Iyengar NCSN (2009) Think-an image based CAPTCHA mechanism (testifying human based on intelligence and knowledge). In: Proceedings of international conference on advances in computing, communication and control (ICAC3–2009), Mumbai IND. ACM, pp 421–424
9. Gupta A, Jain A, Raj A, Jain A (2009) Sequenced tagged captcha: generation and its analysis. In: Proceedings of international advance computing conference (IACC-2009), Patiala IND. IEEE, pp 1286–1291
10. Vimina ER, Areekal AU (2009) Telling computers and humans apart automatically using activity recognition. In: Proceedings of international conference on systems, man and cybernetics, San Antonio US-TX. IEEE
11. Imsamai M, Phimoltares S (2010) 3D CAPTCHA: A next generation of the CAPTCHA. In: Proceedings of international conference on information science and applications (ICISA 2010), Seoul KOR. IEEE, pp 1–8
12. Cui JS, Wang LJ, Mei JT, Zhang D, Wang X, Peng Y, Zhang WZ (2010) CAPTCHA design based on moving object recognition problem. In: Proceedings of 3rd international conference on information sciences and interaction sciences (ICIS-2010), Chengdu CHN. IEEE, pp 158–162
13. Gao H, Yao D, Liu H, Liu X, Wang L (2010) A novel image based CAPTCHA using Jigsaw puzzle. In: Proceedings of 13th international conference on computational science and engineering, Hong Kong CHN. IEEE, pp 351–356
14. Chaudhari SK, Deshpande AR, Bendale SB, Kotian RV (2011) 3D drag-n-drop CAPTCHA enhanced security through CAPTCHA. In: Proceedings of international conference & workshop on emerging trends in technology (ICWET 2011), Mumbai IND. ACM, pp 598–601
15. Raj SBE, Devassy D, Jagannivas J (2011) A new architecture for the generation of picture based CAPTCHA. In: Proceedings of 3rd international conference on electronics computer technology (ICECT-2011), Kanyakumari IND. IEEE, pp 67–71
16. Lee W-B, Fan C-W, Ho K, Dow C-R (2012) A CAPTCHA with tips related to alphabets upper or lower case. In: Proceedings of 7th international conference on broadband, wireless computing, communication and applications (BWCCA-2012), Victoria CAN. IEEE, pp 458–461
17. Goswami G, Powell BM, Vatsa M, Singh R, Noore A (2014) FaceDCAPTCHA: face detection based color image CAPTCHA. Future Gener Comput Syst 31:59–68 (Elsevier)
18. Zhu BB, Yan J, Bao G, Yang M, Xu N (2014) Captcha as graphical passwords-a new security primitive based on hard AI problems. IEEE Trans Inf Forensics Secur 9(6):891–904 (IEEE)
19. Ali FABH, Karim FB (2014) Development of CAPTCHA system based on puzzle. In: Proceedings of international conference on computer, communications, and control technology (I4CT-2014), Langkawi MYS. IEEE, pp 426–428
20. Mehrnezhad M, Bafghi AG, Harati A, Toreini E (2017) PiSHi: click the images and I tell if you are a human. Int J Inf Secur 16(2):133–149 (Springer)
21. Dwivedi U, Ahuja K, Islam R, Barbhuiya FA, Nagar S, Dey K (2017) EyamKayo: interactive gaze and facial expression captcha. In: Proceedings of 22nd international conference on intelligent user interfaces companion (IUI-2017), Limassol CYP. ACM, pp 53–56
22. Kulkarni S, Fadewar HS (2017) Pedometric CAPTCHA for mobile Internet users. In: Proceedings of 2nd international conference on recent trends in electronics, information & communication technology (RTEICT 2017), Bangalore IND. IEEE, pp 600–604
23. Tariq N, Khan FA (2017) Match-the-sound CAPTCHA. In: Information technology-new generations (A Part of Advances in Intelligent Systems and Computing), vol 558. Springer, pp 803–808
24. Uzun E, Chung SPH, Essa I, Lee W (2018) rtCaptcha: a real-time CAPTCHA based liveness detection system. In: Proceedings of network and distributed systems security symposium (NDSS 2018), San Diego CA-US. Internet Society, pp 18–21
25. Al-Hammouri AT, A-Ali Z, Al-Duwairi B (2018) ReCAP: a distributed CAPTCHA service at the edge of the network to handle server overload. Trans Emerging Telecommun Technol 29(4):e3187 (Wiley)
26. Reshef E, Raanan G, Solan E (2005) Method and system for discriminating a human action from a computerized action. Application No 10/790,611, Publication No US 2005/0114705 A1, US Patent

27. Thakur A, Chaware R, Nikhil S, Hafizul Islam SK (2015) A reading oriented overlapping text based CAPTCHA. In: Proceedings of international conference on trends in automation, communications and computing technology (I-TACT-15), Bangalore IND. IEEE, pp 1–6
28. Szydlowski M, Kruegel C, Kirda E (2007) Secure input for web applications. In: Proceedings of 23rd annual computer security applications conference (ACSAC-2007), Miami Beach US-FL. IEEE, pp 375–384)
29. Shirali-Shahreza M, Shirali-Shahreza S (2008) Dynamic CAPTCHA. In: Proceedings of international symposium on communications and information technologies (ISCIT 2008), Lao CHN. IEEE, pp 436–440
30. Shirali-Shahreza MH, Shirali-Shahreza M (2007) Multilingual CAPTCHA, In Proceedings of international conference on computational cybernetics computational cybernetics (ICCC 2007), Gammarth TUN. IEEE, pp 135–139
31. Caruana G, Li M (2012) A survey of emerging approaches to spam filtering. Comput Surv 44(2), Article 9 (ACM)
32. El Ahmad AS, Yan J, Marshall L (2010) The robustness of a new CAPTCHA. In: Proceedings of 3rd European workshop on system security (EUROSEC-10), Paris FRA. ACM, pp 36–41
33. Hidalgo JMG, Alvarez G (2011) CAPTCHAs: an artificial intelligence application to web security. Adv Comput 83:109–181, Chapter 3 (Elsevier)
34. Hernandez-Castro CJ, Ribagorda A (2010) Pitfalls in CAPTCHA design and implementation: the math CAPTCHA, a case study. Comput Secur 29(1):141–157 (Elsevier)
35. Leung C-M (2009) Depress phishing by CAPTCHA with OTP. In: Proceedings of 3rd international conference on anti-counterfeiting. In: Security, and identification in communication (ASID 2009), Hong Kong CHN. IEEE, pp 187–192
36. Chakrabarti S, Singhal M (2007) Password-based authentication: preventing dictionary attacks. Computer 40(6):68–74 (IEEE)
37. Shirali-Shahreza S, Movaghar A (2007) A new anti-spam protocol using CAPTCHA. In: Proceedings of international conference on networking, sensing and control, London GBR. IEEE, pp 234–238
38. Ahmedy I, Portmann M (2010) Using captchas to mitigate the VoIP spam problem. In: Proceedings of 2nd international conference on computer research and development (ICCRD 2010), Kuala Lumpur MYS. IEEE, pp 136–140
39. Chow Y-W, Susilo W, Zhou H-Y (2010) CAPTCHA challenges for massively multiplayer online games: mini-game CAPTCHAs. In: Proceedings of international conference on cyberworlds, Singapore SGP. IEEE, pp 254–261
40. Singh VP et al (2014) Survey of different types of CAPTCHA. Int J Comput Sci Inf Technol (IJCSIT) 5(2):2242–2245

Time Series Analysis on Univariate and Multivariate Variables: A Comprehensive Survey

Satyanarayana Reddy Beeram and Swarna Kuchibhotla

Abstract Time series analysis and forecasting have become an active research area for a couple of years in various domains like signal processing, weather forecasting, earthquake prediction, communication engineering, and any domain which involves temporal measurements. These domains raise important challenges and making it to devise new approaches to accurately predict or forecast into future. This paper surveys the comprehensive studies on different time series analysis, viz., univariate and multivariate and also demonstrates various practical predictions and forecasting models. Furthermore, it proposes some possible research paths that can be explored by active researchers in the area for designing more efficient models for forecasting in time series applications.

1 Introduction

A time series, viewed as a sequence of data points that are collected over a period of time, has wide range of many different domain scientific fields from economics to engineering. In the engineering literature, state-space methods have been developed for the sequential analysis of data [1]. Time series analysis involves to extract the meaningful information or patterns or characteristics from the given raw data to understand the behavior. On the other hand, time series predictions are involved in building a model that can forecast into the future based on available observed data. The fundamental time series model is the regression analysis, which constructs a model by considering one or more independent variables to predict the outcome on continuous variables. The analysis of time series has some pattens involved, which can be discovered. Time series analysis can be further classified into univariate and multivariate based on the number of variables available for observation.

S. R. Beeram (✉)
Kallam Haranadhareddy Institute of Technology, Guntur, AP, India
e-mail: snreddy.beeram@gmail.com

S. Kuchibhotla
Koneru Lakshmaiah Education Foundation (KLEF), Vijayawada, AP, India
e-mail: drkswarna@kluniversity.in

S. C. Satapathy et al. (eds.), *Communication Software and Networks*, Lecture Notes in Networks and Systems 134, https://doi.org/10.1007/978-981-15-5397-4_13

The univariate time series consists of a single observation over a time period. The multivariate time series consists of more than one observations collected over time.

Multivariate time series analysis research is more challenging compared to univariate time series analysis. To design and correlation of multivariate across hierarchical levels vary from "system-to-system". To handle multivariate time series, uses factor analysis which reduces the attribute space from large numbers to smaller numbers of factors. Various prediction and forecasting machine learning models have been proposed in [2–6].

2 State-of-the-Art in Time Series on Various Domains

This section describes the detailed survey of various time series methods proposed by research communities that forecasting various application domains. More than two decades, the BOX-Jenkins method for time series forecasting using ARIMA model treated as benchmark to evaluate some of new methods.

The nonlinear Taylor rule-based exchange rate system has been analyzed by [7], complementing latest studies that discovered that the linear variant of this system outperforms a random walk method. It is found that, with respect to several macroeconomic determinants, the evidence of nonlinearities in the exchange rate suggests that the Taylor rule exchange rate models can in some cases be improved by considering regime changes. Future study in this region might consider alternative nonlinear modeling methods and a broader range of prospective factors for transformation.

Multiple sophisticated time series models have been suggested to forecast forex up to 30 pairs by [8]. Based on the Augmented-Dickey Fuller Test (ADF-test), each forex was converted to be stationary. It is discovered that the enormous benefit of using BCVAR and TVP-BCVAR to estimate EUR-DKK for all forecasting activities compared to the typical Bayesian Auto Regression (AR)(4).

Veraart [9] proposed a new methodology to investigate the connection between the amount of limited order entries and deletions in a limited order book relies on a blended moving average method powered by Levy Noise, called a trawl method. The methodology can be extended to four dimensions to get more accurate predictions.

Phan et al. [10] proposed a new method for filling missing values in the time series which resulted a better prediction accuracy.

The GRU-D model of [11] proved that the model with trainable decays has same time and space complexity to the existing RNN models, but is ahead of non-deep learning methods.

The MTS Networks proposed by [12] facilitates an easy interpretation of dynamics of multivariate time series.

The SFM recurrent neural network developed by [13] is tested on real-time trading data and the accuracy of trading patterns predicted is better than AR model and LSTM.

Ma et al. [14] proposed a new approach for predicting solar flares based on the past data available. Though the approach requires expensive clustering time, guarantees satisfiable prediction accuracy.

Nayak [15] has developed a hybrid system by considering the base system of four higher order neural networks. The performance of these hybrid designs is assessed by predicting some actual stock market's one-step-ahead exchange prices. The efficiency and superiority of the models are established by comparing the results with other models.

Dasgupta and Osogami [16] extended to Gaussian DyBM the dynamic Boltzman Machine (DyBM), which deals with true valued information. The design has also been expanded to a recurrent neural network (RNN) that regulates the DyBM units preference entry. Experiments with synthesized datasets demonstrate that the RNN-Gussian DyBM increases predictive precision by up to 35 percent with normal VAR.

The adaptive inferential model learning design of neuro-complex fuzzy has been expanded by [17] to the multivariate time series data. The writers explored the scheme with design variants of single-input-single-output, multi-input-single-output, and multi-input-multiple-output, testing its efficiency on four multivariate trigger levels. It was evident that the drawings on these datasets are inferior to the published results, and at least as precise as filters based on kernel-based forecasting methods.

The IFS-based technique suggested by [18] utilizes a straightforward max-min synthesis operator for IFSs, making the technique easy and less complicated. Not only does the suggested technique outperform the techniques proposed by different scientists, but it is also sufficiently effective to predict in near line with real-time series data in the hesitation environment.

According to [19], the performance of Functional Link Radial Basis Function method seems better compared to Functional Link method and Radial Basis Function alone.

Opare [20] studied the mortality rate under 5 years using ARIMA model. The study used time series data from 1961 to 2012. The dataset is partitioned into train data from 1961 to 2000 and test data from 2001 to 2012 for each model.

Aboagye-Sarfo et al. [21] presented a paper on the comparison of time series models of univariate and multivarite. The paper demonstrated the comparative performance of VARMA and ARMA models applied to predict the demand for strategic planning and resource allocation from the Emergency Department (ED). The study concluded the superiority of VARMA model.

A mixed regression model for mortality data was proposed by [22].

A hybrid prediction method for the Foreign Exchange Market was proposed by [23]. The findings show that the hybrid technique described is a very helpful and efficient technique for forecasting economic prices and extracting economic patterns.

Adhikari and Agrawal [24] proposed a combined approach to exploit the strengths of the systems Random Walk (RW) and Artificial Neural Network (ANN). The strategy utilizes the RW system to process the linear part of a monetary dataset and the remaining discrete residuals are handled with neural models. Predictive capacity of suggested system is examined in light of three common error stats on four real-world economic time series. The findings acquired obviously show that the combined technique achieves predictive accuracies fairly better than each of the designs RW, FANN, and EANN.

Khashei and Bijari [25] submitted a new hybrid model for time series which is tested on three real datasets. The final extracted results indicate that the model proposed produces higher prediction accuracy.

The authors [26] have proposed an ANN-based naive method to forecast the financial time series.

The time series prediction of electric prices based on asymmetric subset hood product fuzzy neural inference system was proposed by [27]. Authors suggested a better approach for predicting the time series into future with neuro-fuzzy inference system.

Authors proposed in [28], a hybrid approach that combines linear ARMA and Artificial Neural Networks. To reduce the error rate the proposed method used artificial neural network to forecast the time series. This hybrid approach achieves less error rate for real datasets and synthetic time series.

Summary of all the research papaers studied is given in Table 1.

3 Conclusion

The growth of the time series applications enables many services in real time. As the time series data increases, it leads to many challenges. In order to guarantee the accurate forecast need an efficient mechanisms for time series analysis. Even though the AR, MA, ARMA, and ARIMA methods have their advantages and disadvantages that impact to analyze the time series, they can only be applied on univariate time series data.

This paper provides a study of projections of univariate and multivariate time series that closely defines current methods. A comprehensive literature assessment was given, highlighting each study's accomplishments and its primary objectives. Researchers have a strong stake in analyzing the literature proof in time series forecasting over the previous few years, driven by the difficulties generated by the new algorithms. In addition, the evolution of time series demonstrates a growing interest for univariate and multivariate factors. Furthermore, the research continues to explore time series information using sophisticated processes to correctly predict profound training in real time.

4 Suggested Future Research

The comprehensive review suggests the below methods as future research directions.

- It is possible to explore the hybridization of Fuzzy regression and fuzzy support vector regression to see if the stronger time series prediction can be achieved.

Table 1 Summary of time series analysis and fore casting models state of the art

Studies	Approach	Compared with	Winner	Dataset
Lin [29]	Box-Jenkins + Tiao-Box	AR	Box-Jenkins + Tiao-Box	Medical dataset
Watanabe et al. [30]	Rough set + NN	ARIMA models	Rough set theory + NN	Stock price
Rojas et al. [28]	ARMA + NN	ARMA	ARMA + NN hybrid	General time series
Lee and Roberts [1]	Dynamic multi AR	AR	Dynamic multi AR	Global temperature TS data
Narayan et al. [27]	Neuro-fuzzy	General fuzzy inference system	Neuro-fuzzy inference system	Electricity data
Bagheri et al. [23]	ANFIS	Fuzzy models	ANFIS	FOREX dataset
Adhikari and Agrawal [24]	Random walk (RW) + ANN	RW, EANN, FANN	RW + ANN	Real world financial TS data
Ekheden and Hössjer [22]	Mixed regression model	Autoregression models	Mixed-regression model	General TS data
Aboagye-Sarfo et al. [21]	VARMA	ARMA	VARMA	Emergency department demand
Rout and Dash [19]	Functional link RBF neural network	Linear, nonlinear and hybrid neural networks	FLRBFNN	Exchange rates dataset
Zhang and Aggarwal [13]	State frequency memory RNN	AR, conventional LSTM Model	State frequency Memory RNN	Real-time stock price data
Ma et al. [14]	Decision trees	ANN, ARIMA	Decision tree model, but requires extra time for clustering	Sunspots dataset
Yang and Lin [31]	EMD + PSR + ELM	Existing state-of-art models	The Hybrid model of EMD, PSR, ELM	FOREX rates dataset
Nayak [15]	Higher order NN (HONNs)	RBFNN, multilayer perceptron NN, multi-linear regression method	HONNs	Real stock market data
Dasgupta and Osogami [16]	RNN-Gaussian DyBM	DyBM, Gaussian DyBM	RNN-Gaussian DyBM	Synthetic dataset

(continued)

Table 1 (continued)

Studies	Approach	Compared with	Winner	Dataset
Yazdanbakhsh and Dick [17]	Neuro-complex fuzzy inferential system	Kernel-based prediction algorithms (KBPA)	Neuro-complex fuzzy inference system is as accurate as KBPA	General time series data
Che et al. [11]	Deep learning model based on GRU-D	Non-deep learning models	Proposed deep learning model	Synthetic and real-time healthcare dataset
Phan et al. [10]	Novel fuzzy similarity-based measure	General multivariate time series models	Fuzzy similarity-based model	General Non-correlated dataset
Veraart [9]	Mixed moving average model	Moving averages model	The mixed moving averages model	Limit order book dataset
Kumar and Gangwar [18]	Intuitionistic fuzzy time series (IFS) model	Intuitionistic fuzzy set-based model	Intuitionistic fuzzy time series (IFS) model	University of Alabama enrollments, SBI market Price
Wang et al. [7]	Nonlinear Taylor rule-based exchange rate model	Random walk	Nonlinear Taylor rule-based exchange rate model	FOREX dataset
Taveeapiradeecharoen et al. [8]	Bayesian compressed VAR (BCVAR), Time-varying Bayesian compressed VAR (TVP-BCVAR)	Bayesian AR(4)	time-varying Bayesian compressed VAR (TVP-BCVAR)	EUR-DKK EXCHANGE RATE

- The hybrids incorporating Support Vector Machine (SVM) and Evolutionary Computation (EC) can be employed in the future.
- The soft computing constituencies involving Fuzzy Logic (FL) and Evolutionary Computation (EC) need to be investigated to see their efficiency for time series data.
- The evolutionary computation techniques other than Genetic Algorithms (GA) and Particle Swam Optimization (PSO) such as Differential Evolution (DE) in the financial time series analysis context to form new set of hybrid forecasting models.
- Due to their superior exploring capabilities, Quantum-inspired Evolutionary algorithms demonstrated their worth in classification and prediction. A little or no research on the forecast of financial time series is recorded in this area.
- In this context of forecasting financial time series, Deep learning architectures that displayed enormous potential in Image Processing need to be studied and implemented.
- The application of Hidden Markov Model (HMM) is not found in all of the papers that are reviewed. Therefore, the hybrids involving HMM-based models can be developed for forecasting.
- The Quantile Regression (QR)-based hybrids are another alternate method that needs to focus on time series applications.
- The impact of external variables such as sentiment analysis, factors that affect the price of stock need to be properly incorporated into the time series forecasting models.

References

1. Lee SM, Roberts SJ (2008) Multivariate time series forecasting in incomplete environments. Technical report PARG-08-03. www.robots.ox.ac.uk/~parg/publications.html. University of Oxford
2. Cao L, Mees A, Judd K (1998) Dynamics from multivariate time series. Phys D 121(1–2):75–88
3. Chakraborty K, Mehrotra K, Mohan CK, Ranka S (1992) Forecasting the behavior of multivariate time series using neural networks. Neural Netw 5(6):961–970
4. Chen SM, Tanuwijaya K (2011) Multivariate fuzzy forecasting based on fuzzy time series and automatic clustering techniques. Expert Syst Appl 38(8):10594–10605
5. Han M, Wang Y (2009) Analysis and modeling of multivariate chaotic time series based on neural network. Expert Syst Appl 36(2):1280–1290
6. Yazdanbakhsh O, Dick S (2015) Multi-variate timeseries forecasting using complex fuzzy logic. In: 2015 Annual conference of the North American on Fuzzy information processing society (NAFIPS) held jointly with 2015 5th world conference on soft computing (WConSC). IEEE, pp 1–6
7. Wang R, Morley B, Stamatogiannis MP (2019) Forecasting the exchange rate using nonlinear taylor rule based models. Int J Forecast 35(2):429–442
8. Taveeapiradeecharoen P, Chamnongthai K, Aunsri N (2019) Bayesian compressed vector autoregression for financial time-series analysis and forecasting. IEEE Access 7:16777–16786
9. Veraart AE (2019) Modeling, simulation and inference for multivariate time series of counts using trawl processes. J Multivar Anal 169:110–129

10. Phan TTH, Bigand A, Caillault EP (2018) A new fuzzy logic-based similarity measure applied to large gap imputation for uncorrelated multivariate time series. Appl Comput Intell Soft Comput 2018
11. Che Z, Purushotham S, Cho K, Sontag D, Liu Y (2018) Recurrent neural networks for multivariate time series with missing values. Sci Reports 8(1):6085
12. Sun J, Yang Y, Xiong NN, Dai L, Peng X, Luo J (2017) Complex network construction of multivariate time series using information geometry. IEEE Trans Syst Man Cybern: Syst
13. Zhang L, Aggarwal C, Qi GJ (2017) Stock price prediction via discovering multi frequency trading patterns. In: Proceedings of the 23rd ACM SIGKDD international conference on knowledge discovery and data mining. ACM, pp 2141–2149
14. Ma R, Boubrahimi SF, Hamdi SM, Angryk RA (2017) Solar are prediction using multivariate time series decision trees. In: 2017 IEEE international conference on big data (Big Data). IEEE, pp 2569–2578
15. Nayak SC (2017) Development and performance evaluation of adaptive hybrid higher order neural networks for exchange rate prediction. Int J Intell Syst Appl 9(8):71
16. Dasgupta S, Osogami T (2017) Nonlinear dynamic boltzmann machines for time-series prediction. In: AAAI, pp. 1833–1839
17. Yazdanbakhsh O, Dick S (2017) Forecasting of multivariate time series via complex fuzzy logic. IEEE Trans Syst Man Cybern: Syst 47(8):2160–2171
18. Kumar S, Gangwar SS (2016) Intuitionistic fuzzy time series: an approach for handling Non-determinism in time series forecasting. IEEE Trans Fuzzy Syst 24(6):1270–1281
19. Rout AK, Dash P (2016) Forecasting foreign exchange rates using hybrid functional link RBF neural network and levenberg-marquardt learning algorithm. Intell Decis Technol 10(3):299–313
20. Opare PE (2015) Time series models for the decrease in under-five mortality rate in ghana case study 1961–2012. Ph.D. thesis
21. Aboagye-Sarfo P, Mai Q, Sanlippo FM, Preen DB, Stewart LM, Fatovich DM (2015) A comparison of multivariate and univariate time series approaches to modeling and forecasting emergency department demand in western australia. J Biomed Inform 57:62–73
22. Ekheden E, Hössjer O (2015) Multivariate time series modeling, estimation and prediction of mortalities. Insur: Math Econ 65:156–171
23. Bagheri A, Peyhani HM, Akbari M (2014) Financial forecasting using ANFIS networks with quantum-behaved particle swarm optimization. Expert Syst Appl 41(14):6235–6250
24. Adhikari R, Agrawal R (2014) A combination of artificial neural network and random walk models for financial time series forecasting. Neural Comput Appl 24(6):1441–1449
25. Khashei M, Bijari M (2012) A new class of hybrid models for time series forecasting. Expert Syst Appl 39(4):4344–4357
26. Akintola K, Alese B, Thompson A (2011) Time series forecasting with neural network: a case study of stock prices of intercontinental bank nigeria. IJRRAS 3:467–472
27. Narayan A, Hipel KW, Ponnambalam K, Paul S (2011) Neuro-fuzzy inference system (ASuP-FuNIS) model for intervention time series prediction of electricity prices. In: 2011 IEEE international conference on systems, man, and cybernetics (SMC). IEEE, pp 2121–2126
28. Rojas I, Valenzuela O, Rojas F, Guillen A, Herrera LJ, Pomares H, Marquez L, Pasadas M (2008) Soft-computing techniques and Arma model for time series prediction. Neurocomputing 71(4–6):519–537
29. Lin WT (1989) Modeling and forecasting hospital patient movements: univariate and multiple time series approaches. Int J Forecast 5(2):195–208
30. Watanabe H, Chakraborty B, Chakraborty G (2007) Softcomputing approach for stock price trend forecasting from multivariate time series. In: Second international conference on Innovative computing, information and control, 2007. ICICIC'07. IEEE, p 40
31. Yang HL, Lin HC (2017) Applying the hybrid model of EMD, PSR, and ELM to exchange rates forecasting. Comput Econ 49(1):99–116

Basic Review of Different Strategies for Classification of Sentiment Analysis in Social Networks

K. Venkateswara Rao and Mangalampalli Kameswara Subrahmanyam

Abstract The growth of different online networks such as MySpace, Twitter, LinkedIn, and Face book has been increased in recent years, high amount of data outsource via social media into data sources. This huge amount of data analyzed for research on different types of real-time applications. So that analysis of sentiment and mining user opinion is one of aggressive concepts to explore meaning of outsourced data. While different types of approaches are implemented to identifying sentiment and opinion in social networks like pattern-based classification with respect to parts of speech, emotions, and batch model learning while analyzing huge amount of data. So that in this paper description of different machine learning methodologies to describe utilizes sentiment of huge amount data in social networks. We give survey of different approaches with respect to sentiment exploration from online social network. Also describe comparative analysis of different methods used for analysis of sentiment and mining of user opinion in online social networks.

1 Introduction

The utilization of social communication sites, for example, Twitter and Facebook, has been seeing a quick development in recent years. Most likely the purpose of this expansion is that individuals feel good communicating with their conclusions and perspectives calmly exhibit on a subjects with respective relative sites. Then again, our basic leadership process is as a rule affected by other individuals' assessments. The greater part of us would look for our companions', relatives', or colleagues' proposals previously settling on essential buy choices, before eating at a particular eatery, or viewing another motion picture. Here and there we even base our choice

K. Venkateswara Rao (✉)
Developer in IT & Technical Consultant, Computer Science Department, Vijayawada, Krishna, AP, India
e-mail: mailtovenkibujji@gmail.com

M. K. Subrahmanyam
Agile Coach & Development Manager, IT Industry, Hyderabad, Telangana, India
e-mail: haimanyam@gmail.com

S. C. Satapathy et al. (eds.), *Communication Software and Networks*, Lecture Notes in Networks and Systems 134, https://doi.org/10.1007/978-981-15-5397-4_14

exclusively on those feelings. To this end, feeling examination has pulled in an immense research intrigue particularly as of late. Specialists dissected in numerous spaces: motion picture surveys, news articles, online journals, gatherings, audits of item things, and all the more as of late Internet-based life information. Slant investigation of information accessible on the business organization which includes individuals' perspectives is winding up imperative so as to check popular supposition on a specific subject of premium. It can help assess shopper fulfillment about a few items, clients' interests and inclinations, political perspectives, and numerous others. Opinion examination on content is an exceptionally troublesome undertaking without anyone else's input, based on given unstructured best cases badly organized content alongside the setting intricacy [1], not to mention extricating estimation from content as loud as online networking content. There are a few troubles inborn in breaking down assessment from web-based social networking [2]. One precedent is "Negatives False representations" where different words, for example, "poop" and "crying" by and large recommends pessimism, yet they suggest positive feeling when utilized in an analysis of preferred sentiment, for example, "Satisfied with crying referral objects" or "Blessed poo! This is extraordinary". Another precedent is "Contingent estimation, for example," in the event that somebody doesn't get back to me, never work again. Different models define slant investigation of online networking content to be hard. The procedure gets significantly complex with the usage of different things, for example, ("smiley") & probable hash table, for example, "#happy" to explore emotions unexpectedly or snidely.

Notwithstanding the past and specifically to Twitter, content is generally short, a greatest Tweet measure is different characters, and as an outcome, the explored dataset for a particular Twitter data source corpus may have vast element space with few qualities for each Tweeted tweet, bringing about a very scanty dataset that adversely impacts the exactness of the estimation investigation. These inalienable issues in online life message when all is said in done and in Twitter specifically force critical difficulties on the feeling examination process.

So that various types of machine learning related techniques were introduced to explore relations of data present in social networks. We present basic review of different classification-based machine learning approaches to handle analysis of sentiment in social networks. Main contribution of this survey is useful for cataloging of different approaches used in recently published articles according to sentiment analysis. Finally, this survey present in this paper gives more challenging tasks relates to sentiment analysis.

2 Review of Literature

This section describes review of different approaches in social networks. The soonest deal with feeling investigation on Twitter information can be followed back to works of Go et al. [2]. Their way to deal with mockery discovery was to characterize tweets as positive or negative utilizing Naive Bayes, Maximum Entropy, or Support

Vector Machines calculations for characterization also, unigrams, bigrams, unigrams and bigrams, and unigrams with grammatical form labels as highlight extractors. The paper proposed a grouping model with emoji discovery and examination for order and furthermore considered dreary characters. The calculation demonstrated a precision around 80%. In any case, the paper didn't investigate mockery discovery, and limited the arrangement into just positive and negative classes. The subsequent stage around there of research would make the order ongoing. The continuous nature of tweets has propelled numerous inquires about. One among them being created by Bifet et al. [3] which uses the Twitter Streaming API to gain tweets in constant. The paper shows utilization of MOA (Massive Online Examination); it assembles progressing data, uses an amassing of computations, and orchestrates the tweets into two classes, specifically positive and negative tweets. The figuring proposed uses a feature age channel which uses a weighting plan and by then performs change recognizable proof to connect at its results. The work, in any case, is again compelled to just two-class course of action and does not meander into joke area. Moreover, it uses a predefined dataset for getting ready. Another asks about which is comparatively is by Bifet. A. in addition, Straightforward [4] which again uses the twitter spilling Programming interface and a predefined getting ready set to request tweets. This work surveys three figurings specifically Multinomial Naive Bayes, Stochastic Gradient Descent, and Hoeffding Tree for request and shows outcomes of up to 82% on the best fit count. In any case, the proposed work fails to perform joke acknowledgment, anyway uses emoji word reference to help in the game plan.

Most immovably related to the examination done in this paper is created by Aston et al. [5] on Twitter end request in the data stream. The perception figuring and its threw a vote variation with static part assurance were used to foresee thought in a data stream condition for a ton of Twitter data. Their spilling gage achieved relative precision to that of the pack desire for [6]. The words in the best features were uncovered. Also, we hope to apply the Modified Balanced Winnow computation to relational association supposition request in the data stream and to execute online component decision identified with this in order to speak to changing data after some time.

3 Sentiment Analysis (SA)

Sentiment mining, extremity mining, and sentiment mining or on the other hand notion examination is worried about investigation of course based content, content containing sentiments, and feelings. Estimation examination includes numerous undertakings. Four of the essential assignments of notion investigation where most of the exploration exertion is engaged are pre-processing data, class naming, granularity of comment, and distinguishing proof [7]. Information pre-processing is indispensable particularly for the content gathered from web-based life sites since it is unstructured and loaded with spelling botches also, characteristics. All scientists in the zone of notion examination play out a few or the majority of the common dialect pre-processing assignments including: spellchecking, and expulsion of stop words, for example, accentuation marks. Also, a few scientists perform stemming previously arrangement [5, 8].

In class naming procedure (for example, the procedure of commenting on content into marks or classes) some examination centers on ordering content as abstract or goal. In feeling examination, this errand is normally done to start with; on the grounds that it was demonstrated that performing it before extremity grouping enhances the last [1]. In other words, on the off chance that content is recognized as emotional, we can perform extremity grouping to decide if this abstract content is conveying slant or assessment of negative. Then again, a huge collection of research centers on computerizing the procedure of class marking through inaccessible supervision utilizing boisterous names. For example [9] utilized emojis, for example, ":-)" and ":(" to mark +ve or −ve tweets. Be that as it may [10], contended that utilizing boisterous feeling marks may block the execution of opinion classifiers. Proposed different models to abuse Twitter devotee diagram to move forward opinion characterization and developed a diagram that has clients, no. of tweets, unigram related word, bigrams of words, hash tags, what's more, emojis as its hubs to be associated with connection presence (e.g., clients are associated with tweets they made; tweeted data is associated to unigram of words and so on). At that point, they connected a name proliferation technique where slant names were proliferated from a little arrangement of hubs seeded with some underlying name data all through the chart.

Having a pre-handled abstract content with class marks, estimation characterization to be archive [11], sentence of word ambiguity, or state levels (where an expression is a piece of a sentence sequence of words) which we allude to as the characterization of words. At long last, the objective of a feeling is to be considered the difficulties of feeling investigation that was tended to by number of scientists.

a. **Approaches Used for Sentiment Analysis**

After audit of papers referred to in the reference segment, it is discovered that the undertaking of SA can be performed utilizing either the Lexicon/Lexical-based techniques, Machine Learning based strategies, Keyword Spotting, or Concept-based techniques.

Lexical/Lexicon-based strategies

Information is grouped dependent on the quantity of "sentiment words" present in the content [12]. Words communicating either a positive/negative supposition are called conclusion words—"incredible", "splendid", "terrible", "costly" are precedents. The semantic introduction of individual words in an audit is utilized to all in all decide the by and large estimation extremity of the survey [1]. These techniques in this way require a corpus [dictionary of words/notion lexicon], in which each word is commented on with its semantic introduction (positive/negative) [2].

Given a bit of content whose assessment must be resolved, these techniques recognize the supposition words in the content. The corpus is then used to decide the extremity (quality/semantic introduction) of every assessment word. A semantic introduction score of +1 and −1 is relegated to the positive and negative words individually. On the off chance that the all-out extremity score (tps) is certain (i.e., there are more positive supposition words than negative words), the content is delegated positive, else negative. The notion vocabulary is the essential and most critical asset for slant classifiers, which can be made physically/consequently [3]. The slant dictionary comprises conclusion words and expressions [4].

Distinctive strategies that can be utilized to develop an assumption dictionary are—Construction of Manual, Corpus-based, and the Dictionary-based strategies. The upsides of Lexical/Lexicon based strategies are

(1) The technique is straightforward and productive and gives sensible outcomes,
(2) They depend just on named information and needn't bother with any preparation dataset, and
(3) The requirement for a corpus may appear to be a downside. The simple accessibility of dictionaries and their extensible contrasted with preparing set demonstrate something else.

Condensed beneath are the disadvantages of Lexicon-based strategies:

(1) Context subordinate supposition words can't be managed,
(2) Sentences that contain numerous clashing feeling words can't be managed,
(3) A solitary sentence may address numerous substances and the sentiment for every element can differ,
(4) One noteworthy analysis raised against these techniques is that the word references are problematic as they are hand positioned by people [7],
(5) It is difficult to make an exceptional lexical-based word reference to be utilized for various settings [5],
(6) Limited words inclusion—may neglect to perceive words that are not as of now in the dictionary,
(7) They normally perform less precisely than machine learning approaches [2], and
(8) Results could be finished/under broke down prompting a decline in the execution if the word reference is excessively comprehensive/scanty.

Information Machine Learning/Learning-based techniques:

Here, feeling recognition is a double arrangement assignment, i.e., positive or negative. A given bit of content is grouped based on marked models. A preparation dataset is utilized to separate the applicable highlights and at that point used to prepare the calculation (Naive Bayes, Maximum Entropy, SVM, and so on).

Directed, Semi-administered, and Unsupervised strategies are distinctive classifications of machine learning-based SA techniques [8].

i. Directed—utilizing marked prepared information,
ii. Unsupervised—without marked information, and
iii. Semi-directed—blended of marked and unlabeled information.

The upsides of information Machine Learning-based techniques are

(1) Machine Learning technique enhances the exactness altogether [12],
(2) Ability to adjust and make prepared model for explicit purposes and settings [5],
(3) Better outcomes can be acquired utilizing machine learning approaches in limited areas [2], and
(4) Machine learning strategies give more exactness [9].

The downsides of information Machine learning techniques are

(1) Supervised strategies require marked contributions as preparing information—the bigger the better,
(2) If there is an adjustment in the dialect use, these strategies adjust ineffectively,
(3) Availability of named information and thus the low pertinence of the strategy on new information [5],
(4) Labeling information is expensive and restrictive for a few undertakings,
(5) Machine learning strategies depend on the utilization of a vast dataset of marked archives for preparing [9],
(6) When a classifier that has accomplished a high precision is utilized for another area, its execution diminishes [3],
(7) Collecting and commenting on a substantial corpus of labeled preparing information can be both, a testing and costly undertaking.

Keyword Spotting

This strategy incorporates building up a rundown of effect words (catchphrases that identify with a specific feeling). These are typically positive/negative descriptive stop words to be solid pointers of feeling [10]. Impediments of this methodology are

(1) When invalidation is included, influence words are inadequately perceived [10, 11]. For instance, the sentence "the present climate was great" will be effectively named having a positive feelings is an attempting to characterize the sentence "the present climate wasn't great", it is prone to fizzle.
(2) This strategy depends on the nearness of effect watchwords which are surface component of writing [11]. For eq., the sentence "My significant other petitioned for legal separation today" has no influence words yet passes on forceful feelings. In this way stop words spotting is incapable.

Ideal-based approach

Depend on huge semantic learning bases and use web ontology's/semantic systems for semantic content examination [11]. Dissimilar to strategies that utilization keyword(s) and word co-event tallies, they utilize verifiable importance related to normal dialect ideas.

Table 1 Description of different machine learning-related approaches for sentiment analysis

S.no	Technique	Tools	Prediction result	Source of data
1	Prediction based on user	Factorization matrix	Topic-based prediction	Twitter
2	Opinion-based prediction	Sentiment Analysis	Polarity-based classification	Corpos data
3	High amount of matching data	Labeled classification based on	Social media related student data	Social networks & Twitter
4	Classification by sentiment	Corpus based	Sensitive classification w.r.t. cross-domain	Amazon.com
5	Sentiment variations	Latent Deriche	Possible reasons preserving	Twitter
6	Sentiment Detection	Topic-based sentiment analysis	Text with topic selection	Movies related data
7	Hash-based classify	Support vector-based classification	Assessment-based sentiment analysis	Twitter
8	Sentiment analysis detection	Lexical analyzer	Polarity of classification	Movies & other readings

Focal points of this methodology include

(1) Detection of assumptions that are communicated inconspicuously, and
(2) Multi-word calculations can be broke down despite the fact that they don't expressly pass on feelings yet are identified with ideas that do [13]. Overwhelming dependence on the profundity and broadness of the learning base utilized is the real downside of this strategy.

4 Description of Different Approaches used in SA

This section describes basic usage of different approaches used in earlier work regarding sentiment analysis with respect to features used for real-time applications (Table 1).

5 Performance Evaluation Metrics

To evaluate the truth of a classifier, very exercise to make a misunderstandings matrix. Confusion matrix defines table which explains the efficiency of classifier with set of information which are known by user.

The standard conditions showing in the misunderstandings matrix are

(i) True Positives (TP)—the classifier expected yes, and it is actually a yes.
(ii) True Negative (TN)—the classifier expected no, and it is actually a no.
(iii) False Positives (FP)—the classifier expected yes, but it actually is a no (also described as a "Type I error").
(iv) False Negatives (FN)—Classifier expected no, but it actually is a yes (also described as a "Type II error").

Accuracy, Perfection, Remember and F Ranking are four measures/indices used to evaluate the efficiency of the feeling classifiers. These spiders are calculated in accordance with the misunderstandings matrix.

(1) Accuracy—Overall, how often is the classifier:

$$\text{Acc} = (TP + TN)/(TP + TN + FP + FN).$$

(2) Precision—When a classifier predicts yes, how often is it correct? This measure shows how accurately the model makes predictions

$$P = TP/(TP + FP).$$

(3) Recall—it is the basic portion presentation of positive instances

$$\tau = TP/(TP + FN).$$

(4) F-measure describe weighted average of the true positive rate (recall) and precision

$$F_1 = (2 * p * r)/(p + r).$$

6 Major Challenges in SA

The accessibility of a few procedures for SA makes it is hard to state which strategy is better over the other. To enhance the general execution and precision of SA strategies, it is important to address the following issues and difficulties:

(1) An assessment/influence word perhaps viewed as positive in one circumstance and negative in another circumstance,
(2) a similar feeling can be communicated contrastingly by various individuals. They can be conflicting in their announcements,
(3) People may utilize a solitary sentence that joins distinctive suppositions—which is simple for a human to see yet extraordinary for a PC/machine to parse,
(4) When a short bit of content needs setting it winds up troublesome for even human's to comprehend what another person thought,

(5) Handling refutations, polysemy (a word with different implications), slangs and area speculation,
(6) Identifying concealed feelings (e.q., outrage, appall, happiness) is a testing errand, and
(7) Updating/Down-dating Lexicons [11]. Since an assumption analyzer utilizes a vocabulary, its execution depends significantly on the precision of the dictionary. Vocabularies should in this way be refreshed to stifle the words which are never again utilized and to suit new words.

7 Conclusion

Users relate to social network communication based on their emotions defined in public network communications, it defines exploration and monitoring of users feeling. Online social network is one of the important platforms which describes resource utilization and analysis of sentiment analysis and mining of user opinion with respect to stop words relates to various interested concepts like products, words relates to celebrations, etc. Main purpose of this paper gives brief description of different approaches relates to machine learning-based classification approaches based on user emotions taken from twitter and other related social networks. We also discuss description of different lexical, machine learning-related approaches present in social network communication.

References

1. Bollegala D, Weir D, Carroll J (2013) Cross-domain sentiment classification using a sentiment sensitive thesaurus. IEEE Trans Knowl Data Eng 25(8)
2. Tan S, Li Y, Sun H, Guan Z, Yan X, Bu J, Chen C, He X (2014) Interpreting the public sentiment variations on twitter. IEEE Trans Knowl Data Eng 26(5)
3. Lin C, He Y, Everson R, Ruger S (2012) Weakly supervised joint sentiment-topic detection from text. IEEE Trans Knowl Data Eng 24(6)
4. Bing L, Chan KCC, Ou C (2014) Public sentiment analysis in twitter data for prediction of a company's stock price movements. In: 2014 IEEE, 11th international conference on e-business engineering
5. Olensky S (2012) Are brands wielding more influence in social media than we thought? Forbes
6. Xia R, Xu F, Zong C, Li Q, Qi Y, Li T (2015) Dual sentiment analysis: considering two sides of one review. IEEE Trans Knowl Data Eng
7. DeMers J (2013) How social media is changing the SEO industry. Search Engine J
8. Jiang L, Yu M, Zhou M, Liu X, Zhao T (2011) Targetdependent twitter sentiment classification. In: Proceedings of the 49th annual meeting of the association for computational linguistics: human language technologies, Portland
9. Bhuta S, Doshi A, Doshi U, Narvekar M (2014) A review of techniques for sentiment analysis of twitter data. In: Issues and challenges in intelligent computing techniques (ICICT)
10. Yu H, Kim S (2012) SVM tutorial—Classification, regression and ranking. In: Handbook of natural computing. Springer, Berlin, Heidelberg; Saif H, He Y, Alani H (2012) Alleviating data

sparsity for twitter sentiment analysis. In: CEUR workshop proceedings (CEUR-WS. org), Lyon, France
11. Bhuta S, Doshi A, Doshi U, Narvekar M (2014) A review of techniques for sentiment analysis. In: International conference on issues and challenges in intelligent computing techniques (ICICT). IEEE, 978-1-4799-2900-9/14
12. Chen X, Vorvoreanu M, Madhavan K (2014) Mining social media data for understanding students' learning experiences. IEEE Trans Learn Technol 7(3)
13. Ren F, Wu Y (2013) Predicting user-topic opinions in twitter with social and topical context. IEEE Trans. Affect Comput 4(4)

Enhancing the Security, Reliability, and Data Integrity Issues in the Internet of Things by Implementing Blockchain Strategy in Mining: Challenges and Solutions

Biswaranjan Mishra and Neelamadhaba Padhy

Abstract This present and upcoming era are going to endure an unexpected shift transformation from retired system to pervasive network-based enabled entity able to communicate each other and outputting data can be analyzed to extract valuable information. This highly unified global network structure known as Internet of things (IOT). IOT plays a vital role in improvement of government efficiency, easy and smooth communication, and many more. As a consequence of this in coming future an immense of dependents would be there, so it is essential to be reliable by its security and data integrity. Now using the blockchain strategy we can enhance the security by distributing the code into several cluster segments which can be distributed over the host to make IoT effective. In this arrangement, a host is a machine which is used to route the task into packets. Here each packet is associated with the data and address part. The address segment of the packet is also called the head of the packet which holds the address of the other cluster nodes in the route of sink machine. If we take the Conventional security approach, then the privacy of the packet information disfavor for IoT, as it's topology is not centralized. Hence to apply this concept we have to take the distributed architecture of various IoT devices where each of the devices should act as a sink machine and manage the sink we have to create the host machine. Here, our aimed architecture is in chain of command manner and we are using the cloud storage for storing our information data. This structure is also used to coordinating the data transactions between the devices and also with blockchain to make more reliable. Our design uses dispersed trust technique to ensure a suburbansied topology.

B. Mishra
Research Scholar, GIET University, Gunupur, India
e-mail: brmishra89@gmail.com

N. Padhy (✉)
School of Engineering and Technology (CSE Department), GIET University, Gunupur, India
e-mail: dr.neelamadhab@giet.edu

S. C. Satapathy et al. (eds.), *Communication Software and Networks*, Lecture Notes in Networks and Systems 134, https://doi.org/10.1007/978-981-15-5397-4_15

1 Introduction

As we know, IoT is the most recent trends of engineering which not only affect modern science but also the geological science [1].

The most common examples include the Bioinformatics, Physical science, and Geological science. It is a very vast area of research. Here we are going to take the case of underground mines for our case study where we are trying to apply the blockchain strategy for finding the information being obtained from the sensor nodes which are mounted in the underground mines to study out the environmental condition and to aware the miners for safe and secure mining [2].

"As we know that, in the underground mines, the environmental condition is not always identical at every region due to variant soil strength, so the safest mining is becoming a great challenge for the miner".

Hence to make the mining safer we are introducing the sensors in the mines to detect the temperature, moisture, and strength of the mines hole. IoT is the first concept in our era where IoT and blockchain and IoT are best friends where blockchain can integrate sensors. Here we are using the cluster technology to transmit the information from one end to another end [3].

2 Motivation

The Akveren arrangement was created at Umuttepe Region to a grouping of basically white, meager to broad had relations with, calcareous to limy mudrocks and limestone [4].

- The yellowish dim enduring, broadly lied with enormous bioclastic limestone;
- The pink to light red rudistid fix reefs;
- The dim gray to grayish green, and light red mudrocks; and
- The light red limestone aggregate with interlayers of rudistid flotsam and jetsam.

Examination openings showed two layers in this field. The broadness bear by the principal layer is 1.4 m with a thickness of 1.90 gr/cm^3 enclosing to the next layer with a thickness of 2.20 gr/cm^3 [4].

The Izmit arrangement of Triassic age comprising of rock, claystone, sandstone, and shale is shaped in examination block [4]. The rock, sandstone, and shale were seen as more than once included. These are the locations where we can find such type of contaminants which are usually found in the mines.

3 Contribution

Karlen in 1994 expressed that intrinsic cooperations among the five fundamental soil shaping elements [parent material, atmosphere (counting water and temperature impacts), large scale and small scale life forms, geology and time] recognized by Jenny (1941) make a generally steady soil quality that has particular physical, compound, and organic attributes because of winning common or non-anthropogenic elements. In the mines to recognize the dirt quality, we have to contemplate the state of the dirt first [5, 6]. Be that as it may, mankind, the anthropogenic power portrayed as a 6th soil framing factor in the fundamental model for depicting dirt (SSSA 1987), cooperates with the non-anthropogenic factors and impacts soil quality both adversely and emphatically. Soil and yield are the executives' practices forced ashore assets by mankind in this way decide if intrinsic soil quality will be brought down, supported, or improved over moderately brief time intervals [7].

In the mines zone the rooftop falling and soil crumbling are the regular occurrences which happen frequently because of soil quality. The overall significance of anthropogenic or the executives' components contrasted with non-anthropogenic physical, compound, or organic variables will for the most part be controlled by the capacity or application for which a dirt quality appraisal is made. A few natural traits, including microbial biomass, breath, amino acids, soil catalysts, and night crawler action have been proposed as components which impact soil quality.

N. Padhy et al. [8] developed threshold estimation algorithms for software metrics. They have used evolutionary intelligence algorithms for estimation purposes. They have taken a sample of object-oriented examples and derived software metrics. N. Padhy et al. [9] discussed the software reusability metrics algorithms in terms of the software refactor perspective. They have used the bench-marked data set (social media) and extracted software metrics. N. Padhy et.al. [10] discussed how a malicious node can be detected using heterogeneous, cluster-based, secure routing protocols in the broad area of ad hoc sensor networks.

4 Proposed System

In the mines area, the presence of moisture and other variant temperature under the mines are an obvious factor which usually damage the sensors. So if we are going to depend on a single sensor then there is a chance to get erroneous data some times. So we are going to implement the clustered technology where a group of sensors is combined together to form a cluster head which is also known as master node to send and receive the information. In the given cluster if any node gets failed due to any reason then also another sensor can take the initiative to compute the data and process the job. Here the concept is based on source and sink. The numbers of clustered heads are combined in such a manner so that it forms a cloud environment where many, as well as sink nodes, are present. A common arrangement is shown below

5 Proposed Approach

As we know that, whenever the sensor is used to sense the data it requires some amount of energy for its operation. Therefore, cluster heads consume more energy which is relatively expensive when it has a large number of sensor hubs, hence the leftover viatality of the sensor hub is playing as the main factor for the choice of distributed heads [11, 12]. In the general vitality utilization rule, the more hubs we have inside the bunch, the less vitality they require to send information. Considering a sensor organize that is arbitrarily dispersed over a district with a sink hub. The sink hub is set outside having massive amount of vitality. These sensor hubs are isolated into gatherings. All sensor hubs are allotted with a one kind of a unique identification for distinguishing proof, registering force, and preparing power. At first, a portion of those hubs are self-assertively chose as group heads and a level worth 0 is allotted to them with their unique identification [13]. So the identification of the hubs becomes easier in this approach. We are using the blockchain approach to distribute the data being sensed by the sensor node and distributed it over the channel in terms of the packet so that the security of the data can be enhanced. Here we are using the following algorithm which purely depends on the blockchain approach to minimize the energy consumption and secure the data over the channel (Fig. 1).

Applied algorithm driven by the blockchain is given below

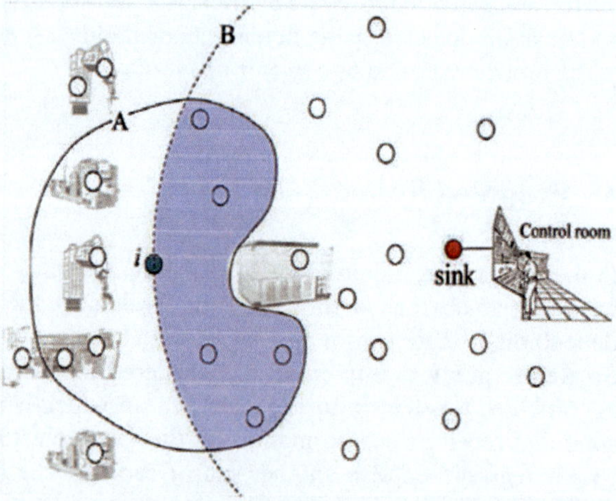

Fig. 1 Proposed system for schematic arrangement of source and sink under the mines

if

$Erd(r) < Erd(i)$

{

$W(j, r) = \gamma\,[Erd(i) / Erd(r)] + (1- \gamma)\,[D(j)/ ((N-1)/K)$

$\beta = Erd(r) / Erd(i)$
$\gamma = 1/(1+ \beta)$

}

γ is vesatile factor

β is the leftover vitality of hub j in round r

$W(j,r)$= influence factor

$Erd(i)$ = The leftover vitality of hub i

$Erd(j)$ = average leftover vitality

N = total number of hubs

K = number of clusters

Else

Hub i surrender group head determination and join a bunch.

6 Experimental Setup

Here in the above algorithm, the component description has been already given. But now the question arises, how it is blockchain based. We know that the blockchain is a bitcoin-based which is mostly used to get a bit of information from a data. In a given sensor node whenever it sense the data, it requires some energy. We used to resemble the energy as a concept of bitcoin to make it distributed over the channel.

At all hoop, every sensor hub figures its own "fixation degree" characterized to be the focus level of hub I, to be specific on the quantity of sensor hubs that it can detect during the rth round. The leftover vitality Erd(i) of a hub I is determined as

Erd(i) = introductory function – (absolute function devoured for channeling and gathering).

Sink center is exceptional for level 1 and transmit a message neighbor hubs to create a directory between bunches are sent with messages of for "Group Heads Arrangement". The pack heads you bounce away from the sink will receive the announcement, then mark the ID as the ID for your father group head ID. Group head getting the announcement will stamp the hub ID, from where the announcement is sent as its dad bunch ID and utilize the level an incentive in the announcement in addition to 1 as next heir worth if its unique level worth is 0 [11]. The directing development procedure will stop in absence to next level 0 bunch head in the sensor organize. In the instatement period of the system, the sink communicates Erd(r) and the normal vitality of sensor hubs in 'group head political decision' messages [09]. At the point when hub I gets the communicate message, it will initially analyze its own remaining vitality Erd(i) with Erd(r).

7 Simulation Result

Here from the above initial value, we are plotting the response graph with respect to the time factor. Here we are trying to record the response of the sensor in a cluster. We have taken the probability of distribution among the sensor with respect to a time interval. In the initial phase, we found the response graph in MATLAB which is shown in Figs. 2 and 3.

After varying the time interval in a large ratio by keeping the probability ratio same we found the deviation is almost negligible. It means the sensor doesn't consume more energy in a large span of time. The final graph so obtained is shown below.

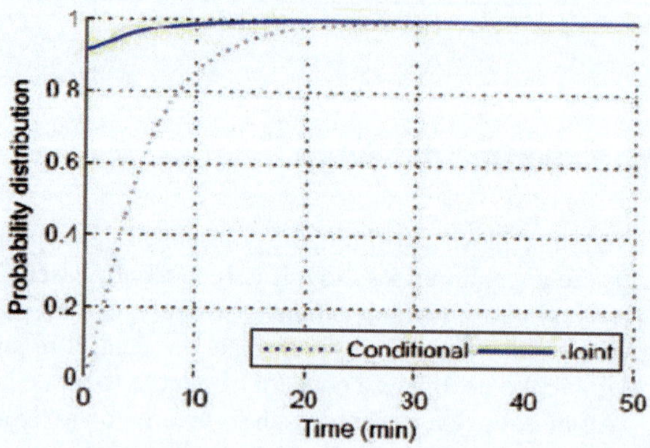

Fig. 2 Initial phase response graph of the sensor in a cluster with respect to the time factor

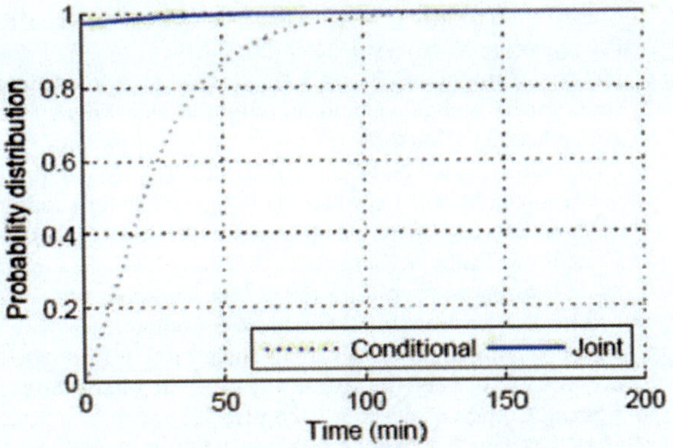

Fig. 3 Final phase response graph of the sensor in a cluster with respect to the time factor

8 Conclusion

We have addressed the problem of information distribution of cluster node in a very given mines region victimization the blockchain thought and that we unfold it over massive areas. Here we appear to become the device's answer in many cases and we premeditate the graph. We continue to consider that the variance of the outcome in the increase in time is unfortunately negligible. This concludes that by applying this logic we are able to save the utmost energy consumption by the device nodes. The results show that, in general, the proximity-based division is that the best among all others. However, this relies on the topology. The enjoy proximity division degrades because the speed of information transfer rates, or equivalently the distances between service centers decreases.

References

1. Outtagarts A (2009) Mobile agent-based applications: a survey. IJCSNS Int J Comput Sci Netw Secur 9(11)
2. Demirkol I, Ersoy C, Alagoz F (2006) MAC protocols for wireless sensor networks: a survey. IEEE Commun Mag Topics Ad hoc Sensor Netw
3. Suriyakala CD, Sankaranarayanan PE (2007) Smart multiagent architecture for congestion control to access remote energy meters. In: International conference on computational intelligence and multimedia applications
4. Bozkurt A, Kurtulus C, Endes H (2009) Measurements of apparent electrical conductivity and water content using a resistivity meter. Int J Phys Sci 4(12): 784–795. http://www.academicjournals.org/IJPS
5. Lu G, Krishnamachari B, Raghavendra CS An adaptive energy efficient and low-latency MAC for data gathering in wireless sensor networks. In: Proceedings of 18th International parallel distributed processing symposium, 26–30 April 2004, p 224
6. Ye W, Heiemann J, Estrin D An energy efficient MAC protocol for wireless sensor networks. In: IEEE proceedings of globecom 2001, pp 2944–2948

7. Marın I et al (2008) LL-MAC: a low latency MAC protocol for wireless self-organised networks. Microprocess Microsyst 32:197–209 Elsvier
8. Padhy N, Panigrahi R, Neeraja K (2019) Threshold estimation from software metrics by using evolutionary techniques and its proposed algorithms, models. Evol Intel. https://doi.org/10.1007/s12065-019-00201-0
9. Panigrahi R, Padhy N, Satapathy SC (2019) Software reusability metrics estimation from the social media by using evolutionary algorithms: refactoring perspective, International Journal of Open Source Software and Processes (IJOSSP) 10(2):21–36 IGI Global
10. Gomathy V, Padhy N, Samanta D, Sivaram M, Jain V, Amiri IS (2020) Malicious node detection using heterogeneous cluster based secure routing protocol (HCBS) in wireless adhoc sensor networks. J Ambient Intell Humanized Comput, 1–7
11. Bouabdallah Fatma, Bouabdallah Nizar (2008) The tradeoff between maximizing the sensor network lifetime and the fastest way to report reliably an event using reporting nodes selection. Comput Commun 31:1763–1776 Elsevier
12. Chen Y, Chen Y (2009) An energy efficient clustering algorithm based on residual energy and concentration degree in wireless sensor networks. In: ISCSCT'09, 26–28 December, pp 306–309
13. Jamieson K et al Sift : A MAC protocol for Event-driven Wireless sensor networks. In: EWSN 2006, LNCS 3868, pp 260–275

Dynamic Hand Gesture Recognition Using 3D-Convolutional Neural Network

Mithun Channayanamath, Akshay Math, Venkat Peddigari, Shilpa Kamath, Kavita Chachadi, Faisal Sabeeh, and Ameen Attar

Abstract Hand gesture recognition in computer science and language translation is the means of recognizing hand gestures through mathematical methods. Gesture recognition has become one of growing fields of research. Hand gesture recognition has ample number of applications including human–computer interaction, sign language and virtual/augmented gaming technology. Users can perform gestures to control or interact with devices without physically touching them. There are many architectures designed in the field of gesture detection, but existing traditional solutions are not robust to detect hand gestures with high accuracy in real time in the presence of complex patterns in performing hand gestures. In this paper, we present a fast and efficient algorithm for classifying different dynamic hand gestures using 3D-convolution neural networks. Unlike 2D-convolution neural networks, 3D-convolution networks extract features along the temporal dimension for analysis of gestures performed in videos. The paper also focuses on improving accuracy and describes data preprocessing and optimization techniques for obtaining the model inference in real time at 30fps. Our method achieves a correct recognition accuracy of 90.7% for the evaluation made on the testing videos in Chalearn LAP Continuous

M. Channayanamath · A. Math (✉) · S. Kamath · K. Chachadi · F. Sabeeh · A. Attar
KLE Technological University, Hubballi, Karnataka, India
e-mail: akshay.math185@gmail.com

M. Channayanamath
e-mail: mithunchannayanamath@gmail.com

S. Kamath
e-mail: shilpakul@gmail.com

K. Chachadi
e-mail: 17kavi17@gmail.com

F. Sabeeh
e-mail: faisalsabeeh98@gmail.com

A. Attar
e-mail: ameen.attar01@gmail.com

V. Peddigari
Samsung Research Institute, Bangalore, Karnataka, India
e-mail: venkatrpeddigari@gmail.com

S. C. Satapathy et al. (eds.), *Communication Software and Networks*, Lecture Notes in Networks and Systems 134, https://doi.org/10.1007/978-981-15-5397-4_16

Gesture dataset [1]. The detection process can be tested on laptops with standard specifications.

1 Introduction

In the beginning, hand gesture recognition for having command over the computers had its genesis with development of glove-based gesture recognition technology. Wired gloves were used as an input device for interaction between computer and humans. The physical data received from these wired gloves are captured using various sensors placed in the gloves. Later, with advancement in computer vision techniques, the glove-based gesture recognition systems started implementing deep learning frameworks for recognizing patterns without using any sensors on the gloves. Various deep learning frameworks have been proposed based on 2D-convolution neural networks [2, 3] which are computationally expensive for dynamic gesture recognition objective. The 3D-convolutional networks [4] extract features along the temporal dimension and are computationally less expensive when it comes to extraction of optimal features from videos for gesture classification. The contributions of this study are as follows.

- Deep Learning model is proposed using 3D-convolution neural networks for classification of large number of dynamic hand gesture classes.
- Optimization of the model for reducing the computational demand of the hyperparameters and to increase the speed of model inference using batch normalization.
- Evaluation of the model on Chalearn continuous gestures dataset achieving recognition of 90.7% accuracy on testing dataset.
- Efficient classification of 39 different classes of dynamic hand gestures from Chalearn LAP continuous gesture dataset.

2 Motivation

To minimize the use of expensive sensors and to avoid the use of wearable devices, the concepts of computer vision techniques and deep learning approaches can be implemented for classification of dynamic hand gestures captured by a camera. The various applications facilitate communication between the users and different smart devices by using dynamic hand gestures as an interface. A lot of research is ongoing in the field of dynamic hand recognition due to its growing applications, it should be noted that the accuracy in classifying the human actions performed in the videos and robustness in the training data pose challenge in building the optimal solution. Our work in this paper is motivated by the traditional methods for recognizing human action in videos [5–8] where either the accuracy or the speed of recognition are compromised while considering the complexity and storage efficiency of the model. This paper mainly focuses on recognizing dynamic hand gestures with high accuracy and with real-time inference based on spatio-temporal features modelled using 3D-Convolutional Neural Networks.

3 Literature Survey

3.1 *Traditional Algorithms for Gesture Recognition*

Many researchers who have worked under this field, have come up with the solutions by defining their own features which are based on motion information for classifying the videos [5, 6]. As in [7, 9] the videos are discriminated based on the spatio-temporal interest points which are described by histogram of gradients and histogram of optical flows as an extension for Harris corner detector. The dense trajectories as in [6] are used to find the interest points along temporal dimension to classify the videos based on action performed. These distinct features which are learned from one dataset cannot be transferred for training the network on other datasets.

3.2 *2D-CNN and RNN-Based Algorithms for Gesture Recognition*

Computer vision algorithms based on deep learning models with convolution neural networks have been implemented for gesture recognition in videos [2, 10]. The fusion of spatial and temporal features from sequence of frames as described in the Slow Fusion model [9] cannot learn the motion information as the temporal features get diminished after the first convolution [11]. While in two-stream model [2], two different networks are implemented to extract spatial and temporal features individually and results are obtained by applying fusion techniques. These models are effective when it comes to learning the spatial features from the frames, but the motion features extracted by these models are not efficient for classifying large number of dynamic gestures in videos.

The Recurrent Neural Networks (RNNs) pass the features extracted from hidden layers to their succeeding layers and therefore the feature maps generated by layers are dependent on features extracted by previous layers. Hence, RNNs are most widely used for extracting temporal features from consecutive frames and classify the action in videos. The pipeline of RNN models for motion recognition consist of convolutional networks at the beginning for extracting features from individual frames and then these features are passed to LSTM units for extracting temporal information, which are used for interpreting the action performed in the videos. Although,

RNNs are more efficient in categorizing the actions when compared to 2D-CNNs algorithms, these models are computationally demanding.

4 Three-Dimensional Neural Networks (3D-CNNs)

Unlike 2D-CNNs, where the convolutions occur along two dimensions (width, height) of frames or the feature maps obtained from consecutive layers of network, 3D-CNNs carry out convolutions along spatial as well as the temporal dimension on stack of frames to extract the motion information in videos [12]. The convolution is carried out by implementing three-dimensional filter kernels to convolve along fixed number of frames arranged contiguously which contain the motion information. Hence, the feature maps obtained from these convolutions are linked to each of the frames in the stack of contiguously arranged frames. The value v at a position denoted by indices x, y and z on jth feature map of the ith layer with size of the 3D filter (A_i, B_i, C_i) and w being the feature map of mth value in preceding layer at position a, b and c is given by Eq. (1) [4].

$$v_{i,j}^{x,y,z} = \tanh\left(b_{ij} + \sum_{m}\sum_{a=0}^{A_i-1}\sum_{b=0}^{B_i-1}\sum_{c=0}^{C_i-1} w_{ijm}^{abc} v_{(i-1)m}^{(x+a)(y+b)(z+c)}\right) \quad (1)$$

5 Proposed Technique

The 3D CNN network is built using Keras deep learning framework. The network is trained for 39 different dynamic hand gesture classes taken from Chalearn continuous gesture dataset, with total of 4823 videos where each class of gesture consists of videos in a range of 90–150, few sample gestures from videos are shown in Fig. 1.

(a) Gesture ID: 3. (b) Gesture ID: 111. (c) Gesture ID: 238.

Fig. 1 Hand gesture samples from Chalearn ConGD dataset

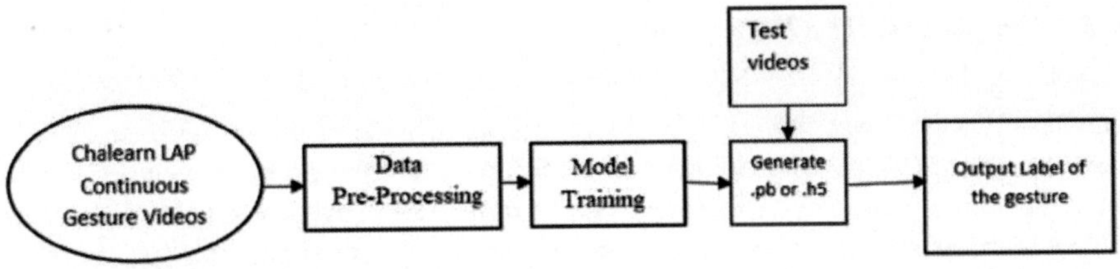

Fig. 2 Functional block diagram

The dataset is rearranged in such a way that all the videos of same gestures are grouped in one folder. Further, the videos are converted to frames using FFMEG tool and the frames are passed through the model for training the network. Finally, the trained network is tested for real-time inference. The functional block diagram of all the processes is shown in Fig. 2.

5.1 Dataset Preprocessing

The videos are converted into frames at 10 fps using FFMPEG tool since frames are labelled at rate of 10 fps in Chalearn-LAP ConGD Dataset. Then a .csv is created using python which holds the information of a video with respect to start frame, end frame and gesture ID. Then, the frames are fetched into particular patch size between start frame and end frame and converted them to grayscale to reduce the number of channels to one and resize them into smaller dimensions with an objective of improving the training process of the network. Since the depth of frames covering the entire gesture may vary in different videos, an average depth size of 30 is considered so that sufficient frames are passed through the network to capture the features of a gesture performed in a video. For gestures with frames greater than depth size of 30, the remaining frames are neglected and importance is given to initial frames and for those gestures with frames less than average depth size of 30 will append the last frame itself to make it a patch size of 30. Choosing the depth size is a very important parameter as the dimension of training data should not be altered. Hence, accommodating the number of frames become important step for training the network. Further, the frames are appended to an array and each set of frames is assigned with class ID. The training dataset which contain the input parameters are normalized by applying mean normalization so that all input weights lie in one common range and hence boosting the speed of training process.

5.2 Proposed 3D-CNN Model

The 3D-Convolutional Neural Network is built using a Keras library with tensorflow backend. The 3D-Convolution layer with required convolution filters is stacked one after the other. The size of feature maps obtained after every convolution in 3D-CNNs is calculated by using the values of filter size (*f*), zero padding (*z*), stride (*t*) and the dimension of feature maps with width (*c*), height (*r*) and depth (*d*). The size of feature map is given by $size = ((c\,f + 2\,z)\,t) + 1$. Hence, the size of input layer is $size = ((30\;5 + 2\;2)\;1) + 1 = 30$ which gets output volume of $30 \times 30 \times 32$, where the size of the input frame is 30×30, $5 \times 5 \times 32$ is the size of the filter *f*; zero padding (*z*) set to 'same' and stride length of 1. We use a patch size of 30 frames as input to the network which describes the entire hand gesture performed in the video.

The input frames with resolution of 320×240 are downscaled to 30×30 for reducing the computational demand. The proposed architecture is shown in Fig. 3, which has $30 \times 30 \times 30$ inputs with fixed size of kernel filters throughout the network. The network has a total of five 3D-convolution layers which uses kernel filter of size of $5 \times 5 \times 32$, the number of filters are increased to 64 and 128 in further convolutions to extract more features in depth. Batch normalization is enabled after every convolution to train the network faster as it converges quickly due to high learning rates [13]. Max pooling layers of size 3×3 are used to downsample the features maps at succeeding levels of network. Dropout of weights below 0.25 is carried at regular intervals to avoid overfitting. All the features passed from the preceding convolutional layers are converted to vectors of dimension 6400 in the fully connected layer. The final softmax layer has 39 output units resulting in vector with the probability of prediction for each gesture class.

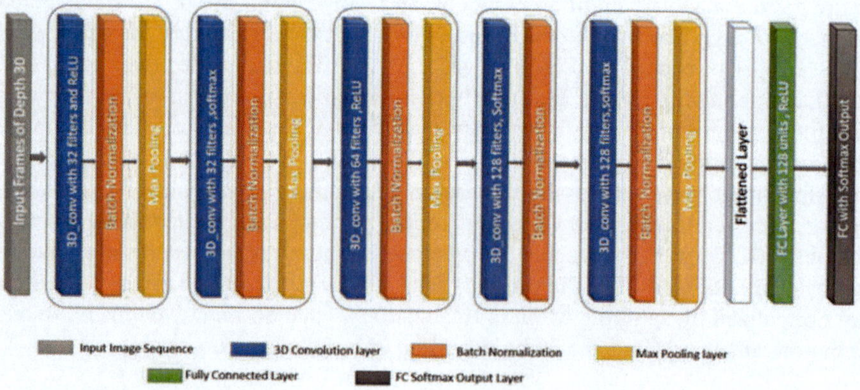

Fig. 3 Proposed 3D-CNN pipeline

6 Training the Network

The entire process is carried out in Python3.5 with OpenCV3.1 and Keras with tensorflow backend in Ubuntu 18.04 OS on system having Intel i5 processor and 8 GB RAM. The model is trained in a DGX server (GPU). Training results can be visualized from Tensorboard. Since training is done in Keras, the weights, network architecture, parameters are saved in .h5 extension. Exponential increase in gesture classes to understand the behaviour of network and improvising it accordingly. Initially, the model was trained from scratch using 20 gestures and later the trained weights are used again to train more number of gesture classes to avoid training from scratch which is time-consuming. The pre-trained network's final softmax layer is replaced with new softmax layer with more number of output units and transfer learning is used for increasing the number of gesture classes without training from scratch.

7 Experimental Results

The model is trained for different number of epochs and batch size with an intent of achieving optimal recognition accuracy. The DGX server took around 8 h to train 39 gestures contained in 4823 videos for 500 epochs with a learning rate of 0.001. The size of network weights after training for 500 epochs and with 39 gestures is 43 megabytes which can be converted into tensorflow model deployable on smart devices. After training the network for 500 epochs with batch size of 3, the results can be visualized in tensorboard. The validation accuracy on Chalearn LAP ConGD dataset for 39 different dynamic hand gestures results to 90.7% accuracy with validation loss of 0.1 and the corresponding graphs are shown in Fig. 4. With OpenCV3.1, the predictions made by the trained model can be visualized by displaying the name of dynamic hand gesture performed in the video. The proposed model can make inference in real

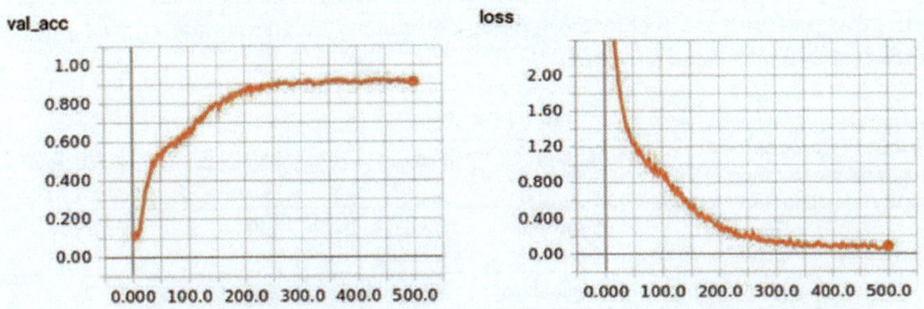

(a) Validation Accuracy. (b) Validation loss.

Fig. 4 Graphical results w.r.t number of epochs

(a) Sample Detection Result 1. (Gesture ID 247).

(b) Sample Detection Result 2. (Gesture ID 79)

(c) Sample Detection Result 3. (Gesture ID 199).

Fig. 5 Predictions made by the proposed model visualized using OpenCV3.1

time at 30 frames per second with high recognition accuracy. The snapshots of predictions made by the trained model on laptop with standard specification of i5 processor are shown in Fig. 5.

8 Conclusion

In this paper, we have proposed dynamic gesture recognition algorithm using 3D convolution neural networks and the experiments are conducted on Chalearn LAP Continuous gesture dataset as benchmark. The work can further be continued with an objective of classifying more number of dynamic hand gesture classes taken from the same dataset with high accuracy. When compared with other state-of-the-art methods based on 3D convolutional neural networks, the proposed technique in this paper performs best while considering more number of dynamic hand gesture classes with high recognition accuracy and low inference time of the model as shown in Table 1. Hence, our work can be implemented in real-time applications with more effective performance where the human–computer interactions are carried out by different dynamic hand gestures.

Table 1 Comparison of recognition accuracies (%) with state-of-the-art methods

Method	Dataset	Number of classes	Accuracy proposed
Proposed	Chalearn LAP ConGD	39	90.71
Liu et al. [5]	Weizmann	10	91.2
Jhuang et al. [8]	KTH	6	91.7

References

1. Wan J et al (2016) ChaLearn looking at people RGB-D isolated and continuous datasets for gesture recognition. In: CVPR workshop
2. Simonyan K, Zisserman A (2014) Two-stream convolutional networks for action recognition in videos. In: Advances in neural information processing systems, pp 568–576
3. Feichtenhofer C et al (2016) Convolutional two-stream network fusion for video action recognition. In: IEEE conference on computer vision and pattern recognition, pp 1933–1941
4. Ji S, Xu W, Yang M, Yu K (2013) 3D Convolutional neural networks for human action recognition. IEEE Trans Pattern Anal Mach Intell 35(1):221–231
5. Liu J et al (2009) Recognizing realistic actions from videos. Comput Vis Pattern Recogn 1996–2003
6. Wang H et al. (2011) Action recognition by dense trajectories. In: IEEE conference on computer vision and pattern recognition, pp 3169–3176
7. Baccouche M et al (2010) Action classification in soccer videos with long short-term memory recurrent neural networks. In: International conference on artificial neural networks. Springer, pp 154–159
8. Jhuang H et al (2007) A biologically inspired system for action recognition. In: 2007 IEEE 11th international conference on computer vision, ICCV2007. IEEE, pp 1–8
9. Laptev I (2005) On space-time interest points. Int J Comput Vis 64 (2–3): 107–123
10. Karpathy A et al (2014) Large-scale video classification with convolutional neural networks. In: IEEE conference on computer vision and pattern recognition, pp 1725–1732
11. Du T et al ((2015)) Learning spatiotemporal features with 3D convolutional networks. In: IEEE international conference on computer vision, pp 4489–4497
12. Baccouche M et al (2011) Sequential deep learning for human action recognition. In: International workshop on human behavior understanding. Springer, pp. 29–39
13. Ioffe S, Szegedy C (2015) Batch normalization: accelerating deep network training by reducing internal covariate shift (Submitted on 11 Feb 2015 (v1). Accessed 2 Mar 2015 (this version, v3))

Evaluating Machine Learning Algorithms for Marketing Data Analysis: Predicting Grocery Store Sales

Deepa Rani Gopagoni, P. V. Lakshmi, and Ankur Chaudhary

Abstract Machine learning is a subset of artificial intelligence that has an ability to automatically learn the data and improve learning through the experience without being programmed. Machine learning is widely employed in numerous applications. There are many machine learning algorithms used to predict the data. Many studies focused on forecasting daily stock market returns using machine learning techniques to perform the analysis. In this paper, the linear regression technique is applied on marketing data, and a model was created to predict the individual customer behaviors for using the given attributes.

1 Introduction

Machine learning techniques enable to extract knowledge by constructing statistical models and predictive analysis on collected datasets from the sources [1–4]. In this analysis, a dataset of direct marketing has been studied. This dataset is from a direct vendor who sells his merchandise solely via email. Catalogs are sent to customers with product characteristics who can order directly using the IDs. Over a period of time, vendor has developed client records to find out what makes some customers spend more than others. Study dataset is obtained from digital learning portal [5].

The dataset has 1000 customer records with information on

D. R. Gopagoni (✉)
Department of CSE, GIT, GITAM (Deemed to be University), Visakhapatnam, Andhra Pradesh, India
e-mail: deepagopagoni7@gmail.com

P. V. Lakshmi
Department of IT, GIT, GITAM (Deemed to be University), Visakhapatnam, Andhra Pradesh, India
e-mail: pvl.7097@gmail.com

A. Chaudhary
Social Prachar, Bengaluru, Karnataka, India
e-mail: socialprachar.ankur@gmail.com

S. C. Satapathy et al. (eds.), *Communication Software and Networks*, Lecture Notes in Networks and Systems 134, https://doi.org/10.1007/978-981-15-5397-4_17

- The age of customers categorized as old, middle and young customers
- Gender of customers categorized as males and females
- Data concerning whether or not the purchasers own a house
- Are they single or married?
- How far do the customers live away from the brick and mortar stores that sell similar kind of products which this direct vendor also sells? So this brick and mortar store is the competitor to the direct vendor.
- The salary of the people or the customers
- How many children they have, ranging from zero to three?
- What has been their past history in terms of volume of purchase and
- How many catalogs the direct seller has sent to the customers?

1.1 Problem Statement

Understanding the customer behavior is very important for the growth of any business. In similar lines a grocery store owner wants to understand the customer distribution who visits their store. They collected some data of customers. The problem statement is to analyze the data of customer shopping behavior, with range of impacting attributes, like age, gender, family size, location of stores, near rival stores, financial status and previous history in a constant store. To evaluate these, we explored a statistical model which might predict the client shopping behavior based on these outlined attributes. The statistical regression model build on store marketing data will enable the understanding of relative customer's distribution.

1.2 Performing Summary Statistics

To begin with, illustrative insights descriptive statistics is performed on the dataset. The obtained marketing dataset is in csv (comma separated values) file format. This provides mean, median, mode, min and max value of the data, and how much data are present in first, second and third quartile [1–3, 6]. The results of descriptive statistics are shown in Fig. 1. Apart from this, descriptive statistics additionally provides information on missing values in each and every column, if there are any. Next, the structure of the dataset is found. This explains which data type belongs to which column. It can be an integer, a numeric, a factor, a character or it can be a date object because we can expect date also.

```
> summary(data)
     Age          Gender      OwnHome        Married       Location
 Middle:508   Female:506   Own :516   Married:502   Close:710
 Old   :205   Male  :494   Rent:484   Single :498   Far  :290
 Young :287


     Salary          Children        History       Catalogs
 Min.   : 10100   Min.   :0.000   High  :255   Min.   : 6.00
 1st Qu.: 29975   1st Qu.:0.000   Low   :230   1st Qu.: 6.00
 Median : 53700   Median :1.000   Medium:212   Median :12.00
 Mean   : 56104   Mean   :0.934   NA's  :303   Mean   :14.68
 3rd Qu.: 77025   3rd Qu.:2.000                3rd Qu.:18.00
 Max.   :168800   Max.   :3.000                Max.   :24.00
  AmountSpent
 Min.   :  38.0
 1st Qu.: 488.2
 Median : 962.0
 Mean   :1216.8
 3rd Qu.:1688.5
 Max.   :6217.0
```

Fig. 1 The results of descriptive statistics on the marketing dataset

1.3 Performing Data Manipulation

Data manipulation is the process of changing data to make it easier to read or be more organized. The next step will be data manipulation. Different libraries are called for plotting. The library dplyr is the useful package for data exploration, library ggplot2 is used for data visualization and the library car is used to provide one function known as VIF that is used for multicollinearity checkup.

These are some of the foremost necessary standard imports used to run a particular project. With this information, it is easy to perform exploratory data analysis (EDA).

2 Exploratory Data Analysis (EDA)

2.1 Performing Univariate Analysis

As per the age variable, initially, univariate analysis is computed. That is the outline statistics of age. Since the amount spent is the target variable, the direct marketer can check the relationship between age and amount spent. As the target variable has continuous data, linear regression is performed. Linear regression indicates the straight-line relationship, that is, $y = m x + c$. Where, if x increases y will also increase. So, direct marketer has developed a relationship between the target variable and the independent variable [5–8]. The target variable is the amount spent and the independent variable indicates age, gender, own home, married, location, salary, children, history and catalogs.

2.2 *Finding Missing Values*

Real-world data are rarely clean and homogeneous. Typically, they tend to be incomplete, noisy and inconsistent, and it is an important task of a data scientist to preposses the data by filling missing values, as they could lead to wrong prediction or classification for any given model being developed. In the current dataset, there are some missing values on history variable. We don't know whether these missing values belong to high, medium or low category. In this case, finding out the average amount spent corresponding within the medium category can be helpful because the missing values belong to medium class. So, we need to find out the relationship between the amount spent with the missing category and the amount spent with the medium category. The missing values and imputation of these values with its mean are described in Fig. 2.

By using ggplot2 function, the relationship is showed clearly because visualization can facilitate allot here instead of doing numerical analysis over that. The visualization shows both are falling under positive skew. It is observed that missing value category has more upper graph compared to the medium category.

From this, it is clear that there are 30% of missing values within the dataset. Keeping this for further analysis, we tend to move to the next step. That is, discard the variables that have the same values. The ggplots of count vs amount spent patterns are shown in Figs. 3 and 4.

```
#Impute Missing values
tapply(data$AmountSpent,data$History,mean)
ind<-which(is.na(data$History))
mean(data[ind,"AmountSpent"])
data%>%filter(History=="Medium")%>%select(AmountSpent)->Amt_M
p<-ggplot(data[ind,],aes(x=AmountSpent))
q<-ggplot(Amt_M,aes(x=AmountSpent))
p+geom_histogram()
q+geom_histogram()
```

Fig. 2 Imputation of missing values with its mean

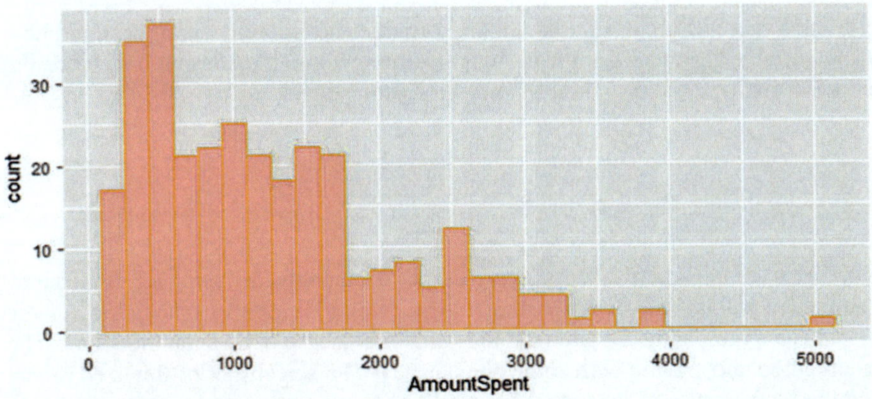

Fig. 3 Plotting amount spent with low category

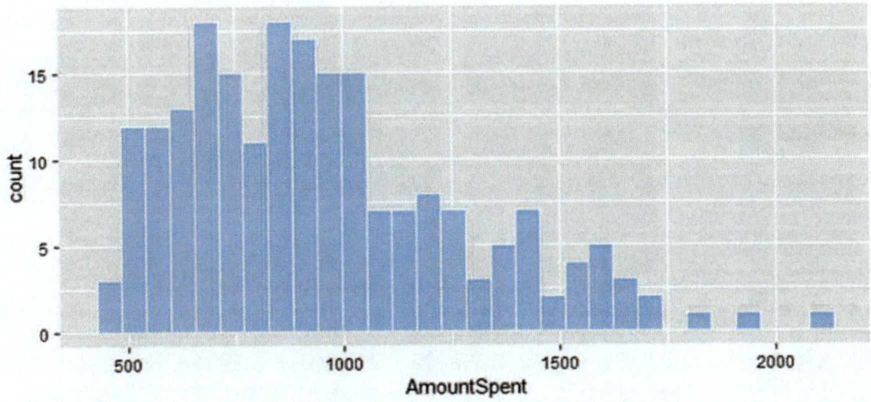

Fig. 4 Plotting amount spent with medium category

2.3 Checking the Significance of Using Linear Model

Let's begin the process. The linear regression is used to find out the relationship between the dependent variable and the independent variable. $Y = b0 + b1 \times 1 + \ldots bnxn$.

First, build the model check whether or not all the independent variables are significant with the dependent variable [6–8]. Our major objective is to find which variables are coming significant by having p-value <0.05.

For this, we need to check all the variables whether they are categorical or continuous in the feature. Because in categorical area, some classes will fall under significant level and some classes will fall under insignificant level. We consider only significant variables. Once we run the model we get the base model. The base model is going to give us an idea on how many variables are actually significant and how many variables become significant by performing some transformations. Once the model is finalized, all the variables will be significant.

2.4 Checking the Accuracy

Now, we need to check the accuracy. Accuracy is computed by using R2. It tells how good the model is fitted. If the accuracy is more, that is closer to 1, suggests that the model is explaining the variation. If the value is closer to 0, then the model must be recomputed and created with some changes [1]. The screenshot of finalized model and the performance of the model are shown in Fig. 5.

```
> summary(mod6)

Call:
lm(formula = sqrt(AmountSpent) ~ Location + Salary + Catalogs +
    Children1 + Med_d + Low_d, data = data1)

Residuals:
     Min       1Q   Median       3Q      Max
-20.0029  -3.7073   0.0778   3.4346  20.9423

Coefficients:
               Estimate Std. Error t value Pr(>|t|)
(Intercept)   1.263e+01  6.444e-01  19.593  < 2e-16 ***
LocationFar   5.399e+00  3.940e-01  13.704  < 2e-16 ***
Salary        2.529e-04  6.565e-06  38.525  < 2e-16 ***
Catalogs      5.953e-01  2.725e-02  21.844  < 2e-16 ***
Children11   -2.137e+00  4.249e-01  -5.030 5.82e-07 ***
Children13-2 -6.060e+00  4.459e-01 -13.590  < 2e-16 ***
Med_d        -4.038e+00  4.556e-01  -8.863  < 2e-16 ***
Low_d        -7.564e+00  5.192e-01 -14.567  < 2e-16 ***
---
Signif. codes:  0 '***' 0.001 '**' 0.01 '*' 0.05 '.' 0.1 ' ' 1

Residual standard error: 5.487 on 992 degrees of freedom
Multiple R-squared:  0.8271,    Adjusted R-squared:  0.8259
F-statistic:   678 on 7 and 992 DF,  p-value: < 2.2e-16
```

Fig. 5 Generating finalized model

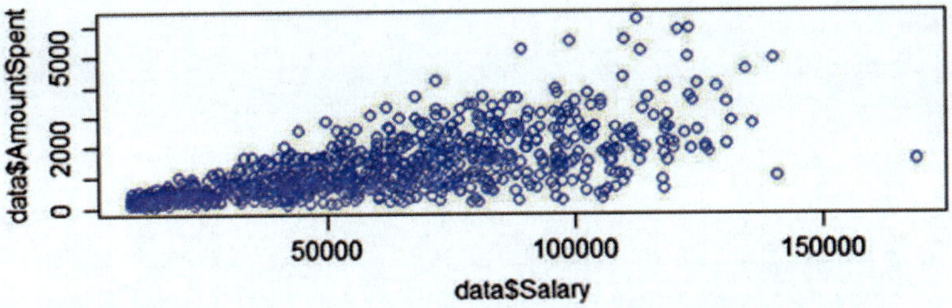

Fig. 6 Heteroscedasticity relationship between amount spent and salary

2.5 Multicollinearity with Homoscedasticity Versus Heteroscedasticity

Once the model is finalized, multicollinearity is checked. It is depicted with variance inflation factor (VIF). VIF can provide the multicollinearity, that is, the relationship between more than two variables with respect to the target variable [5–8]. If the VIF value is <5, it indicates multicollinearity is absent.

Once the VIF procedure is finalized for the training dataset, next is to find out the testing dataset prediction by using predict function. Now, check the residuals, fitted values plot. In visualization this gives the final shape which is heteroscedasticity. That is, there is a pattern where one or more variables have a relationship with the target variable. There should be no heteroscedasticity. To remove this, log transformation is applied. That is applying logarithmic on amount spent. Now, again a plot is generated (Fig. 6).

Here, qqplot is used. It gives the relationship between two continuous variables. From this, a straight line should be obtained. If smoothening is required, it is done by applying a logarithmic transformation or square root transformation. Once the qqplot generates smoothened straight line, this model is finalized. Then this will be the generalized model. Now, check the run chart.

2.6 Run Chart

A run chart is used to check uniform distribution with fitted values. Here ggplot2 function is used with geom function to tell whether the actual data and predicted data are coming in the straight line or not. All the beta coefficients in the data are multiplied with their respective independent variables. From this the direct marketer is in a position to find out the reasons. He will focus on low history and medium history. The marketer also focuses on who are having one child and two children, as this variable is significant in the model (Fig. 7).

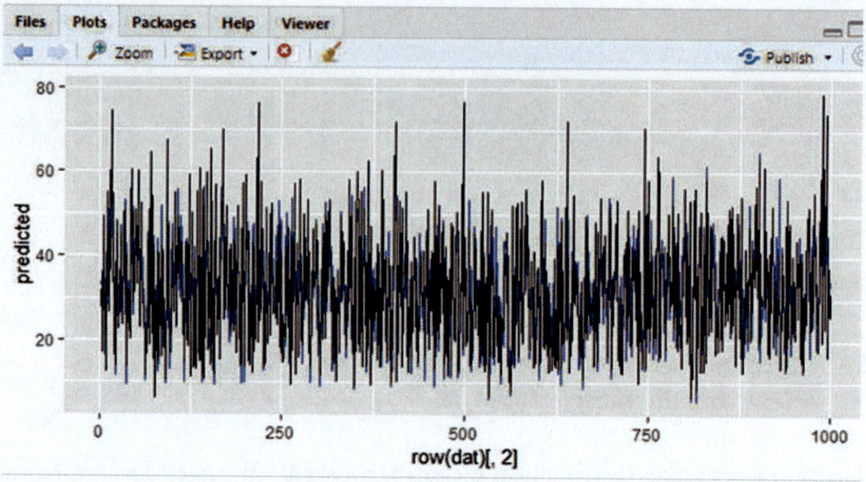

Fig. 7 Linear relationship between actual data versus predicted data

3 Predictions

In order to evaluate the performances of various ML techniques, the prediction results and linear relationship correlated with the actual data. The finalized model predicts beta relationships as follows:

- Increase in location, far, salary and catalogs categories with the amount spent will get increased
- Increase in Children1, Children3-2, Me_d, Low_d categories with the amount spent will get decreased.

4 Conclusion

In this paper, we have analyzed and assessed customer buying behavior in spending more amounts in a supermarket. The supervised machine learning technique and linear regression model suggest the variables that customers are holding positive beta values (estimates) can spend more amount than compared to others. Here, the customers who are living far from the store, the customers who are getting high salary and who receive catalogs in mail will spend more. The customer who has one child will spend more compared to the customers who are having more than one child. Based on these predictions; we are concluding direct super marketer will be able to improve the business by considering the above-mentioned factors which are having positive estimates.

References

1. James G, Witten D, HastieT, Tibshirani R (2015) An introduction to statistical learning with applications in R. Springer Texts in Statistics, pp 26–61
2. Aha DW, Kibler D, Albert MK (1991) Instance-based learning algorithms. Mach Learn 6(1):37–66
3. Debruyne M, Hubert M, Suykens JAK (2008) Model selection in Kernel based regression using the influence function. J Mach Learn Res 9:2377–2400
4. Poggio T, Rifkin R, Mukherrjee S, Niyogi P (2004) General conditions for predictivity in learning theory. Nature 428:419–422
5. Digital Learning Portal. https://digitallearningportal.com/
6. Web based learning link. https://elitedatascience.com/learn-statistics-for-data-science
7. Web based learning link. https://www.analyticsvidhya.com/blog/2016/02/complete-tutorial-learn-data-science-scratch/
8. Web based learning link. https://www.edureka.co/blog/introduction-to-machine-learning/

Review on the Methodologies for Image Segmentation Based on CNN

G. V. Sivanarayana, K. Naveen Kumar, Y. Srinivas, and G. V. S. Raj Kumar

Abstract In computer vision, image distribution is the process of dividing the digital image into several categories. The main reason for the division is that, the image should be divided into regions for better analysis. Today, the distribution method is used for applications such as image classification, facial recognition, object, and video analysis. Several computer vision functions require some knowledge base using which the visual classification is carried out in order to identify the contents. Recent surveys on image distribution techniques often focus on deep learning techniques, to accurately identify real-world objects within each image. Techniques for segmenting several types of objects using numerous algorithms have been presented in the literature, but most are based on a complex strategies and little work has been reported on systematic interpretation. To provide a comprehensive and lucid approach for segmentation of the objects, deep learning approaches are the most appropriate. These methodologies will purposefully study the characteristics of visual aids and help in better identification of the components within the objects more comprehensively. Image Processing can be done using Deep learning design in a Convolutional Neural Network (CNN). These models are distinctly trained and implemented in engrossed processing units like GPUs (Graphical Processing Units), so that the execution time can be better optimized. In the article, other variations of CNN will also be articulated for more elegant object detection.

G. V. Sivanarayana (✉) · K. Naveen Kumar (✉) · Y. Srinivas (✉) · G. V. S. Raj Kumar (✉)
GIT, GITAM, Deemed to Be University, Visakhapatnam-45, Visakhapatnam, India
e-mail: sivan.gayyi@gitam.edu

K. Naveen Kumar
e-mail: naveenkumar.kuppili@gitam.edu

Y. Srinivas
e-mail: srinivas.yarramalle@gitam.edu

G. V. S. Raj Kumar
e-mail: gvsrajkumar.ganapavarapu@gitam.edu

S. C. Satapathy et al. (eds.), *Communication Software and Networks*, Lecture Notes in Networks and Systems 134, https://doi.org/10.1007/978-981-15-5397-4_18

1 Introduction

A real time picture can be transformed with affluence into a digital form using the techniques of image processing, it also aids in performing distinct other functionalities, that includes enhancing the output images from which meaningful interpretations can be drawn about the image. In terms of statistical notes, it can well be interpreted every image generates a signal which is random in nature. The key processing features that can be performed on an image include:

- Acquisition of an image.
- Manipulation and analysis of image the image by using data compression and image enhancement techniques to identify features that are not visible through the bare eye;
- And the concluding step produces a detailed report of the analysis carried out during the image enhancement.

To have a better insight about the image, generally we follow two approaches, i.e., either enhancing the object of interest locally or globally, which approach to be followed depends on the image. The general local enhancement is being adopted by many researchers in order to have accurate identification of objects during the segmentation process, where multiple objects exist. The primary goal to be accomplished before the segmentation process is identifying the objects of interest. To achieve this objective, we divide or partition the image into number of parts known as segments. We will separate and group the set of pixels that have similar attributes with the help of image segmentation [7], so, introduces the deep learning into image segmentation [4]. Object detection builds a bounding box for each and every unique category of images. However, it will not give the information about the shape of an object. Image segmentation considers a pixel-wise mask to be applied against every object within the image.

Deep learning [11] is a branch of machine learning which is based on artificial neural networks, as neural network mimics the human brain, therefore deep learning is also considered to be a part this activity. In the last few decades, the processing power of systems has increased exponentially, as a result the concepts of deep learning and machine learning came into the existence. The concept of artificial neurons function in an exceedingly similar manner as that of the biological neurons exists in our brain, and since deep learning techniques are the sub-branches of artificial learning. Therefore, Deep Learning concepts can be galvanized by the structure and function of the brain known as artificial neural networks. Deep learning fundamentally finds out the features which are more vital for classification by virtue of deep neural networks, whereas in case of Machine Learning techniques, these features are extracted manually.

In the neural network, there is a special collection of electrons, which considers the emission and generates the output. Here, we assign activation functions [3] to neurons to modulate their input. The first column (leftmost column) of the network is known as the input and the last layer is considered output. The middle nodes are called confidential (hidden layers) because their behavior is not detected during training procedures.

In this article, we focus on different types of neural networks and their disparities. The present article portrays a comprehensive view of the various CNN channels as well as its characteristics and the importance of using it. Every neural network has one input layer and that produces an output. For different networks, the number of hidden slots will vary depending on the severity of the problem. There may be different activation functions [3] for each layer that is hidden. The choice of activation function for the device depends on the problem and the type of data being used. By assigning more weight to each layer, the neural network will be able to predict exactly. We can use the backpropagation algorithm to study the weights to be assigned. Neural network has more than one hidden layer often referred to as deep neural network.

Neural network is one of the most widely distributed machine learning algorithms currently. These algorithms perform better alignment and speed compared to other algorithms. The activation mechanism is activated in neurons. The activation functions [3] mainly use the step, sigmoid, tanh, and ReLU (Optimized Linear Unit).

2 Activation Functions

2.1 Step Function

Figure 1a represents a step function [3] and is defined as, if the value of x > = 0, then output is 1 else its value is termed as 0. Calculation of weights at different

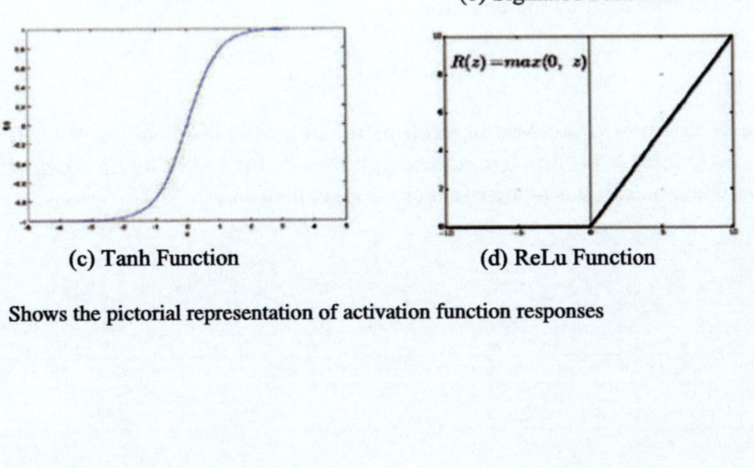

(a) Step Function

(b) Sigmaiod Function

(c) Tanh Function

(d) ReLu Function

Fig. 1 Shows the pictorial representation of activation function responses layers in neural networks can be done using the backpropagation method along with gradient descent. Since the step function is not divisible at zero, therefore, it does not progress with the development of the gradient descent method and impedes the task of improving the parameters. To avoid these problems, sigmoid jobs were brought in place of step functions.

2.2 *Sigmoid Function*

A general form of the sigmoid function [3] or logistic function is depicted in Fig. 1b. When z or the independent variable reaching toward negative infinity, then the value of the function becomes 0, if z tends to infinity then function value is 1. Still, there is a problem of vanishing gradients in a sigmoid function. When the number of layers are increased then this problem also will increase.

2.3 *Tanh Function*

The Tanh(z) [3] function is an upgraded function of the sigmoid, and its yield ranges from -1 to $+1$ rather than 0 and 1. Tanh function is shown in Fig. 1c. The tanh function derivatives are higher and its data is centered on 0, so that we are using tanh in some neural networks instead of sigmoid. A sophisticated having gradient help in an improved learning rate. Nonetheless, still we have a problem of fading gradients in tanh function.

2.4 *ReLU Function*

The Rectified Linear Unit (ReLU) [3] is widely used stimulation function in deep learning models. It is represented in Fig. 1d. If it gets any negative input, then this function output is 0, if any positive value of x is given, then it returns that value back. It is represented by the equation

$$f(x) = \max(0,\ x) \tag{1}$$

One of the most prominent algorithms to achieve is the Leaky ReLU. It is the similar as ReLU for progressive numbers, however, instead of having a value 0 for all negative values, it has a persistent slope (less than 1).

3 Techniques

3.1 Convolutional Neural Network

Convolutional neural networks (CNN) [10] are developed to overcome the problems in Multilayer Perception (MLP) [2] in neural networks, i.e., MLPs uses one perceptron for each input, as the size of image is increased, the amount of weights becomes awkward and uncontrollable and another problematic is that, MLPs respond contrarily to input images and its loosened version; and are not invariant to translation. So, that it is not good idea to use MLPs for image processing. One of the major delinquents is, when images are flattened into an MLP, their spatial information may be lost.

Nodes that are closer together will be of most significance, for that reason they define what are the features within an image? These can be resolved using CNN. We use a filter to analyze the impact of adjoining pixels. We move a filter through the image from top left to bottom right, here the size of filter is described by user (3 × 3 or 5 × 5). A value is calculated for every pixel on the image with the filter using a functionality known as convolution operation. A feature plot is engendered for each filter, once the filters have been distributed over the entire image. These are passed to an activation function, here it decides, is there any feature present at a given location of an image. More operations are done during this stage, those are: we can add number of filtering layers and create more feature maps, which makes CNN as deeper and more complex. We hand-pick the leading values from the feature maps and uses these values as inputs for the next succeeding layers, these operations are performed by pooling layer. The CNN functionality is shown in below Fig. 2.

Step 1(a): In this step, it will detect a feature, with the functionality of the neural network's filters. A set of feature maps are generated and the learning process is done to understand the parameters of such maps, how patterns are detected, the layers of detection, and how these findings are mapped out. The functionality is shown in Fig. 3.

Step 1(b): The second part involves the rectified linear unit or ReLU. We will cover ReLU layers and explore how the functions are linearly combined in the context

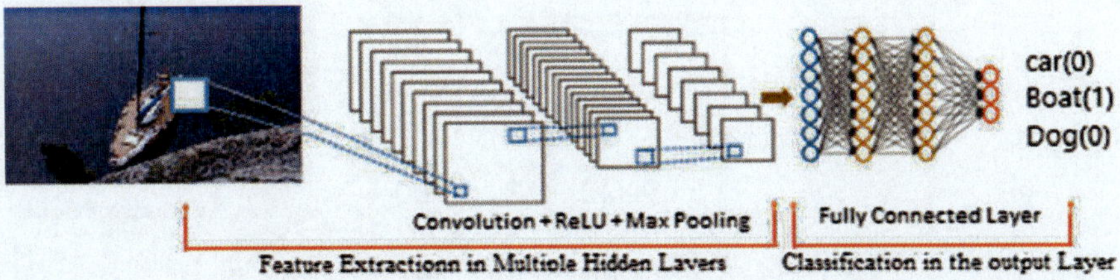

Fig. 2 Shows identifying an object using convolutional neural network layers

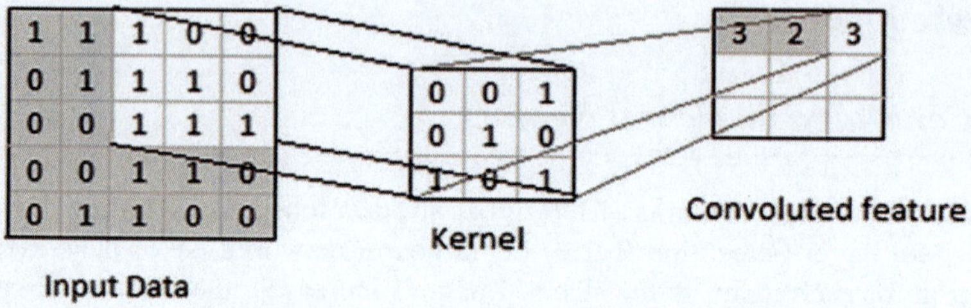

Fig. 3 Convolution operation typically performs a dot product between the filters and local regions of the input

of Convolutional Neural Networks. The negative values will be discarded from an activation map by setting them to zero. Without any effect on the fields of convolution layer, it rises the nonlinear properties of the decision function and of the entire network.

Step 2: Pooling

This part presents the general working principle. The pooling layer combines the output of neuron cluster at one layer into a single neuron in the next layer and decreases the dimensions of data. There are two types of pooling. First one is local pooling, it combines small clusters, generally 2 × 2 as shown in Fig. 4. Second one is global pooling, it operates on all the neurons of the convolutional layer. However, we will be using a specific type of pooling concept, namely, max pooling. There are various approaches, though, including mean (or sum) pooling.

The frequent downsampling operations most considered are max and average, the above figure is depicted by considering a step of 2, where a matrix with four elements is considered (little 2 × 2 square).

Step 3: Full Connection

In an abundantly connected layer, the neurons have comprehensive acquaintances with all stimulations acquired from the previous layers. It connects every neuron from one particular layer to another layer. The activations of neurons can be calculated

Fig. 4 Sampling operations

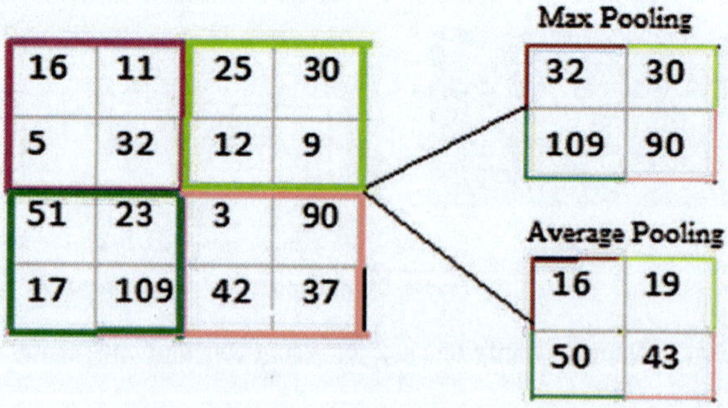

with the matrix multiplication trailed by a bias. The last layer is the output layer called fully connected layer.

3.2 R-CNN

R-CNN is another object detection method in deep learning, this is advanced one to the CNN approach. To perform object detection operation, we must know the details of what class of object it is? And size of bounding box and location of object in image. For each position within the image, we will do the sliding window as shown in Fig. 5.

Depending on the size of the object and distance from the camera, different objects or sometimes same type of objects will have different aspect ratios. When the size of the images is differ, this will effect the size of the window. If we use CNN approach for image classification at each location, this technique is quite slow. So, to avoid this problem we introduce the R-CNN method. The procedure of R-CNN is

i. In the primary step, it uses discriminating search [12] method to produce relative two thousand region proposals that are bounding boxes for image classification.
ii. Then, image classification is processed with the CNN for each and every bounding box.
iii. In the last step, by using regression operation every bounding box is refined.

The procedure to be followed for the Selective search is

1. Firstly process the similarities, i.e., region size, texture similarities, color similarities, and region filling are used as non-object-based segmentation. Then we will attain number of small segmented areas as shown at the bottom left of the image as shown in Fig. 6.
2. Then, we will combine the small segmented areas and obtain the large segmented areas with the approach known as bottom-up approach. Thus, it produces two thousand region proposals as shown in bottom right as shown in Fig. 6.

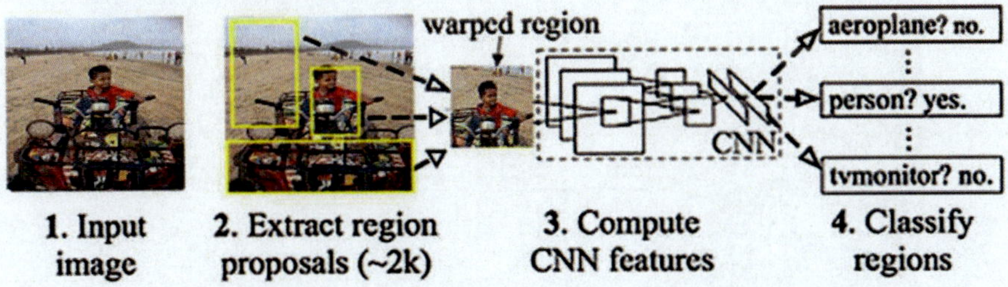

Fig. 5 Represents the regions with CNN features

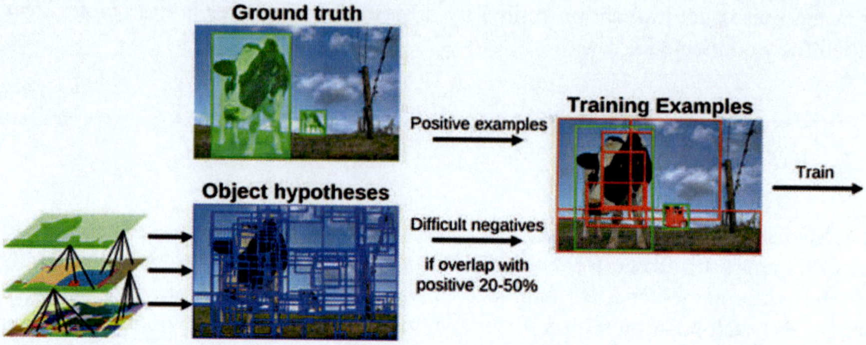

Fig. 6 Selective search operation to generate 2 k region proposals with bounding boxes

3.3 Fast R-CNN

A new-fangled algorithm is proposed which fixes the shortcomings of R-CNN, together aiming at increasing the segmenting speed and accuracy and this technique is considered to be Fast R-CNN [5] because of the relative speed w.r.t training and testing. A Fast R-CNN network grosses input as a complete image and a set of object schemes as shown in Fig. 7.

The processing steps of this method are as follows:

1. In the first step, it generates a convolution feature map by processing an entire image in the network with the convolution and max pooling layers.
2. Then, RoI (Region of Interest) assembling layer will excerpt a fixed-length feature vector from the feature map for each and every object proposal.
3. Each character vector is given by a sequence of fully connected layers that create two relative output layers:

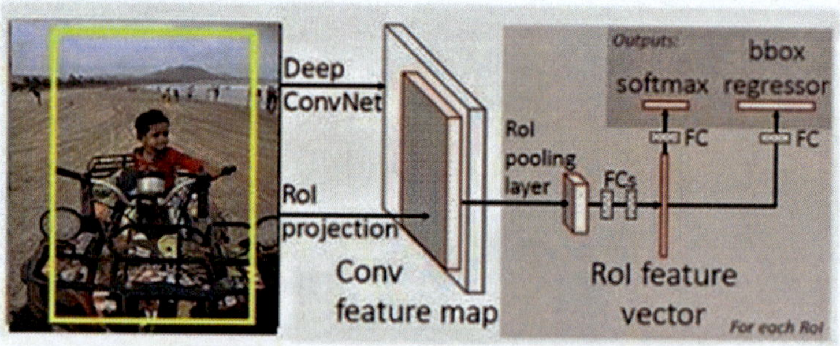

Fig. 7 Shows the flow of operation using fast R-CNN with ROI feature map

i. The first layer generates soft-max probabilities estimates over the K object classes and the background class.
ii. The second layer for each object class K generates four real values. Each set of four values encodes advanced boundary box positions for one of the K classes.

3.4 *Faster R-CNN*

In the previous Fast Region-CNN [5] and Region-CNN [8], by using selective search (SS) [12] region proposals are generated in place of using CNN [10]. In the Faster Region-CNN [6], object detection and region proposal generation are accomplished through the convolution networks. So, by using this type of process object detection can be done much faster. Region Proposal Network (RPN) exists in this approach. Architecture is shown in Fig. 8.

1. Region Proposal Network (RPN)

Selective Search (SS) uses the CNN region and the fast CNN region to invoke region designs initially, classify the object class, and define boundaries using a CNN-based network. (The main discrimination between the CNN region and the fast-CNN region is that the fast R-CNN provides input for region mapping at the map level, whereas the CNN region provides input for pixel-level region suggestions to the CNN for detection.) That a region-CNN and fast region-CNN is performed, a region design approach or a network selection (i.e., a selective search), and the detection network is also dissociated. This department is not a good idea. For example, if there is a false negative result when searching for a selection, this error can directly affect the detection network. It is therefore better to combine them so that they can correlate with each other.

Now in Faster Region-CNN, Network Design Network using CNN replaces RPN using selective search. And the detection network is shared as CNN. The steps in the Network Design Network (RPN) are as follows:

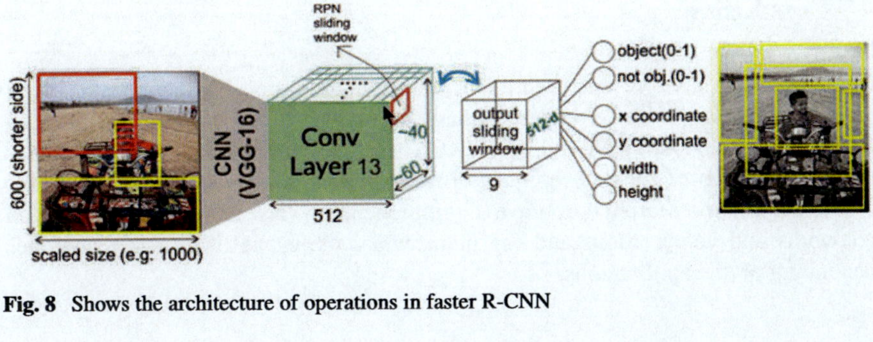

Fig. 8 Shows the architecture of operations in faster R-CNN

- Initially, maps are extracted by passing through the convolution layers.
- Furthermore, the RPN used a sliding window for each location on the feature map.
- The area of the anchor box unit (3 scales 128, 256, and 512 and three aspect ratios 1: 1, 1: 2, 2: 1) is used for (k = 9) to produce region designs for each location Layer.
- The cls layer gives two thousand points if it's for k-pieces an object or not.
- The reg layer provides four thousand coordinates (width, height, center coordinates of the boxes) to the boxes.

There will be a total of WHk anchors for the W × H functional map.

2. Detection Network

Apart from Region Proposal Network, the remaining is same as the fast Region-CNN. First ROI pooling is conducted. Then pooled area goes with convolutional neural network and 2 FC branches for class soft-max and bounding box regressor.

3.5 Mask R-CNN

One improvement for the faster CNN [6] region is the CNN mask-region [1], where a mask prediction branch is added to the class designation and the parallel boundary prediction branch. It includes a small overhead with a faster CNN in the region and runs on fps on the GPU.

There are two main parts in the Mask-RCNN. The first part is the RPN, which generates approximately three hundred regions proposals per picture. During training, all designs (ROI) pass to the object detection network and mask prediction, which is the second part. For each given area of interest, the network predicts masks belonging to all classes because the mask prediction branch runs parallel to the label and box prediction branches.

4 Conclusion

In this work, we have presented a brief insight about the various variations of CNN for effective usage in the area of image segmentation such as Region-CNN, Fast Region-CNN, Faster Region-CNN, and Mask RCNN. The limitations and methodologies involved in each of these techniques are underlined. The work presented in this article helps the researchers to have a comprehensive view about the Convolution Networks and its variations and can plan toward the usage of best-fit techniques depending on the applications.

References

1. He K, Gkioxari G, Dollár P, Girshick R (2018) Mask R-CNN, Facebook AI research (FAIR), arXiv:1703.06870v3 [cs.CV] 24 Jan 2018
2. Ramchoun H, Idrissi MAJ, Ghanou Y, Ettaouil M (2016) Multilayer perceptron: architecture optimization and training. Int J Interact Multimedia Artif Intell 4:26–30
3. Nwankpa C, Ijomah W, Gachagan A, Marshall S (2018) Activation functions: comparison of trends in practice and research for deep learning, arXiv:1811.03378v1 [cs.LG] 8 Nov 2018
4. Badea MS, Felea II, Florea LM, Vertan C (2016) The use of deep learning in image segmentation, classification and detection, arXiv:1605.09612 [cs.CV]
5. Girshick R (2015) Fast R-CNN, microsoft research, arXiv:1504.08083v2 [cs.CV], 27 Sept 2015
6. Ren S, He K, Girshick R, Sun J (2016) Faster R-CNN: towards real-time object detection with region poposal networks, arXiv:1506.01497v3 [cs.CV], 6 Jan 2016
7. Kumar V, Lal T, Dhuliya P, Pant D (2016) A study and comparison of different image segmentation algorithms. In: International conference on advances in computing, communication, & automation (ICACCA). IEEE. 978-1-5090-3481-9 ©2016
8. Girshick R, Donahue J, Darrell T, Malik J (2016) Region-based convolutional networks for accurate object detection and segmentation. IEEE Trans Pattern Anal Mach Intell 38(1):142–158
9. Liu X, Deng Z, Yang Y (2019) Recent progress in semantic image segmentation. Artif Intell Rev 52(2):1089–1106
10. Wu J (2019) Convolutional neural networks, National Key Lab for Novel Software Technology, 20 February 2019
11. Lauzon FQ (2012) An introduction to deep learning. In: 11th International conference on information sciences, signal processing and their applications: special sessions, pp 1438–1439
12. Uijlings JRR, van de Sande KEA, Gevers T, Smeulders AWM (2012) Selective search for object recognition, Technical Report IJCV

A Relativistic Study on Recent Clustering Algorithms

D. Sirisha and S. Sambhu Prasad

Abstract Ever-increasing volumes of data necessitate novel algorithms for extracting inherent information from voluminous data. Generally, clustering is adopted for voluminous and intricate data for determining groups and classifying stimulating categories in the data. Most of the available clustering algorithms are suitable for identifying spherical-shaped clusters and are not capable of handling outliers efficiently. In the present work, a relativistic study of two clustering algorithms, namely BIRCH and CURE algorithms, is carried out. These algorithms are reported to be pertinent for large databases, thus address the problems of traditional clustering techniques. Experimentations are conducted to study the relative efficiency of BIRCH and CURE. Experimental results reveal that CURE is much faster and can efficiently detect the significant outliers and identify non-spherical-shaped clusters with a wide range of size than BIRCH.

1 Introduction

To cluster the huge and dynamic databases, the classical clustering algorithms are not befitting since they presume enough main memory is available to place the data for clustering. Keeping in view of large databases comprising enormous data items, a clustering algorithm must possess the following desired features: one scan (or less) of database is required; provide online status while the algorithm is executing. Moreover, the clustering algorithm must dynamically appraise the results upon the inclusion or exclusion of data from the database [1]. Furthermore, it can be suspended, stopped, resumed, requires less main memory, has an ability to identify diversified data (e.g. sampling), and process each tuple only once [2].

D. Sirisha (✉) · S. S. Prasad
Pragati Engineering College, Kakinada, India
e-mail: sirishad998@gmail.com

S. S. Prasad
e-mail: drssp@gmail.com

S. C. Satapathy et al. (eds.), *Communication Software and Networks*, Lecture Notes in Networks and Systems 134, https://doi.org/10.1007/978-981-15-5397-4_19

In the current work, a comparative study on two clustering techniques, namely balanced iterative reducing and clustering using hierarchies (BIRCH) and clustering using representatives (CURE), is carried out since these two clustering algorithms can be extended for clustering immense databases. The performance of the two clustering techniques, namely BIRCH and CURE, is analyzed and is compared with each other in terms of average execution times, ability to detect significant outliers and identify non-spherical-shaped clusters with varied sizes. The primary aim of the present work is to specify the clustering algorithm befitting for the given data in terms of time and moves.

2 Related Work

Clustering is often defined as grouping large data sets into clusters of smaller sets that exhibit similarity in some way. A good clustering technique ensures high intra-class similarity and low inter-class similarity [3]. Clustering is considered as the most significant unsupervised learning problem as it determines the underlying structure of unlabeled data. The quality of a cluster is usually dependent on the similarity measure employed [4]. Moreover, the efficiency of a clustering algorithm can be gauged by its capability in discovering most of the hidden patterns [5–7].

3 BIRCH

BIRCH [8] is devised aiming at clustering massive data. It presumes meager amount of main memory and scans the database only once attaining linear I/O time. It is an incremental and hierarchical algorithm and handles outlier detection. The key component of this algorithm is that it constructs a tree in order to capture the desirable information for clustering. Subsequently, clustering is done on the tree, and the labeling of nodes in the tree comprises the information for calculating the distance. The chief parameter of this algorithm is its clustering feature (CF), which is a triplet comprising entire cluster information. However, BIRCH is applicable only to numerical data.

BIRCH algorithm constructs a tree called CF tree comprising two parameters B and T, where B is the branch factor which determines maximal child nodes a node may have. T is the threshold which ensures that a sub-cluster in a leaf node possesses diameter that never exceeds the given threshold.

CF: A CF is comprised of information of each cluster and is a triplet (n, s_1, s_2), where n is the number of data points, s_1 is the summation of the data points and s_2 is the sum of the square of the data points.

CF Tree: A CF tree is a balanced tree. Every internal node encompasses a CF triplet for each child node. Moreover, every leaf node denotes a cluster which holds a CF record for each of its sub-cluster. A sub-cluster's diameter will always be lesser than T. A CF tree is dynamically constructed as the data points are added. A data point is placed in the cluster which is embodied in a leaf node much closer to it. Search operation in a CF tree is a top-down approach. Generally, a data point is allocated to a cluster by calculating the distance by Euclidean or Manhattan distance measures. If the diameter of the leaf node exceeds T, the leaf node is split, and the tree is balanced. The algorithm scales to the main memory size by varying T value. Generally, increase in the T value tends to lessen CF tree.

Insertion operations in a CF Tree: The algorithm for performing insertion operation of the new entry E1 in a CF tree is as follows:

- *Identify the leaf node*: The CF tree is searched starting from the root node until a closest leaf node is identified according to a distance metric.
- *Modifying the leaf node*: When the search identifies an appropriate leaf node, say L1 closest to E1, a test is performed to find whether E1 can be placed in L1, by complying with the threshold condition. In such a situation E1 can be placed in L1, and L1 is updated. Otherwise, E1 is included as a new data point provided there is space on the leaf node. If not, the leaf node must be split. A node is split by selecting the far away data points as seeds and reclustering the remaining data points.
- *Updating the leaf node path*: Once E1 is inserted into leaf node, CF information of each non-leaf node on its path must be updated. This can be implemented by adding CF vectors. In situation when leaf split is unavoidable, a new non-leaf node is created and added to the parent node.
- *Refine merging*: The splitting of the nodes causes skewed data that affects the quality of clusters and reduces space utilization. In such situations, merge two clusters. If a leaf node is split, the split is propagated to non-leaf node N_i. Then scan N_i to identify two closer data points. If they are not a pair corresponding to the split, merge them and their children also. If the children comprise more data points which cannot be fit in a page, then merging can be split again.

3.1 Birch Clustering Procedure

Birch clustering process works in four phases.

In phase 1, the input data is scanned, and the CF tree is constructed with the allocated space on main memory and disk. The CF tree reflects the clustering information of the input data. By the end of the phase 1, subsequently, the computations in the further phases will speed up and are accurate since I/O operations are not required. Clustering of sub-clusters is to be performed and outliers are removed.

Phase 2 is not compulsory.

Phase 3: A substantial variation in the size of the clusters resulted after phase 1 and the input size of the clusters in phase 3 can be observed which can be attributed to phase 2. Identical to phase 1, phase 2 also scans the leaf nodes and reconstructs a smaller CF tree by eliminating outliers. Phase 3 addresses skewed input clusters and splitting. Phase 3 results in the clusters reflecting the patterns of data. And trivial errors in clustering may reside as it is applied to data summary.

Phase 4 is not compulsory as it necessitates further data passes to eliminate the trivial errors and improve the quality of clusters.

Advantages of BIRCH: The advantages of BIRCH in comparison to earlier distance-based clustering algorithms are as follows:

- BIRCH makes clustering decisions locally and does not scan all the data. According to the distance metric employed, the clusters are devised on the basis of proximity of data which is an incremental process and further refines the clusters.
- BIRCH algorithm works with a notion that data is spread unevenly and clusters are formed based on the proximity of the data and the farthest data points are identified as outliers and eliminated.
- The running time of BIRCH is linear since it exploits the allocated main memory for deriving quality sub-clusters such that I/O operations are minimal, and the construction of height balanced tree.
- Excluding phases 4 and 5, BIRCH is an incremental approach which does not need the entire data in prior, and data is scanned only once (Fig. 1).

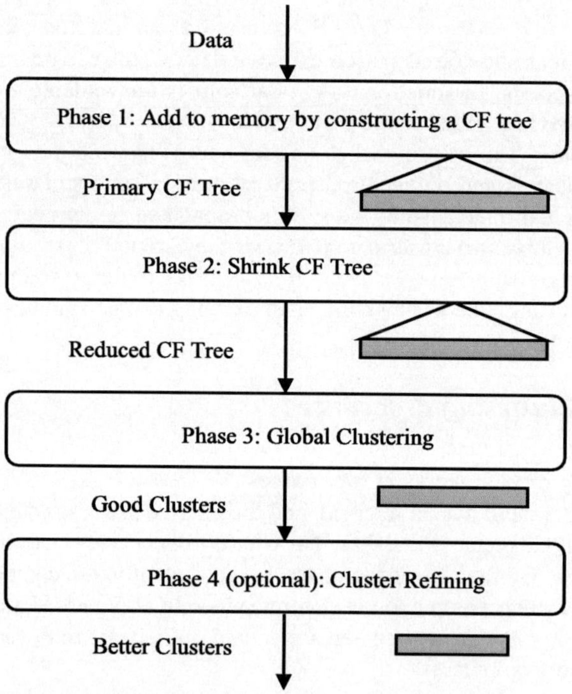

Fig. 1 Phases of BIRCH clustering algorithm

The BIRCH Algorithm

```
Input :
        D = { n1, n2, n3 ...... nn } // Set of data points
        T //  Threshold for constructing CF tree
Output :
        K      // Set of clusters
For each ni ε D do
        Identify closest leaf node and insert ni
        if threshold condition is satisfied then
                insert ti in cluster Ci and update CF, CF triplet;
        else
            if space to add ti then
                add ti as single cluster and update CF triplet;
           else
                split leaf node and redistribute CF features;
```

4 CURE

Most of the earliest clustering algorithms devise spherical-shaped clusters. In contrast, CURE algorithm [9] can efficiently identify outliers, clusters of non-spherical shape with wide range of sizes. First, CURE algorithm chooses constant *c* scattered points in a cluster in order to record shape and size of the cluster. The selected scattered points are narrowed to the cluster centre by α and these points are subsequently designated as representatives of the cluster. CURE is a hierarchical clustering algorithm where the clusters having closest representative points are merged at each step. The clusters of varied shapes can be attained by varying α between 0 and 1. When $\alpha = 1$, CURE clusters as centroid-based algorithm, whereas for $\alpha = 0$, all points are considered for clustering. CURE requires linear space with respect to its size and its worst time for clustering is $O(n^2 \log n)$.

4.1 CURE Clustering Process

First, CURE randomly selects the sample of the data for clustering. Primarily, any clustering approach can be employed, since CURE is devised to identify clusters with unusual shape, and hierarchical approach of clustering is much preferred as it enables to merge the clusters with closest representative points. The Euclidean or Manhattan distance measures can be employed to determine the closest clusters. The merging phase repeats until there are no more sufficiently close clusters (Figs. 2 and 3).

Features of CURE: The salient features of CURE are:

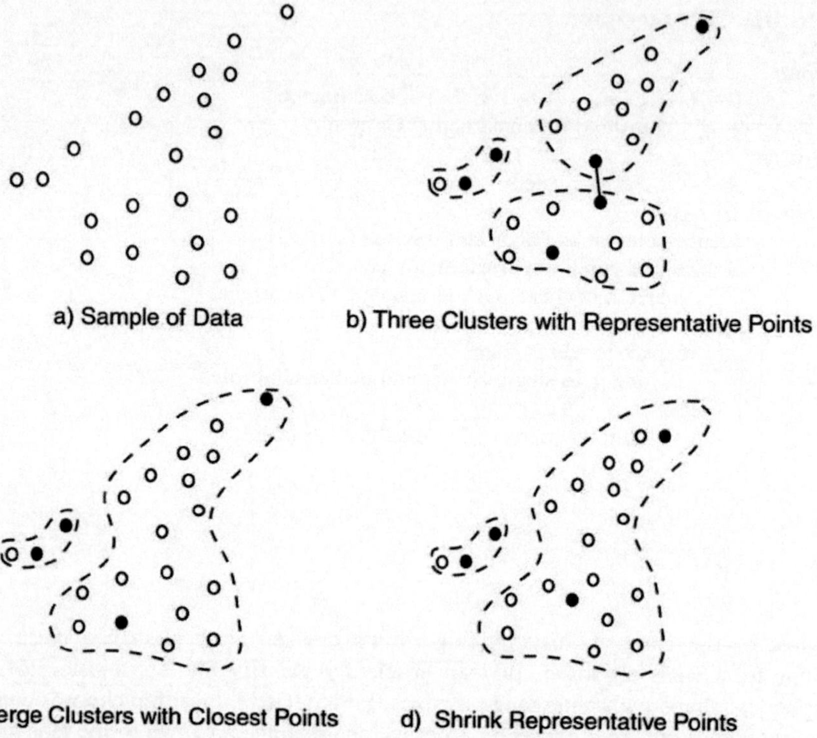

Fig. 2 The CURE clustering process (*Source* [9])

- CURE can identify non-spherical shape clusters with wide variation in size as a consequence of well-scattered representative points and centroid shrinking.
- CURE robust to the outliers.
- CURE needs linear space and time complexity of O(n^2) (Table 1).

5 Results and Discussions

A relativistic study on the performance of the two clustering algorithms BIRCH and CURE is carried out. Two data sets are considered. Data set1 contains big and small clusters while data set2 contains elongated clusters. About 20 experiments were conducted for the two algorithms taking three values for the number of clusters, k and the average execution time is calculated for the two data sets.

Fig. 3 Flowchart of CURE algorithm

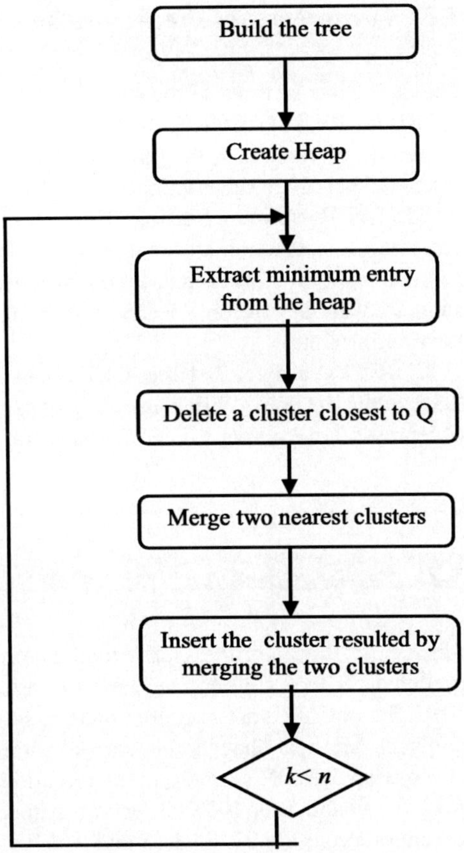

Table 1 Comparison of BIRCH and CURE clustering algorithms

Clustering algorithm	Data type handling	Cluster shape	Complexity	High-dimensional data	Comments
BIRCH	Numeric	Convex	O(*n*)	No	1. Devised for clustering massive numerical data 2. Works better in identifying spherical clusters
CURE	Numeric	Arbitrary	$O(n^2 \log n)$	Yes	1. Uses multiple representatives 2. Improves scalability

5.1 Parameters of the Algorithms

The parameter settings of the two algorithms, like branching factor B, threshold T for BIRCH and for CURE, the sample sizes, the number of representative points c, number of clusters k, the number of partitions p, and α the shrinking factor, are discussed briefly in this section.

BIRCH Parameter Setting. The tree size is a function of the threshold T. The larger the T is, the smaller the tree is. Hence, it is required to set T value to 3, since as the threshold value increases, the number of clusters k decreases. Page size P is suggested to be between P = 256 and 4096 bytes. Larger P values tend to increase the running time.

CURE Parameter Setting. CURE considers the sample size s must be at least 2.5% of the data set. While p the number of partitions is considered as 1, the number of representative points c is 10. The shrinkage factor α value can be between 0.2 and 0.7.

5.2 Performance Analysis of BIRCH and CURE

Since both the algorithms are suitable for large databases, these algorithms are executed on the two data sets, each consisting of 100,000 and 200,000 records. BIRCH and CURE are executed on data set1. The data set1 contains big and small compact spherical-shaped clusters with similar sizes. Also, the variation in the data is very less. BIRCH's average execution time t in seconds are 46.35, 48.25 and 48.2 for $k = 3$, 4 and 5 for 100,000 records, respectively, and for 200,000 records average execution times are 93.4, 94.85 and 94.75 s for $k = 3$, 4 and 5, respectively. Also, it is interpreted that BIRCH can correctly detect the clusters.

While CURE's execution time t, for $k = 3$, t in seconds is 0.0667, and 0.0665 and 0.0664 s for $k = 4$ and $k = 5$ for 100,000 records, respectively. In addition, for 200,000 records the execution times are 0.1008, 0.1015 and 0.1013 s for $k = 3$, 4 and 5, respectively. Moreover, it is observed that CURE is efficient in determining clusters and their shapes. BIRCH and CURE are executed on the data set2. The data set2 consists of clusters with elongated shape. When BIRCH is executed on data set2, the clusters are not correctly being detected as the data set contains non-spherical clusters and there is wide variance in the size of the cluster. CURE is able to accurately determine the unusual-shaped clusters and there is no significant change observed in the execution time of CURE on data set1 and data set2.

Furthermore, BIRCH is able to detect some of the most significant outliers while CURE effectively detected the outliers contained in the data set. Additionally, BIRCH required more memory for execution than CURE since BIRCH builds the CF tree

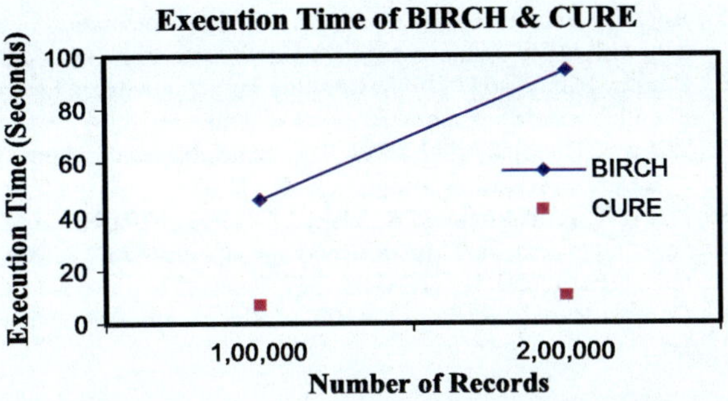

Fig. 4 Execution times of BIRCH and CURE

based upon its T in the RAM, while CURE draws a random sample, partitions the sample and labels the entire database using the sample, and so requires less memory. Figure 4 illustrates the performance of BIRCH and CURE for 100,000 and 200,000 records.

6 Conclusions

In this work, a relativistic study of two clustering algorithms BIRCH and CURE is carried out. These algorithms are reported to be appropriate for large databases and address the problems of the traditional clustering techniques, which do not support large data, and fail to handle clusters with unusual shapes and outliers [9, 8].

Extensive experiments are conducted to study the relative effectiveness of BIRCH and CURE. It is observed that CURE's execution times are much lower than BIRCH's execution time. CURE handles the outliers more efficiently than BIRCH. CURE identifies clusters having non-spherical shapes and wide variances in size, but BIRCH does not. CURE requires two passes on the database, while BIRCH requires only one pass of the entire database.

References

1. Sangalli LM, Secchi P, Vantini S, Vitelli V (2010) K-mean alignment for curve clustering. Comput Statist Data Anal 54(5): 1219–1233
2. Nepolean G, Ganga Lakshmi G (2010) An efficient K-means clustering algorithm for reducing time complexity using uniform distribution data points. In: Trendz in information sciences & computing (TISC2010)
3. Narayanan R, Ozisikyilmaz B, Zambreno J, Memik G, Choudhary A (2006) MineBench: a benchmark suite for data mining workloads. In: 2006 IEEE International symposium on workload characterization, San Jose, CA, pp 182–188
4. Sakthi M, Thanamani AS (2011) An effective determination of initial centroids in K-means clustering using kernel PCA. Int J Comput Sci Inf Technol 2(3): 955–959

5. Anh DT, Thanh LH (2015) An efficient implementation of k-means clustering for time series data with DTW distance. Int J Bus Intell Data Min 10(3): 213–232
6. Bradley P, Fayyad U (1998) Refining initial data items for k-means clustering. In: Proceedings of 15th international conference on machine learning
7. Zhang R, Rudnicky A (2002) A large scale clustering scheme for K-means. In: 16th International conference on pattern recognition (ICPR'02)
8. Zhang T, Ramakrishnan R, Livny M (1996) BIRCH: an efficient data clustering method for very large databases. In: Proceedings of 1996 ACM-SIGMOD international conference on management of data (SIGMOD'96), June 1996, pp 103–114
9. Guha S, Rastogi R, Shim K (2001) Cure: an efficient clustering algorithm for large databases. Inf Syst

Learning Style Recommender System Using VAK Technique and Machine Learning

Srinivas Adapa, Netaji Gandi, Vechalapu Alekya, and Ganthakora Lakshmi Durga

Abstract **We are Teaching, but Are they Learning?**—"This is the main fundamental question every teacher/parent/trainer should ask themselves because there is a huge gap between teaching and learning which resulted in poor learning by students [1]." If the faculty/teachers can adjust their teaching pattern based on the student learning style then surely we can increase the subject interest among students. The learning styles concept has much potential with under-prepared students. Late years have seen an adjustment in the patterns of training from teaching method to andragogy, for example, from an educator focused figuring out how to understudy focused learning. This paper bridges the gap between teaching and learning using VAK self-assessment questionnaire evaluation using different Machine Learning models.

1 Introduction

Each and Every human being is an expert in some or other fields. The human mind works in an altogether extraordinary way from the regular computerized PC. The mind is profoundly perplexing in nature, nonlinear, and parallel PC (data handling framework). It has the capacity to sort out its auxiliary constituents, known as neurons, in order to play out specific calculations (e.g., design acknowledgment, recognition,

S. Adapa (✉) · V. Alekya · G. Lakshmi Durga
Department of Computer Science and Engineering, Coastal Institute of Technology and Management (CITM), Vizianagaram, India
e-mail: a.srinivas@citm.ac.in

V. Alekya
e-mail: v.alekya@citm.ac.in

G. Lakshmi Durga
e-mail: g.lakshmi@citm.ac.in

N. Gandi
Department of Information Technology, Vignan's Institute of Engineering for Women (VIEW), Visakhapatnam, India
e-mail: netaji.gandi@gmail.com

S. C. Satapathy et al. (eds.), *Communication Software and Networks*, Lecture Notes in Networks and Systems 134, https://doi.org/10.1007/978-981-15-5397-4_20

and engine control) commonly quicker than the quickest computerized PC in presence today. By birth, a mind has an unpredictable structure and the capacity to shape its very own standards through their own understanding. To be sure, experience is developed after some time, with the most sensational improvement (i.e., hard-wiring) of the human mind occurring during the initial 2 years from birth, yet the advancement proceeds with well past that stage. During this improvement organize the learning styles of individual is framed which we examine in up-and-coming segment. "The primary assessment is to see whether understudy adapts better by hearing, seeing, or moving data as a piece of preparing it [2]". These learning styles are assembled as VAK (Visual, Auditory, and Kinesthetic (Tactile)). This paper executes Supervised Learning techniques in Machine Learning.

The implementation of this paper is done using offline and online tools through https://colab.research.google.com/.

2 Literature Review

Learning styles are different methodologies or methods for learning. They include instructing techniques, exceptional to every person that is ventured to permit him/her to adapt best. The vast majority incline toward a recognizable strategy for communicating with, taking in, and handling upgrades or data. In view of this idea, individualized "learning styles" began during the 1970s and procured gigantic prevalence. Learning styles are impacted by numerous components, for example, singular experience, various insights, and character factors, for example, an inclination for adapting alone or in a gathering. Our learning style will impact how we adapt to normal errands throughout our life, for example, perusing a guide or preparing a feast. A valuable guide to help comprehend this idea better is the way we figure out how to utilize another bit of innovation. We can move toward it either by sitting alone, perusing directions from start to finish previously or take a "hands-on" approach like squeezing the various controls to find through experimentation or learn by observing others utilizing the equivalent. This VAK idea is taken from the NLP practitioner module [3]. This model reflects about how learning inclinations fluctuate among people. This said be that as it may, conditions may likewise decide how every individual gains some new useful knowledge. Such models help us to consider how we have inclinations for the way in which we learn. In this manner, understanding learning styles approaches causes us to think about a person's predominant or favored perspective thusly helping us to adapt better in lesser time. This paper is also extended from another base paper [4].

3 VAK Learning Style [2]

The learning style suggestions are not exhaustive but are meant to provide instrumental alternatives for teachers in real classrooms. These strategies were developed by instructions as representative of things teachers can do to maximize effective teaching.

A. **Visual Learning Style**

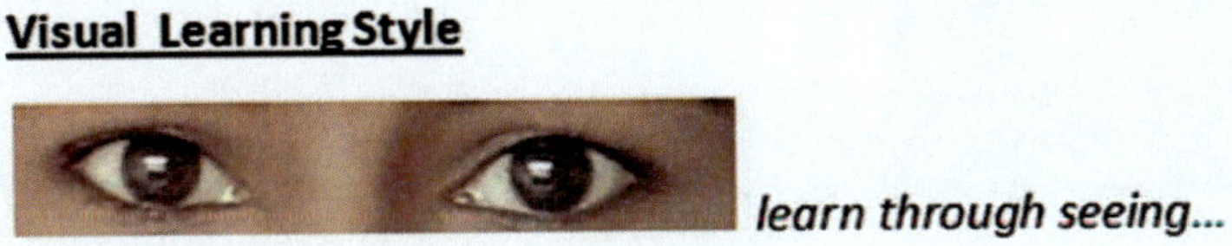

Somebody with a Visual learning style has an inclination for seen or watched things, including pictures, outlines, coordinators, illustrations, displaying, blackboard, stick exercises, instructional TV, non-verbal communication, diagramming, exhibits, shows, freebees, films, flip-graph, and so on [5].

These individuals will utilize expressions, for example, "show me", "we should view that" and will be best ready to play out another errand in the wake of perusing the directions or watching another person do it first. These are the individuals who will work from records and composed headings and guidelines.

B. **Auditory Learning Style**

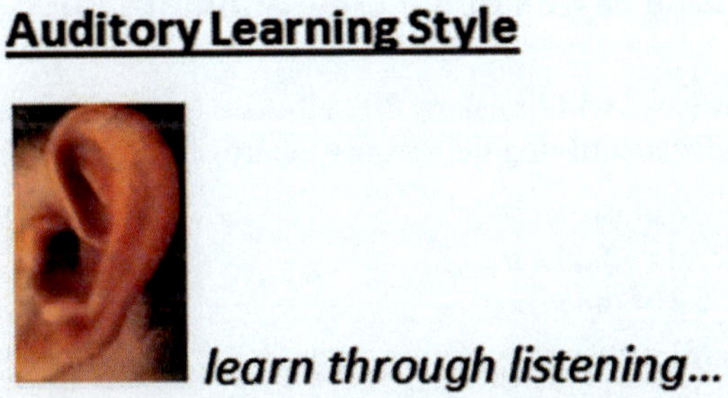

Somebody with an Auditory learning style has an inclination for the exchange of data through tuning in: to the verbally expressed word, of self or others, of sounds and commotions. Recommended procedures for Auditory (Hearing) is tape and read, tune in and look at, address, singing back, talk and tune in, blindfold, sound tape, Socratic addressing, agreeable adapting, orally condense, thoroughly analyze, bunch work [5].

These individuals will utilize expressions, for example, "let me know", "how about we talk it over" and will be best ready to play out another errand in the wake of tuning into directions from a specialist. These are the individuals who are glad being given spoken directions via phone and can recall every one of the words to tunes that they hear!

C. **Kinesthetic/Haptic (Tactile) Learning Style**

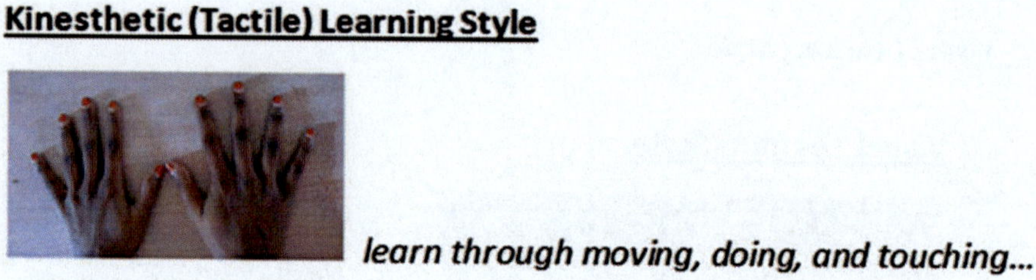

Somebody with a Kinaesthetic learning style has an inclination for physical experience—contacting, feeling, holding, doing, down to earth hands-on encounters. Recommended methodologies for Kinesthetic individuals are Skits, introductions, board work, Projects, Labs, Demonstrations, Investigations, Simulations, Educational Games, bunch work,… etc.

These individuals will utilize expressions, for example, "let me attempt", "how would you feel?" and will be best ready to play out another assignment by feeling free to give it a shot, learning as they go. These are the individuals who like to analyze, hands-on, and never take a gander at the guidelines first!

4 VAK Learning Style Online Questionnaire [6]

In this questionnaire, we will be asking 30 Self-Assessment questions with multiple options to choose for identifying the sensory acuity of the person in terms of:

- Visual (sense of vision).
- Auditory (sense of audio/hearing).
- Kinesthetic (sense of touch).

Images also added for some options to make the user respond according to their understanding and interpretation. The questionnaire is in International language (English) and it is translated to regional language (Telugu) which is available online at https://goo.gl/forms/zsfGIEa7jB87Ie4P2.

We have taken around 900 + user inputs using this online questionnaire and tested it using different machine learning models. The preprocessed data is available at https://tinyurl.com/y5e38epc.

We have converted the user responses to numeric values like:

- For Visual Responses we kept the value **1.**

- For Auditory Responses we kept the value **2.**
- For Kinesthetic Responses we kept the value **3**.

There is demonstrated proof that VAK is a fruitful learning style that advances compelling learning and data input/yield in numerous settings, regardless of whether it is the study hall, a clinical domain, or a lab. In contrast with other learning styles, for example, the Learning Style Questionnaire (LSQ), the VAK learning style model produces results that are progressively positive.

5 Machine Learning Models

5.1 *Training Machine Learning Models: [1]*

The way toward preparing a ML model includes giving a ML calculation (that is, the learning calculation) with preparing information to gain from. The term ML model alludes to the model antique that is made by the preparation procedure. The preparation information must contain the right answer, which is known as an objective or target quality. The learning calculation discovers designs in the preparation information that guide the info information credits to the objective (the appropriate response that you need to foresee), and it yields a ML model that catches these examples.

The preparation procedure includes introducing some arbitrary qualities for every one of the preparation frameworks and endeavor to anticipate the yield of the information utilizing the underlying irregular qualities. Toward the start, the mistake will be huge, however, contrasting the model's expectations and the right yield, the model can modify the loads and predispositions esteems until having a decent foreseeing model as appeared in underneath Fig. 1.

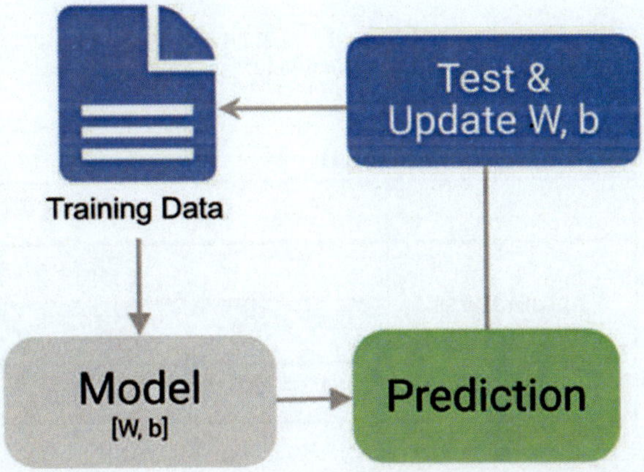

Fig. 1 Machine learning training model [7]

The process is repeated, one iteration (or step) at a time. In each iteration, the initial random line moves closer to the ideal and more accurate one as shown in Fig. 1.

How to Evaluate the Performance of a Model: [8].

This paper demonstrates how to evaluate the performance of a model via Accuracy, Precision, Recall & F1 Score metrics in ML and provides a brief explanation of the "Confusion Metrics" as shown below (Fig. 2).

A **confusion matrix** is a table that is often used to describe the performance of a classification model (or "classifier") on a set of test data for which the true values are known (Table 1) [11].

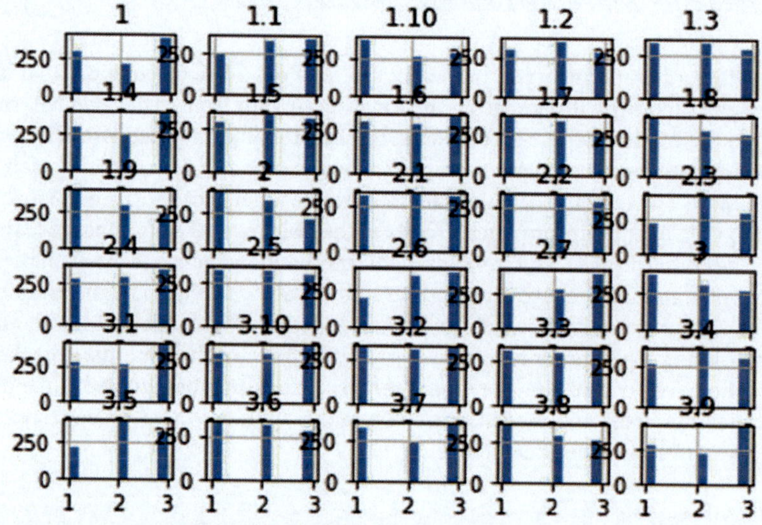

Fig. 2 Histogram plots (Compiled in jupyter notebook using the dataset published in [9] and the questionnaire response published in [10])

Table 1 Confusion matrix [11]

		Predicted class	
Actual Class		Class = Yes	Class = No
	Class = Yes	True Positive	False Negative
	Class = No	False Positive	True Negative

True Positive and **True Negative** are the observations that are correctly predicted and therefore shown in green. We want to minimize false positives and false negatives so they are shown in red color.

Accuracy—Accuracy is the most intuitive performance measure and it is simply a ratio of correctly predicted observation to the total observations. One may think that, if we have high accuracy then our model is best.

Accuracy = TP + TN/TP + FP + FN + TN

Precision—Precision is the ratio of correctly predicted positive observations to the total predicted positive observations. High precision relates to the low false positive rate.

Precision = TP/TP + FP

Recall (Sensitivity)—Recall is the ratio of correctly predicted positive observations to the all observations in actual class—yes.

Recall = TP/TP + FN

F1 score—F1 Score is the weighted average of Precision and Recall. Therefore, this score takes both false positives and false negatives into account. Intuitively it is not as easy to understand as accuracy, but F1 is usually more useful than accuracy, especially if you have an uneven class distribution.

F1 Score = 2*(Recall * Precision)/(Recall + Precision)

Below Table 2 gives the results for this learning style recommender system using several machine learning models.

5.2 *Selecting and Predicting with Best Model: [8]*

We cannot possibly know which algorithm will perform best on our data beforehand. We have to discover it using a process of trial and error. This is also known as spot-checking algorithms. In this paper, we have spot-checked different machine learning algorithms:

- Spot-checking linear algorithms on a dataset [e.g., linear regression (LR), logistic regression, Naive Bayes and Linear Discriminant Analysis (LDA)]
- Spot-checking some nonlinear algorithms on a dataset [e.g., KNN, SVM, and CART]
- Spot-checking some sophisticated ensemble algorithms on a dataset [e.g., Random Forest Classifier (RFC) and stochastic gradient boosting].

The spot-checking algorithm results are shown in Table 3 below and the top two accurate results appeared for Naive Baye's Classifier (**NBC: 63.37%**) and Support Vector Machine (**SVM: 61.34%**).

Table 2 Accuracy, Confusion Matrix, Precision, Recall, F1 Score and support for each model is tabulated below (Based on the questionnaire response [10] and dataset availbale [9])

Model	Accuracy and confusion matrix	Precision, Recall, F1-Score and Support for each model
SVC/SVM	0.7220216606498195 [[44 21 15] [15 77 10] [8 8 79]]	precision recall f1-score support Auditory 0.66 0.55 0.60 80 Kinesthetic 0.73 0.75 0.74 102 Visual 0.76 0.83 0.79 95
LR	0.5379061371841155 [[17 38 25] [20 75 7] [25 13 57]]	precision recall f1-score support Auditory 0.27 0.21 0.24 80 Kinesthetic 0.60 0.74 0.66 102 Visual 0.64 0.60 0.62 95
LDA	0.5884476534296029 [[29 27 24] [29 67 6] [23 5 67]]	precision recall f1-score support Auditory 0.36 0.36 0.36 80 Kinesthetic 0.68 0.66 0.67 102 Visual 0.69 0.71 0.70 95
CART	0.4223826714801444 [[36 30 14] [41 44 17] [39 19 37]]	precision recall f1-score support Auditory 0.31 0.45 0.37 80 Kinesthetic 0.47 0.43 0.45 102 Visual 0.54 0.39 0.45 95
NB	0.6570397111913358 [[34 27 19] [13 79 10] [15 11 69]]	precision recall f1-score support Auditory 0.55 0.42 0.48 80 Kinesthetic 0.68 0.77 0.72 102 Visual 0.70 0.73 0.72 95
RFC	0.48736462093862815 [[38 24 18] [35 54 13] [25 27 43]]	precision recall f1-score support Auditory 0.39 0.47 0.43 80 Kinesthetic 0.51 0.53 0.52 102 Visual 0.58 0.45 0.51 95
KNN	0.40794223826714804 [[68 12 0] [69 27 6] [66 11 18]]	precision recall f1-score support Auditory 0.33 0.85 0.48 80 Kinesthetic 0.54 0.26 0.36 102 Visual 0.75 0.19 0.30 95

6 Result Analysis

The result analysis may not be able to guarantee 100% output unless until the candidates give genuine inputs to the questionnaire based on his past behavior of those scenarios.

Table 3 Spot Checking Algorithm Results (Based on the questionnaire response [10] and dataset availbale [9])

S.No.	Algorithm name	Results_Mean (Results_Standard Deviation)
01.	LR (Logistic Regression)	0.489279 (0.052038)
02.	LDA (Linear Discriminant Analysis)	0.529615 (0.057611)
03.	KNN (K-Nearest Neighbors Classifier)	0.378846 (0.042031)
04.	CART (Classification and Regression Trees)	0.382212 (0.063599)
05.	**NB (Naive Bayes Classifier)**	**0.633774 (0.054632)**
06.	**SVM (Support Vector Machine/Classifier)**	**0.613486 (0.057581**
07.	RFC (Random Forest Classifier)	0.445673 (0.079384)

Histogram Plots:

Below histogram plots show the average values of VAK from average value 1–3.9 having values of

Visual—1

Auditory—2

Kinesthetic—3

Box plots are utilized to show in general examples of reaction for a group [12]. They give a valuable method to imagine the range and different attributes of reactions for an enormous gathering.

From Below box plot Fig. 3, it is seen clearly that Naive Bayes and SVM accuracy is more compared to other models.

Some general observations about box plots from Fig. 3 [12]:

- **The box plot is comparatively short**—This proposes LR and KNN models have a significant level of concurrence with one another. This suggests that LR & KNN models have a high level of agreement with each other.
- **The box plot is comparatively tall**—This proposes CART hold very various suppositions about this viewpoint or sub-part of different models.
- **The four sections of the box plot are uneven in size**—The four areas of the box plot are uneven in size—This CART model shows that numerous applicants have comparative perspectives at specific pieces of the scale, however, in different pieces of the scale up-and-comers are progressively factor in their perspectives as far as learning. The long upper whisker in the CART model implies that students' sees are shifted among the best quartile gathering, and fundamentally the same as for the least positive quartile gathering.

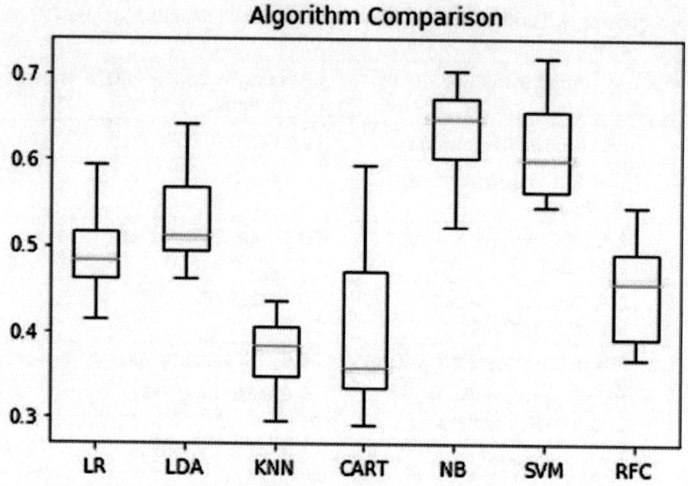

Fig. 3 Box plot for comparison of implementation of different ML models

Comparison of Different Models:

The summary of different Machine Learning models along with the algorithm type and the training dataset accuracy as well as testing dataset accuracy results are shown in Table 4.

From the above summary results in table, it is clear that SVM stands on Top classifier and Naive Bayes Classifier on second position for our Test dataset in terms of accuracy and Naive Bayes Classifier stands on Top and SVM on second position for our Training dataset in terms of accuracy.

Table 4 Comparison of training dataset accuracy and testing dataset accuracy for different models

S. no.	Model/algorithm name	Algorithm type	Training dataset accuracy	Testing dataset accuracy
1.	Logistic Regression	Linear	48.92	53.79
2.	LDA (Linear Discriminant Analysis)	Linear	52.96	58.84
3.	**Naive Bayes Classifier**	**Linear**	**63.37**	**65.70**
4.	Random Forest Classifier	Ensemble	44.87	48.73
5.	KNN (K-Nearest Neighbors)	Nonlinear	37.88	40.79
6.	CART (Classification and Regression Tree)	Nonlinear	38.98	42.23
7.	**SVM (Support Vector Machine)**	**Nonlinear**	**61.34**	**72.20**

The learning styles discussed in this paper have acceptable levels of reliability and validity though much more research in this area is needed. Learning consolidation will be increased if all three of the modalities are used in teaching; visual will reinforce auditory and auditory will reinforce haptic/kinesthetic [5].

The learning styles idea has a lot of potential with under-arranged understudies. Ongoing years have seen an adjustment in the patterns of instruction from teaching method to andragogy, for example, from an educator focused figuring out how to understudy focused learning. In this manner, it isn't just alluring yet additionally fundamental for teachers to perceive that understudies have diverse learning styles and that they should tailor guidelines to the trademark manners by which they want to learn.

7 Conclusion

This paper titled "Learning Style Recommender System using VAK technique and Machine Learning" works good with below two Machine Learning models:

1. Naive Bayes Classifier and
2. SVM (Support Vector Machine)

These models can be further calibrated given we have more information cases. Another way that we can improve the presentation of our models is to consolidate the forecasts from different models. A few models give this ability worked in, for example, irregular woodland for stowing and stochastic slope boosting for boosting.

Individuals ordinarily have a principle favored learning style, yet this will be a piece of a mix of every one of the three. A few people have a very solid inclination; other individuals have an all the more even blend of two or less regularly, three styles. For instance, male formative understudies who have a haptic/Kinesthetic direction ought to likewise be approached to listen cautiously, as when the guidelines are given [4].

The main problem in machine learning is with the number of dimensions in the data known as the **curse of dimensionality**. The curse of dimensionality arises in many places in computer science, especially in machine learning [11]. The future scope of this work can be implemented using deep learning with increased accuracy or we need to increase the dimensionality of data to increase the accuracy.

Acknowledgments With affirmations to Victoria Chislett for building up this assessment [6]. Victoria Chislett spends significant time in execution of brain research and its application inside associations [13–15].

References

1. Training ML models, Amazon web services, Amazon machine learning. https://docs.aws.amazon.com/machine-learning/latest/dg/training-ml-models.html
2. The VAK Learning Style. https://www.southwesterncc.edu/sites/default/files/VAK_Learning_Styles.pdf
3. Jones SG, NLP Trainer (2010) NLP practitioner manual, p 7. www.AUNLP.org
4. Srinivas et al (2017) Recognizing smart people with NLP techniques using AI expert system. Int J Res Eng Sci Technol (IJRESTs) 03(01):14–17. Deepam Publication's
5. Lemire D (1998) Three learning styles models: research and recommendations for development education. 3(2):31–32. Cloud County Community College
6. Chislett V, Chapman A (2005). www.businessballs.com/freematerialsinword/vaklearningstylesquestionnaireselftest.doc
7. Roman V How to develop a machine learning model from Scratch, Date Published: Dec 23, 2018. https://towardsdatascience.com/machine-learning-general-process-8f1b510bd8af. Accessed 27 July 2019
8. Joshi R Accuracy, prevision, recall & f1 score: interpretation of preformance measures, Uploaded on September 9, 2016. https://blog.exsilio.com/all/accuracy-precision-recall-f1-score-interpretation-of-performance-measures/. Accessed 2 April 2019
9. Learning style assessment dataset. https://tinyurl.com/y5e38epc
10. Learning style questionnaire online form. https://goo.gl/forms/zsfGIEa7jB87Ie4P2
11. Goodfellow Ian, Bengio Yoshua, Courville Aaron (2016) Deep learning. The MIT Press, Cambridge, Massachusetts, pp 151–152
12. Wellbeing@School, Understanding and interpreting box plots. https://www.wellbeingatschool.org.nz/information-sheet/understanding-and-interpreting-box-plots. Accessed 26 July 2019
13. Published: August 20 2008. http://knowledge.wharton.upenn.edu/article/the-objective-of-education-is-learning-not-teaching/. Accessed 19 Jan 2019
14. Rizwan M Module 18—machine learning based recommendation systems, Uploaded on Aug 23, 2018. https://engmrk.com/module-18-machine-learning-based-recommendation-systems/?utm_campaign=News&utm_medium=Community&utm_source=DataCamp.com. Accessed 25 July 2019
15. VARK. http://learningstyle2015.weebly.com/flemings-vark-model1.html

Srinivas Adapa is working as an Associate Professor in CSE Department at **ITM Coastal Institute of Technology and Management (CITM)**. He is playing a role as the Head of CSE Department and Heading the Training & Placement Cell. He has a total of 16 years of experience which includes Industry sector, Entrepreneurship, and as an Academician. He has published more than 20 papers and attended more than 20 workshops/FDPs so far. He guided several B.Tech Projects during his career as an Associate Professor. Currently, he is pursuing his Ph.D. at **KL University** and doing his research work in Deep Learning and Computer Vision under the guidance of Dr. E. Vamsidhar, Associate Professor, CSE Dept.—KL University.

Netaji Gandi is working as an Assistant Professor in IT Department at **Vignan's Institute of Engineering for Women**. He is playing a role as an Association Coordinator in IT Department. He has a total 7 years of experience in teaching and he has published 5 papers and attended more than 15 workshops/FDPs. He guided several UG projects in the department.

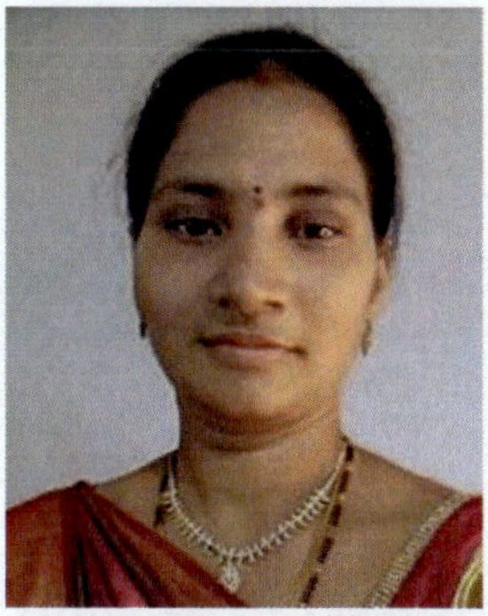

Vechalapu Alekya is working as an Assistant Professor in CSE department at **ITM Coastal Institute of Technology and Management**. She has a total of 8 years of experience in teaching and playing a role as CSE Department Association Coordinator and Project Coordinator. Guided several UG Projects in CSE Department. She also published 7 papers and attended many workshops and FDP's. Her area of interest is machine learning and computer security.

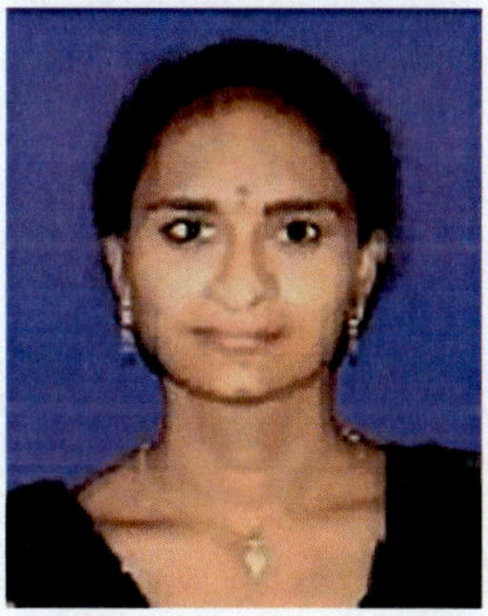

Lakshmi Durga Ganthakora is working as an Assistant Professor in CSE department at ITM Coastal Institute of Technology and Management (CITM). She has a 1 year of teaching experience and also playing a role as class Incharge for Final year B.Tech. She has published 1 Paper and attended many workshops/FDPs. Guided 1 UG Project in CSE Dept. Her area of interest is machine learning and networks.

Advanced Safety Analysis and Modeling of Control Software in Implantable Medical Devices

Umamaheswararao Batta, Jayasri Kotti, and Seetharamaiah Panchumarthy

Abstract Medical Cyber-Physical Systems (MCPS) is an integration of physical system, medical devices, and control software. Software for MCPS should contact with potential hazard of the system identified by safety analysis. Implantable medical devices with embedded software and control software for clinical programming are used in hospital to get safety and high-quality health care. Modeling the error-free and safety software for medical devices is essential and challenging to avoid hazards in MCPS. This paper identifies model for the networked MCPS to overcome some problems related to medical devices. The process starts with modeling the system with the help of some mathematical techniques like timed automata, modeling languages and perform hazards identification process using different hazard analysis techniques. After that derive new safety requirements for system and catalog. Development of an MCPS based on our proposed systems-theoretic software safety approach significantly enhanced the safe operation of the overall system. The proposed model identifies the component and interaction failures within a systems and provides safety to the system.

1 Introduction

Cyber-Physical System (CPS) is a complex system with many interacting components exhibiting aggregate behavior [1, 2]. CPS is used in multiple areas such as medicine, aerospace, automotive, chemical processes, civil infrastructure, and biological systems technology [3]. CPS involves trans-disciplinary techniques,

U. Batta (✉)
Department of CSE, SVEC, Tirupati, India
e-mail: umamaheshbatta@gmail.com

J. Kotti
Department of CSE, LENDI, Vizianagaram, India
e-mail: jayasrikotti@gmail.com

S. Panchumarthy
Department of CSE, Andhra University, Vizag, India
e-mail: psrama@gmail.com

S. C. Satapathy et al. (eds.), *Communication Software and Networks*, Lecture Notes in Networks and Systems 134, https://doi.org/10.1007/978-981-15-5397-4_21

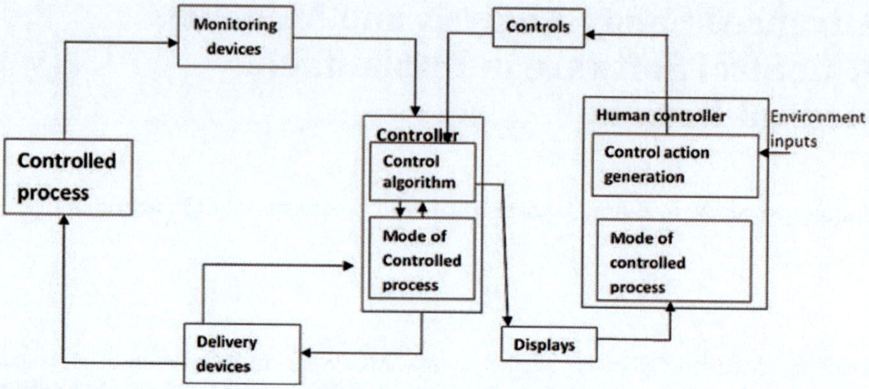

Fig. 1 Medical Cyber-Physical System architecture

combining concept of cybernetics, mechanic design, and process science [4–6]. CPS is safety critical in nature according to Networking and Information Technology Research and Development (NITRD) program [7]. Figure 1 illustrates about the conceptual overview of MCPS.

Safety analysis and modeling of MCPS has been done by Prof. Seetharamaiah in Andhra University [8–11] on medical devices like Impedance Telemetry Monitoring System (ITMS), Digital Speech Processor System (DSPS), and Implantable Receiver Stimulator (IRS) are deployed [8–11]. This paper proposes a hazard-based safety-driven model-based system engineering methodology on MCPS. This paper addresses the core research problem by presenting a hazard-based, Systems-Theoretic approach to modeling software safety, and drawing from research in the fields of systems engineering, software engineering, and safety engineering. These involve additional design, analysis, measurement and verification activities that occur concurrently with the software development activities. To model a process the most critical facet is to catch the dynamic behavior. Different methods for software verification are shown in Fig. 2.

2 Hardware and Software Modules of MCPS

Figure 3 shows the different hardware components of CIS system. The cochlear implant system has four hardware modules such as Digital Speech Processor System (DSPS), Impedance Telemetry Monitoring System (ITMS), HeadSet Cable (HSC), Implantable Receiver Stimulator (IRS) and design and development of a software-controlled Medical Cyber-Physical software system in the medical domain such as

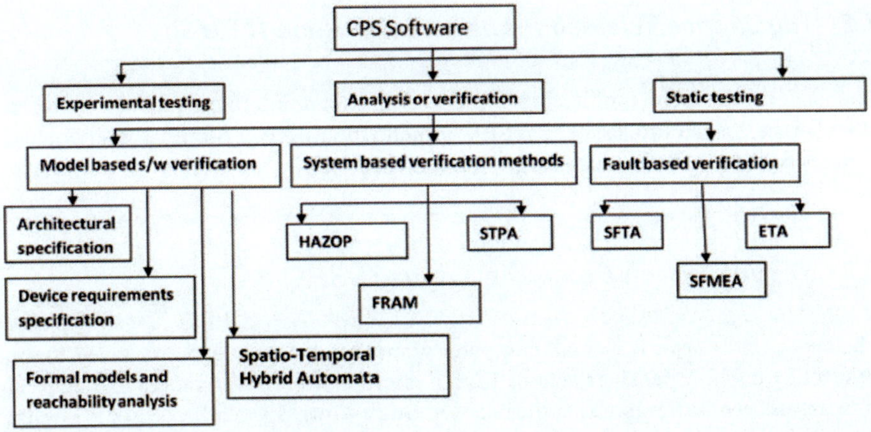

Fig. 2 Safety assurance approaches for CPS software

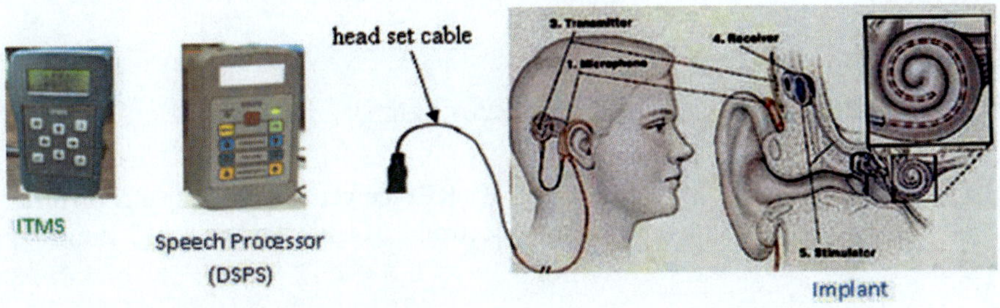

Fig. 3 Hardware components of MCPS

Control Software for Clinical Programming (CSCP) of Cochlear Implant System (CIS). A cochlear implant (CI) is a real-time embedded computing device that can offer a feeling of audio to people who are deaf or profoundly hearing-impaired [12–14]. Significant speech identification can be carried out from the commercial perspective available multichannel cochlear implant systems.

2.1 Impedance Telemetry Monitoring System (ITMS)

Impedance telemetry monitoring system acts as an interface module between Body Warn Speech Processor (BWSP), Implantable Receiver Stimulator (IRS), and Control Software for Clinical Programming (CSCP).

2.2 Digital Speech Processor System (DSPS)

The speech processor is a small computer worn on your body and connected to the headset by cable. It receives sounds from a microphone, converts them to electrical signals, and sends the signals to the headset. Body-Worn Speech Processor is Digital Speech Processing System (DSPS). It is used for capturing incoming sound signal (200–7500 Hz in eight bands), amplifying the sound signal, and compressing the sound signal as per patient's required levels.

2.3 Implantable Receiver Stimulator (IRS)

The IRS unit of Cochlear Implant consists of a RF receiver coil made up of Two turns Pt–Ir, Electrode Array with a Reference Electrode, Magnet, and an ASIC shown as in Fig. 3.

2.4 Control Software for Clinical Programming (CSCP)

The CSCP software is implemented as an interface software for DSPS and ITMS used by an audiologist. The program contains different sub-tasks such as patient registration, UART Settings, ITMS measurements, fitting, and mapping. The data base is developed for sorting patient data, medical record, and evaluation of hearing abilities, evaluation of speech and language status, rehabilitation status, evaluation of psychological status, medical and audiological evaluation, processor programming, and specific training with processor accessories, and so on. The following Fig. 4 shows the partial block diagram of CSCP.

The modules in CSCP software are described as follows:

- **UART settings module** to adjusting the system settings.
- **NEW module** to patient registration.
- **Update module** to update patient data.

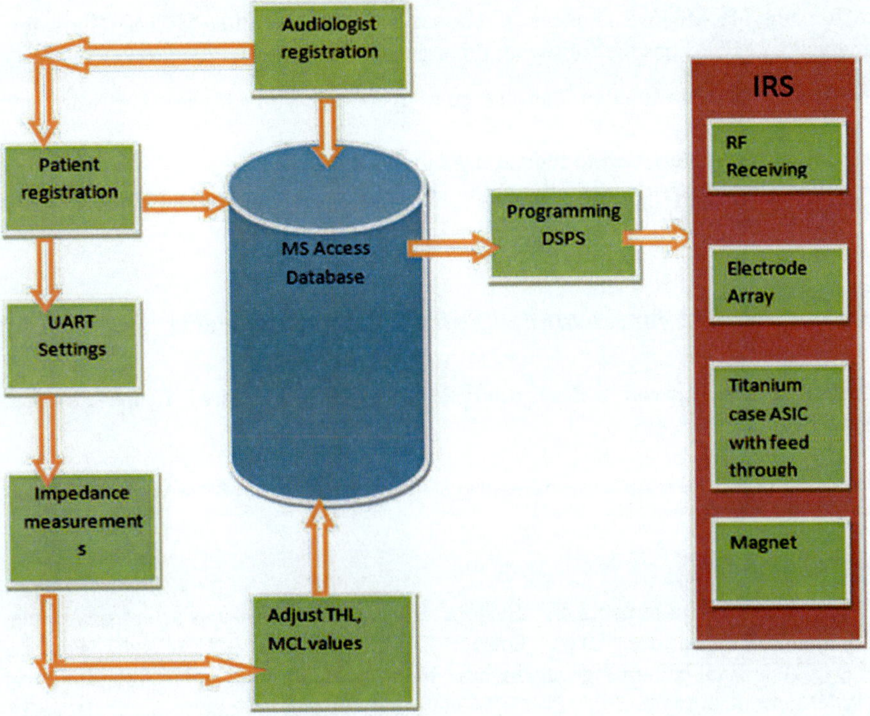

Fig. 4 Database application for CSCP for bionic ear

- **IMP-TELE module** to measure impedance values from ITMS.
- **Fitting module** to Finding THL (threshold level)/MCL (most comfortable level) for each Active Electrode.
- **MAP module** is used to show the patient parameters and to store the parameters in Speech Processor.
- **Search module** used to search patient data.
- **Edit module** is used to edit the patient data.
- **Delete module** is used to edit the patient data.
- **Refresh module** is used to refresh the patient data.

3 Advanced Safety Analysis for MCPS

This proposed approach gives new constraints, requirements, and design decisions of Medical Cyber-Physical System for designers in building a software safety. It develops foundations and techniques for building safe and effective Medical Cyber-Physical System. This Advanced Safety Analysis and Modeling of Control Software in Implantable Medical Devices is recommended in the other MCPS. The safety analysis of MCPS has the following main purposes:

- Hazard Analysis and their causal factors in concert with software, hardware, and human elements.
- Causal factors are used to reduce risks in the system.
- Document the residual safety risk for the delivered system including recommendations for controlling the hazards.

3.1 *Algorithm for Advanced Safety Analysis Process*

Algorithm: Advanced Safety analysis process in Medical Cyber-Physical Systems.

Inputs: functional requirements and system requirements of the Medical Cyber-Physical System.

Variables:

FR: Functional requirements, SR: System requirements, H: Hazards, C: Constraints, CS: Control Structure, UCA: Unsafe Control Actions, CF: Causal Factors, H_TOTAL: Total hazard log information, MSM: Modeling of system for Structural Model in control structure, MCSM: Modeling of system for operating Structure of MCPS control structure, MCM: Modeling of system for MCPS Component Model, DHLSD: Develop High Level System Design, RCS: Requirements & Constraints of subsystems, DDNR: Design Decisions.

BSEM: Behavioral specification on each module, ITMS: Impedance Telemetry Monitoring Systems, DSPS: Digital Speech Processor System

Outputs: Unsafe control actions, causal factors, and new safety requirements and safe design decisions.

Algorithm:

Step 1: *Read the FR, SR of the Medical Cyber-Physical System*
Step 2: *Find the H, C, CS based on FR, SR*
Step 3: *Perform STPA analysis on control structure*
PSMSM←STPA [MSM]: Performing STPA analysis on MSM.
PSMCSM←STPA [MCSM]: Performing STPA analysis on MCSM.
PSMCM←STPA [MCM]: Performing STPA analysis on MCM.
Step 4: *Create hazard log for each hazard, corresponding constraints and casual factors*
HLPMSM←CHL [PMSM]: Create Hazard Log for PSMSM.
HLPMCSM←CHL [PMCSM]: Create Hazard Log for PSMCSM.
HLPMCM←CHL [MCM]: Create Hazard Log for PSMCM.
H_TOTAL= [HLMSM+HLMCSM+HLMCM]
Step 5: *Safety Operations on each Log.*
S1←S [HLMSM]: Safety operations on HLMSM.
S2←S [HLMCSM]: Safety operations on HLMCSM.
S3←S [HLMCM]: Safety operations on HLMCM.
STEP 6: *Perform check operation on each Hazard Log whether safety operation is done or not*

```
C1←C [S1];
C2←C [S2];
C3←C [S3];
C_TOTAL= [C1+C2+C3];
```

Step 7: *Now perform Safety operations on each Hazard Log and obtain S1, S2, and S3.*
Step 8: *Perform Check operation on all Hazard Log i.e., H_total whether the Safety operation is performed on all Hazard Log or not.*

```
If (! H_TOTAL=C_TOTAL)
        {
Do step 9 to step 12;
        }
Else
        {
        /*Develop High Level System Design (DHLSD);
Do R13 to R14;
Return Final Design;
}
```

Step 9: *Get the following RCS, DDNR, and BSEM.*
Step 10: *Perform STA analysis on RCS, DDNR, and BS respectively and obtain information as*

```
SRCS←STPA [RCS];
SDDNR←STPA [DDNR];
SBSEM←STPA [BSEM];
```

Step 11: *Update C_TOTAL ←C_TOTAL+R7;*
Step 12: *Refine c.*
Step 13: *Final Design ←Identify Unsafe Control Actions and change Final Design.*
Step14: *Get new requirements or constraints and design decisions.*
Step 15: *Safe system design.*
Step 16: Get the patient details
Step 17: Measure the impedance values from patient to ITMS and transfer to IMP_TEL module in the CSCP
ITMS ← Patient Impedance data
IMP_TEL ← ITMS values
Step 18: Setting the min and max comfortable levels of electrode in CSCP
Step 19: Store patient parameters into DSPS
Step 20: Using DSPS, Adjust to patient impedance values to hear.

4 Results

This paper identifies 65 scenarios found by FMEA whereas 134 identified by STPA. It uses less time and resources. FMEA results show that it identified only single fault cause of hazard whereas Timed Automata plus STPA identified complex causes of hazards, multiple failures, and no component failure that lead to a hazard (Figs. 5 and 6).

Assessment of MCPS is done by operating each safety analysis technique for different attributes of the system. Assessment of the three safety analysis techniques is done by operating each technique for ten attributes. Our proposed technique performs better than the other two safety analysis techniques. In this technique, the timed automata and STPA analysis are set to different attribute values. This enhances both safety and stability. Proposed technique shows better safety performance for small errors compared to FEMA and FTA since these are used for component failures (Table 1).

5 Conclusion

As a means of addressing the gaps in the understanding of software engineering for safety, a Systems-theoretic functional approach for identification of hazard and proving safety in MCPS was presented. This approach is derived from research in the fields of systems engineering, software engineering, and safety engineering. Improvements in the PC-based Control Software for Clinical Programming are also to

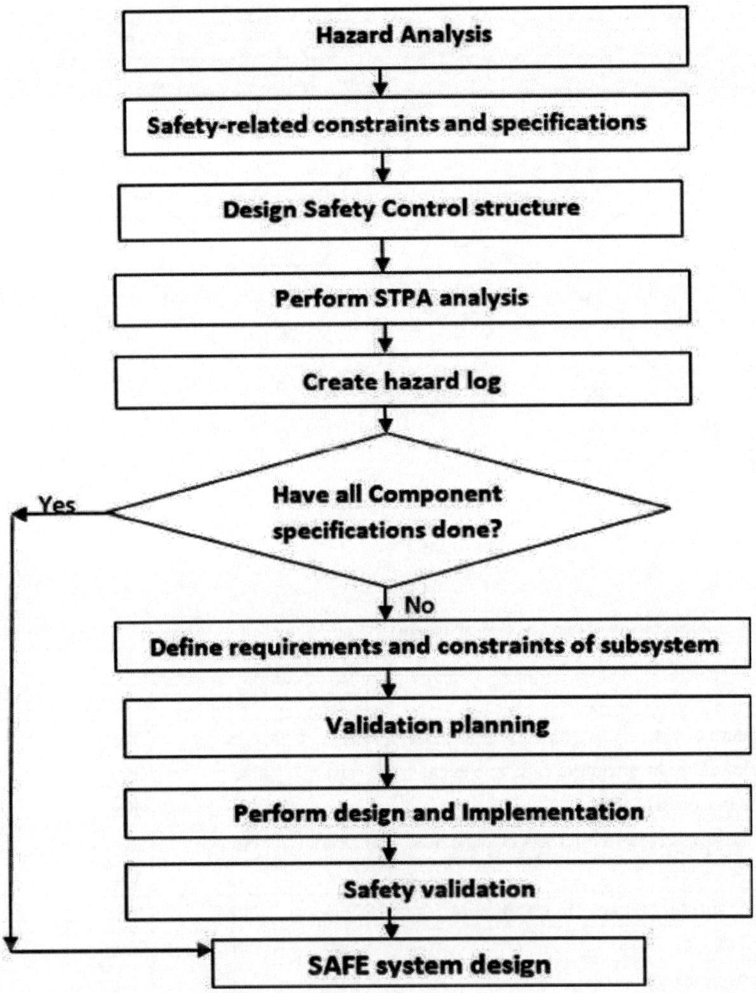

Fig. 5 Safety analysis in MCPS for safe system

be done for ease of usage by the audiologists and also provide options for better patient data fitting and configuring the Body-worn Speech Processor. An upcoming future direction of Cochlear Implant System development is to combine the components of Cochlear Implant System in a single unit.

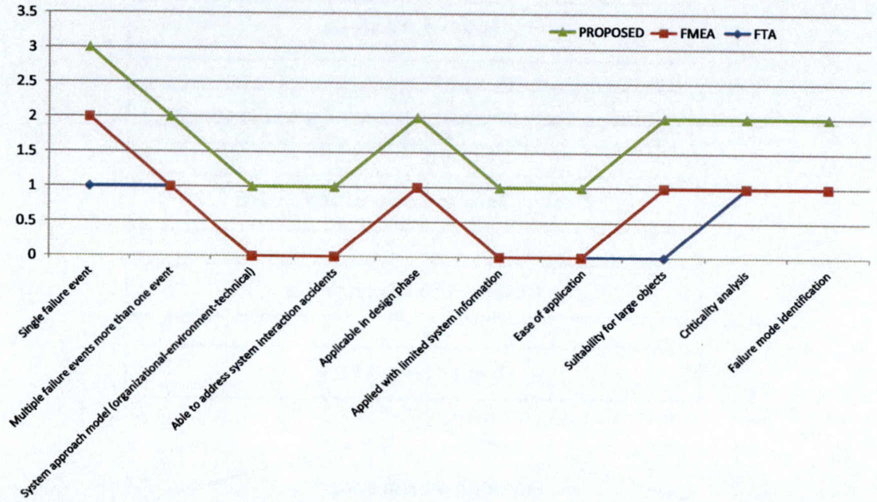

Fig. 6 Assessment of software safety in MCPS

Table 1 Hazard analysis methods comparison

Attributes	Methodology		
	FTA	FMEA	Proposed method
One failure event	Yes	Yes	Yes
Multiple failure events more than one event	Yes	No	Yes
System approach model	No	No	Yes
Able to address system interaction accidents	No	No	Yes
Design phase use	Yes	Yes	Yes
Applied with limited system information	No	No	Yes
Ease of application	No	No	Yes
Suitability for large objects	No	Yes	Yes
Criticality analysis	Yes	No	Yes
Failure mode identification	Yes	No	Yes

References

1. Bar-Yam Y (2003) Dynamics of complex systems. Series Studies in Nonlinearity. Westview Press. http://www.amazon.com/exec/obidos/ISBN=0813341213/newenglandcompleA/
2. Harel D (1987) Statecharts: a visual formalism for complex systems. Sci Comput Program 8: 231–274. http://dl.acm.org/citation.cfm?id=34884.34886
3. Khaitan SK et al (2014) Design techniques and applications of cyber physical systems: a survey. IEEE Syst J
4. Hancu O, Maties V, Balan R, Stan S (2007) Mechatronic approach for design and control of a hydraulic 3-dof parallel robot. In: The 18th International DAAAM symposium, intelligent

manufacturing & automation: focus on creativity, responsibility and ethics of engineers
5. Lee EA, Seshia SA (2011) Introduction to Embedded Systems—A Cyber-Physical Systems Approach, LeeSeshia.org
6. Suh SC, Carbone JN, Eroglu AE (2015) Applied Cyber-Physical Systems, Springer 2014, Rad, Ciprian-Radu; Hancu, Olimpiu; Takacs. Ioana-Alexandra; Olteanu, Gheorghe
7. T. N. I. T. Research and D. Program, Different definition of cyber physical systems. http://www.nitrd.gov/about/blog/white_papers/16-Importance_of_Cyber-Physical_Systems.pdf
8. Raja Kumar K, Seetha Ramaiah P (2006) DSP and microcontroller based speech processor for auditory prosthesis. In: Proceedings of the 14th international conference on advanced computing and communication, ADCOM-2006, Advanced computing and communications society—Bangalore and National Institute of Technology, Surathkal, INDIA, 20–23 December 2006, pp 518–522
9. Raja Kumar K, Seetha Ramaiah P (2008) Programmable digital speech processor for auditory prostheses. In: 2008Proceedings of IEEE international region-10 conference on innovative technologies for societal transformation—TENCON, 18–21 November 2008, Hyderabad, India
10. Raja Kumar K, Seetha Ramaiah P (2010) Development of receiver stimulator for auditory prosthesis. Int J Comput Sci 7(3): 2
11. Raja Kumar K, Seetha Ramaiah P Microcontroller based receiver stimulator for auditory prosthesis. In: 2008 Proceedings of IEEE international region-10 conference on innovative technologies for social transformation—TENCON, 18–21 November 2008, Hyderabad, INDIA
12. Clark GM (2006) The multiple-channel cochlear implant: the interface between sound and the central nervous system for hearing, speech, and language in deaf people—a personal perspective. Phil Trans R Soc B 361:791–810
13. Zeng FG, Rebscher S, Harrison W, Sun X, Feng H Cochlear implants: system design, integration, and evaluation. IEEE Rev Biomed Eng 1: 115–142
14. Wilson BS, Dorman MF (2007) The surprising performance of present-day cochlear implants. IEEE Trans Biomed Eng 54(6): 969–973

Empty Region Detection in an Image Using Deep Convolutional Neural Network

Kavita Chachadi, Vanraj Vala, and Shilpa Kamath

Abstract As Convolutional Neural Networks (CNN) have addressed many computer vision problems like image classification, object detection, object recognition etc. Many object detection algorithms detect an object present in an image by drawing bounding box around that object to localize that particular object using different CNN architectures and have achieved good results. When images are used as background to display text, the arrangement of text in relation to the image can make image not to be so visually appealing. So far object detection and recognition algorithms focused on finding out specific object either in an image or video, but our paper mainly focuses on finding out an empty area in a background image using CNN so that user can enter text inside this empty area without affecting the saliency part of image and also making image to look visually appealing. Finally, we demonstrate the results for empty region detection with accuracy of 85% using IoU as metric.

1 Introduction

Object detection has been found an active area of research which is a challenging problem in computer vision for several decades. The purpose of object detection is to detect if there are any occurrences of objects from the given classes (such as animals, vehicles and humans) in some given image and return spatial location of each object occurrence if present in a given image via a bounding box coordinate [7, 8]. Currently, feature representations can be learnt automatically from data using

K. Chachadi (✉) · S. Kamath
KLE Technological University, Hubballi, Karnataka, India
e-mail: 17kavi17@gmail.com

S. Kamath
e-mail: shilpakul@gmail.com

V. Vala
Samsung Research and Development, Bengaluru, India
e-mail: vanraj.vala@samsung.com

S. C. Satapathy et al. (eds.), *Communication Software and Networks*, Lecture Notes in Networks and Systems 134, https://doi.org/10.1007/978-981-15-5397-4_22

deep learning techniques. In particular, there is a significant improvement for object detection using these techniques [9, 10].

Convolutional neural networks (CNN) have succeeded in visual recognition tasks, such as scene parsing [2], object detection [3], and image classification [1]. Donahue et al. [5] discussed that the features extracted from Krizhevsky's CNN trained on the ImageNet dataset [4] can be reused to generic tasks. Results shown by Razavian et al. [6] concluded that CNN-based deep learning can be used for any visual detection and recognition task.

Object detection can be looked into one of two types: first one is the detection of occurrence of specific object such as Human's face and the building, and second type is the detection of specific classes where different occurrences of predefined object classes such as humans, vehicles, animals, birds can be detected. So far most of the object detection has focused on the detection of single class (such as pedestrians and faces) or a few specific classes.

Our paper focuses on development of CNN model which finds empty area in an image which is the optimal position of stencil for displaying text. Contributions towards this work are as follows:

1. Prepared our own custom dataset for training CNN.
2. Developed CNN model that finds optimal position of stencil which is empty region in an image.
3. Using IoU as a metric obtained average accuracy of 85%.

Finally, the developed model is trained and tested on different images which have empty areas in an image.

2 Motivation

So far, all the existing object detection architectures based on deep learning techniques are used to find a specific object in an image. But in our paper, we have focused on finding empty region in an image using deep learning techniques. So, CNN's can also be used to find region in an image that contains no objects. Finding empty region in an image has also played important role in some of the applications like entering of text in an image, captioning, segmentation, scene understanding, etc. When images are used to display text where the arrangement of text on background image can be such that which can affect saliency part of an image and making image to look distracting. So there is a need to find the empty area where optimal position and size of stencil is decided for displaying text on a given image background.

3 Literature Survey

Recently, deep learning techniques have become strong methods for learning features directly from data and have led to remarkable results in the object detection field. The specific object which is of importance in any given image can be labeled with rectangular bounding box to show the existence of that object can be detected by object detection algorithm and classify them accordingly. ConvNets [11–14] intended for classification of image have realized impressive image features representations, outperforming traditional handcrafted features [15, 16]. There was a significant improvement in accuracy for object detectors adopting these deep ConvNets on various detection benchmarks [17, 19].

Proposal region features are extracted with parameters initialized from a pretrained ImageNet [21] classification model by R-CNN [19] which firstly uses deep ConvNets. Region-of-Interest (RoI) developed by Fast R-CNN [20] in which RoI features are extracted from the convolutional feature maps of the entire image. After that, for generating accurate RoI proposals and sharing computation with detection sub-network, region proposal network(RPN) was introduced by Faster R-CNN [22]. Apart from the proposed one-stage object detection frameworks [23–25], proposal generator and region classifier which is two-stage framework adopted by most of the state of-the-art object detectors. One follows firstly generating region of proposals and classifying generated proposal into different object categories which follows traditional object detection pipeline. The other approach is object detection as a classification or regression problem, implementing a fused framework to achieve final results (categories and locations) directly.

Some of the applications of object detection are discussed below.

Researchers have also extended CNN to salient object detection. Saliency map can be created using high dimensional feature space from AlexNet which was proposed by Kummerer et al. Deep Gaze [26]. Saliency map prediction is integrated into pretrained object recognition DNNs by a similar architecture which was shown by Huang et al. [27]. Saliency maps can also be predicted by taking both global search and local estimation by training two independent deep CNNs (DNN-G and DNN-L) proposed by Wang et al. which captures local information and global contrast [28].

Best Saliency map is selected from a set of saliency maps produced by fast bottom-up saliency approaches is combined with saliency maps from top-down in a weakly supervised top-down saliency detection framework by Cholakkal et al. in his work and results are refined with a multi-scale super pixel-averaging [29]. By combining local and global context with the aid of superpixel segmentation was proposed by Zhao et al. which uses a multi-context deep learning framework for saliency detection [30]. Recurrent Neural Networks (RNNs) and de-convolution networks are used to predict salient objects which are found to be very accurate and efficient method by training CNN architecture. Complete Image saliency detection using deeply supervised recurrent convolutional neural network (DSRCNN) was developed by Tang et al. [31].

As part of pre-processing step, in many face recognition applications face detection plays an important and essential role which uses CNN-based approaches. Without constraint on landmark/pose annotations Multi-view face detection is able to detect faces in a wide range of orientations in a work proposed by Farfade et al. [32, 33] known as Deep Dense Face Detector. To address face detection under unconstrained pose variations and severe occlusions, a complete face detection framework was proposed by Yang et al. based on deep learning, which collects the responses from local facial parts (e.g. eyes, nose and mouth) [34].

Pedestrian detection is essential in applications like tracking of pedestrian, robot navigation and re-identification of person. Zhang et al. made an effort to use Faster R-CNN [20] to pedestrian detection [35].

4 Proposed Technique

We have prepared the image dataset with bounding boxes which tells empty area which is the position of stencil and we have considered shape of stencil as rectangle which is included into the dataset. Initially,we have taken fixed stencil (bounding box) positions like top-left, top-right, bottom-left, bottom-right and centre position as an empty region in an image and developed 4 stage CNN model as shown in Fig. 1 which takes image along with stencil position, with the values bx (bounding box top-left coordinate), by (bounding box top-right coordinate), bh (bounding box height), bw (bounding box width) as its input and predict output which is the optimal position of stencil.

This section describes the results and accuracy of Our Model.

Convolutional neural network efficiency depends to a large extent on the quality of the training dataset; the network will produce good results only if the training data used contain sufficient important characteristics so that they can produce accurate predictions.

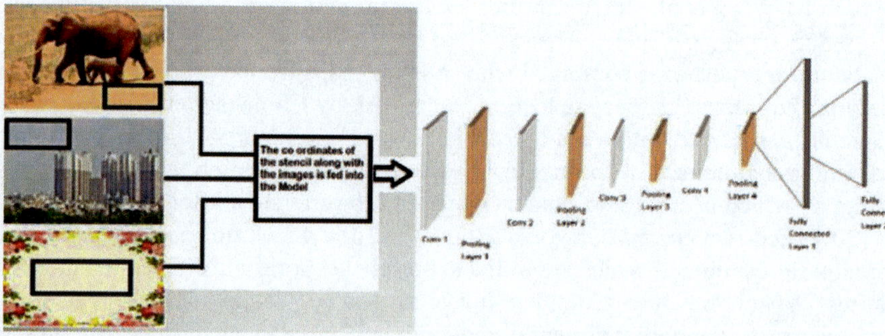

Fig. 1 CNN Model

4.1 Dataset Analysis: Preparation of Custom Dataset

The dataset contains 2500 training images with resolution of 512 × 512 and dividing the dataset among all five fixed position as described above. While selecting images we have identified different varieties of images with empty regions only at top-left, top-right, bottom-left, bottom-right and centre. In some of the images like sky and sea, we have taken sand as empty region for training. We have considered 500 images in each of the fixed positions as empty area in an image and considered 100 test images for validating the model.

4.2 Model Details and Parameters

Our deep learning model is composed of four layers of convolution. The first three convolutional layers use a ReLU activation function. At the output of each of these layers, we add the MaxPooling2D layer of size 2 × 2. We have used two fully connected layers. The size of first fully connected layer being 4096 and second fully connected layer is of size 4. The last fully connected layer gives final output in terms of bounding box coordinates which is the stencil position where model is able to predict empty area. We have considered L2 norm as loss function which tries to minimize the error between ground truth bounding box coordinates and predicted bounding box coordinates. L2-norm loss function is also known as least squares error (LSE). It is basically minimizing the sum of the square of the differences **(S)** between the target value **(Yi)** and the estimated values **(f(xi))**:

$$\mathbf{S} = \sum_{i=0}^{n} (Y_i - (f(x_i))^2 \quad (1)$$

The entire network is trained on Nvidia GTX GPU. Each batch size is considered as 100 with learning rate of 0.0001 and used Adam as optimizer. We have considered Intersection of Union (IoU) as metric to find the accuracy of the model. IoU computes the intersection over the union of the detected bounding box and the ground truth one.

$$\text{IoU} = \text{size of intersection area/size of union area} \quad (2)$$

We have considered an accurate empty area detection if the result of IoU is above some specific value. (i.e. IoU > = 0.5) and calculated precision of the model.

5 Experimental Results

We have tested model for all the fixed five positions which has empty area in a background image and model has performed well in detecting the empty region where user can enter text without distracting other parts of an image. The model has learnt to detect empty region which contains no objects. The model efficiently detects region in upper-right, upper-left, bottom-right, bottom-left and centre position. For our proposed technique, we are getting precision of the model as 85%. Some of the results are shown in Fig. 2.

Fig. 2 Empty region detection in five different positions (Top-Left, Top-Right, Bottom-Left, Bottom-Right and Centre)

The model performs excellent in detecting centre empty region compared to other empty regions in an image.

6 Conclusion

In this paper, we have introduced deep convolution neural network (CNN) for identifying empty region in an image. The proposed model identifies empty area only in five fixed positions where user can enter text without affecting saliency part of an image. In future work, we want to use a pretrained neural network which will allow us to see the performance of model and also, we can make model to predict more than one empty region if present in background image. We can also extend this work to detect empty regions in videos. Many existing architectures does not work on embedded/mobile device, if we have tflite model size info for the proposed method, it can work well for embedded device.

References

1. Krizhevsky A, Sutskever I, Hinton GE (2012) ImageNet classification with deep convolutional neural networks. In: Proceedings of conference NIPS, pp 1097–1105
2. Farabet C, Couprie C, Najman L, LeCun Y (2013) Learning hierarchical features for scene labelling. IEEE Trans Pattern Anal Mach Intell 35(8):1915–1929
3. R. Girshick, J. Donahue, T. Darrell, and J. Malik, "Rich feature hierarchies for accurate object detection and semantic segmentation," in Proc. IEEE Conf. CVPR, 2014, pp. 580–587
4. J. Deng, W. Dong, R. Socher, L.-J. Li, K. Li, and L. Fei-Fei, "ImageNet: A large-scale hierarchical image database," in Proc. IEEE Conf. CVPR, 2009, pp. 248–255
5. J. Donahue, Y. Jia, O. Vinyals, J. Hoffman, N. Zhang, E. Tzeng, and T. Darrell, "Decaf: A deep convolutional activation feature for generic visual recognition." in Proc. ACM Conf. ICML, 2014, pp. 647–655
6. A. Sharif Razavian, H. Azizpour, J. Sullivan, and S. Carlsson, "Cnn features off-the-shelf: an astounding baseline for recognition," in Proc. IEEE Conf. CVPRW, 2014, pp. 806–813
7. Hariharan B, Arbelaez P, Girshick R, Everingham M., Gool L. V., Williams C., Winn J., Zisserman A. (2010) The pascal visual object classes (voc) challenge. IJCV 88(2):303–338
8. Russakovsky O, DengJ., SuH., KrauseJ., SatheeshS., MaS., Huang Z., Karpathy A., Khosla A., Bernstein M., Berg A., Li F. (2015) ImageNet large scale visual recognition challenge. IJCV 115(3):211–252
9. Hinton G., Salakhutdinov R. (2006) Reducing the dimensionality of data with neural networks. science 313(5786):504–507
10. LeCun Y, Bengio Y, Hinton G (2015) Deep learning. Nature 521:436–444
11. A. Krizhevsky, I. Sutskever, and G. E. Hinton. ImageNet classification with deep convolutional neural networks. In NIPS, 2012
12. K. He, X. Zhang, S. Ren, and J. Sun. Deep residual learning for image recognition. in CVPR, 2016
13. K. Simonyan and A. Zisserman. Very deep convolutional networks for large-scale image recognition. In ICLR, 2015
14. C. Szegedy, W. Liu, Y. Jia, P. Sermanet, S. Reed, D. Anguelov, D. Erhan, V. Vanhoucke, and A. Rabinovich. Going deeper with convolutions. in CVPR, 2015

15. N. Dalal and B. Triggs. Histograms of oriented gradients for human detection. in CVPR, 2005
16. Empty region detection in an image using Deep Convolutional Neural Network
17. Empty region detection in an image using Deep Convolutional Neural Network
18. D. G. Lowe. Distinctive image features from scale-invariant key points. IJCV, 2004
19. M. Everingham, L. Van Gool, C. K. Williams, J. Winn, and A. Zisserman. The PASCAL Visual Object Classes (VOC) Challenge. IJCV, 2010
20. T.-Y. Lin, M. Maire, S. Belongie, J. Hays, P. Perona, D. Ramanan, P. Dollar, and C. L. Zitnick. Microsoft COCO: Common objects in context. In ECCV, 2014
21. R. Girshick, J. Donahue, T. Darrell, and J. Malik. Rich feature hierarchies for accurate object detection and semantic segmentation. In CVPR, 2014
22. R. Girshick. Fast R-CNN. In ICCV, 2015
23. O. Russakovsky, J. Deng, H. Su, J. Krause, S. Satheesh, S. Ma, Z. Huang, A. Karpathy, A. Khosla, M. Bernstein, et al. ImageNet large scale visual recognition challenge. IJCV, 2015
24. S. Ren, K. He, R. Girshick, and J. Sun. Faster R-CNN: Towards real-time object detection with region proposal networks. In NIPS, 2015
25. J. Redmon, S. Divvala, R. Girshick, and A. Farhadi. You only look once: Unified, real-time object detection. In CVPR, 2016
26. T.-Y. Lin, P. Goyal, R. Girshick, K. He, and P. Dollar. Focal loss for dense object de-tection. In ICCV, 2017
27. W. Liu, D. Anguelov, D. Erhan, C. Szegedy, S. Reed, C.-Y. Fu, and A. C. Berg. SSD: Single shot multibox detector. In ECCV, 2016
28. M. Kummerer, L. Theis, and M. Bethge, "Deep gaze: Boosting saliency prediction with feature maps trained on ImageNet," arXiv:1411.1045, 2014
29. X. Huang, C. Shen, X. Boix, and Q. Zhao, "Salicon: Reducing the semantic gap in saliency prediction by adapting deep neural networks," in ICCV, 2015
30. L. Wang,H.Lu,X.Ruan, and M.-H. Yang, "Deep networks for saliency detection via local estimation and global search," in CVPR, 2015
31. H. Cholakkal, J. Johnson, and D. Rajan, "Weakly supervised top-down salient object detection," arXiv:1611.05345, 2016
32. R. Zhao, W. Ouyang, H. Li, and X. Wang, "Saliency detection by multi-context deep learning," in CVPR, 2015
33. Y. Tang, X. Wu, andW.Bu, "Deeply-supervised recurrent convolutional neural network for saliency detection," in ACM MM, 2016
34. S. S. Farfade, M. J. Saberian, and L.-J. Li, "Multi-view face detection using deep convolutional neural networks," in ICMR, 2015
35. S. Yang, P. Luo, C.-C. Loy, and X. Tang, "From facial parts responses to face detection: A deep learning approach," in ICCV, 2015
36. J. Yu, Y. Jiang, Z. Wang, Z. Cao, and T. Huang, "Unit box: An advanced object detection network," in ACM MM, 2016
37. L. Zhang, L. Lin, X. Liang, and K. He, "Is faster R-CNN doing well for pedestrian detection" in ECCV, 2016

An Enhanced Approach to Extract Top-K High Productive Itemsets

Preethi Gandreti, Satyanarayana Mummana, and Budimure Ramana Babu

Abstract The main purpose of this work is to develop a superior structure to extract top-K high productive itemsets. Here K is the picked portion of high productive itemsets that is to be established. High productive itemset tunneling is surely a prominent study in data mining but the factors for setting minimum utility margin are definitely a difficult task. In this work, an Enhanced approach to extract top-k high productive itemsets named Enhanced top-K high productive itemset tunneling (ETKU) is proposed. ETKU uses B+ Tree data structure instead of using a Utility Pattern Tree (UP-Tree) data structure that is used in existing Top-K high productive itemsets tunneling (TKU) method. Although TKU helps to reduce the time taken for the process of tunneling by reducing the total number of database scans to two, the complexity lies in the UP-Tree traversal for obtaining potential top-K high productive itemsets. B+ Tree used in ETKU does not have data associated with interior nodes so that more keys can fit into the memory. The leaf nodes of B+ Tree are linearly linked, so a full scan of a tree requires only one linear pass through all the leaf nodes.

1 Introduction

Data mining is stated as the policy to bring out the useful information from vast quantity of data. The technique for uncovering the data to identify unrevealed relationships and project future trends has been a prolonged history. Occasionally data mining is also mentioned as knowledge discovery in databases. Data mining is a

P. Gandreti (✉)
Asst. Professor, Dept. of CSE, Gayatri Vidya Parishad College for Degree and PG Courses (A), Engineering & Technology Program, Visakhapatnam, India
e-mail: preethi@gvpcdpgc.edu.in

S. Mummana
Research Scholar, Department of CS & SE, AUCE(A), Andhra University, Visakhapatnam, India
e-mail: satyam369@gmail.com

B. Ramana Babu
Research Scholar, Pondicherry Engineering College, Pondicherry University, Pondicherry, India
e-mail: budimureramana@gmail.com

S. C. Satapathy et al. (eds.), *Communication Software and Networks*, Lecture Notes in Networks and Systems 134, https://doi.org/10.1007/978-981-15-5397-4_23

process of discovering peculiarities, patterns, and correlations in huge sets of data to forecast the end result. By using a wide variety of techniques, this information could be used to raise revenues, reduce cost, and enhance customer relationships.

Frequent itemset mining is an elementary research issue in data mining. Frequent itemset mining is one of the foremost well known and most accepted data mining procedures. It was actually introduced for market basket analysis but later on it is used for almost any task which needs to identify regularities among variables. It focuses on identifying the regularities in the purchase behavior of the buyer of supermarkets, mail order companies, and online stores. Especially, it attempts to find sets of items that are often purchased together.

Utility mining appears as a key concept in data mining. In utility mining, every item is assigned with a value and count of existence in each transaction. The effectiveness of an itemset shows its importance which can be estimated with respect to weight, worth, volume, or different measurements depending on the user's interest. An itemset is called high profitable itemset if its value is greater than the user indicated minimum utility margin. In most of the cases, discovering a good minimum utility margin by hit and miss method is truly a laborious process. If minimum utility value is inadequate, then great number of high productive itemsets may originate and that will be genuinely ineffective. However, if minimum utility is excessive then no high productive itemsets are going to appear. To overcome the above mentioned consequences, a state-of-the-art scheme to find the top-K high productive itemsets is proposed in this work. Here K is the selected count of high productive itemsets that is to be established.

Using a constant K alternative to minimum utility margin is advisable for numerous applications. For instance, to examine consumer purchase behavior, top-K high productive itemset tunneling serves as an optimistic result for the users who wish to know "What are the top-K sets of products that are bringing huge profits to the company?"

However, reducing the search space for finding high productive itemsets is hard because a superset of low productive itemset can be high productive itemset. In order to face this issue, the transaction weighted utilization (TWU) model was brought to smoothen the functioning of mining task. An itemset is said to be high transaction weighed productive itemset (HTWBI) if its TWU is greater than the minimum utility.

A traditional TWU representation-based algorithm undergoes in two stages. In the first stage, called stage 1, the whole set of HTWBIs is originated. In the second stage, called stage 2, by computing the accurate utility values of HTWBIs with a database scan, the resultant high productive itemsets are acquired.

Two active methods named TKU and TKO are implemented in [1] for drilling such itemsets without setting minimum utility margin. TKU is the two-stage method for tunneling top-K high productive item sets, which integrates five strategies namely pre-evaluation (PE), elevating the margin by node utilities (NU), elevating the margin by minimum utility values of descendants (MD), elevating the margin by considering minimum utility of candidates (MC), and elevating the margin by classifying and computing exact utilities of candidates(SE) to successfully lift the border minimum utility margin and moreover reduces the search space.

The TKO is a one-stage method used for tunneling top-K high productive itemsets, which combines three strategies namely raising the margins by utilities of the candidates (RUC), reducing the approximate utility values by using Z-elements (RUZ), and exploring most promising branches first (EPB) significantly enhance its performance.

2 Related Work

Yin et al. [2] presented an effective method named top-k high utility sequence tunneling (TUS) for digging top-K high productive sequential patterns out of utility-based sequence databases. The method promises that no sequence is missed during the digging process. A new sequence border and a corresponding pruning strategy are introduced to eliminate the unpromising candidates. Pre-insertion and sorting methods are included to lift the minimum utility margin.

Quang et al. [3] designed a new method named ExMiner for tunneling top-K frequent patterns. The seq-BOMA approach is a fusion of seq-ExMiner method and the idea of "build once mine anytime" feature. It permits users to dig top-K frequent patterns where user need not specify any minimum support margin value.

Wu et al. [4] presented a new scenario for tunneling high productive episode in complex event sequences. To find out the utility of episodes, both internal utility and external utility of events are considered. For tunneling high productive episodes containing simultaneous events, the framework of complex event sequences is taken into consideration. An Enhanced method named utility episodes tunneling by spanning prefixes (UP-span) is proposed for tunneling complete set of high productive episodes. The TWU model is extended to episode tunneling and episode weighted utilization (EWU) model is proposed to smoothen tunneling process. The UP-span method is included with two efficient strategies named discarding global unpromising events and discarding local unpromising events for reducing the candidates count in tunneling process and to improve the performance of tunneling task.

Tseng et al. [5] say digging high productive itemsets from a transactional database mentions finding the itemsets with high utility. An efficient algorithm named Utility Pattern Growth is proposed for tunneling high productive itemsets. The information of high productive itemsets is organized in a special data structure called Utility Pattern Tree so that candidate itemsets will be generated with two database scans.

3 Proposed Method

An Enhanced approach to dig top-K high productive itemsets using ETKU is proposed. The selection structured data and policy search will affect the effectiveness of top-K high productive itemset tunneling approach regarding execution time and memory. Crucial contributions for this approach are outlined as follows:

An improved version of TKU named Enhanced top-K productive itemsets tunneling (ETKU) is implemented for tunneling the entire group of top-K high productive itemsets in databases without the requirement to state the minimum utility threshold. Figure 1 represents the structures of the ETKU method. The ETKU method adopts B+ Tree to preserve the information of transactions and utilities of itemsets. ETKU inherits functional characteristics from the TWU model and contains the two stages. In stage 1, potential top-K high productive itemsets (PKHBIs) are produced. In stage 2, top-K high productive itemsets are recognized from the set of PKHBIs identified in stage 1.

The existing TKU method uses UP-tree. Another existing method TKO uses a list-based design named utility list to reserve the utility information of the itemsets in the database. Vertical data representation approach is used to identify the top-K high productive itemsets in a single stage. The proposed ETKU uses B+ tree.

The methods used for high productive itemset tunneling can be commonly classified into two types: two stage and one-stage methods. The chief characteristic of a two-stage method is it contains two stages. In the first stage, a set of candidates called PKHBIs are produced. In the second stage, the precise utility of every candidate identified in the first stage is calculated to discover the high productive itemsets. The characteristic of one-stage method is that high productive itemsets are discovered using a single stage and it does not generate any candidates. High productive itemset digger considers a database in a vertical format and converts it into utility

Fig. 1 Proposed structure

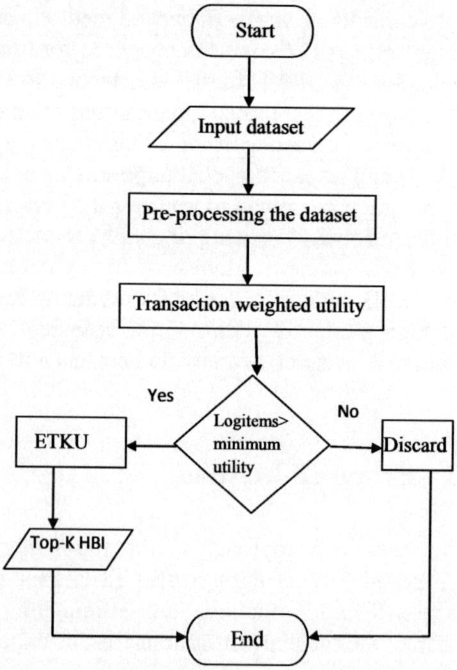

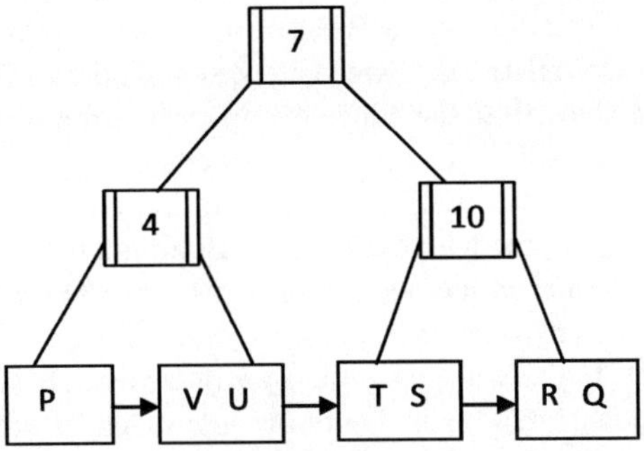

Fig. 2 B+ Tree structure after first database scan

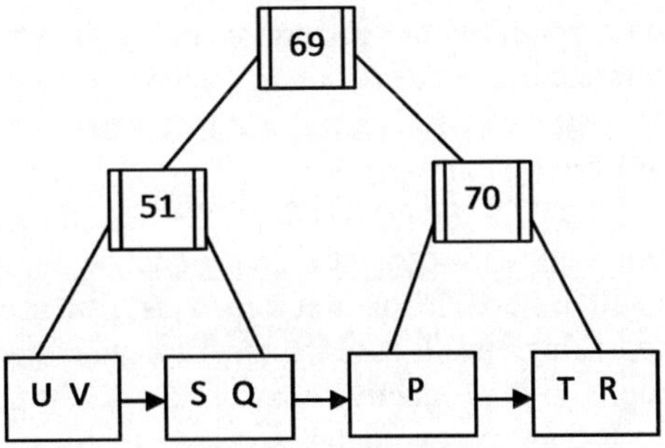

Fig. 3 B + Tree structure after second database scan

lists. The utility list format used in high productive itemset digger (HBI-Digger) permits to directly calculate the utility of produced itemsets in the main memory by eliminating the original database scan. In this paper, a two-stage method is used for ETKU as that in existing TKU but using B+ tree.

3.1 ETKU Structure

TKU is an efficient method to discover top-K high productive itemsets without setting minimum utility. In the existing systems, although the existing algorithms help to speed up the process of rule mining by reducing the total number of database scans to two, the complexity lies in the frequent UP-Tree traversal for obtaining potential high productive item sets. Tree Traversal of UP-Tree structure requires various number of iterations.

In this work, B+ Tree data structure is proposed instead of UP-tree to optimize the solution. Because B+ Tree does not have data associated with interior nodes and more keys can fit on a page of memory. Hence, less store misses are required so as to get to information that is available on a leaf node. The leaf nodes of B+ Tree are connected, so doing a full output of all items in a tree requires only one straight go through all the leaf nodes.

The chief objective of a B+ Tree is rapid traversal and expeditious search. Any record in the B + Tree can be searched quickly since all the nodes are kept at the same distance and the balance of the tree is maintained properly. As the record count in the database increases, the intermediate nodes and leaf nodes are needed to be separated and spread widely to maintain the balance of the tree. Since the nodes are widely spread, the time taken for searching a record becomes faster. As the branches of the tree are widely spread, it takes less I/O on the disk to retrieve the record. Records that are to be retrieved are retrieved in logarithmic fraction of time.

If the file grows in size in the database, the functioning of B+ Tree remains constant. This is due to the maintenance of the records at the leaf nodes and all the nodes are at equal distance from the root. Moreover, if overflow situation arises then the structure of the tree is automatically reorganized. The reorganization of the tree does not affect the performance. B+ Tree has good space utilization because all the intermediate nodes contain pointer to the records and the records are only held by leaf nodes. The space utilized by the pointers is very less when compared to the space utilized by the records.

The ETKU method adopts B+ Tree data structure to retain the information of the transactions and top-K high productive itemsets. The ETKU method is implemented in three steps. In the first step, a B+ Tree is constructed. In the second step, potential top-K high productive itemsets are generated from the B + Tree and in the third step top-K high productive itemsets are discovered from the set of PKHBIs.

It requires two database scans to construct a B+ Tree. The tree construction is explained with an example. Table 1 is an example database which contains five transactions. Let I^* be the set of distinct items $I^* = \{I_1, I_2, \ldots, I_n\}$, here in the considered example, $I^* = \{P, Q, R, S, T, U, V\}$. Let database $D = \{T_1, T_2, \ldots, T_m\}$, here D = {T1, T2, T3, T4, T5} is a set of transactions and each transaction in the database is a subset of I*. Every item is associated with internal utility and external utility. The internal utility of an item is the count of occurrence of that item in a particular transaction. External utility is the profit assigned to that item. Table 2

Table 1 Example database

Tno	Transaction	Transaction utility (TU)
T1	(P, 1) (R, 1) (S, 2)	10
T2	(P, 1) (Q, 4) (R, 1) (S, 6) (T, 1) (U, 4)	25
T3	(P, 2) (R, 6) (T, 2) (V, 2)	34
T4	(Q, 2) (R, 1) (T, 1) (V, 2)	11
T5	(Q, 4) (R, 1) (S, 1) (T, 3)	16

Table 2 Profit table

Item	P	Q	R	S	T	U	V
Unit profit	6	1	2	1	3	1	2

Table 3 Items and their occurrence count

Item	P	Q	R	S	T	U	V
Occurrence count	4	10	10	9	7	4	4

contains the external utilities of all the items in Table 1 example database. Table 3 contains the occurrence count of each item in the database.

During the first database scan, the tree construction is done by calculating the occurrence count of each item in all the transactions. The process of inserting an item into the tree is as follows:

(1) Detect the correct leaf position X1. The search starts at the root node and the key comparisons will direct it to a leaf.
(2) Try to insert the node into the position X1
    (a) If X1 has enough space for new item, then insert the item.
    (b) Else, split X1 (into X1 and a new node X2)
        (i) Reassign X1 entries evenly between X1 and X2.
        (ii) Duplicate the middle key, i.e., recursively insert the middle key into the parent of X1 and add a pointer from X1's parent to X2.
(3) While inserting a new node into an internal node Y1
    (a) If Y1 has enough space, then insert the new item.
    (b) Else, split Y1 (into Y1 and a new node Y2)
        (i) Reassign Y1 entries evenly between Y1 and Y2.
        (ii) Move up the middle key.
(4) Splits spread the tree by making it broad. When the tree splits, the height of the tree gets increased by one.

For Table 1, the B+ Tree structure after the first database scan is shown below:

Before the second database scan, the items with less occurrence count and less unit profit can be discarded in the second database scan. During the second database scan, the tree is reorganized based on the transaction weighted utility (TWU) values of the items in the database. The TWU value of each item is calculated by using the following formulae:

Definition 1 Total utility of an item $I_k I^*$ in a transaction T_s is defined as EU (I_k, T_s) = EX (I_k, D) * IN (I_k, T_s), here EX and IN are external and internal utilities, respectively.

Definition 2 Total utility of an itemset X in a transaction T_s is defined as EU (X, T_s)= $\sum_{I_K \in X} (I_K, T_s)$

Definition 3 Transaction Utility (TU) of a transaction T_s in a database D is defined as TU (T_s) = EU (T_s, T_s).

The Transaction Utility of T_1 in Table 1 is TU (T_1) = EU (P, T_1) + EU (R, T_1) + EU (S, T_1) = (6*1 + 2*1 + 1*2) = 10. The transaction utilities of all the transactions are calculated in the same manner and are mentioned in the Table 1.

Definition 4 The transaction weighted utility (TWU) of an itemset X is the sum of all transaction utilities of the transactions in the database containing X.

Definition 5 An itemset X is called top-K high productive itemset in a database D if there are less than K itemsets whose utility values are greater than EU (X) in f_{HBI}(D, O).

Definition 6 An itemset is called potential top-K high productive itemset (PKHBI) if its TWU and maximum utility are greater than minimum utility margin.

The TWU of an item P in Table 1 is TWU (P) = TU (T1) + TU (T2) + TU (T3) = 10 + 25 + 34 = 69. Table 3 contains the TWU of all the items that are present in the database.

The leaf nodes of the B+ Tree are reorganized based on the TWU values and the reorganized tree for the above example is shown below.

The items are retrieved from the B+ Tree by a doing a full scan and then they must be arranged in the descending order of their transaction utilities. To discover the PKHBIs, ETKU uses an internal variable named as minimum utility margin which is set to 0 initially and it is lifted dynamically after identifying certain count of itemsets with high utilities. Once the PKHBIs are discovered then ETKU computes the utility of PKHBIs by scanning the original database once to discover the top-K high productive itemsets.

4 Conclusion

In this paper, the issue of top-K high productive itemsets tunneling is studied, where K is the picked portion of high productive itemsets that is to be established. A Enhanced method named ETKU is proposed using a B+ Tree data structure. B+ Tree does not have data associated with interior nodes so that more keys can fit on a page of memory. Therefore, fewer cache misses are required to access data that is present on a leaf node. In a B+ Tree, a full scan of all objects in a tree requires just one linear pass through all the leaf nodes. The main objective of a B+ Tree used in EKTU is rapid traversal and expeditious search. The ETKU is a two-stage method for tunneling Top-k high productive itemsets. The proposed ETKU method takes less time and more effective than TKU method for obtaining top-K high productive itemsets.

References

1. Tseng VS, Wu CW, Fournier-Viger P, Philip SY (2016) Efficient algorithms for mining top-k high utility itemsets. IEEE Trans Knowl Data Eng 28(1): 54–67
2. Yin J, Zheng Z, Cao L, Song Y, Wei W (2013) Efficiently mining top-k high utility sequential patterns. In: 13th International Conference on Data Mining (ICDM), pp 1259–1264. IEEE
3. Quang TM, Oyanagi S, Yamazaki K (2006) ExMiner: An efficient algorithm for mining top-K frequent patterns. In: International Conference on Advanced Data Mining and Applications, pp 436–447. Springer, Berlin, Heidelberg
4. Wu CW, Lin YF, Yu PS, Tseng VS (2013) Mining high utility episodes in complex event sequences. In: Proceedings of the 19th ACM SIGKDD international conference on knowledge discovery and data mining, pp 536–544. ACM
5. Tseng VS, Wu CW, Shie BE, Yu PS (2010) UP-Growth: an efficient algorithm for high utility itemset mining. In: Proceedings of the 16th ACM SIGKDD international conference on Knowledge discovery and data mining, pp 253–262. ACM

Study of Algorithms and Methods on Emotion Detection from Facial Expressions: A Review from Past Research

Sasanko Sekhar Gantayat and Swathi Lenka

Abstract Recognizing the emotions of a person's facial expression gives important information about the person's attitude and behavior. Understanding this type of expression exactly is a challenging task. There are two aspects of emotion detection. These aspects of emotion detections are based on facial expression and speech recognition. We have studied and analyzed some algorithms already implemented by the researchers on emotion detection based on facial expression. Emotion detection based on expressions of the face is now a new area in many fields such as education, medicine, computer science, and psychology. Human communication interaction researchers also use facial expressions for emotion classification which gives better results to develop new technologies. Some of the feature extraction methods are developed for recognition of facial expressions for real videos in real time and static images. This article provides a review of research work and various algorithms used for emotion detection based on facial expression recognition.

1 Introduction

Facial expression delivers important information about human emotion and relation and plays crucial role in communication. The main area for facial expression detections comes under pattern recognition and computer vision. Humans recognize emotions quite efficiently and accurately. An automatic emotion recognition system is an element in human–machine interaction. This expression is the most natural, powerful for humans to communicate his/her emotions. It is a computer process for verifying or identifying a person from image or video frame. This process is performed by the computers or humans who contain extracting facial feature, locating faces, and applying facial features.

S. S. Gantayat (✉) · S. Lenka
Department of Computer Science and Engineering, GMR Institute of Technology, Rajam, Andhra Pradesh, India
e-mail: sasankosekhar.g@gmrit.edu.in

S. Lenka
e-mail: swathi050993@gmail.com

S. C. Satapathy et al. (eds.), *Communication Software and Networks*, Lecture Notes in Networks and Systems 134, https://doi.org/10.1007/978-981-15-5397-4_24

Facial emotion part deals with the recognition of expression from the image [3]. This part is used in communication because one can expresses him/her self-basis of mood. Emotions are detected using various approaches, that is, body language, gestures, voice intonation, and complex methods like electroencephalography are used. Six types of universal emotions recognized in various cultures are: disgust, anger, sad, surprise, happy, and fear.

1.1 Facial Expression Recognition System

The facial expression recognition system recognizes the expressions of identifying the human emotions like disgust, fear, surprise, sadness, anger, and happiness. These expressions vary for every individual person. By spoken 7% and by voice 38%, whereas 55% through facial expressions message information is conveyed.

Facial expression recognition system consists of the following four stages.

Image Acquisition

Image sequences or static images are used for recognizing the expressions. For the reorganization of the image 2-D gray-scale image was used popularly. Color images give more information about emotion. Color images are preferred for image recognition because of low cost of equipment. For images we use the phone, camera, or other digital devices.

Pre-processing

It is the key role. This stage enhances mainly input image quality. It locates data of interest by removing the noisy information. It removes redundancy from image without the image detail. Pre-processing also includes filtering and normalization of image which produces uniform size and rotated image.

Feature Extraction

It consists of "interest" part in image. It includes information of shape, motion, color, and texture of facial image. It extracts the meaningful information form image. Compared to actual image, feature extraction is used to reduce information of image, which gives advantage in storage.

Classification

It follows the output of features extraction stage. Classification stage identifies the facial image and groups them according to certain classes and helps in their proficient recognition. Classification is complex because of many factors affected. Classification stage can also be called feature selection stage, deals with extracted information, and groups them according to certain parameters.

2 Background and Literature Survey

2.1 Facial Expression

Outward appearance distinguishing proof is a critical element to know the conduct of a man. It helps the other individual who is in up-close and personal changes with other individuals so that one can take a legitimate measure to talk about or communicate. In any case, the characteristic outward appearance of a man cannot be recognized effortlessly. It requires some master learning to peruse the substance of a man. When we need to peruse the substance of a man by utilizing a camera and investigate utilizing a face acknowledgment framework, it is a troublesome part to a machine to peruse the natural appearance of a man as the machine can break down the noticeable bit of the face, yet not the psyche of the individual.

According to the investigation of Mehrabian [16], among the human correspondence, facial articulations involve 55% of the message transmitted in correlation with the 7% of the correspondence data passed on by semantic dialect and 38% by paralanguage.

2.2 Types of Emotions

Emotion detection through facial expression recognition is becoming quite popular for its increasing scope of applications in human–computer interactive systems. However, there are several methods developed for emotion recognition, including voice, facial expression, posture, and gesture. The emotion recognition mainly consists of two fundamental steps; feature extraction and classification. Feature extraction is defined as the set of features or attributes, which turns to emotional expression. During classification, the features are mapped onto one of the several emotions, such as anger, sad, disgust, and happy.

Figure 2 illustrates the basic seven types of emotions. These are fear, anger, sadness, happiness, disgust, and surprise [18]. Here we have shown only six types of emotions (Fig. 1).

The set of features that are considered for extraction and the classifier that is used for the task of classification are equally important to determine the performance of an emotion recognition system. For a poorly selected set of features, sometimes, even a good classification algorithm cannot give a good result. Thus, selecting good

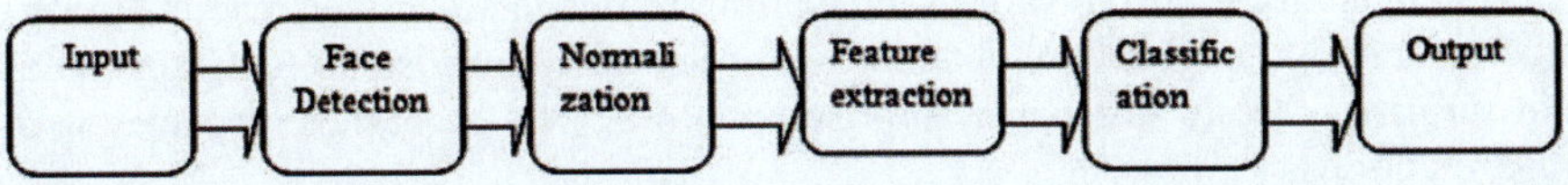

Fig. 1 Facial expression recognition system

Fig. 2 Six basic emotions [19]

features is always a pre-requisite for high classification accuracy and good result. Many algorithms are developed for classifying emotions in the last three decades.

3 Various Methods of Emotion Detection

There are various methods used for detecting emotions from facial expression. Here we have reviewed some methods and algorithms.

3.1 *Fuzzy Rule Method for Expression Recognition*

Mufti and Khanam [1] proposed a fuzzy rule-based algorithm where partial matching of emotion can detect the feature.

Step 1 Input a video which has emotions.
Step 2 Extract frames where input is a continuous video signal. The frame extraction module extracts individual frames from the sequence at a fixed rate to obtain the sampled frames which are essentially 2-D gray-scale images.
Step 3 Extract the feature point for facial expression.
Step 4 Face animation parameter (FAP) extraction module calculates FP movements and decides which particular FAPs have been activated.
Step 5 The crisp inputs are defined by the fuzzification method; by using membership function values are translated into linguistic variables.
Step 6 Detect the expression based on fuzzy parameters

This algorithm is implemented based on the fuzzy logic principle to recognize video image emotions and give high robustness. Because of efficient fuzzy systems, the system is thus robust to various fluctuations in image processing results. Recognition results of emotions are obtained as for joy 60%, surprise 80%, disgust 50%, and surprise 60%. In this paper surprise emotion gives the highest accuracy than other emotions.

3.2 3-D Facial Features Distance Algorithm

To classify emotions on the face, Hamit and Demirel [9] proposed 3-D facial features distance algorithm. It recognizes the basic six expressions. To describe emotions on face, that is, anger, happy, fear, sad, neutral, and surprise, the distance values are used. In this data captured form 60 subjects and each expression is used.

Step 1 Feature extraction of distance vectors characteristics is defined in a table containing the six distances.
Step 2 By using NN the distance vectors are classified and trained using back propagation algorithm.
Step 3 Six distances are used to normalize the five distances

The drawback in this method is that the classification of emotions rate is low for a class named anger because of confusion with anger and neutral classes.

3.3 Algorithm for Feature Fusion to Detect Facial Emotions

In this algorithm, the authors Chen et al. [10] used feature fusion nonlinear dimensions reduction algorithms (HLAC + WPCA), and used a method feature extraction in the face recognition because of the ability to preserve the original geometry. Follow with simple min distance classifier which gives better result than principle components analysis-based method.

Step 1 Expressions are extracted by using integrals projections method.
Step 2 Extract the features by using higher-order auto-correlation method and reduction of dimensions can be done using weighted PCA method. By using facial expression measure system face actions code system weighted values are updated.
Step 3 Minimum distance classifier is used in recognizing the facial expressions for surprise, anger, disgust, happy, sad, and fear

Database used: Image database consists of 336 images and CMU-Pittsburgh Actions Unit Code Face Expression include six persons; 56 images are used per person, which include 7 expressions, and for expression 8 images are used. A total of 168 images are used for training and others are used for testing.

Feature fusion nonlinear dimensions reduction (FFNDR) algorithm is used for features extraction in the expression which recognizes after preserving the original local geometry.

Some researchers are using powerful and complicated classification methods like NN, SVM, and KFDA which are used instead of the minimum distances classifiers to get better performance and improve the system performance with the integration part of a local feature to satisfy the demand for real time and automatic applications, and overcoming the illumination and gesture changes in emotion recognition.

3.4 Automated Emotion Detection Using Facial Expression Recognition

Andrew et al. [7] proposed an algorithm for facial expression recognition. This algorithm is as follows:

Step 1 Video processing: Input video are sequenced into individual frames. Each frame is rated as 25 frames per second. Once the video is sequenced, the frames are moved to the queue.
Step 2 Appearance and shape modeling: A geometric landmark on the face used for automatic face tracking and registration.
Step 3 Classification is carried out on expressions
Step 4 After classification of expressions at runtime, the AFERS offers operators in the real time; output of the expression classification, such as snapshot generation and interrogation reporting, and so on, is done by the analytics engine

The AFERS system is used to recognize the process of interview but it does not detect directly which needs refinement and more research along with contextual models is required.

3.5 Facial Expression Classification Based on Singular Values Decomposition and Principals Component Analysis

In this paper the authors, Kaur et al. [8], have proposed PCA for singular values decompositions for classifying the emotions.

Step 1 Input image is given for pre-processing.
Step 2 Extracted features are sent to the classifier. Image from the expression database is pre-processed and gives knowledge database for training and fed to classifiers.
Step 3 The two images were compared and expressions recognize the emotions to be detected

The database used for emotions recognition system is JAFFE. This database measures the achieved performance of recognition systems. It contains seven fundamental facial expressions for emotion classification, such as sad, neutral, disgust, happy, fear, angry, and surprise. These have gray-scaled images of 213 for expression recognition. Each image size is of 256×256.

The advantage of this algorithm is that excellent classification results are found for all principal emotions along with the neutral emotion from the training database. The image is enhanced, localized, and its features are extracted by using singular value decomposition technique. The algorithm can effectively detect different emotions.

The main drawback of this algorithm is the elimination of errors because of the reflections in the images, that is, persons wearing glasses were not implemented and algorithms used are computationally efficient.

The authors found the results with the highest accuracy. The recognition rate for emotions like angry, disgusts, happy, sad, and surprise along with neutral is obtained. Among them, the expression happy gives 95% accuracy which is a higher rate than existing algorithms. Finally, the network is tested on the real-time dataset with excellent recognition rate.

3.6 Neural Network Algorithm for Expression Recognition

Using this method the authors, Pushpaja et al. [6], have presented a basic facial gesture to recognize the arrangement, and a generic face detection arrangement gives description of emotion classification and proposes system containing the following module or steps:

Step 1 The inputted image is from the webcam.
Step 2 Optical flow method is used for facial detection.
Step 3 Image pre-processing: The image is normalized and noisy or unwanted information in the image is removed. In eigenface library the database images contain two steps: testing dataset and training dataset. Eigenfaces are calculated from the trained dataset. Once the training phase has been done, then these dataset images are matched with the best eigenface which contains the highest eigenvalues.
Step 4 PCA is used for feature extraction.
Step 5 Classification is done using a feed-forward back-propagation network

The advantage of this algorithm is it provides a solution to the problems of recognition in a constraint environment. The drawback is it poses a problem in unconstraint environment.

3.7 Multimodal Emotion Recognition in Response to Videos

Mohan et al. [2] performed the experiment on the extraction of parameters from electroencephalogram and response to multimodel emotions recognition by independent users that is used on pre-existing clips from online resources which when compared with the ground-defined truth gives us the efficiency of the recognition system.

Step 1 Signal setup has been done by recording environment using electroencephalogram.
Step 2 Track real-time expressions obtained by the frame of sequences and apply the algorithm called facetracking.

Step 3 Data acquired from the step above are transferred to other computers and processed using internet communications engine middleware.
Step 4 Candide-3 model was implemented with facial features invariantly.
Step 5 Bayesian network was constructed for the action units (AUs) obtained from the above step.
Step 6 Facial expression was evaluated and emotion was recognized

The advantage of the proposed system is robust detection is possible using this method and it improves classification rate. It provides a good result for the emotion neutral than other states of emotions and also eliminates errors in the user detection by using less data on the trainer at feature extraction state.

The limitation of the system is that the parameters are complex in nature and changes from person to person with age. Light conditions, shadow, and noises can affect papillary responses. The fusion classifiers are hard to find. Different videos and modalities are not directly comparable due to the lack of proper metrics to measure the emotion using eye gaze.

3.8 *Eigenvalue-Based Facial Emotion Classification*

This method uses hardware system known as field program gate arrays that are digital and help in complex computations involving that of finding the largest eigenvalues using power method. The proposed algorithm of Sheily et al. [11] is given as follows:

Step 1 Acquire the image of text form.
Step 2 Crop image to 16 × 16 pixel matrix with every pixel consisting of range 0–255.
Step 3 Apply power method and use Verilog to calculate the largest eigenvalue.
Step 4 A necessary simulation was performed.
Step 5 To get the results, the file (.bit file) format is generated.
Step 6 FPGA ROM maintained the stored files.
Step 7 To detect the corresponding emotions simulated eigenvalue was used

The advantages of the power method implementation using the FPGA kit are it is mobile and flexible, implementation is practical, easily configurable, and easily integrated into other systems. It is found that this methodology involves less mathematical complexity, uses less storage space, and thereby promotes faster recognition.

Limitation of the algorithms is FPGA is expensive, thus making these approaches cost-ineffective. Also, include additional dimensions which involve tedious calculations.

The power method is used to recognize emotions. It is implemented on Xilinx Spartan 3E FPGA kit. The FPGA system consists of size, configurability, flexibility, and mobility. By using this algorithm largest eigenvalues are obtained and then compared with the desired eigenvalues of the stored dataset. It produces results in both MATLAB and Verilog which classify the emotions like angry, sad, and happy. Compared to MATLAB, Verilog provides the results faster.

3.9 *Artificial Neural Network for Emotion Classification*

The authors Siraj et al. [12] proposed the following algorithm:

Step 1: Load video or image

The picture of the face will be taken as input. Then the image process is performed which will convert the image into the desired color and resolution.

Step 2: Frames are detected.

Step 3: Feature extraction: The features are extracted using HAAR method. Contrast values are used instead of HAAR features to detect images. Relative darken and lighten areas are determined from a group of pixels. The image detection varies in various sizes.

Step 4: Train classifier: To train the input against original data which is the target, the method radial basis neural network is used as training classifier. This classifier is trained for different videos and images. After training the train classifier is performed to test target data to identify facial expressions.

From this experiment, the authors found the automated human emotion classification which successfully displays emotion from an image uploaded by the user for matching it with trained dataset. In case when trained dataset is matched with the uploaded image, the system shows the output, otherwise not. Input to the system is an image of human emotion or a video from our database. The output is the emotion of the uploaded image or the frames are taken from the video.

3.10 *Approximated Supervised Learning and Bezier Curve Approach for Facial Recognition*

In this model, Dixit Manish and Silakari Sanjay [4] proposed the facial recognition of an image with human frontal faces using approximation on Bezier curves to ensure curve fitting and smoothing on structural features that are classified by learning from intelligent neural networks. The proposed method was:

Step 1 For converting grayscale, frontal image is taken as input.
Step 2 Applied histogram binarization and equalization thinning of image for pre-processing the inputted image.
Step 3 Harris algorithm was used to extract the corner detection.
Step 4 To use control points, parametric and Bezier curves were used.
Step 5 The curve points were fed to test and train a NN.
Step 6 BP algorithm was used to train the facial images through the network.
Step 7 For recognition of output, testing was done for the accuracy of the network through trained images

The advantage is the facial rate is comparatively higher for this method and it produces a high amount of images when recognition is correct. Use the image-thinning process to reduce redundant pixels and also reduce the cost.

The limitation of this method is images with structural components and angled poses are tough to detect.

The output of the above method is the feature extraction and recognition techniques require longer evaluation and training time if the proper feature points with suitable algorithms are not considered. In this suitable algorithm for edge, corner detections and feature extraction are used with its hybridization in training process. It reduces the error rate.

3.11 Contourlet Transformation and PSO

Li et al. [13] used contourlet transformation and the spatial domain to create feature vector unlike the current working system that works on local binary patterns/steerable pyramid that create feature vector only from transformation and spatial domain. It utilizes properties of directionality which extracts important features. For contour sub-bands, the authors suggested a new coefficient enhancement algorithm which enhances skin region features to make the system more vigorous. They also tested in features level fusion on multiple databases that showed face recognition rate was competitive. Action units are used for recognition and analysis by using classifier in a video.

The first random frog will detect action units and these detected AUs are classified by second random forest which detects expressions. On first frame face landmarks are generated by active appearances model and landmarks are then tracked for the sequence of frames in a video by Lucas-Kanade optical tracker. The vector was created between natural and peak expression. Random forest detects action units from DNNP features, and these AUs are sent to second random forest as an input which then processes these AUs into facial expressions. This method achieves an accuracy rate of 89.37% for the two-fold forest classifier.

Microgeneric algorithms embedded with particles swarm optimizations give the features. It also solves the local optimum problem and premature convergence by introducing non-replaceable memory, a secondary swarm having five participants with a leader and four followers, velocity updating, sub-dimension-based regional facial feature searching and global searching. For recognition, features that are generated from the algorithm are classified with multiclass SVM and ensemble classifier for improved accuracy. Results from the paper LBP-based feature extraction surpassed most recent local binary pattern variants. For expression recognition, 100% accuracy was achieved in case of CK + and 94.66% in case of MMI database for mGA embedded POS and diverse classifier. The assessment was done for 30 trails.

3.12 Faster R-CNN Algorithm

Jiang Huaizu and Learned-Mille Erik [14] proposed an approach that consists of regional proposal network and fast regions of CNN features. The regional proposal network is used instead of using a selective search algorithm because of slow and time-consuming process which affects the quality of the network.

Region proposal network: The output contains boxes/proposals by using regression and classifiers. RPN predicts the possibility of the anchor being processed a foreground or background and refines the anchor.

ROI pooling: Obtains regions with vary in sizes of CNN features. For the reduction of feature maps, the pooling method is used to simplify the problem. This pooling divides feature maps which are an input to a number of regions and applies maximum pool on each region.

The dataset was provided by Chinese Linguistic Data Consortium (CLDC) consisting of multimodal emotional video and audio data. This database has 66,486 images with eight emotions. Among them 12,019 anxious, 9867 happy, 6174 worried, 4977 neutral, 2574 surprise, 10,862 angry, 18,326 sad, and 1687 disgust images.

The model for this algorithm used Pascal_voc.model provided by the py-faster-rcnn, VGG_CNN_M_1024, ZF and VGG16, which are called the faster_rcnn_alt_opt network to fine tune image net model.

For the training set, the sampling was done towards positives, because of compared to background it is extremes to be uncommon.

The result shows that faster R-CNN identifies facial expression easily. The original picture is used as the whole network input. The process of feature extraction in recognizing facial expression is completely avoided. The features are extracted by the network from the training dataset. The region proposal networks (RPNs) generate efficient region proposal. In each image, it locates the face region and recognizes the expression directly.

3.13 Emotion Detection in Virtual Learning Environments

Yang et al. [5] proposed a method for the reorganization of emotions on face virtual learning model, consisting of subset feature, emotion classifier, and feature extraction. HAAR method detects the inputted image. The basic six emotions are classified using NN classifier. The JAFFE database gives high classification performance.

Step 1 Input the frames

Step 2 Acquire the facial image

Step 3 Pre-processing emotion recognition becomes lower when cropping the eye region and grabbing the characteristic values. Here transformation is required to improve accuracy of the system.

Step 4 HAAR cascades method was used for the identification of image frames. If the face does not exist in the system, this algorithm is repeatedly applied. If the face image frames exist in the system, then mouth and eyes were located and cropped.

Step 5 Filter and edge detection: Mean and median filters have been applied to make the image smoother by removing unwanted noisy information. If gradient input images were at the maximum condition, then returns the edge

The authors used the database of 213 images of Japanese women's faces. The actual image and frontal faces are to be readjusted and trim are shown by using the JAFFE database.

This model solves the problems and gives high accuracy and efficiency using emotion classification in virtual learning environments for facial recognition. HAAR method detects the mouth and eyes region. This method identifies the emotions through the network named neural network method. This method is also applied to real distance education.

3.14 Recognition of Facial Expression Using Eigenvector-Based Distributed Features and Euclidean Distance

Jeemoni and Karen [15] proposed a method to recognize facial expression using eigenvector-based distributed features and Euclidean distance to analyze the various parts of the face, like the left eye, right eye, nose, lip, nose, and lip together. They implemented facial expression recognition in three steps. These steps are:

Step 1 To calculate the eigenvectors and eigenvalues of the covariance matrix of the facial images. With the available eigenvectors of expressions, separate subspaces for all the six universal expressions (surprise, happiness, fear, anger, sad, and disgust) are created.

Step 2 Use different Euclidean distances between all the six expressions with the current image that captured.

Step 3 Count the number of different expressions and using voting rule the highest expression is the result

They used real-time facial expressions using a digital camera and analyzed the result and got 95% accuracy. The drawback of this method is the computation time to identify a single facial expression and for more cases, it will take more time.

3.15 Emotion Detection Algorithm Using Frontal Face Image

Hwan et al. [17] proposed a new emotion detection algorithm using a frontal facial image. This algorithm has three stages: the first stage is the image processing stage, the second stage is the facial feature extraction, and the last stage is the emotion detection. They implemented fuzzy classifier to get the emotions. Initially, to extract face region from the facial image, the fuzzy color filter and histogram analysis methods are implemented and later fuzzy classifier is used to classify the emotion from the extracted feature.

They got an accuracy of 82.7% using this new method.

4 Conclusion

In this article, some of the methods and algorithms on emotion recognition are overviewed from the face. These methods involve recognition, extraction, and face categorization. Many approaches give better recognition. Techniques with recognition having higher rate can give greater performance. These approaches provide a better solution of expression recognition and can work in a well-constrained environment. Emotion detection is an issue that causes complexity due to psychological emotions.

We have not considered the emotion detection from the speech as it requires in-depth study and also a complex one. Therefore, research in this field will remain under continuous study for more years to come because many problems have to be solved so that it creates an ideal user-interface and improved recognition of confusion emotions is required.

References

1. Muid Mufti, Assia Khanam: Fuzzy Rule Based Facial Expression Recognition, 2006 International Conference on Computational Intelligence for Modelling Control and Automation and International Conference on Intelligent Agents Web Technologies and International Commerce (CIMCA'06), (2006)
2. Krishna Mohan Kudir P., Abas Md Said, Yunus Nayan M.: Emotion detection using sub-image based features through human facial expressions, International Conference on Computer & Information Science (ICCIS-2012), (2012)
3. Kanade, T., Tian, Y.L., Cohn, J. F.: Hand Book Facial expression analysis. In: Li, S.Z., Jain, "Face Recognition Analysis", pp. 247–276. Springer, New York (2005)
4. Dixit Manish and Silakari Sanjay: Face Recognition Using Approximated Bezier Curve and Supervised Learning Approach, International Journal of Multimedia and Ubiquitous Engineering vol. 10, No.4 (2015)
5. Yang D., Abeer Alsadoon, Prasad P.W.C., Singh A. K., Elchouemic A.: An Emotion Recognition Model Based on Facial Recognition in Virtual Learning Environment, Procedia Computer Science, Vol. 125, pp. 2–10, (2018)
6. Pushpaja V. Saudagare, D.S. Chaudhari: Facial Expression Recognition using Neural Network – An Overview, International Journal of Soft Computing and Engineering (IJSCE) ISSN: 2231–2307, Volume-2, Issue-1, March, (2011)
7. Ryan Andrew, Cohn Jeffery F., Lucey Simon, Saragih Jason, Lucey Patrick, Fernando De la Torre and Rossi Adam, Automated Facial Expression Recognition System, IEEE 43rd Annual 2009 International Carnahan Conference on Security Technology, pp: 172 – 177 (2009)

8. Kaur Mandeep, Vashisht Rajeev, Nirvair Neeru: Recognition of Facial Expressions with Principal Component Analysis and Singular Value Decomposition, International Journal of Computer Applications (0975–8887) Volume 9–36 No.12, November (2010)
9. Soyel Hamit, Hasan Demirel: Facial Expression Recognition Using 3D Facial Feature Distances, International Conference Image Analysis and Recognition (ICIAR 2007), M. Kamel and A. Campilho (Eds.), LNCS 4633, pp. 831–838 (2007)
10. Chen F, Wang Z, Xu Z, Wang Y, Liu F (2008) A Facial Expression Recognition Algorithm Based on Feature Fusion. IEEE Pacific-Asia Workshop on Computational Intelligence and Industrial Application, Wuhan, pp 381–385
11. Verma Sheily, Vilas H. Gaidhane, Asha Rani: Eigen Value Based Facial Emotion Classification, International Journal on Advanced Computer Theory and Engineering (IJACTE), Vol. 2, Issue-5 (2013)
12. Fadzilah Siraj, Nooraini Yusoff, Kee Lam Choong: Emotion Classification using Neural Network, 2006 IEEE International Conference on Computing & Informatics (2006)
13. Li Z, Duan X, Wang C, Yun J, Lu B (2017) An Automatic Deformation for Facial Smile using PSO Detection. Journal of Electronics and Information Science 2:51–55
14. Jiang Huaizu, Learned-Mille Erik: Face Detection with the Faster R-CNN, 2017 12th IEEE International Conference on Automatic Face & Gesture Recognition (FG 2017)
15. Kalita Jeemoni, Das Karen: Recognition of Facial Expression Using Eigenvector Based Distributed Features and Euclidean Distance Based Decision Making Technique, International Journal of Advanced Computer Science and Applications (IJACSA), Vol.4,No.2(2013)
16. Mehrabian A.: Communication without Words, Psychology Today,Vol.2, No. 4, pp. 53–56, September (1968)
17. Kim Moon Hwan (2005) Joo Young Hoon, Park Jin Bae: Emotion Detection Algorithm Using Frontal Face Image, ICCAS2005 June 2-5. KINTEX, Gyeonggi-Do, Korea
18. Ekman Paul (2003) Emotions Reveals. Times Books, New York USA
19. Facial Expressions. www.frontiersin.org
20. Bhaskar A, Gireesh Kumar T (2018) Facial Expression Classification Using Machine Learning Approach: A Review. Data Engineering and Intelligent Computing Advances in Intelligent Systems and Computing 542:337–345. https://doi.org/10.1007/978-981-10-3223-3_32

Data Security Using Matrix Operations

Snehal Bhogan and Ashish Narvekar

Abstract In today's world internet is being widely used for electronic information exchange in one form or the other to the specific receiver using some medium. Secure communication is of major concern while data is being transmitted over the internet. Data security is the study of various methods of protecting data in computer and communication systems from unauthorized disclosure and modification. Security of data can be achieved by a technique called cryptography. This paper attempts to propose a new cipher model using block cipher concept. A matrix is used to store the data. In order to make the data more secure and inaccessible to unauthorized users, various matrix operations can be carried out on the data matrix. Also, a slightly different method is mentioned to share the secret key used during the encryption and decryption mechanisms. This proposed new cipher model enables good diffusion for making a secure cipher.

1 Introduction

Technology today plays an essential role in our daily lives. The world is now making a move toward digital workflow that results in paperless organization, mail messages, online transactions and so on. With the use of internet as the medium for communication, huge amount of data transfer takes place between computers. The communication channel between the computers is open to unauthorized access. Hence, it becomes very important to protect such data during transfer.

Cryptography is the art and science of protecting data from unauthorized users by converting an intelligible message into an unintelligible form and for reconverting the secret form into the intelligible message form while stored and transmitted. Data during communication cannot be changed or misused for any illegal purposes, and

S. Bhogan (✉) · A. Narvekar
Department of Computer Engineering, Agnel Institute of Technology and Design, Goa University, Goa, India
e-mail: sb@aitdgoa.edu.in

A. Narvekar
e-mail: an@aitdgoa.edu.in

S. C. Satapathy et al. (eds.), *Communication Software and Networks*, Lecture Notes in Networks and Systems 134, https://doi.org/10.1007/978-981-15-5397-4_25

this makes security of data very important [1]. Therefore, it is very essential to apply effective encryption/decryption methods to boost data security.

2 Related Works

A design of new block cipher is recommended in this paper. In parallel, multiple algorithms are used to increase the speed of the algorithm. Using security analysis, the recommended algorithm satisfies the correlation coefficient test and considered to be secured and efficient. Using symmetric key cipher approach, in the encryption process the plaintext is converted into appropriate numerical form and generates a key. During decryption process the same key is used to decrypt the message. A matrix is used to store the data. Various operations are carried out on the matrix to protect the data.

In this paper [2] a new encryption algorithm, byte-rotation encryption algorithm (BREA),with parallel encryption model was proposed which makes the encryption process fast and more secure. The BREA works on separate blocks of plaintext and as execution is carried out in parallel the concept of multithreading using single processor is achieved.

Byte-rotation with encryption algorithm is discussed in this paper [3] which guarantees security as well as speed of the encryption method. The key matrix is obtained on application of CBC approach and then the BREA encryption process continues.

A text of small size is replaced by numbers and then stored in a matrix format [4]. Further, the matrix data is summed up with a key matrix which is then trailed by the transpose and rotation operations on the final matrix. Decryption process performs the reverse operation to decrypt the encrypted message.

In paper [5], a new encryption algorithm to convert the plaintext to cipher text and the decryption algorithm to do the opposite are discussed. The encryption and decryption algorithms are performed by applying several bit-rotation procedures with complement, shifting and reordering on key symmetric matrix. The operations make it simple and easy to implement.

In this paper [6], the author presented byte-rotation algorithm for encrypting and decrypting a file within least delay. Using these algorithms the process of file encryption and decryption improves the security and takes less time. BRA was used to provide security for different types of file, like image, text, audio and video.

3 Cryptography

Cryptography is the discipline of secret codes, maintaining the confidentiality of message when moving through an insecure medium. It protects the data from any alteration and misuse of data against unauthorized parties. It uses a cryptographic

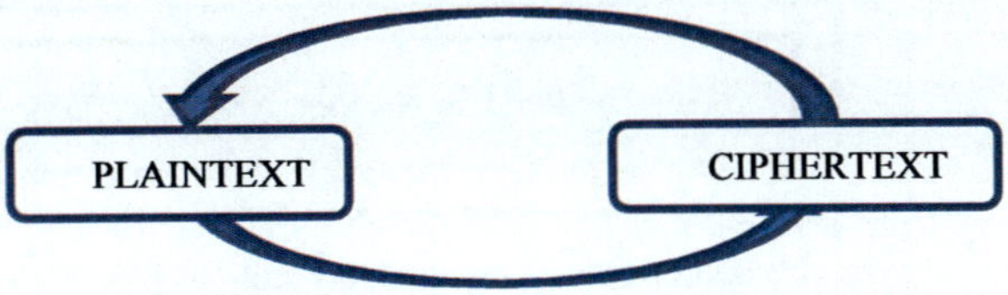

Fig. 1 Process of encrypting and decrypting text

system to transform a plaintext into a cipher text, using most of the time a key. Security is maintained during communication in cryptography.

3.1 Terminologies

Plaintext: The original data that is required to be sent to the receiver is known as the plaintext. This plaintext is in readable and understandable form either by a person or by a computer.

Cipher text: The unclear text which is not readable is called cipher text.

Key: A key is used for encryption and decryption process making cryptography a secured process.

Encryption: The process of converting the plaintext to cipher text using an algorithm and the key is called encryption.

Decryption: The process of converting the cipher text back into its original form that is plaintext with the help of decryption key and decryption algorithm is called decryption (Fig. 1).

3.2 Types of Cryptography

(i) **Private key cryptography**: When the same key is used during the encryption process at the sender side and decryption process at the receiver side, it is called as symmetric key or private key cryptography

(ii) **Public key cryptography**: When two different keys are used, that is one key for encryption process and another key for decryption process, it is called public key or asymmetric key cryptography.

4 Proposed Model

The proposed new algorithm has three parts as follows:

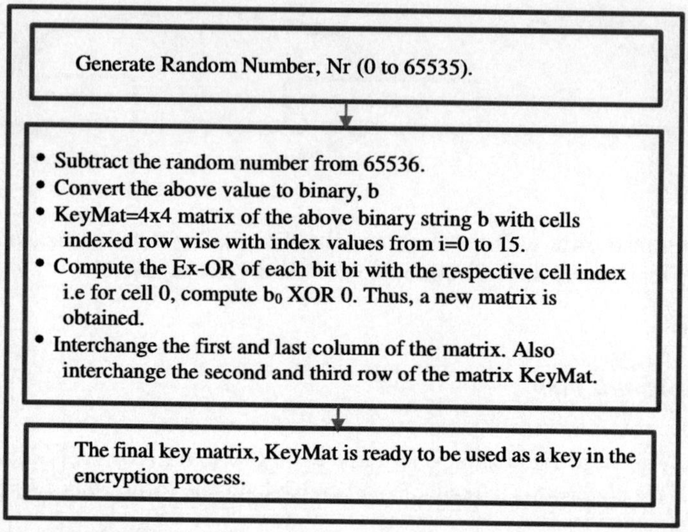

Fig. 2 Key generation process

4.1 Key Generation

The following steps describe the key generation process: (Fig. 2)

1. Generate a random integer Nr 0 to 65535.
2. Subtract the random number from 65536.
3. Convert the above value to binary, b
4. Construct a 4 × 4 matrix called KeyMat of the above binary string b with cells indexed row wise with index values from i = 0 to 15.
5. Compute the Ex-OR of each bit bi with the respective cell index, that is, for cell 0, compute b_0 XOR 0. Thus, a new matrix is obtained.
6. Interchange the first and last column of the matrix. Also interchange the second and third row of the matrix KeyMat.
7. Thus, the final key matrix is ready to be used as a key in the encryption process.

4.2 Encryption

The following steps describe the encryption process: (Fig. 3)

1. Assign numerical values from 0 to 25 to the alphabets A to Z and 26 to 35 to the digits 0 to 9 in sequence.
2. Partition the plaintext consisting of the uppercase alphabets and/or digits into blocks of 16 characters each.
3. Enter the 16 characters of a block in a 4 × 4 matrix Mat1.

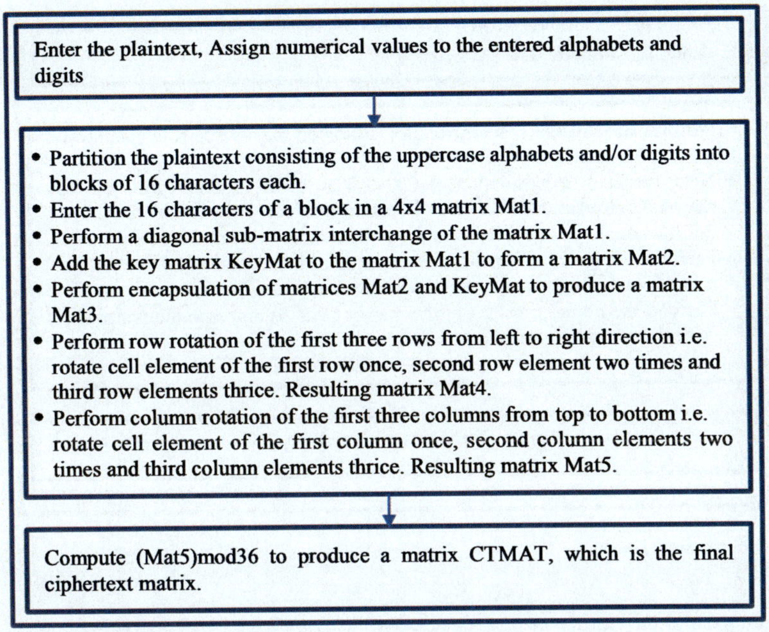

Fig. 3 Encryption process

4. Perform a diagonal sub-matrix interchange of the matrix Mat1.
5. Add the key matrix KeyMat to the matrix Mat1 to form a matrix Mat2.
6. Perform encapsulation of matrices Mat2 and KeyMat to produce a matrix Mat3.
7. Perform row rotation of the first three rows from left to right direction, that is, rotate cell element of the first row once, second row element two times and third row elements thrice. The resulting matrix is Mat4.
8. Perform column rotation of the first three columns from top to bottom, that is, rotate cell element of the first column once, second column elements two times and third column elements thrice. The resulting matrix is Mat5.
9. Compute (Mat5)mod36 to produce a matrix CTMAT, which is the final cipher text matrix.

4.3 Decryption

The following steps describe the decryption process: (Fig. 4)

1. Once the matrix CTMAT is received, perform the reverse column rotation followed by reverse row rotation on it.
2. Next, perform decapsulation of CTMAT in order to disintegrate the key matrix KeyMat and the data matrix called as Mat4.

Enter the ciphertext, CTMAT

- Perform the reverse column rotation followed by reverse row rotation on it.
- Next, perform decapsulation of CTMAT in order to disintegrate the key matrix KeyMat and the data matrix called as Mat4.
- Subtract the key matrix KeyMat from the matrix Mat4 obtained above.
- Perform a diagonal sub-matrix interchange of the matrix Mat4 to obtain a matrix containing weights as defined for A to Z and digits 0 to 9.
- Convert the numeric values in the matrix Mat4 to corresponding alphabets or digits as per the convention.

Finally, the plaintext matrix PTMat is obtained.

Fig. 4 Decryption process

3. Subtract the key matrix KeyMat from the matrix Mat4 obtained above.
4. Perform a diagonal sub-matrix interchange of the matrix Mat4 to obtain a matrix containing weights as defined for A to Z and digits 0 to 9.
5. Convert the numeric values in the matrix Mat4 to corresponding alphabets or digits as per the convention.
6. Finally, the plaintext matrix PTMat is obtained.

5 Implementation of Proposed Algorithm

5.1 Key Generation

1. Generate a random integer Nr in the range of 0–65535. Example: Nr = 5
2. Subtract the random number from 65536.
   $65536 - 5 = 65531$
3. Convert the above value to binary b, 1111 1111 1111 1011
4. Construct a 4×4 matrix called KeyMat of the above binary string

1	1	1	1
1	1	1	1
1	1	1	1
1	0	1	1

5. Compute the Ex-OR of each bit bi with the respective cell index

1	0	3	2
5	4	7	6
9	8	11	10
13	13	15	14

6. Interchange the first and the last column of the matrix. Also interchange the second and third row of the matrix KeyMat.

2	0	3	1
10	8	11	9
6	4	7	5
14	13	15	13

5.2 *Encryption*

Consider the string **"DATA SECURITY1234"**.

D	A	T	A
S	E	C	U
R	I	T	Y
1	2	3	4

The encryption process is illustrated as below:

3	0	19	0
18	4	2	20
17	8	19	24
27	28	29	30

1) Converting the text matrix to its equivalent weight matrix

19	24	17	8
29	30	27	28
19	0	3	0
2	20	18	4

2) Perform sub-matrix interchange(diagonal) to obtain a matrix called Mat1

21	24	20	9
39	38	38	37
25	4	10	5
33	33	33	17

3) Add the matrices Mat1 and KeyMat to form a matrix called Mat2

2	0	3	1
14	13	15	13
21	24	20	9
39	38	38	37
25	4	10	5
16	33	33	17
10	8	11	9
6	4	7	5

4) Perform encapsulation of KeyMat and Mat2 to form a matrix called Mat3

15	21	38	2
9	38	10	13
39	4	33	20
25	33	11	37
16	8	7	5
10	4	1	17
6	3	14	9
0	13	24	5

5) Perform column rotation of the first three columns from top to bottom to generate Mat5

15	21	2	2
9	2	10	13
3	4	33	20
25	33	11	1
16	8	7	5
10	4	1	17
6	3	14	9
0	13	24	5

6) Compute (Mat5) mod36 to produce a matrix CTMAT, which is the final ciphertext matrix

5.3 Decryption

CTMAT as received by the receiver

15	21	2	2
9	2	10	13
3	4	33	20
25	33	11	1
16	8	7	5
10	4	1	17
6	3	14	9
0	13	24	5

The decryption process is illustrated below:

6 Experimental Result

1)Perform the reverse column rotation followed by reverse row rotation on CTMAT

2	0	3	1
14	13	15	13
21	24	20	9
3	2	2	1
25	4	10	5
16	33	33	17
10	8	11	9
6	4	7	5

2) After decapsulation of CTMAT

21	24	20	9
3	2	2	1
25	4	10	5
16	33	33	17

Mat4

2	0	3	1
10	8	11	9
6	4	7	5
14	13	15	13

KeyMat

3) Mat4 after subtracting KeyMat from Mat4

19	24	17	8
29	30	27	28
19	0	3	0
2	20	18	4

4)Mat4 after diagonal sub-matrix interchange of the matrix Mat4

3	0	19	0
18	4	2	20
17	8	19	24
27	28	29	30

5) After converting the cell values of Mat4 to the corresponding alphabets and digits plain text matrix obtained is PTMAT

D	A	T	A
S	E	C	U
R	I	T	Y
1	2	3	4

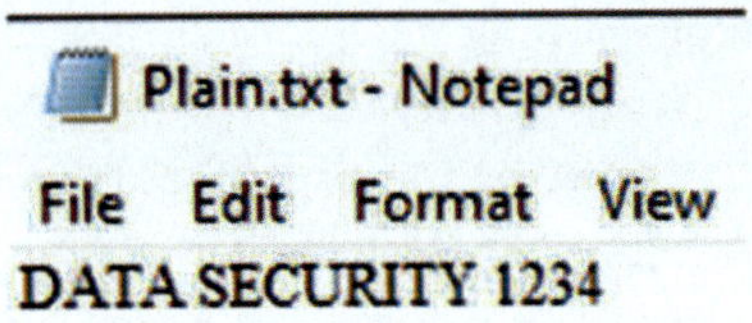

Fig. 5 Plaintext

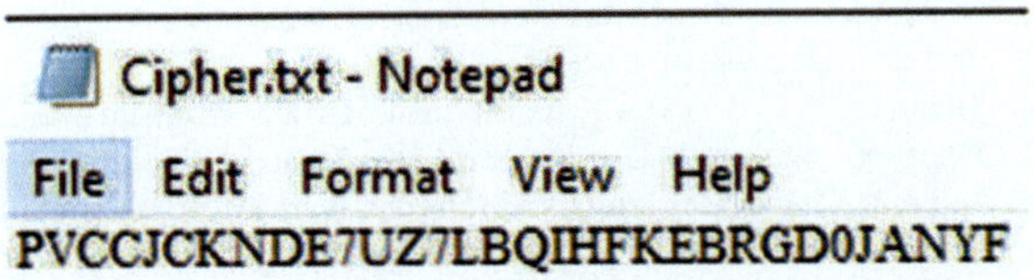

Fig. 6 Encrypted text

Fig. 7 Decrypted text

A new block cipher cryptographic method based on XOR and various matrix operations was proposed in this work. The data undergoes various addition and rotations operations, which make the proposed system more safe and robust. Figure 5 shows the input text file storing the input data to be encrypted. Figure 6 contains the secured data in encrypted form. Figure 7 contains the decrypted form of data. From the figures, it is stated that the proposed method works efficiently.

7 Conclusion

In this paper, the proposed algorithm presents encryption and decryption of text message for protecting of data using XOR operation and various matrix operations like addition and rotation for matrices. In the proposed method, the encryption process using matrix operations will provide security of data that is being transmitted across the network. The proposed method is quite feasible and effective. It is difficult to obtain the original message until the key matrix and construction of matrix which is being transmitted is known. So, the final cipher text obtained is strong and more secure.

References

1. K.Goyal and S.Kinger, Modified Caesar Cipher for Better Security Enhancement, IJCA(0975–8887), Volume 73, July of the World Congress on Engineering and Computer 2013
2. S.Bhati,A.Bhati,S.K.Sharma, A New Approach towards Encryption Schemes: Byte – Rotation Encryption Algorithm, Proceedings Science 2012 Vol II WCECS 2012, October 24-26, 2012, San Francisco, USA
3. Mahendran.R, Byte Rotation with CBC Encryption Algorithm, International Journal of Machine and Construction Engineering. Aug 2014
4. Nirmala.M, C S Pavan Kumar, Ansu Miriam Varghese, Neethu Santhosh, Text Encryption and Decryption Strategy Using Matrix Operations, International Journal of Pharmacy & Technology
5. Mampa Ghosh,Debjani Chakraborty, Various Bit- Rotation Technique On Key-Symmetric Matrix, International Journal Of Engineering And Computer Science ISSN:2319-7242 Volume 4 Issue 5 May 2015, Page No. 11906-11909
6. Punam V. Maitri Rekha V. Sarawade "Secure File Transmission using Byte Rotation Algorithm in Network Security" International Conference for Convergence of Technology – 2014, Pune, India

Impact of Naïve Bayesian Classification on Banking Sector

Sarmistha Saha and G. P. Saradhi Varma

Abstract In our modern era, bank plays an important role for well-being of the world economy, and all the banks are being computerized to handle smoothly ample of data. Hence we should give weightage to the banking sector which in turn can affect global economy in the long run. In this regard, they all need data mining techniques to discover patterns for unknown relationship of customer data and they too need some data mining techniques to classify the unknown variable. Bayesian classifier is the best method and applying its enhanced version, Naïve Bayesian classifier in banking sectors can attract new customers.

1 Introduction

Computer Science is a vast area in the field of Science and Technology. The most emerging subfield in Computer Science is data mining. It consists of various tools for automatic extraction of patterns and associations for large data sets. In the advent of modern Science and Technology, data mining tools help business to select data in the required pattern at first and then to filter and correlate also.

Data analysis consists of two steps: (i) Classification (ii) Prediction.

First of all, application data is to be categorized by classification which in turn gives rise to normalized form and then inference can be drawn using prediction technique.

Here I discuss one prediction technique based on classification. Bayesian classifier is based on Bayes' theorem. Naive Bayesian classifiers assume that the effect of an attribute value on a given class is independent of the values of the other attributes. This assumption is called class-conditional independence. Initially, we used a database composed of 16 independent variables along with their class values.

S. Saha (✉)
Acharya Nagarjuna University, Guntur, Andhra Pradesh, India
e-mail: sarmist@rediffmail.com

G. P. Saradhi Varma
Director, K.L. University, Vijayawada, Andhra Pradesh, India
e-mail: gpsvarma@gmail.com

S. C. Satapathy et al. (eds.), *Communication Software and Networks*, Lecture Notes in Networks and Systems 134, https://doi.org/10.1007/978-981-15-5397-4_26

2 Advantages of I-Banking to the Customer

First of all, advantages or benefits are identified from customers viewpoint. Sixteen benefits have been identified that were asked to obtain for this purpose. These sixteen benefits have been identified based on discussion with customers and review of literature. The sample customers were asked to rank their preference for various I-Banking services and responses have been shown in Table 1.

The analysis of customer's perception is dedicated to present a detailed account of the survey conducted on Internet Banking Practices among the customers of selected branches of ICICI Bank. A questionnaire is administered to the customers of the same branches to extract information related with their perceptions regarding Internet Banking Services.

The data is processed by using MS Excel, SPSS, SAS and have been used according to the requirement. Appropriate tools are used to analyze qualitative data at relevant places.

The primary data collected through questionnaire has its own limitations because some of the sample customers and employees of the bank tend to expose their problems, whereas some others tend to conceal the actual facts. Despite this, efforts were made to elicit correct information to the possible extent.

The above table reveals the fact that there is the highest significance regarding the satisfied customer.

3 Research Methodology

We wish to predict the class label of an unknown sample using Naïve Bayesian classification, given the training data in Table 1. The data samples are described by the attributes, 'Transaction History', 'Linking your Bank/credit card/Demat a/cs through a single user', 'Prepaid mobile charge', 'Send a smart money order', 'Open Fixed deposits and R.D. account', 'Order a Demand draft and pay order', 'Subscribe for mobile banking', 'Request a cheque book', 'Request a change of address', 'Stop Payment request', 'Request a debit card', 'Re-issue/upgrade of ATM/Debit card', 'Duplicate ATM pin', 'Secure mail box', 'E-Instruction', 'Share holding statements'. The class label attributes are Highly satisfied (c1), Satisfied (c1'), neither satisfied nor dissatisfied (c0), Highly dissatisfied (c2), dissatisfied (c2'). But here we consider only two prominent classes, i.e., Highly satisfied (c1), Highly dissatisfied (c2).

My calculations are as follows:

Step 1 Taking Bayes theorem, P(Ci | X) = P(X | Ci)P(Ci) / P(X) where Ci → classes, X → unknown variable, P → Probability Identify a new customer along with its attribute, i.e., X = (transaction history)

Step 2 We need to maximize P(X | Ci)P(Ci) for i = 1, 2

Step 3 We need to compute, P(C1) = P(Highly satisfied)
And P(C2) = P(Highly dissatisfied) from the data of Table 1.

Table 1 Sixteen particulars regarding I-banking services

Particulars	Highly satisfied c1	Satisfied c1'	Neither satisfied nor dissatisfied c0	Dissatisfied c2'	Highly dissatisfied c2	Total
Transaction history	240	96	60	60	24	480
Linking your Bank/Credit card/Demat a/cs through a single user	120	120	60	60	120	480
Prepaid mobile charge	60	120	120	96	84	480
Send a smart money order	84	120	156	24	96	480
Open Fixed deposits and R.D. account	120	135	66	135	24	480
Order a Demand draft and pay order	60	120	36	144	120	480
Subscribe for mobile banking	84	36	60	120	180	480
Request a cheque book	220	96	50	54	60	480
Request a change of address	276	84	72	24	24	480
Stop Payment request	60	84	120	96	120	480
Request a debit card	300	36	24	60	60	480
Re-issue/upgrade of ATM/Debit card	60	120	60	120	120	480
Duplicate ATM pin	192	48	120	60	60	480
Secure mail box	60	36	24	120	240	480
E-Instructions	60	72	108	120	120	480
Share holding statements	72	84	108	96	120	480

Table 2 Result from ANOVA

Particulars		Sum of the squares	Difference	Mean Square	F	Sig
Highly satisfied	Between groups	88940.333	6	14823.389	5.928	0.009
	Within groups	22506.667	9	2500.741		
	Total	111447.000	15			
Satisfied	Between groups	10142.438	6	1690.406	1.247	0.368
	Within groups	12204.000	9	1356.000		
	Total	22346.438	15			
Neither satisfied nor dissatisfied	Between groups	12148.333	6	2024.722	1.674	0.234
	Within groups	10882.667	9	1209.185		
	Total	23031.000	15			
Dissatisfied	Between groups	11504.438	6	1917.406	1.630	0.245
	Within groups	10590.000	9	1176.667		
	Total	22094.438	15			

Step 4 Now, P(Highly satisfied) = 240/480 = 0.5.
P(Highly dissatisfied) = 24/480 = 0.05.

Step 5 Now we have to compute, P(X|Ci) for i = 1, 2
Now P(Transaction history | Highly satisfied) = 240/(240 + 96 + 60) = 240/396 = 0.60
P(Transaction history | Highly dissatisfied) = 24/(24 + 60) = 24/84 = 0.28

Step 6 Using the above probabilities, we get,
P(X | Highly satisfied) = 0.5*0.60 = 0.3
P(X | Highly Dissatisfied) = 0.05*0.28 = 0.014

Step 7 Similarly, we can calculate for P(X | Highly satisfied) and P(X | Highly dissatisfied) for ∀ x
where X ∈ {Linking bank credit card, Prepaid mobile charge, Send a smart money, Open a smart money order, open fixed deposits/R.D. account, Order a Demand draft and pay order, Subscribe for mobile banking, Request a cheque book, Request a change of address, Stop Payment request, Request a debit card, Re – issue/upgrade of ATM/Debit card, Duplicate ATM pin, Secure mail box, E-Instructions, Share holding statements}

Step 8 Observing the above probabilities, i.e., P(Highly satisfied) for ∀ x
or
P(Highly dissatisfied) for ∀ X. Any new customer can take the decision to open account in that corresponding bank

4 Conclusion

This is an efficient and scalable method for evaluating probabilities for an unknown variable and the result has been tallied with the ANOVA method also. With the help of this, so many business organizations including banking sector can predict their risk factor in the present market and depending upon the result they can be alerted and consequently take the precautions too to survive in the present market. It is proved that this method has better performance in predicting risk factor than other methods.

Acknowledgements It is my immense pleasure to express our sincere gratitude to Acharya Nagarjuna University for providing me an opportunity to do our work. I sincerely thank my guide, Prof. Partha Saradhi Varma for guiding and encouraging in carrying out this paperwork.

References

1. Brieman L, Friedman J, Olshen R, Stone C (1994) Classification and Regression Trees. Pacific Grove: Wadsworth
2. Steinberg D, Colla PL (1995) CART: Tree-Structured Non parametric Data Analysis, San Diego, CA: Salford Systems
3. Jiawei H, Micheline K (2008) Data Mining Concepts and Techniques, Morgan Kaufmann
4. Pang Ning Tan (2008) Michael Steinbach. Vipin Kumar, Introduction to Data Mining, Pearson Education
5. Bank Websites
6. Field survey

Smart Door Unlocking Using Face Recognition and Blink Detection

Rajesh K. Yadav and K. M. Karthik

Abstract Internet of Things (loT) has become one of the promising technologies used to connect, manage and control smart objects connected to the Internet. In the development of smart cities, smart homes play a key role among many IoT applications. Smart home also helps physically challenged people in their day-to-day life activities. This paper proposes a solution in providing a smart, secured and easy way of door unlocking. Various concerns involving door unlocking are key lost/forgotten and waiting for someone to open the door. Smart door provides a convenient way to overcome these issues and is a step towards making a smart home. The solution uses face recognition and eye blink detection techniques as the backbone algorithm.

1 Introduction

Smart home is an application of Internet of Things (IoT) where devices interact with other devices present and take actions on their own or these devices are controllable through smartphone, smart watch or through voice and thus making life of people easier. Some examples of smart home involve turning the lights on/off based on the people present in the room, turning on the air conditioner remotely in hot summer weather, monitoring of the appliances and many more. Nowadays people are surrounded by world of automation where they get the work done automatically with little involvement. Smart home is about well-being, security and comfort at your fingertips. It helps to deal with the regular issues, which people generally face daily while working in their home. A layered architecture is presented for smart home in [1, 2]. The concept of smart home would be a great assistance for disabled people. Disabled people completely living alone will have more issues than one living with

R. K. Yadav · K. M. Karthik (✉)
Department of Computer Science & Engineering, Delhi Technological University, Delhi, India
e-mail: karthikkm.dtu@gmail.com

R. K. Yadav
e-mail: rkyadav6711@gmail.com

S. C. Satapathy et al. (eds.), *Communication Software and Networks*, Lecture Notes in Networks and Systems 134, https://doi.org/10.1007/978-981-15-5397-4_28

some non-disabled person. One of the issues that is included in door unlocking is that the person must go physically to open the door, and to handle this issue this paper proposes a smart and secured way of door unlocking.

The structure of the report is as follows. Section 2 gives a summary of related work. Section 3 explains the problem statement that has been addressed and proposes a solution for the addressed problem. Section 4 provides with the implementation of the proposed system. Section 5 gives the result analysis. Section 6 gives conclusion of the paper and last Sect. 7 provides with the future scope of the work.

2 Related Work

In a smart home, home safety is one of the major categories [3]. A smart home system can be categorised into two categories based on controlling: locally controlled and remotely controlled. A locally controlled system involves controlling appliances while staying at home though Bluetooth [4–6] and NFC [7, 8], and remotely controlled involves controlling through Wi-Fi [9, 10]. Bluetooth proves to be the best for short-range communication. An SMS-based solution is proposed in [11], which uses SMS for exchanging data over a GSM network. Exchanging data through SMS will incur additional cost and might be expensive; also, an SMS-based approach lacks in the area of Graphical User Interface (GUI). A door-unlocking mechanism is proposed in [12], which provides with two approaches for unlocking door: speech command and pin input. It is a locally controlled door-unlocking mechanism, which uses Bluetooth for sending and receiving data from smartphone to microcontroller and vice versa. A system based on face recognition [13] uses OTP to provide security to the system. In this system, irrespective of whether the person is authorised or unauthorised, the user receives a notification about the person arrival and must unlock the door after OTP authentication. A face-spoofing technique like 3D mask can be applied to breach the system. Today, facial recognition systems have high accuracy rates and are shown in [14] as a feasible method for secured door unlocking, but lacked automatic capability for image capture. A protocol was developed to transmit visual data to the homeowner for a manual video identification [15]. However, manual identification as well as intrusive identification does not offer visitor and homeowner comfort and convenience. In future, smart door systems should become an automatic identification system, as well as an intrusive identification system.

3 Proposed Solution

A solution that is easy to use and reliable for the users needs to be developed. Face Recognition is used as the principal technology for developing such system. It is currently one of the promising fields of research because of its demand in everyday life. Technology leaders have made great advancement with high accuracy in this

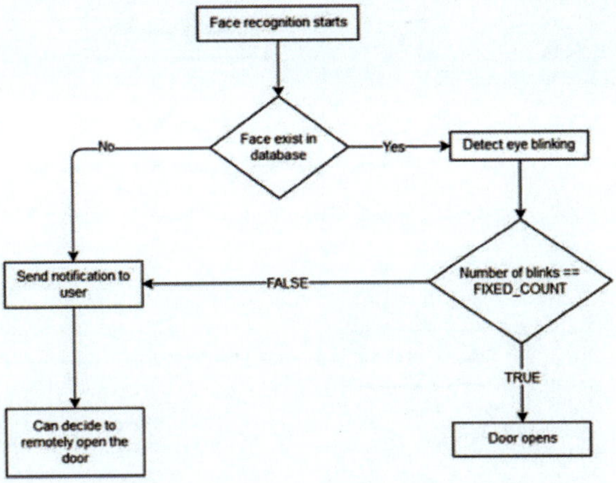

Fig. 1 Face spoofing detection

field. Face ID technology developed by Apple Inc. is being used for unlocking device, making payments, tracking facial expression for Animoji. Facebook uses the face recognition technology for tagging friends in photos. Google Photos allow us to search photos by the people with great accuracy.

The basic flow for the solution is when a person presses doorbell face recognition starts, and the face is matched with the authorised database. If a person is authorised, the door will open automatically; else, a notification will go to an authorised person with the picture of person visited, and the user can decide to open the door remotely. The proposed system sounds easy and convenient for a user but lacks security of face spoofing and authentication at smartphone side. Face spoofing is done mainly in three ways [16]: Photo attack, video attack and 3D mask attack.

To overcome this problem and detect face spoofing, a solution of eye blinking is proposed (Fig. 1). The count will act as a password that will be known to only authorised users.

The person with the smartphone can be someone else and not the actual authorised person. How can the user with the smartphone be authenticated? The flow for authenticating the user is given in Fig. 2. The full proposed system consists of one user acting as a super user who decides to open or not open the door when an unauthorised person visits. The face of the super user is stored in the server, and when the person selects to open the door, the application asks the user to capture his photo, and then this photo is sent to the server where the face recognition takes place. If the authentication is successful, then the door gets unlocked else the user receives an error message.

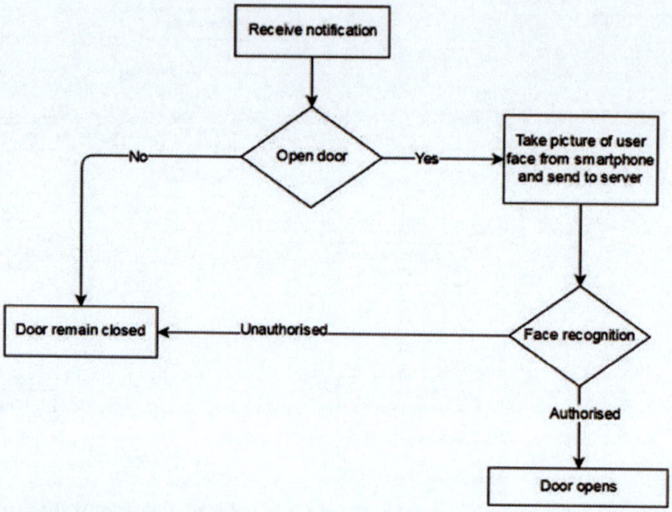

Fig. 2 Smartphone user authentication

4 Implementation

There are three different parts in the proposed system—face recognition, eye blink detection and smartphone application.

4.1 Face Recognition

The implementation of face recognition system is done using a pre-trained model (dlib) based on deep metric learning, which works on ResNet network [17]. The dataset used for training the model is face scrub dataset [18] and the VGG dataset [19]. The model provides with encoding of 128 real-valued numbers for the input face. Face recognition consists of two parts—Face training and Face matching.

Face Training

In this step, image of authorised person is fed to the system one after the other and the vector generated is stored as a JSON file, and whenever a new face needs to be trained (added to the list of authorised users), the image is passed to the system and it is encoded, and the name is added to the existing JSON.

Face Matching

The flow of face matching is calculation of encoding of arrived person and matching it with the database. The calculation of distance between the two vectors (visitor and database) is done by calculating the Euclidean norm, which is given by

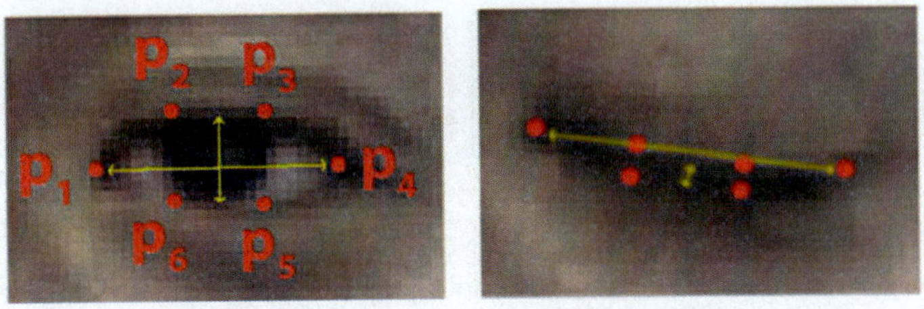

Fig. 3 Points for EAR calculation

$$\|v1 - v2\| = \sqrt{\sum_{1}^{128} (v1 - v2)^{2}} \tag{1}$$

This norm would return the distance between the two vectors (one is the vector of the input image and the other is from the stored JSON). The norm is calculated for each of the encoding stored and the distance with value ≤ 0.5 is returned.

4.2 Eye Blink Detection

Detection of eye blinking is done by calculating Eye Aspect Ratio (EAR) [20]. The EAR method proves to be a fast, simple and straightforward ratio calculation of distances between facial indicators for the eye.

$$EAR = \frac{\|p2 - p6\| + \|p3 - p5\|}{2\|p1 - p4\|} \tag{2}$$

In EAR, p1 to p6 are points present on the eye (Fig. 3). A pre-trained model (dlib) based on [21] is used, which provides with 68 points representing the face landmarks (Fig. 4) for the face which is used to get the six points highlighted within green box which is used to calculate EAR.

The working of EAR is shown in Fig. 5 where the EAR value near to zero for closed state and non-zero for open state can be seen.

4.3 Android

The application is with the super user and decides to act accordingly whenever a notification arrives. On clicking notification, following series of action occurs:

Screen 1 User receives notification
Screen 2 On clicking the notification, the user is shown with the picture of visitor

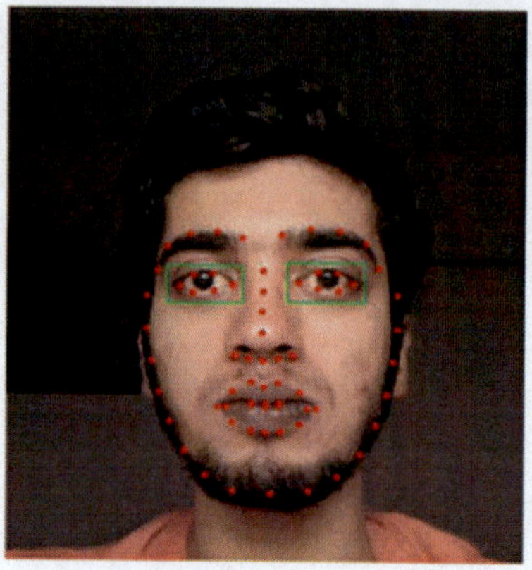

Fig. 4 Facial landmarks

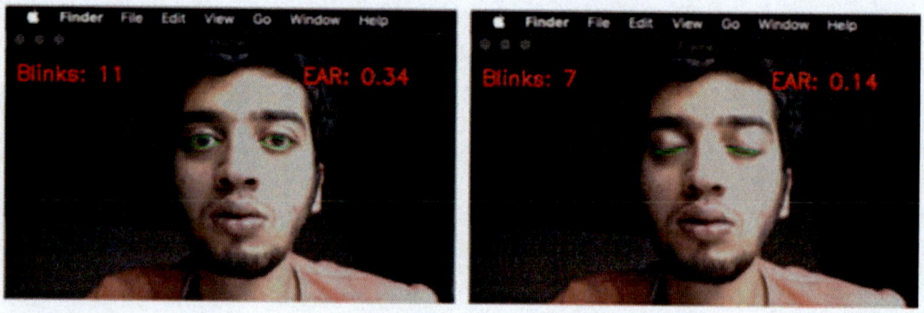

Fig. 5 Working of EAR

Screen 3 If user selects Yes, then he must capture his photo
Screen 4 The photo is sent to server, which authenticates the user
Screen 5 On success, it provides a message saying 'Authentication successful Door opened'
Screen 6 If authentication is not successful, it gives a message saying 'Authentication failed'.

5 Result

The main principal algorithms behind the working of proposed solution are face recognition and eye blink detection. Therefore, these algorithms should provide accurate result in the building of smart, reliable and convenient system.

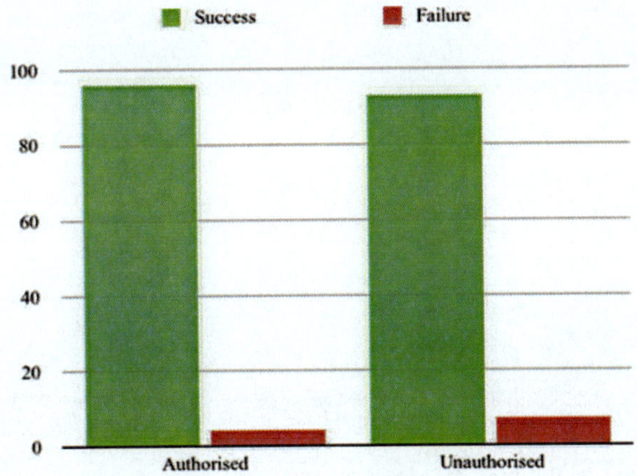

Fig. 6 Face recognition result

5.1 *Face Recognition*

The accuracy of face recognition was tested on FEI face dataset. The system has been trained on 100 different random persons, and for testing, a total of 200 samples were taken with 50:50 ratio for authorised and unauthorised. The accuracy shown is 94.5%, and the analysis is shown in Fig. 6.

5.2 *Eye Blink Detection*

The working of blink detection is based on the calculation of EAR. The EAR value is checked with a threshold value; if it is less than the threshold, then eye is closed else open. If the eye is closed for three consecutive frames, then a blink is detected. The accuracy of blink detection is calculated with different threshold value. For each threshold value, 50 blinks are processed and correspondingly the number of blinks the system detects is noted as output. Each blink is performed at an interval of 1 s. The process is carried for three different users (total 150 blinks), and the average of all three is taken. Figures 7 and 8 show the result of blink detection.

It can be seen through the graph that with the increase in the EAR the system provides with a greater false positive result ($\geq$0.4). Almost accurate number of blinks is shown with EAR of 0.2 having the result of 44 blinks. The result was noted without wearing any spectacles, but if the user is wearing spectacles the result is very inaccurate as shown in Fig. 8 resulting to drawback of the system having constraint of detecting authorised people only without spectacles. The accuracy of the blink detection came out to be 88%.

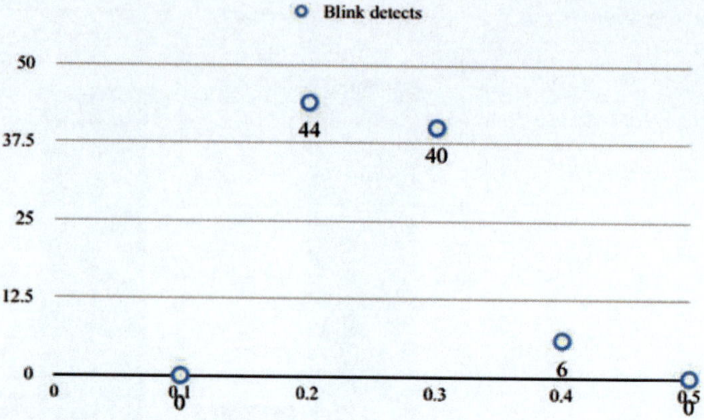

Fig. 7 Result with spectacles

Fig. 8 Result without spectacles

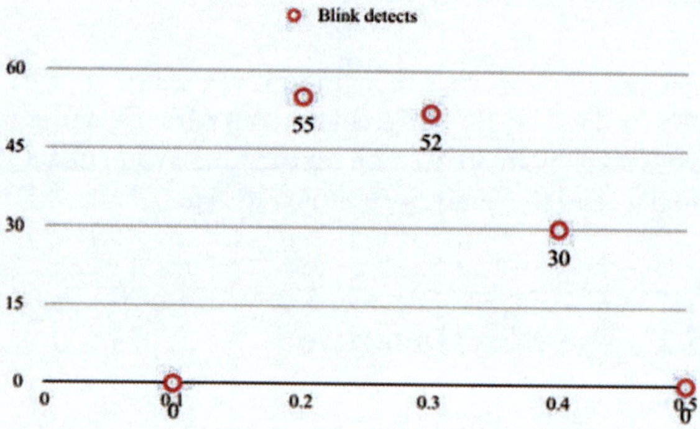

6 Conclusion

IoT has played an important role in making our environment smart, i.e. environment acts on its own based on the conditions. The proposed solution for smart door unlocking is simple and secured. It does not involve any multiple sensors/hardware, it just involves a camera, which captures image, and processing is done on the image, mainly face recognition. The requirement for the proposed solution is a good internet connectivity because when the user receives notification, he captures his photo and sends it to server. If the user has bad network connectivity, then the image would take time to get uploaded at the server and then the computation time and lastly the response from the server. If the system takes time, then visitor might have gone, and the door might get opened later which is unacceptable. The accuracy of the system comes to be same as that of blink detection, which is 91.25%. When a person arrives, there should be enough lighting for eye detection to work accurately.

7 Future Scope

Improving the efficiency of blink detection would be part of future scope. To enhance the security, the password in blink detection, i.e. the number of times the person needs to blink can be changed randomly to a different value every day. Whenever the value changes, a notification would go to the user with the new value. Currently, only one super user exists in the system who will have the application. The system can easily be extended for multiple super users after matching the person image taken from smartphone with all the authorized user list stored in the server. The system of door unlocking can be extended to a full controllable smart home through smartphone, i.e. controlling, monitoring and analysing all appliances through smartphone.

References

1. Gaikwad PP, Gabhane JP, Golait SS (2015) A survey based on smart homes system using internet of-things. In: 2015 International conference on computation of power information and communication
2. Bing K, Fu L, Zhuo Y, Yanlei L (2011) Design of an internet of things-based smart home system. In: The 2nd international conference on intelligent control and information processing, July 2011
3. Alam MR, Reaz MBI, Ali MAM (2012) A review of smart homes—past, present, and future. IEEE Trans Syst Man Cybern Part C: Appl Rev 42(6):1190–1203
4. Ismail NH, Saadon EI (2014) Android-based home door locks application via bluetooth for disabled people. In: Proceedings of the international conference on control system computing and engineering Penang IEEE, November 2014
5. Ramlee RA, Othman MA, Leong MH, Ismail MM, Ranjit SSS (2013) Smart home system using android application. In: 2013 international conference of information and communication technology (ICoICT), Bandung, pp 277–280
6. Kumar S, Lee SR (2014) Android based smart home system with control via Bluetooth and internet connectivity. In: The 18th IEEE international symposium on consumer electronics (ISCE 2014), JeJu Island, pp 1–2
7. Vagdevi P, Nagaraj D, Prasad GV (2017) Home: IOT based home automation using NFC. In: International conference on I-SMAC (IoT in social, mobile, analytics and cloud) (I-SMAC)
8. Chandrakar N, Kaul S, Mohan M, Vamsi CS, Prabhu KR (2015) NFC based profiling of smart home lighting system. In: International conference on industrial instrumentation and control (ICIC)
9. Mao X, Li K, Zhang Z, Liang J (2017) Design and implementation of a new smart home control system based on internet of things. In: IEEE smart cities conference
10. Uddin M, Nadeem T (2012) EnergySniffer: home energy monitoring system using smart phones. In: 8th international wireless communications and mobile computing conference (IWCMC), Limassol, pp 159–164
11. ElKamchouchi H, ElShafee A (2012) Design and prototype implementation of SMS based home automation system. In: International conference on electronics design, system and applications (ICEDSA)
12. Arifin RDH, Samo R (2018) Door automation system based on speech command and PIN using android smartphone. In: International conference on information and communications technology (ICOIACT)

13. Tiwari S, Thakur S, Shetty D, Pandey A (2018) Smart security: remotely controllable door-lock. In: Proceedings of the 2nd international conference on inventive communication and computational technologies
14. Hassan H, Bakar RA, Mokhtar ATF (2012) Face recognition based on autoswitching magnetic door lock system using microcontroller. In: International conference system engineering and technology (ICSET), September 2012, pp 1–6
15. Oh YJ, Paik EH, Park KR (2006) Design of a SIP-based real-time visitor communication and door control architecture using a home gateway. IEEE Trans Consum Electron 52(4):1256–1260
16. Kumar S, Singh S, Kumar J (2017) A comparative study on face spoofing attacks. In: International conference on computing, communication and automation (ICCCA)
17. He K, Zhang X, Ren S, Sun J (2015) Deep residual learning for image recognition
18. Ng H-W, Winkler S (2014) A data-driven approach to cleaning large face datasets. In: Proceedings of IEEE international conference on image processing (ICIP), Paris, France, October 27–30
19. Parkhi OM, Vedaldi A, Zisserman A (2015) Deep face recognition british machine vision conference
20. Soukupova T, Cech J (2016) Real-time eye blink detection using facial landmarks. In: 21st computer vision winter workshop (CVWW2016)
21. Kazemi V, Sullivan J (2014) One millisecond face alignment with an ensemble of regression trees. In: CVPR 2

A Review on Image Compression Techniques

Krishna Marlapalli, Rani S. B. P. Bandlamudi, Rambabu Busi, Vallabaneni Pranav, and B. Madhavrao

Abstract With the continuous advances in technology, the bandwidth requirements of a communication network have been increased. But increased pixel size in and gray level resolution in sensor technology and in digital image representation, the increased bandwidth is not satisfying the requirements. Hence image compression becomes a prominent research area. Image compression decreases the number of bits necessary to symbolize the image without destroying the original quality of the image. This paper briefly discusses different types of image formats to represent the image, types of compression methods used, and their performance measurements.

1 Introduction

Compression is nothing but minimizing the size or bits necessary to symbolize an image, file, or video without much affecting the quality of the input (original) data. One of the significant areas in compression is Image Compression. The primary objective of image compression is to minimize the quantity of data necessary to symbolize an image. Typically a huge number of bits are necessary to signify or store even for an ordinary single band digital (B&W) image. Image compression techniques reduce the redundancy and irrelevancies that exist in the data. The number of bits necessary to symbolize a compressed image is less compared to the original image. Hence the space required for storing and bandwidth required for transmitting the image decreases considerably. So that we can store more number of images and they can be transmitted in a less amount of time, saving the transmitting time and bandwidth. Whatever may be the technique used image compression includes three essential phases Mapper, Quantizer [1], and Encoder. In the first phase, Mapper converts the special domain information of the image to frequency domain. Quantizer

K. Marlapalli (✉) · R. S. B. P. Bandlamudi · V. Pranav · B. Madhavrao
Department of CSE, Sir CRRCoE, Eluru, India
e-mail: marlapallikrishna@gmail.com

R. Busi
Lakireddy Bali Reddy College of Engineering (Autonomous), Mylavaram, India
e-mail: rams1315@lbrce.ac.in

S. C. Satapathy et al. (eds.), *Communication Software and Networks*, Lecture Notes in Networks and Systems 134, https://doi.org/10.1007/978-981-15-5397-4_29

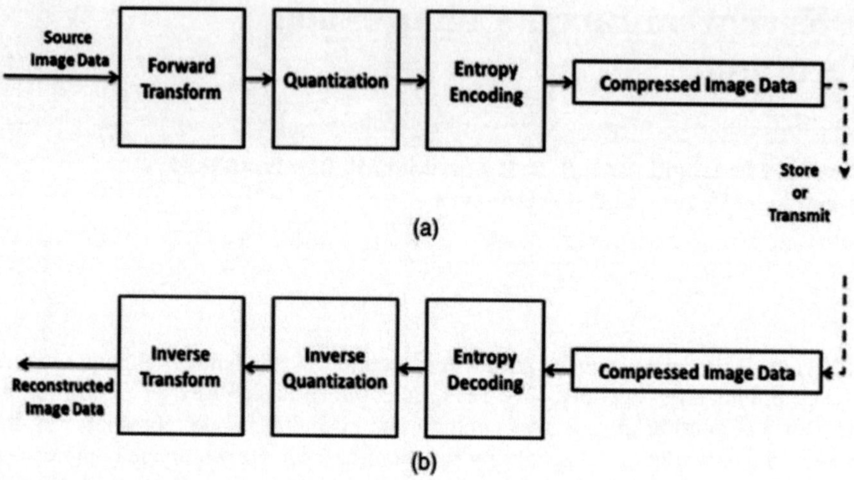

Fig. 1 Block diagram of compression and decompression

works on mapping many to one principle. More values are quantized into smaller value. Hence the numbers of bits necessary to symbolize the image are reduced in the phase. Lastly, in Encoding phase, the resulting data is coded to enhance the compression further. The inverse procedure (Decoding, Dequantization, and Inverse Mapping) is used to recreate the original image back. This process is known as decompression. The entire idea of compression and decompression is depicted in Fig. 1.

In common, three basic redundancies are present in digital images. They are as follows:

Psycho-visual Redundancy: In digital images, there exists some information that has no relative importance in usual visual processing. This unimportant information is called psychovisually redundancy. The removal of psychovisually redundancy results in loss of quantitative information. It doesn't result in the quality of the image that a human eye can identify.

Interpixel Redundancy: Interpixel redundancy is defined as failure to identify and utilize data. If we can predict a pixel value from its adjacent (following/preceding) pixels, then the image is said to have interpixel redundancy. Image resolution influences the interpixel redundancy. It is a redundancy equivalent to statistical dependencies among pixels, particularly between adjoining pixels.

Coding Redundancy: In ordinary image same numbers of bits are used for symbols that have larger probability and for symbols that have lesser probability. We can improve this by assigning fewer bits for the gray levels that appear with more probability and more bits for the gray levels that appear with less probability. There is no relationship between the pixels in Coding redundancy.

2 Image Formats

Majorly image compression techniques can be classified into two types

- Lossy Compression Techniques
- Lossless Compression Techniques

Images are classified into different types such that some image formats support lossy compression and some support lossless. Different image formats and supporting techniques are tabulated in Table 1.

JPG: JPG analyses the image and removes the information which a human eye can't observe. It is used in photographs and images that have continuous tone. The information is stored in 24-bit color. We can adjust the degree of the compression.

JPEG: "Joint Photographic Expert Group" abbreviated as JPEG is a tremendous method to store photographic images of 24-bit. It is widely used in multimedia and web applications. JPEG compresses the image such that there is some loss in the information. JPEG is used for video compression also. It is not suitable for line drawings and graph images.

JPEG 2000: DCT coefficients play an important role in JPEG to compress the image, whereas wavelet-based method is used in JPEG 2000. It was formed by the "*Joint Photographic Experts Group committee*" in the year 2000. JPEG 2000 deals with the lossy techniques as well as lossless.

GIF: GIF full form is Graphics Interchange Format. It supports 8-bit colors. It is also used in gray-scale images and black and white text. The main drawback with GIF is it can't work with images with more than 256 colors, but most of the color images are with more than 256 colors, i.e., 24 bits per pixel. It is widely used in animated images.

PNG: PNG is the short form of Portable Network Graphics. This file format is exploited for lossless image compression. The PNG file format replaces the GIF file format as it achieves 10–30 % more compression compared to GIF. It has two versions PNG-8 (supports $2^8 = 256$ colors) and PNG-24 (supports $2^{24} = 16777216$ colors). PNG generates smaller files and permits more colors. Partial transparency is supported by PNG.

Table 1 Different Image Formats

Lossy compression	Lossless compression
JPG	GIF
JPEG	PNG
JPEG 2000	TIFF
	BMP
	RAW

TIFF: TIFF is the short form for Tagged Image File Format. It is a lossless format. As it has very high quality, the format is principally used in photography and desktop publishing. It saves 8 bits and 16 bits per color (red, green, blue) for 24-bits and 48-bits in total, respectively. The compression in TIFF is relatively less and they are not used in web applications. They are used in high-quality print graphics particularly at massive sizes.

BMP: The BMP is an abbreviation to denote Bitmap Image file. The BMP file is enclosed with raster graphic data. Display devices don't affect this data (independent of display devices). That is a graphics adapter is not necessary to see the BMP file image. Generally, it is used with lossless compression techniques. It is also known as Device Independent Bitmap (abbreviated as DIB) file format or simply a bitmap. It stores bitmap digital images mainly on operating systems OS/2 and Microsoft Windows in large scale. It is able to store monochrome as well as color 2-D digital images.

3 Compression Techniques

In general Image compression techniques [2, 3, 7] are categorized into two types, one is Lossy and the other is Lossless. The following Table 2 shows various techniques under each category.

In lossy compression as the name suggests there will be some loss of the data. The restored image is not identical to the input (original) image. Figure 2 shows the block drawing of Lossy compression.

Whereas in lossless techniques the restored image is completely a duplicate of the input (original) image. There is no loss of data; decompressed image is exactly the same as uncompressed image. Lossless compression methods typically provide about a ten percent decrease in file size for complex images. Medical images or any images are used in courts. Lossless compression methods can provide substantial compression for simple images. The block drawing of lossless compression is shown in Fig. 3.

Table 2 Classification of compression techniques

Lossy techniques	Lossless techniques
Transform coding	Run-length encoding
Vector quantization	Entropy coding
Fractal coding	Prediction coding
Block transform coding	Huffman coding

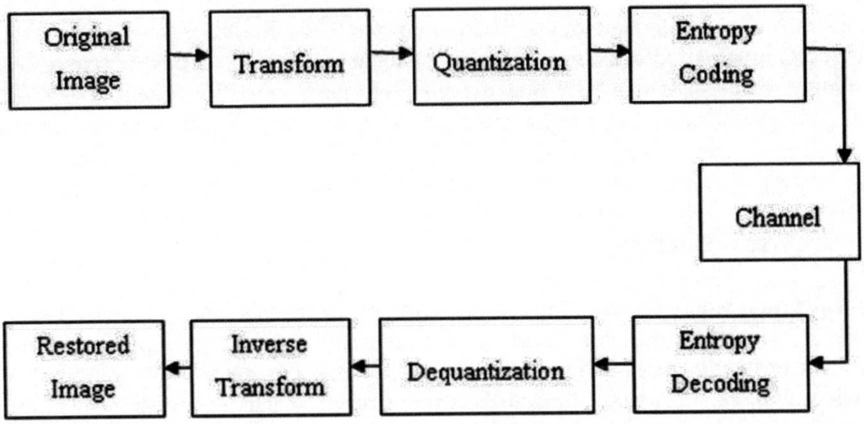

Fig. 2 Lossy image compression

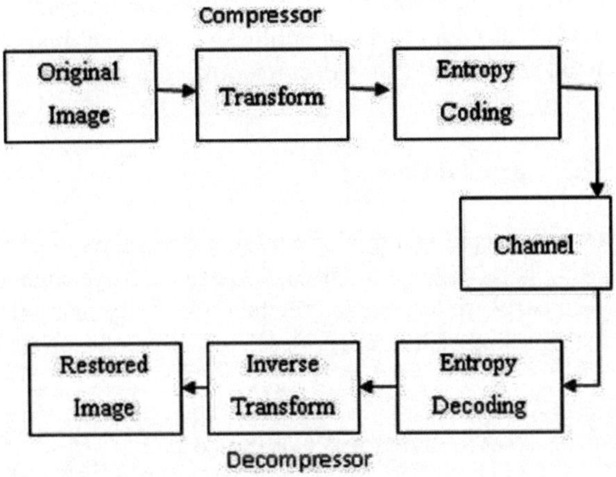

Fig. 3 Lossless image compression

3.1 Lossy Techniques

3.1.1 Transform Encoding

It is a kind of data compression for ordinary data like photographic images or audio signals. It transforms the pixels values of the image from one domain to another domain (spatial domain to frequency domain). More information of the image is represented with very few coefficients and most of the coefficients will have small value or the value tends to zero. A reversible linear transform (such as Discrete Fourier Transform or Discrete Cosine Transform) is used to map the image pixel values into a set of coefficients. These coefficients are further quantized and coded. A good quality transformation groups as much information as possible into a tiny number of transform coefficients. Then to eliminate the coefficients which bear the smallest amount of information, quantization technique is used. In this approach, first an input image of size $N \times N$ is partitioned into number of $n \times n$ nonoverlapping sub-images and then they are transformed to produce $(N/n)^2$ sub-image transform arrays, each array of size $n \times n$. Each of these blocks is transformed separately.

3.1.2 Vector Quantization

Quantization works on map many to one principle. It maps a group of values to one value. Quantization techniques are further divided into two types: they are scalar and vector quantization. Scalar quantization executes a range of single mapping on each value. Vector quantization is the expansion of scalar quantization in multi-dimensions. VQ can be applied in both the spectral or spatial domains. Information theory tells us that better compression can be achieved with vector quantization than with scalar quantization (rounding or truncating individual values). This technique builds up a dictionary of fixed-size vectors known as code vectors. Input image is divided into nonoverlapping blocks labeled as image vectors. Each image vector is encoded with index of the closest matching vector in the dictionary.

3.1.3 Fractal Coding

It works on the assumption that in general a part of an image is similar to its neighboring parts of the input image. These parts are changed into mathematical data and these parts are termed as "fractal codes". By using these fractal codes the encoded image is restored.

3.1.4 Block Transform Coding (BTC)

In BTC, input image is divided into blocks where each block consists of pixels of size 8×8. Block transform coding takes the advantage of correlation among the pixels present in the block. By virtue of this, each block is altered. Finally, each block is quantized and coded individually.

3.2 *Lossless Techniques*

3.2.1 Run-Length Encoding (RLE)

In RLE, the data is replaced by a duo of (length, value) where "length" is the number of repetitions and recurring value is denoted by "value". Particularly in compressing the big-level images, this method is used, because long run of a value is not in common in usual gray-scale images. A resolution to this is to decompose the gray-scale image into bit planes and individual bit planes are compressed separately. Efficient run-length coding method is one of the variations of run-length coding.

3.2.2 Entropy Encoding

In compression framework, entropy means the smallest number of bits needed, on the average, to represent a symbol (the average on all the symbols code lengths). It doesn't depend on the explicit characteristics of the medium. A unique prefix-free code is assigned to every separate symbol in the input. A variable-length prefix-free output codeword is used to replace the every input symbol of fixed-length. The length of codeword is nearly proportional to the –ve logarithm of the probability. As a result, the most commonly used symbols will take shorter codes.

3.2.3 Predictive Coding

It is a technique in which compression is mainly based on the dissimilarity among the original and predicted values. It is also known as DPCM—Differential Pulse Code Modulation. Decomposition of an input image into a group of bit planes is not required. After removing the interpixel similarities only the new information in each pixel is coded. The dissimilarity among the original and predicted pixel value is calculated as new information. The system has two phases an encoder and a decoder, each again consists of a matching predictor. Depending upon the number of precedent inputs the predictor produces the probable pixel value. The predicted value is coded using a variable-length code to produce the next element in the compressed data stream.

3.2.4 Huffman Coding

D. Huffman developed the Huffman code in 1952. It is a minimum length code. This means that given the statistical allocation of the gray levels (the histogram), the Huffman algorithm will produce a code that is as near as possible to the lowest bound, the entropy. The technique outcomes in an unequal (or variable) length code, where the size of the code words can vary.

4 Assessment of Compressed Images

Compression Ratio is the measure of how much the input image is compressed. The quality of compressed image is determined by looking into various factors like Peak Signal to Noise Ratio (PSNR) and Mean Squared Error (MSE) [4–6]. Also some other techniques are existing; PSNR and MSE are frequently used as they are easy to compute.

4.1 Mean Square Error

This is measured by taking the average of difference of squared intensities of compressed and input (original) image pixels. The MSE is given by the equation given below.

$$\text{MSE} = \frac{1}{m * n}\sum_{i=0}^{m-1}\sum_{j=0}^{n-1}\left[g'(p,q) - g(p,q)\right]^2 \tag{1}$$

Here, $g'(p,q)$&$g(p,q)$ *are* pixels values of restored and original images, respectively.

And *m & n are number of rows & columns of the image in spatial domain.*

4.2 Peak Signal to Noise Ratio

PSNR is the usual way to calculate the fidelity [8, 9]. The term PSNR is the ratio between the highest possible signal value (power) and the signal value of deforming noise which changes the value of its representation. The PSNR is usually represented in terms of the logarithmic decibel scale. PSNR is calculated in decibels (dB). Generally, 0.5–1 dB is said to be a noticeable difference.

The mathematical representation of the PSNR is as follows:

$$\text{PSNR} = 10\log_{10}\left(\frac{m * n}{\text{MSE}}\right) \tag{2}$$

where MSE *is* Mean Square Error, *and m & n are image rows & columns in spatial form.*

4.3 Compression Ratio (C_R)

Compression ratio is one more metric of evaluation for measuring the compression. It is the relative quantity between the bits necessary by the input image to the bits necessary by compressed image. The following equation represents the compression ratio.

$$C_R = \frac{\text{Uncompressed (original) image size}}{\text{compressed image size}} \tag{3}$$

In general, low MSE and high PSNR values are desirable.

5 Conclusions

With the rapid growth of internet, the information transmitting through the internet has increased enormously. Image data takes a major share of this information. Compressed images can be transmitted in a simpler way and they use less bandwidth. Taking these points into consideration image compression became a vital research area in computer science. In this review paper, we discussed different types of image formats and different compression methods used for the compression. In lossless compression method, the compressed image is as same as the original but compression ratio is low, whereas in lossy compression methods high compression ratio can be obtained but compressed image is not the same as that of original. There is some loss of data. We conclude the paper by elaborating various quality measuring parameters like Compression ratio, MSE, & PSNR.

References

1. Cosman PC, Oehler KL, Riskin EA, Gray RM (1993) Using vector quantization for image processing. IEEE 81(9):1326–1341
2. Krishna M (2016) The VLIW architecture for real-time depth detection in image processing. Int J Comput Sci Mechatron 2(VI):1–9
3. Krishna M (2016) An efficient multi dimensional view for vehicles by patch memory management in image processing. Int J Comput Sci Mechatron 1(V):1–10
4. Krishna M, Prasad Reddy PVGD, Srinivas G, Ramesh Ch. (2016) A smoothening based JPEG compression for an objective image quality of regular and noisy images. Int J Appl Eng Res 11(6):3799–3804
5. Krishna M, Srinivas G, Prasad Reddy (2016) PVGD Image smoothening and morphological operator based JPEG compression. J Theor Appl Inf Technol 85(3):252–259
6. Krishna M, Rani BSBP, Sri VDS, Karri RR (2017) Filter based JPEG compression for noisy images. J Adv Res Dyn Control Syst 9(18):1233–1248
7. Krishna M, Satyanarayana G, Sri VDS (2017) Digital image processing techniques in character recognition - a survey. Int J Sci Res Comput Sci Eng Inf Technol 2(6):95–101
8. Krishna M, Sri VDS, Rani BSBP, Satyanarayana G (2018) Edge based reliable digital watermarking scheme for authorized ownership. Int J Pure Appl Math 119(7):1291–1299
9. Paravathaneni K, Krishna M (2016) Unadulterated image noises and discrepancy estimation. Int J Technol Res Eng 3(7):1501–1503

A Novel Design of Multiband Monopole Antenna Loaded with Complementary Split Ring Resonator

Y. Sukanya, D. Uma, P. A. Nageswara Rao, and R. P. Das

Abstract A novel patch antenna having Complementary Split Ring Resonator (CSRR) with meandered Coplanar Waveguide (CPW) fed is proposed in this paper. A compact 3 band structure is developed by engraving slots and meandered CPW fed on the radiating element. A triple-band antenna with its reduced size is achieved by etching circles outside the patch along with CSRR slots and with Modified Ground Plane (MGP). The proposed antenna displays acceptable results at all values of resonance, in terms of enhanced bandwidth, return losses, VSWR. The simulated results for different designs are discussed and compared among them. Parametric study is also implemented for different ring widths and etching of circles with different radius on the proposed design. In order to validate the results CSRR permittivity and permeability characteristics are explained. The proposed antenna with greater qualities is suitable for C-band, S-band, Wireless Local Area Network (WLAN), WIMAX applications.

1 Introduction

Owing to consistently developing wireless equipment, a lot of importance is given for multiband antennas. A multiband antenna is worked at several bands of frequency. In order to have different wireless applications, various techniques have been reported for achieving the requirement of multiband antenna. Metamaterials are required for the implementation of antennas [1–5]. Actually, metamaterials do not readily exist in nature; it has an assumed structure which can be obtained from the proposed design [6]. Its principle design is to show negative estimations of μ and ε for

Y. Sukanya (✉) · D. Uma
Vignan's Institute of Information Technology, Duvvada, Visakhapatnam, India
e-mail: sukanyayadlapalli@gmail.com

P. A. Nageswara Rao
Gayatri Vidya Parishad College, Visakhapatnam, India

R. P. Das
Nadimpalli Satyanarayana Raju Institute of Technology, Sontyam, India

S. C. Satapathy et al. (eds.), *Communication Software and Networks*, Lecture Notes in Networks and Systems 134, https://doi.org/10.1007/978-981-15-5397-4_30

different resonators. SRR and CSRR are the prominent strategies to get metamaterial properties [7, 8].

Metamaterial results in best antennas were checked for performance including multiple band activity. Conventional dielectric substrate is not preferred instead this metamaterial substrate is preferred in order to design a miniature antenna [9]. Bandwidth enhancement is achieved by etching multiple circular patches outside and inside of monopole antenna and with modified ground plane [10, 11]. The improvements in amplification and frequency can be achieved by complementary split ring resonator [12]. Because of minimised size of an antenna, matching in impedance became quite difficult and in order to avoid this meandered feed is introduced. Meandered technique is also utilised for reducing the size of antenna. Meandered fed line with hexagonal split-ring resonator is introduced to diminish the geometrical dimensions of antenna and matching impedance [13]. For further improvement, the rectangular monopole can be modified further to corner triangular cut [14].

Multiband operation was achieved without changing the shape and dimensions of basic antenna by inserting an open CSRR on the rectangular patch [15]. The chief aim of the research work is to apply complementary split ring resonator with three openings on the circular monopole antenna for achieving multiband activity and to explore the operation of antenna [16]. Dual elements are present in SRR as well as CSRR. SRR comprises thin metallic rings having split in opposite positions. Several metamaterials exhibit double negative properties at X-band. Double negative is nothing but negative values of permittivity and permeability.

2 Design and Simulation

The complete design evolution process is shown in Fig. 1. The initial step involves monopole structure with coplanar waveguide feeding with Modified Ground Plane (MGP) as shown in design 1, which generates a single band of frequency, resonated at 5.2 GHz with a return loss of −4 dB and a bandwidth of 0.265 GHz as shown in Fig. 2. The second step of the design is CSRR with three slots, and three slits are loaded in the monopole structure with modification in the ground plane as shown in design 2, which generates a triple band of frequency resonated at 3.3 GHz, 5.4 GHz, 7.2 GHz with return losses −19 dB, −12 dB, −18 dB, respectively, as shown in Fig. 3. For additional improvement of loss parameter, a meandered process is introduced in CPW fed with modified ground plane as shown in design 3, which resonates at 3.8 GHz,4.8 GHz, 6.5 GHz with return losses −35 dB, −20 dB, −37 dB, respectively, as shown in Fig. 4. Some other designs are also introduced to improve bandwidth, return losses and gain by considering the design 3. In the fourth step of design, some of the circles are etched outside the monopole antenna with MGP as shown in design 4, which resonates at 3.8 GHz, 4.8 GHz and 6.5 GHz with return losses −34 dB, −19 dB, −36 dB, respectively, as shown in Fig. 5. In the fifth step of design, some of the circles are etched inside the monopole antenna with MGP as shown in design 5, which resonates at 3.8 GHz, 4.8 GHz and 6.5 GHz as shown in Fig. 6. In the last step

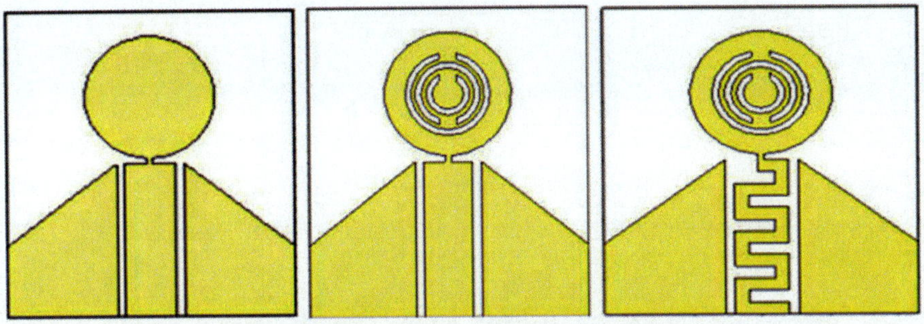

Design 1: Monopole antenna with modified ground plane using CPW fed
Design 2: CSRR loaded monopole antenna with modified ground plane using CPW fed
Design 3: Meandered process CSRR loaded monopole antenna with modified ground plane using CPW fed

Three Different Patterns Followed By Third Design

Design 4: Meander CPW fed CSRR loaded with multiple circles etched at outside the monopole antenna
Design 5: Meander CPW fed CSRR loaded with multiple circles etched at inside the monopole antenna
Design 6: Meander CPW fed hexagonal antenna

Fig. 1 The diagram of antenna structures

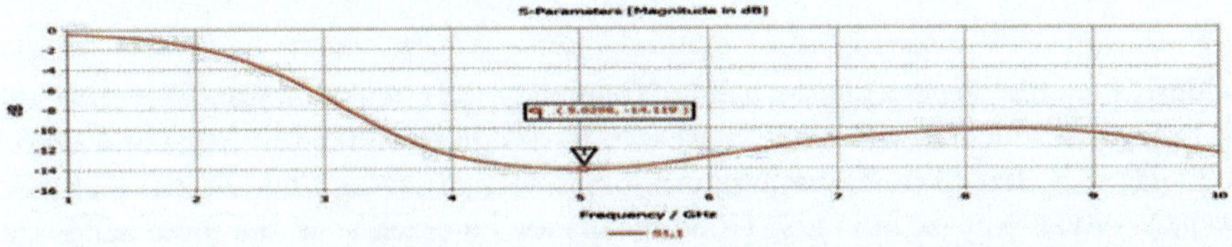

Fig. 2 From design 1 the CPW fed monopole antenna is resonated at 5 GHz with return loss −14 dB

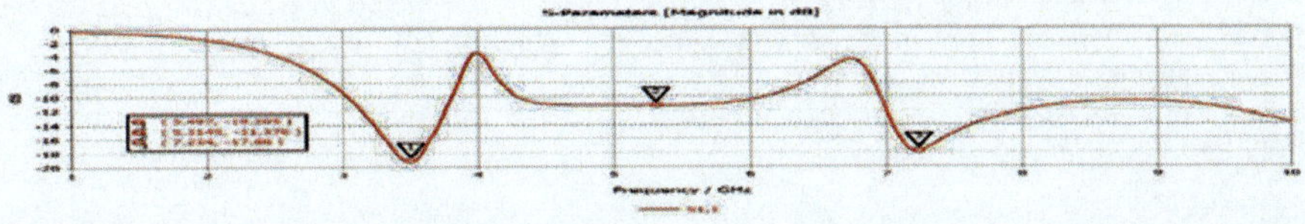

Fig. 3 From design 2 CPW fed CSRR loaded monopole antenna with MGP is resonated at 3.5 GHz, 5.3 GHz, 7.2 GHz with return losses −19 dB, −11 dB, −17 dB

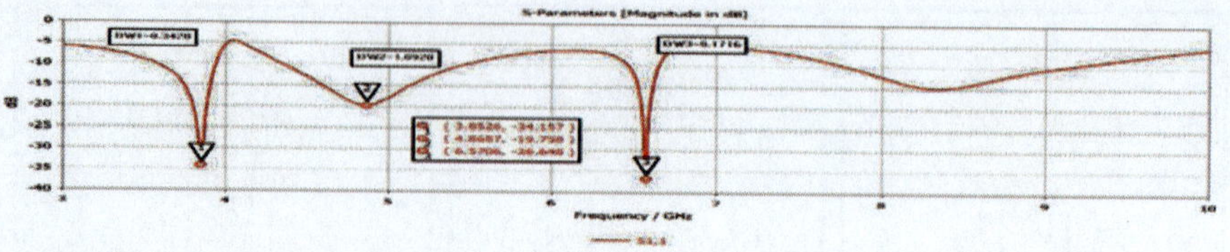

Fig. 4 Meandered CPW fed CSRR loaded monopole antenna with MGP is resonated at 3.8 GHz, 4.8 GHz, 6.5 GHz with return losses −34 dB, −19 dB, −36 dB

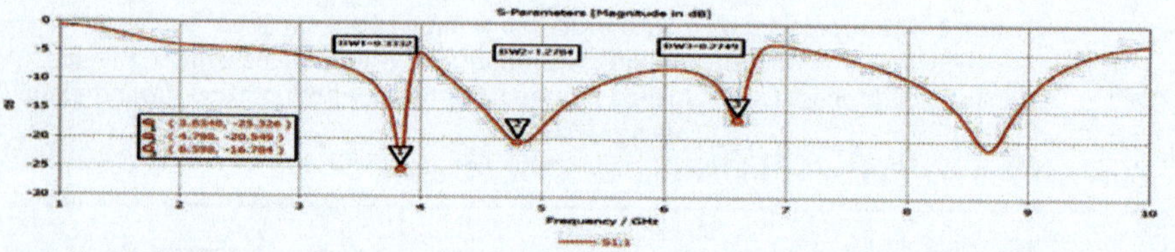

Fig. 5 Meandered CPW fed CSRR loaded with multiple circles etched outside the monopole antenna with MGP is resonated at 3.8 GHz, 4.7 GHz, 6.5 GHz with return losses −25 dB, −20 dB, −16 dB

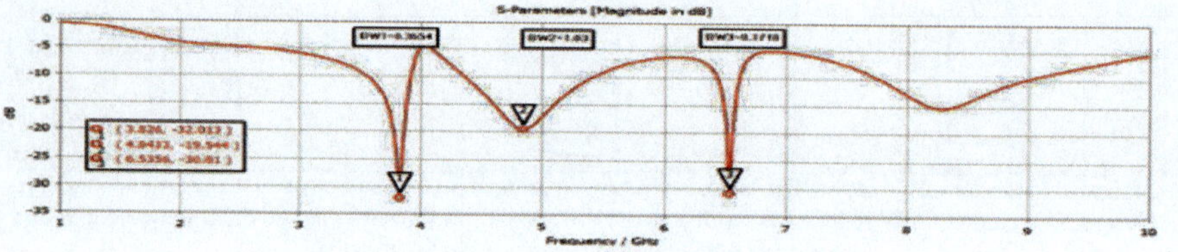

Fig. 6 Meandered CPW fed CSRR loaded with multiple circles etched inside the monopole antenna is resonated at 3.8 GHz, 4.8 GHz, 6.5 GHz with return losses −32 dB, −19 dB, −30 dB

of design in order to get a quartet band of frequency, the circular monopole is changed to hexagonal antenna with MGP as shown in design 6, which resonates at 4 GHz, 4.7 GHz, 7.2 GHz, 8.1 GHz with return losses −42 dB, −20 dB, −24 dB, −25 dB, respectively, as shown in Fig. 7. Good results are obtained at all the three bands by

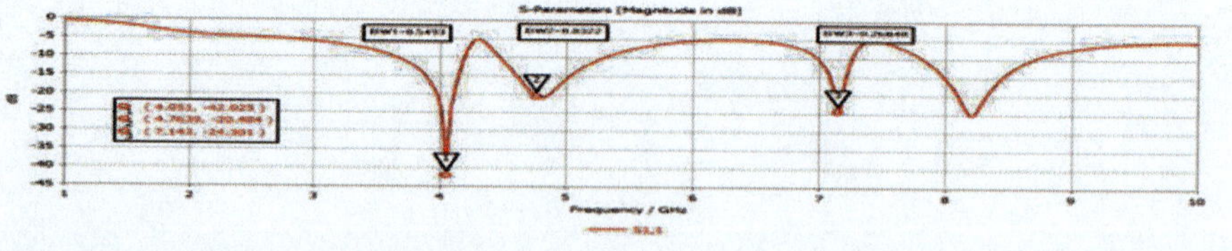

Fig. 7 Meander CPW fed Hexagonal antenna is resonated at 4 GHz, 4.7 GHz, 7.1 GHz, 8.3 GHz, with return losses −42 dB, −20 dB, −24 dB, −25 dB

introducing meandered fed instead of conventional fed. Modified ground plane is used here to improve the return loss at third band. The meandered process is mainly applied for improving bandwidth, return loss characteristics and to lower resonant frequency. All the simulation work of antenna is based on CST software. After the simulation of the results from the obtained S-parameters, the negative characteristics of permittivity and permeability are obtained for complementary split ring resonator. The basic structure suggested in this work is by considering an economical substrate [FR-4 having thickness of 1.6 mm and ε_r4.4]. This structure uses circular monopole of CSRR type with meandered feed process having modification in ground plane.

The exact geometrical dimensions of the suggested structure is shown in Fig. 8 and dimensions are listed in Table 1. A compact size of 26 × 23 × 1.6 mm^3 is used and size reduction is achieved through CSSR and meandered fed. The dimensions

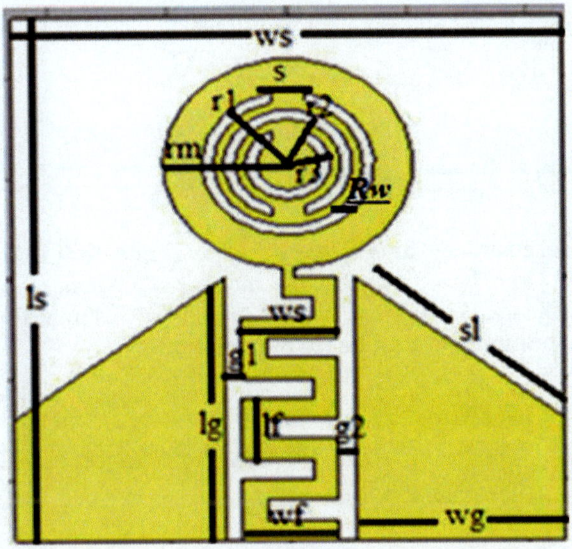

Fig. 8 Geometry of antenna

Table 1 Dimensions of the antenna

Name of parameter	Explanation	Value (mm)
Ls	Substrate length	26
Ws	Substrate width	23
Lg	Ground plane length	13
Wg	Ground plane width	8.8
Wf	Feed line width	4
Lf	Feed length	3
S	Slit width	1.4
Sw	Slot width	0.5

(continued)

Table 1 (continued)

Name of parameter	Explanation	Value (mm)
Rw	Ring width	0.34
*g*1	Gap width 1	0.7
*g*2	Gap width 2	0.7
*S*l	Slant height	4
Rm	Outer ring	5.2
*r*1	Inner ring 1	3.5
*r*2	Inner ring 2	2.7
*r*3	Inner ring 3	1.8

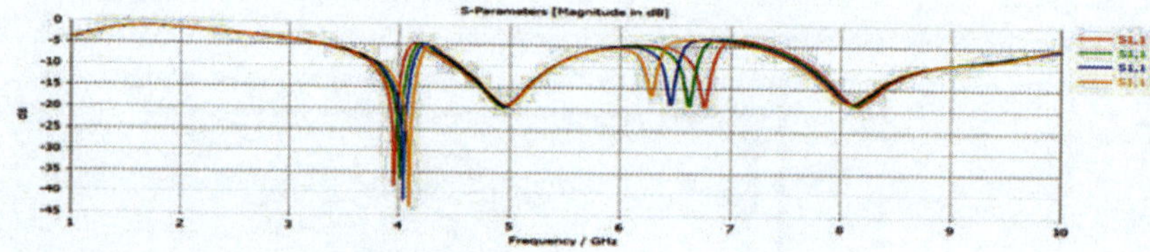

Fig. 9 Parametric study of ring width

are considered as $ls = 26$ mm, $ws = 23$ mm, $lg = 13$ mm, $wg = 8.8$ mm, $g1 = g2 = 0.7$ mm, $lf = 3$ mm, $W1 = 4$ mm, $rm = 5.2$ mm, $r1 = 3.5$ mm, $r2 = 2.7$ mm, $r3 = 1.8$ mm, $S = 1.4$ mm, $rw = 0.34$ mm, $wf = 4$ mm, $sl = 4$ mm.

Simulated Results for S-Parameters

3 Comparison of Parameters Based on CSRR Characteristics

Analysis is carried out for slot width. This structure has three slots and three slits. The parametric study is carried on all the parameters like number of slots (*N*), slit width(*S*) and ring width (*Rw*). In the first study, the effect of number of slots is carried out in order to get better return losses with respect to resonant frequency. In the first stepthe suggested design is placed with single slot that is $N = 1$ slot, it produces a dual-band. Later the suggested design is placed with two slots that is $N = 2$ slot, it produces a triple band of frequencies with return losses. And finally, the suggested design is placed with three slots that is $N = 3$ slot, it produces a triple band of frequency with better return losses. From all the three steps, it concludes that the return losses are improving from 1st slot to 3rd slot. Complementary split ring resonator with three slots is preferred for better results.

In the second step, parametric study is carried out by considering the parameter ring width (*Rw*) and is varied from 0.2 to 0.6 mm with a step of 0.1 mm as shown

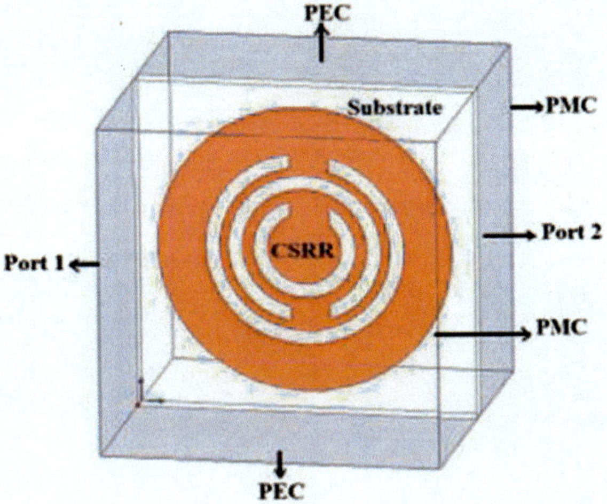

Fig. 10 Waveguide setup for CSRR

in Fig. 9, and also the slit width is varied from 0.2 to 0.4 mm. It is seen that the loss parameter is suitable in case of *Sw* = 0.3 mm and *Rw* = 0.5 mm. The resonance is obtained at frequencies 4.2, 4.9 and 6.5 GHz as shown in Fig. 9.

It is seen that CSSR has negative permittivity characteristics based metamaterial. So these characteristics can be analysed based on the traditional theory of waveguide. The waveguide set up is assigned by applying proper boundary conditions and excitation for the suggested design as shown in Fig. 10. This set up is used to obtain the reflection [S_{11}] and transmission [S_{21}] and VSWR characteristics as shown in Fig. 11. CSRR is complimentary SRR. The CSRR equivalent circuit has LC elements which change the resonance. The required characteristics of CSRR are calculated by using the S-parameters. The negative characteristics of permittivity and permeability determine CSRR characteristics. The negative characteristics are similar for both CSRR and SRR. The negative characteristics of permeability are observed at 5 GHz and 6.1 GHz and negative permeability are observed at 4.2 GHz and 5.1 GHz as shown in Fig. 12. The negative permittivity of CSRR is calculated using Nicolson–Ross–Weir (NRW) approach using equations 1 and 2.

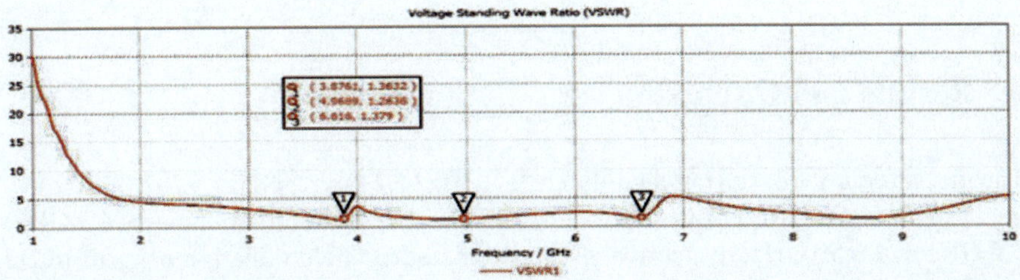

Fig. 11 VSWR characteristics

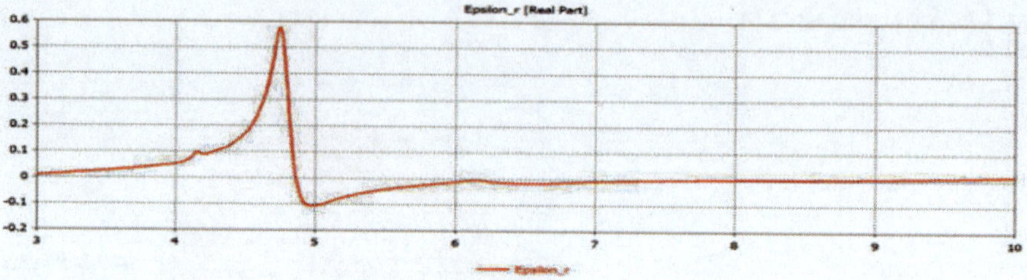

a) Permitivity

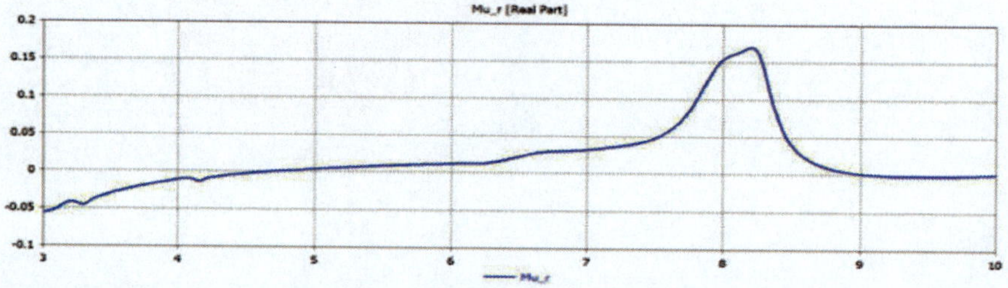

b) Permeability

Fig. 12 Permittivity and Permeability Characteristics of Proposed Antenna

$$\epsilon_r = \frac{2}{jdk_0} * \frac{1 - V_1}{1 + V_1} \tag{1}$$

$$\mu_r = \frac{2}{jdk_0} * \frac{1 - V_2}{1 + V_2} \tag{2}$$

where $V_1 = S_{21} - S_{11}$, $V_1 = S_{21} - S_{11}d$ = substrate thickness, k_0 = wave number of free space.

From all the designs, design 4 Meandered CPW fed CSRR loaded with multiple circles etched outside the monopole antenna is considered as the main design due to its performance in terms of bandwidth and return losses. The design 4 undergoes parametric study by varying the radius from 0.2 to 0.5 in three steps of a circle which is etched outside of the monopole antenna, and parametric study is also carried out by varying number of circles which is etched outside the monopole antenna.

4 Results and Discussion

The measured impedance bandwidths are 90 MHz (4.23–4.32 GHz), 970 MHz (5.15–6.12 GHz) and 230 MHz (7.06 GHz–7.29 GHz) with resonant frequencies at 4.2 GHz, 5.5 GHz and 6.50 GHz, respectively. The simulated far-field radiation patterns in 2D and 3D at a frequency of 5.5 GHz and 10 GHz are shown in Fig. 13a, b, respectively.

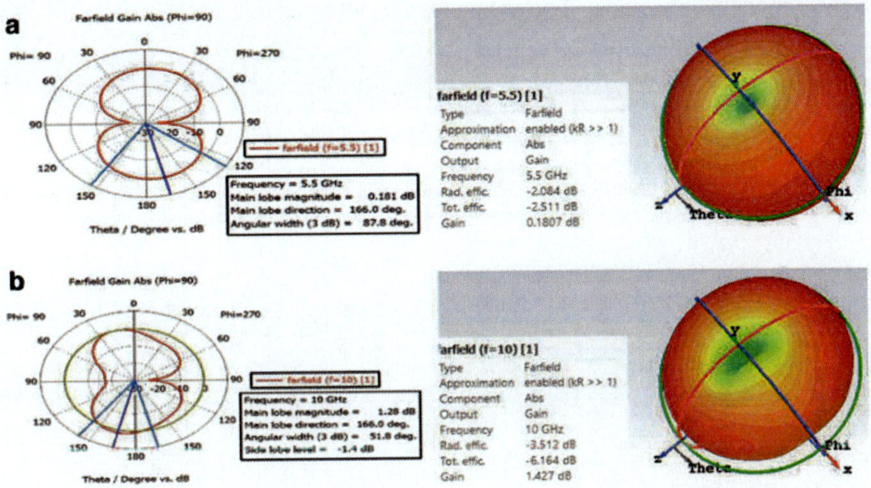

Fig. 13 **a** Radiation pattern of proposed antenna at 5.5 GHz. **b** Radiation pattern of proposed antenna at 10 GHz

5 Conclusion and Future Scope

In the present research endeavour, the compact meandered CPW fed CSRR loaded with multiple circles etched outside the monopole antenna having modification in ground plane is found to be better in performance in all significant characteristics of the antenna by comparing all designs and the comparison table among all the designs is given in Table 2. The proposed antenna structure is suitable for UWB with adequate gain and bandwidth. The pattern of radiation in the band of operation renders this antenna for the desirable area of usage.

The suggested antenna has a good pattern of radiation having augmented bandwidth at all types of resonance. Moreover, the performance measures can be improved in terms of amplification and bandwidth because there is a need for improving the performance of the antenna based on traditional techniques.

Table 2 Performance comparison

Name of the antenna	Monopole antenna with modified ground plane using CPW fed	CSRR loaded monopole antenna with modified ground plane using CPW fed	Meandered process CSRR loaded monopole antenna with modified ground plane CPW fed	CPW Fed CSRR Loaded with multiple circles etched at outside the Monopole antenna	CPW Fed CSRR Loaded with multiple circles etched at inside the Monopole antenna	Meandered CPW fed hexagonal CSRR monopole antenna with modified ground plane
Resonating frequency (GHz)	$F1 = 5$	$F1 = 3.4$	$F1 = 3.8$	$F1 = 3.8$	$F1 = 3.8$	$F1 = 4.05$
		$F2 = 5.5$	$F2 = 4.8$	$F2 = 4.8$	$F2 = 4.8$	$F2 = 4.7$
		$F3 = 7.2$	$F3 = 6.5$	$F3 = 6.5$	$F3 = 6.5$	$F3 = 7.1$
VSWR	$V1 = 1.49$	$V1 = 1.2$	$V1 = 1.04$	$V1 = 1.3$	$V1 = 1.5$	$V1 = 1.13$
		$V2 = 1.75$	$V2 = 1.22$	$V2 = 1.26$	$V2 = 1.24$	$V2 = 1.02$
		$V3 = 1.8$	$V3 = 1.02$	$V3 = 1.37$	$V3 = 1.13$	$V3 = 1.12$
Bandwidth (GHz)	$B1 = 4.78$	$B1 = 0.79$	$B1 = 0.34$	$B1 = 0.33$	$B1 = 0.36$	$B1 = 0.54$
		$B2 = 1.8$	$B2 = 1.09$	$B2 = 1.27$	$B2 = 1.03$	$B2 = 0.83$
		$B3 = 1.06$	$B3 = 1.17$	$B3 = 0.27$	$B3 = 0.17$	$B3 = 0.26$
Gain (dB)	$G = 0.3535$	$G = 0.84$	$G = 1.845$	$G = 1.47$	$G = 1.822$	$G = 1.29$
Return loss (dB)	$R = -14$	$R1 = -47$	$R1 = -34$	$R1 = -25$	$R1 = -32$	$R1 = -42$
		$R2 = -19$	$R2 = -19$	$R2 = -20$	$R2 = -19$	$R2 = -20$
		$R3 = -11$	$R3 = -36$	$R3 = -16$	$R3 = -30$	$R3 = -24$

References

1. Rengasamy R, Kommuri UK (2016) A compact ACS-fed mirrored L-shaped monopole antenna with SRR loaded for multiband operation. Prog Electromagn Res C 64:159–167
2. Arora C, Pattnaik SS, Baral RN (2015) SRR inspired microstrip patch antenna array. Prog Electromagn Res C 58:89–96
3. Li B, Wu B, Liang C-H (2006) Study on high gain circular waveguide array antenna with metamaterial structure. Prog Electromagn Res 60:207–219
4. Datta R, Shaw T, Mitra D (2017) Miniaturization of microstrip Yagi array antenna using metamaterial. Prog Electromagn Res C 72:151–158
5. Pandeeswari R, Raghavan S, Ramesh K (2012) A compact split ring resonator loaded antenna. In PIERS proceedings, Moscow, Russia, August 19–23, 37–40
6. Veselago VG (1968) The electrodynamics of substances with simultaneously negative values of e and μ. Sov Phys Uspekhi 10(4):509–514
7. Caloz C, Itoh T (2005) Electromagnetic metamaterials: transmission line theory and microwave applications. Wiley-IEEE Press, New York
8. Baena JD, Bonache J, Martin F, Sillero RM, Falcone F, Lopetegi T, Laso MAG, Garcfa-Farcfa J, Gil I, Portillo MF, Sorolla M (2005) Equivalent-circuit models for split-ring resonators and complementary split-ring resonators coupled to planar transmission lines. IEEE Trans Microw Theory Tech 53:1451–1461
9. Mookiah P, Dandekar KR (2009) Metamaterial-substrate antenna array for MIMO communication system. IEEE Trans Antennas Propag 57:3283–3292
10. Pandeeswari R, Raghavan S (2014) Broadband monopole antenna with split ring resonator loaded substrate for good impedance matching. Microw Opt Technol Lett 56(10):2388–2392
11. Attia H, Yousefi L, Bait-Suwailam MM, Boybay MS, Ramahi OM (2011) Enhanced gain micro-strip antenna using engineered magnetic superstrates. IEEE Antennas Wirel Propag Lett 10:1198–1201
12. Pandeeswari R, Raghavan S (2015) Microstrip antenna with complementary split ring resonator loaded ground plane for gain enhancement. Microw Opt Technol Lett 57(2):292–296
13. Pandeeswari R, Raghavan S (2015) Meandered CPW-fed hexagonal split ring resonator monopole antenna for 5.8 GHz RFID applications. Microw Opt Technol Lett 57:681–684. Wiley Interscience, USA
14. Pandeeswari R, Raghavan S (2015) A CPW-fed triple band OCSRR embedded monopole antenna with modified ground for WLAN and WIMAX applications. Microw Opt Technol Lett 57:2413–2418. Wiley Interscience, USA
15. Martinez FJH, Zamora G, Paredes F, Martin F, Bonache J (2011) Multiband printed monopole antennas loaded with OCSRRs for PANs and WLANs. IEEE Antennas Wirel Propag Lett 10:1528–1531
16. Boopathi R, Pandey SK (2016) CSRR inspired conductor backed CPW-fed monopole antenna for multiband operation. Prog Electromagn Res C 70:135–143

Gold, Inflation, GDP, and Lending Rate: An Interesting Relation

S. Ravi Kumar Raju and P. S. Avadhani

Abstract The proposed work comprises identifying the relationship between gold, GDP, inflation, and lending rate. The central banks around the globe are constantly in the pursuit of inflation targeting and strengthening their domestic currency for achieving financial stability. In India, gold price plays a major role in determining the currency value. In this context, Beveridge–Nelson Decomposition method was used extensively.

1 Introduction

The transition of the developed countries from a gold-linked currency to a fiat currency led to an increased academic interest in the role of gold in the modern world. Gold and inflation play a major role in the monetary policy of a developing country. The investigation carried out in this endeavor looks at cointegration between the two parameters to understand their interdependency [1].

Gold is back in vogue once again, with purchases made by global central banks pushing up prices to a five year high. Five years ago, the world was going through a rough patch on account of the US Federal Reserve's taper tantrum in 2013, which saw gold prices fall from $1500 an ounce to as low as $1072 in 2015, to subsequently recover some of the lost ground. In 2019, prices are already up 10 percent in dollar terms at $1407 an ounce. For investors, gold is a safe haven every time there is a crisis of confidence. The current rally in gold prices has roots in the ongoing trade tussle between the US and China. US–Iran tensions too have fuelled the rally.

Central banks such as those of Russia, China, and India were heavy buyers of gold in 2018. Central banks' urge for gold soared to a multi-decade high of 651 tons in 2018, a 74 % rise over 2017 [2]. This is the highest since the dissolution of the

S. Ravi Kumar Raju (✉) · P. S. Avadhani
Department of CS&SE, AUCE(A), Andhra University, Visakhapatnam, AP, India
e-mail: ravisayyaparaju@gmail.com

P. S. Avadhani
e-mail: psavadhani@yahoo.com

S. C. Satapathy et al. (eds.), *Communication Software and Networks*, Lecture Notes in Networks and Systems 134, https://doi.org/10.1007/978-981-15-5397-4_31

Bretton Woods System. The RBI now holds the 10th largest gold reserves, at about 613 tons.

The trade war between the US and China, the two largest economies globally, has pushed the world to uncertain territory. If the two do not come to an agreeable solution, there could be bloodbath in financial markets. China held \$1.1 trillion in US government securities as of April 2019. If it decides to offload some of it, the US dollar will likely nosedive. That won't be good for other central banks holding dollar assets. Of India's more than \$400 billion forex reserves, \$155.3 billion was held in US bonds, while the world as a whole held \$6.4 trillion. The stakes are quite high. China is the second-largest economy and will probably displace the US from number one spot by as early as 2030, according to research estimates, but the Yuan is a controlled currency, unlike the dollar.

There is no doubt that the biggest weapon that the US wields now is the dollar, or petrodollar as strategic analysts like to call it. The world runs on fossil fuel and that cannot be bought without dollars. The petrodollar served the world order well for about 50 years, and might continue doing so for another decade or two. But alternatives are emerging fast. India itself is moving to an all-electric vehicle policy by 2030, which is expected to reduce its oil imports [3].

On the other hand, it would increase India's dependence on China for rare earth material needed to make batteries. Besides diversification, there also seems to be an effort to blunt the dollar's potency as an economic weapon. Russia and China have taken the lead in that by accumulating most of the gold available for sale.

US President Donald Trump's haste on hiking tax on everything that hits the US shore is working subtly for countries to move away from a dollar-centric world. In such a scenario, gold is likely to trade with an upward bias in the foreseeable future.

2 Methodology

In this work, we have used the Beveridge–Nelson (BN) Decomposition technique (Figs. 1, 2, and 3).

Fig. 1 Gold versus inflation

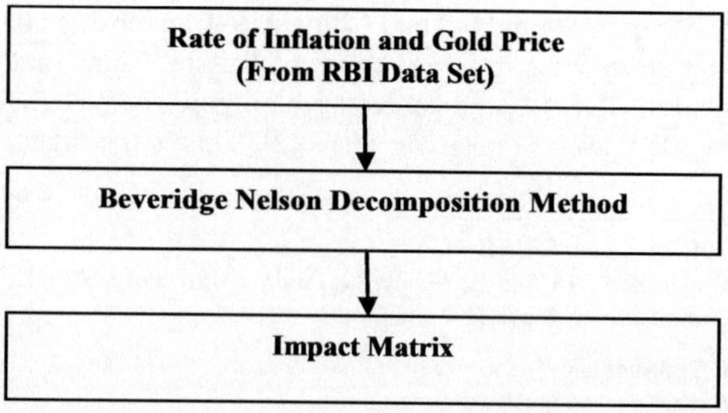

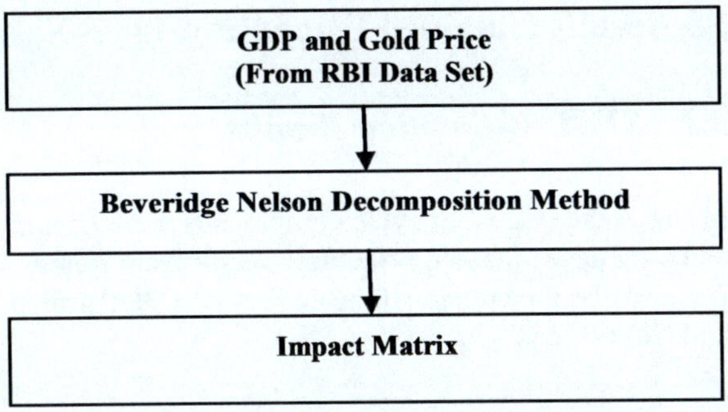

Fig. 2 GDP versus gold

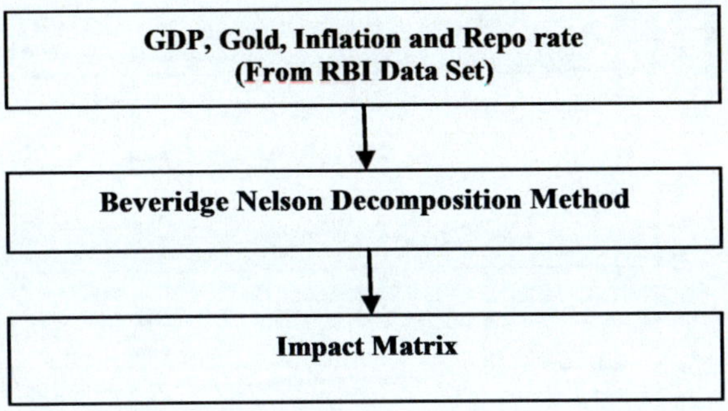

Fig. 3 GDP, gold, inflation, and repo rate

Beveridge and Nelson pioneered trend-cycle decomposition of time series that comprises

(i) The difference-stationary behavior of the data generation and
(ii) Trend discovery.

The two elements coupled together allow the identification of a unique decomposition of any ARIMA(p, 1, q) process into a random walk trend and a stationary transitory component.

The relevance of the decomposition stems from the fact that the extracted trend has an intrinsically stochastic nature, since it is the result of the cumulative propagation effects of random innovations. In this respect, the Beveridge–Nelson decomposition has been deemed to provide a structural interpretation to any ARIMA(p, 1, q) reduced form model fitted according to the traditional Box–Jenkins methodology [4–7].

The BN decomposition has two distinguishing features

(i) The components are correlated and
(ii) The extraction of the components takes into account the current and past observations.

This may actuate a phase shift [8, 9].

3 Results Obtained Using Beveridge–Nelson Decomposition

3.1 SVAR Estimation Results

In Fig. 4 the trend and cycle components were decomposed. In Tables 1, 2, 3, 4, 5, and 6 we have obtained Structural Vector Auto Regression estimates. In this context, we have obtained the relationship between GDP growth, Rate of Inflation, Repo rate, and Gold Price.

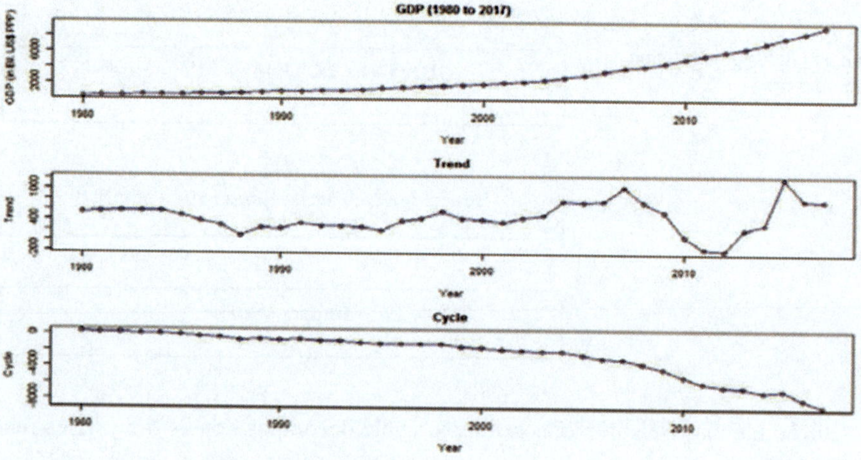

Fig. 4 Trend and cycle after decomposition

Table 1 Estimated contemporaneous impact matrix

Rate of inflation	1.511	−0.1477
Gold price	302.117	−591.3283

Table 2 Estimated identified long-run impact matrix

Rate of inflation	5.906	0.0
Gold price	4506.059	940.4

Table 3 Estimated contemporaneous impact matrix

GDP growth	2.565	−0.1116
Gold price	−516.573	−676.8863

Table 4 Estimated identified long-run impact matrix

GDP growth	2.702	0
Gold price	796.464	1986

Table 5 Estimated contemporaneous impact matrix

GDP growth	0.8148	−0.4201	1.3826	0.5231
Rate of inflation	1.4820	0.5768	−0.4201	−0.1827
Repo rate	−0.4775	0.5669	0.1519	0.0108
Gold price	132.8275	375.6906	−190.4366	−605.2418

Table 6 Estimated identified long-run impact matrix

GDP growth	3.4241	0	0	0
Rate of inflation	10.5775	4.7975	0	0
Repo rate	−0.2772	0.6995	0.6142	0
Gold price	8494.1727	3201.4736	87.974	911.9

4 Conclusion

Governors of the central banks including Reserve Bank of India, United States Federal Reserve, Bank of Japan, and Bank of England have identified that GDP growth rate, Inflation, Repo rate, and the price of gold are crucial to monitor the country's economy. In this paper, an attempt has been made to study the relationship between these parameters.

References

1. Todorova N (2017) The asymmetric volatility in the gold market revisited. Econ Lett 150:138–141
2. Artigas JC (2010) Linking global money supply to gold and to future inflation. World Gold Council
3. Chakravarty S (1985) Report of the committee to review the working of the monetary system. Reserve Bank of India Mumbai
4. Morley JC (2010) The two interpretations of the Beveridge–Nelson decomposition. Macroecon Dyn
5. Low CN, Anderson HM (2008) Economic applications: the Beveridge–Nelson decomposition. In: Forecasting with exponential smoothing. Springer series in statistics. Springer, Berlin
6. Nelson CR (2008) The Beveridge–Nelson decomposition in retrospect and prospect. J Econom 146(2)
7. Vapnik V (1995) The nature of statistical learning theory. Springer, New York
8. Proietti T, Harvey AC (2000) Beveridge–Nelson smoother. Econ Lett 67:139–146
9. Arino MA, Newbold P (1998) Computation of the Beveridge–Nelson decomposition for multivariate economic time series. Econ Lett 61:37–42

A Mechatronics Design Approach of a Low-Cost Smart Reconnaissance Robot

N. Seenu, R. M. Kuppan Chetty, K. M. Adithya Krishna, Taarun Srinivas, and R. G. Pranav Raj

Abstract Mobile robots, now-a-days, are majorly used in various fields such as service robotics, military, education, and surveillance. Human surveillance does have limitations as it is difficult for humans to monitor inaccessible areas and also, they get tired. Recent approaches in surveillance include the use of static cameras to monitor a particular location. But they do not cover the dead spot areas and also people try to avoid the camera's sight. In order to overcome these drawbacks, surveillance is being replaced by mobile robots. In this paper, we presented a low-cost Smart Reconnaissance Robot (SRR) that is able to wander around an environment, capture images, and perform live streaming of video both in daytime and also at night using a camera deployed on the robot. Initially, the robot is modeled and simulated using webots robot simulation software and its performance is analyzed. Then, it is compared with the performance of the robot developed using Arduino microcontroller and other sensors. The user can control the robot's action using a mobile phone or laptop through Internet of Things (IoT) and also can view the video information transmitted by the robot using a wi-fi module. Thus, this robot can be used in domestic or household environments where the user can remotely view any abnormal events or interpretation of human events.

N. Seenu (✉) · R. M. K. Chetty · K. M. A. Krishna · T. Srinivas · R. G. P. Raj
Centre for Automation and Robotics, School of Mechanical Sciences, Hindustan Institute of Technology and Science, Padur, Chennai, India
e-mail: nseenu@hindustanuniv.ac.in

R. M. K. Chetty
e-mail: kuppanc@hindustanuniv.ac.in

K. M. A. Krishna
e-mail: madhu5karigiri@gmail.com

T. Srinivas
e-mail: taarun19@gmail.com

R. G. P. Raj
e-mail: pranav.raj.rg00@gmail.com

S. C. Satapathy et al. (eds.), *Communication Software and Networks*, Lecture Notes in Networks and Systems 134, https://doi.org/10.1007/978-981-15-5397-4_32

1 Introduction

Mobile robots, nowadays, are widely used in different fields such as medical, educational, defense, and even for the house appliances like for security purposes or surveillance [1]. These robots can be wheeled or tracked, walking robots, or flying drones. Generally, human surveillance is done by deploying a person near sensitive areas in order to constantly monitor for changes or any coming dangers. As humans do have limitations in stamina and also monitoring inaccessible places is quite difficult, they get stressed and tired. Moreover there are also the risks of losing the person's life in the act while encountering the enemies. With the help of the advancement in the technologies of the past few decades, it is possible to remotely access the process of surveillance by adding camera to the robots and monitor the area instead of humans [2]. A robot is nothing but an electronic machine that is capable of performing programmed tasks by giving the required input, providing highly accurate results, and easily overcoming the limitations of humans [3]. The advantage of introducing robot for surveillance is having no risk to humans. Even drones are able to monitor the locations that cannot be done by the humans. This paper focuses on developing a surveillance mobile robot for domestic and household environments which is used for aging population, to monitor any human interpretation and abnormal events. The robot is first modeled and simulated using webots, an open-source virtual simulation software which is being developed by cyberbotics Ltd. Its performance is then analyzed and compared with the performance of the robot developed using Arduino Mega microcontroller and various sensors that collect the data and send it to the microcontroller which in turn controls the robot's behavior. A wi-fi module is also deployed to transmit the video information to the user.

The ongoing revolution on IOT is now being blended with robotics in making up a new ear known as Internet of Robotics (IoR) [4]. In order to control the robot's movements and also to view the live streaming of video, an IoT-based application known as Blynk is used which is able to connect any hardware over wi-fi.

Thus, the work of this paper is organized as follows. Section 2 discusses a number of surveillance systems developed for various applications. Section 3 explains the proposed method through block diagram and flow charts. Section 4 presents the modeling and simulation of robot using webots software. Section 5 describes the hardware used and explains the implementation of our prototype. Section 6 discusses the experimental results before we conclude in Sect. 7.

2 Literature Survey

A mobile robot based on the concepts of Internet of Things was proposed by Nayyar et al., for monitoring environmental factors like humidity, gas sensing, and temperature through ESP 8266. ThingSpeak.com delivered the live feed of data in the form of graphs [4]. By utilizing audio and video information Gong et al., developed a household robot to detect any abnormal events. Particle filter algorithms were utilized for detecting the moving target, and the audio information is extracted by using Mel Frequency Cepstral Coefficient (MPCC). They achieved an accuracy of 88.17 % in the detection of abnormal events like gun shooting, falling down, and crying [5].

A cooperative and intelligent surveillance system using a fixed camera on a mobile robot was developed by Chia et al., Dead spot problems in similar systems were resolved using fixed cameras through this method [6]. Similarly, Shaik et al., proposed a surveillance robot which could be controlled using Bluetooth android application. The android application and the robot built on Arduino UNO were interconnected to enable communication between themselves. Thus, it was possible to control the robot remotely and also get the real-time video feed [7].

Kulkarni et al., developed a surveillance robot using Arduino UNO, a video camera, GSM radio, and GPS module. This robot can be controlled remotely using a smartphone application or through PC. Using GPS navigation and mapping software, the best route to a particular location was able to be determined using this robot [8].

From these surveys, it can be understood that surveillance robots are very important in household and domestic environments. They have a wide range of applications such as to detect any abnormal and human interpretation events and to find the shortest route by deploying a GPS module. Also, it is known that IoT plays a major role in developing a surveillance robot, so that that the user can remotely access and control the robot's behavior, also can view the video information transmitted by the robot.

3 Methodology

3.1 Block Diagram

See Fig. 1.

This paper focuses on developing a low-cost wheeled robot which is capable of wandering through an environment and performing the act of surveillance through the camera deployed in the robot. This robot is designed in a circular shape with a diameter of 120 mm and a height of 72 mm. A processor/controller is considered to be a prime unit in any embedded system hardware [9]. So, from the scheme of the proposed system, it can be understood that Arduino Mega is the heart of this robot. Also, sensors such as Ultrasonic, Accelerometer, Microphone, and Camera are used to get the data from the environment and to actuate the robot. L298N motor driver is used to drive the micro metal gear motor. For nighttime surveillance, LED flashlight is utilized. In order to transmit the video information to the user, ESP 8266 wi-fi module is used.

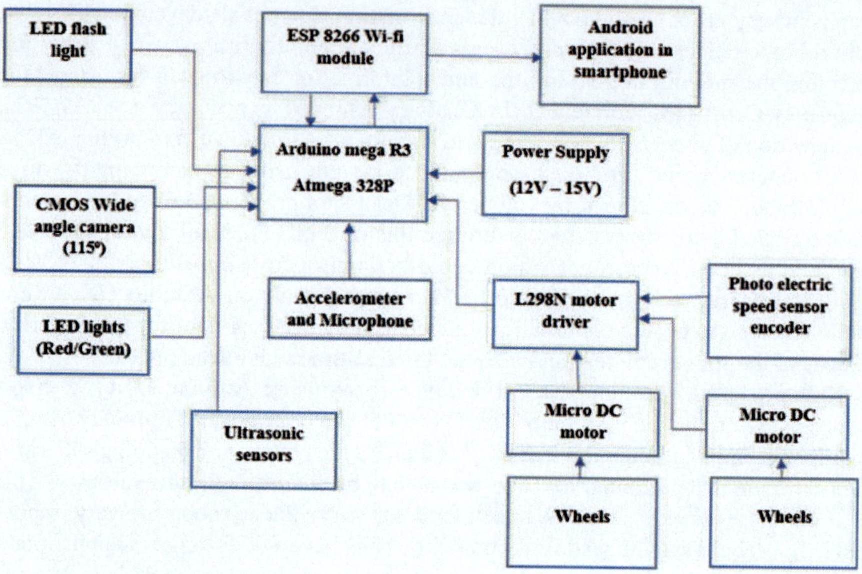

Fig. 1 Scheme of the proposed system

3.2 *Flow Chart*

See Fig. 2.

The robot wanders around the environment freely without colliding any obstacles using an ultrasonic sensor. The flow of control for the motion and the process of transmitting the image captured and the video recorded by the camera to the smartphone application is explained in the below flow chart.

4 Modeling and Simulation of the Robot Using Webots

Webots is an open-source 3D robot simulator software which can be used to model and simulate any robot. This software allows the user to design complex robotic setups, and also the properties of the robot can be chosen by the user. Using this software, the behavior of the robot can be tested in physically realistic worlds. Also, the kinematic modeling of the robotic wheel is discussed in this section. The first step in modeling a robot in webots is to create a robot node. Under the robot node, many sub-nodes can be created using the children's class. The first sub-node is solid node. The solid node is used to create the body of the robot. Under solid node, the geometry of the robot and the appearance has to be given. In our robot the geometry is cylinder. Similarly, distance sensors, camera, hinge joints, and wheels can be created using the other sub nodes (Figs. 3 and 4).

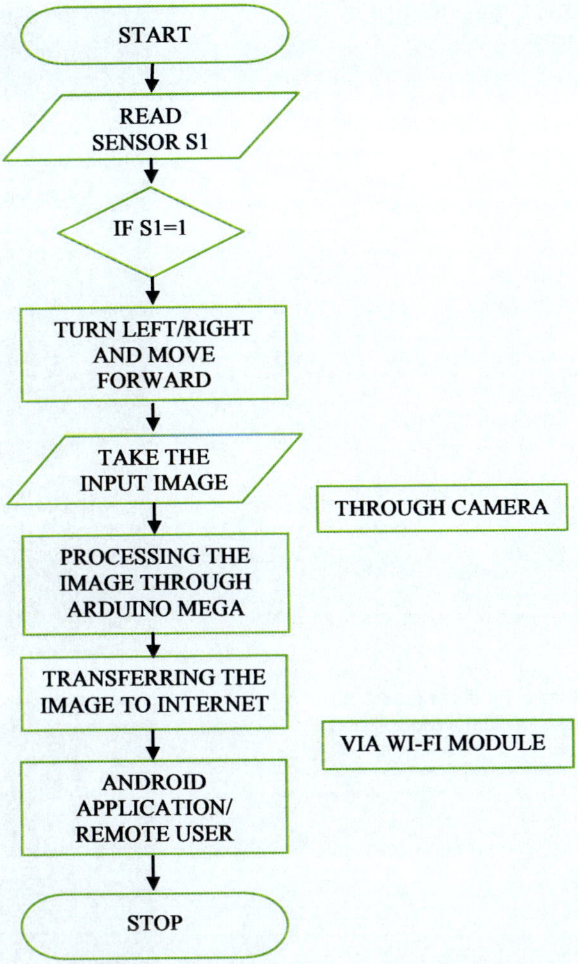

Fig. 2 Working of the proposed system

The first step to stimulate any robot is to design the controller. The controller can be designed using languages like C, C++, Java, Python, and Matlab. The controller is designed using C language for this robot. After designing the controller, the base node and the sub-nodes have to be linked to the controller in order to make the robot move. Braitenberg concept is used as the robot has to avoid the obstacles and perceive the video information.

4.1 Kinematic Design of the Robotic Wheels

See Fig. 5.

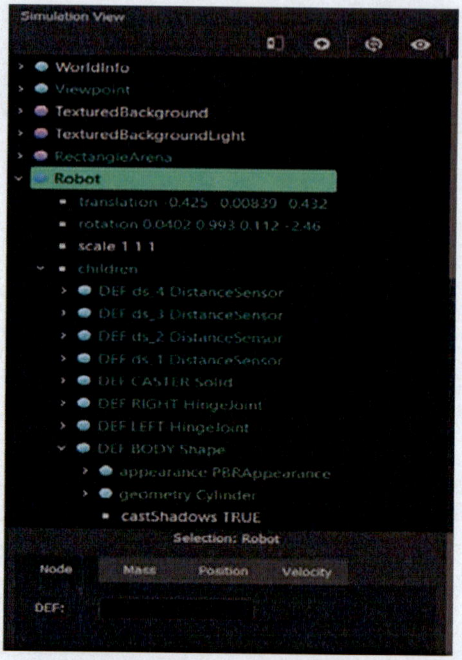

Fig. 3 Scene tree of the robot

Fig. 4 Model of the robot developed in webots

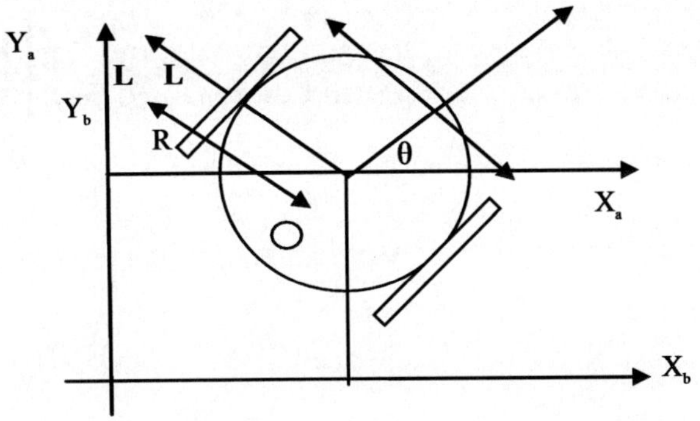

Fig. 5 Differential wheel of the proposed robot

Let,
$\{X_a, Y_a\}$—Moving frame; $\{X_b, Y_b\}$—Base frame
V_r (t)—Linear velocity of right wheel; V_l (t)—Linear velocity of left wheel
ω_r (t)—Angular velocity of right wheel; ω_l (t)—Angular velocity of left wheel
r—Nominal radius of each wheel
R—Instantaneous curvature radius of the robot trajectory, relative to the mid-point axis.

$$ICC = (x - R\sin\theta,\ y + R\cos\theta) \tag{1}$$

$R - L/2$ is curvature radius of trajectory described by left wheel
$R + L/2$ is curvature of trajectory described by the right wheel.

$$\omega(t) = V_r(t)/R + \frac{L}{2} \tag{2}$$

$$\omega(t) = V_l(t)/R - \frac{L}{2} \tag{3}$$

From (2) and (3),

$$\omega(t) = V_r(t) - V_l(t)/L \tag{4}$$

$$R = \frac{L}{2}(V_l(t) + V_r(t)/V_l(t) - V_r(t)) \tag{5}$$

$$V(t) = \omega(t)R = \frac{1}{2}(V_r(t) - V_l(t)) \tag{6}$$

The kinematic model in the robot frame is given by,

$$\begin{Bmatrix} V_x(t) \\ V_y(t) \\ \Theta^\circ(t) \end{Bmatrix} = \begin{Bmatrix} r/2 & r/2 \\ 0 & 0 \\ -r/L & r/L \end{Bmatrix} \begin{Bmatrix} \omega_l(t) \\ \omega_r(t) \end{Bmatrix} \tag{7}$$

The kinematic model in the world frame is given by,

$$V(t) = \omega(t)R = \frac{1}{2}(V_r(t) + V_l(t))$$

$$\omega(t) = V_r(t) - V_l(t)/L$$

Thus,

$$\mathring{X}(t) = v(t)\cos\theta(t) \quad \mathring{Y}(t) = v(t)\sin\theta(t) \quad \Theta^\circ(t) = W(t)$$

$$\begin{pmatrix} \mathring{X}(t) \\ \mathring{Y}(t) \\ \Theta^\circ(t) \end{pmatrix} = \begin{pmatrix} \cos\theta(t) & 0 \\ \sin\theta(t) & 0 \\ 0 & 1 \end{pmatrix} \begin{pmatrix} V(t) \\ \omega(t) \end{pmatrix}$$

$$\text{Thus, } \mathring{q}(t) = S(q)\xi(t) \tag{8}$$

Particular cases:

1. When $V_r(t) = V_l(t)$, Forward linear motion in a straight line (i.e.,) ω is zero
2. When $V_r(t) = -V_l(t)$, the robot rotates about the mid-point of wheel axis [10].

5 Components of SRR

The components used in this robot are selected accordingly so that their function is proper and efficient. The following are the components and their specifications.

Arduino Mega, a microcontroller board based on the ATmega2560, has 54 digital I/O pins, 16 analog inputs, 4 UARTs, a 16 MHz crystal oscillator, a USB connection, a power jack, an ICSP header, and a reset button. The CMOS camera has a resolution of 700TVL and 2.8 mm 90° wide-angle lens. It also supports 7.5–13 V input. The advantage of this camera is that it has a good resolution and consumes very less amount of power. L298N 2A is a motor driver, perfect for driving DC motors. The H–bridge used in this component is used for motor speed and direction control. Ultrasonic sensor detects the presence of any obstacle by emitting ultrasonic waves and receiving the wave reflected back from the obstacle. The detection range of this sensor is available from 20 cm to 20 m.

Light Emitting Diode (LED) is a semi-conductor light source that emits light when current flows through it. It works on the principle of Electroluminescence. They have a voltage of 1.8–2.7 V, and a maximum current flow is about 20 mA. Chassis is the body of the robot. We have designed our robot circular in shape with a diameter of 120 and 72 mm in height. We have selected acrylic material for cutting since it is of lightweight and available at an affordable price.

The ESP8266 wi-fi module is capable of hosting an application or also offloading all wi-Fi networking functions from another application processor. It has a memory of 32 KiB instructions RAM. It is compatible with any microcontroller and also compact in size. A 12 V 600 rpm micro metal DC geared motor is used. The output shaft is 9 mm long and 3 mm in diameter. This motor has a rated torque of 0.18 kg-cm.

Li-ion batteries use an intercalated lithium compound as one electrode material. It is a rechargeable battery. The discharge efficiency of this battery is around 80–90 %. The performance of this battery is high and has a very good capacity. The main advantage of this battery is its lightweight and small size.

The main heart of this robot is its microcontroller, Arduino Mega to which ultrasonic sensors, Accelerometer, Photoelectric speed sensor, L298N motor river, micro DC motors, Microphone, and CMOS wide-angle camera is connected. As discussed earlier, ultrasonic sensors are used for the purpose of obstacle avoidance. Accelerometer is used for determining the position of the robot. Photoelectric speed sensor is used to make the robot operate as a closed-loop system and to improve the performance. L298N motor driver is used to actuate the motors so that the robot can move around. Microphone is used to get the input command from the user and perform operations accordingly. CMOS wide-angle camera is used to perceive the information from the environment and stream it to the user using ESP 8266 wi-fi module. The user can control the operations of the robot using an android application.

6 Results and Discussion

Preliminary investigations of the robot were carried out in webots. As analyzed in Sect. 4.1, the robot was able to travel in a straight line by maintaining the velocities of left and right wheel at the same rate as seen in Fig. 6. Also, the distance sensor which was created within the robot in webots was able to detect the obstacle at a threshold distance of 0.04 m and avoid it as seen in Fig. 7.

The camera deployed on the robot was able to provide the live stream of the video as seen in Fig. 9. The width and height of the image is set as 64 pixels. The spherical field is set to false and the horizontal field of view angle ranges from 0 to π radians. The vertical field of view is determined by a horizontal fieldOfView * height/width. So, this field is set as 0.785 (Fig. 8).

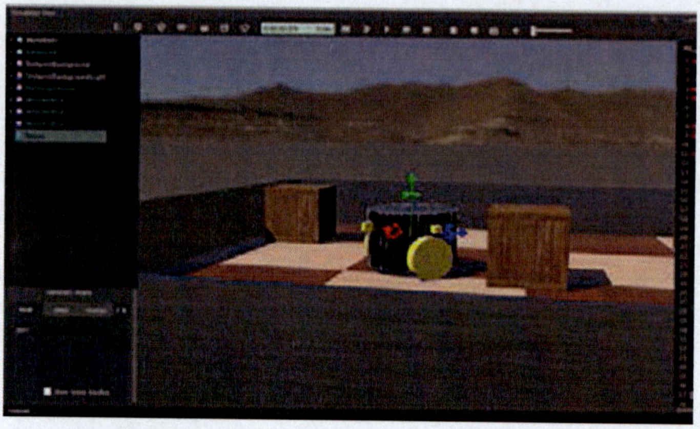

Fig. 6 Robot following a straight path

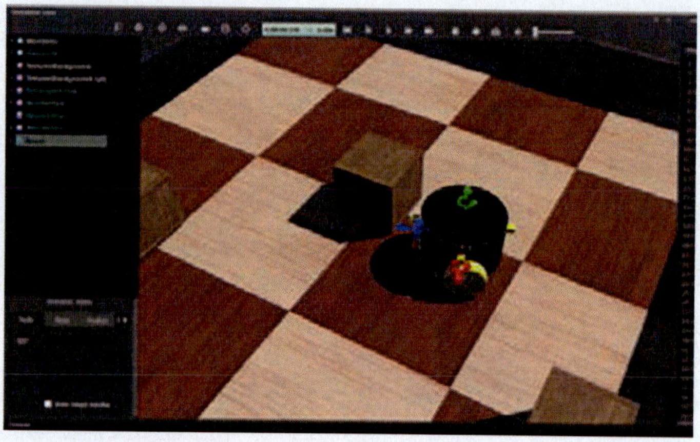

Fig. 7 Robot avoiding an obstacle

7 Conclusion and Future Scope

The usage of mobile robots is increasing day by day in various fields such as medical, industrial, and household environments. In order to overcome the drawbacks of traditional surveillance methods, we proposed a low-cost smart surveillance/reconnaissance mobile robot that can wander around an environment, capture images, and perform live streaming of video. The three main contributions of our paper are (i) an innovative approach to view and monitor an environment remotely using a mobile robot (ii) an easy way to design and simulate any kind of robot for any application using webots and perform kinematic analysis for a differential

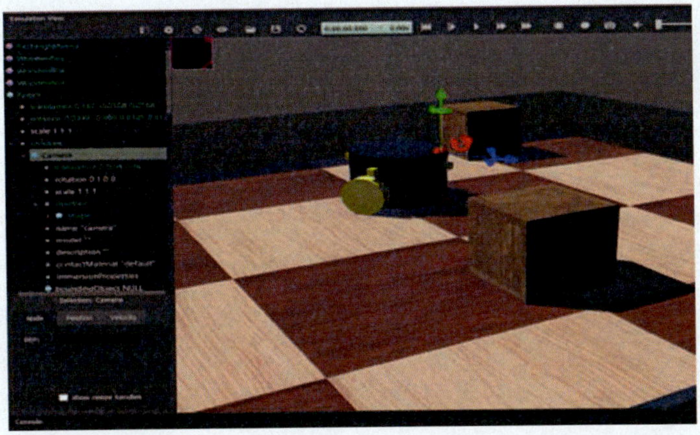

Fig. 8 Robot with camera

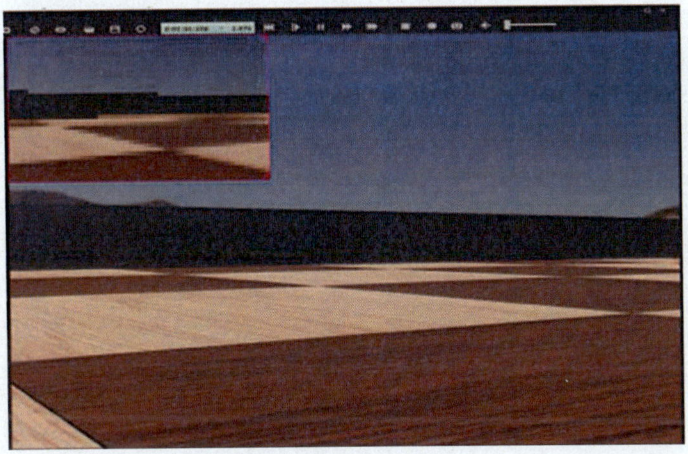

Fig. 9 Live stream provided by the robot in webots

drive wheel (iii) to blend IoT with robotics and highlight its importance. Cooperative system of static cameras is replaced by SRR and also the cost of this robot is affordable. This robot can be used for surveillance in domestic areas and can be used for path planning between a system of similar robots. The future work of this paper is to make this robot to communicate with other robots through swarm intelligence and can be used as cooperative robots, so that they can play a major role in military and defense areas.

Acknowledgments The authors would like to thank Centre for Automation and Robotics, Hindustan Institute of Technology and Science, for providing necessary labs and support to this research work under the Research Incubation Program of Hindustan Institute of Technology and Science.

References

1. Lo CW, Wu KL, Lin YC, Liu JS (2014) An intelligent control system for mobile robot navigation task in surveillance. Adv Intell Syst Comput 274:462–479
2. Adinandra S, Nofriyudi F, Pratama AW, Ratnawati DA (2018) CEPI: A low cost and open source mobile robot platform for research and education. In: AIP conference proceedings 1977, pp 020057-1–020057-8
3. Anandravisekar G, Clinton AA, Raj TM, Naveen L (2018) IOT based surveillance robot. Int J Eng Res Technol 7(3):84–87
4. Nayyar A, Puri V, Nguyen NG, Le DN (2018) Smart surveillance robot for real-time monitoring and control system in environment and industrial applications. Adv Intell Syst Comput 672:229–243
5. Wu X, Gong H, Chen P, Zhong Z, Xu Y (2009) Surveillance robot utilizing video and audio information. Int J Robot Syst 55:403–421
6. Chia CC, Chan WK, Chein SY (2009) Cooperative surveillance system with fixed camera object localization and mobile robot target tracking, vol 5414. Springer, Berlin, pp 886–897
7. Shaik SM, Sufiyan K, Ali A, Ibrahim M, Bodke K (2015) Wireless video surveillance robot controlled using simple bluetooth android application. Int J Adv Res Comput Sci 6(2):100–103
8. Kulkarni C, Grama S, Suresh PG, Krishna C, Antony J (2014) Surveillance robot using arduino microcontroller, android APIs and the internet. IEEE Comput Soc 15(5):83–87
9. Renuka B, Sivaranjani B, Lakshmi AM, Muthukumaran N (2018) Automatic enemy detecting defense robot by using face detection technique. Asian J Appl Sci Technol 2(2):495–501
10. Ribeiro MI, Lima P (2002) Kinematics models of mobile robots. Robótica Móvel

Arrhythmia Recognition and Evaluation of ECG Signal Using Signal Processing Techniques

G. Gayatri and T. Madhavi

Abstract One of the most common causes of death is heart diseases in human beings. To monitor heart activity, Electrocardiogram (EKG or ECG) is a common diagnostic tool used. Most of the clinically important information about the cardiac activities of heart is contained in this signal. Early diagnosis helps devise medical treatment regimes to prevent early death from the disease. This paper explains the study and analysis of an ECG signal processing by using MATLAB tool effectively. It is very difficult to identify heart problems related to P-wave as the information is found in the P wave of the ECG signal, since the P wave is of very less amplitude it is difficult to analyze by simply observing the ECG signal to the physician. So various signal processing techniques such as Frequency-domain analysis (FFT), Time- and Frequency-domain analysis (STFT), and Feature Vector ECG Signal Analysis (FVESA) are applied to extract the data. The signals are recorded from the patients and compared with pre-fed signals to detect the arrhythmias such as atrial fibrillation.

1 Introduction

Early diagnosis helps devise medical treatment regimes to prevent early death from the disease. Most of the clinically useful information about the cardiac activities of heart is contained in the ECG signal. An Electrocardiogram (ECG) is recording the electrical activity of the heart reaching the body surface. The generated electrical activity initiates the contraction of heart's muscles and cardiac cells that pumps blood to the parts of the body. The ECG provides a graphical depiction of electrical

G. Gayatri (✉)
Anil Neerukonda Institute of Technology & Sciences, Electronics & Communication Engineering, Visakhapatnam, India
e-mail: gayatri.ece@anits.edu.in

T. Madhavi
Gandhi Institute of Technology and Management, GITAM University, Electronics & Communication Engineering, Visakhapatnam, India
e-mail: madhavi.tatineni@gitam.edu

S. C. Satapathy et al. (eds.), *Communication Software and Networks*, Lecture Notes in Networks and Systems 134, https://doi.org/10.1007/978-981-15-5397-4_33

forces produced by heart. For each cardiac cycle a series of deflections and waves are produced, by ECG graph.

A typical ECG gives the details about the characteristic waves that correspond to specific events in the cardiac cycle and shown in the intervals and amplitudes. Figure 1 shows a P wave, QRS complex, T wave, and a U wave. The P wave is due to the depolarization of auricles. The P-R interval contains P wave and P-R segment. The interval represents the time of electrical impulse transmission from the atria to the ventricles. Depolarization of both ventricles is reflected in QRS complex. The S-T segment represents the early phase of repolarization of both ventricles. The T wave synchronizes with ventricular diastole or repolarization of ventricles. Repolarization of purkinje fibers is represented by U wave.

The ECG waveform pattern is changed if a disease disrupts the conduction system of heart causing cardiac arrhythmias. Cardiac arrhythmias are generally the irregular heartbeat, abnormal rhythm, or discontinuities in the conduction of an electrical impulse in the heart.

ECG interpretation helps to identify cardiac abnormalities including insufficient blood flow, enlargement, and death of heart muscle. The ECG is also used as primary method to identify problems with heart rate and regularity.

The ECG detection which gives the heart's information and cardiovascular condition is crucial to enhance the living quality of a patient and to provide appropriate treatment. It is an useful tool in diagnosing and analyzing the condition of the heart diseases.

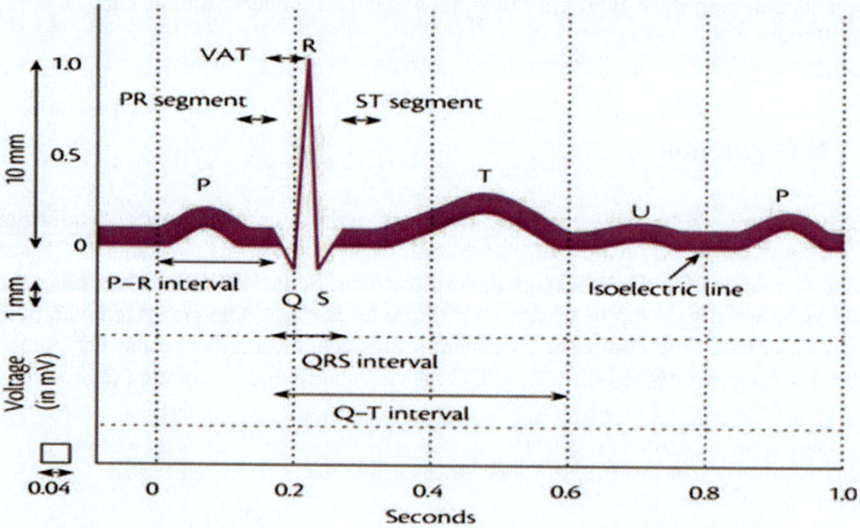

Fig. 1 Typical ECG waveform

Analysis of recorded ECG by manual observation takes longer time for the decision-making. Hence different methods are employed for feature extraction and detection.

2 Generation of ECG Signal

By using real-time data, ECG signal is developed using MATLAB. It produces ECG waveform of different leads and those waveforms can be changed to produce several arrhythmias as possible. This method has several advantages in simulation of ECG signal. Initially, it removes noise and helps in easy detection of abnormalities [1].

2.1 Properties of ECG Waveform

ECG waveform simulated in MATLAB with default specifications is depicted in Fig. 1. The duration of signal and time intervals such as P-R, S-T, and Q-T intervals are the main important properties of the waveforms [2]. The fundamental frequency of the periodic ECG signal is decided by heartbeat. If Fig. 1 is seen carefully, it is recognized that a single period of an ECG signal is a combination of nonlinear and time-varying features [14s]. The shape and duration of each metric of ECG are important. Some values for amplitudes and durations of ECG metrics are shown.

Amplitudes:

P—wave 0.25 mv

R—wave 1.60 mv

Q—wave 25 % of R wave

T—wave 0.5 mv

Durations:

Interval of P-R 0.20 s

Interval of Q-T 0.44 s

Interval of S-T segment 0.15 s

Interval of P wave 0.11 s

Interval of QRS 0.09 s

P wave

Indicates atrial depolarization with duration of <0.11 s

The normal P wave height is <3 mm.

QRS complex: Shows ventricular depolarization

Normal width is less than 0.12 s (sometimes 0.06 s) & 30 mm in height.

T wave

Represents repolarization of ventricles with height <5 mm.

Features of ECG represent shifted and scaled versions of the triangular and sinusoidal waveforms. Such waveform is shown in Fig. 2.

The ECG recording plots amplitude on the vertical axis versus time on its horizontal axis as shown below.

This is the standard ECG signal which is used as reference to compare with recorded ECG signal to detect the cardiac disorders. The default specifications of the standard ECG signal can be modified on the recorded ECG based on user's necessity while MATLAB code simulation.

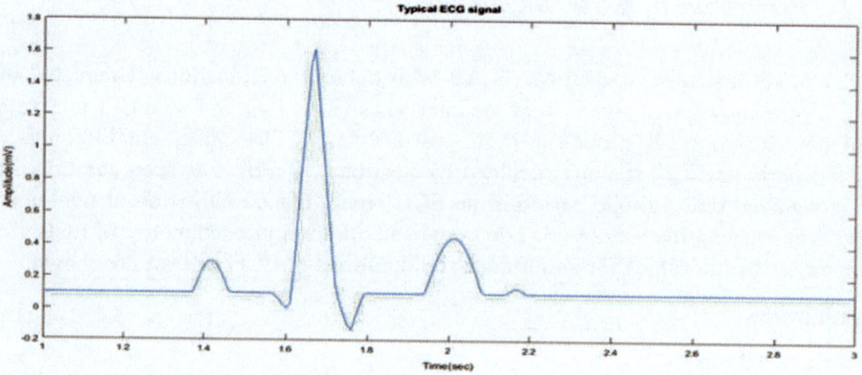

Fig. 2 Typical ECG waveform in ECG

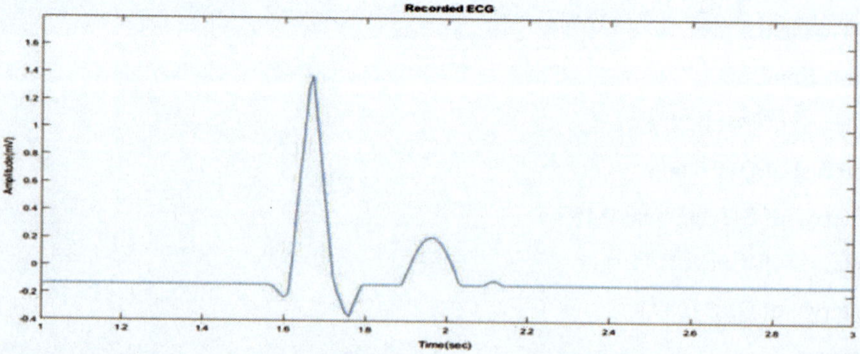

Fig. 3 Recorded ECG with Atrial fibrillation

The waveform with abnormality is shown in Fig. 3. Here the abnormality is atrial fibrillation, i.e., amplitude of P wave is zero. Abnormal ECG signal indicates Arrhythmia [3].

3 ECG Feature Extraction

For diagnosis, extraction process of features picks and stores the relevant information of original signal. It helps to take out diagnostic particulars from the ECG signal and the extracted particulars can be in frequency-domain analysis, time–frequency domain analysis, and feature vector ECG signal analysis.

3.1 Frequency-Domain Analysis

To study all the properties of ECG signal, time-domain method alone is not enough [4]. Therefore, frequency-domain analysis came into existence. Further, a Fourier transform is adopted to perform the spectral analysis that will converts a time-domain signal to frequency domain and frequency coefficients are obtained [5]. For examining the signal in frequency domain, one has to calculate the spectral function. The most used form of this calculation is the Fast Fourier Transform (FFT) [6]. It is a fast algorithm, which implements the discrete Fourier transform.

But this technique has a disadvantage when applied to ECG signal in that although it shows the frequency content of the signal, it failed to provide the accurate location of those frequency components in time.

3.2 Time–Frequency Domain Analysis

The time–frequency is a tool, which allows us to follow the changes in frequencies involved in the signal through time. For this, the visualization of the frequency changes in time is needed. The easy way to do this is to window the signal, see which frequencies are involved at which time window and then plot these vectors as functions in time. The frequency analysis is then the perfect tool to determine the length of such window.

3.2.1 Short-Time Fourier Transform

It consists of information about time and frequency. The phase and sinusoidal frequency information of the signal can be determined from STFT, as it varies with time. The spectrogram based on this technique is simple and fast in comparison to the other time and frequency analysis. It is an easy approach of dividing the given signal into a number of short segments of window size. Then analyzing each segment using standard Fourier transform gives simultaneous analysis of time and frequency.

However, this technique also has limitation in its time–frequency resolution capability in which short windows hardly depict low frequencies, whereas long windows poorly localize short pulses. Hence, the most acceptable method is preferred to overcome this problem.

3.3 Feature Vector ECG Signal Analysis (FVESA)

The FVESA is a tool used to reduce linear dimensionality and forecast the information in the highest variance direction. This method pulls out pertinent properties from the ECG data set [7].

The ECG data consisting of large number of samples over a period of time, and hence FVESA is a highly efficient technique which can be used for ECG signal analysis. Classification of cardiac disorders using ECG analysis is done using covariance method of FVESA by initially providing a set of pre-fed signal and comparing the principal components obtained for the recorded ECG with those of standard ECG [8].

3.3.1 Method of FVESA

The main aim is to analyze ECG signal for disease identification through FVESA to identify cardiac arrhythmia namely Atrial Fibrillation (AF). Hence the FVESA algorithm is provided with a standard ECG as reference and recorded ECG. Standard and recorded ECG data are centered across the origin through data mean corrected so that the DC level in those signals is removed.

The covariance matrix, eigenvectors, and eigenvalues are deliberated. The resulted eigenvalues are sorted along with their corresponding eigenvectors. The feature vector computes the projections of the data sets, which is the transpose of the eigenvector matrix.

4 Results

4.1 Frequency-Domain Analysis

Figure 4 represents the Fast Fourier Transform (FFT) of original ECG signal of a healthy person without any abnormality, and Fig. 5 shows the FFT analysis of recorded ECG with abnormality of atrial fibrillation.

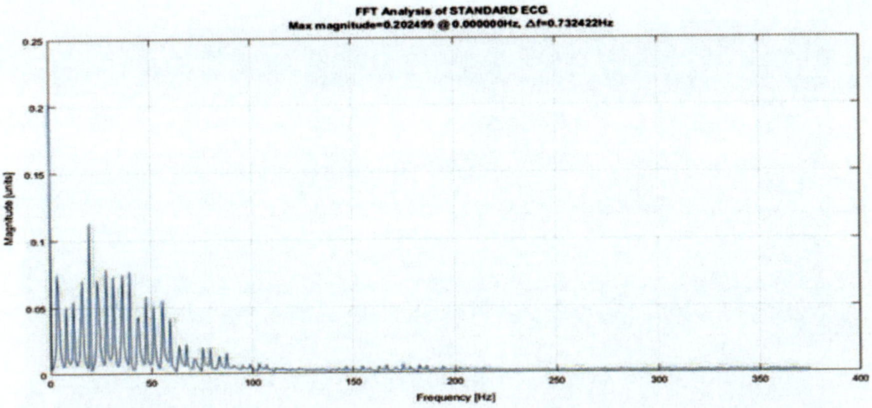

Fig. 4 FFT of standard ECG

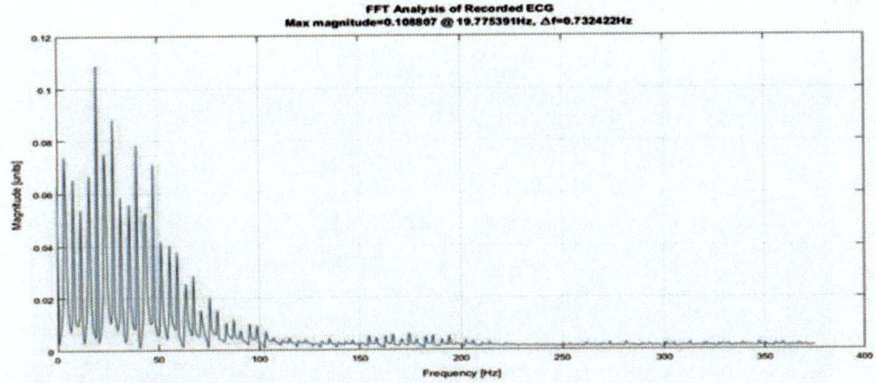

Fig. 5 FFT of Recorded ECG

4.2 *Short-Time Fourier Transform (STFT)*

Even though Fourier transforms indicate the frequency components of the signal, it failed to give the accurate location of those frequency components in time. So to get both the time and frequency components we applied STFT.

Figures 6 and 7 show the STFT analysis of standard ECG and recorded ECG with atrial fibrillation.

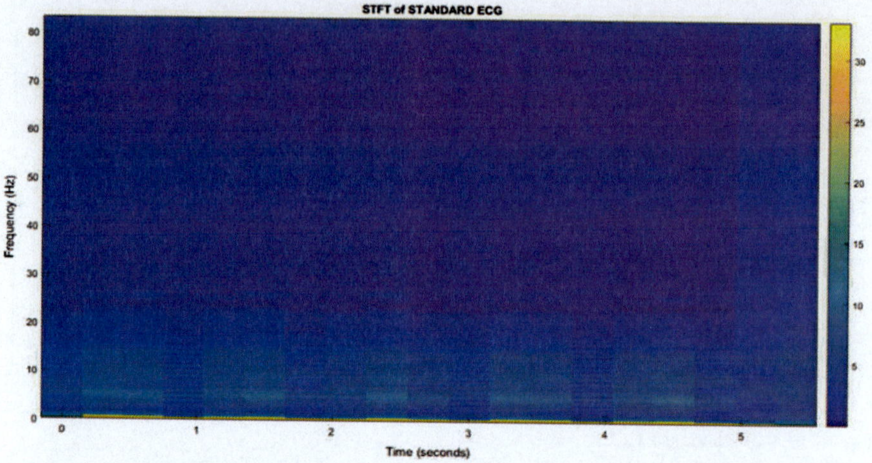

Fig. 6 STFT analysis of standard ECG

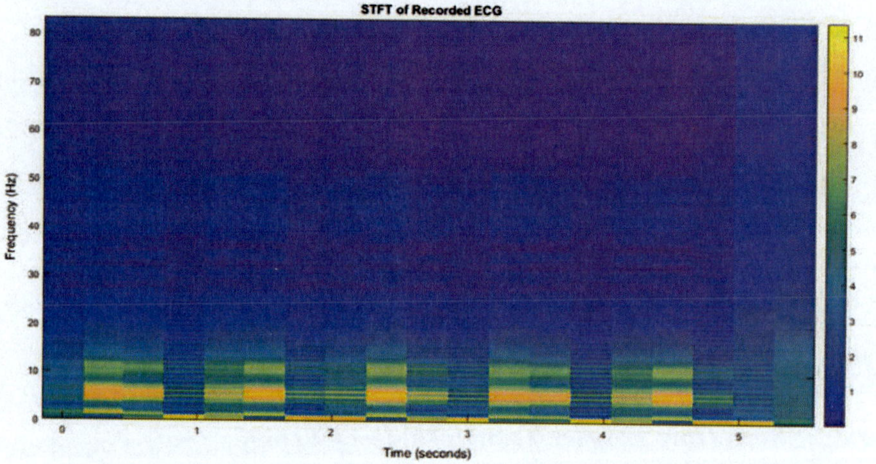

Fig. 7 STFT analysis of recorded ECG

4.3 FVESA

STFT also has limitations in finding time-frequency resolution, due to the uncertainty principle. Short windows hardly depict low frequencies, whereas long windows poorly localize short pulses. Hence to overcome this problem, more suitable technique is used. FVESA is applied to get the principal components of the signal, especially of the P wave. FVESA is a true eigenvector-based multivariate technique.

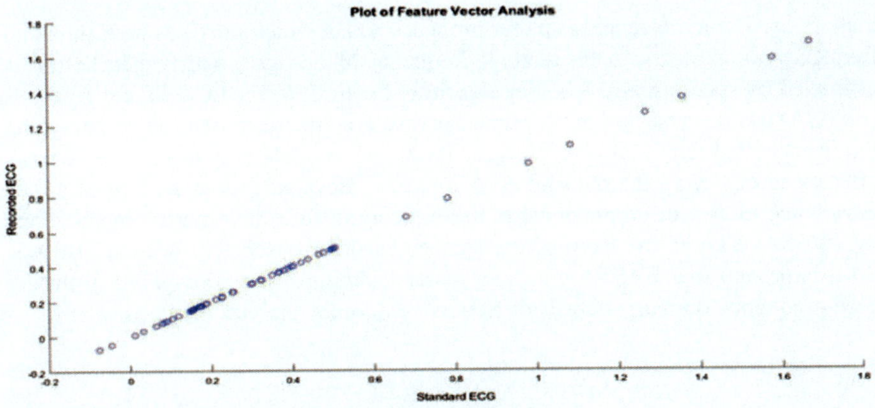

Fig. 8 FVESA analysis of standard ECG without abnormality

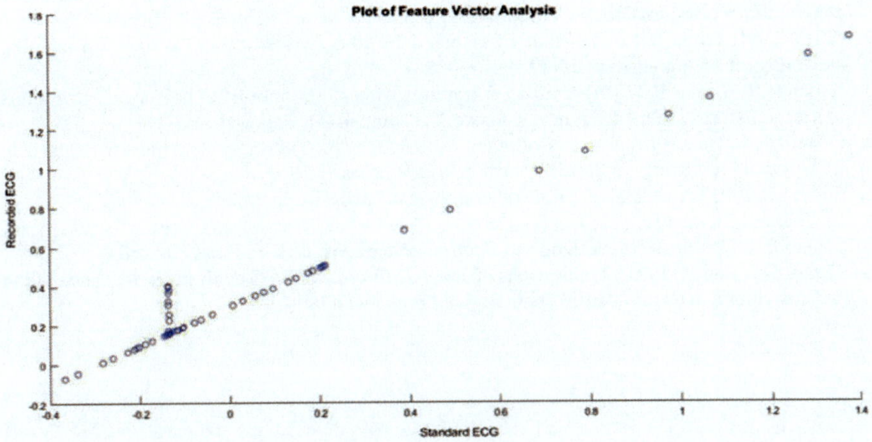

Fig. 9 FVESA analysis of recorded ECG with atrial fibrillation

Figures 8 and 9 represent the FVESA of standard ECG and recorded ECG with atrial fibrillation.

5 Conclusion

Several techniques are used to recognize the abnormalities in ECG signal. Under frequency-domain analysis, FFT has been used to identify the properties and detect atrial fibrillation. The main disadvantage is that the Fourier transform is unable to pick out local frequency content. It has a "hard time" representing functions that are oscillatory. Hence this paper proposes STFT which provides both time and frequency components. In this method, frequency of the signal with respect to time is estimated by spectrogram. Another algorithm Feature Vector ECG Signal Analysis (FVESA) has been carried out for this analysis. It is the most used tool in analyzing the investigated data and also used for predictive models preparation. İn general, after mean centering the data for each metric eigenvalue decomposition of a data covariance matrix or singular value decomposition of a data matrix can be done by FVESA. This is the easy of the true eigenvector-based multivariate analysis. So it is proved that FVESA can be a better technique when compared to that of time–frequency domain analysis or frequency-domain analysis methods.

References

1. Ojha DK, Subashini M (2014) Analysis of electrocardiograph (ECG) signal for the detection of abnormalities using matlab. Int J Med Health Biomed Bioeng Pharm Eng 8(2):120–123
2. (2019) Arrhythmia detection from ECG based heartbeat classification using deep learning networks. Int J Comput Appl 178(26):0975–8887
3. Kedawat S, Kumar R (2011) Cardiac arrhythmias detection in an ECG beat signal using fast fourier transform and artificial neural network. J Biomed Sci Eng 4(4):289–296
4. Castells F, Laguna P, Sörnmo L, Bollmann A, Roig JM (2007) Principal component analysis in ECG signal processing. EURASIP J Adv Signal Process 1–21
5. Smith LI (2002) A tutorial on principal components analysis, pp 2–8
6. IEEE Trans Biomed Eng 37(4):329–342
7. Schmorth L (1986) An introduction to electrocardiography. Oxford Press, Oxford
8. Thakor NV, Zhu Y (1991) Applications of adaptive filtering to ECG analysis: noise cancellation and arrhythmia detection. IEEE Trans Biomed Eng 38(8):785–794

Cloud Virtualized Middlebox: An IDS

K. S. Resma, G. S. Sharvani, and Manisha S. Soneja

Abstract Provisioning services to large data centers in a multitenant environment is a difficult task. It is because of the changing tenant requirements and hence the changing network policies. The above said reasons make the hardware implementation of network functions otherwise called as network middleboxes complicated. The hardware middleboxes make any system highly inflexible, costlier and difficult to maintain. Taking into account all these factors the need for virtualized middleboxes has come up. In this paper, a virtualized middlebox is proposed, designed, implemented and tested. The selected middlebox for virtualization is an Intrusion Detection System which can identify whether a packet coming to a network is an attack or a normal packet. Hence it provides network security. For the sake of virtualization, machine learning techniques are used. The virtualized middlebox is in addition of being flexible and scalable. It also provides better accuracy compared to the existing systems is proved using the obtained results.

1 Introduction

Middleboxes [1] also called network functions or network appliances are a very important part of any network. They include firewalls, intrusion detection systems [2], proxy servers, network address translators, traffic shapers, etc. These are intermediary devices which provide important functionalities in the policy chain. The middleboxes in the past had a hardware implementation. These hardware-based middleboxes come with a lot of disadvantages such as not being flexible, scalable,

K. S. Resma (✉) · G. S. Sharvani
SCOPE, RV College of Engineering, affiliated to VTU - Belagavi, Bangalore, India
e-mail: resma.vin@gmail.com

G. S. Sharvani
e-mail: sharvanigs@rvce.edu.in

M. S. Soneja
SCOPE, Oracle Marathahalli, Bangalore, India
e-mail: manishasoneja.ms@gmail.com

S. C. Satapathy et al. (eds.), *Communication Software and Networks*, Lecture Notes in Networks and Systems 134, https://doi.org/10.1007/978-981-15-5397-4_34

higher cost of maintenance deployment. These disadvantages have lead to the development of virtualized middleboxes. The virtualized middleboxes are a boon to the cloud service providers, as they provide the network with high scalability and flexibility with reduced CAPEX and OPEX. In a cloud network, every tenant will have his own resource needs and the network policies also keep changing. A virtualized middlebox [3] is highly adaptive to these changes. The main objective is to reduce the false alerts and improve the performance of middlebox. It has to be capable of identifying different types of network anomalies or attacks. Machine Learning [4] can be used to implement a virtualized Intrusion Detection System to detect if an alert is truly malicious or not. This IDS can very well act as a middlebox which can provide a secured communicating environment in the cloud environment.

In this paper, a virtualized middlebox an Intrusion Detection System (IDS) is proposed. An Intrusion Detection System has the responsibility of monitoring the network-related activities. It protects the network resources from the intruders. The proposed system works with the support of hybrid classifier which is a combination of decision tree classifier [5] and Naïve Bayes classifier to ensure system security.

The remaining section of the paper is organized as follows: Sect. 2 describes the background study including the descriptions of the algorithms used and the underlying technologies involved. Section 3 describes the system model and Sect. 4 describes the experimental setup. It also describes the data set used and different experimental setups used. Section 5 describes the results and the analysis of the results, followed by the conclusions and references.

2 Background

While using hybrid classifiers collecting every single incoming network request is needed. The basic procedure involves the use of a decision tree algorithm [6] to train the misuse detection model from the available normal and attack training data. Info Gain is calculated for all attributes to understand the contribution of the attributes towards the classification. After training, it is checked against the DT Model to classify if it is an attack or not, if there is no response it is sent to Naive Bayes classifier for anomaly detection.

2.1 Algorithms Used

In the proposed model, a multi-layer Hybrid Classifier is used to identify if an activity contains an attack data or normal data. A misuse detection model is built using Decision tree algorithm by which smaller data subsets are created from the training data, and then multiple one-class Naïve Bayes algorithm models are created for the decomposed subsets. Hybrid Classifier is used as a preprocessor of Intrusion Detection System [7] to reduce the training time and the dimension of feature vectors. Hence minimizes the noise caused by feature differences and elevates the performance of Intrusion Detection System. The proposed hybrid intrusion detection method was experimented with the help of NSL-KDD data set.

2.2 *Hybrid Classifier*

A hybrid classifier is the one in which more than one classification technique is combined to produce the required results. A decision tree algorithm [8] is used to train the misuse detection model from the available normal and attacked training data. Info Gain is calculated for all attributes to understand the contribution of the attributes towards the classification. After training, it is checked against the decision tree model to classify if it is an attack or not, if there is an attack it is sent to improved Naïve Bayes classifier anomaly detection.

2.2.1 Naïve Bayes Classifier

A Naive Bayes classifier is a machine learning model. It is based on probabilities and is used for classification purposes. The essence of this classifier is Bayes theorem [9]. It is a classification technique based on Bayes' Theorem. The assumption is that there is independence among predictors. Naive Bayes classifier assumes that the presence of a particular feature in a class is independent of the existence of any other feature.

We have

$$P\left(\frac{H}{E}\right) = \frac{P\left(\frac{E}{H}\right) \cdot P(H)}{P(E)} \tag{1}$$

where $P\left(\frac{H}{E}\right)$ is the posterior probability, that is how probable is our hypothesis given the observed evidence and it is not directly computable.

$P\left(\frac{E}{H}\right)$ is the likelihood which tells how predictable our affirmation is given that our presumption is true.

$P(H)$ is the prior probability which tells how predictable our hypothesis was before detecting the evidence.

$P(E)$ is the marginal probability which tells how predictable is the new evidence under all possible presumption is.

To apply the classifier in this middlebox, Prior probability $P(Ej)$ can be determined using the training data set in the below equation.

$$P(Ej/X) = (P(X/Ej)P(Ej)/P(X) \tag{2}$$

$$\text{IFFP}(Ej) > P(EiX),\ 1 \le i \le m,\ i \ne j$$

Once the sample has a number of attributes, then *P*(*Ej*) can be determined using the below equation:

$$(Ej) = (X/Ej)/(Ej) \tag{3}$$

$$IFF(EjX) > (EiX),\ 1 \le i \le m,\ i \ne j$$

where *Ej* is the given set of *m* Classes $\{E1, E2, E3\ldots Em\}$, *X* is unknown data sample and *P*(*X*) is a constant for each one category.

To classify records of the test data set or in the online traffic is done as:

$$(Ej) = Sj/S \tag{4}$$

where *Sj* is the training sample size in the class *Ej*, and *S* is the total number of training samples.

2.2.2 Decision Tree Algorithm

To divide the data into classes, the decision tree algorithm has its own training data set. Hence the data gets divided into separate classes. In the decision tree, every node will have edges. And each of those represents values or value ranges corresponding to attribute of the respective nodes. Based on the edge values each of the nodes is again divided. The child nodes are created for each subset of data using the decision tree algorithm [10]. The process iterates for the set of values. Once the stopping rule is reached or when further classification is not possible because of similar datasets or nonexistence of distinguishing attributes leads to the termination of the decision tree algorithm. The created nodes are labelled with the class names. This named node is called a leaf node. Hence, the decision tree algorithm recursively splits the training data set, which creates a tree-like structure. Decision tree algorithm constructs a decision tree based on the training data sets utilizing the concept of information entropy. That is, it is based on the highest attainment of each attribute.

Entropy $= -\sum P(X) \log P(X)$ (5) where *P*(*X*) is a fraction of examples in a given class.

Information Gain (IG) is the measure of information that a particular property gives us about the class. Decision Trees algorithm constantly tries to reach the highest information gain. An attribute having the greatest information gain will test or split with priority.

The Equation of Information gain:

$$GAIN(P, A) = ENGTROPY(P) - \sum(|Pv|\ /\ |P|)ENTROPY(Pv) \tag{5}$$

for *v* element of the values of *A*.

where value of A is the set of positive values for P, and Pv is the subset of P for which attribute A has value v. Once the gain attains the maximum value, depending on that the tree is constructed. This process is carried out until the break down regions become homogeneous followed by the final tree pruning. The tree hence generated has minimum errors as the training set is not general. In this study, the decision tree algorithm is used to train the misuse detection model of the IDS middlebox. The process of training is done using both the attack and normal dataset.

3 System Model

The system model is represented in Fig. 1, the decision tree model is trained using the training data set. The trained algorithm is then used on test data to classify the incoming traffic into normal data and attack data. The classified data is stored in the log file. The suspicious or the attack data is then passed on to Naïve Bayes classifier.

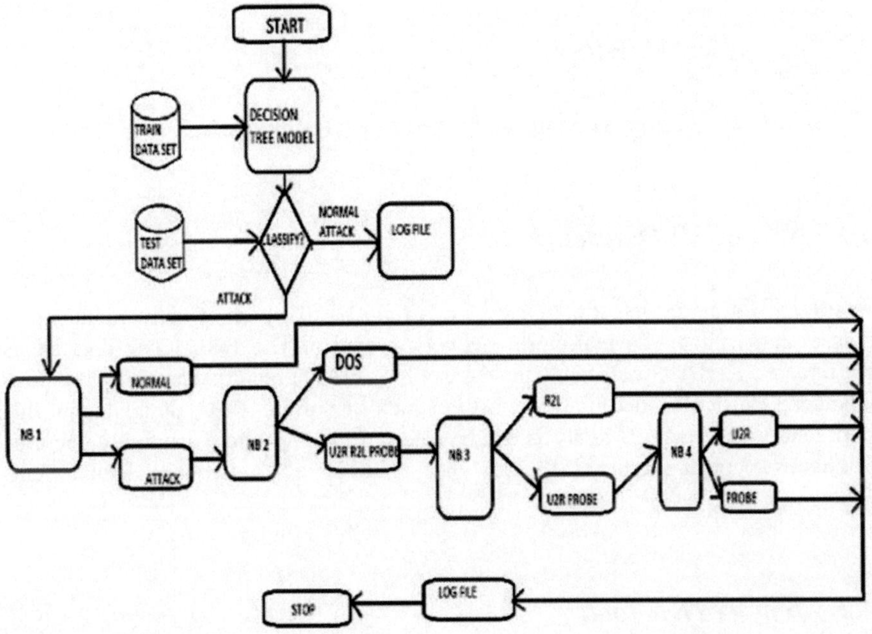

Fig. 1 System architecture

3.1 The Working of Hybrid Model

- Prepossessing data—There were 41 attributes, out of which 3 were categorical, they had to be one-hot encoded for the hybrid model. After one-hot encoding the dimensionality of data set is increased, 50% of top attributes were chosen by select percentile method.
- Decision Tree Model: The data set with normal and anomaly packets was sent to the model. This model classified packets as normal or anomaly.
- Naive Bayes Model: The data with normal and type of attack was sent to the model. Multi-level Naive Bayes classifier was used, where at first level the attack packets are rechecked for normal behaviour, the remaining attack packets are passed to the next level which differentiates dos from other packets, next level differentiates r2l from other attacks and final level which differentiates between u2r and probe.
- The target class is recorded in log file and the resulting accuracy and classification report are recorded.

4 Experimental Setup

The experimental setup is comprised of the following components.

4.1 NSL-KDD Dataset [11]

Intelligent intrusion detection systems can only be using an effective data set. A dataset can imitate the real-world network scenario. The NSL-KDD data set is the updated KDD"99 data set. This dataset has 40+ characteristics which specifies whether a particular incoming request is attack or normal. More details about this particular dataset and its analysis can be referred in [11]. There are four categories of attacks according to the NSL_KDD dataset, they are [11, 12] DOS, Probe, U2R, R2L.

4.2 Jupyter Notebook

The Jupyter Notebook is the application that helped us to create the live code, equations, visualizations, data cleaning, transformation, numerical simulation, statistical modelling and classification for the implementation part of the research.

4.3 *Anaconda Cloud*

The Anaconda Cloud is designed to make software development, release and maintenance supports. It provided package management support environment in elaborate manner. Anaconda Cloud is allowed to have a free public package hosting, as well as package channels, providing a flexible and scalable service.

5 Results and Analysis

Figure 2 represents the accuracy of the decision tree model and the attributes that are selected after one-hot encoding.

Figure 3 represents precision, recall and F1-score for the following categories—Dos, Normal, R2L and U2R. Figure 4 represents precision, recall and F1-score for anomaly and normal packets.

```
Accuracy of descion tree model 0.7792317246273953
43
<class 'numpy.ndarray'>
70
['duration' 'src_bytes' 'dst_bytes' 'wrong_fragment' 'urgent' 'hot'
 'num_failed_logins' 'num_compromised' 'root_shell' 'su_attempted'
 'num_root' 'num_file_creations' 'num_shells' 'num_access_files'
 'num_outbound_cmds' 'count' 'srv_count' 'serror_rate' 'srv_serror_rate'
 'rerro_rate' 'srv_rerror_rate' 'same_srv_rate' 'diff_srv_rate'
 'srv_diff_host_rate' 'dst_host_count' 'dst_host_srv_count'
 'dst_host_same_srv_rate' 'dst_host_diff_srv_rate'
 'dst_host_same_src_port_rate' 'dst_host_srv_diff_host_rate'
 'dst_host_serror_rate' 'dst_host_srv_serror_rate' 'dst_host_rerror_rate'
 'dst_host_srv_rerror_rate' 'service' 'flag' 'land' 'logged_in'
 'is_host_login' 'is_guest_login' 'icmp' 'tcp' 'udp']
```

Fig. 2 Accuracy of the decision tree model

```
<class 'pandas.core.series.Series'>
              precision    recall  f1-score   support

         dos       0.88      0.72      0.79      7460
      normal       0.66      0.97      0.79      9711
       probe       0.75      0.45      0.56      2421
         r2l       0.08      0.02      0.03      2885
         u2r       0.53      0.25      0.34        67

   micro avg       0.71      0.71      0.71     22544
   macro avg       0.58      0.48      0.50     22544
weighted avg       0.67      0.71      0.67     22544
```

Fig. 3 Classification report for Naïve Bayes model

	precision	recall	f1-score	support
anomaly	0.97	0.62	0.76	12833
normal	0.66	0.97	0.79	9711
micro avg	0.77	0.77	0.77	22544
macro avg	0.82	0.80	0.77	22544
weighted avg	0.84	0.77	0.77	22544

Fig. 4 Final classification report

Figure 5 represents a graphical view of the confusion matrix with the false positive rate being reduced to 0.02%. Figure 6 represents the classification report graphically, as shown that the anomaly packets have a very high precision as 97% and standard packets are having a recall of 97%. In this, most of these classifiers have less accuracy, as well as the time complexity is more. In the classifier, the data was fed into the decision tree algorithm and the Naive Bayes classifier, where the decision tree classifier accuracy is 77.7% and the Naive Bayes classifier accuracy is 70.1%. This classifier is fed in the test information and the accuracy obtained is 78%. The summary of the analysis as shown below:

- The first classifier decision tree net shows 77.7% accuracy.
- The second classifier Naive Bayes shows 70.1% accuracy.

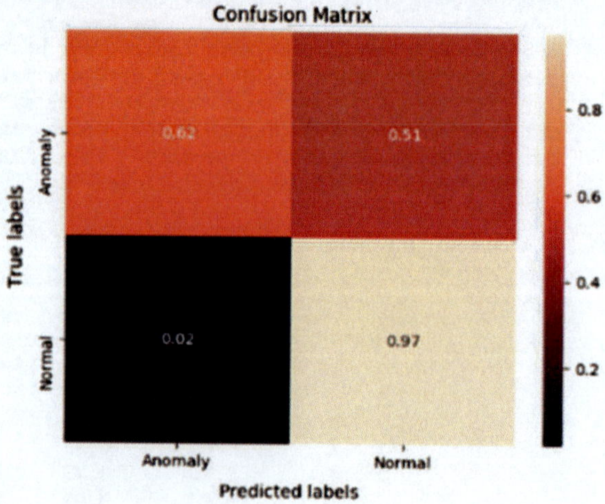

Fig. 5 Final confusion matrix

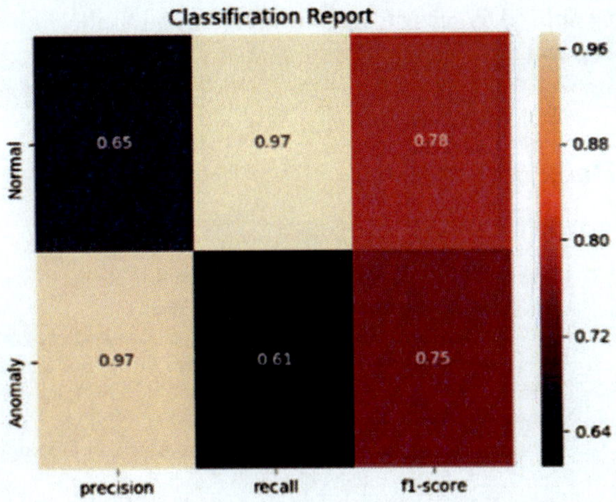

Fig. 6 Graphical view of classification report

- Overall accuracy obtained is 78%.
- As the aim expected was to reduce false-positive errors, it was achieved by reducing to 0.02%.

The remaining classifier shows the moderate accuracy as well as the more time complexity. The final result shows the combination of classifiers such as decision tree algorithm and Naive Bayes that shows the accuracy is 78% which are better than all other classifiers, where decision tree differentiates between normal and attack data, whereas Naive Bayes detects the attack type. As the aim expected was to reduce false-positive errors, it was achieved by reducing it to 0.02%.

6 Conclusion

The aim of our paper was the implementation of a virtualized middlebox. The middlebox selected for the same was IDS. This middlebox is capable of detecting all types of network anomalies or attacks in the available environments. In the implemented model, a multi-layer Hybrid Classifier is adopted to estimate whether the packet is an attack or normal data. This virtualized middlebox detects if the alert is malicious or not. The implementation was done with the help of machine learning algorithms. Our model consists of decision tree algorithm that differentiates between normal and attack data, and the Naive Bayes algorithm that detects the attack type. The experimental results showed that both these selected algorithms give higher accuracy and low time complexity. The developed IDS was tested by installing it on a network to protect this system against real-time attacks. The results are included in the results module. This model is highly flexible and adaptive as it is a virtualized model unlike the traditional hardware middlebox implementations. This model can be utilized by security operation centres, in application intrusion detection, datacenter and cloud infrastructures.

References

1. Resma KS, Sharvani GS (2019) A closer look at the network middleboxes. ICATIECE
2. Kima G, Leeb S, Kim S (2014) A novel hybrid intrusion detection method integrating anomaly detection with misuse detection. 1690/1700
3. Resma KS, Sharvani GS Edge distributed cloud middleboxes. Int J Adv Res Ideas Innov Technol
4. Abd-Eldayem MM (2014) A proposed HTTP service based IDS. Egypt Inf J 15
5. Tsaia C-Y, Hsub Y-F, Linc C-Y, Lind W-Y (2009) Intrusion detection by machine learning: a review. 11994–12000
6. Sivatha Sindhua SS, Geethab S, Kannan A (2012) Decision tree based light weight intrusion detection using a wrapper approach. 129–141
7. An overview of intrusion detection system (IDS) along with its commonly used techniques and classifications. Int J Comput Sci Telecommun 5(2), February
8. Mukhopadhyay I, Chakraborty M, Chakrabarti S (2011) A comparative study of related technologies of intrusion detection & prevention systems. J Inf Secur 2:28–38. Department of Information Technology, Institute of Engineering &Management, Kolkata, India. https://doi.org/10.4236/jis.2011.21003. Published Online January 2011. http://www.SciRP.org/journal/jis
9. Sherry J, Hasan S, Scott C Making middleboxes someone else's problem: network processing as a cloud service. Arvind Krishnamurthy University of Washington Sylvia Ratnasamy UC Berkeley Vyas Sekar Intel Labs, UC Berkeley
10. Lombardo A, Manzalini A, Riccobene V, Schembra G An analytical tool for performance evaluation of software defined networking services. 1DIEEI, University of Catania, Viale Andrea Doria 6, 95125, Catania, Italy
11. Dhanabal L, Shantharajah SP (2015) A study on NSL-KDD dataset for intrusion detection system based on classification algorithms. Int J Adv Res Comput Commun Eng 4(6), June
12. NIST special publication on intrusion detection systems. Mell, Peter; Bace, Rebecca, Standard Form 298 (Rev. 2-89) Prescribed by ANSI Std. Z39-18 298-102. Ashoor AS, Gore S (2011) Importance of intrusion detection system. Int J Sci Eng Res 2(1), January

Novel Approach to New Domain Aspect Identification Using Deep Learning and Word Replacement

Bonson Sebastian Mampilli and Deepa Anand

Abstract Customers of businesses enter a vast amount of data online. The data are analyzed by businesses to measure customer sentiment, allowing businesses to gain insights into what customers feel about their products and services. Sentiment analysis is therefore of prime importance to businesses. Sentiment predictions in sentences can be improved by identifying aspects, also known as Aspect Term Extraction (ATE), and then measuring the polarity of the sentiments associated with the aspects. This detailed measurement of aspect sentiment polarity helps in the overall sentiment classification of sentences. In the area of deep learning using neural networks, embeddings are created across a large corpus of data. These embeddings provide similarity measures that can be used for predicting aspects in unseen domains. In the proposed approach, unseen words of a domain are replaced with words from the training domain. The replacement words are identified using similarity scores between unseen words of a domain and existing words of the training domain. The modified sentences afterword replacement are given to the various domain-specific models for unseen domain aspect prediction. Extensive experiments are carried out across ten datasets to validate our approach. In a majority of the unseen domains, and across the different algorithms used, our approach performs well and gives a higher overall prediction accuracy.

1 Introduction

A lot of data are generated daily by users in many forms, including product reviews, complaints, opinions on events, stock markets, recommendations, and many more. All these different types of text offer some form of insight into a business on what the customer feels about a product, service, or brand. As the number of users of a

B. S. Mampilli (✉)
Research and Development Centre, Bharathiar University, Coimbatore, India
e-mail: bonsonsm@yahoo.com

D. Anand
CMR Institute of Technology, AECS Layout, Bengaluru 560037, India
e-mail: deepa.a@cmrit.ac.in

S. C. Satapathy et al. (eds.), *Communication Software and Networks*, Lecture Notes in Networks and Systems 134, https://doi.org/10.1007/978-981-15-5397-4_35

particular product or service grows, the amount of content generated by them also grows. Businesses grow by understanding what customers feel about their products and services. This includes the positive areas that the customers appreciate their business. It also includes the reasons that the customers are not happy and areas in which the businesses need to improve. However, businesses find it difficult to analyze large amounts of textual data manually to reach these inferences. To understand the details in the text, it is necessary to understand the various sub-topics or details on which the comments have been made. This helps to further understand the sentiment toward specific areas or aspects of the business. These details or sub-topics are known as aspects and form the first step of sentiment identification.

A side challenge is that we need models trained on previously annotated data that can be used for predictions. Unavailability of annotated data in a domain leads to time delays and lots of effort to create such data. Our primary contribution in this paper is to understand how labeled data from one domain can be used to provide better training accuracies on unseen domains. This is done by applying the algorithms and embedding from the labeled data onto the unseen domain. We compare classical machine learning methods of entity and aspect extraction like Conditional Random Fields (CRF) against the deep learning methods like Bi-directional Long Short-Term Memory (BLSTM) networks, Character Embedding networks, and also Embeddings for Language Models (ELMo). Pre-trained GloVe embedding models provide word similarity scores. We have experimented on the impact of replacing words in the testing domain with words from the training domain using the word similarity score. We also analyzed the results of the experiments to see if there might be a relation between the domain of the training data and the domain in which the prediction is being tried out.

The paper is structured as follows: Sect. 2 presents a discussion on related work on ATE using different algorithms. The proposed techniques are presented in Sect. 3. Section 4 discusses the experimental setup and the results of the evaluation and Sect. 5 concludes with a discussion on the novel contributions and directions for the future.

2 Literature Survey

One of the important tasks of NLP is Named Entity Extraction(NEE). NEE is used to automate the extraction of a noun in a sentence to a particular category. For example in the sentence, "Tom went the New York," the noun 'Tom' would be classified as a PERSON and the noun 'NEW YORK' would be classified as a LOCATION. A lot of approaches have been used for the successful implementation of this task with a good F1 score. Initially, the classical machine learning algorithms like CRF, Support Vector Machines (SVM), and Perceptron models were used for this task and also achieved high levels of accuracy [1]. In these approaches, the features needed to be identified and extracted. The features are then fed into the model for training. In comparison, deep learning approaches are able to learn the features from the text directly. Also, they are able to use the learnt features to improve the accuracy of the model predictions by fine-tuning the model predictions in an iterative manner.

Aspect extraction is an exercise very similar to NEE. Aspect extraction has been studied from the syntactical as well as the statistical approaches. In the syntactical approach, it is assumed that a relationship exists between the aspect word and the opinion word. In the statistical approach, CRF has been suggested in [2] to provide good results. This relationship between words can be extracted using the dependency parse tree [3].

The classical machine learning techniques gave good results. However, with the advent of deep learning techniques, it was understood that much higher levels of accuracy can be achieved. While deep learning started getting used extensively in machine vision, it began to also be used on unstructured text. The enormous amount of text available online from Wikipedia and other sites could be used to create embeddings [4]. An example of such an embedding is the GloVe embedding that stores millions of words and their associated contexts as scores. These can then be used by other applications where the word context understanding is helpful.

The above-mentioned approaches used deep learning techniques to understand words and the context of the words. The use of character embeddings in deep learning is useful where there might be new words that have not yet been seen during training [5]. Character embeddings also provide good accuracy levels in identifying named entities or aspects from texts. This is done by automatically extracting the features and using them for model creation and predictions.

In [6], a seven-layer deep neural network architecture to tag a sentence as an aspect or non-aspect word is used. This neural network was used along with a word embedding model for sentiment analysis was used to obtain better accuracy results than the state of the art methods. In this approach, linguistic rules were also identified and then used along with the network and the embeddings.

As neural networks are non-linear in nature, they better fit the aspect data than linear models like CRF, etc. Pre-trained word embeddings also help in the predictions. Along with this, as neural networks are used, there is no need for feature engineering, and because of this, it helps reduce overall development time and cost.

In [7], deep learning techniques have been used to show the abilityto extraction of aspects. The authors have called the aspect the target words and have shown the benefits of tagging the connections between the target words and the context words when building a learning system. The authors use two target-dependent LSTMs to predict aspects without syntactic parsers and lexicons.

3 Approach

3.1 Motivation

ATE across domains has become an important pre-processing step to assist in gathering proper insights from unstructured data. The supervised learning approach to ATE depends on the labeled training data to train on the identification of aspects. This training helps in predicting the aspects of the unseen test domains. In the unsupervised learning approach, different approaches such as linguistic and clustering features are used to understand the characteristics of the data and use it for predicting aspects.

In both the approaches, having knowledge of the vocabulary and how the words are related to each other is useful. If such knowledge can be gathered from the existing domain of words, it is useful. However, it is also possible to learn this from a larger domain of words and this form of learning is provided by word embeddings. Pre-trained Word embeddings help understand the similarity with the context of words in a given vocabulary.

In word embeddings [8], individual words are represented as real-valued vectors in a predefined vector space. Each word is mapped to one vector and the vector values are learned in a way that resembles a neural network. The important aspect of this representation is to use it to create such weighted representation for a word across multiple dimensions and see its importance.

There are three main approaches used for the creation of these contextual word vectors. They are:

1. Count-based methods: These are used to compute some statistical measure of how often words occur with neighboring words in the corpus and then build out dense word vectors for each word.
2. Predictive methods using Word2Vec: These use Neural Network-based Language models. Methods like Continuous Bag of Words (CBOW) and Skip Gram from Google's Word2Vec model fall under this approach [9].
   In CBOW, the words before and after the word in question are used to predict the word. While in Skip-gram, a particular word is taken and on the basis of that the words before and after that are predicted (Fig. 1).
3. Unsupervised learning methods of Glove: This has been developed by Stanford and can be used to obtain dense word vectors similar to the Word2Vec models. Glove model creates a huge word content co-occurrence matrix using the Word to Features relation matrix (WF) and the Feature to Context relation matrix (FC). WF and FC are initialized with random weights and then multiply them to get Word to Content relation matrix (WC) and see how close it is to the actual WC and then improve further by minimizing error using Stochastic Gradient Descent (SGD).

$$\mathrm{WF} * \mathrm{FC} = \mathrm{WC} \tag{1}$$

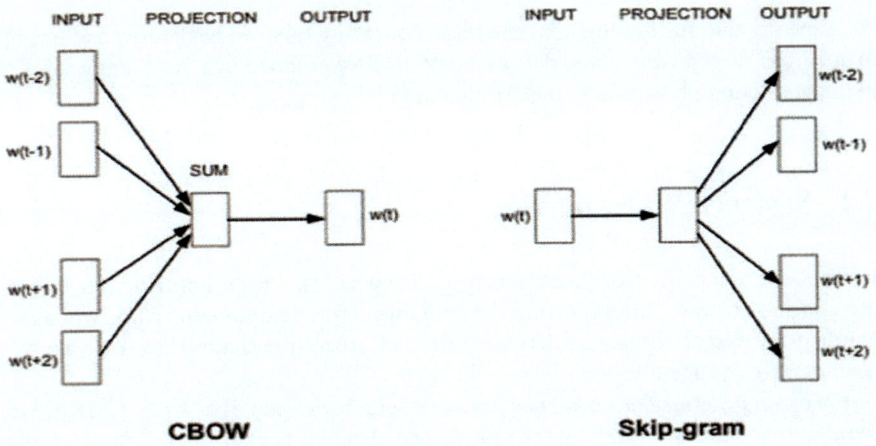

Fig. 1 CBOW and Skip-gram architectures [9]

Large embeddings with words from different domains can be used to find aspect word similarity from different domains.

We also see that Recurrent Neural Networks(RNNs) have been employed to produce promising results in a variety of tasks including language modeling. RNNs maintain a memory based on historical information which enables the model to predict the current output conditioned on long-distance features [10].

However, sentences also contain words to the present word that could help in the proper prediction of the entity type of a word. For this, the Bi-Directional LSTM (BLTSM) is used (Fig. 2).

A BLSTM network can efficiently make use of past features (via forward states) and future features (via backward states) for a specific time frame.

In [1], we see the use of BLSTM for the extraction of named entities from the CoNLL and the OntoNotes5.0 datasets. We see that they are able to achieve an F1 score of 76.29 (±0.29) using this approach. We first used the same approach on ten aspects annotated datasets to see the results of training and testing aspect prediction in the same domain. Then along with BLSTM, we use additional algorithms like CRF, Character Embedding, and ELMo for intra-comparison and see the results on

Fig. 2 A BLSTM network [10]

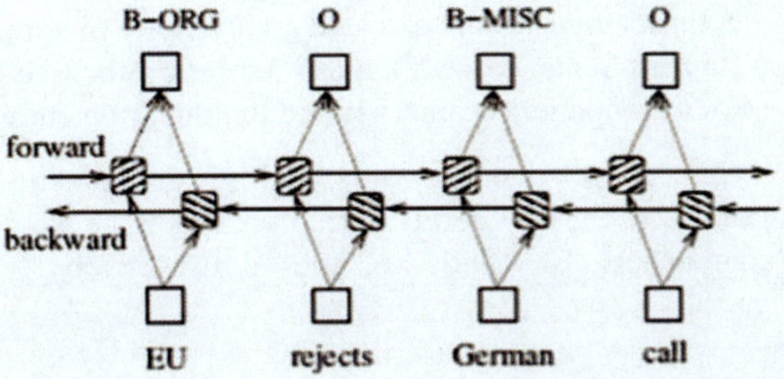

the same domain for training and test data. Following this, we have looked at trying to use cross-domain annotated data along with glove embeddings for training to help in the prediction of aspects in unseen domains.

3.2 Sentence Representation

In cross-domain predictions, lexicons that have been used for training belong to one domain and are very different from the lexicons of another domain. Thus, when the numbers of unseen words are large, entity and aspect predictions do not give the desired levels of accuracy.

For example, consider a sentence from the Camera domain like "It has a 20X zoom capability and allows tweaking the images and also blur backgrounds." Suppose the previously existing models have been trained on laptop data then it would not be able to understand aspect terms such as "zoom capability" and "blur background." Thus, this sentence needs to be represented in a better manner that easily understandable by models created in other domains.

The sentence representation is shown by taking the Laptop domain sentences (the training domain) and taking sentences for aspect prediction from the Camera domain (the testing domain). The sentences from the Laptop domain are taken and all the words are stored as a training domain word list. All the aspect words of the Laptop domain are taken and stored as the Laptop domain aspect list. Now, a sentence of the Camera domain is taken and the words are checked for presence in the Laptop domain. For the words that are not present in the Laptop domain list, the words are compared for similarity and the closet aspect word to this word in the Laptop domain is taken and replaced in the sentence. Below are two examples:

Camera Sentence: The startup time quality of pictures and videos is outstanding.

Camera Sentence with Laptop Aspects: The startup time quality of pictures and videos it performance.

In the sentence, the words 'is' and 'outstanding' have been replaced with 'it' and 'performance,' respectively.

Below is an example of a slightly more complex sentence:

> Camera Sentence: I highly recommend this camera to anybody looking for a compact camera that can take good quality pictures.
> Camera Sentence with Laptop Aspects: I focused options it camera to i looking for a components camera it find to i quality pictures.

Here, we see that the words 'highly,' 'recommend,' 'this,' 'anybody,' 'compact,' 'that,' 'can,' 'take,' 'good' are replaced with the words 'focused,' 'options,' 'it,' 'I,' 'components,' 'it,' 'find,' 'to,' and 'I,' respectively.

The modified sentence is then passed for sentence prediction to extract out the aspect words. These words are then replaced with the original sentence words to give back the aspect words of the new domain.

3.3 Models and Embeddings

For the complete set of experiments with the proposed approach, we have used the following algorithms:

a. Glove Embeddings: Glove [4] is a Global vector that is created from a global corpus from which word context statistics are captured by the model.
b. Conditional Random Fields (CRF): CRF [11] falls in the family of algorithms used for sequence modeling. CRF are a type of discriminative undirected probabilistic graphical models. CRF are able to store the relations that exist between sequential data.
c. Bi-Directional Long Short Term Memory (BLSTM): BLSTM is a modification of the LSTM architecture. LSTM is built on top of Recurrent Neural Networks. BLSTMs are able to look at both forward passes and backward passes across the data to help predict any additional information in sequence modeling. BLSTM has been proven to be very powerful in giving good results in accuracy in Named Entity predictions as mentioned in [1] using the CoNLL-2003 dataset.
d. Character Embeddings [5]: Character Embedding models take in characters as inputs and give back words as outputs. The usage of character embeddings has many advantages that include the ability to form word vectors for out-of-vocabulary words.
e. Embeddings for Language Models (ELMo): ELMo [12] is a novel way to represent words in vectors. It is able to store the syntactic and semantic characteristics of the word and show how the usage varies across contexts. ELMo representations store contextual information.

3.4 Algorithm Implementation

Domain data are used to predict the aspects in three ways:

1. Same-domain testing: The total annotated data of a domain is divided into training and testing data in a 70–30 split. The training data is used to create the aspect prediction model. The accuracy of the model is measured on the test data.
2. Cross-domain testing: The entire data of one domain (training domain) is used to create the aspect prediction model. The accuracy of the model is measured on data of an unseen domain (test domain).
3. Cross-domain testing using replacement with Glove: The algorithm for this approach is given in the algorithm section.

Algorithm 1 Cross-domain testing using replacement with Glove

INPUT: - Set S of all sentences of the testing set
OUTPUT: Aspects of testing domain

1. The entire data of one domain (training domain) is used to create the aspect prediction model.
2. Checks whether the words in the test domain exist in the training domain.
3. If a particular word in the test domain is not present in the training domain, it is marked for checking.
4. Iterate over the marked words and checks for the most similar aspect (based on similarity score) in the training domain aspect list.
5. The word in the test sentence is replaced with the highest similar word of the training domain.
6. Modified sentence is passed to the model for aspect prediction.
7. Words of the original sentence are replaced.
8. These are the new aspect words of the unseen domain.

3.5 Parameters

The features considered for CRF include variables to denote features such as Is the text lower case? Is the text title? Is the text a digit? What are the parts of speech? Is the text in upper case?

The parameters that have been configured for LSTM, Character Embedding, and ELMO for the various algorithms are listed in Table 1:

Table 1 Parameters configured for LSTM, Character Embedding, and ELMO

Parameters	LSTM	ELMO	Character embedding
Preprocessing	Tokenization, Punctuation removal, Lemmentization		
Optimizer	Rmsprop	adam	adam
Number of LSTM layers	1	2	2
Layer 1	BLSTM	BLSTM	LSTM
Layer 2	–	BLSTM	BLSTM
Neurons—Layer 1	100	512	20
Neurons—Layer 2	–	512	50
Dropout—Layer 1	0.1	0.2	0.5
Dropout—Layer 1	–	0.2	0.6
No. of dense layers	1	1	1
Dense layer activation unit	Softmax	Softmax	Softmax
Batch size	32	32	32

(continued)

Table 1 (continued)

Parameters	LSTM	ELMO	Character embedding
Loss type	Categorical cross-entropy	Categorical cross-entropy	Sparse Categorical Cross-entropy
Max. words in sentence	50	50	50
Max. characters in a word:	–	–	10

Table 2 Dataset details per domain

S.No	Domain Name	# Sentences	# Tokens	# Aspect Tokens
1	Computers	530	8569	449
2	Diapers	375	6668	263
3	Antivirus software	380	6777	298
4	Audio/Video players	4656	85160	2053
5	Camera	8123	147893	7387
6	Automobiles	16783	289404	13815
7	Laptop	2981	50190	3246
8	Restaurant	5838	87740	8773
9	Routers	1758	30767	893
10	Web Services	7393	113293	4362

4 Experimental Evaluation

The experimental evaluation is done using four different algorithms, and the details of the evaluation are given below.

4.1 Dataset Preparation

We used ten datasets with a variable number of records and then did cross-domain testing with different data. The dataset information is given in the following Table 2.

The table above gives information on the various domains covered and the data characteristics. Three collections of datasets[1,2,3] were used. Datasets on digital cameras were taken from [1, 4, 13]. The dataset for cellphones was taken from [13]. The datasets for mp3 players were taken from [1, 13] and the dataset for DVD players was taken from [13]. The datasets on routers were taken from [1], and the datasets on diapers were also taken from [1]. Chiu and Nichols [1] also contained the

[1]https://www.cs.uic.edu/liub/FBS/CustomerReviewData.zip.

[2]https://www.cs.uic.edu/liub/FBS/Reviews-9-products.rar.

[3]https://www.cs.uic.edu/liub/FBS/CustomerReviews-3-domains.zip.

Table 3 Results of training and testing using same-domain data for ten domains on the four algorithms BLSTM, CRF, Character Embedding, and ELMO

	BLSTM	CRF	Character embedding	ELMO
Computers	0.44	0.58	0.44	0.63*
Diaper	0.36	0.38	0.38*	0.34
Antivirus software	0.31	0.33	0.72*	0.25
Audio/Video players	0.31	0.31	0.25	0.41*
Camera	0.25	0.24	0.07	0.4*
Automobiles	0.27	0.24	0.32	0.37*
Laptop	0.67	0.76	0.85*	0.76
Restaurant	0.67	0.78	0.86*	0.75
Routers	0.26	0.35	0.43	0.52*
Web services	0.37	0.31	0.27	0.5*

antivirus software dataset. The restaurant datasets from SemEval 2014 and SemEval 2015 and laptop datasets from SemEval 2015 [11] were also used for the evaluations. The web service dataset was used from [14]. The automobiles review was taken from [4].

4.2 Comparison of Various Techniques for Same-Domain ATE

We have trained and tested the same-domain data on the ten domains using the BLSTM algorithm as described in [1]. Along with that, we also applied CRF, Character embedding, and ELMO embeddings, and presented a comparative analysis of F1 scores across the ten domains (Table 3).

It is seen that ELMO performed the best in most of the domains followed by the Character Embedding algorithm. Thus, in same-domain testing, these algorithms proved much better than the BLSTM implementation as suggested in [1].

4.3 Comparison of Cross-Domain ATE with and Without Glove Replacement

We trained the ten datasets using specific algorithms. Next, we applied the approach mentioned in the section on algorithm implementation to study the impact of modifying sentences using Glove similarity on Aspect prediction. The results of using the four algorithms and the results achieved are captured below.

1. Using Conditional Random Fields for training and measuring results of word replacement using the Glove embedding:
   - When training and testing on the same-domain data: Here, we saw that across all datasets, the results were slightly better when the word replacement using Glove is not done. In 51% (38/74) of the cross-domain tests done, the sentence word replacement with embeddings gives a better performance.
   - When testing on the cross-domain data: Here, we saw that the results were more or less divided equally between the approach without using Glove and the approach using Glove.
2. Using BLSTM for training and measuring results of word replacement using the Glove embedding:
   - When training and testing on the same-domain data: Here, across all datasets, we saw that the results were similar in accuracy.
   - When testing on the cross-domain data: Here, the results were much better where words in sentences were replaced using Glove. In 58% (57/97) of the cross-domain tests done, the sentence word replacement with embeddings gives a better performance.
3. Using Character Embedding for training and measuring results of word replacement using the Glove embedding
   - When training and testing on the same-domain data: Here, across all datasets, we saw mostly the results were better without using Glove embedding.
   - When testing on the cross-domain data: Here, the results were much better where words in sentences were replaced using Glove. In 73% (55/75) of the cross-domain tests done, sentence word replacement with embeddings gives a better performance.
4. Using ELMO for training and measuring results of word replacement using the Glove embedding:
   - When training and testing on the same-domain data: Here, across all datasets, we saw the results were more or less the same when working with or without Glove embeddings.
   - When testing on the cross-domain data: Here, it was seen that the results were much better without word replacement. In 76% (66/86) of the cross-domain tests done, predictions without word replacement gave a better performance.

We then did a comparison across datasets and domains to see which algorithm gives the best results. We see that in 5 out of the 10 datasets, ELMo was the best performer. The second best was LSTM with Glove. The highest accuracy was attained using Character Embedding with Glove. However, this was not generic across all datasets. The details of the results are given below (Fig. 3).

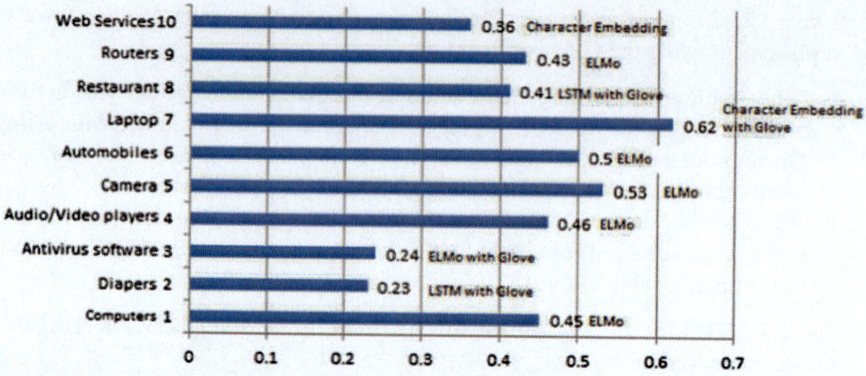

Fig. 3 Best accuracies per domain. The *Y*-axis depicts the ten domains and the *X*-axis depicts the F1 scores. The bars legends denote the training data used in cross-domain prediction

5 Conclusion

We have proposed an integrated approach of using a combination of algorithm implementations and a word matching and replacement using Glove to identify aspects in unseen domains. The proposed technique has been used to train on one domain and predict aspects across ten domains. It is noticed that training using the data from the computer domain provided the best cross-domain accuracy in seven of the ten domains. The ELMO algorithm gave the best accuracy in five of the ten domains while using Glove with replacement gave good results in four of the ten domains. There seem to be high benefits in the ELMo algorithm usage as it is able to learn the context of the words much better during the training process.

In the future, we plan to experiment with combinations of other embeddings and also use it in the combination of other rules-based extraction methods.

References

1. Chiu J, Nichols E (2016) Named entity recognition with bidirectional {LSTM}-{CNN}s. Trans Assoc Comput Linguist 357–370
2. Lafferty J, McCallum A, Pereira FCN (2001) Conditional random fields: Probabilistic models for segmenting and labeling sequence data. In: ICML'01, pp 282–289
3. Liu Q, Gao Z, Liu B, Zhang Y (2015) Automated rule selection for aspect extraction in opinion mining. In: Twenty-Fourth international joint conference on artificial intelligence, 2015
4. Pennington J, Socher R, Manning C (2014) {G}love: global vectors for word representation. In: Proceedings of the 2014 conference on empirical methods in natural language processing ({EMNLP}), pp 1532–1543
5. Yoon K, Yacine J, David S, Alexander M (2015) Character-aware neural language models. http://arxiv.org/abs/1508.06615
6. Poriaa S, Cambriab E, Gelbukhc A (2016) Aspect extraction for opinion mining with a deep convolutional neural network. Knowl-Based Syst 108

7. Tang D, Qin B, Feng X, Liu T (2015) Target-dependent sentiment classification with long short term memory, DBLP:journals/corr/TangQFL15
8. Goldberg Y (2017) Neural network methods for natural language processing (Synthesis Lectures on Human Language Technologies)
9. Tomas M et al (2013) Efficient estimation of word representations in vector space, arXiv preprint. http://arxiv.org/abs/1301.3781
10. Huang Z, Xu W, Yu K (2015) Bidirectional LSTM-CRF models for sequence tagging, CoRR
11. Charles S, Andrew M (2010) An introduction to conditional random fields
12. Peters ME et al (2018) Deep contextualized word representations. http://arxiv.org/abs/1802.05365
13. Xiang Z, Yann L (2015) Text understanding from scratch. http://arxiv.org/abs/1502.01710
14. Hu M, Liu B (2004) Mining and summarizing customer reviews. In: Proceedings of the tenth ACM SIGKDD international conference on Knowledge discovery and data mining, ACM, pp 168–177
15. Joshi V, Peters M, Hopkins M (2018) Extending a parser to distant domains using a few dozen partially annotated examples. http://arxiv.org/abs/1805.06556
16. Ding X, Liu B, Yu PS (2008) A holistic lexicon-based approach to opinion mining. In: Proceedings of the 2008 international conference on web search and data mining, ACM, pp. 231–240
17. Pontiki M, Galanis D, Papageorgiou H, Manandhar S, Androutsopoulos I (2015) Semeval-2015 task 12: Aspect based sentiment analysis. In: Proceedings of the 9th international workshop on semantic evaluation (SemEval 2015), pp 486–495
18. Kessler JS, Eckert M, Clark L, Nicolov N (2010) The ICWSM 2010 JDPA sentiment corpus for the automotive domain. In: 4th International AAAI conference on weblogs and social media data workshop challenge (ICWSM-DWC), Washington, DC, Citeseer
19. Jakob N, Gurevych N (2010) Extracting opinion targets in a single-and cross-domain setting with conditional random fields. In: Proceedings of the 2010 conference on empirical methods in natural language processing, Association for Computational Linguistics, 2010, pp 1035–1045

Power Quality Improvement of a Grid-Connected System Using Fuzzy-Based Custom Power Devices

V. Anantha Lakshmi and T. R. Jyothsna

Abstract Generally, the power systems are mainly affected by the continuous changes in operational requirements and the increasing amount of distributed energy systems because this causes the effect of deregulation. This paper proposes a new concept, i.e. power-control strategies for a micro-grid generation system for better transferring of power. This paper proposes a single-stage converter-based micro-grid to reduce the number of converters in an individual AC or DC grid. The proposed micro-grid concept can work in both stand-alone mode and also in a grid-interfaced mode. The distortions occurred in the power system due to the changes in load or the usage of non-linear loads can be eliminated by using control strategies designed for shunt active hybrid filters such as series and shunt converters. Conventional Proportional Integral (PI) and Fuzzy Logic Controllers are used for power quality enhancement by reducing the distortions in the output power. The MATLAB simulation results were compared among the two control strategies, with fuzzy logic controller and PI controller.

1 Introduction

Over the past few years, the growth in the use of non-linear loads has caused many power quality problems like high-current harmonics, low-power factor, and excessive neutral current. Non-linear loads appear to be current sources injecting harmonic currents into the supply network through the utility's Point of Common Coupling (PCC). This results in a distorted voltage drop across the source impedance, which causes voltage distortion at the PCC. Other customers at the same PCC will receive a distorted supply voltage, which may cause overheating of power factor correction

V. Anantha Lakshmi (✉) · T. R. Jyothsna
Department of Electrical Engineering, Andhra University, Visakhapatnam, Andhra Pradesh, India
e-mail: anusark2001@yahoo.co.in

T. R. Jyothsna
e-mail: thummalajyothsna@gmail.com

S. C. Satapathy et al. (eds.), *Communication Software and Networks*, Lecture Notes in Networks and Systems 134, https://doi.org/10.1007/978-981-15-5397-4_36

capacitors, motors, transformers and cables, and mal-operation of some protection devices [1].

Distributed energy resources are one of the power generation systems in small-scale range such as renewable energy resources examples of photovoltaic cell, wind energy generation system, or hydro energy. Placing the micro-grid concept near to the load centers has the advantage of improving efficiency by reducing the transmission line losses or voltage drops.

By increasing the domestic and commercial appliances and increasing demand for critical or sensitive loads causes the growing electricity consumption. In this paper, a micro-grid concept-based single-stage AC–DC converter is proposed for reducing processes of multiple reverse conversions in an individual AC or DC grid and to facilitate the connection of various renewable sources and loads to power system. The coordination control scheme such as maximum power point tracking converters is proposed for obtaining maximum power from the renewable energy sources under variations in input or any demand conditions [2–10]. This type of micro-grid systems is even generated electrical power under normal abnormal conditions such as if it is solar, it operates at room temperature or if it generates energy at normal speed, i.e. in plain surface area. However, power electronic-based converters are proposed in this paper for controlling purposes.

Generally, harmonics and reactive power are two of the serious problems associated with the grid. They are caused by non-linear loads, including saturated transformers, arc furnaces, and semiconductor switches. The presence of harmonics and reactive power in the grid is harmful because it will cause additional power losses and malfunctions of the grid components [2–6]. To prevent the inflow of harmonic and reactive currents and to improve the operating ability of the transmission systems, a kind of Flexible AC Transmission System (FACTS) has been proposed [7–11]. The static var compensator (SVC) is an important component of FACTS.

Reviewing recent studies shows the necessity for a concrete controller, which fits between the utility grid and the distributed energy resources at the load end to enhance and maintain the balance in the overall system. The contribution in this article is more toward the control mechanism of the UPQC device since the control technique plays a significant role in getting the maximum benefits out of any device, especially if it is a power electronic device. So far, many control techniques have been proposed to this UPQC, the work done in this particular paper is a combination of acclaimed evolutionary schemes focused on the control schema of the actual device. For obtaining better control of harmonics in a system, the conventional PI controller is replaced with a custom-designed combination of Neuro-Fuzzy and a PSO controller. The work in this article follows in the direction of providing a solution for

1. Maintaining Power Quality at the distribution end during switching from grid-connected to isolated mode.
2. Compensation of harmonics in grid voltage and in load currents.
3. Controlling real and reactive powers in both grid-connected and -isolated modes.
4. Expediting both real and reactive powers to the micro-grid in isolated mode.

2 Grid-Interfacing System

In the present scenario, the integration of grid with the renewable energy sources such as photovoltaic system is the most important application. These advantages include the favorable incentives in many countries that impact straightforwardly on the commercial acceptance of grid-connected PV systems. This condition imposes the necessity of having good quality designing tools and a way to accurately predict the dynamic performance of three-phase grid-connected PV systems under different operating conditions in order to make a sound decision on whether or not to incorporate this technology into the electric utility grid (Fig. 1).

2.1 *Photovoltaic Cell*

The photovoltaic energy system is one of the methods for generating electrical power by the direct conversion of solar irradiation into electrical energy using power semiconductor devices. These photovoltaic power generation systems consist of series and parallel connected solar cells. The power generated from the photovoltaic system is varied as changes occur in the input of solar panel, i.e. sun radiation; in order to get maximum output at any instant of time irrespective of time, we are going to use maximum power point tracking controller (Fig. 2).

The increasing commercial or domestic appliances such as sensitive or non-linear loads cause the changes in the transmission system parameters such as voltage or current unbalances or harmonics in current and other power quality problems. In order to meet the requirement of power quality standards, it is necessary to use some sort of compensating technique. Basically, from the first generation, the reactive elements are used for compensating purposes. Later, power electronics control-based devices are implemented and also called as flexible AC transmission systems.

3 Unified Power Quality Controller

One of the compensating devices forms the FACTS family called as Unified Power Quality Conditioner that is the efficient method improving for the power quality [5]. The unified power quality controller is a combination of series and shunt controller separated by a common DC-link for exchanging reactive power.

A shunt device is one of the compensated equipments, which is connectedto the transmission system. This shunt compensated system has the capability of either to absorb or generate active power at the point of connection, thereby controlling the voltage magnitude. As the bus voltage magnitude can only be varied within certain limits, controlling the power flow in this way is limited and shunt devices mainly serve other purposes. A device that is connected in series with the transmission line is

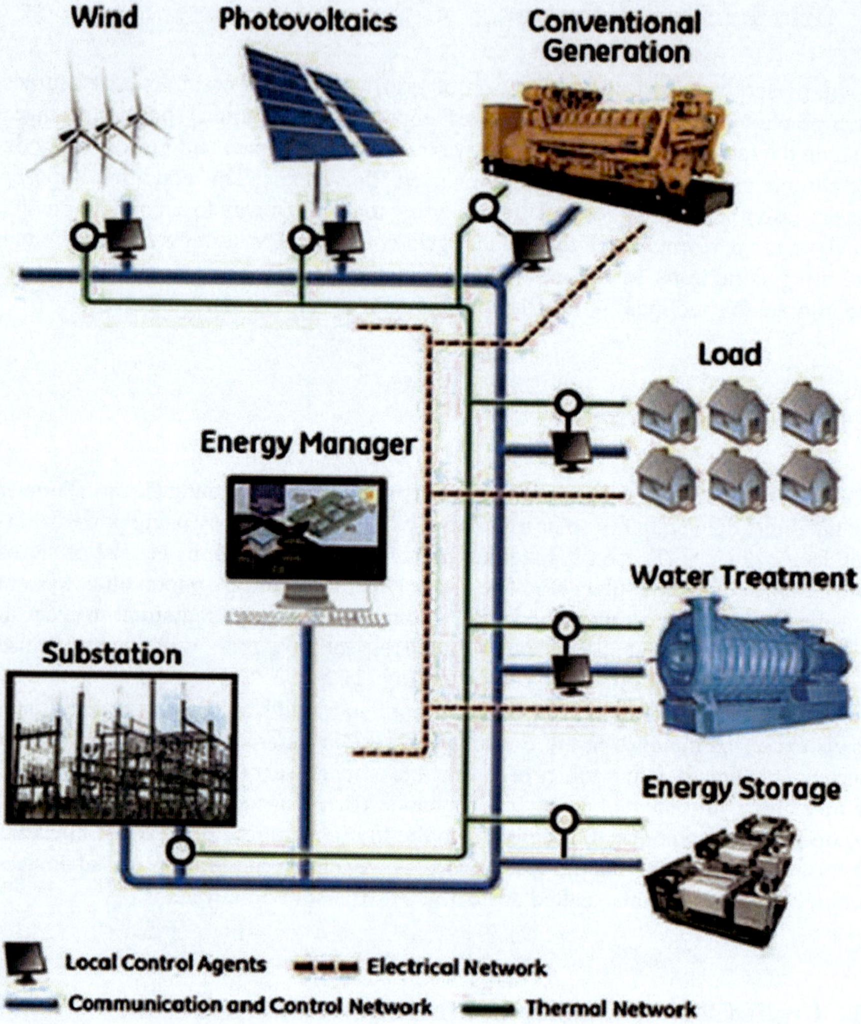

Fig. 1 Microgrid power system

referred to as a 'series device'. Series devices influence the impedance of transmission lines. The principle is to change (reduce or increase) the line impedance by inserting a reactor or capacitor. To compensate for the inductive voltage drop, a capacitor can be inserted in the line to reduce the line impedance.

The series compensated device is connected in series with the line for controlling the transmission parameters such as transmission impedance by controlling reactance, fluctuations in system voltage (Fig. 3).

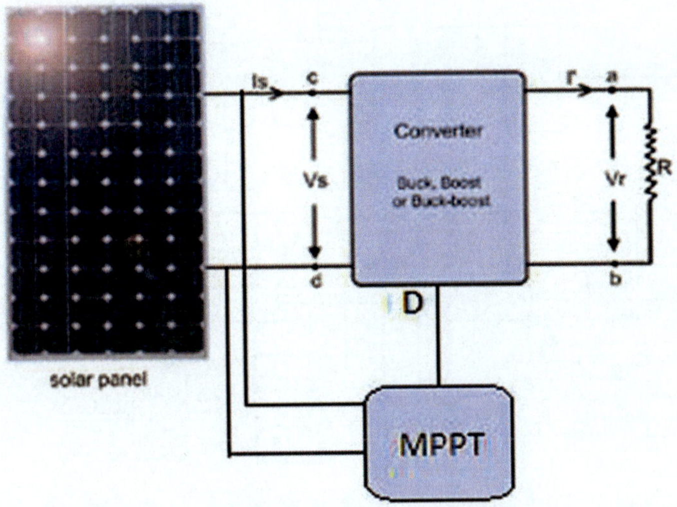

Fig. 2 PV system with power converter

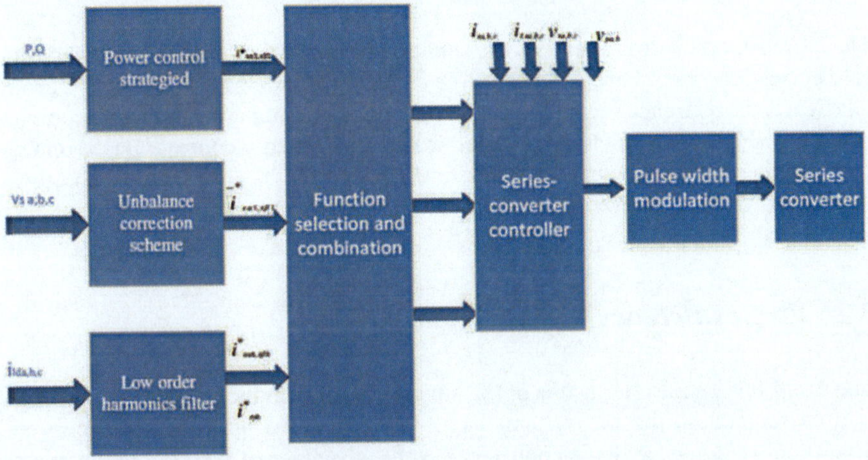

Fig. 3 Block diagram of overall control structure with series converter

The series controller, explained in the previous section, is used for compensating the power quality problems. It is a controller with the help of a three-phase converter. The gate pulses for this series converter are generated with the help of closed-loop control diagram as shown in Fig. 4. In this control technique, first, the voltages are compared and the error obtained for this is converted to two-phase orthogonal vectors. With the help of these vectors, the pulses for the series converter are calculated.

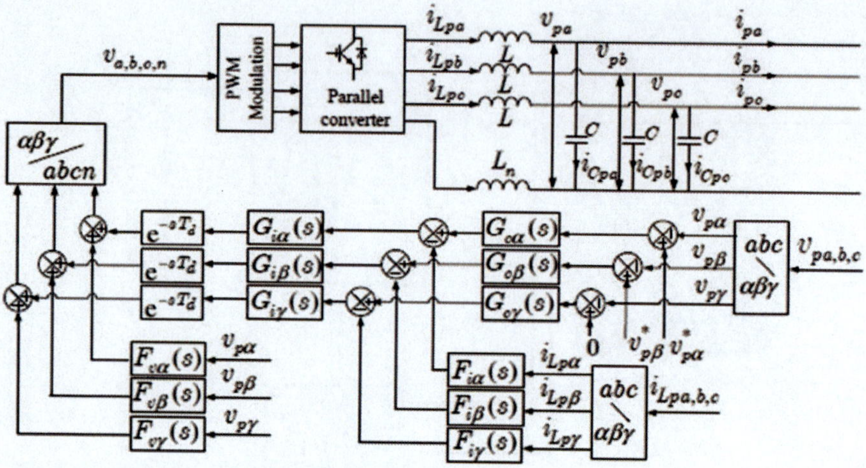

Fig. 4 Closed loop control diagram for shunt converter

3.1 Closed-Loop Control Diagram for Shunt Converter

The closed-loop control circuit of the shunt converter which is used for generating gate pulses for three-phase voltage source converter is as shown in Fig. 4 [7]. From this figure, the reference vector currents are obtained by comparing the actual currents with normal currents which are obtained from the PQ theory concept. Based on this error obtained from this concept, the gate-triggering pulses are generated are given to the VSC of shunt converter (Fig. 5).

3.2 Fuzzy Inference System

The fuzzy logic controller is one of the advanced soft computing controllers, which is used for controlling the system output. As compared with the other conventional controllers, fuzzy logic controller has the advantage of fast computing, better response, low settling time, and high running response. The fuzzy logic controller operation can be explained in mainly four ways, i.e. 1. Fuzzification, 2. Membership function, 3. Rule-base formation, and 4. Defuzzification.

The basic block diagram for the fuzzy logic controller is shown in Fig. 6. The rules taken for this system are shown in the table given below. The input variables such as error and error rate are expressed in terms of fuzzy set with the linguistic terms VN, N, Z, P, and VP. In this type of Mamdani fuzzy inference system, the linguistic terms are expressed using triangular membership functions (Table 1).

L(e, ce) = {VN, N, Z, P, VP}

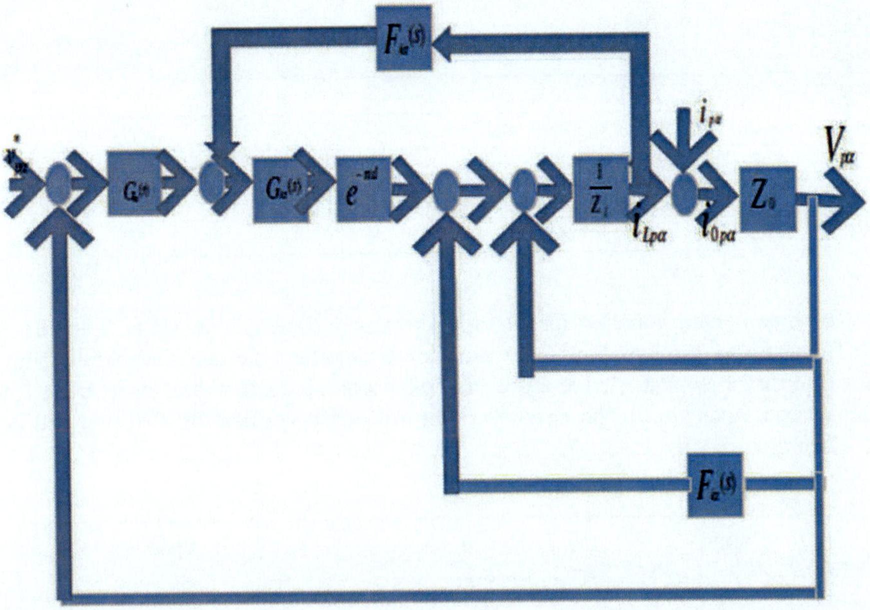

Fig. 5 Closed loop block diagram for α

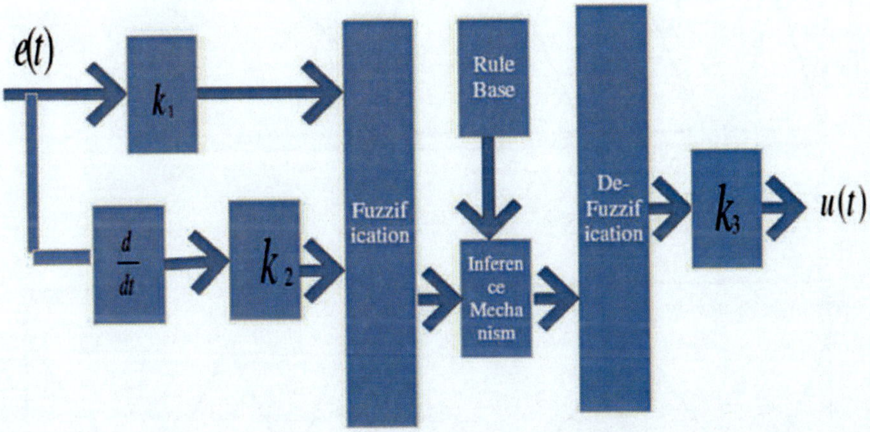

Fig. 6 Block diagram of fuzzy logic controller

Ex: Let us choose one rule with the terms for error and error rate as Zero and Zero. In this, the rules are formed with the help of if–then rule base formation. For this example, these two inputs are related to the logical operator AND (means Min value) and then the output is derived.

If input 1 e is Z and input 2 ce is Z, then the output is Z.

Table 1 Seven variable rule base

e/ce	VN	N	Z	P	VP
VN	VN	N	Z	P	VP
N	VP	P	Z	N	VN
Z	N	Z	P	VP	VN
P	N	N	VN	P	P
VP	N	Z	P	P	VP

The membership function for the input error is as shown below (Figs. 7 and 8).

The membership functions here used are triangular type and the membership function range is −1 to 1, i.e. a universe of discourse. Defuzzification is done by using centroid method. The seven-variable rule base consists of 49 rules and is If–Then type.

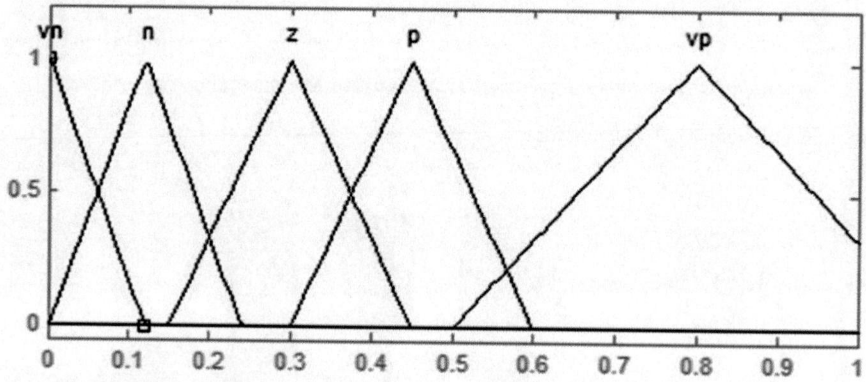

Fig. 7 Membership function representation for input 1

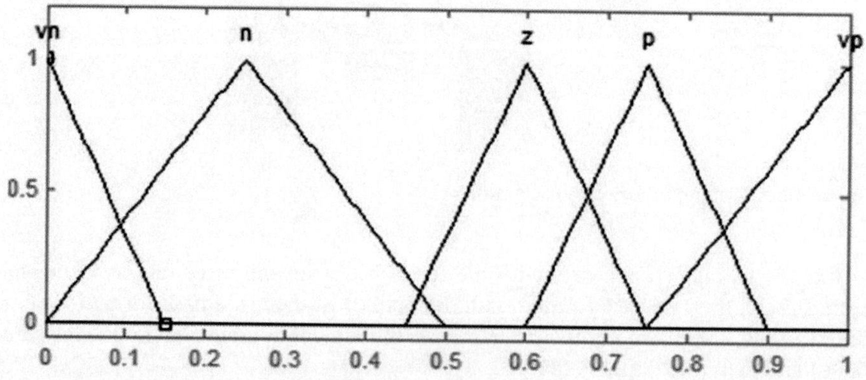

Fig. 8 Membership function representation for input 2

4 Simulation and Results

In order to study the performance of the proposed control scheme indicated in Fig. 1, it is executed in Matlab/Simulink environment. The micro-grid system employs two DG units, which can work in coordination with the utility grid, or in isolated mode when the main grid is unavailable to serve consumer load. The control scheme, as depicted in Fig. 1, makes it possible for the DG unit to support the grid in different scenarios (Fig. 9).

In this paper, the simulation is done for micro-grid, and the results are compared with two cases.

Case 1: with PI Controller

In this, the conventional PI controller is used for series and shunt controllers. The obtained results are shown in the figures below.

Figure 10 shows the simulation results of the system feeder currents. Due to the usage of non-linear loads, these currents are affected by distortions with a magnitude change of 0.5 p.u of its final value. With the help of the shunt converter, connected in parallel with micro-grid system compensates current harmonics. Figure 10a shows the simulation result of micro-grid current, and the simulation result of

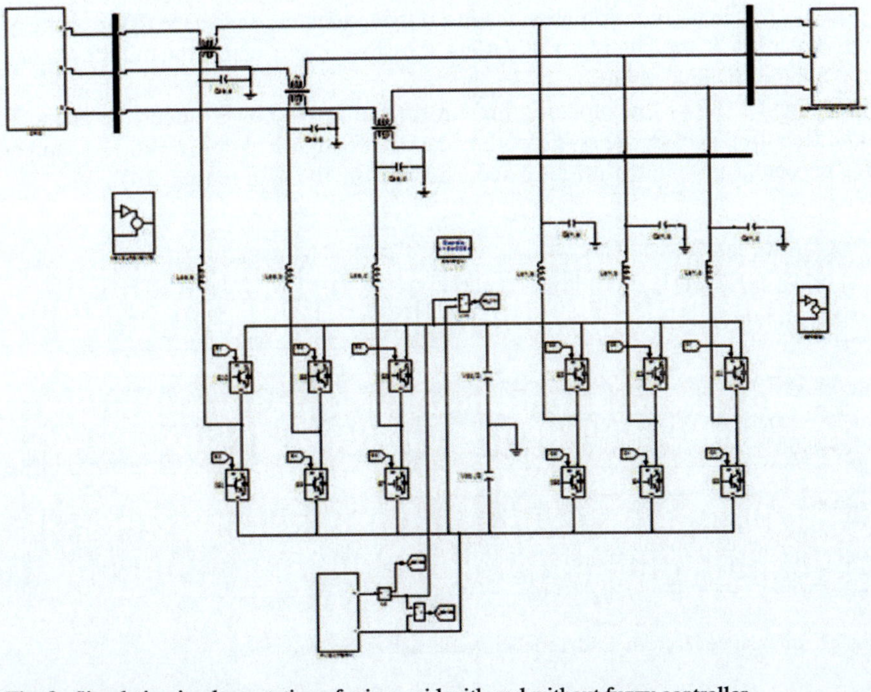

Fig. 9 Simulation implementation of micro grid with and without fuzzy controller

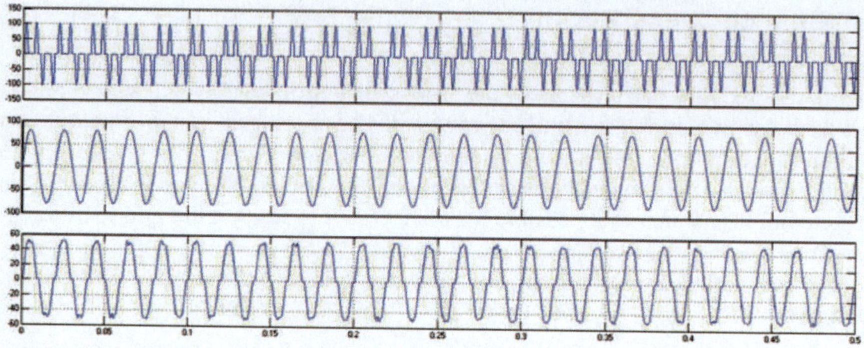

Fig. 10 Simulation result for Feeder currents 1, 2 and 3

shunt converter-injected current in Fig. 10b, c shows the simulation results of the compensated microgrid current.

The three-phase source voltages are demonstrated in Fig. 11, which shows the simulation results for the system grid voltages. In this system, grid voltage affects its variations with 0.5 P.U in magnitude as shown in Fig. 11a. The series converter-injected voltage during the variations in Fig. 11b, c shows the simulation result of the compensated micro-grid voltage.

Figure 12 shows the simulation results of the system currents. Due to the usage of non-linear loads, these currents are affected by distortions with a magnitude change of 0.5 p.u of its final value.

Figure 13 shows the simulation result for the active power under the islanded condition. In this case, we consider the islanded condition at time $t = 0.17$ s, and at this particular instant, the grid is disconnected from the system (Fig. 14).

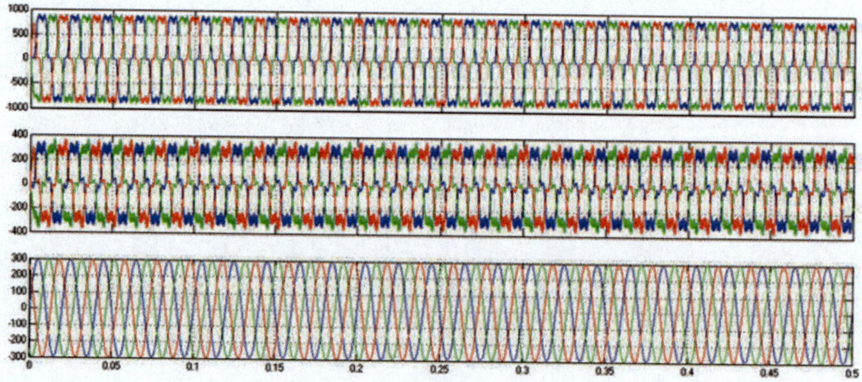

Fig. 11 Simulation result for grid, series converter and micro-grid voltage

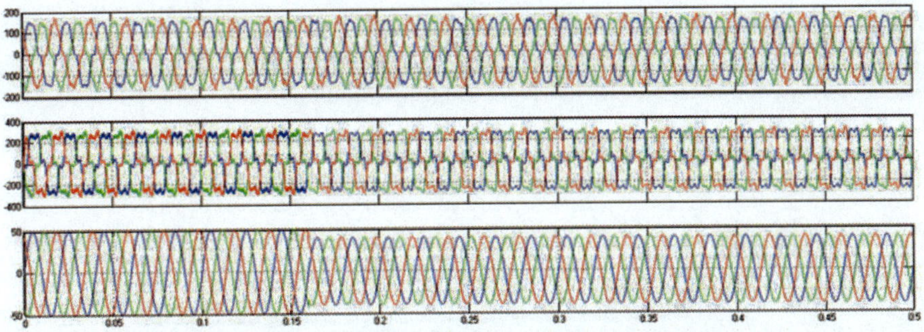

Fig. 12 Simulation result for grid, series converter and micro-grid current

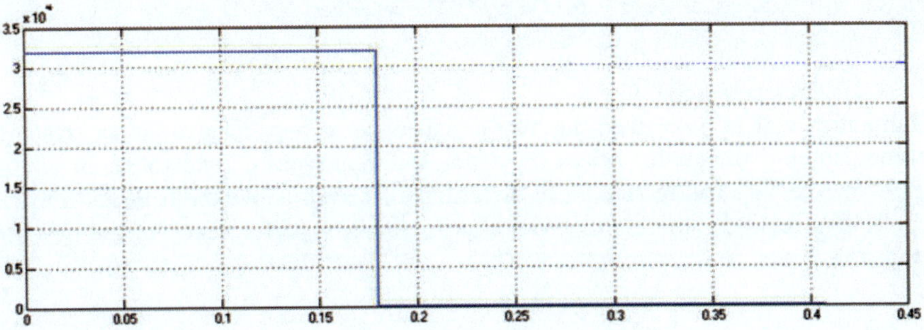

Fig. 13 Simulation result for active power under islanded condition

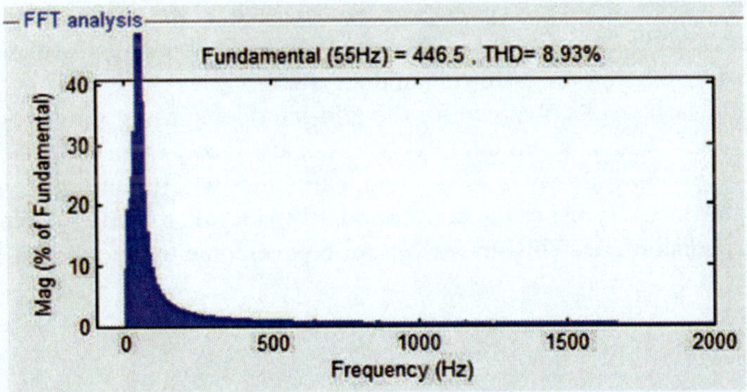

Fig. 14 FFT analysis for feeder current

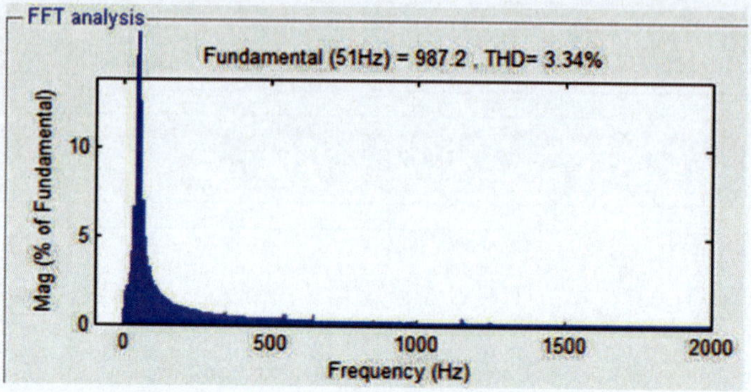

Fig. 15 FFT Analysis for feeder current with FUZZY controller

Case 2: with fuzzy controller

Harmonics will be generated due to the existence of non-linearity in the system accompanying voltage distortions. By using different types of available controllers in the market, we can reduce these distortions. As seen above, both PI and Fuzzy controllers seem to stumble to provide prime control performance of the system (Fig. 15).

5 Conclusion

This paper has successfully implemented the micro-grid-based unified power quality conditioner along with the fuzzy logic controller. Generally, the micro-grid concept mainly concentrates on the reduction of transmission losses, and the power quality problems that occurred in the system are compensated by the unified power quality conditioner. The fuzzy logic controller is used for getting better performance by the reduction of total harmonic distortion in the system.

The simulation results obtained for the grid-interfacing using series and parallel converter system with conventional PI controller and Fuzzy logic controller. Due to the presence of non-linearity in the system, harmonics will produce which leads to voltage distortions. By using the conventional PI controller in the system, we can reduce these distortions. This drawback can be overcome by adopting a fuzzy set theory.

References

1. Tey LH Member, IEEE, So PL Senior Member, IEEE, Chu YC Member, IEEE (2005) Improvement of power quality using adaptive shunt active filter. IEEE Trans Power Deliv 20(2), Apr
2. Wang F, Duarte JL, Hendrix MAM (2011) Grid-interfacing converter systems with enhanced voltage quality for microgrid application concept and implementation. IEEE
3. Wang F, Duarte JL, Hendrix MAM (2010) Pliant active and reactive power control for grid-interactive converters under unbalanced voltage dips. IEEE Trans Power Electron. in press
4. Farhangi H (2010) The path of the smart grid. IEEE Power Energy Mag 8(1):18–28
5. Fujita H, Akagi H (1998) The unified power quality conditioner: the integration of series- and shunt-active filters. IEEE Trans Power Electron 13(2):315–322 Mar
6. Silva S, Donoso-Garcia PF, Cortizo PC, Seixas PF (2002) A three phase line-interactive ups system implementation with series-parallel active power-line conditioning capabilities. IEEE Trans Ind Appl 38(6):1581–1590
7. Han B, Bae B, Kim H, Baek S (2006) Combined operation of unified power-quality conditioner with distributed generation. IEEE Trans Power Deliv 21(1):330–338 Jan
8. Tao H (2008) Integration of sustainable energy sources through power electronic converters in small distributed electricity generation systems. PhD dissertation, Eindhoven University of Technology
9. Guerrero JM, Vicuna LGD, Matas J, Castilla M, Miret J (2004) A wireless controller to enhance dynamic performance of parallel inverters in distributed generation systems. IEEE Trans Power Electron 19(5):1205–1213 Sept
10. Li YW, Kao C-N (2009) An accurate power control strategies for power-electronic-interfaced distributed generation units operating in a low-voltage multi bus micro grid. IEEE Trans Power Electron 24(12):2977–2988 Dec
11. Wang F, Duarte JL, Hendrix MAM (2008) Reconfiguring grid interfacing converters for power quality improvement. In: Proceedings of IEEE Benelux young researchers symposium \ in electrical power engineering, pp 1–6

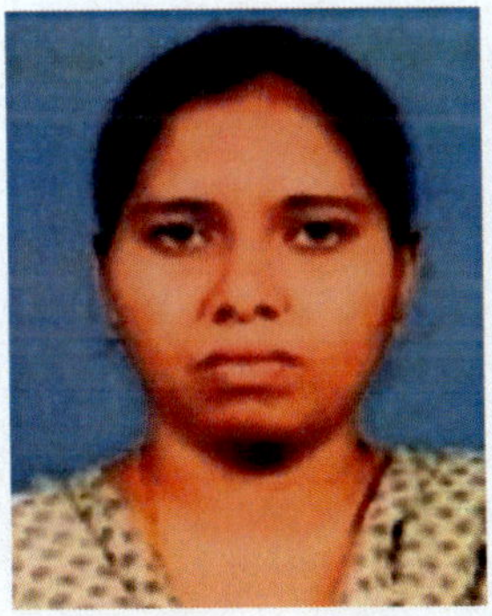

Mrs. V. Anantha Lakshmi received the B.Tech degree in Electrical and Electronics Engineering and M.Tech. Degree in power electronics and power systems from Nagarjuna University, Guntur, Andhra Pradesh, India in 2003 and 2008 respectively. She is currently pursuing Ph.D. degree at Andhra University, Visakhapatnam, India, and working as Asst. Professor in department of EEE, Andhra Loyola Institute Of engineering and technology, Vijayawada. Her research interests are FACTS, Smart grids, micro grids, and Power quality.

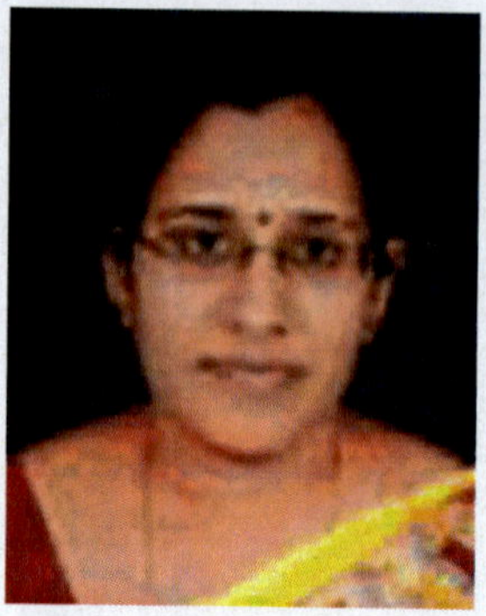

Dr. T. R. Jyothsna received her B.Tech, M.Tech, and Ph.D. from Andhra University, Visakhapatnam, India. She published a number of research papers in various national and international journals and presented papers in various conferences in and out of India. Presently, she is working as a Professor in the Department of Electrical Engineering, Andhra University. Her research interests are Smart grids, Power Quality, Power system stability, and FACTS controllers.

Detection and Prediction of Land Use and Land Cover Changes Using Deep Learning

Eliza Femi Sherley, Ajay Kumar, Revathy, and Divyashree

Abstract Our planet Earth is endowed with plenty of natural resources that sustain life for millennia. Nature provides us with five elements such as water, land, air, sky, and fire. Of these, land is an important element in nature to analyze the change in Earth. The land is covered with trees, crops, wetlands, water, and buildings. However, land use/land cover changes contribute to the major environmental challenges in various parts of the world. Globally, there has been an increase in the exploitation of land by humans due to industrialization and globalization, and it poses a major threat to the environment. Therefore, analyzing the land use/land cover (LU/LC) change has become an important and crucial issue to be solved all around the world. Identifying the physical aspect of the Earth's surface (Land cover) as well as how we exploit the land (Land use) is a challenging problem in environmental monitoring and many other subdomains. This can be done through field surveys or analyzing satellite images (Remote Sensing). While carrying out field surveys is more comprehensive and authoritative, it is an expensive project and mostly takes a long time to update. So to overcome these hurdles, we have proposed a deep learning-based approach to analyze the historical land use and land cover changes in a particular region from satellite images and use it to predict land use and land cover in the future. This prediction will be very useful for urban planning and environmental management of rapidly growing cities. The use of Geo-informatics is immensely helpful in accomplishing this task saving a huge amount of time and energy.

E. F. Sherley · A. Kumar (✉) · Revathy · Divyashree
Department of Information Technology, Anna University (MIT Campus), Chennai, India
e-mail: ajaykumar1998.ak29@gmail.com

E. F. Sherley
e-mail: elizafemis@gmail.com

Revathy
e-mail: susirevathy@gmail.com

Divyashree
e-mail: divyaanjushree@gmail.com

S. C. Satapathy et al. (eds.), *Communication Software and Networks*, Lecture Notes in Networks and Systems 134, https://doi.org/10.1007/978-981-15-5397-4_37

1 Introduction

Land cover and land use refer to the pattern of ecological resources and human activities dominating different areas of the Earth's surface. The main motivating factor of this study is to see the effect of human activities on nature over time. The importance of land use and land cover information for global land change and environmental sustainability research is realized worldwide, and there have been numerous efforts aiming to derive accurate land cover information at various scales mostly by using remote sensing technology. The spectral, temporal, and spatial information present in satellite image data collected over a particular time period offers unique opportunities for monitoring and managing land cover changes from local to global spatial scales. Therefore, our objective is to leverage the deep learning and machine learning techniques to analyze the satellite data and predict the land use and land cover changes in the near future. Land use is the human use of land for different purposes, and land cover refers to physical and biological cover on the surface of land. The land use and land cover pattern of a region is a result of natural and socio-economic factors and their utilization by man in time and space. Knowledge of land cover and land use change is important for many planning and management activities. Satellite remote sensing, in conjunction with Geographic Information Systems, has been widely applied and been recognized as a powerful and effective in detecting land use and land cover change. The significance of the proposed method to detect the land use and land cover classes for a geographic region and to predict the land use and land cover change in the near future using deep learning techniques lies in urban planning and natural resources management.

2 Related Work

In recent years, many image-processing techniques have been developed to classify different objects from a normal RGB image. But here, we propose a technique using multispectral images to classify the various land use and land cover classes based on remote sensing indices. Chen and Gao [1] proposed a method for object-based classification of Airborne Lidar data. Airborne Lidar data are traditionally used to classify ground and non-ground points and some methods have been designed to detect vegetation or building features extraction. The Airborne Lidar data for Cambridge city has been taken as input and pre-processing was performed to remove the noise and elevation outliers. Kamga Guy et al. [2] proposed a new hybrid system for satellite image classification that combines the deep features which has distinct information and generates a discriminative representation by preserving the essential information of original feature space. Features are extracted via transfer learning by the use of a pre-trained convolutional neural network. For this purpose, a single strategy of using fully connected layers to represent different levels of image features is applied. Secondly, an approach called entropy-controlled neighborhood component analysis is then proposed to optimize the fusion of multiple layers of different architectures in a unified hierarchical manner. Zhao et al. [3] proposed an object-based deep learning method to accurately classify the high-resolution imagery. These images are taken from Beijing, Pavia, and Vaihingen cities under the complex urban conditions. The proposed method has been built on a combination of a deep feature learning strategy and an object-based classification for the interpretation of high-resolution images.

3 Proposed Work

The proposed methodology as given in Fig. 1 introduces a new approach for detection and prediction of land use and land cover changes using deep learning. Advancements in remote sensing provide us satellite data in the form of multi-temporal images. These time-series data help us to predict the land changes in the near future. In order to process the time series data, we have used recurrent neural network (RNN), namely long short-term memory (LSTM). A recurrent neural network is a special type of network, which are able to remember the last states and can process the long sequences of input because of its recurrent connections. LSTM network is an extension of the recurrent neural network, which extends their memory and enables RNN to remember their inputs over a long period of time. So, we have proposed a new method for the prediction of land changes using LSTM.

The satellite images are acquired from Bhuvan website of ISRO and are pre-processed for sharpening the images and to get reliable data that will give appropriate information for further processing. From the pre-processed image, the unsupervised

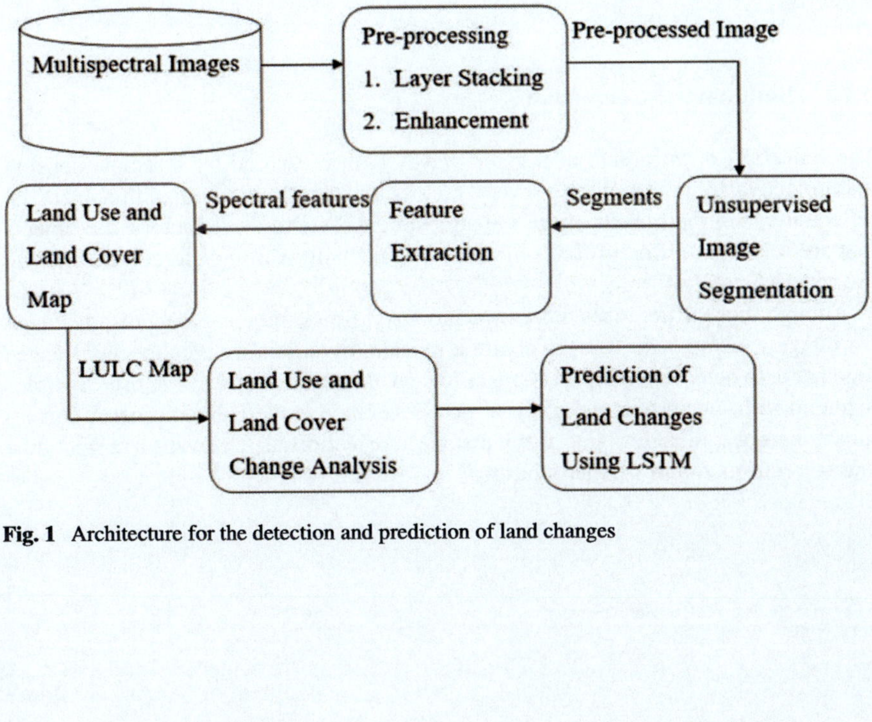

Fig. 1 Architecture for the detection and prediction of land changes

image segmentation has been performed and the spectral features are extracted using remote sensing indices such as Normalized Difference Vegetation Index (NDVI), Normalized Difference Water Index (NDWI), and Normalized Difference Built-up Index (NDBI). Then these segments are classified into water area, built-up area, vegetation, and sparse vegetation. The classified regions of multi-temporal images are analyzed to find the rate of change and it is used for the prediction using LSTMs.

3.1 Pre-Processing

Pre-processing of satellite images prior to image classification and change detection is essential. Pre-processing commonly comprises a series of sequential operations, including atmospheric correction or normalization, image registration, geometric correction, and masking (e.g., for clouds, water, irrelevant features). The goal conveniently should be that following image pre-processing, all images should appear as if they were acquired from the same sensor. It is achieved using ENVI software. The following are the image pre-processing methods applied.

3.1.1 Layer Stacking

Layer stacking is done by combining the different band images acquired from the Bhuvan website [4] as given in Fig. 2a. The four band images are in separate TIF formats, which are then stacked one behind another to form a single TIF file. So, we will be able to obtain the pixel information for each and every band easily for clustering.

3.1.2 Radiometric Correction

The reflection of an object at a given wavelength measured by a remote sensing instrument varies in response to several factors like the illumination of the object, its reflectivity, and the transitivity of the atmosphere. Due to these factors, the objects that are sensed at different times of a day or year will exhibit different radiometric characteristics.

Though these differences are advantageous at times, they can be a disadvantage for image analysts who want to create a mosaic, by adjoining neighboring images together or to detect meaningful changes in land use and land cover over time. In order to put up with such problems, analysts have developed numerous radiometric correction techniques, including Earth–sun distance corrections, sun elevation corrections, and corrections for atmospheric haze.

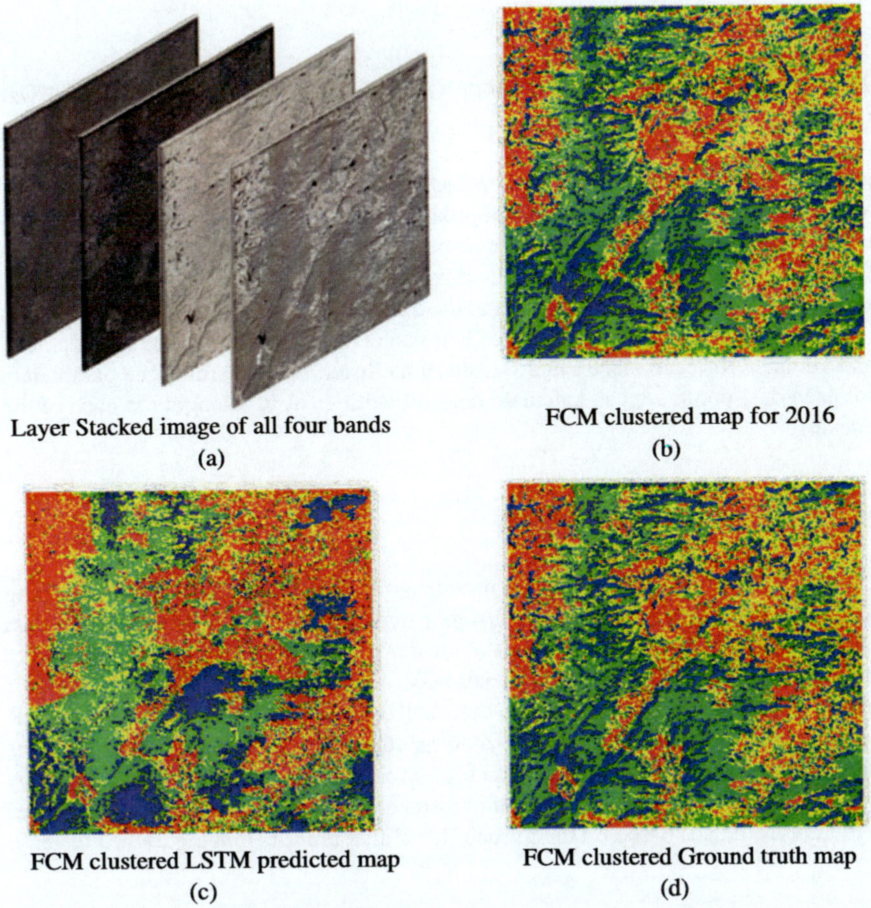

Fig. 2 FCM clustering and LSTM prediction performed on the multispectral images

3.2 Unsupervised Image Segmentation

In this architecture, unsupervised segmentation is considered as a clustering problem, i.e. given an image, produce a pixel-wise prediction that segments the image into coherent clusters corresponding to objects in the image due to the unavailability of ground truth data. K-means is a conventional clustering algorithm for data clustering, the data point is deterministically assigned to only one cluster, but in reality, there may be overlapping between the clusters. To overcome this problem, we have used the gaussian mixture model and fuzzy c-means clustering algorithm. We have then compared the two clustering algorithms under some performance measures and used the best one for further processing.

3.2.1 Gaussian Mixture Model

To address the overlapping problem of k-means, the Gaussian mixture model was introduced. Gaussian mixture models are a probabilistic model for representing normally distributed subpopulations within an overall population. Mixture models, in general, do not require knowing which subpopulation a data point belongs to, allowing the model to learn the subpopulations automatically. Since the subpopulation assignment is not known, this constitutes a form of unsupervised learning. In this approach, each cluster is described by its centroid, covariance, and the size of the cluster. We fit a set of k-Gaussian to the data, instead of identifying clusters by its "nearest" centroids. We also estimate Gaussian distribution parameters such as mean and variance for each cluster and weight of a cluster. After learning the parameters for each data point, we can calculate the probabilities of it belonging to each of the clusters.

3.2.2 Fuzzy C-Means Clustering

Fuzzy C-Means Clustering assigns membership to each data point corresponding to each cluster center based on the distance between the cluster center and the data point. The more the data is near to the cluster center more its membership toward the particular cluster center. Mathematically, an n-dimensional vector and an $m * n$ matrix are formed. Let us assume that matrix is $A[m][n]$. Let us say, we have K clusters. So, the problem becomes dividing the data into k clusters such that the distance between the centroid and each point is minimum. Here, the distance may be Euclidean distance or the Manhattan distance. Each point in a cluster is assigned a membership (uij) based on the probability of that point belonging to the cluster.

3.3 *Feature Extraction*

Feature extraction is the extraction of the necessary information from the pre-processed data to use for further analysis. Here, we extract the features from the band images using remote sensing indices. In satellite images, spectral band values are important for classification. So, the Spectral features are extracted from the images to calculate indices values. Based on the values the classes are labeled according to the maximum membership to a particular cluster and Land Use and Land Cover Map is generated for further processing. The various indices in remote sensing are, NDVI, NDWI, and NDBI as given in Eqs. (1), (2), and (3) which gives a value in the range of -1 to $+1$, where -1 signifies very less amount of presence of that particular class, and $+1$ refers to the abundance in that land use land cover class.

3.3.1 Normalized Difference Vegetation Index

NDVI is a Vegetation Index to monitor the vegetation or vegetation health in a region.

$$\text{NDVI} = \frac{\text{NIR} - \text{RED}}{\text{NIR} + \text{RED}} \tag{1}$$

3.3.2 Normalized Difference Water Index

Normalized Water Index is an index to extract water bodies from satellite imagery.

$$\text{NDWI} = \frac{\text{GREEN} - \text{NIR}}{\text{GREEN} + \text{NIR}} \tag{2}$$

3.3.3 Normalized Difference Built-up Index

Normalized Difference Built-up Index is used to extract built-up regions.

$$\text{NDBI} = \frac{\text{SWIR} - \text{NIR}}{\text{SWIR} + \text{NIR}} \tag{3}$$

3.4 *Land Use and Land Cover Map*

The band images are collected from RESOURCESAT [4] satellite which includes four bands (Visible Green, Visible Red, Near Infrared, Short Wave Infrared) collected from 2012 to 2016. Pre-processing tasks were performed. Unsupervised image segmentation, i.e. Gaussian mixture model clustering was performed and is compared with fuzzy C-means clustering, and the best clustering algorithm is chosen based on lesser intra-cluster distance. The remote sensing indices such as NDVI, NDBI, and NDWI are calculated for each of the clusters and the cluster with the maximum index is assigned as its respective class. For example, the cluster with the maximum water index is assigned as water. The land use and land cover map is then generated by giving colors to the pixels of different clusters in the image as shown in Fig. 2b.

Table 1 Percentage change of land use and land cover

Year\Class	Sparse vegetation	Built-up	Vegetation	Water
2012	31.61	20.28	40.54	7.57
2013	21.64	33.25	40.34	4.77
2014	24.16	28.26	43.39	4.19
2016	23.95	32.38	32.12	11.56

3.5 Prediction of Land Changes Using LSTM

In order to process the time-series data, we have used LSTM which is a recurrent neural network. All the four bands captured at a particular time are separately predicted for the future using LSTMs. Each band at different time periods is given as input to LSTM architecture which predicts the respective band image in the near future. After predicting the four bands, they are again given to the FCM clustering algorithm for separating the region into different clusters and the remote sensing indices are used to find the land class for every cluster as given in Fig. 2c.

3.6 Results and Discussion

The FCM clustering technique is compared with GMM clustering using the intra-cluster distance (i.e. the sum of average Euclidean distance between each of the clusters with their respective cluster centers), and it is shown that the intra-cluster distance of FCM is less than GMM leading to better clustering. The percentage of the land classes are then calculated as shown in Table 1 (by taking the ratio of the pixels belonging to each class to the total number of pixels in the image) to analyze their changes over the years and help us in making future decisions in urban planning, etc.

The entire programming is done in MATLAB running on a Windows Computer. The classified band images from the predictions given by LSTMs shown in Fig. 2(c) are compared pixel-wise with that of the original classified band images as shown in Fig. 2d in the near future. Comparing both the LULC maps, the percentage of similarity of the predicted map with the original map and is found to be 78.6% for that geographic region.

3.7 Conclusion

In this work, we have detected, analyzed, and predicted the land use and land cover changes using deep learning techniques and their performances are evaluated. The acquired satellite images are pre-processed using ENVI software to get the data required for analysis. We have compared Gaussian mixture Model and fuzzy C-Means Clustering for unsupervised segmentation. The FCM algorithm performs better than GMM because it has a lesser intra-cluster distance. Land use and land cover classification is carried out by calculating remote sensing indices for each cluster to label them according to the indices. Prediction is performed using LSTM neural network and the accuracy of the prediction is calculated by comparing the number of accurately classified pixels in the predicted LULC map with that of the pixels in the original LULC ground truth. Our experimental result gives 78.6% accuracy for LSTM. In the future, we plan to reduce the computation time of the algorithm from hours to minutes using mathematical optimization and also incorporate other techniques like DenseNet5 and W-Net in the project.

References

1. Chen Z, Gao B (2014) An object-based method for urban land cover classification using airborne lidar data. IEEE J Sel Top Appl Earth Obs Remote Sens 7(10):4243–4254, Oct
2. Kamga Guy F, Akram T, Laurent B (2018) A deep heterogeneous feature fusion approach for automatic land use classification. Inf Sci 467:199–218, Oct
3. Zhao W, Shihong D, Emery WJ (2017) Object-based convolutional neural network for high-resolution imagery classification. IEEE J Sel Top Appl Earth Obs Remote Sens 10(7):3386–3396
4. Bhuvan, NRSC. https://bhuvan-app3.nrsc.gov.in/data/download/index.php

A Study of Machine Learning Approaches to Detect Cyberbullying

Subbaraju Pericherla and E. Ilavarasan

Abstract Social media networks like Facebook and Twitter create a great platform to share public views, opinions, feelings by text message, image, video. The public is very much interested to use these networks because of the comfortable Graphical User Interface (GUI) by a single click and taps to share content from their electric gadgets, gizmos, and mostly by their smartphones. On the other hand, some people performing cyberbullying activities like aggressive comments, abusing, trolling. Sometimes, these negative activities lead to cyberbullying victims to attempt suicide. In this paper, the authors are presenting essential approaches to recognize cyberbullying over social media using advanced machine learning and deep learning algorithms.

1 Introduction

Cyberbullying can be defined as an aggressive behavior activity such as posting negative comments, sharing annoying information in online social media networks (OSN) [1]. OSN are the right podium for cyberbullies to target victims [2]. The main reason for choosing OSN is its ability to spread maximum content in a short period, which is not possible in traditional approaches like emails and phone text messages. A recent report was given by cyberbullying research organization that the average victimization rate is 27.8% for the past 30 days [3]. Cyberbullying is riskier than the traditional approach because it can occur irrespective of time and place online. Most of the cases, cyberbullying victims cannot express their problems to others [4, 5]. Recently, few studies were performed to improve cyberbullying detection [6, 7]. The different forms of cyberbullying are sharing aggressive comments, abusing, writing negative comments, trolling unwanted information among the group [8].

The remaining sections in this paper are as follows: Sect. 2—Need for developing Cyberbullying Prediction Models. Section 3—Methodology for Constructing

S. Pericherla (✉) · E. Ilavarasan
Department of CSE, Pondichery Engineering College, Pondicherry, India
e-mail: Raju.pericherla74@pec.edu

S. C. Satapathy et al. (eds.), *Communication Software and Networks*, Lecture Notes in Networks and Systems 134, https://doi.org/10.1007/978-981-15-5397-4_38

Cyberbullying Models. Section 4—Gaps in Cyberbullying Detection. Section 5—Conclusion and Future Scope.

2 Need for Developing Cyberbullying Predictions Models

Absconding from cyberbullying activities is very difficult because these types of activities occur irrespective of time and place. The impact of cyberbullying causes health-related problems in children and adolescents [9]. Research proved that cyberbullying affects public psychologically and physically. Some studies reported that victim takes the major chance of attempting suicide, and there is a correlation between victims and suicide attempts [9, 10]. Cyberbullies target social media users like Twitter, Facebook, Instagram, etc., because mass spread by sharing comments and negative opinions among the public [11]. In the era of OSN, there is a desperate need for automatic monitoring and prediction models for cyberbullying activities [12].

3 Methodology for Constructing Cyberbullying Prediction Models

Data collection: The major source for data collection is social networks like Twitter, Face book, and Instagram. Data can be extracted from these websites using authorized application program interfaces (APIs) by supplying keywords, hashtags, and user profiles [13]. The drawback of extracting data from SM's APIs is a limited amount of data returned [14].

Data Wrangling: Data wrangling is the process of structuring and cleaning the metadata into a meaningful format. It is one of the crucial steps in building the models because the extracted data may be in an unstructured format, which affects the performance of the model. It consists of tasks like finding missing values, identifying outliers in the raw data, and removal of unnecessary data. Data wrangling is also being referred to as the pre-processing of data (Fig. 1).

Feature Engineering: The performance of the predictive models depends on choosing the right feature set, which is used to train the model [15]. This task plays a vital role in building cyberbullying prediction models by machine learning algorithms [16, 17]. The right combination of feature set gives the effective classifier for the prediction of cyberbullying activity from machine learning algorithms [18]. Figure 2 displays some of the features used for cyberbullying prediction models [19].

Feature selection Techniques: Feature selection techniques are very useful to increase the performance of cyberbullying prediction models. Chi-square test and PCA are used to extract significant features from the set [20]. Information Gain explains the relation between training datasets and class labels [21]. Pearson correlation could be utilized to decrease the feature dimensionality in classification models

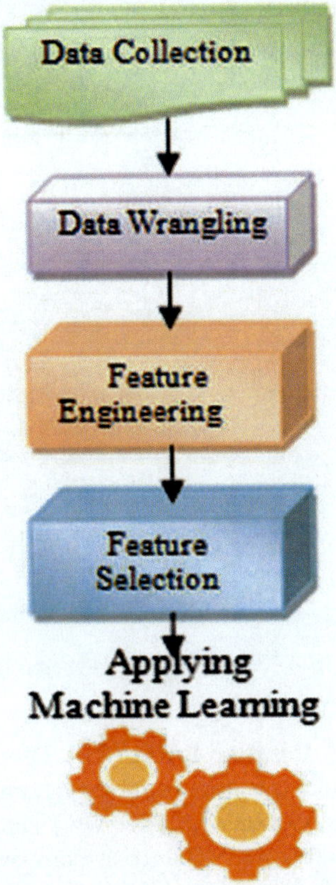

Fig. 1 Workflow of constructing cyberbullying prediction mode

[22]. The chi-square test is used to test independence between a feature and its class [23].

Machine learning Algorithms: Most of the machine learning algorithms follow three kinds of approaches for cyberbullying detection: (1) Supervised learning approach, (2) Lexicon-based approach, and (3) Rule-based approach.

Support Vector Machine (SVM): It is a kind of supervised classifier mostly used in the classification of text [24]. Chen et al. developed a prediction model for the offensive text in social media [25]. Mangaonkar et al. implemented the SVM model for the detection of cyberbullying on Twitter [26]. Dinakar et al. considered comments from YouTube to predict cyberbullying using SVM and concluded that SVM is more accurate than Naïve Bayes and J48 [27].

Naïve Bayes (NB): NB classifiers can be implemented by Bayes' theorem between features. It is mostly used for the classification of text. NB is used to build the prediction model for cyberbullying [28]. Shreyas Kumar et al. used the NB algorithm in Twitter Bullying Detection [29]. NB is a very frequently used algorithm in machine learning.

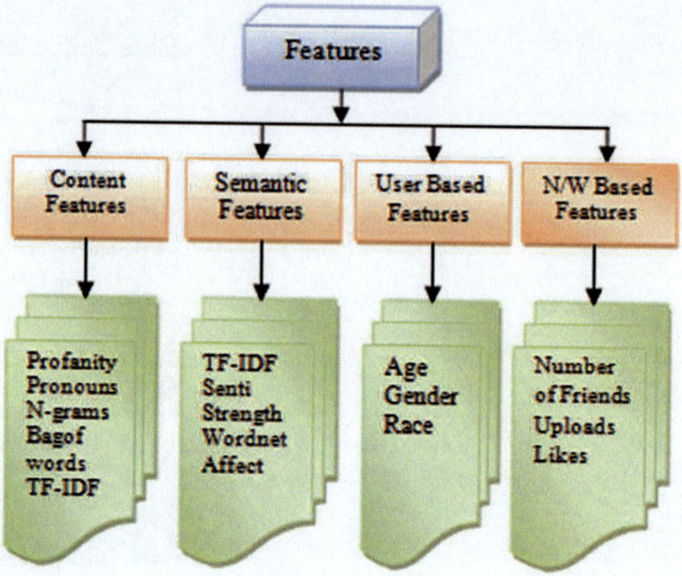

Fig. 2 Features used for cyberbullying detection

Random forest (RF): RF combines the decision tree and ensemble process. RF selects feature variable randomly for classification. RF was used to build cyber prediction models [30]. RF works based on the principle of bagging and improves classification performance as compared to other models. The advantages of RF are proper handling of missing values and the classifier model will not overfit.

Decision Tree (DT): C 4.5 is the most improved algorithm in DT, which is used in cyberbullying prediction models [31]. This initial step of C 4.5 follows the divide and conquers method. DT can be used for both regression and classification problems. Information gain and entropy are the two measures for the decision trees. DT can be used for both categorical and numerical data. Overfitting problem occurs due to noise data in decision trees.

K-nearest neighbor (KNN): KNN is a basic non-parametric classification algorithm in machine learning. It uses Euclidean distance as a parameter. KNN classifier is used for the prediction of cyberbullying messages in Turkish [32]. KNN is also called as a lazy learner because it doesn't involve in any training process. The disadvantage of KNN is the processing time becomes slow when the volume of data increases.

Logistic regression (LR): LR is a static technique used for classification. It uses the logistic sigmoid function to transform its output and return a probability value. LR separates the hyperplane between two datasets using the logistic function. LR takes features (variables) as inputs and generates probability value in the output. If the probability value is >0.5, the classification is positive class, otherwise negative class. Figure 3 shows the frequency of algorithms used in cyberbullying (Fig. 4).

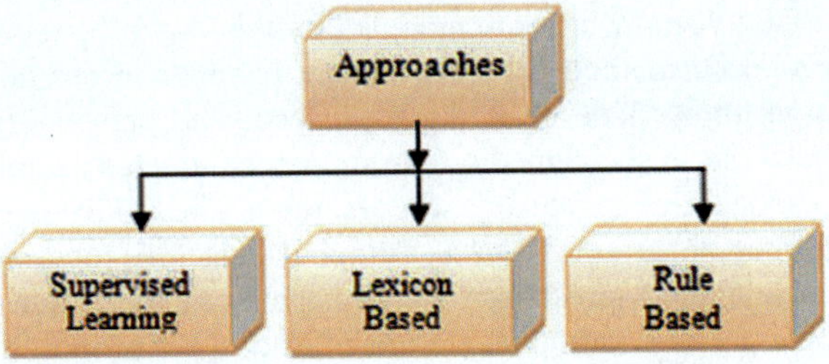

Fig. 3 Cyberbullying prediction models

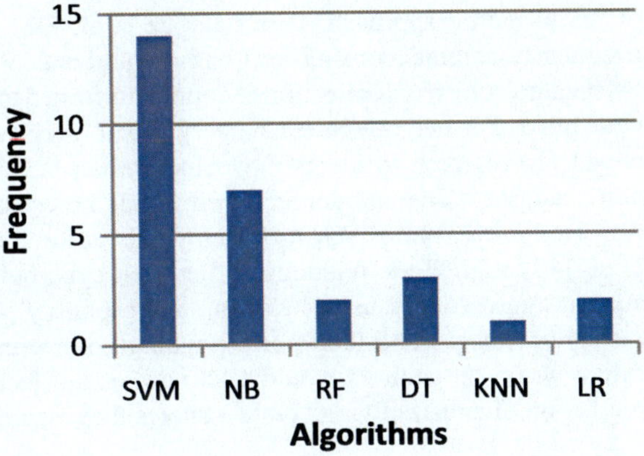

Fig. 4 Frequency of ML algorithms [34]

4 Research Graphs in Cyberbullying Detection

Subramani et al. [33] considers Domestic Violence(DV) as one kind of cyberbullying activity and applied a deep learning approach for identification of DV victims who are in desperate need based on their Facebook posts. The authors tried to classify the posts into two categories called "Critical" and "Non-Critical." The proposed model achieved an accuracy of 94%. In the same way, Michalska et al. [34] implemented advanced deep learning techniques such as LSTMs, GRUs, and BLSTMs to classify the DV posts in multi-level categories. The authors considered only 1655 posts due to manual annotation. They can achieve 92% accuracy. Mahata et al. [33] implemented a deep learning classifier for the identification personal intake of medicine from Social networks such as Twitter and Facebook.

Authors' main objective is to classify the tweets into three categories named Class 1 (personal meditation intake), Class 2 (possible medication intake), and Class 3 (non-intake). They proposed an ensemble of five CNN architectures that produce F1-score of 0.693. Galan Garcia et al. [35] implemented a methodology to identify troll profiles on Twitter. Authors collected twitter profiles and different features such as tweets, geo-location, language, and tweeted time from Twitter API, and were compared with different classification algorithms such as Random Forest, J48, K-Nearest Neighbor, and Sequential Minimal Optimization using the WEKA tool. The authors collected

1900 tweets, which are modeled by the Vector Space Model (VSM) to evaluate the proposed method. The experiment results showed that SMO and DT are the best algorithms. Tadesse et al. [28] investigated personality predictions based on user's behavior in social media. The authors conducted a comparative study to know user's personalities on the same set of features. The XG Boost classifier outperforms with the highest accuracy of 74.2%. Tyshchuk et al. [36] focused to study human behaviors on social media in response to significant events such as natural disasters, terror attacks, emergency conditions, etc.. Fazil et al. [37] proposed a hybrid approach for spam detection in Twitter by adding 19 special features. The authors used three classifiers, Random Forest, Decision Tree, and Bayesian network, to study the performance of the proposed approach. Pastor Lopez et al. [38] proposed a method to detect irregularity comments using troll-based comments. Authors gathered dataset from 'Meneame' which is a collection of comments from 5 to 12 April 2011. The comments were labeled as two categories, namely "Troll" and "Not Troll." Margono et al. [39] conducted research to identify two Indonesian bullying words called as "anjing" and "bangsat." The authors analyzed trends between the Indonesian words using Association rule and FP-Growth techniques in the RapidMiner tool. Hosseinmadri et al. [40] studied the relationship between cyberbullying and its related features such as aggression, abusive behavior, and profanity [41–43]. The authors collected Instagram images and their corresponding comments as dataset. As part of their future work, the authors would take new image features and sensor data for the prediction of cyberbullying. Table 1 shows the research gaps in Cyberbullying.

5 Conclusion and Future Scope

Cyberbullying provokes most of the netizens for suicide attempts. In this connection, this research focuses on the study of various methods and approaches used to detect cyberbullying activities in social media posts through machine learning. Most of the existing techniques are not considering the sarcastic text, considers only a limited number of features in the content for detecting the cyberbullying activity. In our future work, existing algorithms will be improved to get rid of these limitations for the effective detection of harmful and provoking posts for cyberbullying.

Table 1 Summarization of research gaps in Cyberbullying

S.N	Authors	Techniques/Tools/Datasets used	Limitations
1	Subramani et al. [33]	**Techniques**: Deep Learning (DL), tenfold cross-validation to finds accuracy **Datasets**: (a) word2vec (b) Glove	(1) Limited size data set (2) Manual labeling of post (3) Applied only on Facebook posts (4) Feature extraction algorithm not proposed
2	Michalska et al. [35]	**Techniques**: DL threefold cross-validation used to finds performance	(1) Small corpus dataset (2) Manual annotation process
3	Mahata et al. [28]	**Techniques**: DL fivefold cross-validation finds accuracy **Dataset**: AMIA 2017 site	(1) Pre-trained word embedding
4	Galan-Garcia et al. [36]	**Techniques**: Machine learning approach, **Dataset**: Twitter API, **Tools**: WEKA,	(1) Analysis of specific features on troll profile also language features in tweets (2) Write a style of the abusive user
5	Taddesse et al. [37]	**Techniques**: Machine learning approach, XGBoost classifier **Dataset**: myPersonality	(1) Small dataset size
6	Tyshchuk et al. [38]	**Techniques**: Cross-lingual trigger clustering approach **Dataset**: Twitter API,	(1) Framework designed only for social media
7	Fazil et al. [39]	**Technique**: SMOTE for balance the dataset **Dataset**: Twitter Weka tool	(1) No specific feature set is defined to detect spam
8	Pastor-Lopez et al. [40]	**Techniques**: VSM approach, fivefold cross-validation **Dataset**: Meneame new site	(1) Labeling of text with a machine learning algorithm is a very time-consuming process
9	Margono et al. [41]	**Algorithm**: FP-Growth **Dataset**: Indonesian Twitter **Tools**: RapidMiner	(1) Limited to identify only specific words
10	Hosseinmadri et al. [42]	**Algorithm**: SVM **Techniques**: Snowball sampling for data collection **Dataset**: Instagram comments	(1) Implement only for highly negative media sessions (2) Crowd-source labeling

References

1. Smith PK et al (2008) Cyberbullying: its nature and impact in secondary school pupils. J Child Psychol Psychiatry 49(4):376–385
2. Hinduja S et al (2019) Cyberbullying research center. Kowalski RM et al (2007) Electronic bullying among middle school students. JAH 41(6):S22–S30
3. Mason KL (2008) Cyberbullying: a preliminary assessment for school personnel. Psychol School 45(4):323–348
4. Huang Q, Singh VK, Atrey PK (2014) Cyber bullying detection using social and textual analysis. SAM 14. ACM, New York, USA, pp 3–6
5. Squicciarini A, Rajtmajer S, Liu Y, Griffin C (2015) Identification and characterization of cyberbullying dynamics in an online social network. In: Proceedings of the 2015 IEEE/ACM international conference, ACM, pp 280–285
6. Watters PA et al (2012) Detecting illicit drugs on social media using automated social media intelligence analysis in Cyberspace safety and security., pp 66–76
7. O'Keeffe GS, Clarke-Pearson K (2011) The impact of social media on children, adolescents, and families. Pediatrics 127(4):800–804
8. Kowalski RM, Limber SP (2013) Psychological, physical, and academic correlates of cyberbullying and traditional bullying. J Arts Humanit 53(1):S13–S20
9. Sampasa-Kanyinga H et al (2014) Associations between cyberbullying and school bullying victimization and suicidal ideation, plans and attempts among Canadian schoolchildren. PloS one 9(7)
10. Whittaker E et al (2015) Cyberbullying via socialmedia. J Sound Vib 14(1):11–29
11. Van Royen K et al (2016) Harmonizing freedom and protection: Adolescents' voices on automatic monitoring of social N/W sites. Youth Serv Rev 64:35–41
12. Bellmore A et al (2015) The five W's of "bullying" on Twitter: who, what, why, where, and when. Comput Hum Behav 44:305–314
13. González-Bailón S et al (2014) Assessing the bias in samples of large online networks. Social Netw 38:16–27
14. Ghahramani Z (2015) Probabilistic machine learning and artificial intelligence. Nature 521(7553):452–459
15. Dadvar M et al (2012) Improved cyberbullying detection using gender information
16. Hosseinmardi H et al (2015) Detection of cyberbullying incidents on the instagram social network. arXiv preprint. http://arxiv.org/abs/1503.03909
17. Dong W et al (2016) Leading effect of social media for financial fraud disclosure: a text mining based analytics
18. Semiu Salawu et al (2017) Approaches to automated detection of cyberbullying: a survey. IEEE Trans Effective Comput
19. Chavan VS, Shylaja S (2015) Machine learning approach for detection of cyber aggressive comments by peers on social media network. In: Advances in computing, communications and informatics (ICACCI), 2015 International Conference on. IEEE
20. Qabajeh I et al (2014) An experimental study for assessing email classification attributes using feature selection methods. In: ACSAT, 3rd international conference on. IEEE
21. Benesty J et al (2009) Pearson correlation coefficient. In Noise reduction in speech processing. Springer, pp 1–4
22. Zheng Z, Wu X, Srihari R (2004) Feature selection for text categorization on imbalanced data. ACM SIGKDD Explor Newsl 6(1):80–89
23. Joachims T (1998) Text categorization with support vector machines: learning with many relevant features. In European conference on machine learning. Springer
24. Chen Y et al (2012) Detecting offensive language in social media to protect adolescent online safety. In Privacy, security, risk and trust (PASSAT), 2012 international conference and international conference on social computing (SocialCom). IEEE
25. Mangaonkar A et al (2015) Collaborative detection of cyberbullying behavior in Twitter data. In International conference on electro/information technology (EIT). IEEE

26. Dinakar K, Reichart R, Lieberman H (2011) Modeling the detection of textual cyberbullying
27. Galán-García P et al (2014) Supervised machine learning for the detection of troll profiles in twitter social network: application to a real case of cyberbullying. In International joint conference SOCO'13-CISIS'13-ICEUTE'13. Springer
28. Tyshchuk Y et al (2018) Modeling human behavior on social media in response to significant events. IEEE Trans
29. Van Hee C et al (2015) Detection and fine-grained classification of cyberbullying events. In International conference recent advances in natural language processing (RANLP)
30. Reynolds K et al (2011) Using machine learning to detect cyberbullying. In: ICMLA, 2011 10th international conference on. IEEE
31. Selma Ayse Ozel et al (2017) Detection of cyberbullying on social media messages in Turkish. In: 2017 International conference on computer science and engineering, October
32. Ali M et al (2019) Predicting cyberbullying on social media in the bigdata era using ML Algorithms: Review of Literature and Open challenges. IEEE
33. Galan-Garcia P et al (2015) Supervised machine learning for the detection of troll profiles in twitter social network: application to a real case of cyberbullying, vol. 24. OU
34. Mahata D et al (2018) Detecting personal intake of medicine from Twitter. IEEE Intell Syst, August
35. Taddesse MM et al (2018) Personality predictions based on user behaviour on the facebook social media platform. IEEE Access
36. Fazil M et al (2018) A hybrid approach for detecting automated spammers in Twitter. IEEE Trans Inf Forensics Secur
37. Pastar Lopez I et al (2016) Anomaly-based user comments detection in social news websites using troll user comments as normality representation. OU Press
38. Hendro et al (2014) Mining Indonesian cyber bullying patterns in social networks. In: 37th Australasian computer science conference, Auckland, New Zealand
39. Homa et al (2015) Detection of cyberbullying incidents on the instagram social sites
40. Founta AM et al (2019) A unified deep learning architecture for abuse detection. In: 10th ACM conference on web science, pp 105–114, July
41. Sanchez H et al (2011) Twitter bullying detection. UCSCISM245 DM course report
42. Subramani S et al (2016) Domestic violence crisis identification from facebook posts based on deeplearning, vol 4. IEEE Access
43. Michelska S et al (2019) Deep learning for multi-class identification from domestic violence online posts, vol 7, IEEE Access

Dynamic and Collaborative Group Key Generation with Quadtree-Based Queue-Batch Algorithm

V. Srinadh and P. V. Nageswara Rao

Abstract Secure communication is very much needed. Maintaining secrecy in the communication of members of a group is a must. Without Security, group communication is not preferable. In general, A secret key is used to communicate the members of any Group [1]. This secret key is called Group Key. In group communication, if we keep the same group key for a long time hackers can discover it. In this context, the secret group key has to be altered frequently. This can be done for each join operation as well as for each leave operation. We can maintain a key tree so that we can easily calculate the group key after each join or leave the event. Height of the key tree can be reduced if we use a quadtree to represent the key tree instead of a ternary tree. Rekeying time depends totally on the height of the key tree. For the same number of group members, the key tree height of quadtree is less than the key tree height of the ternary tree. We propose quadtree representation of group members so that the height is less rekeying operations will be done quickly. Here, we are going to implement quadtree-based Queue-Batch Algorithm.

1 Introduction

Security is basic employment in any correspondence system, especially in Group arranged correspondence. In social event-centered correspondence structure, entire correspondence will occur with the help of one key which is called Group Key. This social affair key must be changed at whatever point another part joins into the get-together. With the objective that the joined people can't get to the past conferred data. Similarly, as social affair key must be balanced at whatever point a present

V. Srinadh (✉)
Computer Science and Engineering, GMR Institute of Technology, Jawaharlal Nehru Technological University, Kakinada, Andhra Pradesh, India
e-mail: srinadh.v@gmrit.edu.in

P. V. Nageswara Rao
Computer Science and Engineering, GITAM Institute of Technology, GITAM University, Visakhapatnam, Andhra Pradesh, India
e-mail: nagesh@gitam.edu

S. C. Satapathy et al. (eds.), *Communication Software and Networks*, Lecture Notes in Networks and Systems 134, https://doi.org/10.1007/978-981-15-5397-4_39

part leaves from the get-together with the objective that the left people can't get to the further correspondence data. Changing the social occasion key at whatever point another part joins into the get-together and a present part leaves from the get-together is known as rekeying. This social occasion correspondence can be addressed using a key tree. In case we use ternary tree, height of the tree will be extended if the people are more, so that rekeying action takes extra time. As opposed to ternary tree, in case we use the quadtree the tree height will be less, so that the rekeying movement takes less time. To achieve this, we need an ensured scattered assembling key comprehension and approval show with the objective that people can set up and affirm a run of the mill gathering key for secure and private correspondence. The first key understanding show was proposed by Diffie–Hellman. It can guarantee the security of correspondence between the members [2].

2 Quadtree-Based Group Diffie–Hellman Protocol

Tree-based group Diffie–Hellman (TGDH) protocol is used to maintain group key in a group communication [3]. Each member of the group should maintain two keys: secret key and blinded key. These are maintained using a quadtree. We allocate a vertex digit v to all vertices of the tree. We can represent K_v as secret key and BK_v is an open key of vertex v. Public key is computed as follows:

$$\mathrm{BK}_v = \alpha^{K_v} \bmod p \tag{1}$$

The last level vertex of the tree relates to the secret and public keys of gathering part Mi. Each part keeps all the mystery keys next to its keyway beginning from leaf vertex up to the root vertex. Consequently, the mystery key held by the root vertex is shared by every one of the individuals and is viewed as the group key [2]. Zero will be set as a vertex ID of the root vertex. Each nonleaf vertex comprises of either four youngster hubs whose vertex IDs are $4v$+1, $4v$+2, $4v$+3, and $4v$+4 for its left kid, center 1, center 2, and right kids separately; or three hubs whose vertex IDs are $4v$+1, $4v$+2, and $4v$+3 for its left, center, and right kids separately; or three hubs whose vertex IDs are $4v$+1, $4v$+2, and $4v$+3 for its left, center, and right kids individually; or two kid hubs whose vertex ID's are given by $4v$+l and $4v$+2 for its left and center kids separately. In view of the Diffie–Hellman convention [4], the mystery key of a nonleaf vertex can be produced, in three ways: as pursues:

Case 1: Nonleaf node with two child nodes

Give v a chance to be the nonleaf vertex whose two youngsters are $4v$+l and $4v$+2, then the mystery key of v can be determined by the mystery key of one kid vertex of v and blinded key of another kid vertex of v. Logically, we have

$$\begin{aligned} K_v &= (\mathrm{BK}_{4v+1})^{K^{4v+2}} \bmod P \\ &= (\mathrm{BK}_{4v+2})^{K_{4v+1}} \bmod P \\ &= \alpha^{K_{4v+1}K_{4v+2}} \bmod P \end{aligned} \tag{2}$$

Case 2: Nonleaf node with three child nodes

Give v a chance to be the nonleaf vertex whose three youngsters are $4v$+l, $4v$+2 and $4v$+3, then the mystery key of v can be determined by the mystery key of one kid vertex of v and blinded key of other two kid hubs of v [5]. For example, with the help of the secret key of node $4v$+l and blinded key of node $4v$+2, we can generate the secret key according to the Eq. (2); using this secret key and blinded key of node $4v$+3, we can generate the secret key of the node v using the same Eq. (2). Applying Eq. (2) for two times can be replaced by Eq. (3). Out of three child nodes, we can use any child node's private key and public key of the other two child nodes. All possible cases can be represented mathematically as follows:

$$\left.\begin{array}{c} K_v = (\mathrm{BK}_{4v+3})^{((\mathrm{BK}_{4v+2})^{K_{4v+1}} \bmod P)} \bmod P \\ \text{OR} \\ K_v = (\mathrm{BK}_{4v+2})^{((\mathrm{BK}_{4v+3})^{K_{4v+1}} \bmod P)} \bmod P \\ \text{OR} \\ K_v = (\mathrm{BK}_{4v+1})^{((\mathrm{BK}_{4v+3})^{K_{4v+2}} \bmod P)} \bmod P \\ \text{OR} \\ K_v = (\mathrm{BK}_{4v+3})^{((\mathrm{BK}_{4v+1})^{K_{4v+2}} \bmod P)} \bmod P \\ \text{OR} \\ K_v = (\mathrm{BK}_{4v+2})^{((\mathrm{BK}_{4v+1})^{K_{4v+3}} \bmod P)} \bmod P \\ \text{OR} \\ K_v = (\mathrm{BK}_{4v+1})^{((\mathrm{BK}_{4v+1})^{K_{4v+3}} \bmod P)} \bmod P \end{array}\right\} \tag{3}$$

Case 3: Nonleaf node with four child nodes

Give v a chance to be the nonleaf vertex whose three youngsters are $4v$+l, $4v$+2 = $4v$+3 and $4v$+4, then the mystery' key of v can be determined by the mystery key of one kid vertex of v and blinded key of other three kid hubs of v. For example, with the help of secret key of node $4v$+l and blinded key of node $4v$+2, we can generate secret key according to the Eq. (2); using this secret key and blinded key of node $4v$+2, we can generate the secret key of the node v using the same Eq. (2). Using this secret key and blinded key of node $4v$+3, we can generate the secret key of node v using the same Eq. (2). Applying Eq. (2) for three times can be replaced by applying Eq. (3) for two times. Same thing can be done by applying Eq. (4) only one time. Out of four child nodes, we can use any child node's private key and public key of the other two child nodes. All possible cases can be represented mathematically as follows:

$$K_v = \left.\begin{array}{l} \left(\mathrm{BK}_{4v+1}{}^{K_{4v+2}} \bmod P\right)^{((\mathrm{BK}_{4v+3})^{K_{4v+4} \bmod P)} \bmod P \\ \text{OR} \\ = \left(\mathrm{BK}_{4v+1}{}^{K_{4v+2}} \bmod P\right)^{((\mathrm{BK}_{4v+4})^{K_{4v+3} \bmod P)} \bmod P \\ \text{OR} \\ = \left(\mathrm{BK}_{4v+1}{}^{K_{4v+1}} \bmod P\right)^{((\mathrm{BK}_{4v+3})^{K_{4v+4} \bmod P)} \bmod P \\ \text{OR} \\ = \left(\mathrm{BK}_{4v+1}{}^{K_{4v+1}} \bmod P\right)^{((\mathrm{BK}_{4v+4})^{K_{4v+3} \bmod P)} \bmod P \end{array}\right\} \quad (4)$$

3 Quadtree-Based Queue-Batch Algorithm

We have found that before rekeying calculations like ternary tree-based Rebuild calculation and ternary tree-based Batch calculation play out all rekeying ventures toward the start of each rekeying interim [6]. It prompts high preparing overhead during update occurrence thus the beginning of the gathering correspondence will be delayed. That is the reason we propose a progressively powerful calculation which we call quadtree-based Queue-Batch calculation. The idea in ternary tree-based Queue-Batch computation is to diminish rekeying load by preprocessing the joining people during the inactive rekeying between time.

Quadtree-based Queue-Batch calculation is isolated into two stages:

(i) Queue-subtree
(ii) Queue-consolidate

3.1 *Queue-Subtree Phase*

Queue-subtree phase happens when another part joins the gathering during the rekeying interim. During this stage, we add this recently joining part in an impermanent key tree T′. Algorithm for Queue-subtree stage is as follows:

```
Algorithm Queue-subtree (T')
  { if (a new member is there to join)
  { if (T'!= NOT NULL)
Build a new quad tree T' which consists of this newly joining member; else /*Already there exists some
members in T' */{ Discover the node where to be inserted; Now include this newly joining member to
ternary tree T';Choose the furthest right part under the subtree established at the child of the joining
vertex to be a support;
if(support)
Rekey the reestablished nodes and communicate new open keys;
}// end of else
}// end of if
}// end of algorithm
```

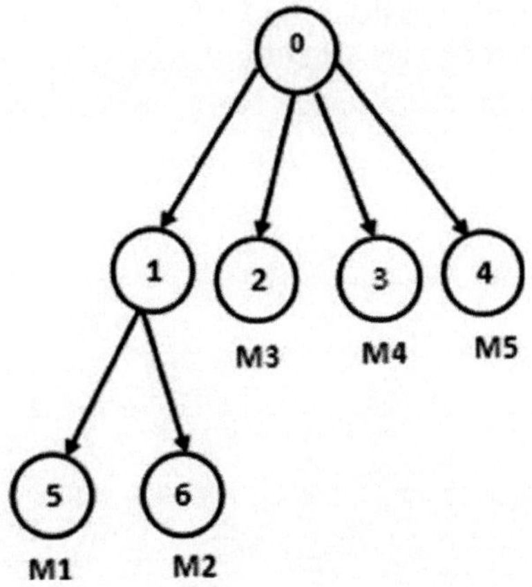

Fig. 1 Subtree when 5 members join the group

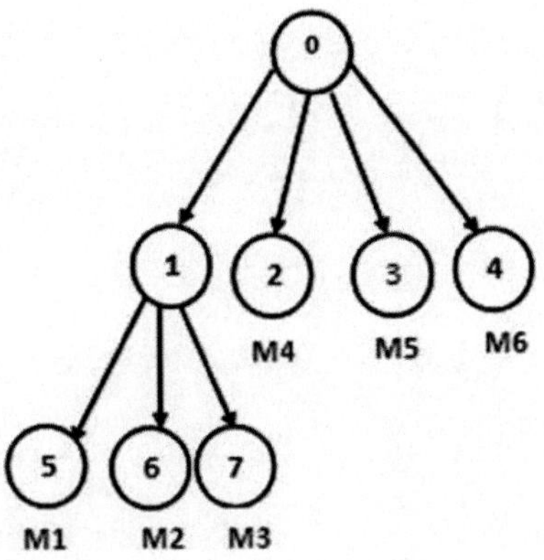

Fig. 2 Subtree when 6 members join the group

While generating group key if 2 or 3 or 4 members join the group, then subtree is created simply with these 2 or 3 or 4 members as children, respectively. But if 5 children join the group, then the subtree is as shown in Fig. 1. If 6, 7, and 8 children join the group, then the subtrees are as shown in Figs. 2, 3, and 4, respectively.

3.2 *Queue-Merge Phase*

Queue-merge phase happens toward the start of each rekeying interim. In this stage, we consolidate the brief tree T′ with the current key tree T. Algorithm for Queue-merge phase is as follows:

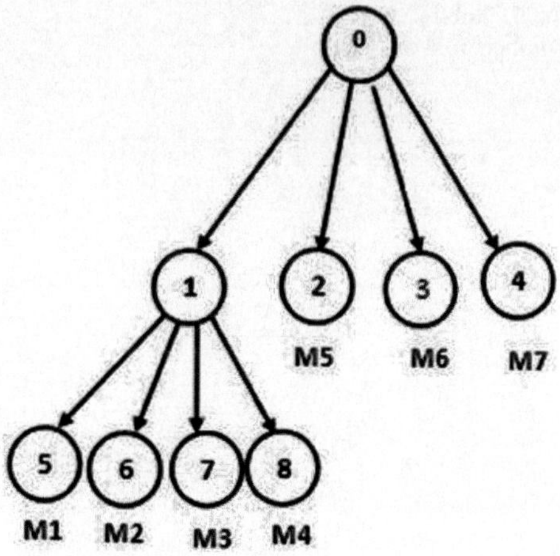

Fig. 3 Subtree when 7 members join the group

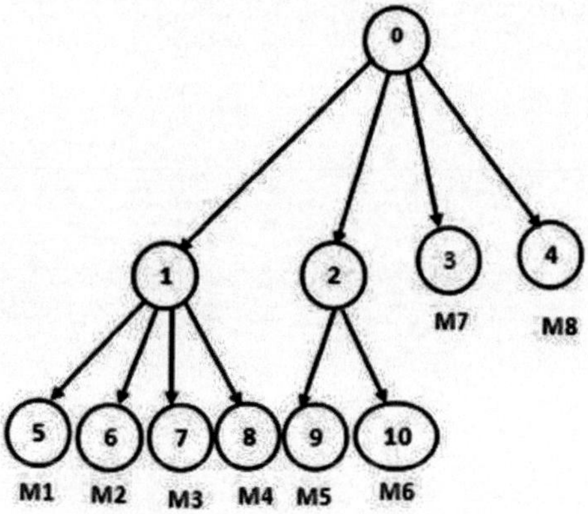

Fig. 4 Subtree when 8 members join the group

```
Algorithm Queue-merge(T, T', Ml, L)
{
  if (L==0) /* number of members want to leave is zero */
    {
          Add T' to either the shallowest node of quad tree T such that the merge will not
          increase the height of resulting tree, or the root node of T if the merge to
          any locations will increase the height of the resulting tree.
    }
  else
    {
          Add T' to the highest leave position of the key tree T;
          Remove remaining L-1 leaving leaf nodes and promote their siblings;
    }
```

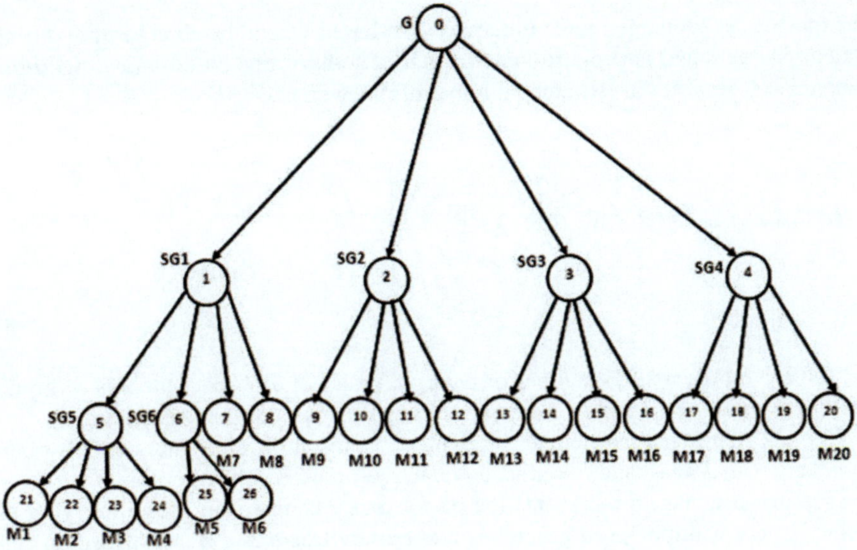

Fig. 5 Key tree with 20 members

Consider the key tree with 20 members as shown in Fig. 5.

Consider the situation where six members (M21, M22, M23, M24, M25, M26) join the group, while four members (M2, M5, M14, M19) leave the group. In this case, the resultant tree will become as shown in the following figure. Here, except for SG2, remaining all subgroup keys will be updated. These subgroup keys will be calculated by using Eqs. (2), (3), and (4). Finally, the secret key at root node 0 will be calculated. The secret key at root node will become a new group key. Here onwards, any correspondence will be finished utilizing this new gathering key until there is no join or leave occasion occurs.

4 Mathematical Analysis and Experimental Results

The primary thought of the ternary tree-based Queue-batch calculation misuses the inert rekeying interim to pre-process some rekeying activities. At the point when we contrast its presentation and ternary tree-based Batch calculation, we have to consider just the rekeying activities happening toward the start of each rekeying interim. When there are no joining individuals, ternary tree-based Queue-Batch algorithm is equal to the unadulterated leave instance of ternary tree-based Batch algorithm. For $J > 0$, the quantity of recharged hubs in ternary tree-based Queue-batch algorithm during Queue-merge stage is proportionate to that of ternary tree-based batch algorithm when $J = 1$. Hence, the estimated number of renewed nodes is

$$E[R_{\text{Queue - batch}}] = \begin{cases} 0, & if\ J = 0,\ L = N \\ \sum_{i=0}^{h-1} 4^{l}\left[1 - \frac{C(N-\frac{N}{4^{l}},L)}{C(N,L)}\right] - L & if\ J = 0, 0 \leqslant L < N \\ \sum_{i=0}^{h-1} 4^{l}\left[1 - \frac{C(N-\frac{N}{4^{l}},L)}{C(N,L)}\right] - (L-1), & \text{Otherwise} \end{cases} \tag{5}$$

Table 1 represents the number of changed nodes at different numbers of group members with $J = 17$ and $L = 3$ using ternary tree-based Batch Algorithm:

Table 2 represents the number of changed nodes at different numbers of group members with $J = 17$ and $L = 3$ using quadtree-based Batch Algorithm:

The graph in Fig. 6 represents the performance comparison of the rekeying of nodes using quadtree-based group key and ternary tree-based Batch algorithm and our proposed algorithm (quadtree-based Batch Algorithm). Here, we have considered $J = 17$ and $L = 3$ (Fig. 7).

5 Conclusion

To speak to group individuals in the event that we utilize the ternary tree, stature of the tree will be expanded when the individuals are more, so that rekeying activity takes additional time. Rather than ternary tree in the event that we use quadtree the tree tallness is less contrasted with ternary tree, so that rekeying activity takes less time. So, the correspondence will be progressively secure and quicker.

Table 1 No. of transformed nodes at different number of group members with $J = 17$ and $L = 3$ using Ternary tree-based Batch Algorithm

N	J	L	l	$W = 3^{\wedge}l$	$N/3^{\wedge}l$	N-$(N/3^{\wedge}l)$	$X = \mathrm{cmb}(N\text{-}N/3^{\wedge}l,L)$	$Y = \mathrm{cmb}(N,L)$	$Z = 1-X/Y$	$W*Z$	$\sum(W*Z)$
20	17	3	1	3	6.67	13.33	286	1140	0.75	2.25	4
20	17	3	2	9	2.22	17.78	680	1140	0.4	3.6	
30	17	3	1	3	10	20	1140	4060	0.72	2.16	9
30	17	3	2	9	3.33	26.67	2600	4060	0.36	3.24	
30	17	3	3	27	1.11	28.89	3276	4060	0.19	5.13	
40	17	3	1	3	13.33	26.67	2600	9880	0.74	2.22	8
40	17	3	2	9	4.44	35.56	6545	9880	0.34	3.06	
40	17	3	3	27	1.48	38.52	8436	9880	0.15	4.05	
50	17	3	1	3	16.67	33.33	5456	19600	0.72	2.16	7
50	17	3	2	9	5.56	44.44	13244	19600	0.32	2.88	
50	17	3	3	27	1.85	48.15	17296	19600	0.12	3.24	
60	17	3	1	3	20	40	9880	34220	0.71	2.13	7
60	17	3	2	9	6.67	53.33	23426	34220	0.32	2.88	
60	17	3	3	27	2.22	57.78	29260	34220	0.14	3.78	
70	17	3	1	3	23.33	46.67	15180	54740	0.72	2.16	7
70	17	3	2	9	7.78	62.22	37820	54740	0.31	2.79	
70	17	3	3	27	2.59	67.41	47905	54740	0.12	3.24	
80	17	3	1	3	26.67	53.33	23426	82160	0.71	2.13	6
80	17	3	2	9	8.89	71.11	57155	82160	0.3	2.7	
80	17	3	3	27	2.96	77.04	73150	82160	0.11	2.97	
90	17	3	1	3	30	60	34220	117480	0.71	2.13	13

(continued)

Table 1 (continued)

N	*J*	*L*	*l*	*W* = 3^*l*	*N*/3^*l*	*N*-(*N*/3^*l*)	*X* = cmb(*N*-*N*/3^*l*,*L*)	*Y* = cmb(*N*,*L*)	*Z* = 1−*X*/*Y*	*W***Z*	∑(*W***Z*)
90	17	3	2	9	10	80	82160	117480	0.3	2.7	
90	17	3	3	27	3.33	86.67	102340	117480	0.13	3.51	
90	17	3	4	81	1.11	88.89	109736	117480	0.07	5.67	
100	17	3	1	3	33.33	66.67	45760	161700	0.72	2.16	14
100	17	3	2	9	11.11	88.89	109736	161700	0.32	2.88	
100	17	3	3	27	3.7	96.3	142880	161700	0.12	3.24	
100	17	3	4	81	1.23	98.77	152096	161700	0.06	4.86	

Table 2 Number of transformed nodes at different number of group members with $J = 17$ and $L = 3$ using Quadtree-based Batch Algorithm

N	J	L	l	$W = 4^l$	$N/4^l$	N-$(N/4^l)$	$X = \text{cmb}(N\text{-}N/4^l,L)$	$Y = \text{cmb}(N,L)$	$Z = 1-X/Y$	$W*Z$	$\sum(W*Z)$
20	17	3	1	4	5	15	455	1140	0.6	2.4	5
20	17	3	2	16	1.25	18.75	816	1140	0.28	4.48	
30	17	3	1	4	7.5	22.5	1540	4060	0.62	2.48	4
30	17	3	2	16	1.88	28.12	3276	4060	0.19	3.04	
40	17	3	1	4	10	30	4060	9880	0.59	2.36	4
40	17	3	2	16	2.5	37.5	7770	9880	0.21	3.36	
50	17	3	1	4	12.5	37.5	7770	19600	0.6	2.4	5
50	17	3	2	16	3.13	46.87	15180	19600	0.23	3.68	
60	17	3	1	4	15	45	14190	34220	0.59	2.36	4
60	17	3	2	16	3.75	56.25	27720	34220	0.19	3.04	
70	17	3	1	4	17.5	52.5	22100	54740	0.6	2.4	9
70	17	3	2	16	4.38	65.62	43680	54740	0.2	3.2	
70	17	3	3	64	1.09	68.91	50116	54740	0.08	5.12	
80	17	3	1	4	20	60	34220	82160	0.58	2.32	8
80	17	3	2	16	5	75	67525	82160	0.18	2.88	
80	17	3	3	64	1.25	78.75	76076	82160	0.07	4.48	
90	17	3	1	4	22.5	67.5	47905	117480	0.59	2.36	8
90	17	3	2	16	5.63	84.37	95284	117480	0.19	3.04	
90	17	3	3	64	1.41	88.59	109736	117480	0.07	4.48	
100	17	3	1	4	25	75	67525	161700	0.58	2.32	8
100	17	3	2	16	6.25	93.75	129766	161700	0.2	3.2	
100	17	3	3	64	1.56	98.44	152096	161700	0.06	3.84	

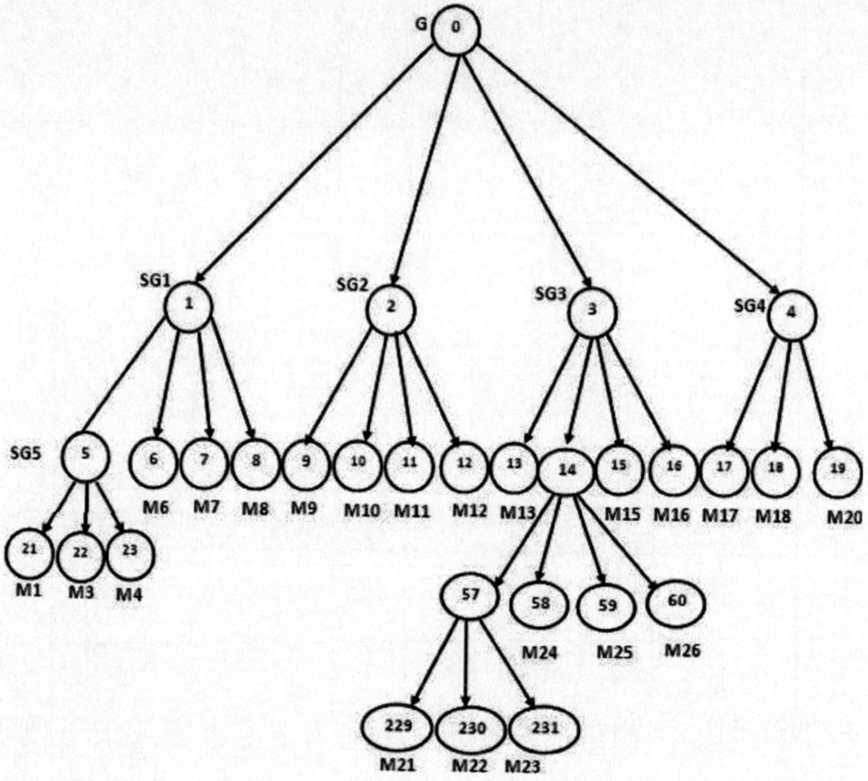

Fig. 6 Key tree after 6 members (M21, M22, M23, M24, M25, M26) join and 4 members (M2, M5, M14, M19) leave the group

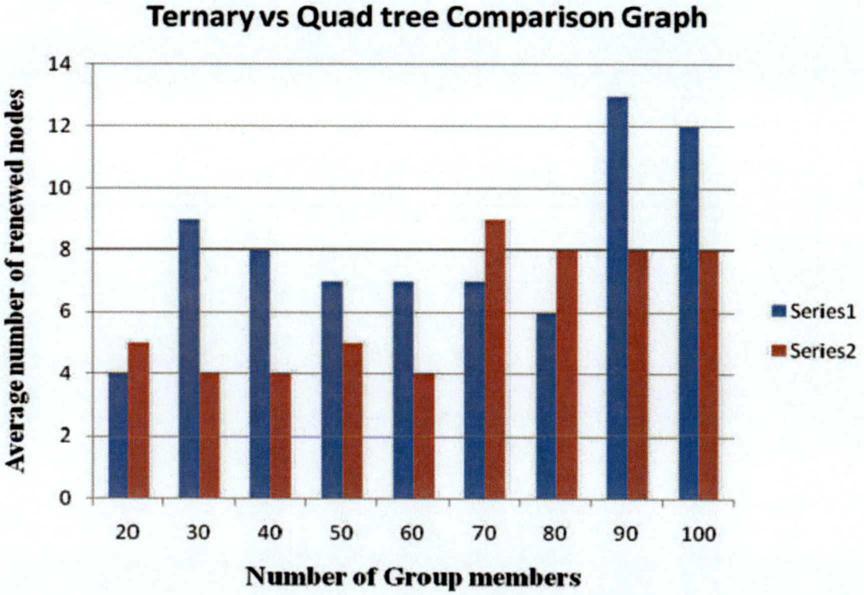

Fig. 7 Mathematical analysis: average numbers of renewed nodes at different numbers of joins

References

1. Diffie W, Hellman M (1976) New directions in cryptography. IEEE Trans Inf Theory 22(6):644–654
2. Patrick PCL, Lui JCS, Yau DKY (2006) Distributed collaborative key agreement and authentication protocols for dynamic peer groups. IEEE/ACM Trans Netw 14(2), April
3. Srinadh V, Nageswararao PV (2015) Ternary tree based group key generation. IJETM Int J Eng Technol Manag 2(5), Sept–Oct
4. Renuga devi N, Mala C (2014) Ternary tree based group key agreement for cognitive radio MANETS. Int J Comput Netw Inf Secur 10:24–31
5. Srinadh V, Nageswararao PV (2017) Ternary tree based batch algorithm for group key generation. J Adv Res Dyn Control Syst 13 Special Issue September:505–511
6. Steiner M, Tsudik G, Waidner M (2000) Key agreement in dynamic peer groups. IEEE Trans Parallel Distrib Syst 11(8):769–780 Aug

Performance Evaluation of Tandem Communication Network Model with DBA Having Direct Weibull Inter Arrival Times and Phase-Type Transmission

G. Sita Ratnam, P. Srinivasa Rao, K. Srinivasa Rao, and Mugada Swetha

Abstract Communication network models play a significant role in designing and modeling the performance of communication systems. Due to the rapid development of sophisticated technological advancements, a wide variety of communication network models is developed in order to utilize the resources more effectively. Generally, the analysis of communication system is mainly concerned with the problem of allocation, distribution of data/voice, packetization, statistical multiplexing, flow control, link capacity assignment, delay, and routing. In this paper, we proposed a new communication network model with the assumption of the arrival process follow the non-homogeneous Duane process and the transmission process follow the Poisson process. From this analysis, we observed, the nature of Duane arrivals has a significant influence on the performance measures of the network, and it can be predicted closer to the practical situations. The proposed communication network model is useful for analyzing and designing communication systems like telecommunications, satellite communications, computer communications, and the internet.

1 Introduction

To improve the quality of service, the differentiated communication services have been considered asscalable traffic management. In this paper, we proposed a new communication network model with dynamic bandwidth allocation, having non-homogeneous Duane arrivals studied and analyzed. This model is useful for various

G. Sita Ratnam · M. Swetha (✉)
Lendi Institute of Engineering & Technology, Visakhapatnam, India
e-mail: bepalaswetha@gmail.com

G. Sita Ratnam
e-mail: Sitagokuruboyina@gmail.com

P. Srinivasa Rao (✉) · K. Srinivasa Rao (✉)
Andhra University College of Engineering, Visakhapatnam, India
e-mail: peri.srinivasarao@yahoo.co.in

K. Srinivasa Rao
e-mail: ksraoau@gmail.com

S. C. Satapathy et al. (eds.), *Communication Software and Networks*, Lecture Notes in Networks and Systems 134, https://doi.org/10.1007/978-981-15-5397-4_40

places like telesatellite and computer communications. However, in telecommunication and internet services, two nodes are connected in tandem having direct arrival to both buffers having intermediary departure of packets. This type of arrival strategy is known as two-stage direct arrivals. For example, in telecommunications, the service station may receive packets from domestic users as well as ISD users also [1]. In this situation, the network may create congestion in packet transmission. The idea here is reduced to the congestion in buffers and delays in transmission while maintaining the satisfactory quality of service in transmission. Here, it is assumed that the messages are packetized and stored in buffers for forward transmission. The transmission is carried with dynamic bandwidth allocation (dependent on the content of the buffer). Further, it is assumed that the number of transmission completions follows the Poisson process. The packets are transmitted through first in– first out discipline. The proposed communication network model also describes the arrival of packets are time-dependent and transmission of packets follows a non-homogeneous Duane process, which has an efficient utilization of resources it is also needed to have dynamic bandwidth allocation strategy for transmission [2–4].

2 Related Work

Recently, the dynamic bandwidth allocation or load-dependent transmission strategy is developed for improving the quality of service by utilizing the idle bandwidth [3]. The authors assumed that the arrival of packets to the buffers is homogeneous and time-independent. Hence, they used the Poisson process for characterizes arrivals. Modeling and analysis of network traffic based on the Poisson process assume the arrival of packets is smoother and less burst [5, 6]. However, the present-day increases intensity on communication systems, the aggregate traffic may not remain smooth and bursty. It is known that the actual traffic in the Ethernet, LAN exhibits the property of self-similar (burstness) and long-range similarity [7, 8]. The self-similar or burstness is due to the time-dependent nature of arrivals. Very little work has been reported with respect to the communication network models with time-dependent arrivals, which described communication network models with the non-homogeneous Poisson process [4, 9].

They considered that the mean arrival rate is linearly dependent on time and inter-arrival times of packets follow an exponential distribution. But, the linear dependence of the time of the mean arrival rate has certain drawbacks in approximating the performance measure of the communication network. Hence, variable time-dependent arrival rate can characterize the inter-arrival times of packets with a Weibull distribution and the arrival process can be Duane process.

The Weibull inter-arrival time distribution was used for modeling the G/M/1 queuing system [10]. Hence, we analyzed the performance of WWW servers using Weibull distribution for inter-arrival times assuming that there is a single server in each queue and queues are independent [11]. In many communication systems, such as satellite communications and telecommunications, the output of one queue is the input to the other queue and is not independent. There is no work reported regarding tandem communication network models with Weibull inter-arrival times, which model the self-similarity network traffic. With this motivation in this paper, we proposed a new communication network model with dynamic bandwidth allocation having time-dependent Duane arrival process with different patterns of transmission flows studied and analyzed.

3 Communication Network Model and Transient Analysis

The communication network model having two nodes in tandem with direct arrivals to the buffers connected to node 1 node 2 is studied. Here, it is assumed that the data/Voice packets arrive to the buffers connected at node 1 and node 2 and follow Duane processes with mean arrival rates $\lambda(t)$ and $\varepsilon(t)$, respectively. After being transmitted from the first transmitter, the packets may join the buffer connected to the second node with probability θ or may get terminated with probability π such that $\theta + \pi = 1$. The number of transmission completions in each node follows Poisson processes with parameters μ_1 and μ_2. The transmission time of each packet in each node is adjusted depending on the content of the buffer connected to it [9, 12]. The schematic diagram representing the communication network is shown in Fig. 1.

The general solution of the probability generating function of the number of packets in the first buffer and the number of packets in the second buffer at time t as follows:

$$P(s_1, s_2; t) = \exp\left[\left[a_1 \cdot b_1\left((s_1 - 1)e^{-\mu_1 \cdot t} \int_0^t e^{\mu_1 \cdot v} \cdot v^{b_1 - 1} dv - \left(\frac{e^{-\mu_1 \cdot t}}{\mu_1}\right)\right)\right.\right.$$

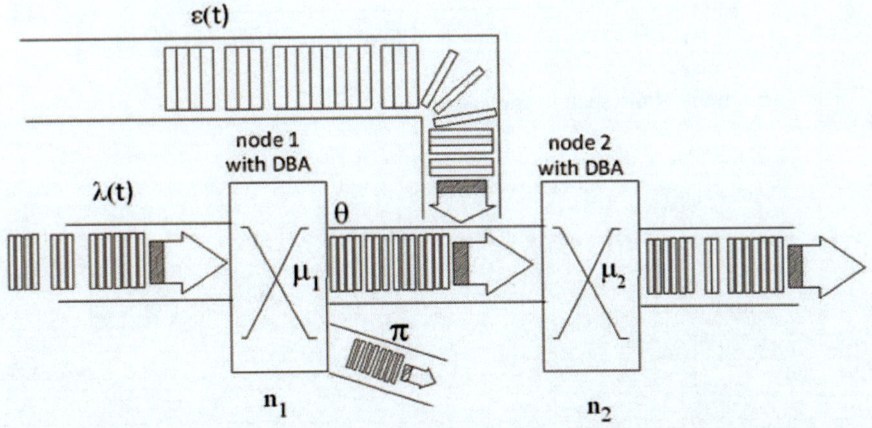

Fig. 1 Communication network model with two-stage direct arrivals

$$+\theta\cdot\frac{(s_2-1)}{(\mu_2-\mu_1)}\cdot\left(e^{-\mu_2\cdot t}-e^{-\mu_1\cdot t}\right)+\Bigg(\theta\cdot\frac{\mu_1\cdot(s_2-1)}{(\mu_2-\mu_1)}\cdot$$
$$\left(e^{-\mu_1\cdot t}\int_0^t e^{\mu_1\cdot v}\cdot v^{b_1-1}dv-e^{-\mu_2\cdot t}\int_0^t e^{\mu_2\cdot v}\cdot v^{b_1-1}dv\right)-\left(\frac{e^{-\mu_2\cdot t}}{\mu_2}\right)\Bigg)\Bigg]$$
$$+a_2\cdot b_2\cdot\left[(s_2-1)e^{-\mu_2\cdot t}\int_0^t e^{\mu_2\cdot v}\cdot v^{b_2-1}dv-\left(\frac{e^{-\mu_2\cdot t}}{\mu_2}\right)\right]\Bigg] \tag{1}$$

4 Performance Measures of the Network

We derive and analyze the performance measures of the communication network under transient conditions. Expanding $P(S_1, S_2; t)$ given in Eq. (1) and collecting the constant terms, we get the probability that the network. The mathematical (1) describes here the performance measures of the network.

The throughput of the first transmitter is

$$Thp_1(t)=\mu_1(1-P_{0.}(t))$$
$$=\mu_1\left[1-\exp\left[-a_1\cdot b_1\left(e^{-\mu_1\cdot t}\int_0^t e^{\mu_1\cdot v}\cdot v^{b_1-1}dv-\left(\frac{e^{-\mu_1\cdot t}}{\mu_1}\right)\right)\right]\right] \tag{2}$$

The mean delay in the first buffer is

$$W_1(t)=\frac{L_1}{Thp_1}=\frac{E[N_1]}{Thp_1}$$
$$=\frac{\left[a_1\cdot b_1\left(e^{-\mu_1\cdot t}\int_0^t e^{\mu_1\cdot v}\cdot v^{b_1-1}dv-\left(\frac{e^{-\mu_1\cdot t}}{\mu_1}\right)\right)\right]}{\mu_1\left[1-\exp\left[-a_1\cdot b_1\left(e^{-\mu_1\cdot t}\int_0^t e^{\mu_1\cdot v}\cdot v^{b_1-1}dv-\left(\frac{e^{-\mu_1\cdot t}}{\mu_1}\right)\right)\right]\right]} \tag{3}$$

The throughput of the second transmitter is

$$Thp_2(t)$$
$$=\mu_2(1-P_{.0}(t))$$
$$=\mu_2\Bigg[1-\exp\Bigg[-a_1\cdot b_1\Bigg(\frac{\theta}{(\mu_2-\mu_1)}\cdot\left(e^{-\mu_2\cdot t}-e^{-\mu_1\cdot t}\right)\Bigg)$$
$$+\frac{\mu_1\cdot\theta}{(\mu_2-\mu_1)}\cdot\left(e^{-\mu_1\cdot t}\int_0^t e^{\mu_1\cdot v}\cdot v^{b_1-1}dv-e^{-\mu_2\cdot t}\int_0^t e^{\mu_2\cdot v}\cdot v^{b_1-1}dv\right)-\left(\frac{e^{-\mu_2\cdot t}}{\mu_2}\right)\Bigg]$$
$$-a_2\cdot b_2\cdot\left[e^{-\mu_2\cdot t}\int_0^t e^{\mu_2\cdot v}\cdot v^{b_2-1}dv-\left(\frac{e^{-\mu_2\cdot t}}{\mu_2}\right)\right]\Bigg] \tag{4}$$

The mean delay in the second buffer is

$$W_2(t) = \frac{L_2(t)}{Thp_2(t)} = \frac{E(N_2)}{Thp_2(t)}$$

$$= \frac{\begin{array}{c}\left[a_1 \cdot b_1\left[\frac{\theta}{(\mu_2-\mu_1)}\cdot\left(e^{-\mu_2\cdot t}-e^{-\mu_1\cdot t}\right)+\frac{\mu_1\cdot\theta}{(\mu_2-\mu_1)}\cdot\right.\right. \\ \left.\left(e^{-\mu_1\cdot t}\int_0^t e^{\mu_1\cdot v}\cdot v^{b_1-1}dv - e^{-\mu_2\cdot t}\int_0^t e^{\mu_2\cdot v}\cdot v^{b_1-1}dv\right)-\left(\frac{e^{-\mu_2\cdot t}}{\mu_2}\right)\right] \\ \left.+a_2\cdot b_2\cdot\left[e^{-\mu_2\cdot t}\int_0^t e^{\mu_2\cdot v}\cdot v^{b_2-1}dv-\left(\frac{e^{-\mu_2\cdot t}}{\mu_2}\right)\right]\right]\end{array}}{\begin{array}{c}\mu_2\left[1-\exp\left[-a_1\cdot b_1\frac{\theta}{(\mu_2-\mu_1)}\cdot\left(e^{-\mu_2\cdot t}-e^{-\mu_1\cdot t}\right)+\frac{\mu_1\cdot\theta}{(\mu_2-\mu_1)}\cdot\right.\right. \\ \left(e^{-\mu_1\cdot t}\int_0^t e^{\mu_1\cdot v}\cdot v^{b_1-1}dv - e^{-\mu_2\cdot t}\int_0^t e^{\mu_2\cdot v}\cdot v^{b_1-1}dv\right)-\left(\frac{e^{-\mu_2\cdot t}}{\mu_2}\right) \\ \left.\left.-a_2\cdot b_2\cdot\left[e^{-\mu_2\cdot t}\int_0^t e^{\mu_2\cdot v}\cdot v^{b_2-1}dv-\left(\frac{e^{-\mu_2\cdot t}}{\mu_2}\right)\right]\right]\right]\end{array}} \tag{5}$$

For different values of t, a_1, a_2, b_1, b_2, μ_1 and μ_2, the probability of emptiness of each buffer, the utilization of both buffers, the throughput of the two transmitters, the mean delay in the buffers, the mean number of packets in the network, and variability of the number of packets in the network are computed.

5 Performance Evaluation of the Network

The performance of the developed communication network is discussed through numerical illustration [13, 14]. Different values of the parameters are considered for bandwidth allocation and arrival of packets. After interacting with the technical staff at internet-providing station, it is considered that the packet arrival rate parameters (a_1) and (a_2) varies from 10×10^4 packets/s to 13×10^4 packets/s and 6×10^4 packets/s to 9×10^4 packets/s, and b_1, b_2 varies from 1.25 to 2, 0.8 to 1.1, respectively, with an average packet size of 102 bytes.

After transmitting from node 1, packets may reach node 2 with probability θ, ($\theta = 1 - \pi$) and some packets may be terminated with probability (π). The transmission rate μ_1 varies from 11×10^4 packets/s to 14×10^4 packets/s. The rate of transmission rate μ_2 varies from 19×10^4 packets/s to 22×10^4 packets/s. In all these nodes, dynamic bandwidth allocation strategy is considered i.e., the transmission rate of each packet depends on the number of packets in the buffer connected to it at that instant.

From Table 1, we observed the analysis of communication network, as time (t) increases the throughput of first and second nodes are increasing from 53,051 packets to 99,893 packets and from 84,069 packets to 155,700 packets, respectively, for fixed values of the other parameters.

Table 1 Values of the mean number of packets and mean delay of the communication network model with dynamic bandwidth allocation

t*	a_1	b_1	a_2	b_2	$\mu_1$$	$\mu_2$$	π	Θ	$Thp_1(t)$	$Thp_2(t)$	$W_1(t)$	$W_2(t)$
0.7	9	2	5	2	10	18	0.5	0.5	5.3051	8.4069	0.1425	0.0844
1	9	2	5	2	10	18	0.5	0.5	8.0210	11.0168	0.2020	0.0967
1.5	9	2	5	2	10	18	0.5	0.5	9.7718	13.8807	0.3868	0.1192
2	9	2	5	2	10	18	0.5	0.5	9.9893	15.5700	0.6847	0.1442
1.5	**10**	2	5	2	10	18	0.5	0.5	9.8500	14.1772	0.4264	0.1234
1.5	**11**	2	5	2	10	18	0.5	0.5	9.9015	14.4523	0.4666	0.1276
1.5	**12**	2	5	2	10	18	0.5	0.5	9.9353	14.7076	0.5073	0.1319
1.5	**13**	2	5	2	10	18	0.5	0.5	9.9575	14.9446	0.5483	0.1361
1.5	9	**1.3**	5	2	10	18	0.5	0.5	8.4033	11.0052	0.2183	0.0937
1.5	9	**1.5**	5	2	10	18	0.5	0.5	9.0877	12.0200	0.2635	0.1010
1.5	9	**1.8**	5	2	10	18	0.5	0.5	9.5212	12.9839	0.3192	0.1095
1.5	9	**2**	5	2	10	18	0.5	0.5	9.7718	13.8807	0.3868	0.1192
1.5	9	2	**6**	2	10	18	0.5	0.5	9.7718	14.4915	0.3868	0.1253
1.5	9	2	**7**	2	10	18	0.5	0.5	9.7718	15.0117	0.3868	0.1316
1.5	9	2	**8**	2	10	18	0.5	0.5	9.7718	15.4548	0.3868	0.1383
1.5	9	2	**9**	2	10	18	0.5	0.5	9.7718	15.8322	0.3868	0.1451
1.5	9	2	5	**0.8**	10	18	0.5	0.5	9.7718	11.3331	0.3868	0.1036
1.5	9	2	5	**0.9**	10	18	0.5	0.5	9.7718	11.5953	0.3868	0.1047
1.5	9	2	5	**1**	10	18	0.5	0.5	9.7718	11.8472	0.3868	0.1058
1.5	9	2	5	**1.1**	10	18	0.5	0.5	9.7718	12.0891	0.3868	0.1070
1.5	9	2	5	2	**11**	18	0.5	0.5	10.653	13.8994	0.3246	0.1361
1.5	9	2	5	2	**12**	18	0.5	0.5	11.504	13.9149	0.2771	6.1585
1.5	9	2	5	2	**13**	18	0.5	0.5	12.323	13.9280	0.2398	0.1897
1.5	9	2	5	2	**14**	18	0.5	0.5	13.109	13.9391	0.2102	0.2364
1.5	9	2	5	2	10	**19**	0.5	0.5	9.7718	14.3147	0.3868	0.1031
1.5	9	2	5	2	10	**20**	0.5	0.5	9.7718	14.7237	0.3868	0.0905
1.5	9	2	5	2	10	**21**	0.5	0.5	9.7718	15.1096	0.3868	0.0804
1.5	9	2	5	2	10	**22**	0.5	0.5	9.7718	15.4743	0.3868	0.0721
1.5	9	2	5	2	10	18	**0.4**	**0.6**	9.7718	14.3989	0.3868	0.1268
1.5	9	2	5	2	10	18	**0.3**	**0.7**	9.7718	14.8579	0.3868	0.1344
1.5	9	2	5	2	10	18	**0.2**	**0.8**	9.7718	15.2479	0.3868	0.1421
1.5	9	2	5	2	10	18	**0.1**	**0.9**	9.7718	15.5942	0.3868	0.1499

* = Seconds, $ = Multiples of 10,000 packets/s

As the arrival rate parameter (a_1) varies from 10×10^4 packets/s to 13×10^4 packets/s, the throughput of the first buffer and the second buffer are increasing from 98,500 packets to 99,575 packets and the second buffer varies from 141,772 packets to 149,446 packets, respectively, when other parameters remain fixed.

Similarly, the arrival rate parameter (a_2) varies from 6×10^4 packets/s to 9×10^4 packets/s, the throughput of the first buffer remains fixed at 97,718 packets and the second buffer is increasing from 144,915 packets to 158,322 packets, respectively, when other parameters remain fixed. When the arrival rate parameter (b_1) varies from 1.3 to 2, the throughput of the first buffer and the second buffer are increasing from 84,033 packets to 97,718 packets, 110,052 packets to 138,807 packets, respectively, when the other parameters remain fixed. As the arrival rate parameter (b_2) varies from 0.8×10^4 packets/s to 1.1×10^4 packets/s, the throughput of the first buffer remains fixed at 97,718 packets and the second buffer is increasing from 113,331 packets to 120,891 packets, respectively, when other parameters remain fixed.

When the transmission rate (μ_1) varies from 11×10^4 packets/s to 14×10^4 packets/s, the throughput of the first buffer and the second buffer are increasing from 106,538 packets to 131,096 packets, 138,994 packets to 139,391 packets, respectively, when other parameters remain fixed. Similarly, the transmission rate (μ_2) varies from 19×10^4 packets/s to 22×10^4 packets/s, the throughput of the first buffer remains at 97,718 packets and the throughput of the second buffer increases from 143,147 packets to 154,743 packets, respectively, when the other parameters remain fixed. When the parameter (θ) varies from 0.6 to 0.9, the throughput of the first buffer remains at 97,718 packets and the throughput of the second buffer increases from 143,989 packets to 155,942 packets when the other parameters remain fixed.

It is observed that as the time (t) and the arrival rate parameter (a_1) are increasing, the mean delay in buffers is increasing from fixed values of the other parameters. As the arrival rate parameter (a_2) increases, the mean delay in buffer 1 remains fixed at 3868 packets. The mean delay in buffer 2 is increasing from 1253 packets to 1451 packets, when fixed values of the other parameters.

It is also observed that as the arrival rate parameter (b_1) is increasing, the mean delay in buffers is increasing from 2183 packets to 3868 packets and 937 packets to 1192 packets, respectively, when other parameters are fixed. As the arrival rate parameter (b_2) increases, the mean delay in buffer 1 remains fixed at 3868 packets. The mean delay in buffer 2 is increasing from 1036 packets to 1070 packets, when fixed values of the other parameters.

When the transmission rate (μ_1) increases, the mean delay in the first buffer decreases, when the other parameters remain fixed. Similarly, the transmission rate (μ_2) increases, the mean delay in the second buffer decreases, when the other parameter remains fixed.

When the parameter (θ) increases, the mean delay in the second buffer increases when other parameters are fixed.

From this analysis, it is observed that the dynamic bandwidth allocation strategy has a significant influence on all performance measures of the network. It is further observed that the performance measures are highly sensitive toward smaller values of time. Hence, it is optimal to consider dynamic bandwidth allocation under and non-homogenous Poisson arrivals and evaluate the performance under transient conditions. It is also to be observed that the congestion in buffers and delays in transmission were reduced to a minimum level by adopting dynamic bandwidth allocation. This phenomenon has a vital bearing on the quality of transmission.

6 Comparative Study

The comparative study between the model with Poisson arrivals (time-independent) and the proposed communication network models with Duane arrival processes (time-dependent). The time-dependent nature of traffic will occur a particular time period like in the telecommunication network, the more International/Domestic calls are connected early in the morning, night, and festival seasons. At that time, the congestion will occur in buffers delay will increase and quality of service will loss users. So here, we try to compare these two models and analyzed the computed performance measure of both models are presented in Table 2, for different values of time $t = 1, 1.5, 2, 2.5$ s. This comparison will help us to users better understand how the time-dependent arrivals (network traffic) will influence communication systems.

As t increases, the percentage variation of performance measures between the models also increases. The model with non-homogenous Poisson arrivals with dynamic bandwidth allocation has more utilization compared to that of the model with homogenous Poisson arrivals with dynamic bandwidth allocation. From this analysis, it is observed that the assumption of non-homogenous Poisson arrivals has a significant influence on all the performance measures of the network.

7 Conclusions

This paper contributes to the descriptive modeling and performance evaluation of some communication network models with Duane arrival process and dynamic bandwidth allocation. Due to technological innovations and exploration of scientific skills, the perceptions of the customers and service providers are changing rapidly in this modern era because the communication network models play a significant role in providing high performance. The communication network model is a mathematical model that represents the physical flow of information in a communication system. These models can approximate the performance measures of the system more close to reality. Conducting laboratory experiments under different load and transmission conditions is time-consuming and costly. Several authors have developed various communication network models with various assumptions on the constituent process of the networks.

Table 2 Comparative study of homogeneous and non-homogeneous Duane arrivals

Time (t) (s)	Parameters measured	Model with Poisson arrivals	Proposed model	Difference	% of variation
$t = 1$	$L_1(t)$	0.7000	1.2600	0.5600	80.00000
	$L_2(t)$	0.1944	0.3284	0.1340	68.93004115
	$U_1(t)$	0.5034	0.7163	0.2129	42.2924116
	$U_2(t)$	0.1767	0.2799	0.1032	58.4040747
	$Thp_1(t)$	5.0340	7.1635	2.1295	42.30234406
	$Thp_2(t)$	3.1805	5.0386	1.8581	58.42163182
	$W_1(t)$	0.1390	0.1759	0.0369	26.54676259
	$W_2(t)$	0.0611	0.0652	0.0041	6.710310966
$t = 1.5$	$L_1(t)$	1.0500	2.9400	1.8900	180.0000
	$L_2(t)$	0.1944	0.5228	0.3284	168.9300412
	$U_1(t)$	0.6501	0.9471	0.2970	45.68527919
	$U_2(t)$	0.1767	0.4072	0.2305	130.4470855
	$Thp_1(t)$	6.5006	9.4713	2.9707	45.69885857
	$Thp_2(t)$	3.1807	7.3290	4.1483	130.4209765
	$W_1(t)$	0.1615	0.3104	0.1489	92.19814241
	$W_2(t)$	0.0611	0.0713	0.0102	16.69394435
$t = 2.5$	$L_1(t)$	1.7500	8.4000	6.6500	380.00000
	$L_2(t)$	0.1944	0.9117	0.7173	368.9814815
	$U_1(t)$	0.8262	0.9998	0.1736	21.01186153
	$U_2(t)$	0.1767	0.5982	0.4215	238.5398981
	$Thp_1(t)$	8.2623	9.9978	1.7355	21.00504702
	$Thp_2(t)$	3.1807	10.7671	7.5864	238.5135348
	$W_1(t)$	0.2118	0.8402	0.6284	296.6949953
	$W_2(t)$	0.0611	0.0847	0.0236	38.62520458

8 Scope for Further Research

In this paper, we developed and analyzed the descriptive modeling of single and multiple tandem communication network models with Duane arrival process having dynamic bandwidth allocation. It is also possible to develop and analyze the perspective modeling of communication network models with Duane arrival process and dynamic bandwidth allocation using various cost considerations and control policies. The models developed in this thesis can also be further generalized with general service time distribution, which requires further investigations.

Communication network modeling is one of the prime areas of communication systems that utilize the resources more efficiently and optimally. Due to technological innovations and advancements, it is possible to develop many more communication network models with plausible conditions in order to utilize the resources more effectively and efficiently and to develop advanced communication systems.

References

1. Lea C-T, Alyatama A (1995) Bandwidth quantization and states reduction in the broadband ISDN. IEEE Trans Netw 2(3):352–360
2. Guodong H, Jianhu W, Jiangxing W (2006) Dynamic bandwidth allocation algorithm utilizing full band. J Electron 23(1):48–53
3. Padmavathi G, Srinivasa Rao K, Reddy KVVS (2009) Performance evolution of parallel and series communication network with dynamic band width allocation CIIT. Int J Netw Commun 1(7):410–442
4. Rama Sundari MV, Srinivasa Rao K, Srinivasa Rao P, Suresh Varma P (2011) Three node tandem communication network model with dynamic bandwidth allocation and non-homogeneous Poisson arrivals. Int J Comput Appl (0975-8887) 31(1)
5. Abry P, Baraniuk R, Flandrin P, Riedi R, Veitch D (2002) Multi-scale nature of network traffic. IEEE Signal Process Mag 28–27
6. Cappe O, Moulines E, Pesquet JC, Petropulu A, Yang X (2002) Long-range dependence and heavy-tail modeling for teletraffic data. IEEE Signal Process Mag 14–27
7. Fowler H, Leland W (1991) Local area network traffic characteristics, with implications for broadband network congestion management. IEEE J Select Areas Commun 9:1139–1149
8. Singhai R, Joshi SD, Bhatt RKP (2007) A novel discrete distribution and process to model self-similar traffic. In: 9th IEEE international conference on telecommunications—Con TEL 2007, Zagreb, Croatia, pp 167–171
9. Suhasini AVS, Srinivasa Rao P, Reddy PRS (2013) Transient analysis of tandem queuing model with non-homogeneous Poisson bulk arrivals having state dependent service rates. Int J Adv Comput Math Sci 3(3):272–284
10. Fisher MJ, Gross D, Masi D, Shortle JF (2001) Analyzing the waiting time process in internet queuing systems with the transform approximation method. Telecommun Rev 12:21–32
11. Araikthamazian, Bogachev M (2015) The telecommunication reviews, pp 21–32
12. Parthasarathy PR, Selvaraju N (2001) Transient analysis of a queue where potential customers are discouraged by queue length. Math Probl Eng 7:433–454
13. Reiser M (1982) Performance evaluation of data communication systems. Proc IEEE 70(2):171–196
14. Gundale AS, Yardi AR (2008) Performance evaluation of issues related to video over broadband network. In: Proceedings of world academy of science, engineering and technology, vol 36, pp 122–125

A Novel Approach for Detection of Driver Drowsiness Using Behavioural Measures

D. Venkata Subbaiah, Pushkal Padala, and K. Venkata Rao

Abstract Among many reasons, drowsiness is a significant cause of a road accident, especially driver who tends to drive continuously for long distances or during long night drive is mostly subjected to drowsiness. Hence, the existing drowsiness detection system requires a counter-measure device for sleepiness-related accident prevention. This paper aims to perform drowsiness detection with the help of driver's facial expressions. At first, the input data was collected to detect and track the driver's face, eyes and mouth regions from the input video. The extracted features from the input video of the eye region were combined with the threshold value of mouth region and head pose angle. IFS algorithm was utilized for selecting the best features and these were given as the input for SVM binary classification algorithm to classify the stages of drowsiness detection. The experimental result shows that the recommended system improves the efficiency from 84.85 to 88.97% as compared to the existing systems.

1 Introduction

Drowsiness is a powerful cause of a road accident, specifically driver who tends to drive constantly for long distances or during long night drive is mostly subjected to drowsiness. The NHTSA, 2011 announced drowsiness adds roughly 3% of all traffic-related destructiveness [1]. The drowsiness detection assesses different measures like vehicle behaviour, visual features, physiological features, etc. In vehicle-based measures, a number of metrics are included for detecting the driver drowsiness such

D. Venkata Subbaiah (✉) · K. Venkata Rao
Department of CS&SE, Andhra University, Visakhapatnam, India
e-mail: vsdesanamukula.auce@gmail.com

K. Venkata Rao
e-mail: drkvrcsse.auce@gmail.com

P. Padala
Department of CSE, National Institute of Engineering, Mysore, India
e-mail: pushkalpadala.niem@gmail.com

S. C. Satapathy et al. (eds.), *Communication Software and Networks*, Lecture Notes in Networks and Systems 134, https://doi.org/10.1007/978-981-15-5397-4_41

as movement of steering, lane departure and pressure on acceleration pedal [2]. Drowsiness, especially in its subsequent stages, is likewise connected with eyelid closures and micro-sleeps, which may obstruct with the driver's capacity to identify the outset of a forward crash event [3, 4]. The main issue with this technique is the efficiency depending on specific properties of the vehicle as well as driver. So, the techniques based on visual features like yawning, facial expressions, head movement and eye state showed an effective performance in driver drowsiness detection because of its non-intrusive nature [5, 6]. Besides expanding crash, drowsy driving crashes are regularly more serious than different accidents since they as often as possible happen on high-speed expressways, and are frequently run-off-the-road crashes with no braking prior to impact [7]. Eye tracking and head movement detection are broadly examined as elective interface methods [8]. This paper attempts drowsiness detection by assessing the facial expressions or features on the basis of machine learning methods for classifying the levels of drowsiness detection.

2 Related Work

Various researches are developed by the researchers in different stages of driver drowsiness detection. In this section, a brief review of some important contributions of the literature is presented.

de Naurois et al. [9] used Artificial Neural Networks (ANNs) for detecting the driver's fatigue level or for predicting the outset of an defective driving state. Here, two ANN-based approaches were utilized for predicting the level of fatigue state and also for detecting the time period, how long driver takes to reach moderately drowsy state. In this research study, the drowsiness detection performance of developed methodology was improved approximately 80% in the detection and 40% in prediction compared to the other existing systems. The subject-specific adaptation of driver's data delivers a better response. Besides, the developed research work did not concentrate on different road conditions and time.

Mittal et al. [10] had proposed driver fatigue was one of the most widely recognized purposes behind deadly road accidents around the world. The similar examination of these procedures demonstrates that behavioural measures are very easy to acquire and does not irritate the driver as they are non-invasive. Among different behavioural measures, head tilt measure was observed to be most exact and effective.

Guo and Markoni [11] developed an effective system for real-time driver drowsiness prediction using hybrid classifier: CNN and LSTM. Here, the developed hybrid CNN-LSTM model was performed with better accuracy and low computational cost. The output of the developed model was tested on an online available dataset: public drowsy driver database. The experimental end result shows the effectiveness of the evolved paradigm in terms of accuracy. In this experimental model, the classification was quiet complex when the size of the attribute values was very large.

In order to resolve the above issues, a new automated system is developed here for improving detection of driver's fatigue.

3 Proposed System

In this conference, the designed model consists of five phases: data collection, object detection and object tracking, feature extraction, feature selection and classification. Figure 1 Shown below describes the workflow of the system and the detailed explanation about the proposed system.

3.1 Data Collection

At first, the input data is gathered from a dataset: public drowsy driver dataset from ACCV 2016 competition [12]. Here, the video is captured for several people with different activities such as yawning, talking, still condition, etc. In this dataset, it also

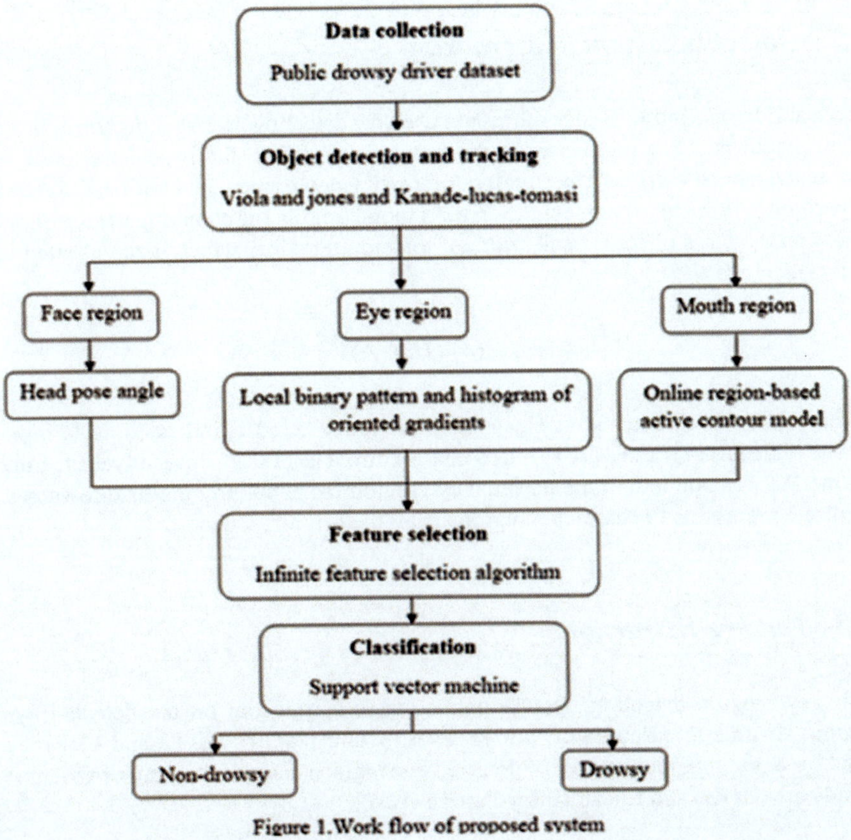

Fig. 1 Workflow of proposed system

Fig. 2 Sample images of public drowsy driver dataset

comprises of two sets, which are training set and testing set. The training set ranges from 1 to 1.5 min of the video length. Similarly, the length of the video testing set ranges from 1 to 10 min. The sample images of public drowsy driver dataset are shown in Fig. 2.

3.2 *Object Detection and Tracking*

Viola and Jones, and KLT algorithms are used for detecting and tracking the driver's face, mouth and eye regions from the input video. Then, the head pose angle is estimated from the driver's face region for every video frames. In addition, ORACM algorithm is utilized to segment the driver's mouth region for obtaining the threshold value on the basis of height and width of the mouth region, which is represented in Eq. (1).

$$\frac{\partial \varphi}{\partial (x, y)} = H(spf(I(x, y))) \times \varphi(x, y) \tag{1}$$

where $H(.)$ is represented as Heaviside function, $I(x, y)$ is indicated as an input image (video frame) and $\varphi(x, y)$ is denoted as current level set. Consecutively, feature extraction was applied on the detected eye region for extracting the feature values. Figure 3 represents the extracted regions from the original video.

3.3 *Feature Extraction*

After eye region detection, feature extraction is carried out on the detected eye regions. In this research study, a high-level texture features (HOG and LBP) are used for extracting the features of detected eye regions using Histogram of Oriented Gradients (HOG) and Local Binary Pattern (LBP).

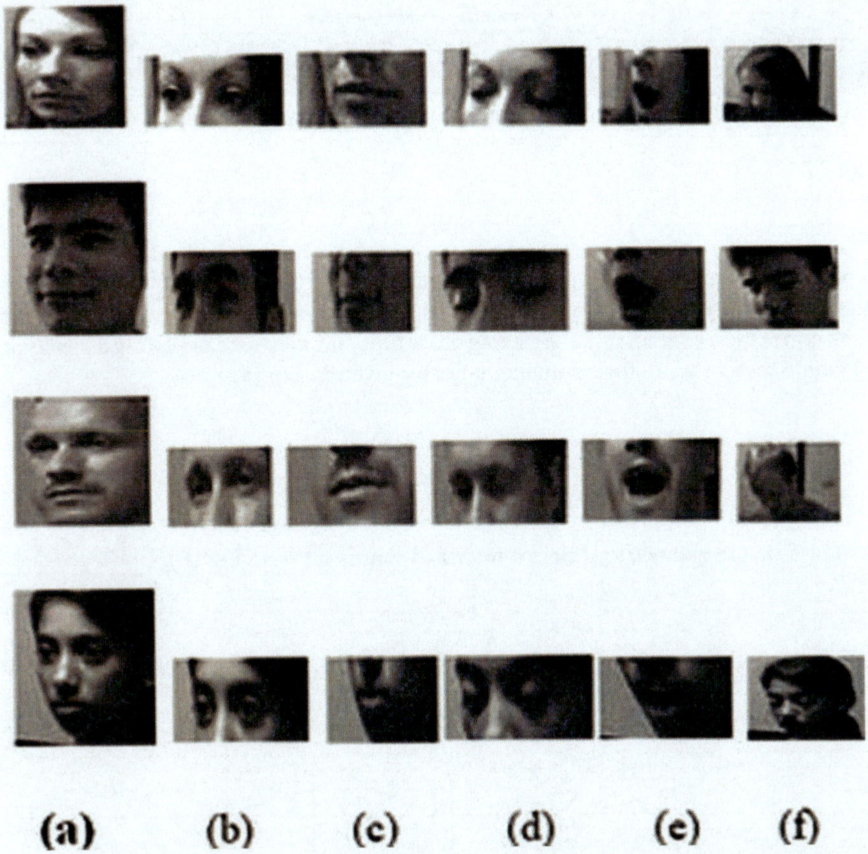

Fig. 3 **a** Extracted face region, **b** extracted eye region (open), **c** extracted mouth region (normal), **d** extracted eye region (closed), **e** extracted mouth region (yawn) and **f** head pose (bending)

3.4 Feature Selection Using Infinite Feature Selection Algorithm

In this exploration study, Infinite Feature Selection (IFS) algorithm is employed to choose the most favourable feature vectors. Given a set of features, $F = \{f^{(1)}, \ldots, f^{(n)}\}$ and $x \in R$ signify a sample of generic distributions f. Then, an indirected fully connected graph $G = (V;\ E)$; V is developed for each feature distributions. E codifies the edges for pairwise feature distributions and A specifies the nature of weighted edges: for every element a_{ij} of A, $1 \leqslant i;\ j \leqslant n$ states a pairwise energy term. Energies are represented as a weighted linear combination of two simple pairwise measures linking $f^{(i)}$ and it is defined in Eqs. (2), (3) and (4).

$$a_{ij} = \alpha\sigma_{ij} + (1 - \alpha)c_{ij} \tag{2}$$

where

$$\sigma_{ij} = \max\left(\sigma^{(i)} - \sigma^{(j)}\right) \tag{3}$$

$$c_{ij} = 1 - \left|\text{Spearman}\left(f^{(i)},\ f^{(j)}\right)\right| \tag{4}$$

where α is denoted as loading coefficient that ranges from [0; 1], $\sigma^{(i)}$ is indicated as standard deviation of the samples $\{x\} \in f^{(i)}$ and *Spearman* is denoted as spearman's rank correlation coefficient. The passage to infinity implies for calculating a new type of single feature score that is mathematically given in Eq. (5).

$$s\ (i) = \sum_{l=1}^{\infty} s_l(i) = \sum_{l=1}^{\infty}\left(\sum_{j\in V} R_l(i, j)\right) \tag{5}$$

Let S be the geometric series of matrix A that is given in Eq. (6).

$$S = \sum_{l=1}^{\infty} A^l \tag{6}$$

where S is utilized to obtain $s(i)$ as given in Eq. (7).

$$s\ (i) = \sum_{l=1}^{\infty} s_l(i) = \left[\left(\sum_{l=1}^{\infty} A^l\right)e\right]_i = [S_e]_i \tag{7}$$

where r is represented as real-valued regularized factor, r^l is the paths weight for length l, $\check{s}(i)$ is effectively computed by using Eq. (8) and then energy scores are obtained by using Eq. (9).

$$\check{S} = (\boldsymbol{I} - rA)^{-1} - \boldsymbol{I} \tag{8}$$

$$\check{s}(i) = [\check{S}_e]_i \tag{9}$$

A rank for the feature is selected by decreasing the order of $\check{s}(i)$ energy scores. The ranking is utilized for determining the number of features m to be selected by adopting classifier. The achieved feature values are incurred as the input for binary classifier: SVM.

3.5 Classification Using Support Vector Machine

SVM is a special classification approach, which is represented by an isolated hyperplane. In recent decades, SVM classification method is extensively utilized in many applications such as signal processing, bioinformatics, computer vision fields, etc. because it has the ability to perform in high-dimensional data. The formula to calculate the linear discriminant function is represented as $w{\cdot}x{+}b = 0$. In SVM classifier, an optimum hyperplane is used between the two classes (drowsy and drowsiness) in order to distinguish the samples without noise, which is mathematically represented in Eq. (10).

$$pi[w \cdot x + b] - 1 \geqslant 0, i = 1, 2, \ldots N \tag{10}$$

Then, w^2 in Eq. (14) is reduced, so the optimization problem is resolved by the saddle point of a Lagrange function with multipliers α_i. The ideal segregate function is given in Eqs. (11) and (12).

$$f(x) = \text{sign}\{(w^* x) + b^*\} = \text{sign}\left\{\sum_{i=1}^{N} \alpha_i^* \cdot pi\left(x_i^* - x\right) + b^*\right\} \tag{11}$$

$$f(x) = \text{sign}\left\{\sum_{i=1}^{N} \alpha_i^* \cdot pi \cdot k(x, x_i) + b^*\right\} \tag{12}$$

Finally, the interior product $\left(x_i^* - x\right)$ is interchanged by a linear kernel function $k\left(x, x'\right)$ in Eq. (12) for diminishing the computational complexity in higher dimensional data.

4 Experimental Result and Discussion

The performance of designed system was compared with an existing algorithm hybrid CNN-LSTM [11] on a reputed database: public drowsy driver dataset. The proposed system performance was evaluated by means of classification accuracy, Precision Predictive Value, False Positive Rate and F1-Score.

4.1 Performance Metrics

The relationship between the input and output variables of the designed system is understood by using the performance metrics like classification accuracy, Precision Predictive Value (PPV), False Positive Rate (FPR) and F1-Score. The general formula to evaluate the False Positive Rate (FPR), Precision Predictive Value (PPV) and F1-Score are given in Eqs. (13), (14), (15) and (16).

$$FPR = \frac{TNT}{TNT + FPT} \times 100 \tag{13}$$

$$PPV = \frac{TPT}{TPT + FNT} \times 100 \tag{14}$$

$$F1 - Score = \frac{2TPT}{2TPT + FPT + FNT} \times 100 \tag{15}$$

$$Accuracy = \frac{TPT + TNT}{TPT + TNT + FPT + FNT} \times 100 \tag{16}$$

where FPT is indicated as false positives, TNT is specified as true negatives, TPT is stated as true positives and FNT is represented as false negatives.

4.2 Result Analysis Using Public Drowsy Driver Database

Here, public drowsy driver database is used for evaluating the achievements of the designed and existing system (hybrid CNN-LSTM [11]). The public drowsy driver dataset comprises 22 subjects. In that, four subjects (004, 022, 026 and 030) are used for testing evaluation and the remaining subjects (001, 002, 005, 006, 008, 009, 012, 013, 015, 020, 023, 024, 031, 032, 033, 034, 035 and 036.) are used for training evaluation.

The average accuracy of SVM classifier is 88.97% and the comparative classification methodology Naïve Bayes delivers 65.47% of average accuracy. The average Precision Predictive Value (PPV) of SVM classifier is 89.83% and the comparative classification method attains 73.66% of average PPV. The average False Positive Rate (FPR) of SVM classifier is 88.72% and the comparative classification method delivers 78.22% of average FPR. In addition, the average F1-score of SVM classifier is 87.85% and the existing classification methods deliver 73.56% of average F1-score. Tables 1 and 2 clearly show that the SVM classifier with infinite feature selection algorithm performs adequately related to other classification methods on ACCV 2016 dataset. Graphical comparison of proposed system with dissimilar classifier is denoted in Fig. 4, and graphical representation of the proposed model performance is denoted in Fig. 5

Table 3 represents performance of the designed system with IFS algorithm and without IFS algorithm, and Table 4 presents the cross verification of designed system, which is verified on a database: public drowsy driver dataset. In with infinite feature selection, the classifier (SVM) averagely improves the accuracy in driver drowsiness detection up to 1.46% compared to without infinite features selection. The performance of proposed system worked well compared to all existing systems in the

Table 1 Performance evaluation of proposed system with dissimilar classifiers

Classifier	Subjects ID	PPV (%)	FPR (%)	F1-score (%)	Accuracy (%)
Naïve Bayes	004	73.4	77.30	73.47	65.69
	022	76.70	78.15	76.29	74.82
	026	68.30	76.12	71.54	62.80
	030	76.25	81.32	72.95	58.60
SVM	**004**	88.62	87.21	85.34	87.93
	022	90.56	88.59	89.26	87.49
	026	90.81	89.10	87.92	90.59
	030	89.35	90.00	88.91	89.87

Table 2 Mean value of proposed system with different classifiers and performance measures

Classifier	PPV (%)	FPV (%)	F1-score (%)	Accuracy (%)
Naïve Bayes	**73.66**	**78.22**	**73.56**	**65.47**
SVM	89.83	88.72	87.85	88.97

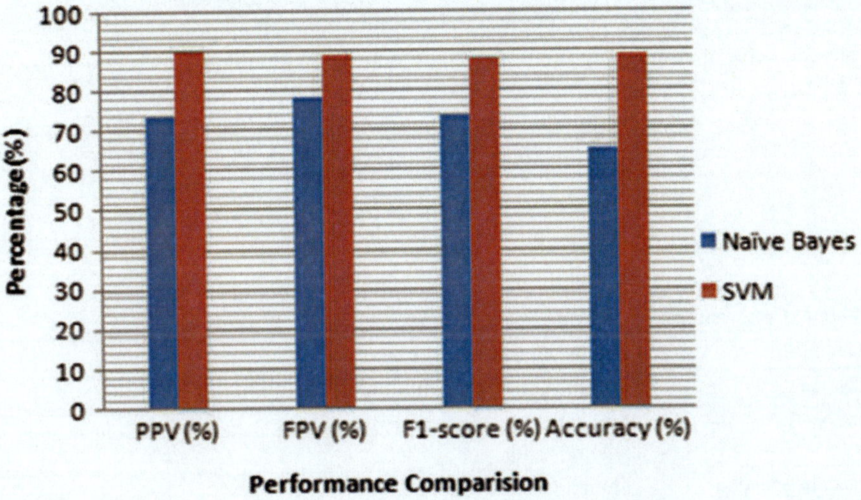

Fig. 4 Graphical comparison of proposed system with dissimilar classifiers

view of metrics like accuracy, Precision Predictive Value, False Positive Rate and F1-Score.

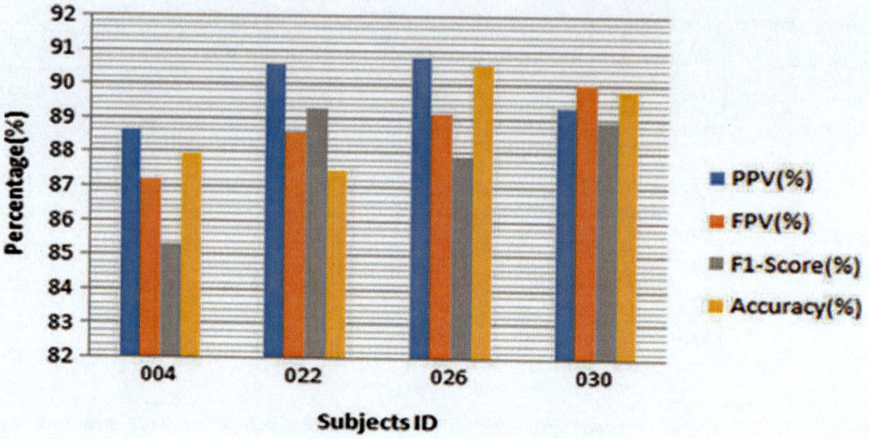

Fig. 5 Graphical representation of proposed system performance

Table 3 Performance evaluation of designed arrangement/system using with IFS and without IFS

Feature extraction	Feature selection	Classifier	Subjects ID	Accuracy (%)
Combination of LBP, HOG, head pose angle and threshold value of mouth region	Without infinite feature selection	SVM	004	87.22
			022	86.11
			026	89.78
			030	86.96
	With infinite feature selection		004	87.93
			022	87.47
			026	90.58
			030	89.91

Table 4 Cross verification of the designed system

Parameter	Verification
Dataset	Public drowsy driver dataset
Feature extraction	Combination of HOG, LBP, head pose angle and threshold value of mouth region
Feature selection	Infinite feature selection
Classification method	SVM
Total number of samples	22 samples
Training samples	18 samples
Testing samples	4 samples
Average accuracy	88.97

Table 5 Comparative analysis of proposed and existing system

Methodology	Database	Accuracy (%)
Hybrid CNN-LSTM [11]	Public drowsy driver dataset	84.85
Proposed system (SVM using infinite feature selection algorithm)	Public drowsy driver dataset	88.97

4.3 Comparison of Results

The results compared with existing and developed system are represented in Table 5. Guo and Markoni [11] developed a concept for managing driver fatigue detection using hybrid CNN-LSTM. Performance of the developed system was tested on an online dataset: public drowsy driver dataset. Compared to the existing paper with accuracy 84.85%, the designed work achieved 88.97% of accuracy and that is higher than in the available paper.

5 Conclusion

In this research paper, to detect the driver's drowsiness, a new approach was developed. The main objective of this experiment is to classify stages of drowsiness either drowsy or non-drowsy. Here, in feature selection process, Infinite Feature Selection (IFS) algorithm is used to select the optimized features, and support vector machine is used for classification purpose. From the experimental investigation, proposed approach accomplished an accuracy of 88.97%, but the present system obtained an accuracy of 84.85% only on public drowsy driver dataset. In future, a new stack of deep learning techniques will be used to extract best features and increase the accuracy of driver's drowsiness detection.

Acknowledgments The author would like to thank NTHU Computer Vision Lab for providing driver drowsiness dataset.

References

1. McDonald AD, Lee JD, Schwarz C, Brown TL (2018) A contextual and temporal algorithm for driver drowsiness detection. Accid Anal Prev 113:25–37
2. Flores MJ, Armingol JM, de la Escalera A (2010) Real-time warning system for driver drowsiness detection using visual information. J Intell Rob Syst 59(2):103–125
3. Gaspar JG, Schwarz J-CW, Brown TL, Kang J (2017) Gaze position modulates the effectiveness of forward collision warnings for drowsy drivers. Accid Anal Prev 1–6
4. Wang X, Xu C (2016) Driver drowsiness detection based on non-intrusive metrics considering individual specifics. Accid Anal Prev 95:350–357

5. Wang X, Xu C (2015) Driver drowsiness detection based on non-intrusive metrics considering individual specifics. Accid Anal Prev XXX:1–8
6. de Naurois CJ, Bourdin C, Bougard C, Vercher J-L (2018) Adapting artificial neural networks to a specific driver enhances detection and prediction of drowsiness. Accid Anal Prev 121:118–128
7. Mbouna RO, Kong SG, Chun M-G (2013) Visual analysis of eye state and head pose for driver alertness monitoring. IEEE Trans Intell Trans Syst 14(3):1462–1469
8. Al-Rahayfeh A, Faezipour M (2013) Eye tracking and head movement detection: a state-of-art survey. IEEE J Trans Eng Health Med 1:1–12
9. de Naurois CJ, Bourdin C, Bougard C, Vercher JL (2018) Adapting artificial neural networks to a specific driver enhances detection and prediction of drowsiness. Accid Anal Prev 121:118–128
10. Mittal A, Kumar K, Dhamija S, Kaur M (2016) Head movement-based driver drowsiness detection: a review of state-of-art techniques. In: IEEE international conference on engineering and technology, pp 1–6
11. Guo JM, Markoni H (2018) Driver drowsiness detection using hybrid convolutional neural network and long short-term memory. In: Multimedia tools and applications, pp 1–29
12. Weng C-H, Lai Y-H, Lai S-H (2016) Driver drowsiness detection via a hierarchical temporal deep belief network. In: Asian conference on computer vision workshop on driver drowsiness detection from video, Taipei, Taiwan, Nov. 2016

Interpretation of Statistical Errors for Precise GPS Receiver Positioning Using Recursive Least Squares and Extended Kalman Filter Approaches

P. Sirish Kumar and V. B. S. Srilatha Indira Dutt

Abstract The Global Positioning system prevalently known as GPS is a range based positioning scheme that gives a 3D position of an obscure object on top of the earth. Object location accuracy commonly depends on the satellite clock error, atmospheric delays, multipath, poor satellite geometry, and receiver measurement noise, etc. Mostly, none of the above parameters do have constant behavior throughout the world and should be inspected territorially to give an exact solution. This paper mainly concentrates on the statistical analysis of the Recursive Least Squares (RLS) and Extended Kalman Filter (EKF) algorithms. Both of these algorithms were examined in this article by determining the most frequently used positional accuracy parameters in 2D and 3D spaces for statistical error analysis in a specific region, which expose their relationships and clarify several prevalent misinterpretations about precision. For the optimization of analytical outcomes, statistical attributes can be used. To evaluate these parameters, IISC, Bangalore, GPS receiver data were possessed with RLS and EKF methods.

1 Related Study

This paper aims to involve iterative and recursive navigation algorithms, including linear algebra, to comply with ideas and system information on positioning systems accuracy analysis techniques. The literature also offers a wide array of critical solutions for positioning systems, which show an enhancement in need and demand in positioning systems [1]. Various algorithms of positioning for these measures, such as linear, non-linear, iterative, closed-form alternatives, are also researched and studied [2]. Although these algorithms provide accurate position estimates, their use is restricted to offline GPS information-handling implementations owing to their

P. Sirish Kumar (✉)
Department of ECE, Aditya Institute of Technology and Management, Tekkali 532201, India
e-mail: sirishdg@gmail.com

V. B. S. Srilatha Indira Dutt
Department of ECE, GITAM, Visakhapatnam 530045, India
e-mail: srilatha06.vemuri@gmail.com

S. C. Satapathy et al. (eds.), *Communication Software and Networks*, Lecture Notes in Networks and Systems 134, https://doi.org/10.1007/978-981-15-5397-4_42

iterative nature and time-consuming process [3]. These algorithms include complicated mathematical calculations such as Jacobian computing and the inverse matrix. Algorithms in the recurrent and direct form [4] are necessary for predictions for the position of a GPS receiver in a real-time setting. At last, to evaluate the outcome of the proposed method, GPS position accuracy measures like DRMS, 2DRMS, CEP can be utilized as 2D measures. Likewise, statistical analysis can be performed for 3D position accuracy observations by taking into account the SEP, MRSE, 3DRMS [5, 6].

2 Introduction

Global Positioning System gathers estimations or range data from realized emanating sources to decide an obscure objects position and the system is likewise called a satellite-based positioning system [7, 9]. Twenty-four-satellite constellation groups created the structure of the Global Positioning System that can reveal to you where you are in three dimensions. These 24 satellites are positioned in six orbital planes. Each orbital plane comprises of four satellites that are placed on over the earth at a height of 20,200 km altitude and 55° inclination [7]. GPS navigation and position calculation depend on estimating the separation from the user position to the exact location of the GPS satellites as they circle. By estimating the separation to four GPS satellites, it is conceivable to set up three directions of a user position (latitude, longitude, and altitude) [10] just as GPS time. Every GPS made of three major important segments [7]. They are Space: Satellites or Space Vehicles (SV) orbiting the Earth twice a day at 20,200 km. Control: Ground stations provide navigation information update and SV control. User: GPS receiver. To find the user position, a minimum of four satellite signals are required. Commonly, satellite GPS signal consists of three main components [7]. They are Carrier wave: 1575.42 MHz (LINK 1) and 1227.60 MHz (LINK 2). Navigation message: 50 bit/s contains ephemeris data (detailed orbital information for the transmitting satellite) and almanac data (more general orbital information for all satellites).

3 Recursive Least Squares Algorithm

Linearization is the prevalent way of solving the four nonlinear concurrent balances. However, generally, more than four satellites are noticeable at any point in the time of moment. If there are more than four satellites, all satellites in perspective are more common in solving the user location [7]. In this scenario, there are first linearized arrangements of nonlinear equations and then more equations than unknown. For the solution [8], you can use the commonly recognized least square approximation algorithm.

3.1 Navigation Solution with More Than Four Satellites

At the point when multiple satellites are in view from a given recipient area, an increasingly well-known way to deal with tackle the user position is to utilize every one of the satellites [11]. Then the navigation equation with more than four satellites can be given as in Eq. 1.

$$S_i = \sqrt{(a_i - a_u)^2 + (b_i - b_u)^2 + (c_i - c_u)^2} + E_u \tag{1}$$

where,

S_i is measured pseudo-range between a satellite antenna and GPS receiver
a_u, b_u, c_u are the coordinates of user position
a_i, b_i, c_i are the coordinates of satellite position
E_u is receivers clock bias.

The navigational solution can be achieved and expressed in the form of linearization of Eq. 1 [7].

$$\Delta S_i = \beta \Delta A_u \tag{2}$$

where ΔS_i is the measured pseudo-range error, β is the observation matrix, ΔA_u is the receiver's state vector error.

Equation 2 could be extended as follows:

$$\begin{bmatrix} \Delta P_1 \\ \Delta P_2 \\ \Delta P_3 \\ \Delta P_4 \\ \vdots \\ \Delta P_n \end{bmatrix} = \begin{bmatrix} \beta_{11}\,\beta_{12}\,\beta_{13}\,1 \\ \beta_{21}\,\beta_{22}\,\beta_{23}\,1 \\ \beta_{31}\,\beta_{32}\,\beta_{33}\,1 \\ \beta_{41}\,\beta_{42}\,\beta_{43}\,1 \\ \vdots \\ \beta_{n1}\,\beta_{n2}\,\beta_{n3}\,1 \end{bmatrix} \begin{bmatrix} \Delta a_u \\ \Delta b_u \\ \Delta c_u \\ \Delta E_u \end{bmatrix} \tag{3}$$

where

$$\beta = \begin{bmatrix} \beta_{11}\,\beta_{12}\,\beta_{13}\,1 \\ \beta_{21}\,\beta_{22}\,\beta_{23}\,1 \\ \beta_{31}\,\beta_{32}\,\beta_{33}\,1 \\ \beta_{41}\,\beta_{42}\,\beta_{43}\,1 \\ \vdots \\ \beta_{n1}\,\beta_{n2}\,\beta_{n3}\,1 \end{bmatrix}, \quad \beta_{i1} = \frac{a_i - a_u}{S_i - E_u} \beta_{i2} = \frac{b_i - b_u}{S_i - E_u} \beta_{i3} = \frac{c_i - c_u}{S_i - E_u}$$

Since the grid β is certifiably not a square lattice, it cannot be reversed straightforwardly. Anyway, Eq. 3 is as yet a linear equation. In several linear equations, the least square approximation can be used to find the solution when there are more equations than unknowns. To acquire the desired answer, you can use the pseudo-inverse of the β matrix. Therefore, Eq. 3's navigation solution [8, 11] was

$$\Delta A_u = [\beta^T \beta]^{-1} \beta^T \Delta S_u \tag{4}$$

Using Eq. 4, we can obtain the values of Δa_u, Δb_u, Δc_u, and ΔE_u. Generally, speaking, the least square approximation provides a better alternative than only four satellites, as more information is used. The traditional way in which a GPS receiver is determined needs an iterative algorithm, which involves original estimates. The navigation solution in the recursive least squares approach originates from the estimated location of the user and is then predicted using that iterative approach to the actual position of the user.

4 Extended Kalman Filter

In EKF, the noisy measurement sequences overtime is contemplated to determine unknown state vector residuals. These validations will give some additional accuracy than having a single measurement [4, 8]. System model and measurement of linearity are utilized in EKF to predict system residuals which makes EKF exactly suitable for dynamic GPS receivers as it gives an accurate navigational solution. EKF method was implemented in two phases [4, 11, 12]. They are time update phase and measurement update phase.

$$J_G = F a_G + W_G \tag{5}$$

$$a_G = \phi_G a_{G-1} + \omega_G Y_{G-1} + \gamma_G \tag{6}$$

where φ_G, ω_G, F are the state transition, control input, and observation matrices respectively; γ_G, W_G are the process and measurement noises, respectively; J_G is the required measurement at time G. The matrix F in Eq. 3 is comparable to β and is indicated in Eq. 2. EKF [7, 11] uses pseudo-range as the observable vector, with its state vector a_G as position feature $[a_u\ b_u\ c_u]$, and the state vector estimate is set by the following equations. In reality, there is no real state and the real state a_G is therefore substituted by its estimate $\hat{a}_G$.

Phase 1: Time Update:

$$\hat{a}_G^- = \phi \hat{a}_{G-1} + \gamma_G \tag{7}$$

$$S_G^- = \phi_{G-1} S_{G-1} \phi_{G-1}^T + H_{G-1} \tag{8}$$

$$K_G = S_G^- F^T (F S_G^- F^T + W_G)^{-1} \tag{9}$$

Phase 2: Measurement Update:

$$\hat{a}_G = \hat{a}_G^- + K_G (J_G - F \hat{a}_G^-) \tag{10}$$

$$S_G = (I - K_G F) S_G^- \tag{11}$$

where $\hat{a}_G^-$ is the estimate of state vector at time G before update, S_G^- is the estimate of covariance at time G before update, K_G is the EKF gain, $\hat{a}_G$ is the state vector estimate at time G after update, S_G is the covariance at time G after update, I is the Identity matrix, and H_{G-1} is the system noise covariance matrix. After observation–prediction, we constantly calculate the prediction process.

5 Results

The RLS and EKF algorithm performances were analyzed in a particular zone (South zone of India) with the navigational raw data collected from IISc, Bangalore, over for 24 hours. Table 1 provides information on RLS and EKF's three-error pdfs over GPS receiver data from IISc, Bangalore. These parameters are utilized in figuring the 2D and 3D position accuracy estimates appeared in Tables 2 and 3, respectively. Accuracy measures are the factual strategies for portraying the execution of GPS

Table 1 X, Y, Z co-ordinate error analysis with RLS and EKF for 24-h data

Parameter	Recursive least squares			Extended Kalman filter		
	X (m)	Y (m)	Z (m)	X (m)	Y (m)	Z (m)
Error (max.)	63.3	55.9	13.5	46.5	46.2	18.3
Error (min.)	22.0	19.6	0	23.1	1.46	0
Mean error (μ)	32.9	32.9	5.10	33.6	27.7	4.60
Standard deviation (σ)	6.24	7.66	2.76	5.10	10.2	4.49
Variance (σ^2)	38.938	58.676	7.618	26.01	100.400	20.160

Table 2 2-dimensional statistical error analysis of RLS and EKF

Statistical parameter	RLS (m)	EKF (m)
DRMS	9.8829	11.465
2DRMS	19.765	22.931
3DRMS	29.648	34.396
R95	17.0974	19.8345
CEP	8.2462	9.2250

Table 3 3-dimensional statistical error analysis of RLS and EKF

Statistical parameter	RLS (m)	EKF (m)
SEP	8.5015	10.133
MRSE	10.262	12.316
90% spherical accuracy standard	13.8778	16.4851
99% spherical accuracy standard	18.6925	22.2044

position estimation calculations. It is very well seen from Tables 2 and 3 that half of the assessed horizontal point positions (i.e., x, y) by RLS will be inside 8.25 m of the genuine position and half of the evaluated 3D point positions will be inside 8.5 m, though it is 9.22 and 10.1 m for EKF individually. Even though that the CEP and SEP estimations of RLS indicate it is exact by 1–2 m than EKF. The estimated positions of x, y, z co-ordinates compared with the original receiver position by using both algorithms, shown in Figs. 1, 2, and 3.

This paper utilizes one progressively factual measure ‘confidence level’ to break down the exhibition of these algorithms. The zone of the curve inside which the estimations or assessed parameters are well on the way to be is called the confidence level. The confidence level or probability that the error will be inside 40 m is appeared both the calculations in Table 4.

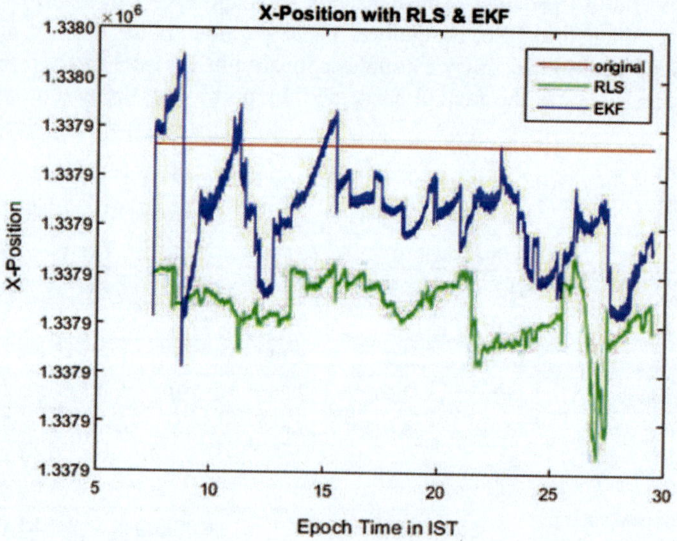

Fig. 1 X-position with RLS and EKF

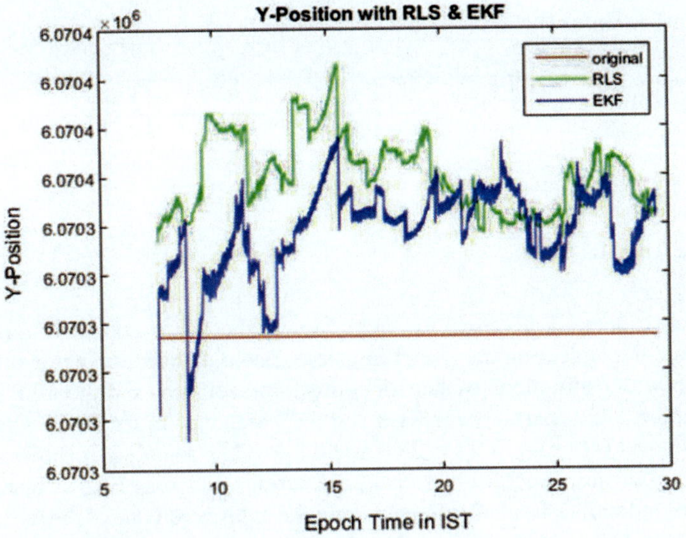

Fig. 2 Y-position with RLS and EKF

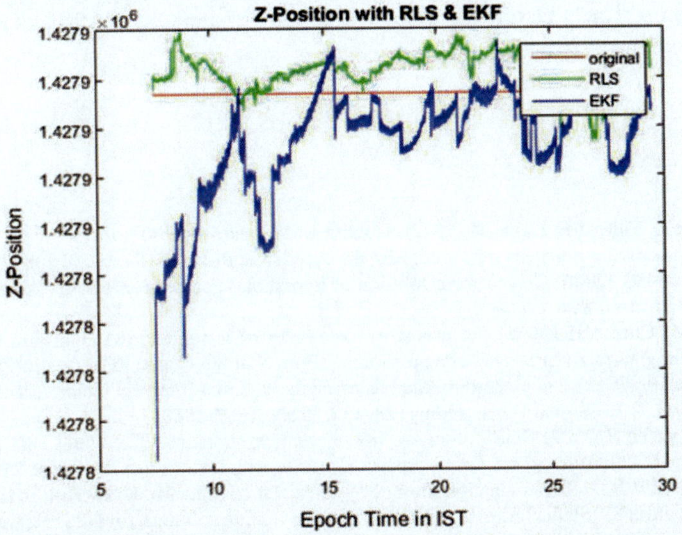

Fig. 3 Z-position with RLS and EKF

Table 4 Confidence level of 35 and 40 m for RLS and EKF

Method	CL 35			CL 40		
	X %	Y %	Z %	X %	Y %	Z %
RLS	63	60	100	90.4	78.4	100
EKF	62	76	100	90.2	90.2	100

6 Conclusion

This paper especially focuses to break down the RLS and EKF algorithms performance-dependent on the IISc, Bangalore, raw data over some undefined time frame 24 h with estimations of statistical position accuracy, and the CEP and SEP are 8.25 and 8.5 m separately for RLS and 9.23 and 10.1 m for EKF, whereas the confidence level (40 m) of 0.9 for EKF and 0.8 for RLS is noted. It shows over this area (IISC, Bangalore), RLS gives exact evaluations, whereas EKF is having great precision of more than 40 m. Contingent upon the application one of the two methodologies may demonstrate helpful as RLS assumes a significant job in evaluating the raw quality of the data, while EKF gives extra learning about system concealed parameters (Velocity, Direction, etc.,) those can't be observed by RLS. In this paper, the position accuracy factors can also be used for Russian GLONASS and European Galileo systems.

References

1. Dardari D, Falletti E, Luise M (2012) Satellite and terrestrial radio positioning techniques: a signal processing perspective, 1st ed. Elsevier Academic Press, ISBN: 9780123820846
2. Bakhoum EG (2006) Closed-form solution of hyperbolic geolocation equations. IEEE Trans Aerosp Electron Syst 42(4)
3. Yang M, Chen KH (2001) Performance assessment of a noniterative algorithm for global positioning system (GPS) absolute positioning. Proc Natl Sci Counc ROC (A) 25(2):102–106
4. Faragher R (2012) Understanding the basics of the Kalman filter via a simple and intuitive derivation. IEEE Signal Process Mag 128–132, ISSN: 1053-5888
5. van Diggelen F (2007) GNSS accuracy: lies, damn lies, statistics. GPS World 18(1):26–32
6. Rutledge D (2010) Accuracy versus precision: a Primer on GPS truth. GPS World 21(5):42–49
7. Rao GS (2010) Global navigation satellite systems, 1st ed. Tata Mc-Graw Hill, India
8. Sirish Kumar P, Srilatha Indira Dutt VBS (2019) Navigation solutions for GPS receiver position estimation over the southern region of India. Int J Recent Technol Eng (IJRTE) 7(6), ISSN: 2277-3878
9. Conley R (1994) An overview of the GPS standard positioning service signal specification. In: Proceedings of the 7th international technical meeting of the satellite division of the institute of navigation (ION GPS 1994), September 20–23, 1994, Salt Palace Convention Center, Salt Lake City, UT, pp 179–188
10. Srilatha Indira Dutt VBS, Sasibhushana Rao G, Swapna Rani S, Goswami R (2009) Investigation of GDOP for precise user position computation with all satellites in view and optimum four satellite configurations. J Indian Geophys Union 13(3):139–148

11. Laveti G, Sasibhushana Rao G et al (2016) Modified Kalman filter for GPS position estimation over the Indian subcontinent. Elsevier Procedia Comput Sci J 87:198–203, ISSN: 1877-0509
12. Hassibi A, Boyd S (1998) Integer parameter estimation in linear models with applications to GPS. IEEE Trans Signal Proces 46:2938–2952

Robust Blood Vessels Segmentation Based on Memory-Augmented Neural Network

K. Arunabhaskar and R. Kiran Kumar

Abstract Retinal blood vessel segmentation is used in various areas and, for instance, it can be used as an indicator of diseases such as hypertension, diabetes, and arteriosclerosis. In this research, Memory-Augmented Neural Network (MANN) is used to segment the blood vessel from the fundus image of the eye. The fundus images are obtained from the two databases such as DRIVE and STARE; those are considered to be a standard database. The preprocessing techniques such as Global Contrast Normalization and Zero-phase Component, used by the classifiers to function on reliable features in the image. The feature is extracted from the input and it is used by the MANN to classify the blood vessel in the database images. The simulation result showed that the proposed MANN gave a better performance compared to existing methods. The accuracy of this method attained up to 97% in the STARE database and also achieved higher values in other parameters. MANN refers to the type of neural network equipped with external memory, which helps in achieving high accuracy classification for learning long-term dependency.

1 Introduction

Different growth in the retinal vascular formation in the eye is the indication of some diseases, including diabetes, cardiovascular disease, hypertension, and stroke [1, 2]. Study of these vessels is one of the important tasks in the medical analysis because it can be used for diagnosis several medical pathologies [3, 4]. Age-related Macular Degeneration and Diabetic retinopathy are two major reasons for the cause of blindness and vision impairment [5, 6]. Diabetic retinopathy occurs mainly on working-age people across the world and cause of modification in blood vessel structure. So, finding the changes in the blood vessel of retina helps in diagnosing

K. Arunabhaskar (✉) · R. Kiran Kumar (✉)
Department of Computer Science, Krishna University, Machilipatnam, Andhra Pradesh, India
e-mail: letter2arunbhaskar@gmail.com

R. Kiran Kumar
e-mail: kirankreddi@gmail.com

S. C. Satapathy et al. (eds.), *Communication Software and Networks*, Lecture Notes in Networks and Systems 134, https://doi.org/10.1007/978-981-15-5397-4_43

the diseases. Some features in the retinal vessels such as branching pattern, width, and length help in diagnosis some diseases [7]. The blood vessels of the eye appear in the fundus images and this can be used to identify the changes in blood vessels. Blood vessels of humans are directly visible from the backside of the eye in a non-invasive way. Color fundus images are taken by making photography of the backside of the eye and these images are largely available in the database for research purposes [8].

The blood vessel of retina appeared to have low contrast in the images, which makes it difficult to segment the blood vessel from the background. This makes the needs of image processing technique to accurately segment the blood vessels from dataset image and these techniques depend on features such as uniform edges, cross-sectional profiles, and intensity regions [9]. Every year, the number of diabetic retinopathy patients is increasing and causes blindness in patients. In order to minimize the clinician workload and resource utilization, extraction of retinal blood vessels with high accuracy is needed [10]. In this research, blood vessels are classified using the Memory-Augmented Neural Network (MANN). MANN uses external memory, while other networks have internal memory architecture. This method has achieved the highest classification accuracy compared to the other existing methods. The DRIVE and STARE are the two different datasets used in this method for evaluating the performance of the proposed method. The experimental result is measured with two types of preprocessed images and plain images, shown in the experimental result section.

2 Proposed System

The proposed method involves three processes, namely: (1) preprocessing, (2) Feature extraction, and (3) Classification. The MANN is used for the classification, which achieves higher classification for long-term learning due to its external memory. The Global Contrast Normalization and Zero-phase Component analysis are the two preprocessing techniques used in this method.

2.1 *Processing*

The deep neural network has the capacity to process the raw fundus images and, using the pre-processing technique, the accuracy of the method will be increased. Two kinds of pre-processing techniques carried out in this method. The two methods of preprocessing are (1) Global Contrast Normalization and (2) Zero-phase component analysis.

2.2 Global Contrast Normalization

The brightness of Figs. 2 and 3 clearly indicates that brightness may vary across the field of view. This will provide abstract from these fluctuations and focus on vessel detection, then performs local brightness and contrast normalization. The patch of the images is subtracted from the average and dividing with a standard deviation of its elements. These operations are made independently in R, G, and B channels.

2.3 Zero-Phase Component Analysis

The neighboring pixel in the natural images is highly correlated and they are represented as the same structure in the scene. While learning a statistical model of images, it has to neglect the universal characteristics of the image and focus on the higher order correlations. The universal characteristics are neglected by multiplying the data matrix with a whitening matrix. This makes it difficult to measure the value of one pixel from giving only one other value pixel in the deep learning. Formally, feature data stored in the centered matrix X, features in columns, and data points in rows. The value of X is taken as $3 \times 27 \times 27 = 2187$ columns and number of rows similar to the number of examples. The covariance matrix $\Sigma = \frac{1}{\sum_{i=1}^{m} X \times X^T}$ has eigenvectors in columns of U and eigenvalues on the diagonal of Λ, so that $\Sigma = U \Lambda U^T$. Then, U denotes the orthogonal rotation matrix and U^T provides the rotation needed to de-correlate the data. The equation of classical PCA whitening transformation is given as:

$$W_{\text{PCA}} = \Lambda^{\frac{-1}{2}} U^T \tag{1}$$

Each component W_{PCA} contains variance provided by the respective eigenvalue.

However, whitening is not unique and the data of whitening remain whitened under rotation. That shows $W = RW_{\text{PCA}}$ with an orthogonal matrix R is the whitening transformation. When the zero-phase component whitening happens, U is considered as an orthogonal matrix

$$W_{\text{ZCA}} = U \Lambda^{\frac{-1}{2}} U^T = \Sigma^2 \tag{2}$$

These preprocessed patterns are used by the feature extraction technique to extract the features to perform the blood vessel classification.

2.4 *Memory-Augmented Neural Network*

The controller network is developed for updating the dynamic state of the recurrent layer from the time step $t-1$. This is different from long short-term memory having the dynamic state $Z_t = z_t$ rejected to access the external memory M_t. This kind of network learns from long-term dependencies in a de-mixing system and stores in external memory. The MANN has the tendency of storing longer sequential temporal data compared to LSTM. There are two basic components in MANN. The first one is the controller network, which operates differently from the LSTM with the specialized configuration for interactions with memory. The second one is addressing mechanism depending on the content-based soft attention and the real-valued de-mixing measures in sequential read and write. These two components are discussed in this section.

2.4.1 Local Connectivity

Local connectivity refers that a given unit gets the data only form its receptive field (RF), which is the pixel of the small rectangle in the image or units in the previous layer. The neighboring unit's layers of RFs are typically offset by stride. The dimensions of a layer depending on the image size, stride, and RF size together. For example, a unit layer with 3×3 RFs with one-pixel stride needs only nine units when applied to a 5×5 monochromatic image, because three RFs of width 3 each, overlapping by two pixels, span the entire image. The larger stride and larger RF create the smaller layer and local connectivity that reduces the number of weights in comparison to the fully connected conventional networks. This is reliable to the spatial nature of information related to the visual of image and mimics certain aspects of the natural visual system.

2.4.2 Parameter Sharing

The measure of sharing of weight across units in the same layer is the parameter sharing and when the unit in the layer shares the same weight vectors to others, then the feature map is created. Each of them calculates the same local feature, albeit from a different part of the image. This will help to minimize the number of parameters even further and makes the extracted features equivalent. If the single-channel images with a unit layer 3×3 RFs have only ten parameters, nine channels for the RF pixels, and one for neuron threshold independent of the number of units.

2.4.3 Pooling

The pooling collects multiple units' outputs using convolution. Max-pooling is the most popular aggregation pooling, each unit in pooling returns the highest values in its RFs. The function of pooling is like the local sharing, which reduces the resolution of the previous layer and gives translation invariance. The architecture of MANN consists of several max-pooling layers and finalized with one fully connected layer that allows only the excitations in output neurons, where each neuron relates to one decision class. The layers tend to be smaller due to sliding RFs by the number of pixels defined in stride across the input images and finally grid given as input into the fully connected is generally much smaller than input images. Often numerous feature maps working in parallel and each of the maps are responsible for extracting one feature. In the case of multi-channel images, separate feature maps are connected to all channels. These features are utilized by the classifier MANN for the segmentation of the blood vessel from DRIVE and STARE image.

2.4.4 Controller Networks

The LSTM variant is used to design the controller networks

$$i_t = \sigma\left(W_{ix}z_t^{(l-2)} + W_{iz}z_{(t-1)}^{l-1} + W_{ir}r_t - 1 + b_i\right)$$

$$f_t = \sigma\left(W_{fx}z_t^{l-2} + W_{fz}z_t - 1^{l-1} + W_{fr}r_t - 1 + b_f\right)$$

$$o_t = \sigma\left(W_{ox}z_t^{l-2} + W_{oz}z_t - 1^{l-1} + Wr_t - 1 + b_O\right)$$

$$u_t = \tanh\left(W_{ux}z_t^{(l-2)} + W_{uz}z_{t-1}^{(l-1)} + W_{ur}r_{t-1} + b_u\right)$$

$$c_t = f_t \odot c_t - 1 + i_t \odot u_t$$

$$z_t^{(l-1)} = o_t \odot \tanh(c_t) \tag{3}$$

where W_{ix}, W_{fx}, W_{ox} and W_{ux} denote the weight parameters from the input $z_t^{(l-2)}$ at current time t to input gate, forget gate, output gate, and cell, respectively.

2.4.5 Addressing Mechanism

The four steps of addressing the process are controlled by addressing parameters in recurrent layers. The external memory at the time t is expressed as a matrix $M_t \in R^{N \times M}$ consists of N memory slots with an M-dimensional vector in each

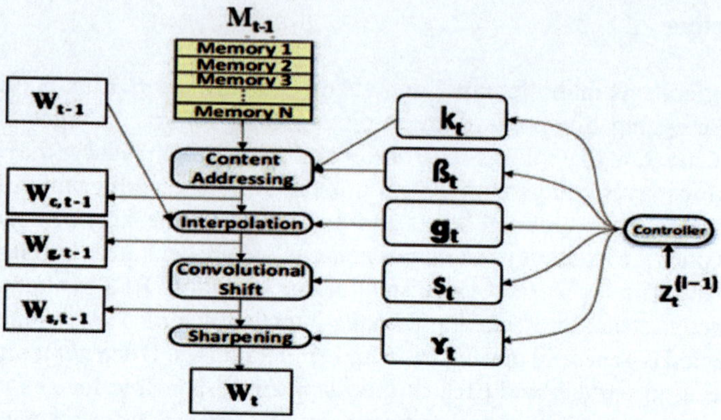

Fig. 1 Four steps of addressing procedure which is driven by parameters $\{k_t, \beta_t, g_t, s_t, \gamma_t\}$

slot. The ith memory slot is denoted by $M_t(i)$. The soft attention of content-based is fulfilled and in the first step, the content-based addressing is performed by measuring initial attention (Fig. 1).

$$\omega_{c,t}(i) = \frac{\exp(\beta \cos(M_t(i), k_t))}{\sum_j \exp(\beta_t \cos(M_t(j), k_t))} \tag{4}$$

A common procedure in examining the eye is retinal imaging and an optical camera is used to view the rear inner surface of the eyeball through pupil of the eye. The picture is captured and the optic nerve, fovea, the retinal layer, and surrounding vessels can be seen in the images. The images are analyzed and referred to while considering any observed findings.

Where the cosine similarity $\cos(M_t(i), k_t)$ measures the strength of a memory slot $M_t(i)$ accessed by a key $K_t \in R^M$. This similarity is weighted by β_t to find the normalized attention $0 \leq \omega_{c,t}(i) \leq 1$. Secondly, an interpolation step is performed to update the attention as $w_{g,t} = (\omega_{g,t}(i))$ by using scalar interpolation gate $0 \leq g_t \leq 1$ which balances between the content-based attention $w_{c,t} = \omega_{c,t}(i)$ and the read or write head $wt - 1 = w_{t-1}(i) \in (w_{r,t-1}, w_{\omega,t-1})$ at consecutive time steps $t - 1$ and t

$$\omega_{g,t}(i) = g_t\omega_{c,t}(i) + (1 - g_t)\omega_{t-1}(i) \tag{5}$$

A convolution shift is performed in the third step for location-based attention by blurring the attention over S points instead of spotting at a single point or location. The attention is then updated to $w_{s,t} = (\omega_{s,t}(i))$, where $\omega_{s,t}(i) = \sum_{j=0}^{S-1} \omega_{g,t}(j)s_t(i - j)$. Here, $s_t = \{s_t(i)\}$ denotes a set of S shift weights. In the fourth step, the read or write head is updated by

$$\omega(i) = \frac{\omega_{s,t}(i)^{\gamma t}}{\sum_j \omega_{s,t}(j)^{\gamma t}} \quad (6)$$

where the dispersed vector $w_{s,t}$ due to the convolution shift is sharpened by using $\gamma_t \geq 1$.

2.4.6 Reading and Writing

Reading and writing operations are carried out at every time t to retrieve and store the de-mixing information in recurrent layers for separating monaural source, respectively. In the reading phase, a read vector of length M is calculated by $r_t = \sum_{i=1}^{N} \omega_{r,t}(i) M_t(i)$ which is obtained by attending over N memory slots using the weights of read head $w_{r,t} = \{\omega_{r,t}(i)\}$. This read vector r_t is incorporated into the controller network for updating the recurrent variables z_{t+1}^{l-1} at the next time step $t + 1$. In addition, the writing process consists of two operations, erase and add, which are controlled by write head by using the erase vector e_t and add vector a_t, respectively. The ith memory slot is updated by erasing and adding

$$M_t(i) = M_{t-1}(i) \odot \left[1 - \omega_{\omega,t}(i) e_t\right] + \omega_{\omega,t}(i) a_t \quad (7)$$

Algorithm 1 Recurrent layers in MANN

Input: $\mathbf{z}_t^{(l-2)}, \mathbf{z}_{t-1}^{(l-1)}, \mathbf{r}_{t-1}, \mathbf{w}_{t-1} = \{\mathbf{w}_{t-1,r}, \mathbf{w}_{t-1,w}\}$

$\mathbf{z}_t^{(l-1)} = \text{LSTM}(\mathbf{z}_t^{(l-2)}, \mathbf{z}_{t-1}^{(l-1)}, \mathbf{r}_{t-1})$

$\mathbf{w}_t = \text{Address}(\text{Controller}(\mathbf{z}_t^{(l-1)}), \mathbf{w}_{t-1}, \mathbf{M}_{t-1})$

$\{\mathbf{e}_t, \mathbf{a}_t\} = \text{Controller}(\mathbf{z}_t^{(l-1)})$

$\mathbf{r}_t = \text{Read}(\mathbf{w}_{r,t}, \mathbf{M}_{t-1})$

$\mathbf{M}_t = \text{Write}(\mathbf{w}_{w,t}, \mathbf{e}_t, \mathbf{a}_t, \mathbf{M}_{t-1})$

Output: $\mathbf{z}_t^{(l-1)}, \mathbf{r}_t, \mathbf{w}_t = \{\mathbf{w}_{r,t}, \mathbf{w}_{w,t}\}$

Algorithm 2 Memory augmented source separation

Input: $\mathbf{x}_t^{\text{mix}}$: mixed signal

Feature extraction: $\mathbf{z}_t^{(l-1)} = \text{MANN}(\mathbf{x}_t^{\text{mix}})$

Mask prediction: $\mathbf{y}_t^1, \mathbf{y}_t^2$

Source signal estimation: $\widehat{\mathbf{x}}_t^i = \mathbf{y}_t^i \odot \mathbf{x}_t^{\text{mix}}$ for $i = 1, 2$

Output: $\widehat{\mathbf{x}}_t^1, \widehat{\mathbf{x}}_t^2$: estimated source signals

MANN classifies the blood vessel from the database images and the outcome is evaluated using the evaluation metrics. These parameters are compared with existing systems to analyze the function of the proposed method (Figs. 2, 3, and 4).

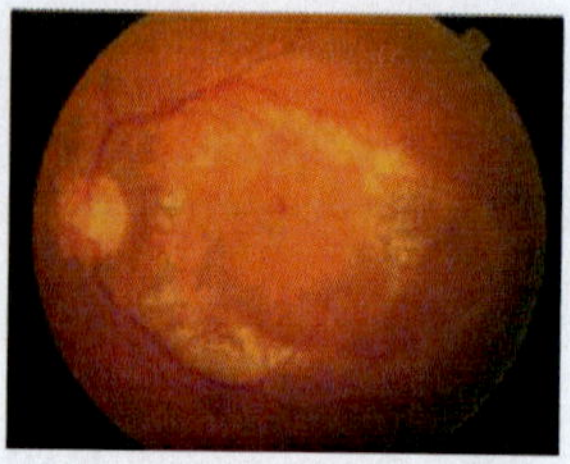

Fig. 2 Sample image of STARE database

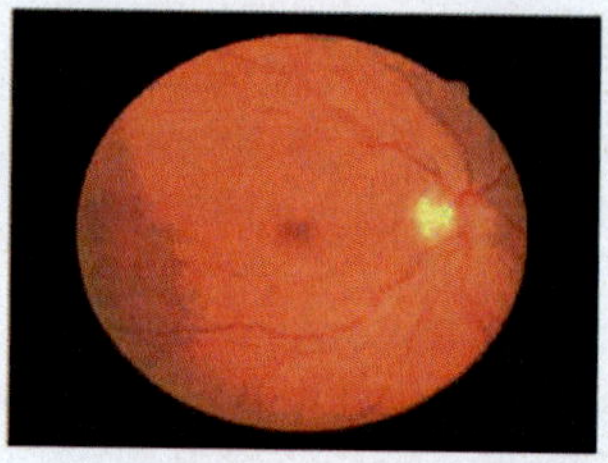

Fig. 3 Sample image of DRIVE database

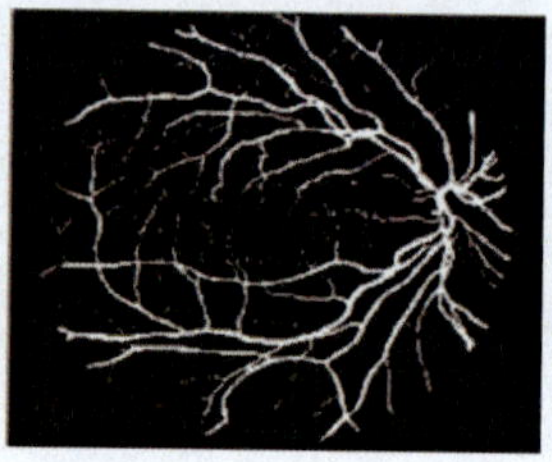

Fig. 4 Manual segmentation of blood vessels

3 Conclusion

MANN classifiers helped in classifying the blood vessel from the fundus image, which was taken from the two databases. A number of images from DRIVE and STARE database were taken and the blood vessels were segmented. The experiment results were evaluated from the plain image and preprocessed image. This research work made a comparison with the value of the proposed method and existing methods based on two different databases and this clearly showed that the proposed method has higher accuracy compared to the other methods. The accuracy of the MANN achieved up to 97% and some parameters were also measured. The manual segmentation of the blood vessel is also available in the database. The classifiers used these images to learn the features and help in classifying the blood vessels from the images. The specificity of some methods is high compared to the proposed method and some feature needs to be used to improve the specificity of this MANN.

References

1. Nguyen UT, Bhuiyan A, Park LA, Ramamohanarao K (2013) An effective retinal blood vessel segmentation method using multi-scale line detection. Pattern Recogn 46(3):703–715
2. Rodrigues LC, Marengoni M (2017) Segmentation of optic disc and blood vessels in retinal images using wavelets, mathematical morphology and Hessian-based multi-scale filtering. Biomed Signal Process Control 36:39–49
3. Aslani S, Sarnel H (2016) A new supervised retinal vessel segmentation method based on robust hybrid features. Biomed Signal Process Control 30:1–12
4. You X, Peng Q, Yuan Y, Cheung YM, Lei J (2011) Segmentation of retinal blood vessels using the radial projection and semi-supervised approach. Pattern Recogn 44(10–11):2314–2324
5. Colomer A, Naranjo V, Engan K, Skretting K (2017) Assessment of sparse-based inpainting for retinal vessel removal. Sig Process Image Commun 59:73–82
6. Javidi M, Pourreza HR, Harati A (2017) Vessel segmentation and microaneurysm detection using discriminative dictionary learning and sparse representation. Comput Methods Programs Biomed 139:93–108
7. Franklin SW, Rajan SE (2014) Computerized screening of diabetic retinopathy employing blood vessel segmentation in retinal images. Biocybern Biomed Eng 34(2):117–124
8. Zhou L, Yu Q, Xu X, Gu Y, Yang J (2017) Improving dense conditional random field for retinal vessel segmentation by discriminative feature learning and thin-vessel enhancement. Comput Methods Programs Biomed 148:13–25
9. Singh NP, Srivastava R (2016) Retinal blood vessels segmentation by using Gumbel probability distribution function based matched filter. Comput Methods Programs Biomed 129:40–50
10. Zhu C, Zou B, Zhao R, Cui J, Duan X, Chen Z, Liang Y (2017) Retinal vessel segmentation in colour fundus images using extreme learning machine. Comput Med Imaging Graph 55:68–77

Mutual Information and Machine Learning Based Distinguishers for Pseudo Random Bit Sequences

Bharat Lal Jangid

Abstract In cryptology, a strong pseudo random bit sequence must be indistinguishable from pure random bit sequence or other strong pseudo random bit sequence. In this paper, distinguishers based on machine learning and mutual information are presented to distinguish various pseudo random bit sequences (generated by AES, 3DES, IDEA, Blowfish, RC4, and '*rand*' function of C) from the strong pseudo random bit sequence of AES in CBC mode. Significant features are selected by using mutual information from the extracted features. SVM, Resilient Propagation ANN, and Multinomial Naïve Bayes classifiers are used to distinguish. It is observed that pseudo random bit sequences of four block ciphers in ECB mode are distinguishable with more than 99% accuracy as compared to CBC and OFB mode. Bit sequences generated by RC4 stream cipher and '*rand*' function of C are also significantly distinguishable.

1 Introduction

In present era of technology, communication security in presence of adversary is a challenging task. Confidentiality can be achieved through various cryptographic approaches [1]. Pseudo random bit sequences are generated by algorithmic process by using secret key [2, 3]. Cipher text/keystream generated by any strong block/stream cipher must possess randomness characteristics and it must be indistinguishable from a true/pure random bit sequence (generated by physical means) or other strong pseudo random bit sequence. The property of indistinguishability does not reveal any clue to adversary about the source of generator for cryptanalysis. Distinguishing the pseudo random bit sequence is an important area of research in cryptanalysis and some of the recent developments are presented in next section.

B. L. Jangid (✉)
Scientific Analysis Group, Defence R&D Organization, Delhi, India
e-mail: bharatlal_ce@yahoo.com

S. C. Satapathy et al. (eds.), *Communication Software and Networks*, Lecture Notes in Networks and Systems 134, https://doi.org/10.1007/978-981-15-5397-4_44

2 Related Work

Golic et al. presented statistical distinguisher for irregularly decimated linear recurring sequences, and it was based on the correlation analysis of clock-controlled Linear Feedback Shift Registers (LFSR) [4]. This approach consists of a probabilistic reconstruction of bits in the regularly clocked LFSR sequence. Martin et al. suggested a distinguisher for stream ciphers, which has capability to exploit certain weaknesses of feedback polynomials [5]. It generalizes that polynomials having tap points in several groups and far away may lead to distinguish the generated bit sequence from random one. Chang et al. developed distinguishers for four-branch and eight-branch Generalized Feistel Network (GFN) in a known key attack model to conclude that eight-branch type GFN is weaker than four-branch type GFN [6]. Grassi et al. proposed a distinguisher for five-round AES based on structural differential property [7]. All of these discussed distinguishing approaches generally exploited the generator-specific weaknesses or properties to distinguish.

Some development based on machine learning approaches is also there in this area. Dileep et al. suggested an approach for identification of encryption method for block ciphers using document vector with Support Vector Machine (SVM) and presented their performance on five block ciphers [8]. Sharif et al. developed a technique to classify encryption algorithms using pattern recognition approaches [9]. In this approach, four block ciphers were distinguished by using eight pattern recognition techniques in Electronic Code Book (ECB) [1] mode.

In this paper, mutual information and machine learning based distinguishers are presented to distinguish a pseudo random bit sequence from the strong pseudo random bit sequence. Here, cipher text generated by Advance Encryption Standard (AES) [10] in Cipher Block Chaining (CBC) [1] mode is taken as reference strong pseudo random bit sequence. Significant features are selected from the extracted features of bit sequences by using mutual information approach. Three machine learning models (SVM, Resilient Propagation Artificial Neural Network (RPROP-ANN), and Multinomial Naïve Bayes (MNB)) are used to distinguish them. Four block ciphers (AES, 3DES [11], IDEA [12], and Blowfish [13]) in ECB mode and bit sequences from RC4 and '*rand*' function of C language are distinguishable.

Three machine learning techniques are discussed in Sect. 3, which are used here. Feature extraction and selection are discussed in Sect. 4. In Sect. 5, proposed distinguishing approach is discussed. Data generation and experimental setup is discussed in Sect. 6. Results and their discussion are described in Sect. 7. The paper is concluded in Sect. 8.

3 Brief Discussion of Machine Learning Techniques

Following machine learning approaches are used in the proposed distinguisher.

3.1 Support Vector Machine

Support Vector Machine is an optimal classifier with maximum margin on both side of discriminatory hyperplane [14]. Let $\{X_i, d_i\}_{i=1}^{N}$ represent training samples where X_i is the input pattern (feature vector) of the ith example and d_i is the corresponding class label which can be -1 or $+1$. The discriminatory hyperplane is $g(X) = W_0^T X + b_0$ where X is an input feature vector, W_0 is an adjustable weight vector, and b_0 is bias. It is an optimization problem to find weight vector W and bias b that minimize $\emptyset(W) = (1/2)W^T W$ subject to constraints $d_i\left(W^T X_i + b\right) \geqslant 1$ where $i = 1, 2, 3, \ldots, N$. Its dual representation finds α that maximizes the objective function $Q(\alpha) = \sum_{i=1}^{N} \alpha_i - (1/2) \sum_{i=1}^{N} \sum_{j=1}^{N} \alpha_i \alpha_j d_i d_j K\left(X_i, X_j\right)$ subject to constraints $\sum_{i=1}^{N} \alpha_i d_i = 0$ and $\alpha_i \geqslant 0$ where $i = 1, 2, 3, \ldots, N$. Here, $K\left(X_i, X_j\right)$ is kernel function that computes dot product of feature vectors into higher dimensional space. A polynomial kernel function is chosen in our experiment.

3.2 Resilient Propagation Artificial Neural Network

Resilient propagation is a variant of backpropagation model of ANN that improved the drawback of unequal and slow learning due to inherent disadvantages of pure gradient [15]. It uses direct adaptive learning method that uses sign of gradient instead of magnitude for fast adaption. A four-layer architecture including two hidden layers is used in the proposed distinguisher. The number of neurons in input layer is equal to number of features. Number of neurons in hidden layer 1, hidden layer 2, and output layer are 100, 25, and 1, respectively. Random initial weights and sigmoid activation function is used. Threshold mean square error and number of epochs are chosen as 0.0001 and 500, respectively, to terminate training.

3.3 Multinomial Naïve Bayes Classifier

In the proposed distinguisher, all features are positive integers so that Multinomial Naïve Bayes statistical classifier is chosen [16]. It is a variant of Bayes classifier which assumes that each feature is independent of others and sample space follows multinomial distribution. Let $X = \{x_1, x_2, x_3, \ldots, x_d\}$ be d-dimensional input feature vector of a sample that represents value of features $f_1, f_2, f_3, \ldots, f_d$, respectively, and C_i is class label, then posterior probability $p(C_i/X) = p(X/C_i)p(C_i)/p(X)$ can be computed for each class where $p(X/C_i)$ is likelihood and $p(C_i)$ is prior probability of ith class. In multinomial distribution, class conditional probability $p(X/C_i) \propto$

$\prod_{j=1}^{d} p(f_j/C_i)^{x_j}$. So, posterior probability $p(C_i/X) \propto p(C_i) \prod_{j=1}^{d} p(f_j/C_i)^{x_j}$. For classification of input sample, posterior probability for each class is computed and sample is assigned to the class which has maximum posterior probability.

4 Feature Extraction and Selection

Extraction of features and selection of significant features is the foremost activity in distinguishing problem. The proposed feature extraction technique is as follows:

4.1 Feature Extraction Approach

The input pseudo random sequence is binary which has to be converted into feature vectors (called samples). The features should cater class-dependent characteristics so that frequencies of all possible j-bit (where $j = 1$ to 10) words are computed in overlapping manners in blocks of input sequence. Proposed algorithm is as follows:

i. **Input:** Pseudo random bit sequence of length N bits.
ii. Divide input bit sequence of length N bits into blocks of length M bits, so that total number of blocks are $F = \lfloor N/M \rfloor$.
iii. Initialize *feature_count =0*;
iv. For *i =1 to F* do
    For *j = 1 to 10* do
        For *k = 0 to 2^j-1* do
            Compute *Feature[feature_count]* = Frequency of *j*-bit word having decimal value equal to *k* in current *i^th* M bits block in overlapping manner.
            feature_count=feature_count+1;
        End do
    End do
  End do
v. **Output:** F × 2046 matrix of *F* samples, where each sample has 2046 features.

4.2 Feature Selection Based on Mutual Information

All the extracted features may not be significant so that selection of significant features is important. The feature set S that exhibits maximum average mutual information between each feature $f_i \in S$ and class label c is the maximum relevant feature set as given below [17].

$$\max Rel(S, c), \quad Rel = (1/|S|) \sum_{f_i \in S} I(f_i; c) \tag{1}$$

where $I(X; Y)$ is the mutual information exhibited by two discrete random variables X and Y that are defined in terms of their probability mass functions as follows:

$$I(X; Y) = \sum_{x \in X} \sum_{y \in Y} P_{X,Y}(x, y) \log\left(\frac{P_{X,Y}(x, y)}{P_X(x) P_Y(y)}\right). \tag{2}$$

The features selected based on maximum relevance may exhibit redundancy (high dependency on other features). So, instead of using all redundant features, only one of them may be sufficient. The set of features S that exhibits minimum average mutual information by all possible pairs of features is the set of minimum redundant features. Minimum redundancy is defined as follows:

$$\min Red(f_i, f_j), \quad Red = (1/|S|^2) \sum_{f_i \in S} \sum_{f_j \in S} I(f_i; f_j). \tag{3}$$

The mutual exclusive set of features is selected by combining (2) and (3) as follows:

$$\max \emptyset(Rel, Red), \quad \emptyset = Rel - Red. \tag{4}$$

By using (4), features are selected incrementally. Suppose S_{m-1} is the set of $m-1$ selected features from the set S, then mth feature is selected as follows:

$$\max_{f_j \in S - S_{m-1}} \left[I(f_j; c) - (1/(m-1)) \sum_{f_i \in S_{m-1}} I(f_j; f_i) \right]. \tag{5}$$

5 Proposed Approach of Distinguisher

The aim of proposed distinguisher is to distinguish a given pseudo random bit sequence from a strong pseudo random bit sequence taken from cipher text generated by AES in CBC mode. The steps of distinguisher are as follows:

i. **Input**: θ_1 is an input pseudo random bit sequence that is to be distinguished from the reference pseudo random bit sequence θ_2.
ii. Let θ_1 and θ_2 have class labels +1 and −1, respectively, and their length is N bits.
iii. Features are extracted from both θ_1 and θ_2 as described in Sect. 4.1.

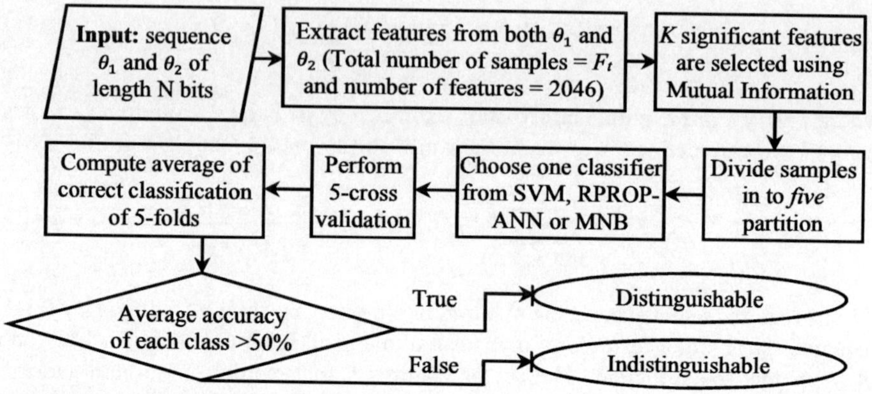

Fig. 1 Flow diagram of proposed distinguisher

iv. Let $FEAT_1$ be feature matrix of θ_1 and $FEAT_2$ be feature matrix of θ_2. Size of $FEAT_1$ and $FEAT_2$ is $F_1 \times 2046$ and $F_2 \times 2046$, respectively.
v. Combine samples of $FEAT_1$ and $FEAT_2$ in a matrix $FEAT$ of size $F_t \times 2046$ where $F_t = F_1 + F_2$.
vi. Select significant features from $FEAT$ as described in Sect. 4.2. Now, size of feature matrix $FEAT$ becomes $F_t \times K$ after selection of K significant features (i.e., there are F_t samples, and each one has K features).
vii. Divide total number of samples in $FEAT$ matrix into five equal parts, and each partition must have almost equal number of samples of both classes.
viii. Choose one classifier from SVM, RPROP-ANN, and MNB as described in Sect. 3.
ix. Perform five-cross validation on five possible training and testing samples set (i.e., train the classifier on samples of four partition and test it on samples of remaining one partition).
x. Find average correct classification accuracy of five-cross validation.
xi. **Decision Criterion**: if average correct classification accuracy of each class is more than 50%, then input pseudo random bit sequence is said to be distinguishable from reference one.

Flow diagram of proposed approach of distinguisher is as shown in Fig. 1.

6 Data Generation and Experimental Setup

The distinguisher is implemented in C++ on RHEL 7.3 WS 64-bit platform. Length of each pseudo random bit sequence (input as well as reference) is taken as $N = 10^6$ bits and size of block for feature extraction is chosen as $M = 5000$ bits as discussed in Sect. 4.1. 100 files of N bits are generated by encrypting plain text file by each of the four block ciphers (AES, 3DES, IDEA, and Blowfish (BF)) in ECB mode. Different key is used to encrypt each plain text block. Same amount of data is generated in CBC as well as OFB mode. But, here, different key is used to encrypt each set of 100 plain text blocks. 100 files of N bits are generated by encrypting plain text by keystreams of RC4 stream cipher and '*rand*' function of C language.

7 Experimental Results

Experimentation has been done on all data sets generated as described in Sect. 6 to distinguish them from reference strong pseudo random bit sequence. First, experimentation is done on cipher data set in ECB mode, and their achieved average five-cross validation accuracies of correct discrimination are listed in Table 1. There are a total of 800 samples of both classes in one experimentation, and average five-cross validation accuracy of them is computed. Like this, 100 experiments are performed to distinguish each one of the 100 files. So, average accuracies in Tables 1, 2 and 3 are the average of accuracies achieved for 100 files.

It is observed from Table 1 that pseudo random bit sequences of four block ciphers in ECB mode are clearly distinguishable. Impact of mutual information-based feature selection approach to enhance the accuracy is also clearly observable from Table 1. It is recommended that ECB mode of block ciphers should be avoided for security of highly sensitive information.

Second experimentation is done on data sets of three block ciphers (AES-CBC is used as the reference class, so it is excluded here) in CBC mode. Experimentation is also done on cipher data of RC4 and '*rand*' function of C language. Achieved average five-cross validation accuracies are listed in Table 2.

It is observed from Table 2 that most of the distinguishing accuracies are less than 50% for pseudo random bit sequences of four block ciphers in CBC mode so that CBC mode of block ciphers is considered very strong. But pseudo random bit

Table 1 Average five-cross validation accuracies of correct discrimination of pseudo random bit sequences generated in ECB mode of four block ciphers

Classifiers		SVM	RPROP-ANN	MNB
Accuracies without feature selection (no. of features = 2046)	AES-ECB (%)	97.82	80.25	75.55
	3DES-ECB (%)	98.11	81.42	78.64
	IDEA-ECB (%)	97.45	83.74	80.41
	BF-ECB (%)	98.23	82.58	80.85
Accuracies with feature selection (used only 1023 features)	AES-ECB (%)	**99.65**	83.88	77.60
	3DES-ECB (%)	**99.72**	85.65	80.05
	IDEA-ECB (%)	**99.80**	85.26	82.36
	BF-ECB (%)	**99.62**	83.50	82.18

Table 2 Average five-cross validation accuracies of correct discrimination of pseudo random bit sequences generated in CBC mode of three block ciphers along with RC4 and '*rand*'

Classifiers		SVM	RPROP-ANN	MNB
Accuracies without feature selection (no. of features = 2046)	3DES-CBC (%)	48.96	48.09	47.52
	IDEA-CBC (%)	48.90	48.95	47.12
	BF-CBC (%)	48.65	47.80	46.82
	RC4 (%)	48.49	48.21	47.09
	'*rand*' (%)	**63.18**	**61.48**	**58.64**
Accuracies with feature selection (used only 1023 features)	3DES-CBC (%)	50.78	48.56	47.89
	IDEA-CBC (%)	49.21	49.45	48.08
	BF-CBC (%)	51.21	50.34	47.14
	RC4 (%)	**51.73**	**51.20**	**51.01**
	'*rand*' (%)	**68.02**	**57.20**	**50.21**

Table 3 Average five-cross validation accuracies of correct discrimination of pseudo random bit sequences generated in OFB mode of four block ciphers

Classifiers		SVM	RPROP-ANN	MNB
Accuracies without feature selection (no. of features = 2046)	AES-OFB (%)	48.95	48.02	47.15
	3DES-OFB (%)	49.18	48.20	47.82
	IDEA-OFB (%)	48.85	48.38	46.67
	BF-OFB (%)	48.96	48.01	47.62
Accuracies with feature selection (used only 1023 features)	AES-OFB (%)	49.87	49.28	47.96
	3DES-OFB (%)	49.68	49.78	48.55
	IDEA-OFB (%)	50.16	49.88	47.42
	BF-OFB (%)	51.02	50.15	49.48

sequences of RC4 and '*rand*' function are distinguishable with more than 51% and 68% accuracies, respectively, when feature selection approach is applied. So, '*rand*' function is considered as a weak generator, and use of RC4 may also be avoided in high security applications. The SVM is outperforming the other two classifiers. The accuracies listed in Table 2 on cipher data of block ciphers in CBC mode are better than quoted 27.25 and 37.25% in [8] on cipher data generated by different keys. Cipher data in this paper is also generated by using different keys for each set of 100 plain text blocks in a file of 10^6 bits.

Third experimentation is done on data sets of block ciphers in OFB mode, and their achieved average five-cross validation accuracies of correct discrimination are listed in Table 3.

It is observed from Table 3 that OFB mode of block ciphers is also very strong against distinguishing attack.

If execution time is considered on standalone PC (Intel Xeon E5 2.3 GHz), then SVM takes approximately 37 s with feature selection for one file. MNB takes approximately 7 s but RPROP-ANN takes approximately 285 s due to multiple epochs/iterations to minimize error during training phase.

8 Conclusion

This paper describes mutual information and machine learning based distinguishers to distinguish a pseudo random bit sequence from other strong pseudo random bit sequence. SVM with mutual information-based feature selection is performing better than RPROP-ANN and MNB classifiers. Pseudo random bit sequences of block ciphers in ECB mode are clearly distinguishable but bit sequences of block ciphers in CBC and OFB mode are not distinguishable. Pseudo random bit sequences generated by RC4 and '*rand*' function of C language are significantly distinguishable, so uses of these should be avoided in high security applications. Accuracies on cipher data of described block ciphers in CBC mode by using different keys are better than quoted in [8]. It is a generalized distinguisher not for a particular kind of generator. It can also be used to distinguish a pseudo random bit sequence from a true random bit sequence (generated by physical quantities).

References

1. Schneier B (1996) Applied cryptography, 2nd ed. Wiley
2. Schindler W (2009) Random number generators for cryptographic applications. In: Cryptographic engineering. Springer, Berlin, pp 5–23
3. Blum L, Blum M, Shub M (1986) A simple unpredictable pseudo-random number generator. SIAM J Comput 15(2):364–383
4. Jovan D, Golic Renato M (2006) Statistical distinguishers for irregularly decimated linear recurring sequences. IEEE Trans Inform Theory 52(3):1153–1159
5. Martin H, Thomas J, Lennart B, Hakan E (2012) Improved distinguishers on stream ciphers with certain weak feedback polynomials. IEEE Trans Inf Theory 58(9):6183–6193
6. Chang D, Kumar A, Sanadhya SK (2017) Distinguishers for 4-branch and 8-branch generalized Feistel network. IEEE Access Int J 5:27857–27867
7. Grassi L, Rechberger C, Ronjom S (2017) A new structural-differential property of 5-round AES. In: Advances in cryptology-Eurocrypt 2017, pp 289–317
8. Dileep AD, Sekhar CC (2006) Identification of block ciphers using support vector machines. In: International joint conference on neural networks, pp 2696–2701
9. Suhaila OS, Kuncheva LI, Mansoor SP (2010) Classifying encryption algorithms using pattern recognition techniques. In: IEEE international conference on information theory and information security, pp 1168–1172
10. Advanced Encryption Standard. Federal Information Processing Standards Pub. 197 (2001)
11. Data Encryption Standard. Federal Information Processing Standards Pub. 46-3 (1999)
12. Basu S (2011) International data encryption algorithm (IDEA): a typical illustration. J Glob Res Comput Sci 2(7):116–118
13. Bruce S (1993) Description of a new variable-length key, 64-bit block Cipher (Blowfish). In: Cambridge security workshop proceedings, Springer-Verlag, pp 191–204
14. Haykin S (2009) Neural networks and learning machines/Chapter 6, 3rd ed. Pearson Prentice Hall

15. Martin R, Braun H (1993) A direct adaptive method for faster backpropagation learning: the RPROP algorithm. In: IEEE conference on neural networks, pp 586–591
16. Mohana R, Sumathi S (2014) Document classification using multinomial Naïve Bayesian classifier. Int J Sci Eng Technol Res (IJSETR) 3(5):1557–1563
17. Hanchuan P, Fuhui L, Chris D (2005) Feature selection based on mutual information: criteria of max-dependency, max-relevance and min-redundancy. IEEE Trans Pattern Anal Mach Intell 27(8):1226–1238

Rectangular Patch with Optimized Circular Slot and Perturbed Rings for Multiband Application

Ribhu Abhusan Panda, Md Danish Iqbal, Abhishek Panigrahi, G. Sai Varun, and Debasis Mishra

Abstract This artefact provides a novel design, which includes a rectangular patch having an optimized circular slot. The patch is supported by modified rings to enhance the antenna gain. The unique patch structure is designed using HFSS software on a substrate of dimension 70 mm × 70 mm × 1.6 mm. The ground plane has the same dimension as that of substrate but only differs in height. Height of ground plane is 0.01 mm. The optimization has been done for the circular slot in HFSS by using the Optimetrics tool. The perturbed rings having a width of 2 mm act as the supporting element for enhancement of antenna gain. The proposed antenna is designed with a simple line feed of 3 mm width. Parameters like S-Parameter (<10 dB), surface current distribution, standing wave ratio, antenna gain and directivity are determined from the simulation at the optimized value of the circular slot. The antenna is designed for multiband application in the frequency range from 1 to 6 GHz, which also includes the 5.9 GHz WLAN application.

R. A. Panda (✉) · D. Mishra
Department of Electronics and Telecommunication Engineering, V.S.S. University of Technology, Burla, Odisha, India
e-mail: ribhupanda@gmail.com

D. Mishra
e-mail: debasisuce@gmail.com

M. D. Iqbal · A. Panigrahi · G. Sai Varun
Department of Electronics and Communication Engineering, GIET University, Gunupur, Odisha, India
e-mail: danishiqbal227@gmail.com

A. Panigrahi
e-mail: abhishekpanigrahi46@gmail.com

G. Sai Varun
e-mail: saivarun1425@gmail.com

S. C. Satapathy et al. (eds.), *Communication Software and Networks*, Lecture Notes in Networks and Systems 134, https://doi.org/10.1007/978-981-15-5397-4_45

1 Introduction

Augmented antenna gain and bandwidth can be achieved by different techniques that have been developed in recent years [1–5]. The implementation of diverse slots on the disconcerted patches also leads to bandwidth and gain enhancement [6]. Metal rings around the patch designed on the substrate play a vital role for the increment of antenna gain [7–9]. Superstrates and hybrid substrates also play a vital role for the increment of antenna gain [10, 11]. Some more modification has been considered with effective design for different applications [6, 12–14]. In this paper, a conventional rectangular patch has been designed with a circular slot. The radius of the circular slot has been optimized using a specific tool that is included in high frequency structure simulator, which uses finite element method for simulation.

2 Antenna Design with Specific Design Parameters

The proposed antenna has basically three parts that are ground plane, substrate and patch. The ground plane has the dimension 70 mm × 70 mm × 0.01 mm, substrate has the dimension 70 mm × 70 mm × 1.6 mm and the conventional rectangular patch has the dimension 38.01 mm × 29.42 mm × 0.01 mm. The optimized value of the radius of the circular patch has been found out to be 2 mm. Width of the feed is 3 mm. Perturbed rings have been implemented along with patch on the substrate which not only supports the multiband resonance but also improves the antenna gain. The symmetric rings have the uniform width of 2 mm. The complete design has been shown in Fig. 1.

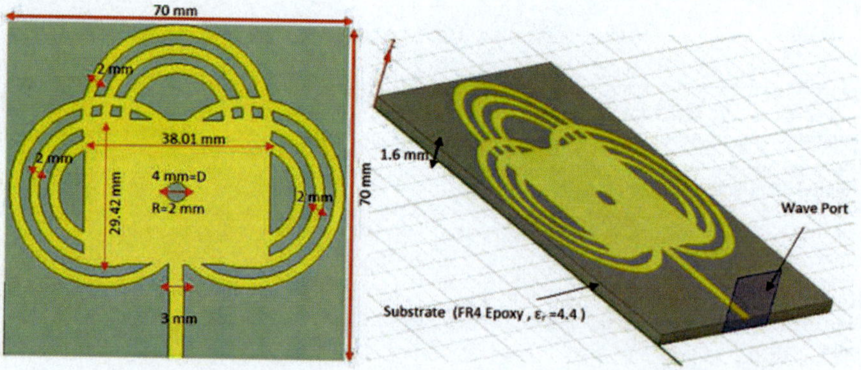

Fig. 1 Pictorial representation of the geometry associated with the antenna

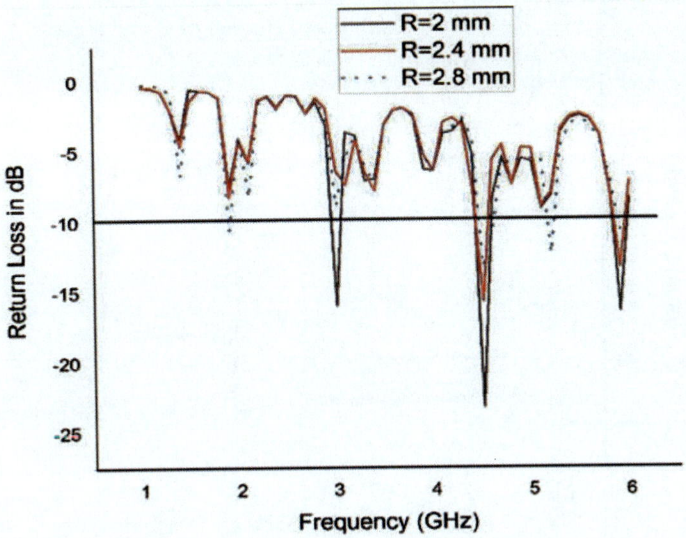

Fig. 2 Return loss of three optimized values of the radius of circular slot

3 Outcomes from the Simulation

3.1 *Return Loss and Resonant Frequency*

Focus has been on the return loss and resonant frequency to acquire the clear observation on the frequency ranges where the designed antenna can be operated. The proposed antenna has been designed from 1 GHz to 6 GHz frequency range, and it has been observed from the simulation result that there are three resonant frequencies at 3 GHz, 4.5 GHz and 5.9 GHz, respectively. The return loss has significant value at the optimum radius of the circular slot. Figures 2 and 3 indicate the S_{11} (<10 dB) versus Frequency plot highlighting the resonant frequencies.

3.2 *Peak Antenna Gain and Directivity*

The peak Antenna gain of the designed antenna has a value of 4.55 dB and the peak directivity is 7.11 dB. These are shown in Figs. 4 and 5, respectively. At 5.9 GHz frequency, the resonance occurs with a return loss of −16.39 dB and the antenna gain at 5.9 GHz is 4.55 dB. This is illustrated in Fig. 6.

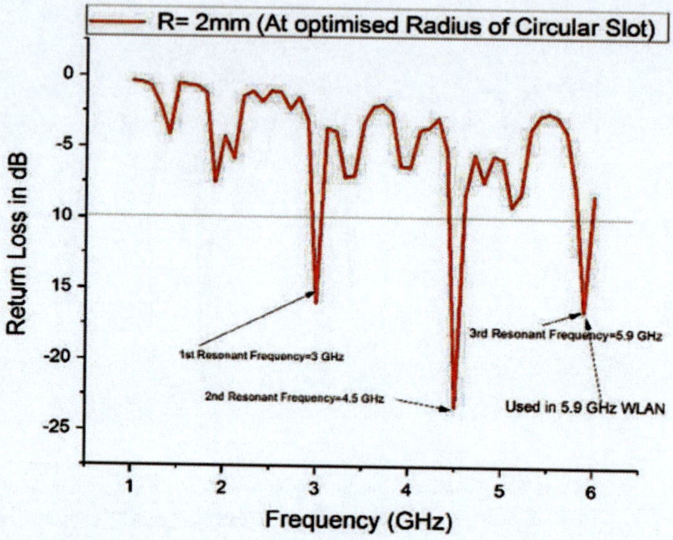

Fig. 3 The best result highlighting the resonant frequencies

Fig. 4 Peak antenna gain

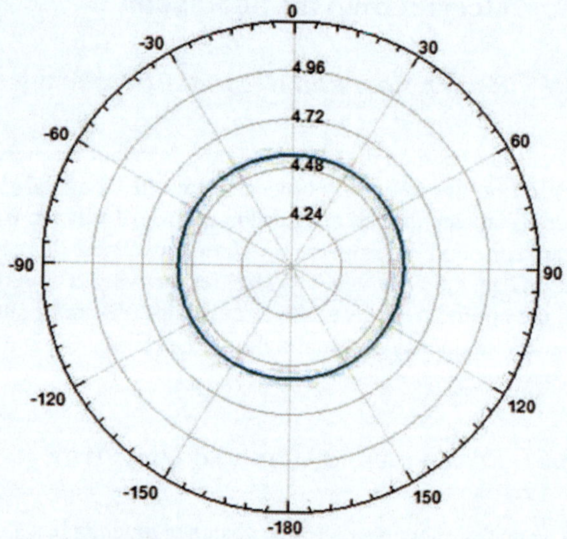

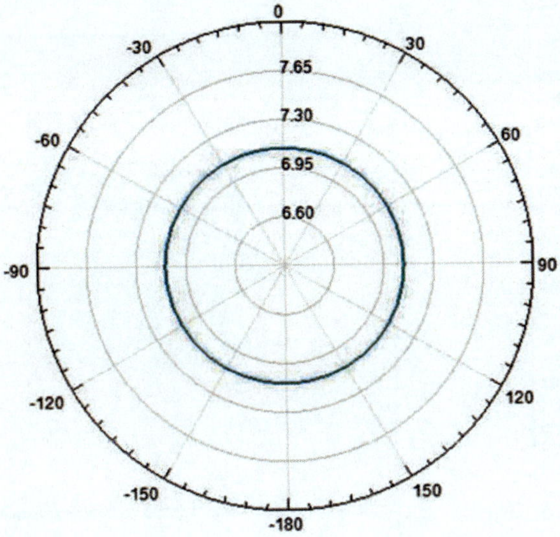

Fig. 5 Peak directivity

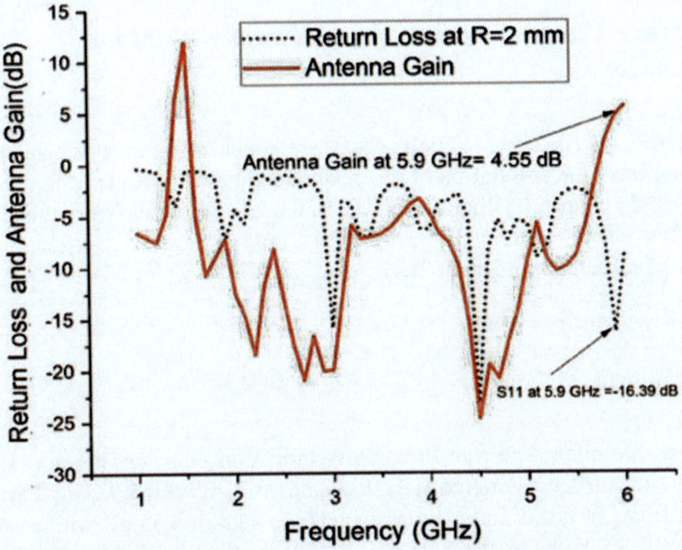

Fig. 6 Antenna gain and return loss at 5.9 GHz

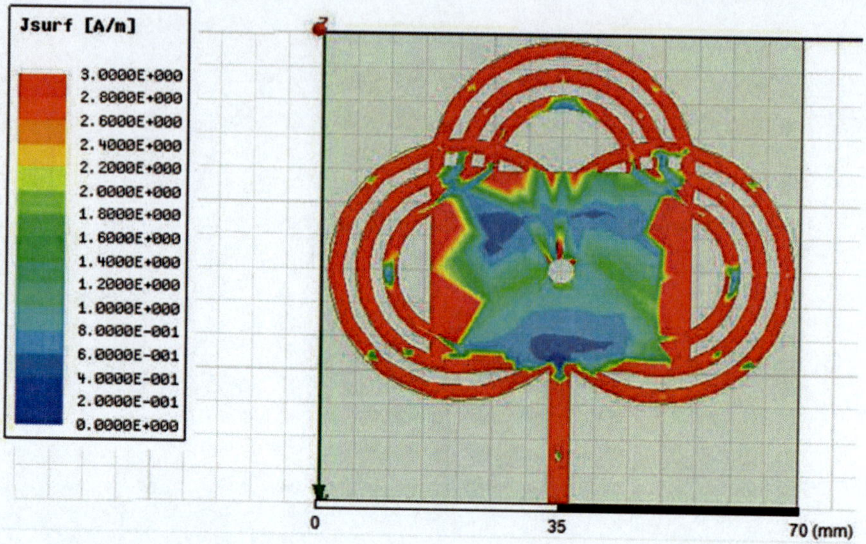

Fig. 7 Surface current distribution

3.3 Surface Current on the Altered Patch with Metal Boundary

Surface current distribution is a vital aspect while measuring the performance of the antenna, and from the simulation of the designed antenna, it is clear that the current is waving inside the patch which means the radiation is effective from the patch with the metal boundary. Figure 7 illustrates the surface current distribution, and Fig. 8 illustrates the radiation pattern in 2D.

4 Conclusion

With appreciable antenna gain and directivity, the proposed patch antenna can be used efficiently as a multiband antenna at the frequencies 3, 4.5 and 5.9 GHz. Particularly, at 5.9 GHz frequency, this antenna has an antenna gain of 4.55 dB with a return loss of −16.39 dB. So, it can be used for 5.9 WLAN application (Table 1).

Fig. 8 Radiation pattern in 2D

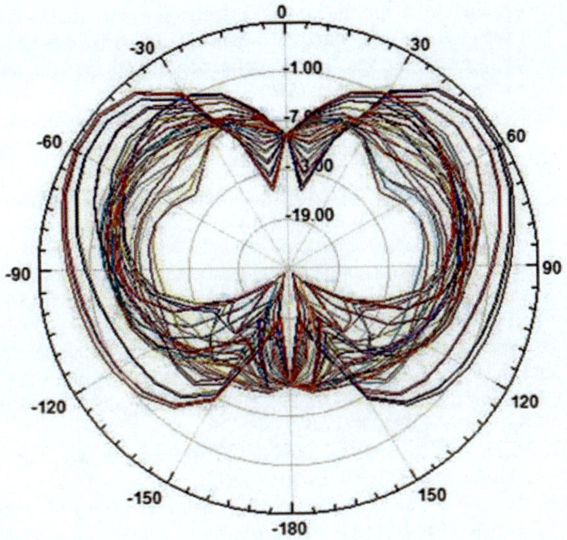

Table 1 Simulated parameters

Parameters	Value
Resonant frequency 1 (fr1) (GHz)	3
Resonant frequency 2 (fr2) (GHz)	4.5
Resonant frequency 3 (fr3) (GHz)	5.9
Return loss at fr1 (dB)	−15.9209
Return loss at fr2 (dB)	−23.2234
Return loss at fr3(dB)	−16.3965
Peak antenna gain (dB)	4.55
Peak directivity (dB)	7.11

References

1. Razi ZM, Rezaei P, Valizade A (2015) A novel design of Fabry-Perot antenna using metamaterial superstrate for gain and bandwidth enhancement. Int J Electron Commun (AEÜ) 69:1525–1532
2. Vettikalladi H, Lafond O, Himdi M (2009) High-efficient and high-gain superstrate antenna for 60-GHz indoor communication. IEEE Antennas Wirel Propag Lett 8
3. Saravanan M, Beslin V, Geo, Umarani SM (2018) Gain enhancement of patch antenna integrated with metamaterial inspired superstrate. J Electr Syst Inform Technol 5:263–270
4. Asaadi M, Sebak A (2017) Gain and bandwidth enhancement of 2 × 2 square dense dielectric patch antenna array using a holey superstrate. IEEE Antennas Wirel Propag Lett 16
5. Mollaei MSM, Zanganeh E, Farahani MF (2017) Enhancement of patch antenna gain using cylindrical shell-shaped superstrate. IEEE Antennas Wirel Propag Lett 16
6. Panda RA, Panda M, Nayak S, Das N, Mishra D (2019) Gain enhancement using complimentary split ring resonator on biconcave patch for 5G application. In: International conference on

sustainable computing in science, technology and management (SUSCOM-2019), pp 994–1000
7. Panda RA, Dash P, Mandi K, Mishra D (2019) Gain enhancement of a biconvex patch antenna using metallic rings for 5G application. In: 6th International conference on signal processing and integrated networks (SPIN), pp 840–844
8. Meena D, Meena RS (2015) Gain and directivity enhancement of microstrip patch array antenna with metallic ring for WLAN/Wi-Fi applications. In: Communication, control and intelligent systems (CCIS), pp 9–11
9. Kumar A, Kumar M (2014) Gain enhancement in a novel square microstrip patch antenna using metallic rings. In: International conference on recent advances and innovations in engineering (ICRAIE-2014), Jaipur, pp 1–4
10. Attia H, Yousefi L (2011) High-gain patch antennas loaded with high characteristic impedance superstrates. IEEE Antennas Wirel Propag Lett 10:858–861
11. Rivera-Albino A, Balanis C (2013) Gain enhancement in microstrip patch antennas using hybrid substrates. IEEE Antennas Wirel Propag Lett 12:476–479
12. Panda RA, Mishra D, Panda H (2018) Biconcave lens structured patch antenna with circular slot for Ku band application. Lecture notes in electrical engineering, vol 434. Springer Nature, Singapore, pp 73–83
13. Panda RA, Mishra D, Panda H (2016) Biconvex patch antenna with circular slot for 10 GHz application. In: SCOPES 2016. IEEE, pp 1927–1930
14. Panda RA, Panda M, Nayak PK, Mishra D (2020) Log periodic implementation of butterfly shaped patch antenna with gain enhancement technique for X-band application ICICCT-2019. In: System reliability, quality control, safety, maintenance and management. Springer Nature, Singapore, pp 20–28

A Hybrid Imbalanced Data Learning Framework to Tackle Opinion Imbalance in Movie Reviews

Salina Adinarayana and E. Ilavarasan

Abstract Opinion Mining is an important buzzword in recent times for research and industry for data science applications. Many concerns in opinion imbalance particularly in movie reviews were analyzed and handled for efficient recommendations. Opinion imbalance in terms of binary class, which often compromises the classifier prediction results, was scarcely studied. In this work, a Hybrid Imbalanced Data Learning Framework (HIDLF) is proposed to handle the opinion imbalance in the movie review dataset and then classify the movie reviews through the proposed HIDLT-SVM algorithm, which is a part of HIDLF for effective movie review classification. Experimental comparisons of the proposed work are done on movie reviews with Logistic Regression, CART, and REP Tree. Different evaluation metrics are used for capable classification of opinions from the movie reviews. The results recommend that the planned HIDLT-SVM framework performs better than the competent algorithms.

1 Introduction

Data sources from different social media sites like movie review sites, blogs, and Twitter having reviews as dataset are with class imbalance nature; total numbers of instances in the dataset are in the higher ratio for one class than the other one. The majority subset also contains many different subpatterns, representing the different group of instances. These different subpattern representations are known as with-in class imbalance subpatterns, which led to the subset majority and minority, patterns. As the classifiers try to learn knowledge from majority and minority sets to build the model, the class imbalance also exists in real world applications like product recommendations. Class imbalance nature in the majority class is due to noisy minority

S. Adinarayana (✉)
Shri Vishnu Engineering College for Women, Bhimavaram, India
e-mail: s.suhasini2k9@gmail.com

E. Ilavarasan
Department of CSE, Pondicherry Engineering College, Pondicherry, India

S. C. Satapathy et al. (eds.), *Communication Software and Networks*, Lecture Notes in Networks and Systems 134, https://doi.org/10.1007/978-981-15-5397-4_46

class samples, which are misclassified or understated concepts. Due to many representations for majority class data, this occurrence is rare. The noise in the minority class may influence class imbalance nature so that the strength of the groups corresponding to the class imbalance becomes an important problem, i.e., whether these samples denote actual concept, or are just attributed to noise. The existing machine learning algorithms are efficient to discover the hidden information from the balanced movie review dataset, but they are not that much successful in discovering the relevant information from the imbalance movie review dataset. To tackle the opinion imbalance problem in the movie reviews, Hybrid Imbalanced Data Learning Framework (HIDLF) is proposed to handle the imbalance nature in the movie review dataset and classify the opinions for movie recommendation. The remaining paper is organized as follows. The literature review of the opinion mining and class imbalance is presented in Sect. 2. The proposed hybrid framework for movie recommendation is presented in Sect. 3. Section 4 presents the planned framework for movie review classification. The results are conferred in Sect. 5. At last, the conclusion and the future work is presented at the end.

2 Literature Review

Gilad [1] has done analysis on different feature selection schemes for mood classification in web blogs by using a diversity of features for the classification process with content and non-content features. Blei et al. [2] have developed Latent Dirichlet Allocation (LDA) for feature extraction to mine the latent aspects, but this method is less successful in identifying specific aspect terms. Bollegala et al. [3] have discussed document-level opinion analysis techniques, which reveal whether the opinion expressed is positive or negative in the entire document. Chen and Zimbra [4] presented that, for opinion classification, different algorithms in Machine learning, like NB, Maximum Entropy, and SVM, are used. Among these, NB is a simple probabilistic classifier based on the conditional probability used in the Bayesian theorem, with similar assumptions on the variables used as features. Wang and Xia [5] have compared features with semantic lexicons to determine the feature-level polarity of the document, which is a semantic-oriented approach for opinion mining. Sun et al. [6] have presented a collection of opinion words of the Chinese online product review document with polarity details. Malof et al. [7] have proposed an algorithm to provide a solution for the class imbalance issue in the clustering process using under-sampling techniques. Ahmed et al. [8] have done the research on sampling techniques like hybrid, under and over-sampling. Siriseriwan and Sinapiromsaran [9] have proposed a Safe-Level Synthetic Minority Oversampling Technique (Safe-Level–SMOTE) by generating synthetic instances away from possibly surrounding the majority instances and handling the minority outcast with the 1-nearest neighbor model. Cao and Zaïane [10] have presented a wrapper approach for improving the performance of accurate classification by finding the best feature subset pair and misclassification cost sensitive technique, which is effective under different networks.

Fragoso et al. [11] have proposed one-class slab Support Vector Machine-based classifiers, for the idealized one-class solution for open-set recognition. Seiffert et al. [12] have proposed hybrid algorithms based on sampling techniques, namely, Random Under-Sampling Boosting (RUSBoost) and SMOTE Boost to minimize the problem of class imbalance using data sampling techniques.

3 Proposed Hybrid Imbalanced Data Learning Framework

This framework defined in this section is used to tackle opinion imbalance and classify the opinion from the movie reviews shown in the Fig. 1. It is a three-step process and is presented in the below subsections.

3.1 Collection of Reviews

The movie reviews are collected from different websites, namely filmskerecenzije.com, happynovisad.com, 2kokice.com, kakavfilm.com, filmskihitovi.blogspot.com, mislitemojomglavom.blogspot.com, popboks.com, and yc.rs. To handle the imbalance nature in dataset, purposefully, a subset of reviews with imbalanced nature in terms of binary polarity is considered as dataset for experimental analysis. The individual reviews of users may mislead the opinion of the new users, so it is recommended to go through several reviews before summarizing the reviews.

3.2 Feature Selection

The reviews collected are processed for removal of erroneous data. A Two-Level Feature Selection (TLFS) process is used here; in level 1, each opinion in the movie review is converted into fixed-length vectors with length equal to the vocabulary size using the Continuous Bag of Words (CBoW) model, and then skip gram model is used to predict the source context words. Here, two models finally produce a set of feature vectors. In level 2 of TLFS, the correlation-based feature selection process is used to find the relevant features from these feature vectors.

Fig. 1 Hybrid imbalanced data learning framework for movie reviews

3.3 Sentiment Classification Using HIDLT-SVM

The sentiment classification is the process of classifying review instances into positive or negative review depending upon the text message in the review. To classify the sentiment in each preprocessed IMBMR dataset collected from TLFS algorithm, Hybrid Imbalanced Data Learning through Support Vector Machine (HIDLT-SVM) algorithm is proposed which is a hybrid framework and is a modified version of the well-known classifier Support Vector Machine (SVM) to balance the dataset and classify the reviews. The first stage of the algorithm will have balanced the dataset, and stage 2 of the algorithm uses the functionality of SVM classifier to classify the movie reviews. After reviewing literature on different classifiers, SVM is chosen as base classifier. The classified opinions from HIDLT-SVM are used for movie recommendation.

4 Hybrid Imbalanced Data Learning Through Support Vector Machine (HIDLT-SVM)

IMBMR dataset contains a slight amount of movie review instances in one class as compared with the other class. A minority class is considered as a positive class, whereas a majority class is regarded as a negative class. To overcome this imbalance nature in the dataset, instances under the minority class should be improved so that the resultant dataset with the improved minority set gets balanced when combining with the majority opinion instances. To improve minority instances for balancing the dataset, HIDLT-SVM algorithm is used to balance and classify the opinions. During the balance phase of HIDLT-SVM, it improves the minority opinions; thus, the improved minority opinions count is more or less close to the majority opinions count. Once dataset is balanced, phase 2 of HIDLT-SVM algorithm with this balanced dataset will produce classified movie reviews. The framework of the proposed Hybrid Data Learning approach on movie reviews is shown in Fig. 2.

4.1 Balancing Phase of HIDLT-SVM Algorithm

IMBMR dataset is divided into two subsets, the majority and minority opinion movie review subsets. This balancing process is an over-sampling technique to generate improved opinion instances; minority opinion subset is further analyzed. The pseudocode of the algorithm is presented below:

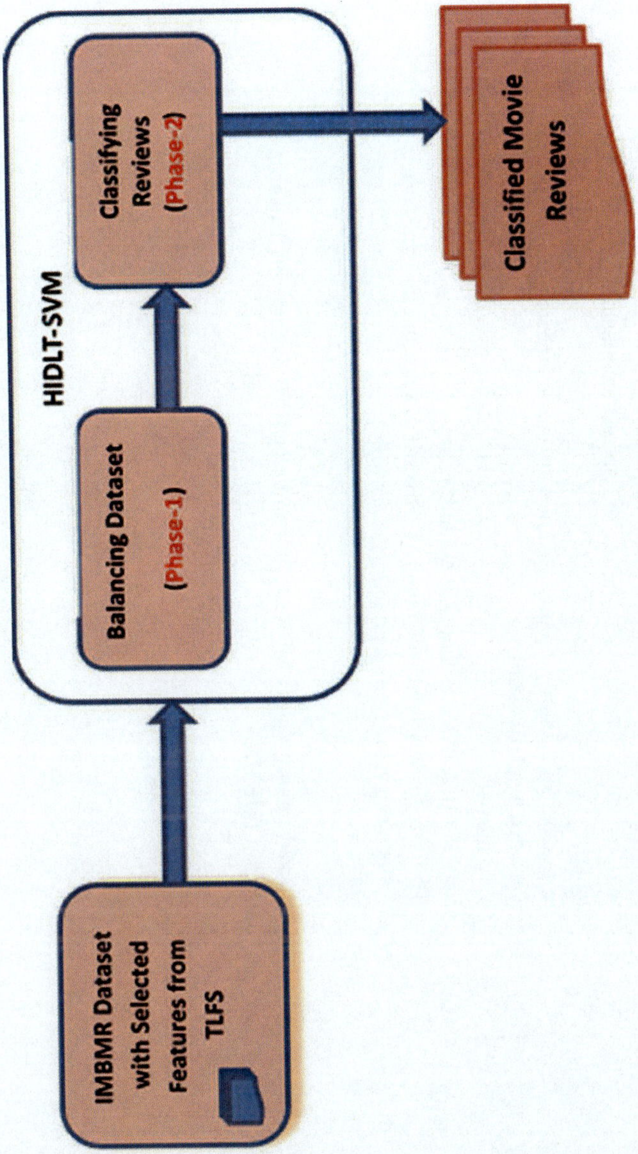

Fig. 2 The hybrid data learning approach for IMBMR movie reviews

Input: Minority Sample set MP and Majority Sample set NP of IMBMR dataset
Output: Balanced IMBMR Dataset
Procedure:

1. *for* every pi' in the minority-class MP
2. // pi represents ith opinion instance in the minority class MP
3. From the entire training set T, its 'm' nearest neighboring samples is calculated
4. The number of minority opinion samples in the 'm' nearest-neighbors is represented by p' (0 ≤p'≤p).
5. *If* p'= p then
6. Let PR= {p'1, p'2 ..., p'dnum}, 0 ≤dnum≤pnum
7. To form the new improved minority-class MP1, substitute missing values with the mean value from the minority-class MP
8. *endif*
9. *endfor*
10. Improved minority-class MP1 and majority-class NP to form the balanced dataset of IMBMR

The balancing phase of HIDLT-SVM algorithm uses only existing pure opinion cases in the current minority opinion subset for producing unreal opinion cases. So, this balancing process in HIDLT-SVM algorithm gives better performance over the existing over-sampling approaches. The training set size is increased in the result to balance the dataset; with this balanced dataset, HIDLT-SVM algorithm will generate classified opinions effectively.

4.2 *Classifying Reviews*

During phase 2 of the HIDLT-SVM algorithm, it adopts the functionality of SVM classifier to find a separating line between majority and minority samples by finding the points closest to the line from both the samples. This HIDLT-SVM is trained with IMBMR dataset and adjusts the tuning parameters to control the compromise between the smooth decision borderline and classifying training opinions correctly. The training and test sets in the IMBMR dataset contain a disjoint set of movies, so no substantial performance is gained by remembering movie-unique terms and their relation with observed labels.

5 Result Analysis

The details of the IMBMR dataset are provided in Table 1. The investigational procedure used for simulation is 4-fold cross validation (4CV). In 4CV, the IMBR set is distributed into four identical partitions. In every trip, one of the folds is used for testing and the remaining folds are used for training the model. The mean of 4CV trips is used to calculate the performance of the model. Accuracy, Precision, Recall, and F-score are used as evaluation measures for the proposed hybrid learning approach for movie recommendation. HIDLT-SVM algorithm of the proposed framework is implemented in open-source tool WEKA.

Accuracy (A) defined as the ratio of correctly predicted opinions from Twitter reviews to the total number of opinions predicted is given in Eq. 1.

$$A = \frac{TP + TN}{(TP + FP) + (TN + FN)} \tag{1}$$

Table 1 Movie Review dataset and their opinion classes

S. no	Dataset	Instances	Missing	Majority	Minority	IR
1	IMBMR	2834	No	2417	417	5.796

where True Positives (TP) is the number of reviews, which were correctly classified to be positive by the classifier. False Positives (FP) is the number of reviews, which were incorrectly classified as positive by the classifier. True Negatives (TN) is the number of reviews, which were correctly classified to be negative by the classifier. False Negatives (FN) is the number of cases, which were incorrectly classified as negative by the classifier. The Precision (Prc) is the measure of accuracy provided so that a specific class has been predicted and is calculated by using Eq. 2.

$$Prc = \frac{TP}{(TP) + (FP)} \tag{2}$$

The Recall (Re) is the ratio of amount of positive reviews (TP) acceptably predicted to the actual number of positive reviews (TP + FN) in the corpus, and it is calculated using Eq. 3.

$$Re = \frac{TP}{(TP) + (FN)} \tag{3}$$

F-score (Fs) is calculated by using the Eq. 4 given below:

$$Fs = \frac{2 \times Prc \times Re}{Pr + Re} \tag{4}$$

5.1 Opinion Mining with IMBMR Dataset

The IMBalanced Movie Review (IMBMR) dataset shown in Table 1 is with 2834 opinions; out of which, 2417 are positive opinions and 417 are negative opinions. It is a subset of a group of 4725 movie reviews in Serbian Dataset [13]. The opinion Imbalance Ratio (IR) of the dataset is 5.796.

5.2 Performance Analysis

Performance of the HIDLT-SVM algorithm on the IMBMR dataset is compared against the existing classifiers such as Logistic regression, CART, and REP Tree classifiers. The percentage of improvement of HIDLT-SVM algorithm trends with respect to accuracy over other classifiers such as Logistic Regression, CART, and REP Tree are 4.7%, 8%, and 9.1%, respectively. The percentage of improvement of HIDLT-SVM algorithm trends with respect to Precision over other algorithms such as Logistic Regression, CART, and REP Tree are 3.9%, 1.6%, and 10.6%, respectively. The percentage improvement of Recall values of HIDLT-SVM algorithm over other

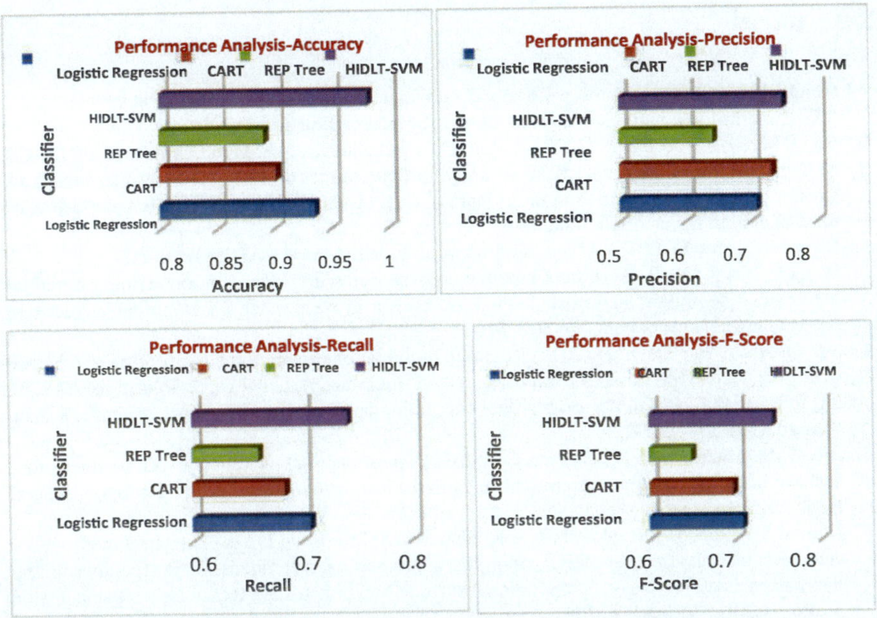

Fig. 4 Trends in accuracy, precision, recall, and F-score values on IMBMR dataset

algorithms, namely, Logistic Regression, CART, and REP Tree are 3.4%, 5.7%, and 8.1%, respectively, and the corresponding graph is shown in Fig. 4. The percentage of improvement of HIDLT-SVM algorithm with respect to F-score over other algorithms such as Logistic Regression, CART, and REP Tree are 3.5%, 4.6%, and 9.4%, respectively. The performance analysis graphs of the proposed HIDLT-SVM algorithm for each measure are shown in Fig. 4. The improved values of the evaluation measures recommend that the planned approach is superior to the competent approaches on the imbalanced movie opinion datasets in terms of performance.

6 Conclusion

In this work, a framework for hybrid imbalanced data learning approach is proposed for the imbalanced movie reviews classification. The investigational results indicate that the approach presented in this work is a competitive one when compared with other well-known algorithms. The proposed approach is validated with performance measures accuracy, precision, F-score, and recall. In future work, this hybrid imbalanced data learning framework can be adjusted to implement on multiclass, complex, and heterogeneous datasets.

References

1. Gilad M (2005) Experiments with mood classification in blog posts. In: Proceedings of the SIGIR workshop on stylistic analysis of text for information access
2. Blei DM, Ng AY, Jordan MI (2003) Latent Dirichlet allocation. J Mach Learn Res 3:993–1022
3. Bollegala D, Weir D, Carroll J (2011) Using multiple sources to construct a sentiment sensitive thesaurus for cross-domain sentiment classification. In: Proceedings of 49th annual meeting of the association for computational linguistics, pp 132–141
4. Chen H, Zimbra D (2010) AI and opinion mining. IEEE Intell Syst 25(3):74–80
5. Wang L, Xia R (2017) Sentiment Lexicon construction with representation learning based on hierarchical sentiment supervision. In: Proceedings of conference on empirical methods in natural language processing, pp 502–510
6. Sun Q, Niu J, Yao Z, Qiu D (2016) Research on semantic orientation classification of Chinese online product reviews based on multi-aspect sentiment analysis. In: Proceedings of IEEE/ACM 3rd international conference on big data computing applications and technologies (BDCAT), Shanghai, pp 262–267
7. Malof JM, Mazurowski M, Tourassi GD (2012) The effect of class imbalance on case selection for case-based classifiers: an empirical study in the context of medical decision support. Neural Netw 25(1):141–145
8. Ahmed S, Mahbub A, Rayhan F, Jani R, Shatabda S, Md. Farid D (2017) Hybrid methods for class imbalance learning employing bagging with sampling techniques. In: Proceedings of 2nd international conference on computational systems and information technology for sustainable solution (CSITSS), pp 126–131
9. Siriseriwan W, Sinapiromsaran K (2016) The effective redistribution for imbalance dataset: relocating safe-level SMOTE with minority outcast handling. Chiang Mai J Sci 43(1):234–246
10. Cao P, Zaïane OR (2013) Measure optimized wrapper framework for multi-class imbalanced data learning: an empirical study. In: Proceedings of international joint conference on neural networks (IJCNN), August 2013
11. Fragoso V, Scheirer W, Hespanha J, Turk M (2016) One-class slab support vector machine. In: Proceedings of international conference on pattern recognition (ICPR)
12. Seiffert C, Khoshgoftaar TM, Van Hulse J, Napolitano A (2010) RUSBoost: a hybrid approach to alleviating class imbalance. IEEE Trans Syst Man Cybern Part A Syst Humans 40(1):185–197
13. Batanović V, Nikolić B, Milosavljević M (2016) Reliable baselines for sentiment analysis in resource-limited languages: the serbian movie review dataset. In: Proceedings of the 10th international conference on language resources and evaluation (LREC 2016), Portorož, Slovenia, pp 2688–2696

Design and Implementation of an Efficient VLSI Architecture for 10T Full Adder Used in Ultra Low Power Applications

Vagu Radha Haneesha, Vemu Srinivasa Rao, and S. Adinarayana

Abstract Full adder is an important digital design for development of many digital systems. For an efficient digital system design, an area efficient and high-speed full adder is very much needed. This full adder is very much needed for ultra low power applications. In this paper, an efficient VLSI Architecture is proposed for the 10T full adder. The proposed VLSI architecture consumes less power, less area, and operates at higher speed. This VLSI architecture has been implemented using 180 nm technology using Generic Process Design Kit (GPDK) with the help of Cadence Design suite. The VLSI architecture can be synthesised with reduction in power and high speed. The proposed VLSI architecture has been compared with different full adder cells. The proposed full adder architecture performance can be analyzed by using parameters such as energy, propagation delay, leakage power, and total power.

1 Introduction

The full adder is an important building block in any digital system. Full adder is a basic element in circuits such as comparators, multipliers, shifters, etc. [1]. Adder is also useful in the design of digital signal processors, microprocessors to implement arithmetic operations. Any digital design performance can be analyzed depending on several parameters such as delay, speed, power consumption, area, estimation of critical path, etc. In this paper, the full adder performance can be analyzed by estimating the delay and power consumption. The main purpose of this work is to reduce the delay so as to achieve high speed and minimize the power consumption [2]. This work is carried out under cadence EDA environment in 90 nm technology.

V. Radha Haneesha (✉) · V. Srinivasa Rao · S. Adinarayana
Raghu Institute of Technology, Visakhapatnam, India
e-mail: radha.haneesha@gmail.com

V. Srinivasa Rao
e-mail: vemu1974@gmail.com

S. C. Satapathy et al. (eds.), *Communication Software and Networks*, Lecture Notes in Networks and Systems 134, https://doi.org/10.1007/978-981-15-5397-4_47

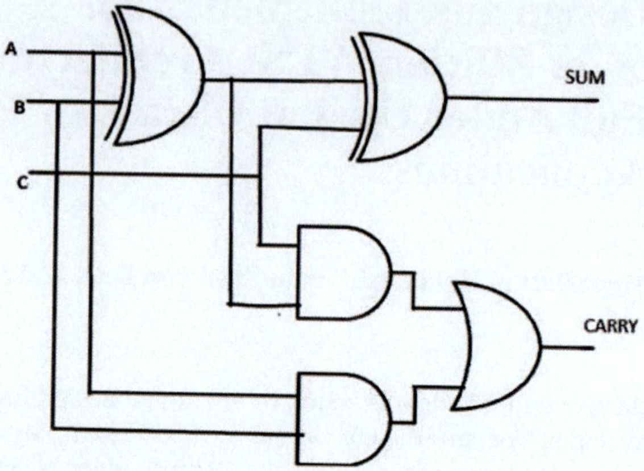

Fig. 1 Circuit diagram of full adder

The objective of this work is to design an efficient full adder, which can be used for ultra low power applications [3].

Power consumption is an important parameter, which depends on the supply voltage VDD. Reduction in power consumption can be obtained by scaling VDD using scaling factor 1/β. Power reduction can also be done by using power down mode concept [4]. Delay is another important parameter to assess the performance of any digital circuit design [5]. During the measurement of delay, estimation of critical path delay is very important. This delay is entirely responsible to estimate the speed of the digital systems. In this paper, both delay and power [6] are estimated using cadence EDA environment. The typical circuit diagram of one-bit full adder is shown in Fig. 1.

$$\mathrm{SUM} = \mathrm{A\ xor\,B\ xor\,C} \tag{1}$$

$$\mathrm{CARRY} = \mathrm{AB} + \mathrm{BC} + \mathrm{CA} \tag{2}$$

The truth table for one-bit full adder is shown in Table 1.

This paper is organized as follows: Sect. 1 describes introduction, Sect. 2 discusses the related work, Sect. 3 covers the proposed work, Sect. 4 demonstrates the result, and finally Sect. 5 concludes the work.

2 Related Work

In this section, existing full adder circuits [2, 3, 5] are discussed. In the full adder design [3], it exhibits poor functionality in sub-threshold region. If they are operated in 1.8 V as supply voltage, then they are exhibiting different output voltage levels. For example, Fig. 2 shows the full adder 1 schematic diagram with a, b, c as inputs

Table 1 Truth table for one-bit full adder

A	B	C	Sum	Carry
0	0	0	0	0
0	0	1	1	0
0	1	0	1	0
0	1	1	0	1
1	0	0	1	0
1	0	1	0	1
1	1	0	0	1
1	1	1	1	1

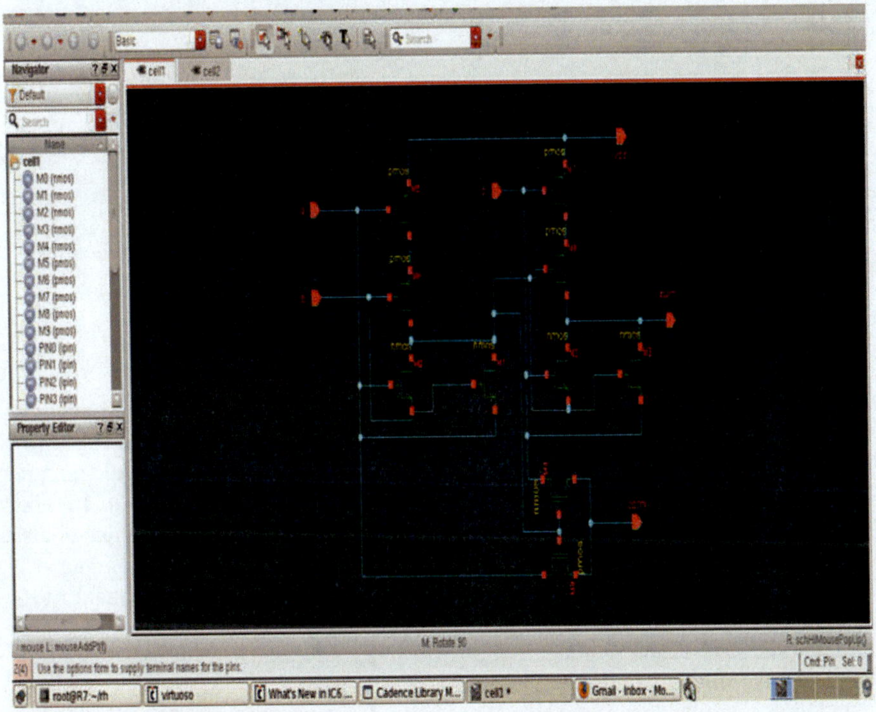

Fig. 2 Full adder 1 schematic diagram

and sum, carry as outputs. In Fig. 2, it has the output of sum as HIGH when the vector input ABC is equal to 001, but the output shows 1.2 V instead of 1.8 V in Fig. 9. This shows the worst-case output of the full adder 1 [7].

And then, next in Fig. 3, full adder 2 is shown. It is designed to reduce the delay with same number of transistors. Here, the sum output is LOW when the vector combination of inputs ABC is 110. But the output in Fig. 10 shows the sum output

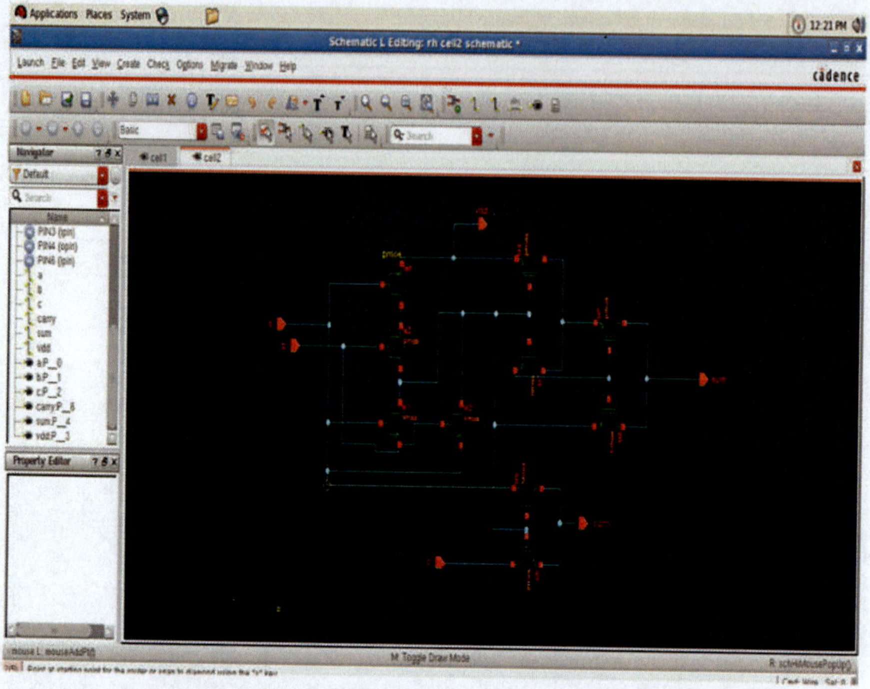

Fig. 3 Full adder 2 schematic diagram

as 1 V instead of gnd. Hence, this shows the worst output case in the full adder 2. To reduce the worst-case outputs, full adder 3 is designed and is shown in Fig. 4 [8]. In Fig. 4, the full adder 3 has ten transistors as same as the above full adders, and the carry output has to show HIGH when the vector input ABC is equal to 101, but the carry output in Fig. 11 shows 1.3 V instead of 1.8 V. Hence, this case is also considered as worst output case for the full adder 3 [9]. In Fig. 5, full adder 4 is designed by using an inverter for the input c. Then, for the vector ABC equal to 110, the sum output has to show 0 and carry output as 1. Instead of these values, carry output is showing 1.2 V. Hence, the full adder 4 is also considered as non-functional adder or unacceptable adder.

In Fig. 6, the full adder 5 is shown. This full adder shows 2 V as carry output for the vector input case ABC equal to 010. Hence, this case is also treated as unacceptable [10]. Thus, full adder 5 is completely non-functional at deep sub-threshold levels [11]. Hence, our normal full adder with the expressions shown above is designed with 28 transistors as shown in Fig. 7 with the name as full adder 6. Full adder 6 takes more silicon area to design because it has more number of transistors than the above full adders. But the worst-case outputs compared to the above full adders are reduced as shown in Fig. 14 [12]. But the main disadvantage of this circuit is high consumption of silicon area. Hence, the proposed architecture is designed in the next section.

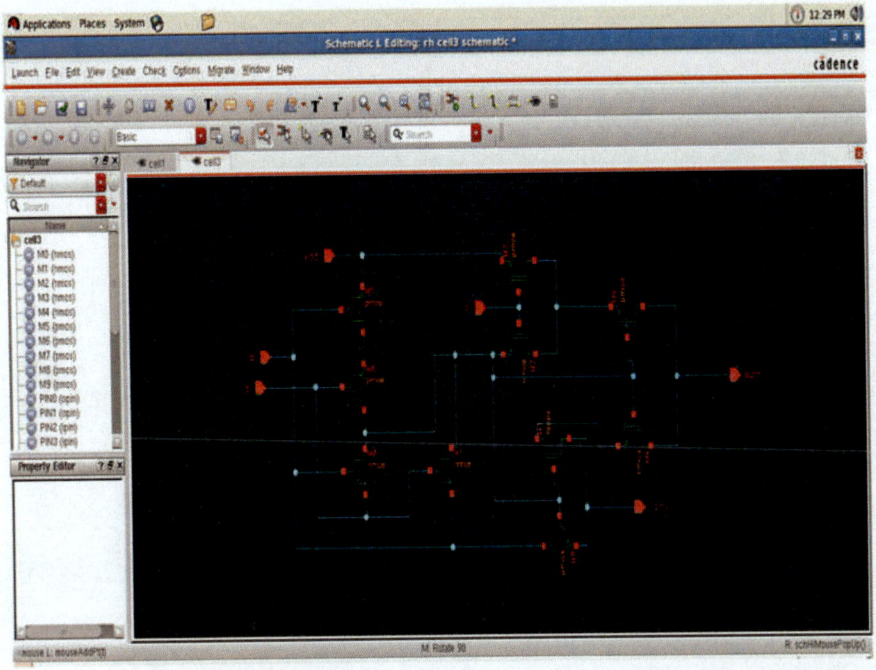

Fig. 4 Full adder 3 schematic diagram

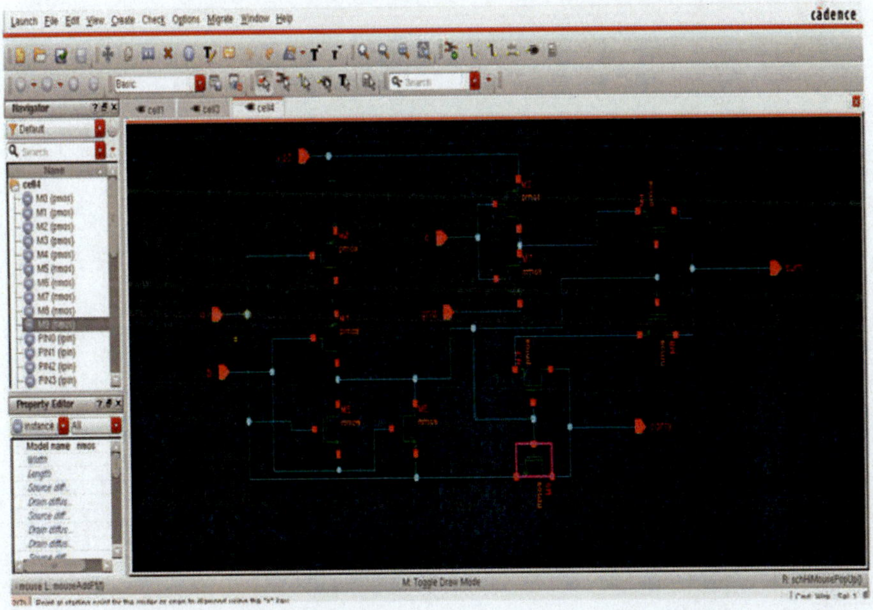

Fig. 5 Full adder 4 schematic diagram

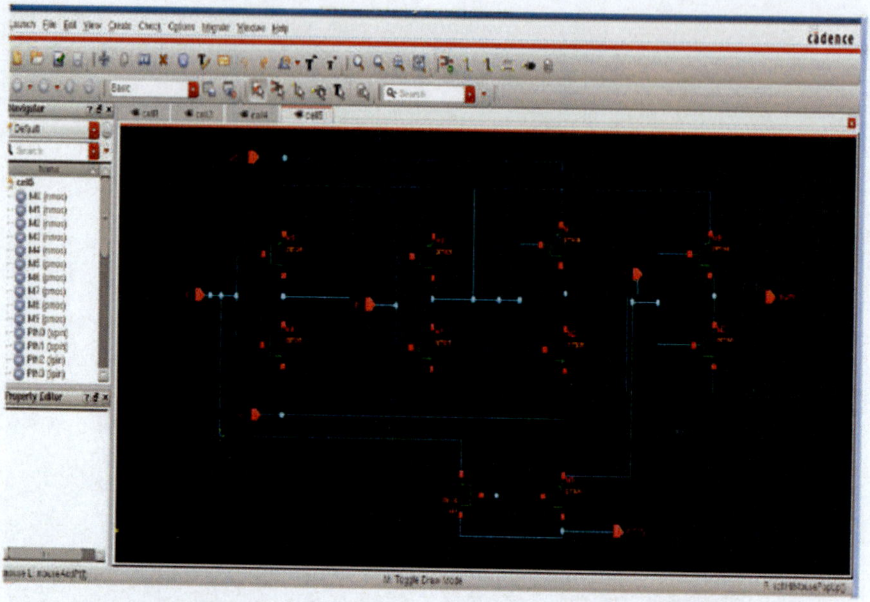

Fig. 6 Full adder 5 schematic diagram

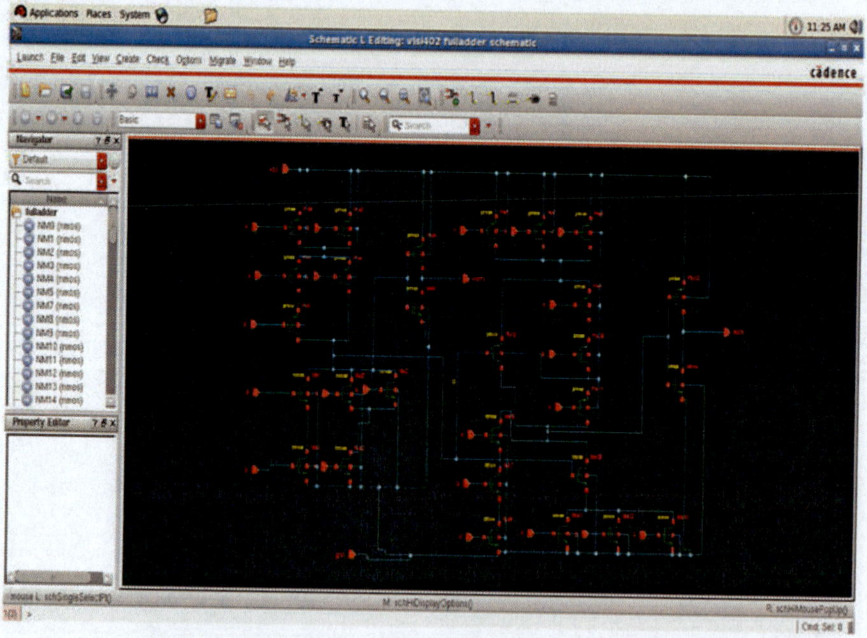

Fig. 7 Full adder 6 schematic diagram

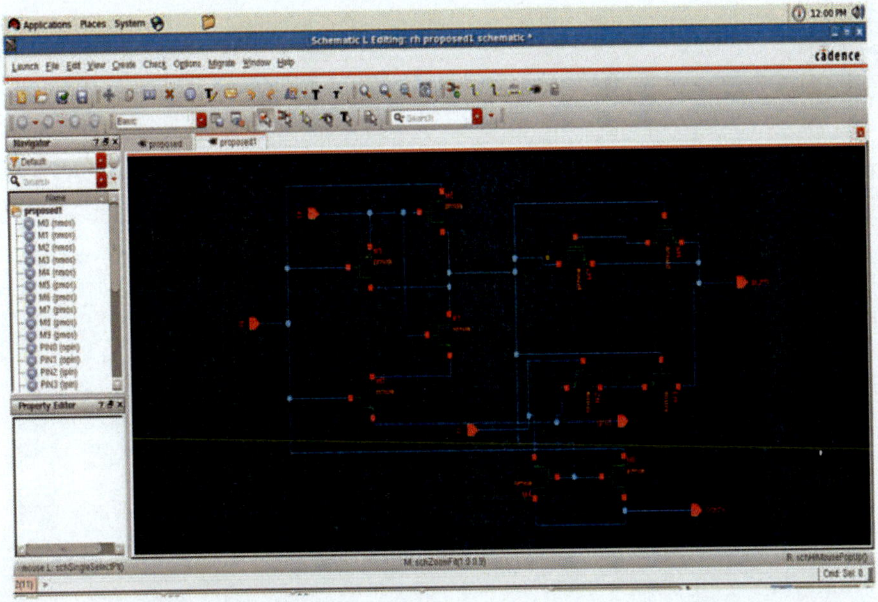

Fig. 8 Proposed schematic diagram of full adder

3 Proposed Work

The existing designs have no perfect output level at 1.8 V for the case of HIGH and no perfect output level of 0 V for the case of LOW. This is called degradation of output voltage levels. In order to overcome this drawback, the proposed work incorporates decreasing of W/L ratios of ten transistors. The technology also changed from 180 to 90 nm to overcome the existing drawback [3]. In order to restore the logic levels, the proposed work introduces device aspect ratio reduction and also reduces the value of β. For ultra low power applications, full output voltage swing is required, which has been proposed in this work. The proposed full adder circuit is shown in Fig. 8 consisting of ten transistors in which xor gates are used. The first four transistors are considered as xor gate 1 and the next four transistors are considered as xor gate 2, and the input c is constructed as back-to-back transistors. If A xor B is 1, then the output is shown as A, or else if A xor B is equal to 0, then the output is C. The Output of the proposed full adder is shown in Fig. 15.

4 Results

The existing full adder simulated in cadence EDA tool under 90 nm technology. The simulation results for various full adders are shown in Figs. 9, 10, 11, 12, 13, and

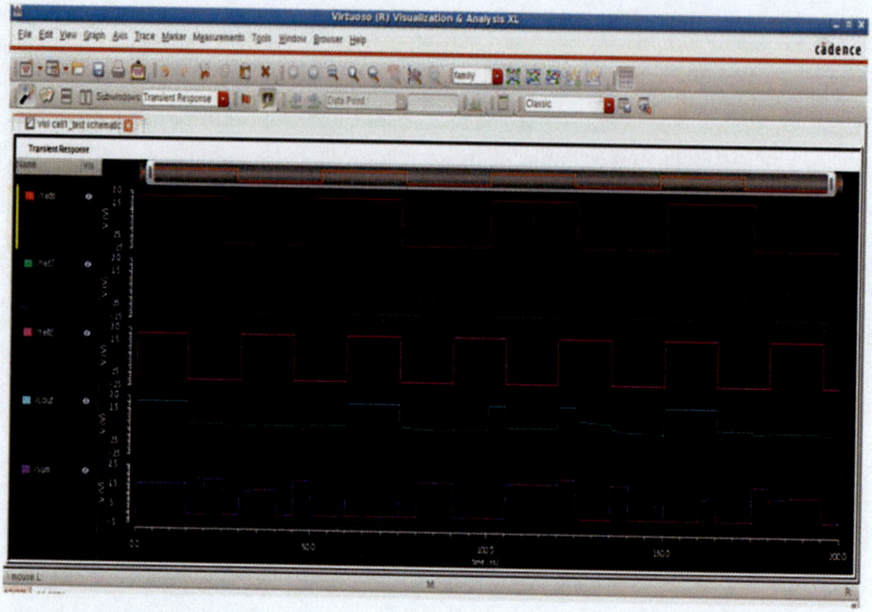

Fig. 9 Output of full adder 1 [3]

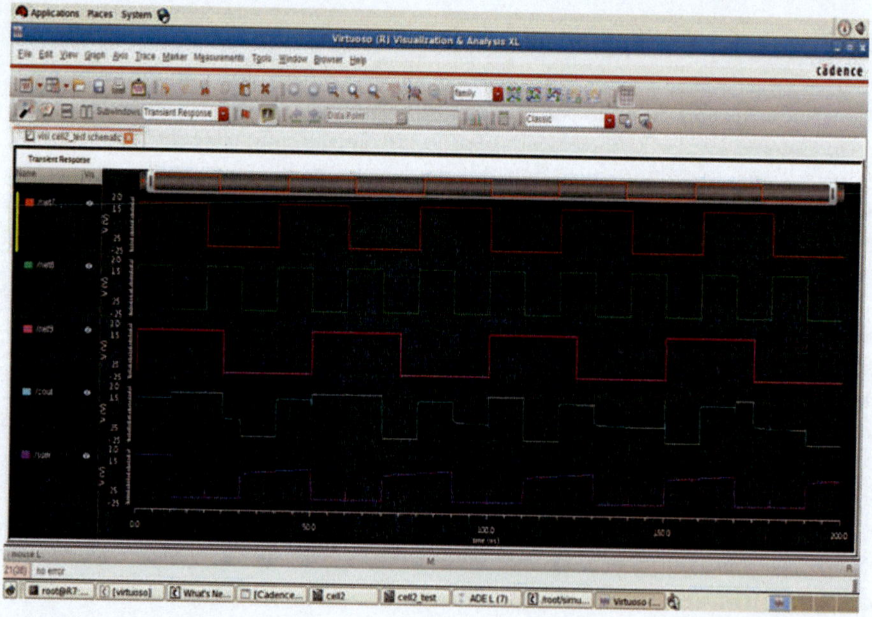

Fig. 10 Output of full adder 2 [3]

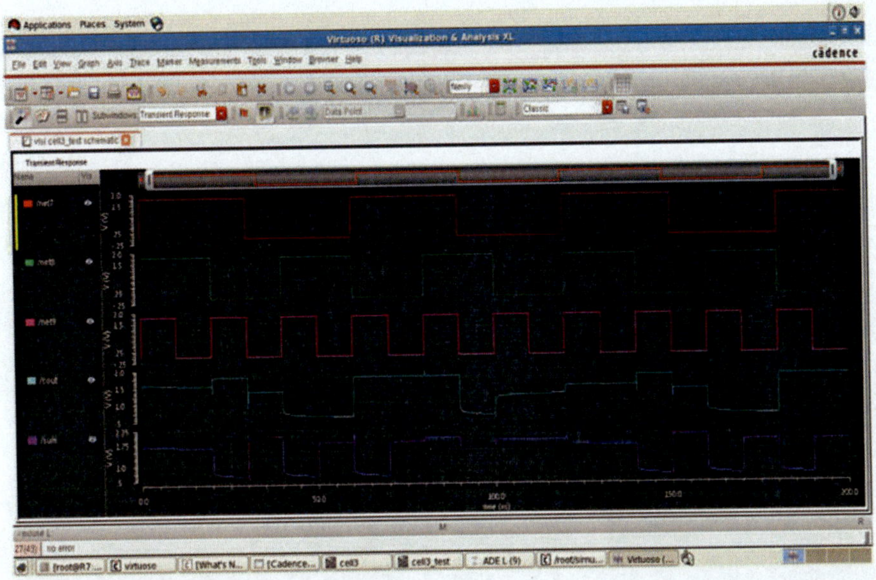

Fig. 11 Output of full adder 3 [3]

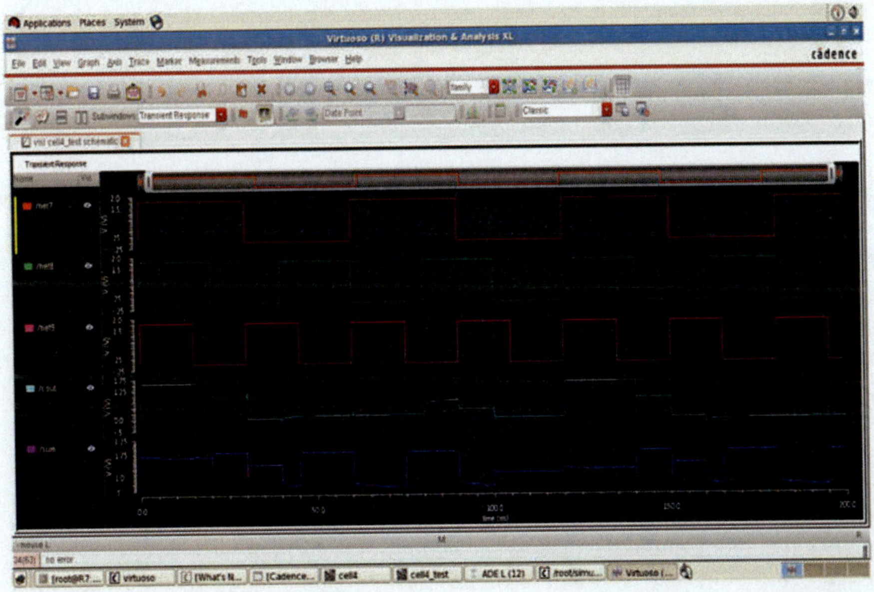

Fig. 12 Output of full adder 4 [3]

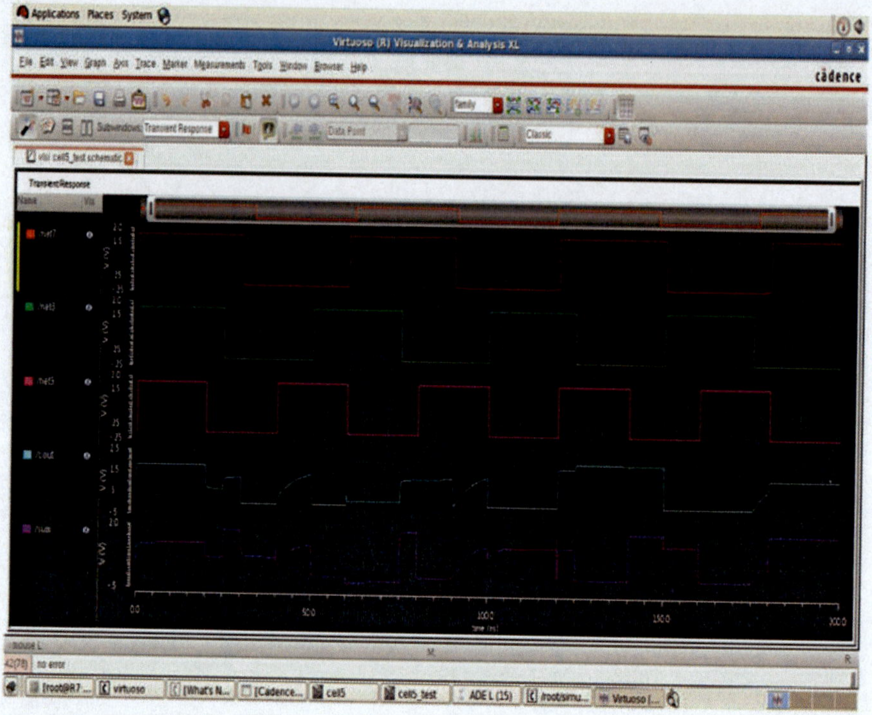

Fig. 13 Output of full adder 5 [3]

14. The existing results show that there is a degradation of the output levels of all the full adders.

The proposed work also simulated in cadence EDA tool under 90 nm technology. The simulated results are shown in Fig. 15 for 10T full adder. The simulation results show that there is no degradation in the output levels, which are suitable for ultra low power applications. The simulation results are also free from glitches, which reduces the power consumption and also avoids unnecessary transitions in the logic levels. Due to avoiding of unnecessary transitions, the power consumption also drastically reduced. The simulation results also show that the signal travels through the shortest critical path from input to the output of the full adder circuit. This shortest critical path will result in less delay so as to achieve highest speed for full adder, which is required to implement efficient digital systems. The delay also reduced due to reduction in the width of the transistors, which are there in the critical path [13–15].

The average power is to be considered as an efficient parameter for estimation of power consumption instead of dynamic power consumption in cadence EDA environment.

The comparison has been performed between the existing method and proposed method and shown in Table 2. The proposed results show that delay and average power proved to be better compared to existing results.

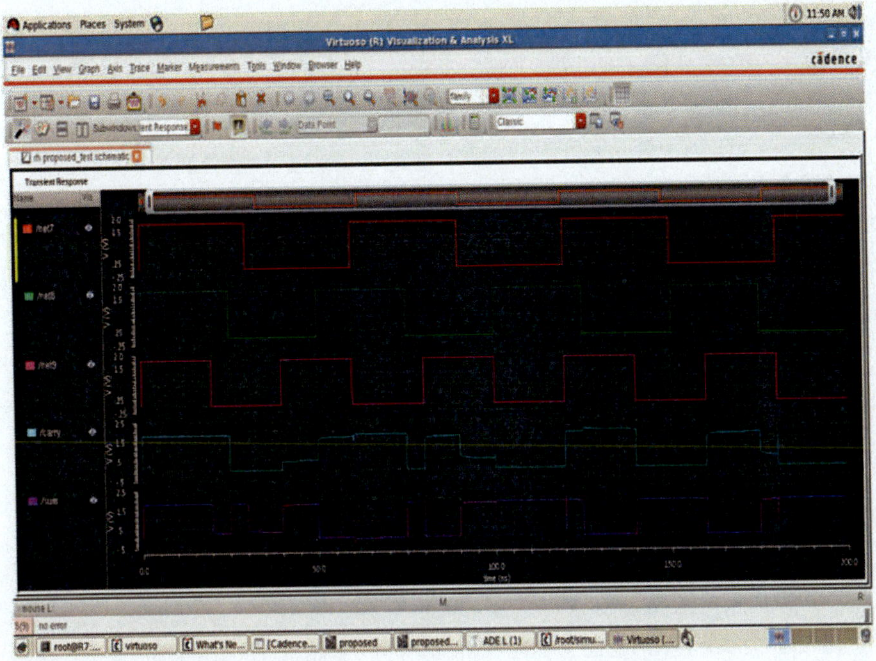

Fig. 14 Output of full adder 6 [3]

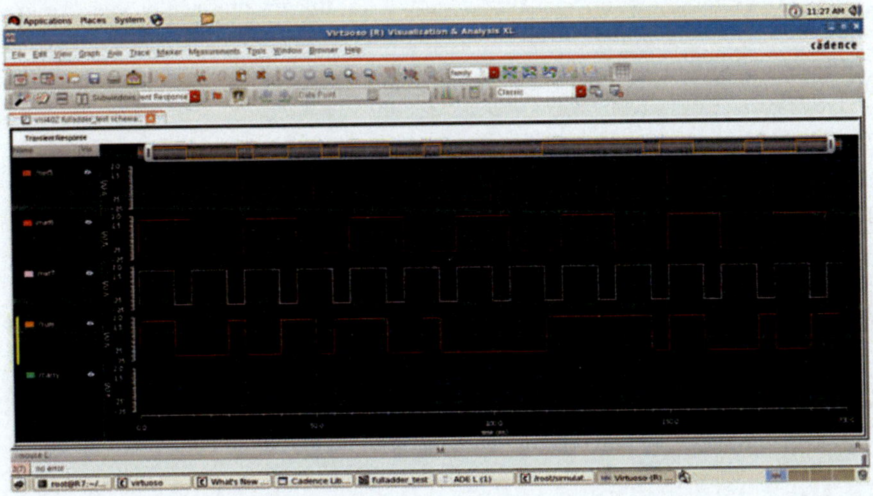

Fig. 15 Output of proposed method

Table 2 Comparison of existing method and proposed method for delay and average power

Full adder	Existing [3]						Full adder (proposed)
	Full adder 1	Full adder 2	Full adder 3	Full adder 4	Full adder 5	Full adder 6	
Delay (10^{-8} s)	10.94	10.68	8.50	9.23	9.89	10.60	7.79
Average power (10^{-10} W)	4.874	4.923	4.762	3.869	5.667	3.678	2.096

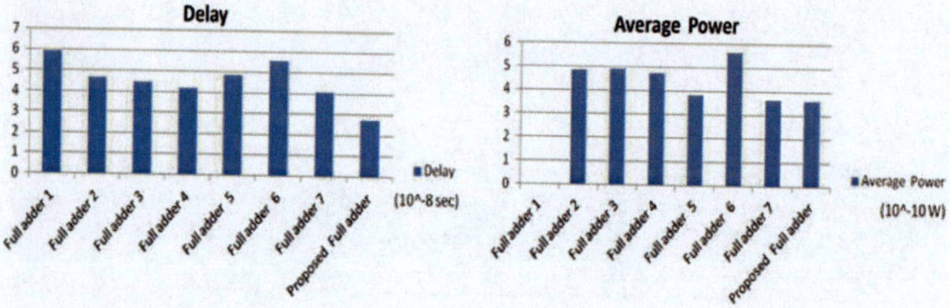

Fig. 16 Comparison of delay and average power for both proposed and existing full adders

The comparison of delay and average power for the existing and proposed work is shown in graphical representation in Fig. 16.

5 Conclusion

The proposed full adder has been implemented in 90 nm technology under cadence EDA environment. The proposed design proved that it consumes less power and it operates at higher speed by 10% compared to the existing architectures have been developed and standardized collaboratively by using the VLSI architecture. In this paper, a one-bit full adder using 10T is proposed, which uses optimum energy point having transistors W/L ratio equal to 11.1 at 90 nm technology. This design is suited for ultra low power applications such as IoT devices, energy harvesting, etc.

References

1. Dev A, Nimade S (2017) A efficient technique for low-power high-speed adder circuit design in DSM technology
2. Hassoune I, Flandre D, O'Connor I, Legat JD (2010) ULPFA: a new efficient design of a power-aware full adder. IEEE Trans Circuits Syst I Reg Papers 57(8):2066–2074

3. Bhattacharyya P, Kundu B, Ghosh S, Kumar V (2016) Performance analysis of a low-power high-speed hybrid 1-bit full adder circuit
4. Nigam A, Singh R (2016) Comparative analysis of 28T full adder with 14T full adder using 180 nm
5. Dokania V, Verma R, Guduri M, Islam A (2018) Design of 10T full adder for ultra low power applications. Ain Shans Eng J 9:2363–2372
6. Singh S, Sharma T, Sharma KG, Singh BP (2011) 9T full adder design in subthreshold region
7. Guduri M, Islam A (2015) Design of hybrid full adder in deep subthreshold region for ultralow power applications. In: 2015 2nd international conference on signal processing and integrated networks (SPIN), Noida, pp 931–935
8. Blair GM (1994) Designing low power CMOS. IEEE Electron Commun Eng J 6:229–236
9. Dokania V, Islam A (2015) Circuit-level design technique to mitigate impact of process, voltage and temperature variations in complementary metal-oxide semiconductor full adder cells. IET Circuits Devices Syst 9(3):204–212
10. Predictive Technology Model. Nanoscale Integration and Modeling (NIMO) Group. Arizona State University (ASU). Available Accessed on 18 September 2016
11. Safaei Mehrabani Y, Eshghi M (2016) Noise and process variation tolerant, lowpower, high-speed, and low-energy full adders in CNFET technology. IEEE Trans Very Large Scale Integr (VLSI) Syst 24(11):3268–3281
12. Rajaei R, Bakhtavari MS (2017) Ultra-low power, highly reliable, and nonvolatile hybrid MTJ/CMOS based full-adder for future VLSI design. IEEE Trans Device Mater Reliabil 17(1):213–220
13. Bhattacharyya P, Kundu B, Ghosh S, Kumar V, Dandapat A (2015) Performance analysis of a low-power high-speed hybrid 1-bit full adder circuit. IEEE Trans Very Large Scale Integr (VLSI) Syst 23(10)
14. Agrawal P, Raghuvanshi DK, Gupta MK (2017) A low-power high-speed 16T 1-bit hybrid full adder. In: Proceeding international conference on recent innovations is signal processing and embedded systems (RISE-2017) 27–29 October 2017
15. Naseri H, Timarchi S (2018) Low-power and fast full adder by exploring new XOR and XNOR gates. IEEE Trans Very Large Scale Integr (VLSI) Syst

Location Based Scalable Recommendations Using Social Relationships

Joydeep Das, Abhishek Agarwal, Subhashis Majumder, and Kalyani Mali

Abstract Over the last few years, Location Based Services (LBS) has gained huge popularity due to the tremendous advancement of mobile handheld devices. Some of the applications of LBS include recommending nearby restaurants or shopping malls, local news, concerts or events, and local advertisements offering prizes or discounts. These applications are dependent on the location of the users. In this paper, we propose a location based recommendation framework that not only considers the location of the users but also utilizes their social connections in order to improve the recommendation accuracy and scalability. The proposed work recommends restaurants to users using Collaborative Filtering (CF) techniques and while recommending to a target user, it will consider only the ratings of those users who are also her friends in a social network. For this, we compute the similarity among the target user and her friends. Typically the most costly step of any CF algorithm is the similarity computation and since we are calculating similarity only with the friends of a target user bypassing the entire user set, this should definitely enhance the scalability of the system. In this work, we further classify the friendship relations present in social network into *immediate friends* and *distant friends*. Experimental evaluations using Foursquare dataset verify that better recommendation accuracy can be achieved considering the opinions of the friends rather than considering the entire set of users.

J. Das (✉)
The Heritage Academy, Kolkata, WB, India
e-mail: joydeep.das@heritageit.edu

A. Agarwal · S. Majumder
Department of Computer Science and Engineering, Heritage Institute of Technology, Kolkata, WB, India
e-mail: agarwlabhishek@gmail.com

S. Majumder
e-mail: subhashis.majumder@heritageit.edu

K. Mali
Department of Computer Science and Engineering, University of Kalyani, Kalyani, WB, India
e-mail: kalyanimali1992@gmail.com

S. C. Satapathy et al. (eds.), *Communication Software and Networks*, Lecture Notes in Networks and Systems 134, https://doi.org/10.1007/978-981-15-5397-4_49

1 Introduction

The increasing popularity of smartphones combined with enhancement in location-aware technologies permit us to add location as one of the dimensions to the traditional user-item-rating-based recommender systems [6]. In fact, location plays a major factor while recommending to the users engaged in a location-based social networks like Foursquare,[1] GeoLife [9], Yelp,[2] etc. For example, when recommending restaurants to a user present in a particular geographic location, we can utilize the ratings and/or reviews provided by the users of the social network to the nearby restaurants of that user and suggest the most popular ones.

Location-based recommender systems typically provide users with a set of venues like a restaurant, a movie theater, or a shopping mall which matches her preferences and also located within a limited distance. These systems generally employ Collaborative Filtering (CF)-based recommendation algorithms, and nowadays the velocity with which data gets added to these systems often lead to severe scalability problems. The primary reason that hinders the scalability is the similarity computations which are needed between every pair of users/items in the training part of a CF algorithm. To overcome this problem, we propose a recommendation system in the restaurant domain that will recommend restaurants to a user on the basis of the opinions of her social connections (friends) while avoiding the opinions of the other users on those restaurants. This will certainly reduce the number of similarity computations up to a great extent and subsequently lessen the recommendation time.

In this work, we consider the following factors as the most influential ones that affect the quality of our location-based recommender system. (a) location of the user: people normally prefer to visit locations (restaurants) which are nearer to them, and therefore we need to identify the restaurants within a geographical span convenient to the user. (b) preferences of the user: preferences (e.g., Chinese food or Italian food) of a target user must be elicited using her historical data in order to recommend the most popular restaurants serving those foods. (c) opinions (ratings/reviews) of the other users about the target restaurants: it is probably the decisive factor that influences the target user's decision as people normally tend to rely more on the social opinions given by the nearby users. However, sometimes the most popular locations can not be recommended due to the distance preferences of the user. Thus our objective is to build a restaurant recommendation system that recommends the most suitable restaurants to the users by incorporating their distance preferences as well as social relationships. Experimental results using Foursquare dataset show that our social connections-based model is more effective and efficient than the existing baselines.

---

[1] https://foursquare.com/city-guide.

[2] https://www.yelp.com/.

2 Related Work

Location-based recommendations are generally of two types: generic and personalized. In generic location-based recommender systems, the mass opinions on the locations such as restaurants, shopping malls, or travel routes are considered for providing recommendations without taking into account the individual users preferences. These kinds of systems either use the GPS trajectories generated by the users [11] or find correlation among the locations [3] in order to provide suitable recommendations. On the other hand, personalized location-based systems [1] suggest locations on the basis of the individual preferences of a user and her location history.

The concept of location-based ratings was utilized by Sarwat et al. [8] for providing scalable and effective recommendations. They proposed a framework that includes the following two modules: (i) preference locality, which implies people in nearby areas tend to have similar preferences, and (ii) travel locality, which implies people generally prefer to visit venues (shopping mall) which fall within a limited distance. On the contrary, Zheng et al. [10] defined a user similarity measure on the basis of hierarchical graph and proposed a location-aware recommender algorithm that outperforms the traditional cosine similarity-based recommendation models. Our model is closely related to the work proposed by Bao et al. [2]. They developed a location-based recommendation system by integrating the personal preferences of the users as well as their social networking data. In this work, we also take help of the social relationships found in a typical social network to elicit the opinions of the people (friends) who are connected with the target user. In addition, to filter the friends further, we classify them as either immediate or distant friends.

3 Our Framework

In this paper, we investigate the use of location and the impact of friends on generating recommendations for a user. To evaluate our proposed system, we use Foursquare dataset [8]. It contains 2,153,471 users, 1,143,092 venues (e.g., restaurants), 1,021,970 check-ins, 27,098,490 social connections, and 2,809,581 ratings that were assigned by the users to the venues. The location of each user in the dataset is identified by a geospatial coordinate (latitude and longitude) that specifies the home town of the user. Similarly each venue is also represented by a geospatial location (latitude and longitude). We go by the intuition that the impact of the ratings of friends on a particular restaurant will be more than that of the other users concerned. To prove this, we find out the cosine similarity between the users and their friends and also with the other users who are not her friends. To compute the cosine similarity between two users a and b, we first need to represent them as rating vectors whose dimension is equal to the total number of restaurants both the users have rated together. The similarity is measured by calculating the dot product between the two vectors and dividing the value by the product of their L_2 (Euclidean) norms as given below [4].

$$sim(a, b) = \cos(a, b) = \frac{a \cdot b}{\|a\|_2 * \|b\|_2} = \frac{\sum_{i=1}^{n} a_i b_i}{\sqrt{\sum_{i=1}^{n} a_i^2}\sqrt{\sum_{i=1}^{n} b_i^2}}, \tag{1}$$

where n is the total no of restaurants, a_i and b_i are the ratings of user a and user b on restaurant i.

We found that for friends the average cosine similarity is around 0.9 and for other users it is coming around 0.7. This motivates us to design our recommendation framework only using the friends while leaving out the opinions of the other users of the system.

3.1 Classification of Friends

Let us analyze the social connections present in the social network shown in Fig. 1. First we classify the friendship relations into two types: (a) immediate friends and (b) distant friends. Given a user u, the immediate friends of u consists of all the users who are directly connected to u, and the distant friends of u comprises the users who are at least one hop away from u. In Fig. 1, we have a user named Barry, and he has got three immediate friends Bruce, Oliver, and Tony. Those immediate friends of Barry also have got few friends. Immediate friends are represented in red, while distant friends are one hop away are represented in green and distant friends who are two and three hops away are represented in blue and pink, respectively. While recommending restaurants to a user considering her social connections, typically the

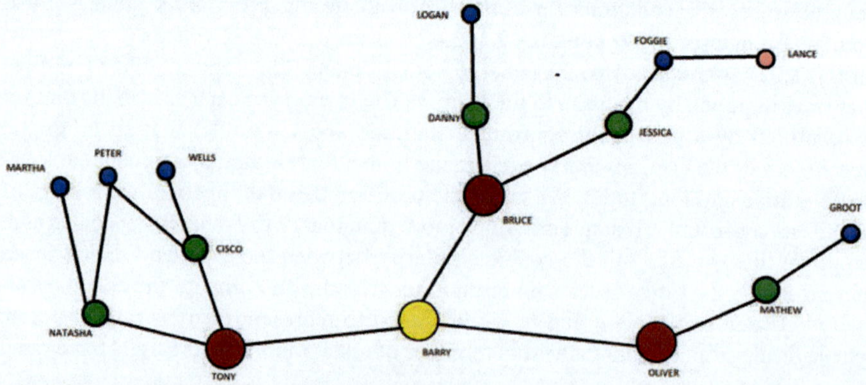

Fig. 1 Friendship relations in a social network

weights of all the ratings from both immediate and distant friends are considered to be the same. We believe that this might not be true. Some immediate friends might influence a user more than the rest of its immediate friends. Similarly, some distant friends may have a more pronounced influence on a user in comparison to other distant friends. Also, in cases where we do not have enough ratings from immediate friends we need to rely on the ratings of the distant friends. Now to increase the coverage even more, we may need to consider the rating of the distant friends who are 2 or 3 hops away from a given user. To solve the abovementioned issues we propose a friend selection strategy in the next section.

3.2 Friend Selection Scheme

Let us draw a radius around the target user and name it as User Radius (UR). UR will consist of the immediate and distant friends living around the target user who have visited the restaurant we want to recommend. Similarly draw a radius around the restaurant and name it as Restaurant Radius (RR). RR will consist of the immediate and distant friends of the target user who have visited and also live around the restaurant we want to recommend. Now let us consider the following possibilities. It can happen that Bruce, the immediate friend of the target user, Barry, might be present in the UR. So we should treat the rating of this friend with higher priority as it is more likely that a friend living in the same area of the target user will have more influence in comparison to a friend who is living in a different area. It is also possible that Tony, another immediate friend of Barry, is present in the RR. In this case, we should take the review of Tony more into consideration in comparison to others as he has already visited the restaurant we want to recommend as well as he lives nearer to the restaurant and thus has a better idea about the restaurant than the other friends. A third case where Oliver, another immediate friend of Barry, is present in both RR and UR which means that he lives in the same area where Barry lives and the restaurant we want to recommend is located. In such cases, the weightage of the ratings of friends like Oliver should be the highest.

Similarly, we can have different possibilities for distant friends of a given user. For example, it might be possible that Jessica who is an immediate friend of Bruce and distant friend of Barry is present in both UR and RR. In this case, we can take the rating of Jessica more into consideration while recommending a restaurant to Barry in comparison to the rating/review of other distant friends who are only present in UR or RR and also those distant friends who are neither present in UR nor RR. Thus, we form a priority list for immediate and distant friends who might or might not be present in UR or RR or both.

Priority list:

P1: Immediate Friends present in both RR and UR.
P2: Distant Friends present in both RR and UR.
P3: Immediate Friends present in RR only.
P4: Immediate Friends present in UR only.
P5: Distant Friends present in RR only.
P6: Immediate Friends not present in both RR and UR.
P7: Distant Friends present in UR only.
P8: Distant Friends not present in both RR and UR.

3.3 *Recommendation Using Friends*

As already discussed above, we will only consider the ratings of the friends of a target user, and not the entire rating data while providing recommendations. In the previous section, we have assigned different priorities to the immediate and distant friends and there are 8 priority types. In the recommendation phase, we only consider the first five priority types P1, P2, P3, P4, and P5. Our main emphasis is on the friends who are part of both UR and RR, and also on those friends who live near the vicinity of the restaurant we want to recommend. Therefore we leave out P6, P7, P8. Next, we assign weightage to the ratings of the friends belonging to the considered priority types. Then we predict the rating for a target user u on an unknown restaurant r using the following formula.

$$Predicted\_Rating_{(u,r)} = \sum_{i=1}^{m} W_{Pi} * \left(\sum_{j=1}^{n} \frac{Rat_j}{n} \right), \tag{2}$$

where $\sum_{i=1}^{m} W_{Pi} = 1$, m is the total number of priority types, W_{Pi} is the weight associated with the priority type i, n is the number of friends in the priority type i, and Rat_j is the rating of the jth friend belonging to the priority type i to restaurant r. The detailed recommendation process is depicted in Algorithm 1.

Algorithm 1: *Top-N* Restaurant Recommendation Algorithm

Input: Target user t_u, Friendship relations of t_u, the number of recommended restaurants, N.
Output: A list of *Top-N* recommended restaurants.

1 Find the immediate and distant friends who have visited the restaurants we want to recommend to t_u.
2 Segregate these immediate and distant friends into UR, RR, and others.
3 Assign weightage to the ratings of these friends present in UR and RR in accordance with the priority list.
4 Predict the rating for t_u on a restaurant on the basis of the weighted ratings of her friends.
5 Recommend *Top-N* restaurants to t_u according to descending order of predicted ratings.

---

4 Experiments and Results

We evaluate our framework by conducting experiments on Foursquare dataset [8]. The description of this dataset is already detailed in Sect. 3. The dataset is divided into five disjoint sets for fivefold cross-validation. One of these sets is used for testing while the remaining four are used for training, thus forming 5 independent training/testing sets. To evaluate the prediction accuracy of our algorithm, we use Mean Absolute Error (MAE) and Root Mean Squared Error (RMSE) [4] while Precision, Recall, and F1 metrics [4] are used to assess the performance of the *Top-N* recommended restaurants. Precision is defined as the ratio of relevant restaurants recommended to the total number of restaurants recommended. It indicates the probability that a recommended restaurant is relevant to the user. Recall is defined as the ratio of relevant restaurants recommended to the total number of relevant restaurants. It measures the probability that a relevant restaurant is recommended to the user. $F1$ score is widely used to test the recommender quality and it treats both Precision and Recall with equal priorities.

$$Precision = \frac{true_p}{true_p + false_p} \tag{3}$$

$$Recall = \frac{true_p}{true_p + false_n} \tag{4}$$

$$F1 = \frac{2 \times Precision \times Recall}{Precision + Recall} \tag{5}$$

In the above equations, $true_p$ (True Positive) indicates the case when a relevant restaurant is recommended by the system while $false_p$ (False Positive) implies that an irrelevant restaurant is suggested. $false_n$ (False Negative) implies the case when a restaurant of users's choice is not recommended. The aim of any recommendation algorithm is to minimize the values of MAE and RMSE while retaining higher values for Precision, Recall, and $F1$.

We compare our framework by implementing two memory-based CF models: User-based and Item-based [4], and two model-based CF methods of Singular Value Decomposition (SVD) [7] and Non-negative Matrix factorization (NMF) [5]. Pearson's correlation coefficient is used as the similarity metric for User-based method and for Item-based method, we use a cosine-based similarity measure. For SVD and NMF, we fix the dimension of the latent feature as 6. We have conducted three sets of experiments: (a) considering all the users of the dataset (base), (b) considering only the friendship relations present in the dataset but without using any priority, and (c) Considering friendship relations with priority. Finally we compare their performances.

4.1 Recommendation Performance on Foursquare Dataset

We have presented the experimental results performed on Foursquare dataset in terms of the different evaluation metrics in Table 1. In the table, we have used MAE, RMSE, *Precision*@10, *Recall*@10, and *F*1@10 to evaluate the quality of the *Top-10* recommended restaurants. Here, time reports the overall time needed to perform the recommendation task. Experiments were conducted using a computer with Core i7-8550U CPU@1.80 GHz and 8 GB RAM. The bold numbers in Table 1 indicate the results which are better than the base. In Table 1, base indicates the case where we use the rating data of all the users of the dataset. Using Friends Without Priority (UFWOP) represents the case where only the friends of a target user are considered for the purpose of recommendation. However, here we do not consider the priorities associated with the friends. Using Friends With Priority (UFWP) is the case where friendship relations are leveraged along with their priorities in order to generate recommendations.

From the results reported in Table 1, we clearly observe that in most of the cases our approach has better results than base. Further note that, UFWP-based method has advantage over both UFWOP and base, which suggests that considering priority classes makes sense. Regarding recommendation time, we can observe that both UFWOP and UFWP methods run faster than the corresponding base cases. For example, in case SVD-based CF algorithm, total recommendation time for base is 801.4 s while the corresponding time for recommending all the users of the dataset

Table 1 Recommendation performance on foursquare dataset

	MAE	*RMSE*	*Precision*@10	*Recall*@10	*F*1@10	*Time* (s)
User-based						
Base	0.664	0.864	0.905	0.416	0.569	780.5
UFWOP	**0.660**	0.868	0.876	**0.615**	**0.737**	**401**
UFWP	0.671	**0.853**	0.864	**0.624**	**0.712**	**424.65**
Item-based						
Base	0.655	0.849	0.914	0.432	0.586	775.9
UFWOP	**0.648**	**0.832**	**0.919**	**0.621**	**0.740**	**376.3**
UFWP	0.662	**0.828**	0.893	**0.628**	**0.737**	**382.5**
SVD						
Base	0.618	0.804	0.950	0.446	0.607	801.4
UFWOP	0.632	**0.795**	0.933	**0.639**	**0.758**	**350.3**
UFWP	**0.588**	**0.783**	0.910	**0.657**	**0.762**	**354.2**
NMF						
Base	0.624	0.824	0.948	0.448	0.608	797.5
UFWOP	0.638	**0.814**	0.923	**0.628**	**0.747**	**372**
UFWP	**0.602**	**0.795**	0.942	**0.652**	**0.770**	**378.66**

using UFWOP and UFWP techniques are 350.3 and 354.2, respectively. Thus the runtime reduces by about 56% and 55%, respectively. Similar results can be observed for the other cases too. Therefore we can finally conclude that our approach reduces the runtime as well as produces quality recommendations.

5 Conclusion and Future Work

In this paper, we have attempted a social relationship-based recommendation approach to improve the performance of existing CF algorithms. We have designed a recommendation system in the restaurant domain where recommendations to a user are provided not only considering her personal preferences but also using the preferences of her friends present in a social network. Moreover, we classified the friends into two types: immediate and distant, and also assigned priorities to them in order to further enhance the recommendation accuracy. Experimental analysis using Foursquare dataset demonstrates that our approach is efficient and scalable. In future, we have plans to consider the effect of hops while considering the distant friends and accordingly revise the priority list in order to further enhance the performance of our system.

References

1. Bao J, Zheng Y (2017) Location-based recommendation systems. In: Encyclopedia of GIS, pp 1145–1153
2. Bao J, Zheng Y, Mokbel MF (2012) Location-based and preference-aware recommendation using sparse geo-social networking data. In: Proceedings of the 20th international conference on advances in geographic information systems, SIGSPATIAL'12, pp 199–208
3. Cao X, Cong G, Jensen CS (2010) Mining significant semantic locations from GPS data. Proc VLDB Endow 3(1–2):1009–1020
4. Ekstrand MD, Riedl JT, Konstan JA et al (2011) Collaborative filtering recommender systems. Found Trends Human-Comput Interact 4(2):81–173
5. Lee D, Seung H (2001) Algorithms for non-negative matrix factorization. Adv Neural Inform Process Syst 13:556–562
6. Ricci F, Rokach L, Shapira B (2015) Recommender systems: introduction and challenges. In: Ricci F, Rokach L, Shapira B (eds) Recommender systems handbook. Springer, Boston, MA, pp 1–34
7. Sarwar B, Karypis G, Konstan J, Riedl J (2000) Application of dimensionality reduction in recommender systems: a case study. In: Proceedings of the ACM WebKDD workshop
8. Sarwat M, Levandoski JJ, Eldawy A, Mokbel MF (2014) Lars*: an efficient and scalable location-aware recommender system. IEEE Trans Knowl Data Eng 26(6):1384–1399
9. Zheng Y, Xie X, Ma WY et al (2010) Geolife: a collaborative social networking service among user, location and trajectory. IEEE Data Eng Bull 33(2):32–39
10. Zheng Y, Zhang L, Ma Z, Xie X, Ma WY (2011) Recommending friends and locations based on individual location history. ACM Trans Web 5(1):1–44
11. Zheng Y, Zhang L, Xie X, Ma WY (2009) Mining interesting locations and travel sequences from GPS trajectories. In: Proceedings of the 18th international conference on world wide web, WWW'09, pp 791–800

A Survey on Identity-Based Security in Wireless Sensor Networks

Lakshmanarao Kalabarige and Hima Bindu Maringanti

Abstract This paper presents the literature on identity-based security management approaches in Wireless Sensor Networks (WSNs). With limited communication, storage and computational capabilities of motes, it is difficult to ensure Authentication, Integrity and Confidential distribution of data and secret keys in WSNs. The open nature of wireless channels used in WSNs may easily lead to eavesdrop on conversations between nodes by an adversary. Hence, authentication of wireless motes, confidentiality of data to be transmitted and no modification of data during its transmission are the important security services of WSNs. The identity-based protocols of WSNs ensure security services like Confidentiality, Integrity and Authentication during data transmission. In literature, various identity-based security approaches have been proposed in recent years to ensure the above-mentioned security services during data transmission in WSNs. The study of literature on identity-based security approaches is classified into four categories and presents the proposed literature on different protocols belonging to each category.

1 Introduction

Nowadays, the WSNs are used in wide variety of high sensitive applications [1] like battlefield surveillance, military and some other applications, which require high security to the data to be transmitted [1, 2]. However, the wireless sensors are insecure to different passive and active attacks [1, 3] such as eavesdropping, masquerading, impersonating, Sybil, interception and physical capture of a node due to the unattended deployment. Hence, the security of WSNs is highly important.

L. Kalabarige (✉)
Department of Computer Science and Engineering, GMR Institute of Technology, Rajam, India
e-mail: lakshmanarao.k@gmrit.edu.in

H. B. Maringanti
Department of Computer Applications, North Orissa University, Baripada, Mayurbhanj 757003, Odisha, India
e-mail: profhbnou2012@gmail.com

S. C. Satapathy et al. (eds.), *Communication Software and Networks*, Lecture Notes in Networks and Systems 134, https://doi.org/10.1007/978-981-15-5397-4_50

The traditional cryptographic approaches [4] are infeasible to resource-constrained sensor nodes due to its high memory, communication and computational overheads. The security strength of the protocol which provides security to WSNs is inversely proportional to the implementation complexity in security evaluation benchmarks like Communication, Computation and Storage. All these security evaluation benchmarks of a proposed protocol are inversely proportional to the implementation complexities in security services like CIA triad, cryptographic techniques, key generation and sharing. Hence, implementation of security services should be lightweight.

The organization of the remaining paper is as follows: The related work is discussed in Sect. 2. The Identity-Based Cryptography (IBC) schemes which provide Authentication to the WSN are discussed in Sect. 3. Section 4 covers the IBC approaches for secure key distribution. Section 5 presents IBC schemes which establishe signature for secure communication. Section 6 describes the IBC approaches to address Sybil attack. Section 7 presents research challenges and future scope, and lastly, concluding remarks are presented in Sect. 8.

2 Related Work

The study of proposed cryptographic methods for WSNs says that the Symmetric Key Cryptography (SKC) is lightweight, encrypts huge amount of data and suitable to the WSNs. However, it suffers from key distribution problem. The Public Key Cryptography (PKC) algorithm like RSA is suitable to WSNs when a sensor is equipped with dedicated hardware accelerator for cryptographic operations. On the other hand, the Elliptic Curve Cryptography(ECC) is similar to PKC and its security strength depends on the strength of Elliptic Curve Discrete Logarithm Problem (ECDLP), and the Elliptic Curve Diffie–Hellman (ECDH) is used for the exchange of secret keys between pair of nodes which does not ensure authenticity as well as is exposed to man-in-the-middle attack. In this context, the researchers concentrated on design of lightweight IBC schemes for WSNs. The assurance of security services and distribution of secret keys or session keys made easy and lightweight in IBC. Recently, many IBC schemes for WSN have been proposed. In this state of art, the study on IBC is classified into four types based on the working style of the proposed method as shown in Fig. 1.

3 Identity-Based Authentication

Authentication is a most important security need of WSNs. It controls attacks formulated based on identity of motes. Various protocols under authentication are discussed below. Mahmud et al. proposed identity-based authentication and access control to WSN [5]. The authors Oliver et al. proposed TinyPBC [6]. The authors Mohad

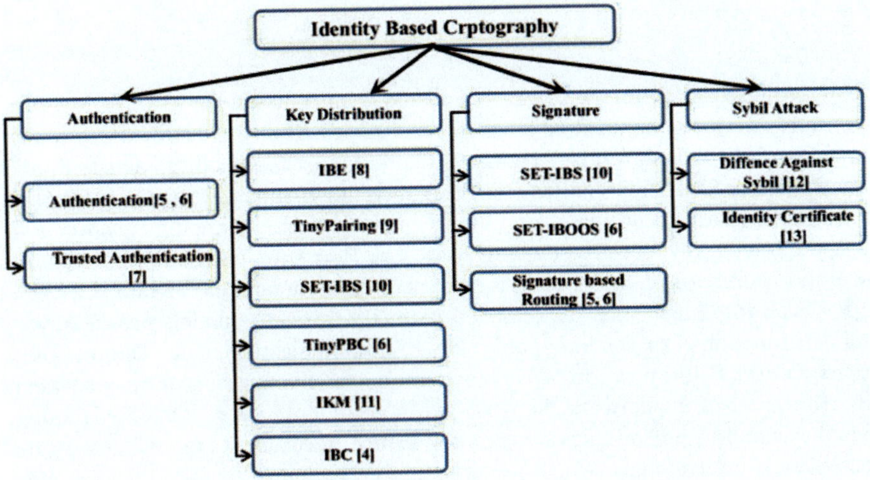

Fig. 1 Classification of IBC based on proposed methods

et al. proposed identity-based trust authentication [7] for WSNs. Following this, the comprehensive survey on these protocols has been discussed.

3.1 *Authentication and Access Control Scheme*

In [5], the Base Station (BS) acts as private key generator (PKG) where the motes and users take their corresponding private key from BS. Motes authenticate the user to grant access on a requested data. This approach has five phases such as 1. System initialization, 2. Authentication and establishment of session key, 3. Data access, 4. Update of user information and 5. User revocation. The first phase performs operations such as 1. BS initializes master secret key, public key, public system parameter and data access structure on it's own, 2. Stores Identity, Timestamp, Group identity and Access structure of both users and sensors in its database and 3. The sensor nodes of the network stores details of each user broadcasted by BS. The second phase performs authentication as well as generates session key between mote and user after successful completion of initialization phase. The third phase validates session key and authenticates user while accessing data from a respective sensor node. The *Msg* (*NO_ACCESS*) will be delivered to the user who requested for unprivileged data. In fourth phase, BS updates about the list of compromised and newly added users in the network. The announcement of compromised list of users preserves network security. In fourth phase, the compromised list of users is detained by the sensor node from accessing network.

3.2 TinyPBC

In [6], proposed a non-interactive authenticated identity-based key sharing in WSNs to ensure secure communication among sensor nodes. The single node compromise should not lead to compromise of entire network. In this approach, BS acts as Public Key Generator (PKG), BS builds S (master secret key) and converts the identity of sensor node as x, y coordinates on elliptic curve through a mapping function ϕ to compute private key of corresponding node. The $P_i = \phi(id_i)$ is the x, y coordinates on elliptic curve for ith node; then, the private key of ith node is calculated as $S_i = [s]P_i$. Each ith sensor node is preloaded with (i) id_i, (ii) the node's private key S_i and (iii) function ϕ to generate public key of communicating node. The proposed approach uses Bilinear pairing to achieve non-interactive sharing of keys between any two motes of the network. Suppose two nodes i and j know IDs of each other, then they should agree on a same secret key. Hence, according to the Bilinearity, the secret key is calculated as $e(S_i, P_j) = e([s]P_i, P_j) = e(P_i, P_j)^s = e(P_i, [s]P_j) = e(P_i, S_j) = e(S_j, P_i)$. Note that i with S_i can calculate $P_j = \phi(id_j)$ since it knows ID of jth node. Likewise, j with S_j can calculate $P_i = \phi(id_i)$. Hence, it is proved that both i and j are able to calculate common secret key $K_{i,j} = e(S_i, P_j) = e(S_j, P_i)$. Hence, it is proved that the proposed method provides authentication and non-interactively distributes common secret key in secure manner.

3.3 Identity-Based Trusted Authentication (IBTA)

The IBTA [7] proposed Trust Authentication to WSNs using the concept of IBE [8]. The Authenticated Key Exchange (AKE) through ID-based one-pass [7] implements exchange of shared secret and session key. Each sensor node of the network consists of trusted list of sensor nodes. The destination is authenticated based on sender ID. The AKE is implemented using symmetric bilinear pairings as stated below: The ith sensor node randomly generates a number $r \in Z_q$, calculates shared secret random number $R = rQ_i$ where Q_i is a public key and sends R to receiver j.

Both sender and receiver commonly calculate $h = H_2(R, ID_i || ID_j)$, and private keys S_i, S_j of both parties are stored securely in their respective On-SoC ROMs. The sender i and receiver j calculate shared secret key as $K_{ij} = e((r + h) S_i, Q_j)$ and $K_{ji} = e(R + hQ_i, S_j)$, respectively. Lastly, ith and jth nodes calculate session key as $\boldsymbol{K}(Kij)$ and $\boldsymbol{K}(Kji)$, respectively, where $\boldsymbol{K}$ is derivation function.

4 Identity-Based Key Exchange

The secure key exchange plays an important role in WSNs to share secret keys, pairwise keys, cluster keys, group keys and session keys. Various protocols under identity-based key exchange are listed below. The authors Boneh et al. proposed

IBE using Weil pairing [8]. The authors Lu et al. proposed Secure and Efficient data Transmission through Identity-Based Signature (SET-IBS) [10]. The authors Xiong et al. proposed TinyPairing [9]. The authors Oliveria et al. proposed TinyPBC [6]. The authors Boujelben et al. proposed pairing identity-based key management protocol for Heterogeneous WSNs (HWSNs) [14]. The [6, 9, 14] are pairing-based cryptography approaches. The authors Boujelben et al. proposed IKM—An identity-based Key Management Scheme for HWSNs [11].

The authors Szczechowiak et al. proposed an IBC approach [4] for secure WSNs. The authors Lakshmanarao et al. [15] proposed Hashed Identity-based Secure Key and Data Exchange (HISKDE). Following this, the comprehensive study on these protocols has been discussed.

4.1 *Identity-Based Encryption (IBE)*

The IBE [8] proposes secure key sharing to WSNs. It has four phases. 1. Setup: BS generates public parameters and its master key s by taking security parameter k as input. The master key and public parameters are preloaded into all sensor nodes. 2. Extract: In this phase, mote ID is mapped as x, y coordinates on elliptic curve E/F_q: $y^2 = x^3 + 1$ using $Q_{IDx} = H_1(ID_x)$, and Q_{IDx} is multiplied by master key s as $d_x = sQ_{IDx}$ to obtain the private key d_x. 3. Encrypt: The common parameters, ID of destination and message M are considered as input to this process to calculate ciphertext C. 4. Decrypt: The common parameters, private key d_x and ciphertext C are considered as input to this process to calculate plain text M.

4.2 *SET-IBS*

The SET-IBS [10] proposed a secure key exchange technique. The BS performs pre-distribution of key to all motes as described in Sect. 5.1. The strength of this approach relies on the hardness of Diffie–Hellman key pairing technique. Key management for security [10, 16] performs tasks like authentication of intermediate nodes and provides security to the secret key during transmission from source to destination. The key exchange method of SET-IBS is used for secure routing in [16], whereas [10] uses it for the exchange of key through authenticated nodes of CWSNs. The only one difference between [10] and [16] is that the digital signature is generated on ciphertext C_j of jth sensor node [10], whereas [16] generates digital signature on the message M of jth sensor node.

4.3 Identity-Based Key Pairing

The Bilinear Pairing grabs the attention of research community since the introduction of Pairing-Based Cryptography (PBC). In [9], lightweight and fast pairing-based cryptography library TinyPairing for WSN is proposed. It has additional library for IBE [8] scheme proposed by Boneh and Franklin. TinyPBC [6] is used to generate secret key and non-interactively share a common secret key between two motes in secure manner as described in Sect. 3.2. The library of TinyPairing is used in TinyPBC to implement PBC primitives for the sensor nodes which use 8, 16 and 32-bit processors.

A Pairing Identity-Based Key Management Protocol [14] for HWSNs generates pairwise keys as well as other keys such as cluster key common to all nodes belonging to same cluster and a group key which is common to the entire network. The cluster keys are used to secure the messages that are broadcast locally within the cluster.

The group key is used by BS to secure the broadcast messages, which are common to entire network. The key establishment and sharing in pairing identity-based cryptography [14] has two phases (i) Pre-Deployment phase: preloads all nodes with public parameters, private and public keys. (ii) After deployment phase: performs tasks such as Cluster formation, Neighbourhood Discovery and generation of pairwise, cluster and group keys.

4.4 IKM

The IKM [11] for HWSNs is extended [14] by adding two issues: 1. Node addition and 2. Re-keying as extra contribution. The Node addition adds new nodes in two cases: 1. Whenever an existing node fails and 2. To extend coverage of the network. The usage of same keys for long time may lead to network compromise by an adversary. Hence, Re-keying is applied periodically to update the existing pairwise and cluster keys to avoid threats from compromised nodes.

4.5 IBC

A practical identity-based key agreement [4] scheme reduces the required number of keys by using bilinear paring concept and does not require prior communication between communicating sensor nodes to establish keys. It has two phases: 1. *Pre-Deployment phase* preloads sensor node with secret key, public key and public parameters and 2. *Key agreement* generates pairwise session key $K_{A,B}$ and $K_{B,A}$ between nodes A and B where $K_{A,B} = KDF\ (\hat{e}(sA, B)) = KDF(\hat{e}\ (A, B)^s) = KDF(\hat{e}(A, sB) = KDF\ (\hat{e}(sB, A)) = K_{B,A}$.

4.6 HISKDE

The proposed HISKDE [15] is a lightweight protocol which combines both IBC and SKC methods to provide confidentiality and authentication to the transmitted data and secret key. This approach generates random prime number as secret key K based on key generation rules and extracts the key value k_v from spiral arrangement using K and encrypts key using hashed ID of a sensor node and encrypts data using k_v.

5 Identity-Based Signature

The Signature is a technique which keeps identity of a sensor node in the digital data to be transmitted, and it is verified by the receiver to check the authenticity and integrity of sender. Three protocols under signature are listed below. The authors Lu et al. [10] proposed SET-IBS and SET-IBOOS. The SET-IBS is discussed in Sect. 4.2. The authors Oliveira et al. proposed secure routing [6] through signature. Following this, the comprehensive study on these protocols has been discussed.

5.1 SET-IBOOS

The proposed Secure and Efficient data Transmission (SET) Identity-Based online/offline digital signature (SET-IBOOS) [10] is an extension to [17]. There are Cluster Header (CH), BS and sensor nodes in CWSNs. The BS preloads all sensor nodes with (i) an encryption key $\boldsymbol{k}$ where $\boldsymbol{k} \in [m\ 1]$, m is large integer, (ii) pairing parameters $(p, q, E/F_p, G_1, G_2, e)$ and selects $P \in G_1$ randomly, (iii) two hash functions H for the point mapping on a curve and h is for producing fixed length output for a given input and (iv) a random integer $\tau \in Z_q$ as master key msk sets $P_{pub} = \tau P$ as network public key. The implementation of identity-based signature in SET-IBS has four phases such as setup, extraction, signature and verification. Specifically, setup is performed by BS; the extraction, signature and signing were performed by sender, and verification is performed by receiver. The signature generation approach proposed in Secure Routing Protocol for CWSNs using ID-based Digital Signature [6] is an extension to [5].

6 Identity-Based Solution to Sybil Attack

The WSNs use public and open wireless medium for communication which is vulnerable to identity-based attacks such as Spoofing and Sybil. These attacks degrade the performance of network. Sybil attack grabs identity of nodes and creates multiple redundant fake sensor node identities to breach the security of WSNs. The authors Chen et al. proposed Detecting and localizing identity-based attacks in WSNs [12]. The authors Zhang et al. proposed an approach to defend against Sybil attacks in WSNs [13]. The following subsections present study on these protocols.

6.1 Identity-Based Defence Against Sybil Attack

In [12], the mathematical statistics on physical property such as Received Signal Strength (RSS) of wireless medium is used to identify and mitigate Sybil attack in WSNs. The strength of RSS is measured from a known location of a static node to detect Sybil attack based on identity of motes. The proposed approach establishes relationship between the distance of signal and physical spaces of a node based on RSS to develop analytical expression to find location of nodes, detection rate and false-positive rate.

6.2 Identity Certificate

In [13], proposes a lightweight identity-based certificate to defeat Sybil attack. This certification process avoids public key cryptography and provides authentication to all the messages by using one-way key chains and Merkle hash trees. Each node i is set with unique information I_i and information commitment C_i where computation of C_i from I_i is computationally infeasible. Each leaf vertex V_i of Merkle hash tree is represented as one sensor node where $V_i = H(C_i || I_i)$. Certificate $IDCert_i$ consists of the label of leaf and root nodes on the corresponding $Authentication\text{-}Path_i$. The ith node should show its certificate at jth on demand to prove its authenticity.

7 Research Challenges and Future Scope

The evaluation criteria of security protocols of WSNs depend on the implementation complexity of security services such as confidentiality, authentication, integrity, key generation and cryptographic technique. This study observes that the security approaches which use IBC [4] reduce computation, communication and storage complexity when compared with SKC and PKC. The strengths and weaknesses of the proposed Identity-Based Cryptography approaches are listed in Tables 1 and 2. From the comparison, it can be observed that most of the proposed IBC approaches are vulnerable to node capture attack and incur more computational complexity. The implementation of all the proposed techniques combines IBC with either ECC or PKC. Hence, all proposed methods are more or less equal to the implementation of PKC. In this context, the combination of IBC with SKC gives better results to the

Table 1 The comparison of [4–7, 9, and 10] IBC schemes for WSNs

Criteria	[4]	[5]	[6]	[7]	[9]	[10]
Simulator	TinyOS	NS2	TinyOS	Hardware	TinyOS	Omnet++
Authentication	Maintained	Mutual	Maintained	Maintained	Maintained	Maintained
Access control	No	Maintained	No	No	No	No
Session key agreement	Available	Available	Available	Available	Available	Available
Confidentiality	Maintained	Maintained	Maintained	Maintained	Maintained	Maintained
Integrity	No	Maintained	No	No	No	Maintained
Key management	Pairing based	No	Non-interactive	Trust based	Maintained	Signature based
Cryptographic technique	Bilinear pairing	IBS	Bilinear pairing	Weil pairing	Weil pairing	Bilinear pairing
Vulnerability	Node capture	Node capture	Node capture	Energy Consumption and computational complexity	Computational complexity	Node capture
Advantage	Defend intruder	Node revocation	Secure non-interactive sharing of key	Secure non-interactive sharing of key	Secure non-interactive sharing of key	Defend passive attacks

Table 2 The comparison of [11–14, 16, 17] IBC schemes for WSNs

Criteria	[11]	[12]	[13]	[14]	[16]	[17]
Simulator	TinyOS	NS2		AVISPA tool	Omnet++	Omnet++
Authentication	Maintained	Maintained	Maintained	Maintained	Maintained	Maintained
Access control	No	No	No	No	No	No
Session key agreement	Available	No	No	Available	No	Available
Confidentiality	Maintained	No	Maintained	Maintained	Maintained	Maintained
Integrity	No	No	No	No	Maintained	Maintained
Key management	Yes	No	No	Yes	Yes	Asymmetric
Cryptographic technique	Bilinear pairing		One-way key chain, hash tree	IBC and Tate pairing	IBS	Bilinear pairing
Vulnerability	Node capture	Communication overhead	Node capture	Node capture	Node capture	Node capture
Advantage	Prevents Sybil	Adversary	Defend man-in-the-middle attack	Security to routing protocols	Size of message is smaller than Tesla	Defend passive and active attacks

WSNs since it is proved that the SKC and IBC are lightweight in their implementation and IBC can be used to address the key sharing problem of SKC.

8 Conclusion

In this paper, the literature on IBC-based security approaches is presented. The existing IBC-based protocols are analysed and classified based on the proposed cryptographic method, and it is observed that the proposed identity-based security approaches ensure CIA—Confidentiality, Integrity and Authentication triad to the WSNs. As discussed earlier, the protocols which implement CIA triad ensure secure communication in WSNs.

References

1. Simplicio Jr MA, De Oliveira BT, Margi CB, Barreto PS, Carvalho TC, Naslund M (2013) Survey and comparison of message authentication solutions on wireless sensor networks. Ad Hoc Networks 11(3):1221–1236
2. Swain G, Sahu AK (2019) Dual stego-imaging based reversible data hiding using im-proved lsb matching. Int J Intell Eng Syst 12(5):63–73
3. Rani A, Kumar S (2017) A survey of security in wireless sensor networks. In: 2017 3rd international conference on computational intelligence & communication technology (CICT). IEEE, pp 1–5
4. Szczechowiak P, Scott M, Collier M (2010) Securing wireless sensor networks: an identity-based cryptography approach. Int J Sens Networks 8(3/4):182–192
5. Al-Mahmud A, Morogan MC (2012) Identity-based authentication and access control in wireless sensor networks. Int J Comput Appl 41(13)
6. Oliveira LB, Aranha DF, Gouvêa CP, Scott M, Camara DF, Lopez J, Dahab R (2011) Tinypbc: pairings for authenticated identity-based non-interactive key distribution in sensor networks. Comput Commun 34(3):485–493
7. Yussoff YM, Hashim H, Baba MD (2012) Identity-based trusted authentication in wireless sensor network. arXiv preprint arXiv:1207.6185
8. Boneh D, Franklin M (2001) Identity-based encryption from the weil pairing. In: Advances in cryptology | CRYPTO 2001. Springer Berlin Heidelberg, Berlin, Heidelberg, pp 213–229
9. Xiong X, Wong DS, Deng X (2010) Tinypairing: a fast and lightweight pairing based cryptographic library for wireless sensor networks. In: 2010 IEEE wireless communication and networking conference. IEEE, pp 1–6
10. Lu H, Li J, Guizani M (2013) Secure and efficient data transmission for cluster-based wireless sensor networks. IEEE Trans Parallel Distrib Syst 25(3):750–761
11. Boujelben M, Youssef H, Mzid R, Abid M (2011) Ikm: an identity based key management scheme for heterogeneous sensor networks. JCM 6(2):185–197
12. Chen Y, Yang J, Trappe W, Martin RP (2010) Detecting and localizing identity-based attacks in wireless and sensor networks. IEEE Trans Veh Technol 59(5):2418–2434
13. Zhang Q, Wang P, Reeves DS, Ning P (2005) Defending against sybil attacks in sensor networks. In: 25th IEEE international conference on distributed computing systems workshops. IEEE, pp 185–191

14. Boujelben M, Cheikhrouhou O, Youssef H, Abid M (2009) A pairing identity based key management protocol for heterogeneous wireless sensor networks. In: 2009 international conference on network and service security. IEEE, pp 1–5
15. Lakshmanarao K, Maringanti H (2015) Hashed identity based secure key and data exchange in wireless sensor networks using IEEE 802.15.4 standard. Int J Appl Eng Res 10(10):23231–23241
16. Lu H, Li J, Kameda H (2010) A secure routing protocol for cluster-based wireless sensor networks using id-based digital signature. In: 2010 IEEE global telecommunications conference GLOBECOM 2010. IEEE, pp 1–5
17. Liu JK, Zhou J (2009) An efficient identity-based online/offline encryption scheme. In: International conference on applied cryptography and network security. Springer, pp 156–167

Analysis of Crimes Against Women in India Using Machine Learning Techniques

K. Ravi Teja, K. R. M. Ajay Kumar, G. S. S. Praveen, and D. N. D. Harini

Abstract Data mining and Machine learning, both, have become the crux phrases in search engine optimization, web suggestions, predictions, and serve multiple options to the end user based on his search pattern. India is the second most populous country which flourishes in the rate of growth of the economy and also leads in the rate of crimes. Complexity and diversity which have occurred due to the huge population make it an arduous task for the crime investigators to restrict their frequent occurrences. In this paper, an approach has been discussed such that the pattern of crimes and crime characteristics in India, such as rape, sexual assault, abduction, etc., are analyzed by the inclusion of techniques of machine learning in the investigation process. To analyze, a powerful python library, pandas has been used to study and work on the datasets gathered from each state in the country. The scope of the paper not only analyzes the pattern of crimes, but also emphasizes the primary root cause and measures to be taken to prevent them in the future through the implementation of machine learning algorithm of the decision tree.

1 Introduction

Crime is an inevitable force concomitant with integrity in society. Crime leads to incongruity in the social contract. Poverty and unhealthy social conditions are primordial causes of crime [1]. A person can either be directly or indirectly responsible for

K. Ravi Teja (✉) · K. R. M. Ajay Kumar · G. S. S. Praveen · D. N. D. Harini
Department of CSE, Gayatri Vidya Parishad College of Engineering (A), Madhurawada, Visakhapatnam, Andhra Pradesh, India
e-mail: ravijd77@gmail.com

K. R. M. Ajay Kumar
e-mail: ajay.prince555@gmail.com

G. S. S. Praveen
e-mail: praveengatti123@gmail.com

D. N. D. Harini
e-mail: hariniauphd@gmail.com

S. C. Satapathy et al. (eds.), *Communication Software and Networks*, Lecture Notes in Networks and Systems 134, https://doi.org/10.1007/978-981-15-5397-4_51

the crime. If a person commits a crime by himself, then he is considered to be directly responsible for the crime. If a person or group of people is the causal of crime by another person, then the latter is considered to be indirectly responsible for the crime. Crime usually follows an irregular trend and involves analysis of humongous data. The Indian government has an organized format of this discrete data. Intelligible analysis and prediction of this data can be achieved through machine learning algorithms [2]. Although the government provides an analysis of crime data, it showed a trivial effect in curbing crimes. We propose an efficient analysis to detect crime patterns and crime solving.

Data Mining [3], also known as knowledge mining from data, is the process of unveiling hidden patterns, extracting useful information from humongous raw data, and paving the way for the generation of new information, mostly from unstructured data. It involves Data Cleaning, Data Integration, Data Selection, Data Transformation, and Data Evaluation [4]. Its interdisciplinary nature-incorporated methods from diverse domains are Statistics, Machine Learning, Database Systems and Data Warehouses, and Information Retrieval. Out of these, Machine Learning plays a vital role in bridging the characteristic rift between input and processing by using intricate algorithms such as Decision Tree, Support Vector Machine, Naive Bayes, ID3, Random Forest, etc. Implementation of these algorithms facilitates the generation of productive analysis which is beneficial to track and understand the flux of any real-world situation.

2 Related Work

Several researchers and authors have used Machine Learning Techniques in solving real-world problems. Machine Learning Techniques are proved to be efficient in solving real-world problems like crime. Some of the researchers used these efficient techniques in detecting patterns and solving crimes.

The authors Motwani et al. [10] studied the crime data against the women by analyzing the patterns and its relations between the crimes and predicted a model using time series algorithm; the dataset contains information from 2003 to 2012 for the state Andhra Pradesh.

Oil Sarkar et al. [11] analyzed the crime in a systematic pattern and clustered by using DB SCAN and Agglomerative clustering to observe and prevent crime adversity in the state West Bengal. The authors Shiju Sathya Devan and Surya Gangadharan [3] predicted the crime using Naive Bayes Classifier and integrated the concept named entity recognition in the crime articles to get details like victim name, location, and time of the crime.

The crime location is segregated into 200 m square clusters and labeled as the center location, and prediction models are developed using global, pooled, hierarchical, and multitasking learning models. Identify non-homogeneous crime patterns and share the information of different areas within this developed model.

2.1 *Pandas*

Pandas is a powerful python data analysis toolkit. It reads files and stores them into python objects such as data frames. Data frames in pandas are two-dimensional size-mutable, heterogeneous tabular structures with labeled rows and columns. Analysis can be done on these objects as per the user's requirements. It should be installed explicitly using pip. Any type of files such as .csv, .xls, json, etc., that are used to represent statistical data, can be loaded into data frames.

2.1.1 Classification

One form of data analysis is classification [5]. The classification task can be seen as a supervised technique where each instance belongs to a class. In the process of classification, there is a predetermined set into which a given data are grouped, based on its attributes. This could help in detecting high-yielding patterns. For example, an email-labeling program might attempt to classify an email as legitimate or spam. Decision Tree is one of the most widely used methods for classification. It classifies instances by sorting, from the root node to the leaf node. The process begins with the root node, testing the attribute specified by the node, and then moving down the branch corresponding to the value of the attribute. This is repeated recursively for the subtree rooted at the new node. The ultimate goal of classification is to accurately predict the target class for each case in data.

2.1.2 Decision Tree: ID3 Classification Algorithm

Decision trees are implemented using numerous algorithms whose primary focus is on factors such as splitting criterion, elimination of overfitting, and handling incomplete data. One of the most widely used among these is the ID3 (Iterative Dichotomizer 3) algorithm [6]. It is a precursor to the C4.5 algorithm. Invented by Ross Quinlan in the year 1986, it aims to build the smallest possible decision tree. In the scope of this project, the algorithm builds the tree heuristically bestowing the most optimal decision combination for a set of given attributes. The fundamental decision would be the choice of an attribute at each node in the tree.

The algorithm's basic roots are built on two parameters: Entropy, to calculate the homogeneity and Information, and Gain, which determines the decrease in entropy after the division of data on an attribute (dominant attribute).

Dichotomization is the process of dividing into two opposite/obverse classes or groups. The algorithm iteratively divides the dataset into two groups, the most dominant attribute and the remaining attributes that fall under the dominant attribute. It calculates the entropy and information gain of each attribute in the next step. The most dominant attribute is found through the attribute estimation and is put on the tree as a decision node.

With the root of the tree determined, entropy and gain scores are estimated in the next iteration for the other attributes to find the next most dominant attribute. The process continues as long as a decision is reached for that branch.

2.1.3 Random Tree

Random tree [7], an extensively used Machine Learning Classification Algorithm, is used to classify data rooted on a salient feature called "votes."

Disparate datasets are taken, directed to be trained and on a predefined set of parameters yielding a cluster of variant trees. The datasets are generated using a procedure called bootstrap procedure. In this, each set to be trained is considered and vectors are chosen with replacement in an unordered way, leading to selection of random vectors. However, this method does not ensure the constant frequency of a vector, on the whole, rising count disparities. This follows that, at each instance, all the variables would not be used because of the stochasticity. Best split is obtained from these random vectors at each node instance with its size being constant. This does not require error-correction or error-estimation techniques unlike other methodologies as its key factor of building is randomness. Hence, any errors occurring in this procedure could be internally estimated utilizing the vector subset property.

3 Dataset

This dataset provides detailed information about the different types of crimes (with sections from the Indian Constitution) against women and the number of cases registered in the police station for each subset of crimes in all the states and union territories in the year 2014. Each of the crimes is subdivided based on various categorical classifications in the view of making accurate predictions. This imparts primary discernment regarding the ascending delinquencies with statistical evidence.

4 Proposed Methodology

The state-wise analysis of crimes in this paper is done in python programming language with the help of pandas module. The dataset chosen to conduct our work on is obtained from authenticated source, i.e., data.gov.in [8] which is the latest release under National Data Sharing and Accessibility Policy and contributed from the Ministry of Home affairs, department of states, and National Crime Records Bureau (NCRB). The data refers to different states and union territories' crime against women during the year 2013–2014. The most vulnerable crimes covered in this dataset are assault on women, kidnapping and abduction, insult to the modesty of women, cruelty by husband or relatives, immoral traffic, dowry deaths, rapes, etc.

- First step in performing efficient analysis is data preprocessing.
- In data preprocessing, we convert the raw data into nominal data that assists in efficient extraction of data patterns.
- We upload the nominal data into Weka tool [9] and build a plot matrix using this nominal data and find the correlation between all the attributes.
- Using J48 in Weka, a decision tree classifier is constructed which is used in the analysis of causes of a particular crime in all the states and union territories.
- To enhance the scope of analysis, random tree classifier is constructed by considering all the attributes as the decision tree accepts a single attribute.

Once the preprocessing of the dataset has been completed, the user can do the analysis on the newly formed structure. Pandas provide various plots that aid in data plotting for visualization and drawing inferences from them. The input to the ID3 algorithm should be a nominal dataset. This mandates the need to form a new data frame corresponding to the requirement. The steps to perform classification are as follows:

- Define a method that performs the required operation and returns the categorial value of the cell. Create a new data frame bypassing this function to an existing data frame.
- Import the above-formed dataset to obtain the graphical representation of the attributes.
- Use the DecisionTreeClassifier from sklearn with the criterion set to "State/UT" attribute in the dataset.
- The DecisionTreeClassifier thus formed can be visualized using pydotplus module in python.
- We derive a tree analysis, and the decision is made from the tree obtained.

5 Analysis

5.1 Analysis of the Crime Rate

From Fig. 1a, data obtained as a result of the ID3 classification algorithm gives the most chronic crime that is likely to occur in that particular State or Union Territory. For example, in states like Andhra Pradesh, Assam, Gujarat, and Haryana, "Cruelty by Husband or his relatives" is the most frequently committed crime. Similarly, "Rape" in Meghalaya, Mizoram, Nagaland, and Sikkim is the most recurring crime. Considering the country as a whole, it can be deduced that one of the basic reasons for the crime against women is the cold and cruel nature of her husband or her relatives. The very same conclusion can be drawn from the data of 29 states in the country. "Assault with intent to outrage women's modesty" has been the most prevalent crime in the Union Territories.

The pictorial representation of the statistical data in the form of pie graph (Fig. 1b) shows the discrepancy in the intensity levels of major crimes that are taking place

a

	State/UT	Prominent Crime
0	Andhra Pradesh	Cruelty by Husband or his Relatives
1	Arunachal Pradesh	Assault on Women with intent to outrage her mo...
2	Assam	Cruelty by Husband or his Relatives
3	Bihar	Kidnapping and Abduction
4	Chhattisgarh	Assault on Women with intent to outrage her mo...
5	Goa	Assault on Women with intent to outrage her mo...
6	Gujarat	Cruelty by Husband or his Relatives
7	Haryana	Cruelty by Husband or his Relatives
8	Himachal Pradesh	Assault on Women with intent to outrage her mo...
9	Jammu & Kashmir	Assault on Women with intent to outrage her mo...
10	Jharkhand	Cruelty by Husband or his Relatives
11	Karnataka	Assault on Women with intent to outrage her mo...
12	Kerala	Cruelty by Husband or his Relatives
13	Madhya Pradesh	Assault on Women with intent to outrage her mo...
14	Maharashtra	Assault on Women with intent to outrage her mo...
15	Manipur	Kidnapping and Abduction
16	Meghalaya	Rape

b

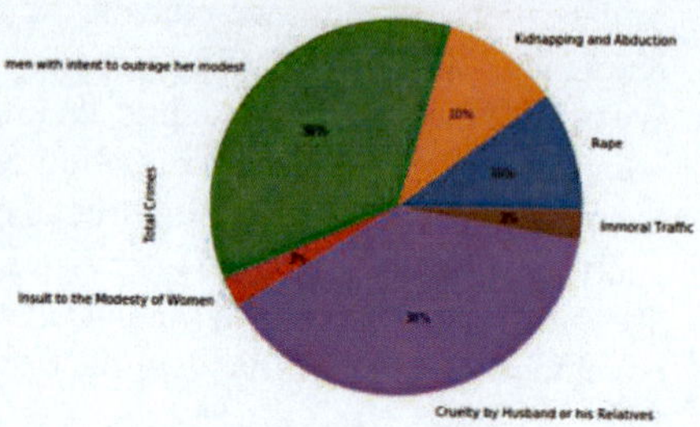

Fig. 1 **a** States and most frequently occurred crimes. **b** Percentages of individual crimes across India

in India. Thereby, we infer that "Cruelty by Husband or his Relatives" has been the crime with the greatest intensity of all the crimes, followed by "Assault on Women with intent to outrage her modesty" and so on.

It is uncanny to notice that the assault on a woman is predominantly done by her friends, relatives, or acquaintances rather than strangers. Individuals entrusted to protect their own women are becoming an inevitable threat to women's freedom and safety, which are the most basic aspects needed in a society to live in. Failing to meet this elementary bar would jeopardize the lives of millions of women in the country.

Laws must be austere toward crimes so that criminals must be intimidated to commit the crime. Punishments levied on felons must be enduring. When women are subjected to crime, they must try to buy some time and find a way to communicate the rescue forces or neighbors who could provide her with some help. She should be brave and try to defend herself.

5.2 *Analysis of Crime-Prone States*

Figure 2a provides information about the list of states that recorded the greatest number of crimes with respective crime heads. For example, Madhya Pradesh tops the table with the greatest number of crimes with "Rape" as the crime head followed by Uttar Pradesh and Maharashtra.

This pie graph (Fig. 2b) depicts the state-wise crime rate in India. It is comprehensible that southern states of India, as a whole, are more prone to crime than northern states. Moreover, we can gather that Uttar Pradesh and Maharashtra have the highest percentage of crime occurrence, i.e., 42% of crimes against women tend to eventuate in these states. Furthermore, Andhra Pradesh and West Bengal, each with 14% crime rate, are the second most menacing states in the country.

Grading of zones based on the intensity and frequency of crimes would aid in precise implementations of security acts. We classify the states and union territories

a

	Crime	Proned State
0	Rape	Madhya Pradesh
1	Attempt to commit Rape	West Bengal
2	Kidnapping and Abduction	Uttar Pradesh
3	Dowry Deaths	Uttar Pradesh
4	Assault on Women with intent to outrage her mo...	Maharashtra
5	Sexual Harassment	Uttar Pradesh
6	Voyeurism	Maharashtra
7	Insult to the Modesty of Women	Andhra Pradesh
8	Cruelty by Husband or his Relatives	West Bengal
9	Importation of Girls from Foreign Country	Bihar
10	Abetment of Suicides of Women	Maharashtra
11	Indecent Representation of Women	Rajasthan
12	Sati Act	Andhra Pradesh
13	Immoral Traffic	Tamil Nadu

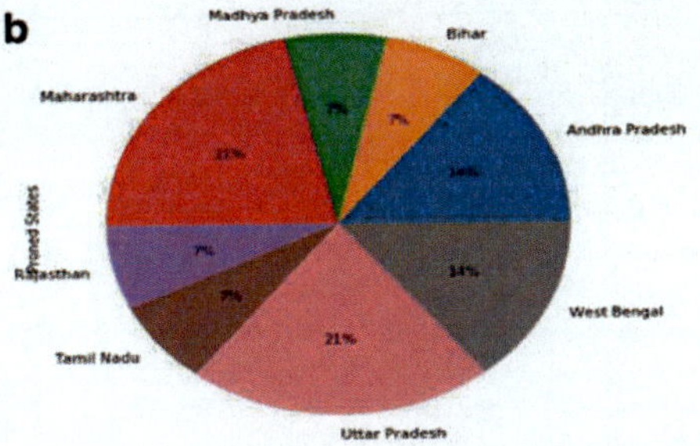

Fig. 2 **a** States with their crime heads. **b** Prone states

into a hot zone, warm zone, and cold zone. The territories where the crime rate is greater than 20% are classified under the hot zone. Similarly, the territories having a crime rate between 10 and 20% are classified under warm zone, and the territories having a crime rate under 10% are classified under cold zone. This classification helps the government to legislate laws and deploy security forces accordingly.

5.3 Analysis of Literacy Rate

The graph (Fig. 3) shows the percentage of literates and crimes across the country. The blue line portrays the percentage of crimes across all the states and union territories in India. The orange line portrays the literacy rates of the states and union territories.

General opinion of law and public is that if the literacy rate in a region is high then there would be less chance of committing crimes in that region. So, we analyzed the data of literacy rate and crimes that are being committed in each state and arrived at the following surprising conclusions.

Here, we observe the consistency in the trend of literacy rate across all the states in the country, whereas the crimes in respective states follow irregular trends. We cannot infer a chronic relation between the literacy rate and the crimes in respective states. This deliberately states that literacy has little effect on the crimes that are happening. Hence, we can conclude that even though people are educated, there is a certain dearth of morality in society. Providing security to women is an ephemeral solution. There is no perfect solution that could eradicate crimes from this world, but building a society with good moral values will have significant declivity in crime rate. Building a society with such probity is an onerous task as long as the incorporation of morals into the people does not instigate from schooling.

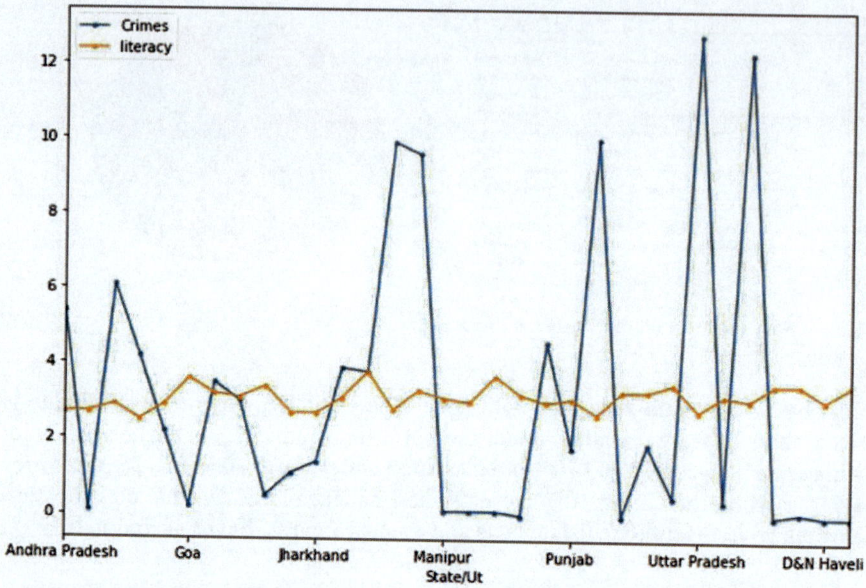

Fig. 3 Graph showing crimes and literates in different states

5.4 Inter-crime Correlation

In order to realize how and to what degree the crimes are dependent on one another, we construct a plot matrix using Weka tool. This plot matrix gives the inter-crime correlation. We can infer how much a crime is linearly correlated with other crimes. For example, consider the crimes "Assault on Women" and "Dowry Deaths," we clearly infer from the plot matrix that these two crimes are strongly correlated, i.e., the probability of committing "Dowry Deaths" increases with "Assault on Woman" and vice versa. Furthermore, we can also analyze the crimes that are not strongly correlated, such as "Dowry Deaths" and "Attempt to rape," i.e., the probability of committing "Attempt to rape" has a negligible effect on "Dowry Deaths." If two crimes are strongly correlated, it clearly purports that if a particular crime is committed, then the correlated crime is more likely to happen. So, the Police Department will have a clear analysis of what crime is more likely to happen and take the corresponding precautions in curbing the latter crime.

5.5 Analysis of Causes of Crime

The decision tree is constructed using J48 in Weka which is an extension of ID3 Algorithm. The decision tree (Fig. 4) depicts the flow of causes of crime narrowed down

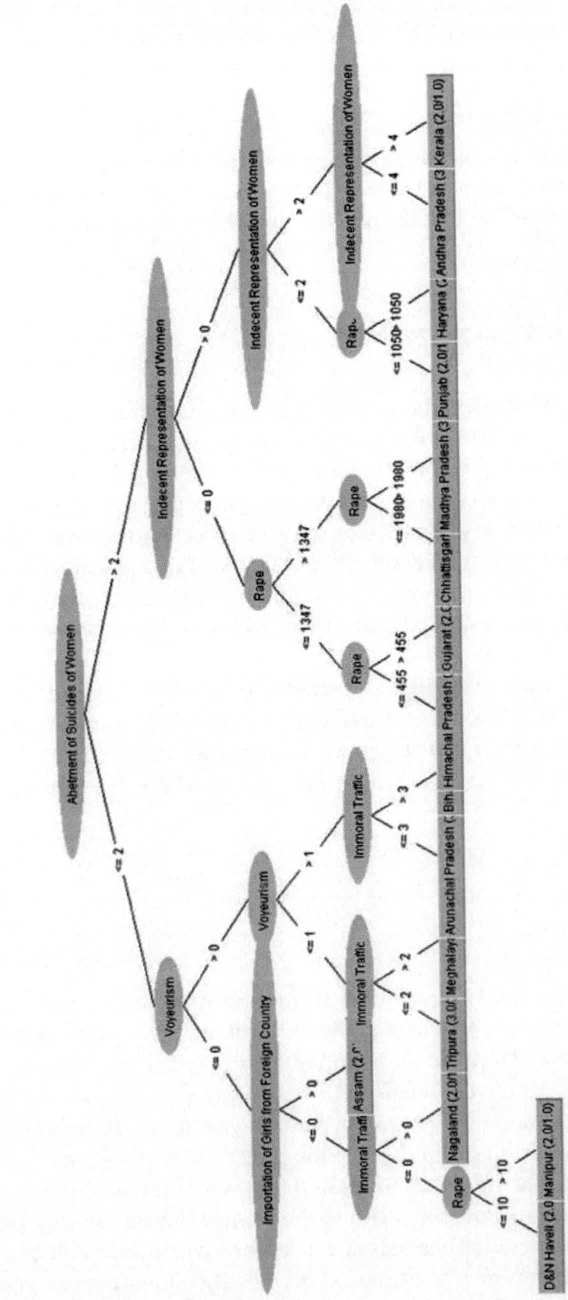

Fig. 4 Decision tree depicting the flow of causes of a crime

to individual states. The root node and the internal nodes represent the crimes. The crime that is at the root node is the most probable crime in all the states. We start with leaf node (states) and we revert to the root (crime) following the branches connecting the internal nodes. This path from leaf node to the root node includes all the prominent crimes that are being committed in each state. For example, in the state of Tripura, the crimes such as "Immoral traffic," "Voyeurism," and "Abetment of Suicide of Women" are prominent crimes. The approach stated above, reduces the effort in finding the most probable crimes in every state that are likely to happen. Therefore, the Police Department can devise the necessary precautions in avoiding such crimes.

5.6 Analysis Using Random Tree

As all the attributes are not taken into consideration in classifying the states by using decision tree classifier, we extend our analysis using Random Tree. Figure 5 shows a fully classified random tree representation. Random tree enables us to take all the crimes into consideration and classify the states using crimes. For example, "Rape" is a factor that classifies the states into two categories further followed by other factors. We have a piece of brief information on crimes that are taking place in all the states. This makes easy to predict the most probable crimes if we have the information about the place of crime. We can also predict the most probable places of the crime if we have the information about the crime.

If there is a situation where an unprecedented crime took place in a state, with the help of random tree classification, we can alert the state that criminals from other states might have migrated to this state and committing similar crimes as they did in their home state. We can also discern the states from where these criminals have migrated.

6 Conclusion

There has been an inexplicable increase in the number of crimes against women in the past five years. The effect on women inflicted by these delinquencies is insanely high and sometimes suicidal. It would be forbidden to predict the pattern of crimes from such humongous data without using the Data Mining techniques and Machine Learning algorithms. Plotting pie graph for percentages of crimes helps to analyze the intensity of various crimes. Plotting the pie graph for crime-prone states helps us to classify the states and union territories of the country into three different zones making it easier for the government to devise the commensurate security measures. Decision tree assists in the quick analysis of causes for a particular crime in any state or union territory. Random tree presents a clear classification of states and crime. Investigation can be more efficient if we use random tree classification.

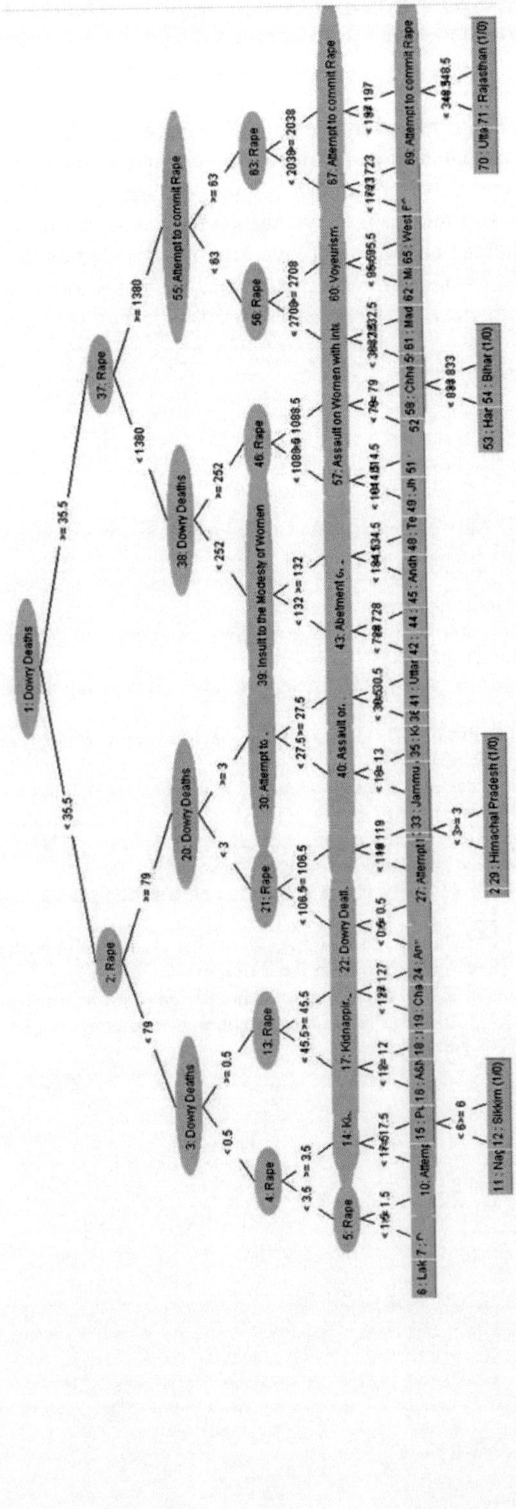

Fig. 5 Random tree depicting all the crimes and territories in India

Succumbing to social conditions, pressure, and the fear of humiliation, crimes are not being reported or acknowledged in most of the rural regions in the country. The data provided gives statistical information after a productive analysis of the raw crime data which would help the government, society, and each of the individuals to comprehend that the women of our country are stranded in a static, miserable situation which could be improved by the collective action of the society as a whole.

Awareness, public acceptance, and eradication of conventional and harmful social norms could put an end to the seamless silence we are in now. This drives the immediate necessity to formulate strict legislations against the heinous crimes against women. Anti-crime squads could be deployed in areas graded as the hot zones, to eliminate the possibilities of a crime to take place. Women should be cognizant of how to play bold in exigency and to be able to find their own way out or try to find help.

References

1. Al Boni M Gerber MS (2016) Area Specific Crime Prediction Models. In: Machine learning and applications (ICMLA). IEEE, pp 671–676
2. Michalski RS, Carbonell JG, Mitchell TM (eds) Machine learning: an artificial intelligence approach. Tioga Publishing Company, Palo Alto
3. Sathyadevan S, Gangadharan S (2014) Crime analysis and prediction using data mining. In: Networks & Soft Computing (ICNSC). IEEE
4. Sangeeta Rani A, Sonika A (2014) Effectiveness of data preprocessing for data mining. Int J Curr Eng Technol 4
5. Gorade SM, Deo A, Purohit P (2017) A study of some data mining classification techniques. Int Res J Eng Technol (IRJET) 4
6. Colin A (1996) Building decision trees with the ID3 algorithm. Dr Dobbs J
7. Mishra AK, Kesari Ratha B (2016) Study of random tree and random forest data mining algorithms for microarray data analysis. Int J Adv Electr Comput Eng (IJAECE) 3
8. https://data.gov.in/resources/crime-against-women-during-2014
9. Yashoda P, Kannan M (2011) Analysis of a population of diabetic patients databases in Weka tool. Int J Sci Eng Res 2
10. Motwani M, Purwar R, Madhur R, Jamshed A (2018) An efficient approach towards crime against women using Time Series algorithm. Int J Comput Appl 179(34):22–26
11. Sarkar O, Datta S, Kannan K (2018) Analysis and prediction of crime rate against women in west Bengal by clustering. In: IADS international conference on computing communications and data engineering, 7–8 Feb 2018

Comparative Analysis of Genomic Personalized Cancer Diagnosis by Machine Learning Approaches ROC Curve

V. Kakulapti, P. Lalitha Bhavani, K. Swathi Reddy, and P. Nissar Ahmed

Abstract Incredible developments are happening in the healthcare domain during the past decades for how precision medication and even more concretely, how hereditary testing is obtainable to interrupt the line of attack sickness like malignancy are identification. However, this is only to some extent, happen appropriate to the enormous quantity of human intervention at a standstill needed. Once progression, a malignant growth tumor can have a huge number of hereditary transformations. Another than confront is distinctive the metamorphosis that gives to cancer intensification from the unbiased metamorphosis. This analysis of hereditary transformation is completed manually, which is the time-consuming task where a medicinal pathologist needs to physically analysis and characterize every hereditary transformation dependent on evidence from text-based medical prescription. To overcome this, we develop a machine learning classifiers and also analyzing the comparative analysis of different classifiers with ROC curve. After calculating probabilities for each class, the class with the highest probability taken. Depending on the result, pathologist distinguishes between transformations that confer to cancer intensification and the unbiased transforms.

V. Kakulapti (✉) · P. Lalitha Bhavani · K. Swathi Reddy · P. Nissar Ahmed
Sreenidhi Institute of Science and Technology, Yamnampet, Ghatkesar, Hyderabad, Telangana 501301, India
e-mail: vldms@yahoo.com

P. Lalitha Bhavani
e-mail: lalithabhavani22@gmail.com

K. Swathi Reddy
e-mail: swathireddy.k.14@gmail.com

P. Nissar Ahmed
e-mail: pnissarahmednemo@gmail.com

S. C. Satapathy et al. (eds.), *Communication Software and Networks*, Lecture Notes in Networks and Systems 134, https://doi.org/10.1007/978-981-15-5397-4_53

1 Introduction

One of the leading causes of death is cancer is also called as malignancy throughout worldwide; approximated 15.5 million citizens have cancer according to American cancer society report 2018. Cancer growth is a lot of infections portrayed by cell adjustments, the unpredictability of which is characterized by various dimensions of cell association [1, 2]. The consummation of the progression of human genes, significant advancement occurs in portraying the being epigenome, proteome, and metabolome. An enhanced perceptive of pharmacogenomics is increasing, and the possibilities for adapting health care for the being develop enormously. Personalized medicine tries to merge an enduring genomic and medical trait to develop a healing approach that utilizes existing perceptive of the genetic methods of cancer [3, 4].

Nowaday, personalized medicine has included the methodical utilization of hereditary or other data about an enduring individual choose or increase that enduring anticipatory and healing concern. Subatomic profiling in healthy and persistent examples may take into consideration a more noteworthy level of personalized medicine than is recently offered. The patient data includes proteinaceous, hereditary, and metabolic report can be utilized to alter health care to that patient's requirement. The main aspect of this healthcare technique is the improvement of companion investigations, whereas molecular evaluates that quantify levels of proteins, genes, or precise transformations are utilizing to give a precise treatment for a patient state by stratifying status of the disease, choosing the appropriate prescription, and alter prescribed amount to those enduring precise requirement.

Hereditary anomaly, either somatic or inherited, may direct to cancer. Cancer occurs due to hereditary is a most important branch of health genetics, which can be implicitly subsequent malignancy hereditary. This type of cancer growth spread 10–15% of entire cancers and the remaining cancers persuaded by ecological features, contaminations, and the standard of living, i.e., smoking, etc. This data enables researchers to decide the danger of malignant growth improvement in a person's lifetime [5]. Though, only a small number of malignancy-disposing patterns in which an allele isolated in an outsmart-leading fashion, thus causal to an elevated threat of cancer improvement. Additionally, non-hereditary features provide metamorphosis or further hereditary changes. Malignancy is detected to increase in patients who have no ancestor's history of cancer.

Apart from hereditary disparities in cancer, attained hereditary variations in qualities that utilize and process drugs additionally impact reaction to treatment. These variations may amplify the toxicity of precise medicines; which facilitated the advance of the discipline of "pharmacogenomics." It recognizes patients because of their genotype information, will react to a precise treatment [2]. The adapted medicine aim is to utilize accurate medicine at the accurate dose, with negligible or no toxicity, for the patient.

One of the popular techniques of artificial intelligence is machine learning, which plays a significant role in the analysis of numerous diseases like cancer since electronic healthcare data. The ROC curve is a visualization representation of the counterfeit-optimistic rate and the accurate-optimistic rate from various categorization imperatives. In general, the accurate-optimistic rate is on the *x*-axis, and the counterfeit-optimistic rate is on the *y*-axis. Every categorization rule creates a position on the chart. The conventional ROC curve increases when a constant value determined in every topic, and the categorization is optimistic if the value is higher than a threshold. Since the threshold fluctuates, a new categorization rule generated and the consequential graph is a solitary curve.

2 Related Work

In recent years, researchers investigated numerous machine learning methods in the field of the healthcare system. One of the significant and valuable methods for assessing the performance of investigative healthcare assessments is the ROC (the receiver operating characteristic) curve [6, 7] by machine learning approaches. Generally, the execution of a categorization imperative is frequently précis by two measures communicated to the two kinds of faults: accurate-optimistic rate and counterfeit-optimistic rate. In this perspective, the accurate-optimistic tempo is the possibility that an issue with malignancy appropriately categorized as containing malignancy. The counterfeit-optimistic tempo is the possibility that an issue exclusive of malignancy in accurately categorized as containing malignancy. The accurate-positive tempo called as sensitivity, and counterfeit-positive tempo called as specificity.

In mammography screening of cancer, the positive outcome stringently characterized like a proposal for the surgery, the counterfeit-optimistic rate was 0.01, and the genuine optimistic rate was 0.80 [8, 9]. Several features concerning the calculation of malignancy effect based on genetic expression signatures are conferred [9, 10]. These are the prospective and the restrictions of microarrays for the calculation of the malignancy effect. Although gene signatures could effectively enhance the capability for prediction in malignancy enduring, deprived progress has been preparing for their function in the services. Though before genetic articulation profiling can be utilized in medical practice, learning with huge volumes of data, i.e., big data, illustrations, and additional satisfactory confirmation is required.

At this time, it is most favorable to utilize AI techniques to determine an explicit diagnosis. The ultimate aspire is to acquire prepared algorithms which enumerate category and expanding nature of malignant enlargement by convention of one or numerous categorization aspects. All these algorithms can be utilized by physicians as supplementary implements to progression large quantities of enduring information for applying analysis [11, 12].

If there should be an occurrence of expected tumors inspected, it is conceivable to utilize methods based on gene activity sample in influenced cells [13, 14]. As a consequence, gene articulation stages can utilize as categorization attributes which describe the fabrication tempo of protein in lung cancer cells evaluated with normal cells [15]. For attaining the assignment of malignancy type categorization, algorithms usually applied, for example, SVM (Support Vector Machines), RF (Random Forests), DT (Decision Tree), Boosting, KNN (K-Nearest Neighbor), LASSO, NN (neural networks) [16, 17]. The efficiency of a variety of algorithms is different relying on examined datasets. To calculate the efficiency of the algorithms and evaluate them it is recognized to utilize the ROC curve and MCC (Matthews Correlation Coefficient) as an assessment of the value of binary as well as non-binary categorizations [18].

3 Methodology

The dataset taken from the Kaggle database and divided into training and testing data. This work utilized two data files. One of the data files contains the genetic mutations information, and another file contains the clinical evidence (textual information) which is generated by human/pathologists employ to categorize the genetic metamorphosis. These files have ID contains data files information is a common column (ID, Gene, Variations, Class) training_text (ID, **Text**). The hereditary metamorphosis is there in the training set as test_variants and test_text gives the medical evidence in the form of text. The identification of attribute is common in both. Hence the hereditary mutation row with precise ID in training_variants was categorize utilizing the medical evidence from the row with ID (Fig. 1).

3.1 *Preprocessing*

The dataset is preprocessing by removing noise, missing values, and any anomalies. The dataset is cleaning by removing stop words, propositions, and other irrelevant data.

3.2 *Univariate Analysis*

It is the uncomplicated structure of investigating the data with solitary attribute at a time. It is commonly utilizing to discover which include in the entire dataset gives more to predict the proper output as indicated by the information given to the model. It typically focuses on depicting and shortening the data to retrieve meaningful patterns from the dataset.

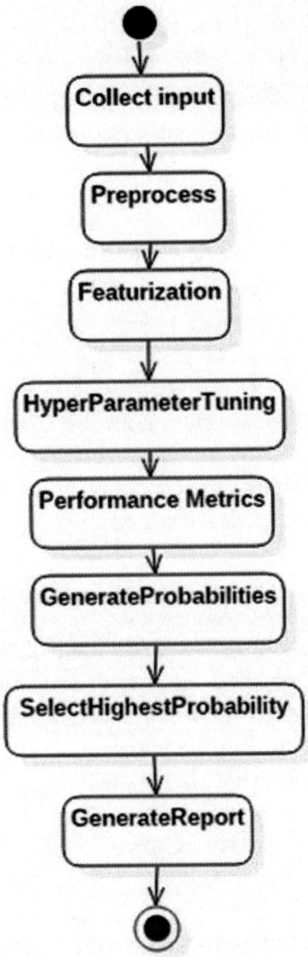

Fig. 1 Workflow of the proposed system

3.3 Featurization

It is the most widely used approach and works very well on categorical data taking a not very large number of values. It generally creates new binary columns in addition to the columns that are already present in the dataset. These binary columns indicate the presence of all possible values from the original data. An array of strings or integers given as the input which returns a sparse matrix or a dense array as an output

3.4 *Bag of Words*

These generally utilized in the categorization of documents where the probability of all words considered as an aspect of the training phase.

3.5 *Performance Metrics*

3.5.1 Log Loss

This is the performance measure of a categorization model where the frequency of occurrence between 0 and 1 signifies the prediction of given participation. By utilizing machine learning techniques is to reduce the log loss value to the extent that to make the model perform well. If the log loss value increases, the performance of the model decreases.

3.5.2 Accuracy

This is the number of expectations where the anticipated value equivalent to the real value. Accurateness isn't generally a superior pointer since it can even function well with "dumb" models.

3.5.3 Roc Curve

This is the performance measure attained by visualizing the proper positive rate and counterfeit-positive rate against each other. It demonstrates the capability of a binary classifier as its intolerance is diverse.

3.6 *Random Forest*

It is one of the popular machine learning algorithms which generates elevated outcomes, and this utilized for categorization and regression exclusive of hyper-parameter tuning. It is a supervised learning approach which generates a forest arbitrarily collection of collaborative decision trees. These decision trees are well trained to utilize a bagging technique which integrates numerous machine learning approaches to enhance the performance and produce better outcomes. Random forest is an ensemble technique.

4 Experiment Analysis

Proof (text) that pathologists utilize to categorize hereditary transformations (Tables 1 and 2).

Both these data files have a common column called ID.

Data file's information: training_variants (ID, Gene, Variations, Class).

Number of data points: 3321.

Number of features: 4.

Features: (ID, Gene, Variations, Class).

4.1 Univariate Analysis on Gene Feature

4.1.1 Features of Gene

Train data contains 252 dissimilar classes which distributed as

Where ID: row identification utilized to connect the transformation to the medical evidence

Gene: the gene where this genetic mutation located

Variation: the aminoacid change for these mutations

Table 1 After preprocessing of dataset the features and their class

	ID	Gene	Variation	Class
0	0	FAM58A	Truncating mutations	1
1	1	CBL	W802*	2
2	2	CBL	Q249E	2
3	3	CBL	N454D	3
4	4	CBL	L399V	4

Table 2 After preprocessing of text data

	ID	Gene	Variation	Class	Text
0	0	FAM58A	Truncating mutations	1	Cyclin-dependent kinases cdks regulate variety...
1	1	CBL	W802*	2	Abstract background non-small cell lug cancer...
2	2	CBL	Q249E	2	Abstract background non-small cell lug cancer...
3	3	CBL	N454D	3	Recent evidence demonstrated acquired uniparen...
4	4	CBL	L399V	4	Oncogenic mutations monomeric casitas b lineage

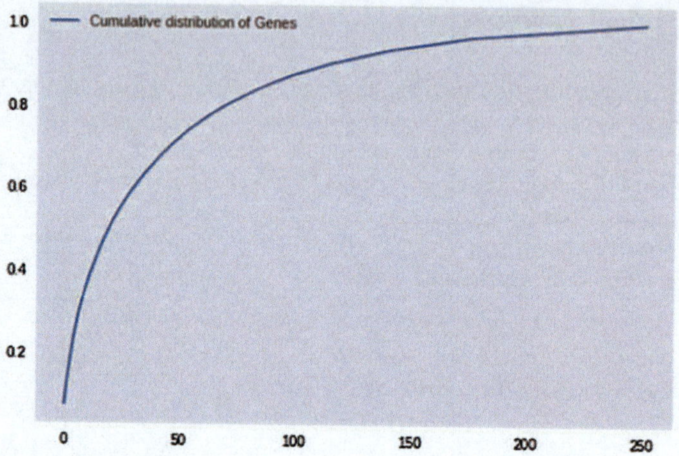

Fig. 2 The cumulative distribution of genes

Class: 1–9 the class the genetic mutation has classified on (Fig. 2).

4.1.2 Hyperparameter Tuning

One of the supervised algorithms is Naïve Bayes, which is fast and correctly classified and reliable. It is an efficient algorithm for large volumes of data. It is one of the simplest supervised learning algorithms. **Naive Bayes** classifier is the fast, accurate, and reliable algorithm. It is a baseline model for text data and applied the training data to the model and utilized the CV data for finding best hyperparameter and fitted the model. Model is examined with test data and found out the log loss value is 1.27 whereas random model is higher than this. In this, calculate the no of miss classified to be 39.8% and the probabilities of every class interpreted every point, which predicting particular class randomly (Figs. 3, 4, 5, and 6).

4.2 Logistic Regression

In this work, the LR model worked efficiently with univariate analysis and with a thorough analysis on LR by taking both imbalanced data and balanced data (Figs. 7, 8, and 9, 10).

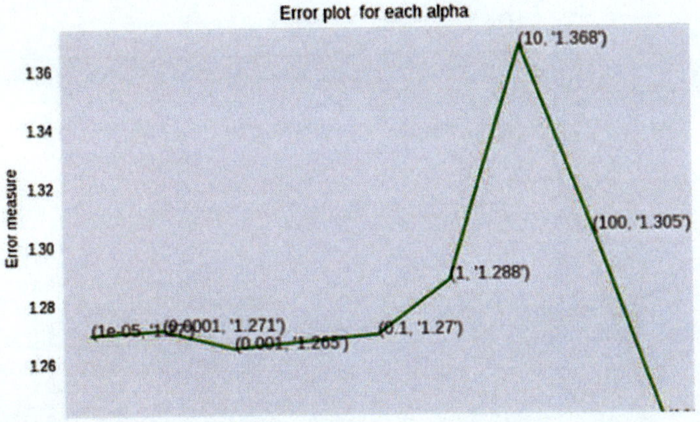

Fig. 3 Cross-validation by hyperparameter tuning

Fig. 4 Naïve Bayes precision matrix of gene mutation

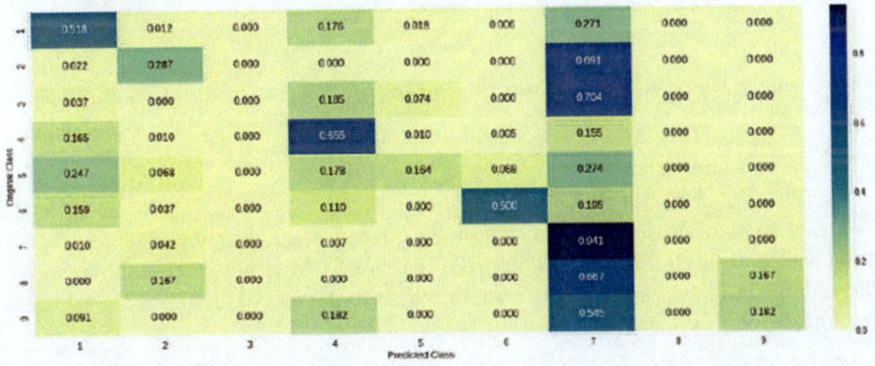

Fig. 5 Naïve Bayes recall matrix of gene mutation

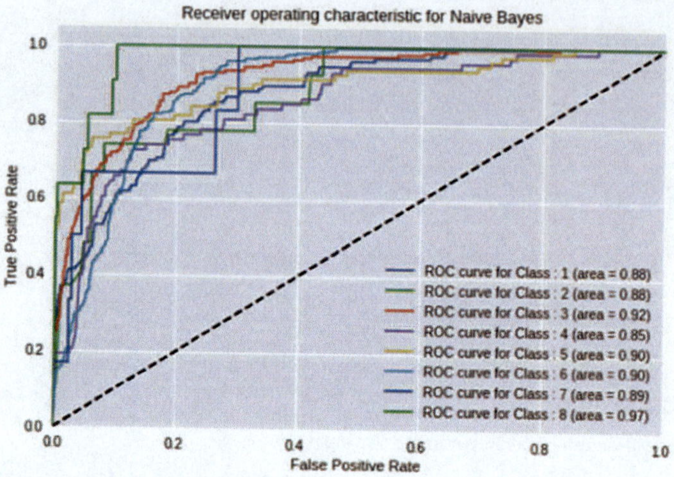

Fig. 6 Naïve Bayes ROC curve of gene mutation

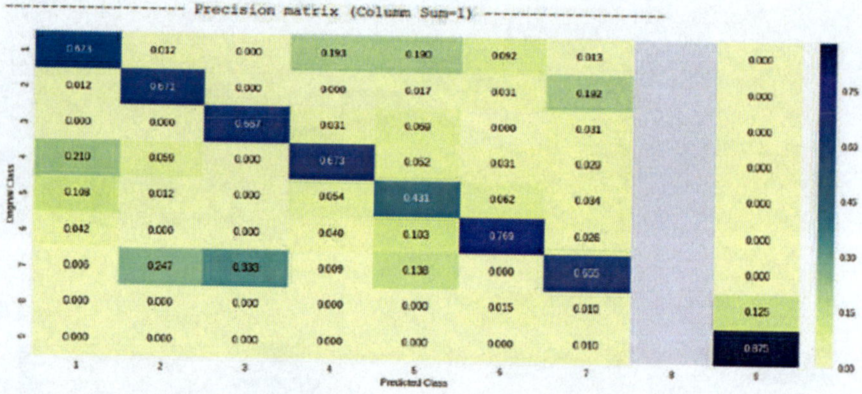

Fig. 7 Logistic regression precision matrix

4.2.1 SVM Classifier

We used Linear SVM (with class balancing) because it is interpretable and works very well with high dimension data. RBF Kernel SVM is not interpretable, so we did not use it. Now we applied the training data to the model and used the CV data for finding best hyperparameter (C) (Figs. 11, 12, 13).

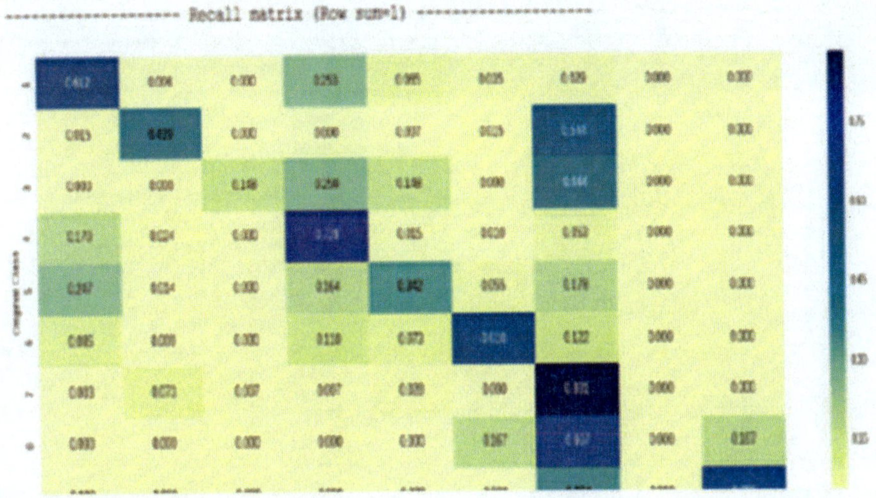

Fig. 8 Logistic regression recall matrix

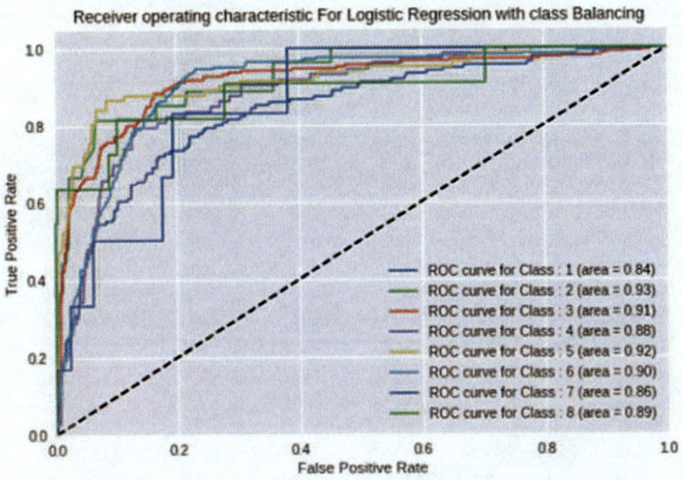

Fig. 9 Logistic regression ROC curve with class balancing

4.2.2 Decision Tree Classifier

DT works well with low-dimension data. It is also interpretable (Figs. 14, 15, and 16).

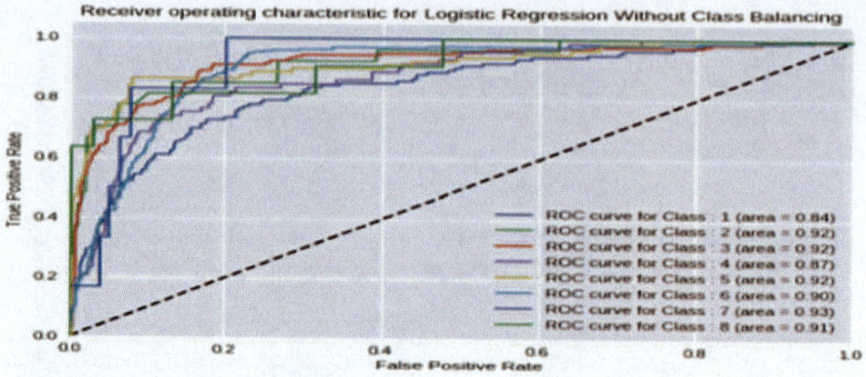

Fig. 10 Logistic regression ROC curve without class balancing

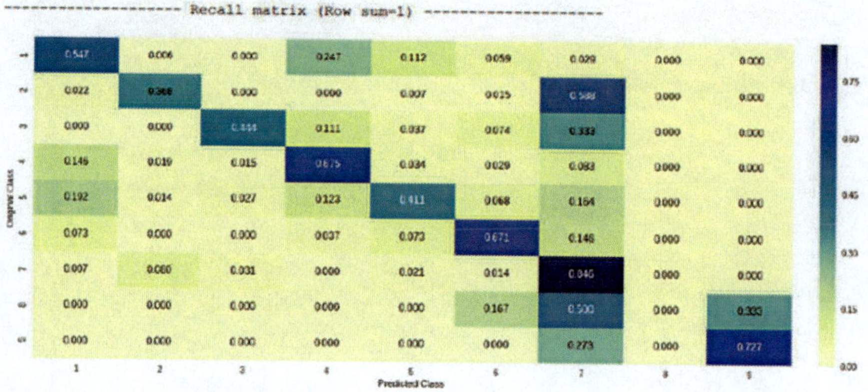

Fig. 11 SVM classifier precision matrix

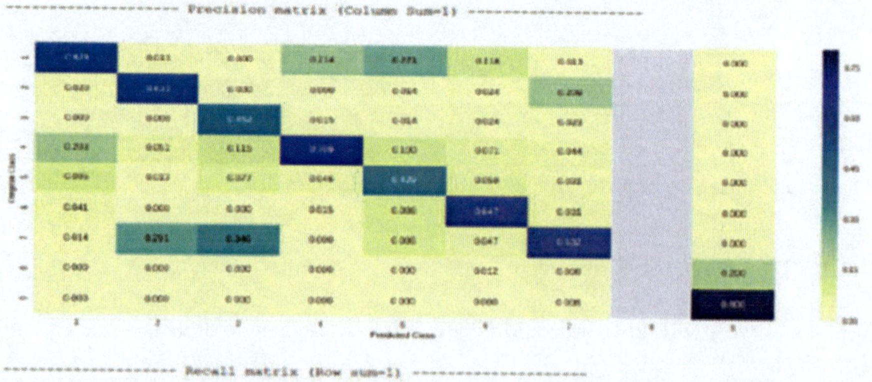

Fig. 12 SVM classifier recall matrix

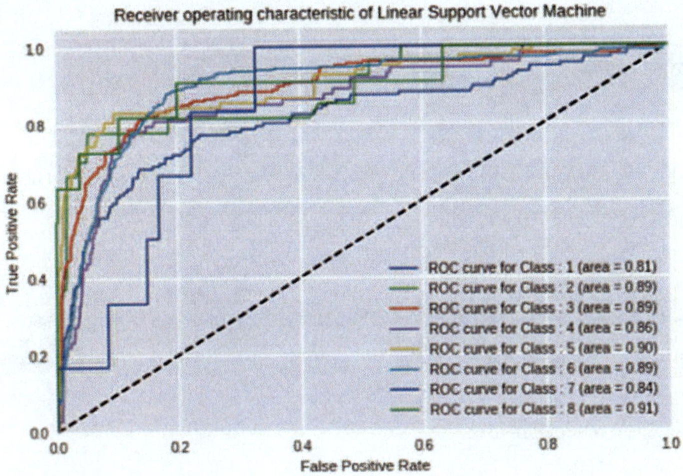

Fig. 13 SVM classifier ROC curve

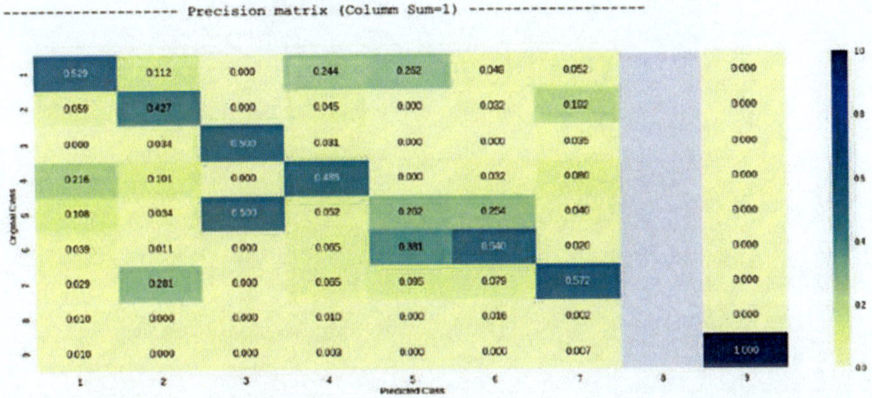

Fig. 14 Decision tree precision matrix

4.2.3 Random Forest Classifier

By changing the no of base learners and max depth in Random Forest Classifier, we found that best base learners = 2000 and max depth = 10. Then we fitted the model with best hyperparameters, and test data is applied to it and found out that log loss value is 1.097 (near to LR) and total no of miss classified points 36.84% (more than LR).

Response Coding: By changing the no of base learners and max depth in Random Forest Classifier, we found that best base learners = 100 and max depth = 5. Then we fitted the model with best hyperparameters and found that train log loss

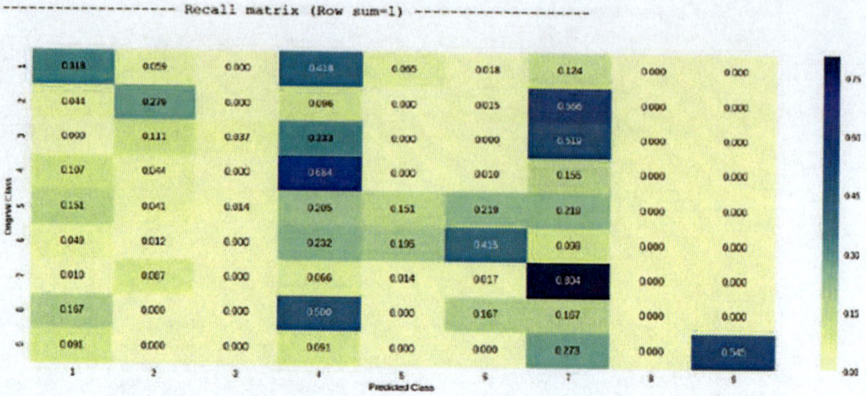

Fig. 15 Decision tree recall matrix

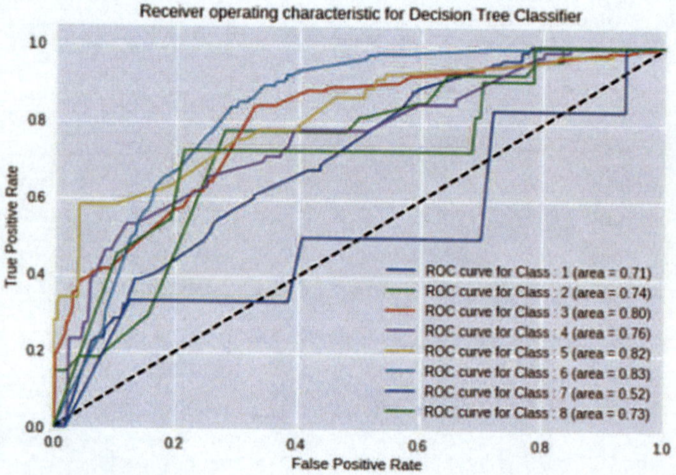

Fig. 16 Decision tree ROC curve

is 0.052 and CV log loss is 1.325 which says that model is over fitted even with best hyperparameters. So, we don't use RF + Response Coding (Figs. 17, 18, and 19).

5 Performance Analysis

Table 3 shows a performance analysis of ROC curve for the six different networks, i.e., Decision Tree Linear Support vector machines, Logistic regression, Naïve Bayes, SGD classifier, and random forest. The results shown in Table 3, the random forest

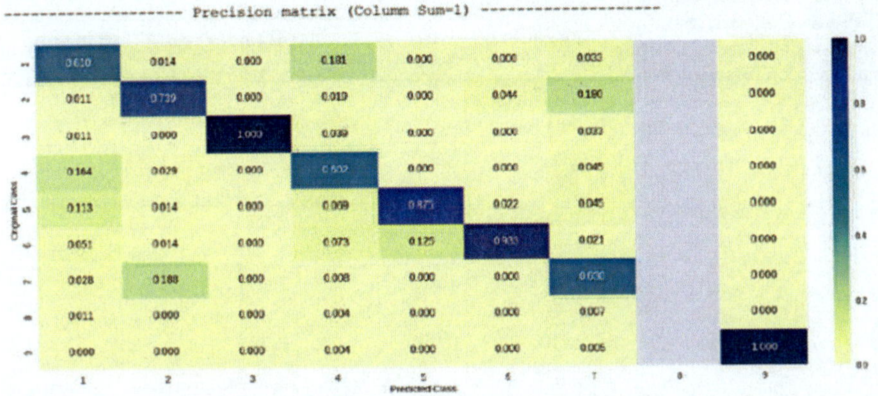

Fig. 17 Random forest classifier precision matrix

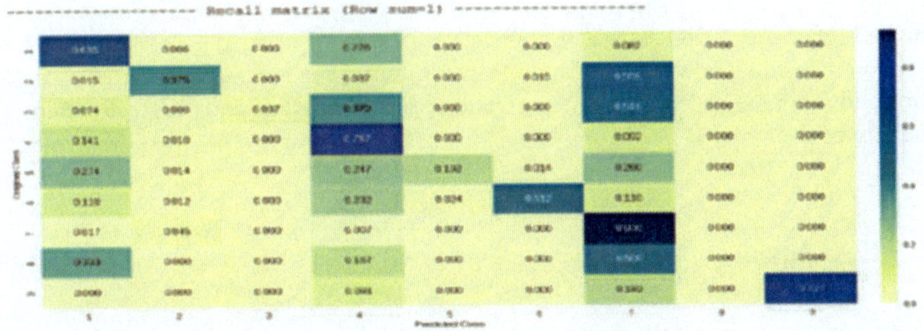

Fig. 18 Random forest classifier recall matrix

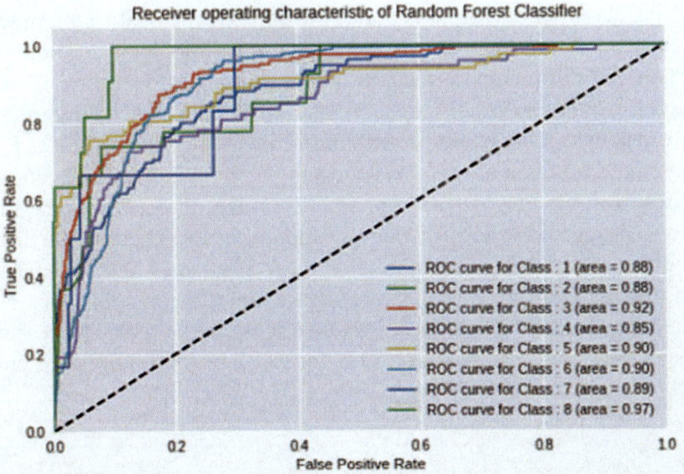

Fig. 19 Random forest classifier ROC curve

Table 3 Comparative analysis of different classifiers and their log loss values

Algorithm	Best hyperparameter	Train log loss	Test log loss
Decision tree	100	1.03	1.46
Linear SVM	0.001	0.74	1.11
Logistic regression	0.001	0.59	1.03
Naive Bayes	1000	1.055	1.23
SGD classifier	0.001	0.735	1.11
Random forest	2000	0.714	1.11

had a higher prediction accuracy, followed by the SGD using the ROC curve. The proposed results show that machine learning techniques can give precise prediction and facilitate appropriate categorization of patient's state and progress patient's quality of life.

6 Conclusion

After applying some machine learning algorithms, by comparing the test log loss of each algorithm, observe that logistic regression is performing well compared to any other algorithm. The ROC curve consequences should constantly locate in perception, as an effective classifier does not assurance the ultimate medical result, in scrupulous for instance-dependent incidents in transmission, prediction, and prediction learning wherever particular numerical provisions and techniques are required. Finally, such data may proficiently and consistently originate from randomized organized examinations. Even for the most complex problems, logistic regression works fairly well. As it has sigmoid function along with squashing technique, it gives even more optimized models which in turn lead to comparably good results. It uses L2 regularization to reduce overfitting and underfitting and can handle outliers better. The data training time of logistic regression is approximately O (nd) where n is the number of data points and d is the number of features, but space and time complexity at run time is approximately around O (d), which is very low. In the end, logistic regression is both time and space-efficient and performs well for low latency requirement problems.

7 Future Enhancement

In this work, logistic regression is performing well than another algorithm, further extend it by applying deep learning algorithms, to make it work better by reducing the performance metrics like log loss less than 0.7, which is considered to be an optimal value. As deep learning models that currently exist are developed mostly by taking the ideas of logistic regression, the proposed model can be easily extending those algorithms without any major changes. To extend this research by apply Metaheuristic algorithms to optimize the model. Optimizing the model, in turn, leads to increasing the performance of the model. If performance is increasing, there would be fewer errors occur by the model which would be useful for the prediction of cancer with as fewer mistakes as possible.

References

1. Reimand J, Wagih O, Bader GD (2013) Themutational landscape of phosphorylation signaling in cancer. Sci Rep 3:article 2651
2. Hanahan D, Weinberg RA (2011) Hallmarks of cancer: the next generation. Cell 144(5):646–674
3. Simon R (2010) Clinical trial designs for evaluating the medical utility of prognostic and predictive biomarkers in oncology. Person Med 7(1):33–47
4. Bedard PL, Hansen AR, Ratain MJ, Siu LL (2013) Tumour heterogeneity in the clinic. Nature 501(7467):355–364
5. Stricker T, Catenacci DV, Siewert SY (2011) Molecular profiling of cancer—the future of personalized cancer medicine: a primer on cancer biology and the tools necessary to bring molecular testing to the clinic. Semin Oncol 38:173–185
6. Schroth et al (2009) Association between CYP2D6 polymorphisms and outcomes among women with early stage breast cancer treated with tamoxifen. JAMA 302:1429–1436
7. Shapiro DE (1999) The interpretation of diagnostic tests. Stat Methods Med Res 8:113–134
8. Baker SG (2000) Identifying combinations of cancer biomarkers for further study as triggers of early intervention. Biometrics 56:1082–1087
9. Baker SG, Pinsky P (2001) A proposed design and analysis for comparing digital and analog mammography: special ROC methods for cancer screening. J Am Stat Assoc 96:421–428
10. Koscielny S (2010) Why most gene expression signatures of tumors have not been useful in the clinic. Sci Transl Med 2:14 ps12–14 ps12
11. Sun T, Wang J, Li X et al (2013) Comparative evaluation of support vector machines for computer aided diagnosis of lung cancer in CT based on a multi-dimensional data set. Comput Methods Programs Biomed 111:519–524
12. Yu Z, Lu H, Si H et al (2015) A highly efficient gene expression programming (GEP) model for auxiliary diagnosis of small cell lung cancer. PLoS One 10:125517
13. Han Y, Wang XB, Xiao N, Liu ZD (2013) mRNA Expression and Clinical Significance of ERCC1, BRCA1, RRM1, TYMS and TUBB3 in postoperative patients with non-small cell lung cancer. Asian Pac J Cancer Prev 14:2987–2990
14. Liu M, Pan H, Zhang F et al (2013) Screening of differentially expressed genes among various TNM stages of lung adenocarcinoma by genomewide gene expression profile analysis. Asian Pac J Cancer Prev 14:6281–6286
15. Cheng P, Cheng Y, Li Y et al (2012) Comparison of the gene expression profiles between smokers with and without lung cancer using RNA-Seq. Asian Pac J Cancer Prev 13:3605–3609

16. Lei Win S, Htike ZZ, Yusof F, Noorbatcha AI (2014) Gene expression mining for predicting survivability of patients in earlystages of lung cancer. Int J Bioinf Biosci 4:1–9
17. Cai Z, Xu D, Zhang Q et al (2015) Classification of lung cancer using ensemble-based feature selection and machine learning methods. Mol BioSyst 11:791–800
18. Baldi P, Brunak S, Chauvin Y, Andersen CAF, Nielsen H (2000) Assessing the accuracy of prediction algorithms for classification: an overview. Bioinformatics 16:412–424
19. Michiels S, Koscielny S, Hill C (2005) Prediction of cancer outcome with microarrays: a multiple random validation strategy. Lancet 365:488–492

DNA Playfair Cryptosystem Based on 16 × 16 Key Matrix Using DNA ASCII Table

P. Bharathi Devi, R. Kiran Kumar, and P. Ravindra

Abstract DNA Cryptography contributes significantly in the field of network security. Data is the crucial aspect which has to be secured during transmission. Several traditional cryptographic algorithms are developed with variant cryptographic approaches. Among all of them, DNA Cryptography is prevalent because of its inherent property of storing huge information in small amount of DNA. In the present scenario, everything is done online. All the services, infrastructure or platforms are provided on demand to the clients. In such a situation, there is a need to impart the trust to the client in the storage of a cloud. In the present paper, the authors implemented DNA-based Cryptosystem in a Cloud using DNA ASCII Table and extended the playfair key matrix from 5 × 5 to 16 × 16. The key matrix developed in this paper contains only the DNA Sequences instead of alphabets. The key aspect of this algorithm is the DNA ASCII Table and the Key matrix is randomly generated so that it becomes a challenging task for the intruder to perform bruteforce attack.

1 Introduction

Everything is digitized. The digitized data is traveled through various social networking applications or other internet applications. Most of the people are not interested to buy the products permanently. They are interested in on-demand services and pay on utility services. In this aspect, cloud computing plays a vital role. In

P. Bharathi Devi (✉)
Department of Computer Applications, K.B.N.College, Vijayawada, Andhra Pradesh, India
e-mail: bharathipatnala@gmail.com

R. Kiran Kumar
Department of Computer Sciene, Krishna University, Machilipatnam, Andhra Pradesh, India
e-mail: kirankreddi@gmail.com

P. Ravindra
Department of Computer Science, K.B.N.College, Vijayawada, Andhra Pradesh, India
e-mail: ponugumatiravindra@gmail.com

S. C. Satapathy et al. (eds.), *Communication Software and Networks*, Lecture Notes in Networks and Systems 134, https://doi.org/10.1007/978-981-15-5397-4_54

Table 1 DNA digital coding

DNA sequence	DNA digital coding
A	00
C	01
G	10
T	11

Cloud, the resources are sharable from one cloud to other. One cannot trust the information that is stored in a cloud and that is being transmitted. In order to provide security to the information stored in a cloud many security applications are developed and regulatory issues are also discussed [1–3]. In order to give privacy to our data, service providers have identified new algorithms. Though many cryptography algorithms exist, its time to implement DNA-based cryptographic algorithms due to the enormous size of the data. By nature the DNA has the capable of storing petabytes of information in single gram by DNA and it was proved by Harvard University Research Team [4]. DNA cryptography is a part of natural science, which has extensive information storage capacity. It stores information of living organisms. Living life forms has unique DNA information. It is characterized as information storage, even parallel handling and highly secured data transmission. DNA cryptography depends on one-time-pad scheme. Cryptography needs to combine with atomic science for increasingly secure information transmission and information covering up. A plaintext message is encoded in DNA sequences. DNA sequences get powerful, when combined with nucleotide base A–T and C–G. DNA cryptography innovation is required in data security to protect and hide data. In conventional cryptography (like as DES, RSA), encrypted messages are detectable by attacker [5–7]. DNA has an ability to store enormous data instead of existing algorithms. DNA is presented as another technology for whole data. Genetic data is encoded as a grouping of nucleotides Guanine-G, Adenine-A, Thymine-T and Cytosine-C. Adenine, Thymine and Guanine, Cytosine are base pairs, which are attached to a sugar and a phosphate to keep up helical structure [8, 9]. DNA strands are joined with hydrogen bond. A and T DNA groupings are joined with twofold hydrogen bond while C and G are joined with triple bond. Each nucleotide comprises the accompanying three components, A Nitrogenous Base, A five-carbon Sugar and A Phosphate Group. Each DNA base is coded with binary value. The complement of A should be assigned to T because those two are complementary bases. The complement of C should be assigned to G because those two bases are complementary to each other [10]. The Digital Coding of DNA Sequence is as follows (Table 1).

2 Literature Survey

A great deal of work is going on DNA Cryptography and many researchers implemented various algorithms. The playfair method which is implemented as DNA structure by many authors has some limitations. Sabry et al. [11] proposed a play fair cipher usage utilizing DNA Amino Acids. In this paper, they separated the procedure into two stages. The initial step is pre-handling, in which they arranged the secret key by evacuating spaces, rehashed characters and afterward changed over the key into capitalized. The plaintext is likewise pre-prepared by removing the spaces to stay away from the aggressor's follow. At that point there comes the handling in the second step. Here, the plaintext is changed over into double and after that changed over into DNA bases. From the DNA sequence, think about three bases as a codon and supplant the codon into its comparable amino acids. In protein synthesis, 20 amino acids are shaped from 64 codons. As 26 letter sets exist in English language, these 20 amino acids are changed in accordance with these 26 letters in order. Further, ambiguity number is likewise noted to eliminate the ambiguity of the amino acids. Arrange the key an incentive into 5 × 5 play reasonable grid and apply the play fair cipher procedure. The resultant code is DNA type of Cipher Text. Ambiguity number is additionally changed over into DNA base. The last cipher content is the connection of the DNA type of cipher content and ambiguity DNA base. In this algorithm, building the scattering of codon table to get all the English letter set letters and furthermore monitor the ambiguity estimation of every amino acid letter set takes high time unpredictability. The dimension of encryption process is high since it takes two dimensions to perform. It likewise had the weakness of putting away key size up to 25 characters simply because the game plan of key should be possible in 5 × 5 matrix .

Atito [12] created playfair strategy utilizing insertion method by taking the Mona Sabry playfair technique. He expanded the work by executing the algorithm in two dimensions. In the primary level, he connected the method of playfair proposed by Mona Sabry. In the following dimension, he connected addition system to the encoded information of the primary dimension. The cipher content is as faked DNA grouping, however, it contains concealed information. In this likewise, the creator utilized 5 × 5 key grid.

Reddi et al. [13] proposed new DNA-based playfair cryptosystem, which depends on the DNA LOOKUP table and furthermore dependent on the course of action of key network of size 8 × 8 which contains all the 64 codons (Fig. 1).

3 Proposed Algorithm

The proposed model is based on the rules of using traditional playfair cryptosystem [14] and it was an extension work of the model "A Novel Encryption Scheme to Secure the Data Using DNA Based Playfair Cipher Technique" by Kiran K Reddi et al. The traditional playfair cryptosystem has so many limitations and also some DNA-based cryptosystems earlier discussed in Sect. 2 also has limitations. To overcome all these limitations, the proposed model has been introduced. In this model, the ASCII Table proposed in the "A Novel Text Encryption Algorithm using DNA ASCII Table with Spiral Approach" is used for processing the key value of the cryptosystem [15]. Generally, in traditional playfair the key value contains only 26

Amino acid	DNA codons				
Alanine	GCT	GCC	GCA	GCG	
Arginine	CGT	CGC	CGA	CGG	AGA	AGG
Asparagine	AAT	AAC				
Aspartic acid	GAT	GAC				
Cysteine	TGT	TGC				
Glutamic acid	GAA	GAG				
Glutamine	CAA	CAG				
Glycine	GGT	GGC	GGA	GGG		
Histidine	CAT	CAC				
Isoleucine	ATT	ATC	ATA			
Leucine	CTT	CTC	CTA	CTG	TTA	TTG
Lysine	AAA	AAG				
Methionine	ATG					
Phenylalanine	TTT	TTC				
Proline	CCT	CCC	CCA	CCG		
Serine	TCT	TCC	TCA	TCG	AGC	AGT
Threonine	ACT	ACC	ACA	ACG		
Tryptophan	TGG					
Tyrosine	TAT	TAC				
Valine	GTT	GTC	GTA	GTG		
Start (CI)	ATG					
Stop (CT)	TAA	TAG	TGA			

Fig. 1 Structured codon table

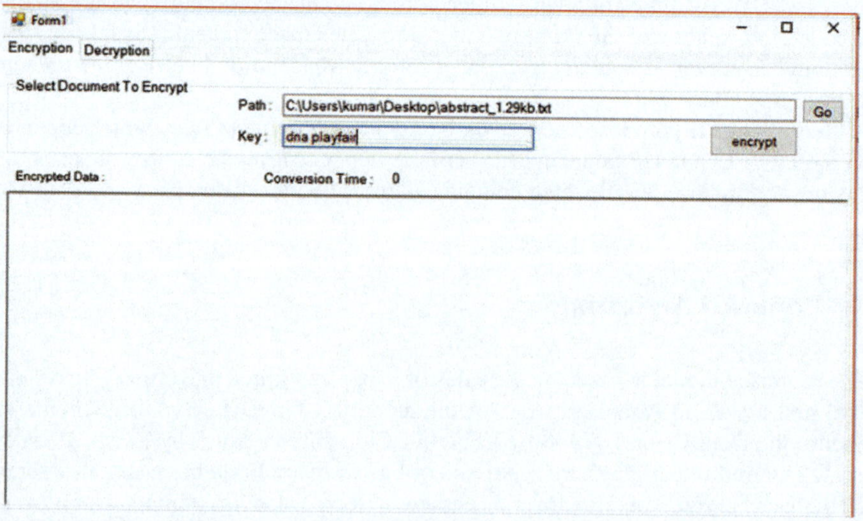

Fig. 2 Encryption of text document with the keyword “dna playfair”

alphabet values. But in this cryptosystem, the authors used all the 256 ASCII characters. These characters mapped with the value defined in the DNA ASCII Table. In the proposed model, the key matrix is arranged by using all 256! permutations DNA basses into 16 × 16 matrix. The first step in this algorithm is processing the key. Identify the repeated characters in the keyword and eliminate. Then map these characters into their equivalent DNA bases from the DNA ASCII Table. Arrange all these DNA bases into row wise and then place the remaining DNA bases of the Key matrix. Later, process the plaintext. Convert the plaintext into its equivalent ASCII and then in turn convert into DNA bases. Divide the DNA strand into four bases each. If two consecutive bases are same insert a dummy DNA sequence which is of length 4. After that apply playfair rules. That is, if the two pairs are in same row, then replace them with their upward pairs. If the two pairs are in same columns, then replace them with their downward pairs. If the two pairs are not in same row and column, then form a rectangle and replace with the opposite pairs. The decryption is the reverse process of the same above.

Algorithm DNAEncrypt (P, Key)

Begin

1. Process the Key, i.e., remove the repeated characters from it and convert the characters into its equivalent ASCII Values and map these values with DNA sequences by using DNA ASCII Table and then arrange them into Key matrix (Table 2).
2. Process the Plaintext, i.e., convert the characters into its equivalent ASCII Values and then convert them into DNA bases.
3. Divide the bases into four bases each and consider the pairs and then apply playfair technique.

End

Algorithm DNADecrypt (C, Key)

Begin

1. Process the Key, i.e., remove the repeated characters from it and convert the characters into its equivalent ASCII Values and map these values with DNA sequences by using DNA ASCII Table and then arrange them into Key matrix (Table 2).
2. Divide the bases in the ciphertext into four bases each and consider the pairs and then apply playfair technique
3. After applying, the output which is in the form of DNA can be converted into its equivalent binary and then convert into its equivalent ASCII character.

End

Table 2 Key matrix of proposed algorithm

	0	1	2	3	4	5	6	7	8	9	10	11	12	13	14	15
0	CGTG	CGCA	CGTT	CACA	CATG	CGCT	TTTA	TTTC	TTTG	TTTT	TTCA	TTCC	TTCG	TTCT	TCTA	TCTC
1	TCTG	TCTT	TTGA	TTGC	TTGG	TTGT	TTAA	TTAC	TTAG	TTAT	TCCA	TCCC	TCCG	TCCT	TCAA	TCAC
2	TCAG	TCAT	TCGA	TCGC	TCGG	TCGT	TATA	TATC	TATG	TATT	TACA	TACC	TACG	TACT	TGTA	TGTC
3	TGTG	TGTT	TAGA	TAGC	TAGG	TAGT	TAAA	TAAC	TAAG	TAAT	TGCA	TGCC	TGCG	TGCT	TGAA	TGAC
4	TGAG	TGAT	TGGA	TGGC	TGGG	TGGT	ATTA	ATTC	ATTG	ATTT	ATCA	ATCC	ATCG	ATCT	ACTA	ACTC
5	ACTG	ACTT	ATGA	ATGC	ATGG	ATGT	ATAA	ATAC	ATAG	ATAT	ACCA	ACCC	ACCG	ACCT	ACAA	ACAC
6	ACAG	ACAT	ACGA	ACGC	ACGG	ACGT	CATA	CATC	CATT	CACC	CACG	CACT	CGTA	CGTC	CAGA	CAGC
7	CAGG	CAGT	CAAA	CAAC	CAAG	CAAT	CGCC	CGCG	CGAA	CGAC	CGAG	CGAT	CGGA	CGGC	CGGG	CGGT
8	CTTA	CTTC	CTTG	CTTT	CTCA	CTCC	CTCG	CTCT	CCTA	CCTC	CCTG	CCTT	CTGA	CTGC	CTGG	CTGT
9	CTAA	CTAC	CTAG	CTAT	CCCA	CCCC	CCCG	CCCT	CCAA	CCAC	CCAG	CCAT	CCGA	CCGC	CCGG	CCGT
10	AATA	AATC	AATG	AATT	AACA	AACC	AACG	AACT	AGTA	AGTC	AGTG	AGTT	AAGA	AAGC	AAGG	AAGT
11	AAAA	AAAC	AAAG	AAAT	AGCA	AGCC	AGCG	AGCT	AGAA	AGAC	AGAG	AGAT	AGGA	AGGC	AGGG	AGGT
12	GTTA	GTTC	GTTG	GTTT	GTCA	GTCC	GTCG	GTCT	GCTA	GCTC	GCTG	GCTT	GTGA	GTGC	GTGG	GTCT
13	GTAA	GTAC	GTAG	GTAT	GCCA	GCCC	GCCG	GCCT	GCAA	GCAC	GCAG	GCAT	GCCA	GCGC	GCGG	GCGT
14	GATA	GATC	GATG	GATT	GACA	GACC	GACG	GACT	GGTA	GGTC	GGTG	GGTT	GAGA	GAGC	GAGG	GAGT
15	GAAA	GAAC	GAAG	GAAT	GGCA	GGCC	GGCG	GGCT	GGAA	GGAC	GGAC	GGAT	GGGA	GGGC	GGGG	GGGT

4 Results and Discussion

The proposed model is developed using.NET environment and the simulations were tested on various text document files. The screenshots and the simulations are tabulated in the following table (Table 3). From the results, it is observed that the proposed model takes lesser time when compared to the existing models proposed by Kiran K. Reddi et al., It is a herculean task to perform bruteforce attack because of randomization of DNA ASCII Table and the Key matrix. Applying randomization always increases the prediction accuracy (Figs. 2, 3, 4, 5, and 6).

Table 3 Time analysis of proposed and existing models

Text document size (in terms of MB)	Encryption time (in terms of ms) proposed algorithm	Encryption time (in term of ms) existing algorithm	Decryption time (in terms of ms) proposed algorithm	Decryption time (in terms of ms) existing algorithm
1	0.011	3.35	0.10	2.98
5	3.45	8.52	6.32	9.05
10	5.23	12.25	8.23	11.56
50	10.84	19.26	11.54	18.25
100	18.36	29.25	17.59	28.56

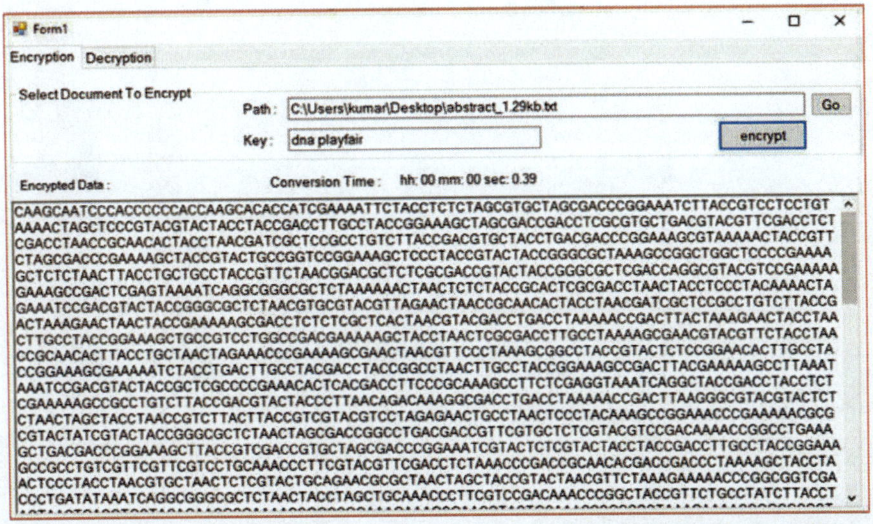

Fig. 3 Resultant DNA sequence after encryption of text document with the keyword "dna playfair"

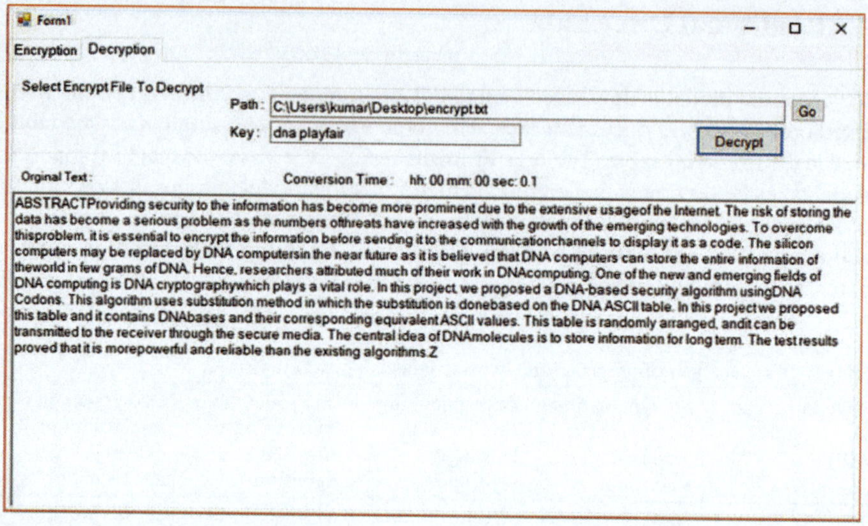

Fig. 4 Reverting back the original text

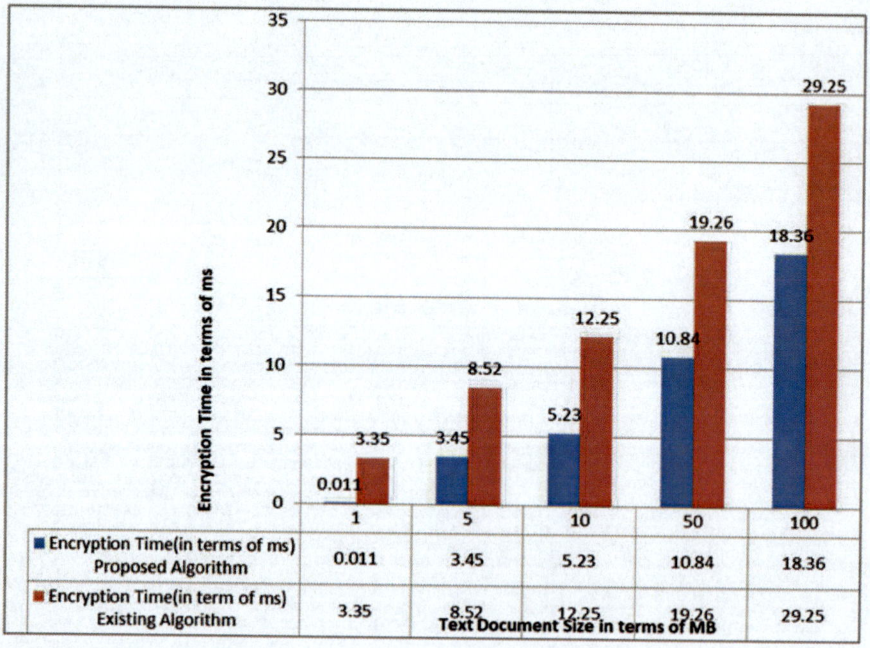

	1	5	10	50	100
Encryption Time(in terms of ms) Proposed Algorithm	0.011	3.45	5.23	10.84	18.36
Encryption Time(in term of ms) Existing Algorithm	3.35	8.52	12.25	19.26	29.25

Fig. 5 Comparison of time taking to encrypt the text document of proposed and existing model

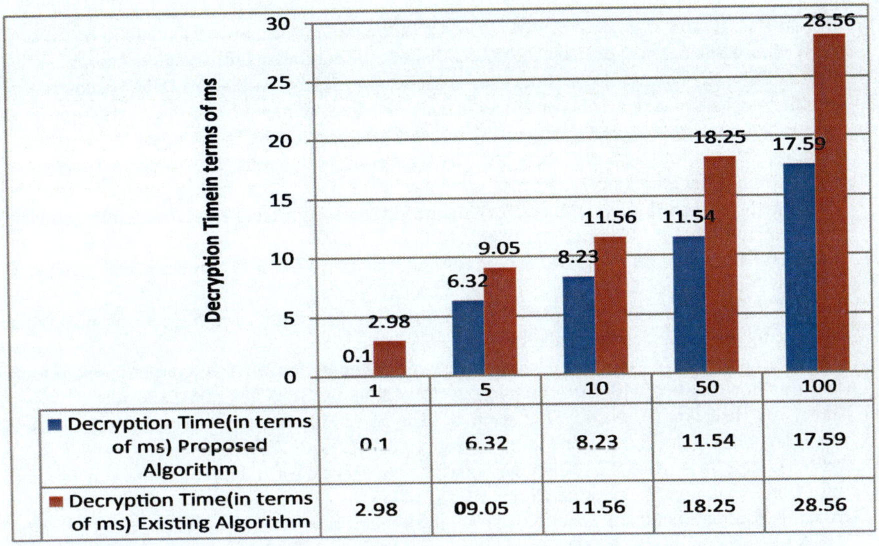

Fig. 6 Comparison of time taking to decrypt the text document of proposed and existing model

5 Conclusions

This model ensures the privacy of the information either stored in a cloud or in transmission. The key aspect of this model is the key matrix that is extended to 16 × 16 so that total 256! permutations were needed to identify the text and also the recovery of original plain text is impossible for the intruders because in each transmission the key matrix is randomly arranged. Another Key aspect of this model is arrangement of DNA ASCII Table. This table is also randomized in each transmission. The arrangement of DNA ASCII Table also can be taken 256! permutations. By considering earlier implementations, this model provides better security because of increasing size of Key matrix and DNA ASCII Table.

References

1. Kiran Kumar R, Suneetha D (2019) A novel approach for data security in cloud environment using image segmentation and image steganography. In: Information systems design and intelligent appliations. Advances in intelligent systems and computing, vol 862. Springer, Singapore
2. Cloud Security Alliance (2013) The notorious nine: cloud computing top threats in 2013 Retrieved from https://downloads.cloudsecurityalliance.org/initiatives/top_threats/The_Notorious_Nine_Cloud_Computing_Top_Threats_in_2013.pdf
3. Rao RV, Selvamani K (2015) Data Security Challenges and Its Solutions in cloud computing. Procedia Comput Sci 48:204–209

4. Robert F (2017) Service, Mar 2, 2017, DNA Could store all of the world's data in one room, https://www.sciencemag.org/news/2017/03/dna-could-store-all-worlds-data-one-room
5. Suresh Babu E, Nagaraju C, Krishna Prasad MHM (2015) Light-weighted DNA based cryptographic mechanism against chosen cipher text attacks. In: 2nd international doctoral symposium on applied computation and security systems, Kolkata, India. LNCS, Springer
6. Gao, Q(2010), Biological alphabets and DNA-based Cryptography, in American Society for Engineering Education
7. Chauhan SS, Singh H, Gurjar RN (2014) Secure key exchange using RSA in extended playfair cipher technique. Int J Comput Appl, 0975–8887
8. Leonard M. Adleman (1994) Molecular computation of solution to combinatorial problems. Science 266(5187):1021–1024
9. Gehani A, LaBean T, Reif J (2004) DNA-based cryptography, In: Aspects of molecular computing. Springer, Berlin Heidelberg, pp 167–188
10. XueJia Lai, MingXin Lu, Qin Lei, JunSong Han, XiWen Fang (2010) Asymmetric encryption and signature method with DNA technology. Sci China Inf Sci 53(3):506–514
11. Sabry M, Hasheem M, Nazmy T, Essam Khalifa M (2010) A DNA and amino acids-based implementation of playfair cipher. Int J Comput Sci Inf Secur 8(3):129–136
12. Atito A, Khalifa A, Rida SZ (2012) DNA-based data encryption and hiding using play fair and insertion techniques. J Commun Comput Eng, 44–49
13. Kiran K Reddi, Bharathi P Devi (2018) A novel encryption scheme to secure the data using DNA based playfair cipher technique, i-mangers. J Inf Technol 7(3):1–9
14. Stallings W (2003) Cryptography and network security: principles and practices, 3rd edn. Prentice Hall, NJ
15. Devi Bharathi P, Kumar Kiran K (2018) A novel text encryption algorithm using DNA ASCII table with spiral approach. Int J Recent Sci Res 9(1):23588–23595

Non-negative Matrix Factorization Procedure for Characteristic Mining of Mathematical Formulae from Documents

K. N. Brahmaji Rao, G. Srinivas, P. V. G. D. Prasad Reddy, and B. Tarakeswara Rao

Abstract Non-negative matrix factorization (NMF) has been one of the substantial progresses in the use of various clustering tasks in the recent times. NMF solves challenging data mining and machine learning problems. Our proposed technique solves this precarious problem of mining mathematical keywords from textual documents using NMF clustering based on a training set of mathematical keyword and formulae pairs. The recital of the proposed method is calculated using the metrics such as retrieval time, sensitivity, efficiency, and FNR are used for appraisal of the proposed technique.

1 Introduction

In technology chastisement, the search for allied mathematical expressions by the researchers is operational and cannot be ended in effect with text-based search except apposite text keywords are identified. The ability of accessible text search engine to search mathematical expression vanquished with a math-aware search engine [1–3, 7].

The search for mathematical formulae is significant and tricky as they restrain both constitutional and interpretation information. In general, the theoretical basis

K. N. Brahmaji Rao (✉) · B. Tarakeswara Rao
Department of CS&SE, AUCE(A), Andhra University, Visakhapatnam, A.P., India
e-mail: brahmaji77@gmail.com

B. Tarakeswara Rao
e-mail: prasadreddy.vizag@gmail.com

G. Srinivas
Deparment of CSE, GITAM Deemed to Be University, Visakhapatnam, A.P., India
e-mail: srinivas.gitam@gmail.com

P. V. G. D. Prasad Reddy
Deparment of CSE, Kallam Haranadhareddy Institute of Technology, Guntur, A.P., India
e-mail: tarak7199@gmail.com

S. C. Satapathy et al. (eds.), *Communication Software and Networks*, Lecture Notes in Networks and Systems 134, https://doi.org/10.1007/978-981-15-5397-4_55

of the knowledge in numerous technical documents is usually represented with mathematical formulae [5].

The reliable search engines like Google and Yahoo are familiar for text-based information. But they are inefficient in searching data with mathematical formulae. Let us consider an example $\cos x + e^{\mathrm{Sinx}}$, this formula contains three symbols e, cos, and sin signify exponential and trigonometric functions. In this equation, the sin and cos terms contain some semantic meaning and cos and sin terms are structurally related to exponential.

As already quite a few methods were proposed for the extraction of mathematical formulae, in this paper we intended two newfangled methods for retrieving repetitive and non-repetitive formulae occurred in the given text.

The later part of this paper is designed as follows. Section 2 reviews the related non-negative matrix factorization clustering method used for the reclamation mathematical formulae. Section 3 includes the relative study of the projected methods. As a final point, the conclusion is accessible in Sect. 4.

2 Motivation and Contribution

Ma et al. [1] presented that in order to represent scientific knowledge mathematical formulae or expressions are essential. It is very complex to reclaim related mathematical formulae, as mathematical formulae contain both semantic and structural information. Three prevalent clustering algorithms *k*-Means, Self-Organizing Map (SOM), and Agglomerative Hierarchical Clustering (AHC) have been used in the proposed model for mathematical formulae retrieval.

Samarasinghe et al. [2] proposed document retrieval to solve mathematical problems. Kohonen's Self-Organizing Maps have been used in this approach for data clustering. The efficiency of the planned approach with other clustering techniques was presented.

Rao et al. [10] implemented Levenshtein distance and Sequence matcher to check string similarity on mathematical texts and keywords. An evaluation of their performance was carried out based on the time taken for the retrieval of keywords from mathematical texts. Sequence matcher performed with lesser false negatives when compared with Levenshtein distance.

Rao et al. [11] considered three different algorithms, viz., Sequence matcher, Levenshtein Distance, and Fuzzy-Wuzzy for mathematical formulae retrieval. Two different variants of Fuzzy-Wuzzy are found applicable to this study out of four variants. Efficiency measure, sensitivity analysis, and time series are the metrics used to calculate the performance of these. Fuzzy-Wuzzy partial ratio algorithm scored better over the other variants on efficiency measure and sensitivity analysis.

3 Proposed Approach

The matrix factorization method Non-negative Matrix Factorization (NMF) is a method which confines the matrices to be nonnegative. In order to comprehend NMF, we should elucidate the core insight between matrix factorization. The NMF method is suitable if the essential factors can be construed as non-negative for a particular task. When there are numerous attributes which are uncertain and are not sturdy predictors, then NMF is the best choice for feature extraction. With NMF technique the user can break down the multivariate data with a predefined number of features [2, 18, 19]. The threshold value is not fixed as 10 it can be any value. Depending upon the formulae in the text document we can change the threshold value. As there is no predefined or benchmark dataset available with math formulae for formulae retrieval. So, we have trained the NMF algorithm with synthesized dataset prepared by us contains more than 800 documents including text documents which contains a culmination of mathematical symbols, formulae or expressions, and literals from the natural languages such as alphabets and words. In one document, one particular formula is dominated and in another document, another document another formula is dominated, likewise we have prepared formula dominating documents dataset [20].

3.1 NMF Algorithm

Step 1: Load the text file into our trained program.
Step 2: Remove the stop words from the program.
Step 3: Calculation of Tf-idf Tokenizer terms.
Step 4: Load feature data into NMF Algorithm.
Step 5: Specify the number of clusters the given dataset to be clustered into.
Step 6: Run the NMF Algorithm.
Step 7: Retrieve the math formulae based on the threshold value.
Step 8: Display and count the number of math formulae and non-math formulae retrieved in each cluster.
Step 9: Repeat same for every cluster and display the results.
Step 10: Calculate the total retrieval time.

3.2 Flow Chart of the Actual Process

See Fig. 1.

Fig. 1 The procedure to retrieve math formulae with NMF clustering

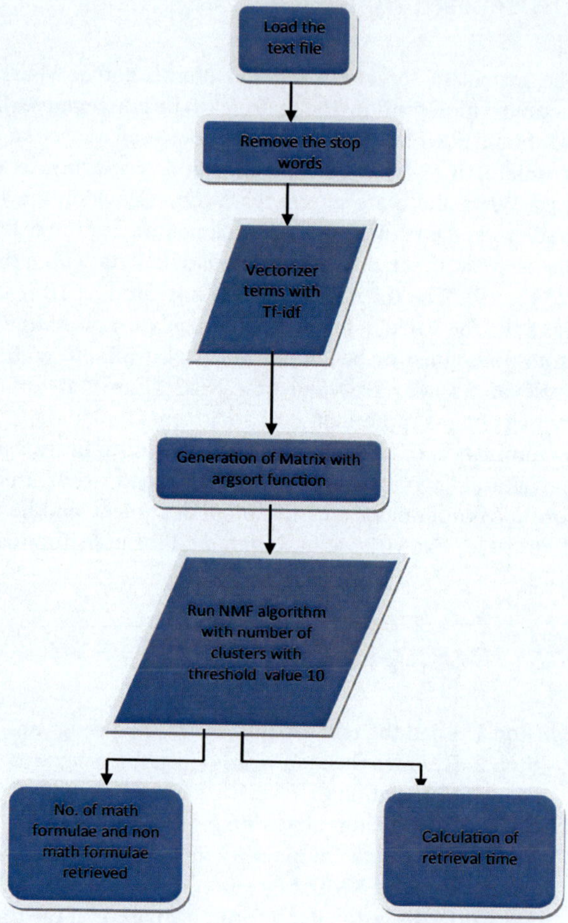

4 Results and Discussion

Time analysis is the time taken to retrieve the matched formula. The RAM and processor speed plays a major role in the retrieval time. In this paper, the experimentation for calculating the retrieval time is performed with 4 GB RAM and I3 Processor system. If the number of clusters increased, the number of matched formulae retrieved will also be increased, as a result the retrieval time is more. The number of formulae retrieved from the number of formulae in the training document is termed as efficiency. The efficiency of proposed approach increases with increase in number of clusters (Table 1).

N_{f} = Number of formulae in a sample
N_{c1} = Number of Formulae identified in Cluster 1

Table 1 Efficiency and retrieval time of math formulae with 2 and 3 clusters

$N_f = 40$	No. of clusters = 2						No. of Clusters = 3							
	Match		Mismatch		E_f	C_t	Match			Mismatch			E_f	C_t
	N_{c1}	N_{c2}	N_{c1}	N_{c2}			N_{c1}	N_{c2}	N_{c3}	N_{c1}	N_{c2}	N_{c3}		
Permutations	5	7	5	3	30	0.036	7	6	7	3	4	3	50	0.037
Limit	5	6	5	4	27.5	0.035	7	7	7	3	3	3	52.5	0.0355
Sigma	6	5	4	5	27.5	0.045	7	6	6	3	3	4	47.5	0.0454
Addition	7	7	3	3	35	0.086	6	7	7	4	3	4	50	0.104
Trigonometric	7	6	3	4	32.5	0.049	6	6	7	4	4	3	47.5	0.051
Factorial	5	4	5	6	2.5	0.037	7	7	7	3	3	3	52.5	0.038
Square	6	6	4	4	30	0.044	8	7	7	2	3	3	55	0.046
Exponential	6	7	4	3	32.5	0.036	6	6	6	4	4	4	45	0.0371
Integral	6	5	4	5	27.5	0.0357	7	6	7	3	4	3	50	0.0367
Logarithmic	5	6	5	4	27.5	0.0385	7	7	6	3	3	4	50	0.039

N_{c2} = Number of Formulae identified in Cluster 2
C_t = Run Time for Clustering and retrieval in ms.
E_f = Efficiency
S = Sensitivity
FNR = False-Negative Rate

Initially, we started experimentation by taking 2 and 3 clusters and number of formulae in the test document is 40. Here, efficiency increasing but run time for clustering and retrieval is increasing when the no. of clusters increased from 2 to 3 (Figs. 2, 3, 4, and 5, Table 2).

Here also, the no. of clusters increased from 4 to 5 the efficiency increases with increased time complexity (Figs. 6 and 7, Table 3).

As the no. of clusters increased from 2 and 3, the sensitivity is increasing with the decreasing of false-negative rate (Table 4).

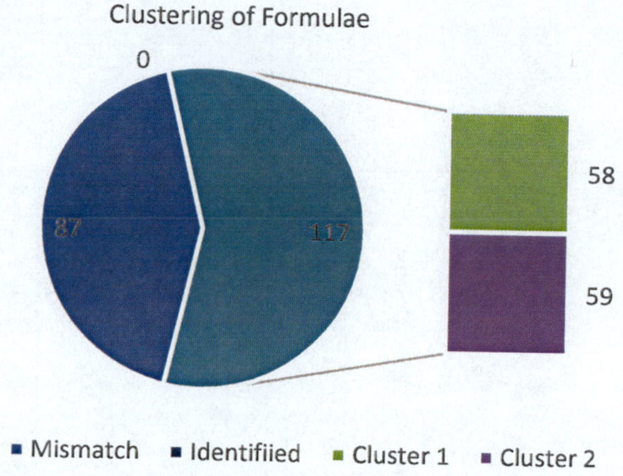

Fig. 2 Number of formulae matched and mismatched 2 clusters

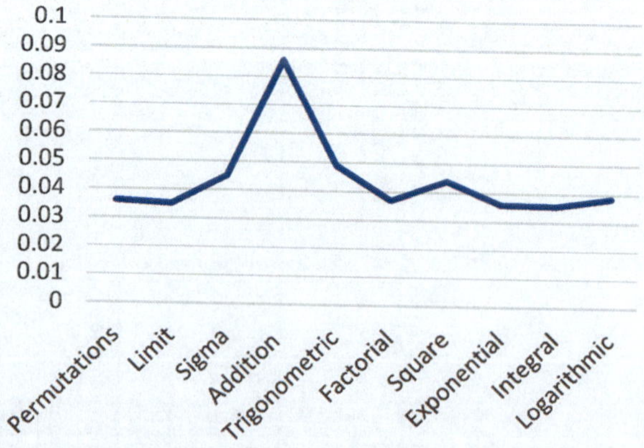

Fig. 3 Time required to ran and retrieve

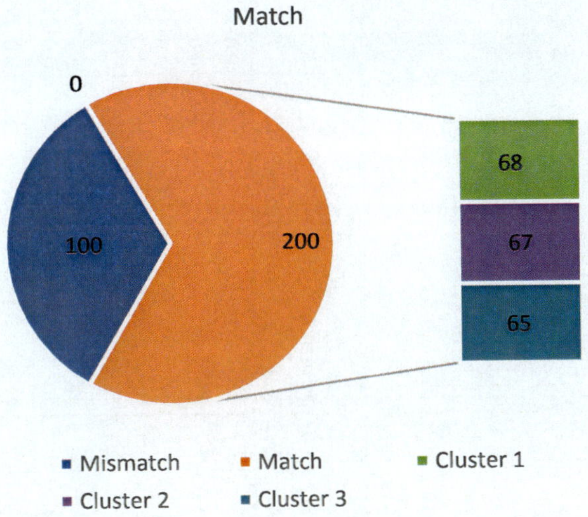

Fig. 4 Number of formulae matched and mismatched with 3 clusters

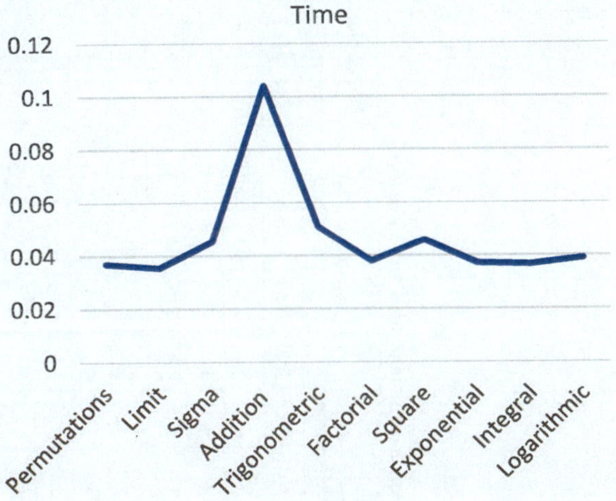

Fig. 5 Time required to run and retrieve

Here the sensitivity increases and false-negative rate decreases when the no. of clusters increased from 4 to 5.

Hence, the efficiency, sensitivity increases with increase in number of clusters. But the time required to retrieve math formulae also increases with increase in number of clusters. Whereas, the false-negative rate decreases with number of clusters.

Table 2 Efficiency and Retrieval time of math formulae with 4 and 5 clusters

$N_f = 40$	No. of clusters = 4										No. of clusters = 5											
	Match				Mismatch				E_f	C_t	Match					Mismatch					E_f	C_t
	$Nc1$	$Nc2$	$Nc3$	$Nc4$	$Nc1$	$Nc2$	$Nc3$	$Nc4$			$Nc1$	$Nc2$	$Nc3$	$Nc4$	$Nc5$	$Nc1$	$Nc2$	$Nc3$	$Nc4$	$Nc5$		
Permutations	7	6	7	6	3	4	3	4	65	0.0375	7	6	5	7	7	3	4	5	3	3	80	0.038
Limit	7	7	6	5	3	3	4	5	62.5	0.0361	7	6	6	7	7	3	4	4	3	3	82.5	0.037
Sigma	7	7	5	7	3	3	5	3	65	0.0472	7	7	7	6	6	3	3	3	4	4	82.5	0.0461
Addition	6	6	6	6	4	4	4	4	60	0.107	7	8	7	7	7	3	2	3	3	3	90	0.109
Trigonometric	7	6	6	6	3	4	4	4	62.5	0.0518	6	6	7	7	6	4	4	3	3	4	80	0.0525
Factorial	6	6	7	6	4	4	3	4	62.5	0.0386	7	7	7	6	7	3	3	3	4	3	85	0.0391
Square	6	7	6	6	4	3	3	4	62.5	0.047	7	7	6	6	7	3	3	4	4	3	82.5	0.0479
Exponential	5	7	6	7	5	4	3	4	62.5	0.0372	8	7	7	7	8	2	3	3	3	2	92.5	0.0373
Integral	5	7	6	7	5	3	4	3	62.5	0.037	6	6	6		8	4	4	4	3	2	82.5	0.0378
Logarithmic	7	6	7	6	3	4	3	4	65	0.0375	7	6	7		7	3	4	3	4	3	82.5	0.044

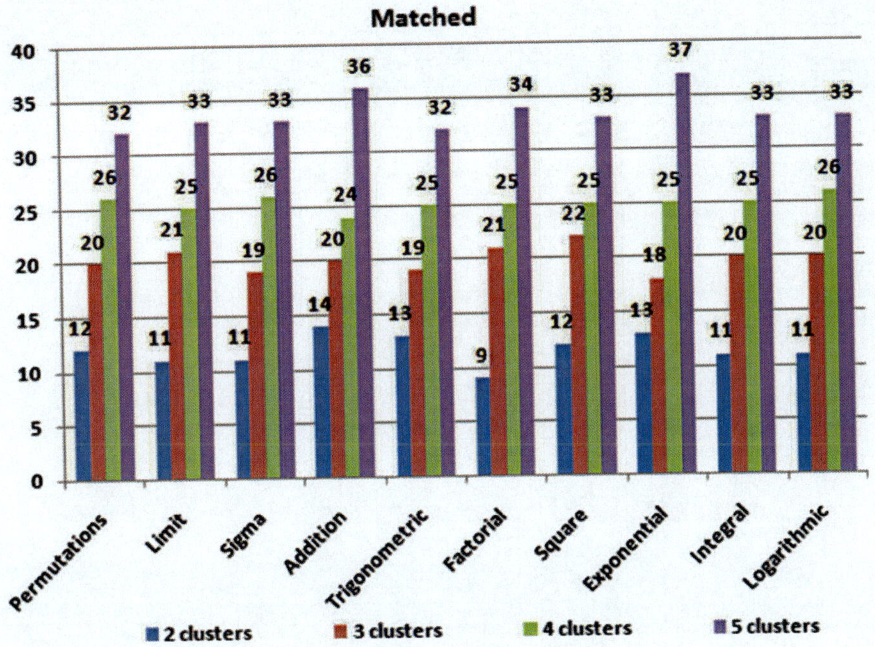

Fig. 6 Number of formulae matched with 2, 3, 4, and 5 clusters

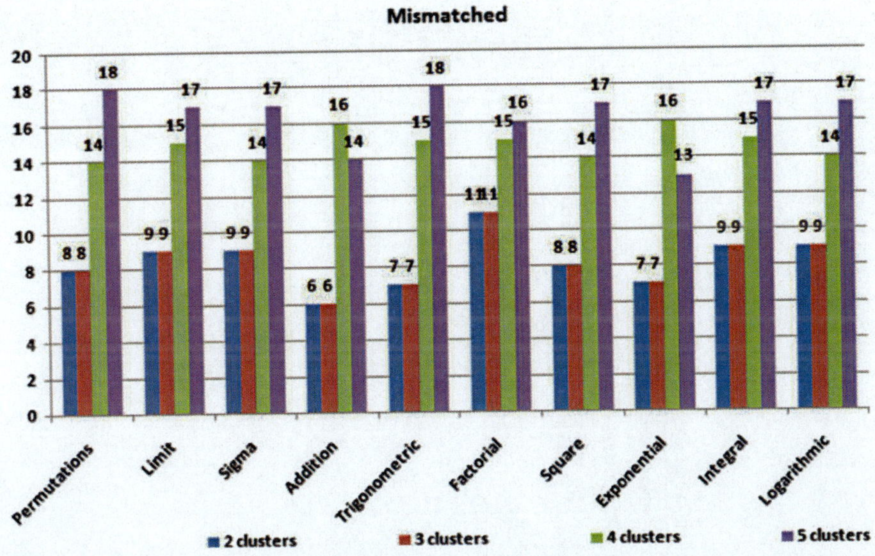

Fig. 7 Number of formulae mismatched with 2, 3, 4, and 5 clusters

Table 3 Sensitivity and false-negative rate of math formulae with 2 and 3 clusters

$N_f = 40$	No. of clusters = 2						No. of Clusters = 3							
	Match N (T_p)		Mismatch N (F_n)		S	FNR	Match N (T_p)			Mismatch N (F_n)			S	FNR
	N_{c1}	N_{c2}	N_{c1}	N_{c2}			N_{c1}	N_{c2}	N_{c3}	N_{c1}	N_{c2}	N_{c3}		
Permutations	5	7	5	3	30	40	7	6	7	3	4	3	50	33.33
Limit	5	6	5	4	27.5	45	7	7	7	3	3	3	52.5	30
Sigma	6	5	4	5	27.5	45	7	6	6	3	3	4	47.5	34.48
Addition	7	7	3	3	35	30	6	7	7	4	3	4	50	35.48
Trigonometric	7	6	3	4	32.5	35	6	6	7	4	4	3	47.5	36.67
Factorial	5	4	5	6	2.5	55	7	7	7	3	3	3	52.5	30
Square	6	6	4	4	30	40	8	7	7	2	3	3	55	26.67
Exponential	6	7	4	3	32.5	35	6	6	6	4	4	4	45	40
Integral	6	5	4	5	27.5	45	7	6	7	3	4	3	50	33.33
Logarithmic	5	6	5	4	27.5	45	7	7	6	3	3	4	50	33.33

Table 4 Sensitivity and false-negative rate of math formulae with 4 and 5 clusters

$N_f = 40$	No. of clusters = 4										No. of clusters = 5											
	Match N (T_p)				Mismatch N (F_n)				S	FNR	Match N (T_p)					Mismatch N (F_n)					S	FNR
	N_{c1}	N_{c2}	N_{c3}	N_{c4}	N_{c1}	N_{c2}	N_{c3}	N_{c4}			N_{c1}	N_{c2}	N_{c3}	N_{c4}	N_{c5}	N_{c1}	N_{c2}	N_{c3}	N_{c4}	N_{c5}		
Permutations	7	6	7	6	3	4	3	4	65	35	7	6	5	7	7	3	4	5	3	3	80	20
Limit	7	7	6	5	3	3	4	5	62.5	37.5	7	6	6	7	7	3	4	4	3	3	82.5	17.5
Sigma	7	7	5	7	3	3	5	3	65	35	7	7	7	6	6	3	3	3	4	4	82.5	17.5
Addition	6	6	6	6	4	4	4	4	60	40	7	8	7	7	7	3	2	3	3	3	90	10
Trigonometric	7	6	6	6	3	4	4	4	62.5	37.5	6	6	7	7	6	4	4	3	3	4	80	20
Factorial	6	6	7	6	4	4	3	4	62.5	37.5	7	7	7	6	7	3	3	3	4	3	85	15
Square	6	7	6	6	4	3	3	4	62.5	37.5	7	7	6	6	7	3	3	4	4	3	82.5	17.5
Exponential	5	7	6	7	5	4	3	4	62.5	37.5	8	7	7	7	8	2	3	3	3	2	92.5	7.5
Integral	5	7	6	7	5	3	4	3	62.5	37.5	6	6	6	7	8	4	4	4	3	2	82.5	17.5
Logarithmic	7	6	7	6	3	4	3	4	65	35	7	6	7	6	7	3	4	3	4	3	82.5	17.5

5 Conclusion

In this article, an approach which retrieves mathematical formulae was projected. The efficiency non-negative matrix factorization-based math classification retrieves all the math formulae that are matched with data in the training document. The efficiency, sensitivity, and false-negative rate with proposed NMF always depends on the number of clusters. The time of retrieval with the proposed approach with different numbers of clusters is calculated. Presently experiments are being performed on much enormous set of training documents. Currently, we are working with a method called TensorFlow for characteristic mining of mathematical formulae.

References

1. Ma K, Cheung Hui S, Chang K (2010) Feature extraction and clustering-based retrieval for mathematical formulas. In: The 2nd international conference on software engineering and data mining. pp 372–377
2. Harshanath Samarasinghe S, Cheung Hui S (2009) Mathematical document retrieval for problem solving. Int Conf Comput Eng Technol, 583–587
3. Misutka J, Galambos L (2008) Mathematical extension of full text search engine indexer. In: Proceedings of 3rd international conference on information and communication technologies: from theory to applications (ICTTA 08), April 2008, pp 1–6
4. Web Link to TensorFlow. Accessed 15 May 2019
5. https://sourcedexter.com/tensorflow-text-classification-python/
6. Miller BR, Youssef A (2003) Technical aspects of the digital library of mathematical functions. In: Annals of mathematics and artificial intelligence. Springer, Netherlands, pp 121–136
7. Zhang H, Ho TB, Lin MS (2004) An evolutionary K means algorithm for clustering time series data. In: Proceedings of the international conference on machine learning and cybernetics, pp 1282–1287
8. Munavalli R, Miner R (2006) MathFind: A math-aware search engine. In: Proceedings of the annual international ACM SIGIR conference on research and development in information retrieval, pp 735–735
9. Kohlhase M (2006) Markup for mathematical knowledge. In: An open markup format for mathematical documents, Ver. 1.2. Lecture notes in computer science. Springer, Berlin, pp 13–23
10. Rao GA, Venkata Rao K, Prasad Reddy PVGD, Lava Kumar T (2018) An efficient procedure for characteristic mining of mathematical formulas from document. Int J Eng Sci Technol (IJEST) 10(3):152–157
11. Rao GA, Srinivas G, Venkata Rao K, Prasad Reddy PVGD. Characteristic mining of mathematical formulas from document—a comparative study on sequence matcher and Levenshtein distance procedure. Int J Comput Sci Eng 6(4):400–403
12. Rao GA, Srinivas G, VenkataRao K, Prasad Reddy PVGD (2018) A partial ratio and ratio based fuzzy-wuzzy procedure for characteristic mining of mathematical formulas from documents. IJSC—ICTACT J Soft Comput 8(4):1728–1732
13. Peng M, Chen D, Xie Q, Zhang Y, Wang H, Hu G, Gao W, Zhang Y (2018) Topic-net conversation model. Int Conf Web Inf Syst Eng, 483–496
14. Peng M, Xie Q, Zhang Y, Wang H, Zhang X, Huang J, Tian G (2018) Neural sparse topical coding. In: Proceedings of the 56th annual meeting of the association for computational linguistics, vol 1: long papers. pp 2332–2340

15. Peng M, Xie Q, Wang H, Zhang Y, Tian G (2018) Bayesian sparse topical coding. IEEE Trans Knowl Data Eng, 1080–1093
16. Wang H, Zhang Y (2018) An efficient framework for the analysis of big brain signals data. In: Australasian database conference. pp 199–207
17. Peng M, Shi H, Xie Q, Zhang Y, Wang H, Li Z, Yong J (2018) Block Bayesian sparse topical coding. In: 2018 IEEE 22nd international conference on computer supported cooperative work in design (CSCWD), Nanjing. pp 271–276
18. Peng M, Zhu J, Wang H, Li X, Zhang Y, Zhang X, Tian G (2018) Mining event-oriented topics in microblog steam with unsupervised multi-view hierarchical embedding. ACM Trans Knowl Disc Data (TKDD) 12(3), Article No. 38
19. Web Link to NMF. Accessed 15 May 2019. http://mlexplained.com/2017/12/28/a-practical-introduction-to-nmf-nonnegative-matrix-factorization/
20. Web Link to NMF. Accessed 15 May 2019. https://datascience.stackexchange.com/questions/10299/what-is-a-good-explanation-of-non-negative-matrix-factorization
21. Brahmaji Rao KN, Srinivas G, Prasad Reddy PVGD, Surendra T (2019) A Heuristic ranking of different characteristic mining based mathematical formulae retrieval models. Int J Eng Adv Technol (IJEAT) 9(1), ISSN: 2249-8958

A Comparative Study on Coverage-Hole Detection Improvement with Inner Empty Circle over Delaunay Triangulation Method in Wireless Sensor Networks

Smita Das and Mrinal Kanti Debbarma

Abstract Wireless Sensor Network (WSN) is the collection of tiny sensor nodes, competent of transmitting data throughout the Region of Interest (RoI) and is very essential in surveillance-related applications. The most essential quality of WSN is coverage which can be at a stake due to the occurrence of coverage-hole in the RoI. In this paper, coverage-hole is identified based on the concept of Delaunay Triangulation (DT) and Empty Circle (EC) property. Then hole-area is estimated using Inner Empty Circle (IEC) property which is an improvement of Empty Circle (EC) property. Mathematical proofs and simulation results show that the IEC method is much accurate in comparison to traditional EC method in hole-identification in terms of number of coverage-holes, hole-area estimation, and hole-discovery time. Also the identification of coverage-hole concerning the size of sensing radius, size of RoI and no. of sensor nodes is promising.

1 Introduction

WSN is composed of small, cheap sensor nodes requiring very less power. The reason behind choosing sensors for data collection is that sensors can be deployed very close to the incident to be monitored and thus precise data can be collected. Due to their efficiency in surveillance, WSNs are very useful in habitat monitoring [1], disaster management [2], endangered species tracking [3], waste management [4], health monitoring [5], intrusion detection [6], etc. For all these applications, it is essential to recognize how much area is being covered and hence area coverage [7] becomes a hot field of research in WSN. Practically, there may be quite a few coverage-holes which are the area inside the RoI not covered by any sensor and thus incapable of sensing and transmitting data. These coverage-holes are created on various grounds

S. Das (✉) · M. K. Debbarma
Department of Computer Science & Engineering National Institute of Technology, Agartala 799046, Tripura, India
e-mail: smitadas.nita@gmail.com

M. K. Debbarma
e-mail: mkdb06@gmail.com

S. C. Satapathy et al. (eds.), *Communication Software and Networks*, Lecture Notes in Networks and Systems 134, https://doi.org/10.1007/978-981-15-5397-4_56

[8]: random deployment of nodes in harsh environment, obstacle in the sensing field, destruction of a sensor node due to power depletion, ambiguous network topology, etc. The proper hole-detection can also clarify whether a sensor node is in working condition or not which improves quality of coverage. In this paper, we have analyzed different computational geometry-based algorithms to identify whether coverage-hole is present or not in the RoI of a large scale WSN. Derived from the EC property, first we have defined the network topology by using DT. Then, based on the concept of DT and by using Inner Empty Circle (IEC) property, we refined the process of hole-detection. Finally, we have compared the IEC-based hole-detection algorithms with DT-based hole-detection in terms of number of coverage-holes, hole-area estimation, and hole-discovery time. For this comparison, we have concerned the size of sensing radius, size of RoI and no. of sensor nodes in the WSN.

The rest of the paper is prepared as follows: Sect. 2 describes the related work about the recent works done in coverage-hole-detection. Section 3 consists of some preliminary definitions and explains the proposed method along with theoretical proofs. Simulation, result analysis and comparison between IEC and DT-based hole-detection methods are done in Sect. 4. Finally, we have concluded the paper in Sect. 5.

2 Related Work

The coverage-hole detection algorithms roughly can be divided [9] into three types-Topology based, Computational Geometry based, and Statistical approach based. Topology-based approach depends on the topological properties like connectivity information and they do not necessitate any extra location identification device such as GPS. Computational Geometry (CG) based approach is based on node location information. In node location identification, an alternate solution to GPS may be geometry-based node localization methods. CG-based approach is much accurate in boundary detection in comparison to topological and statistical approaches. Statistical approach is based on statistical properly such as network model or probability of node distribution. Here node location information is not necessary but the accuracy is quite less in comparison to topology and CG-based approaches. Since the Computational Geometry-based approach is widely used in Hole-detection; hence, in this section we have discussed only this approach.

In 2009, J. Kanno et al. proposed a novel method [10] for detecting number of coverage-holes along with their location in a coordinate-free sensor network. In addition, they have detected the hole-boundary by processing information embedded in a non-planar communication graph. In 2014, P. Ghosh et. al proposed two novel distributed hole-detection algorithms [11] as DVHD (distance vector hole-determination) and GCHD (Gaussian curvature-based hole-determination). Using distributed Delaunay Triangulation; DVHD calculates the shortest distance path between any pair of nodes and utilizes path distance as metric for identifying the hole-boundaries. W. Li and W. Zhang in 2015 proposed a distributed algorithm [12] for hole-detection based on DT and empty circle property. The centers of the empty circles, containing the coverage-holes, are connected with lines in each pair of neighbor DT. If the line does not cross the common side of the two triangles or it intersects the common side whose length is greater than 2RS (where RS is the sensing radius), then the two empty circles, generated by the pair of neighbor DTs, incorporate in the same coverage-hole. In 2016, Li and Wu in [13] proposed a distributed coverage-hole detection and tree-dissect-based hole-healing algorithm (THH). In the tree, the Inscribed Empty Circles are stored in a list according to their sizes and the largest from the list is removed and is set as the source to construct a new sub-tree. In 2017, G. Zhang et. al proposed computational geometry-based approach [14] for hole-detection and boundary identification in a hybrid sensor network. At first, the static sensors estimate the size of coverage-hole based on DT construction. Then the static sensors conduct the mobile sensors move to the optimal location to heal the coverage-holes. Using computational geometry, bound of assisted sensor numbers is determined. In 2018, S.M. Koriem and M.A. Bayoumi proposed a hole-detection algorithm [15] by dividing the RoI cells based on Grid algorithm. The algorithm stores the exact location of the four edges of each cell to find out the nearest three nodes to each cell's edge points for hole-identification.

3 Proposed Method

For the proposed method to be executed, we have assumed that the deployment of the sensor nodes in the RoI is of random in nature, containing homogeneous, static nodes. All nodes within the WSN have the same sensing radius, R_S and communication radius, R_C. Besides, we would like to note the following definitions used in hole-detection algorithms.

Empty Circle Property The empty circle (EC) property says that the circumcircle of a triangle is the unique circle that passes through the three vertices of the triangle, as shown in Fig. 1.

Delaunay Triangle (DT) A triangulation of a finite point set S is called a Delaunay triangulation, if the circumcircle of every triangle must be empty, i.e., there must not be any point from S in its interior. The DT will be created based on the node positioning.

Inner Empty Circle Advancement in EC is the Inner Empty Circle (IEC) which is concentric with its corresponding EC. If R_{E} is the radius of an EC and $R_{\mathrm{E}} > R_{\mathrm{S}}$, then radius of IEC, $R_{\mathrm{IEC}} = (R_{\mathrm{E}} - R_{\mathrm{S}})$. EC property can identify the coverage-hole but the size of EC and Coverage-hole are not the same. Since $R_{\mathrm{E}} > R_{\mathrm{S}}$, the size of an EC can be much larger than the subsequent coverage-hole and that is why IEC is basically used to calculate the hole-area much accurately. In Fig. 2, red circles are showing the ECs and the green circles represent the IECs.

In this paper, we have tried to identify the presence of coverage-hole in the RoI. For this, we have done the following: (1) Construction of DT, (2) Identify the presence of Coverage-Hole in the RoI, and (3) Comparing the detection of Coverage-Hole.

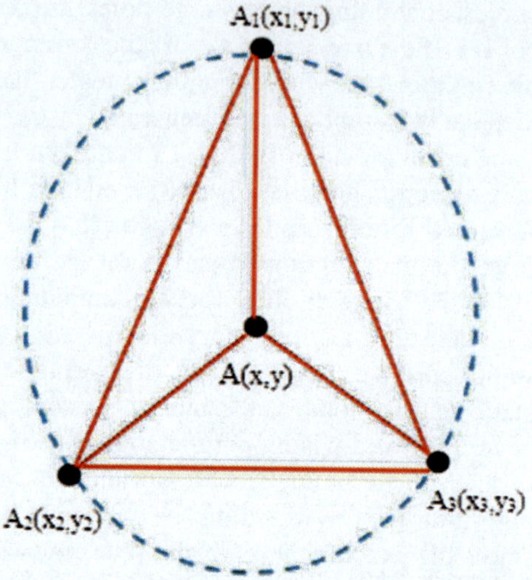

Fig. 1 Empty circle property representation

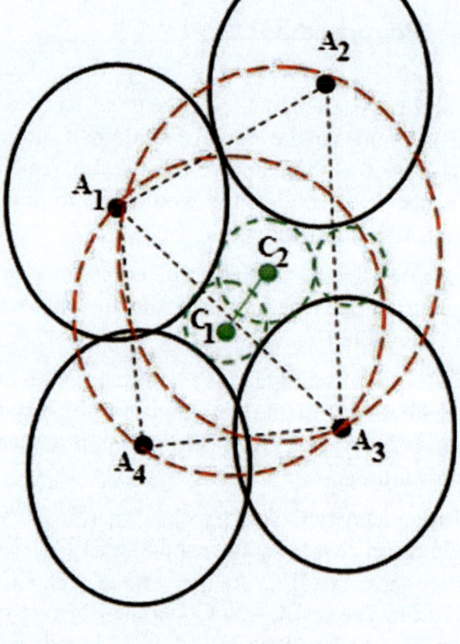

Fig. 2 Identification of coverage-hole based on IEC

3.1 Construction of DT

After the random deployment of the N no. of nodes in the RoI, the node locations are identified based on RSSI profiling. Then the DT is formed using those nodes. For each node, the circum-center and the circum-radius w.r.t the DT, to which the node belongs, is calculated. As per definition, circum circle of the DT is nothing but the empty circle itself.

3.2 Identify the Coverage-Holes

Now in the RoI, the existence of coverage-hole is identified based on the following two theorems:

Theorem 1 *If R_S and R_C are the sensing and communication radius of sensor node respectively, then there exist some coverage-hole in the RoI for the relation $R_C > R_S$*

Explanation: From Fig. 3 [16], it is obvious that $R_C > R_S$. Now R_C is the radius of the EC and O_C is its center. Therefore, O_C is definitely out of coverage of R_S when radius of EC is greater than the sensing radius. As per definition, this out of coverage area produces the coverage-hole. Hence, coverage-hole exists when $R_C > R_S$. Also we can comment that, if there is some coverage-hole in the RoI, then the relation $R_C > R_S$ must hold.

Theorem 2 *If the line segment, joining the centers of the two ECs of two neighboring DTs, intersects the common side of the two DTs and the length of the common side is greater than (2*sensing radius), then the two ECs intersect the same coverage-hole.*

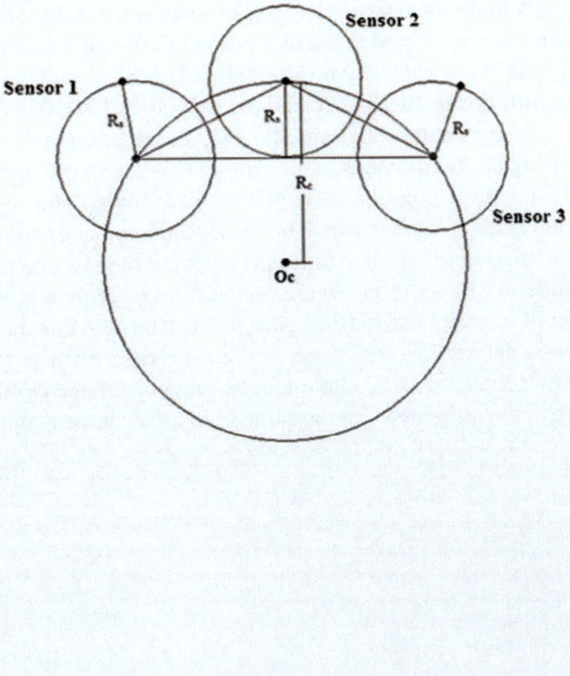

Fig. 3 Coverage-hole in RoI when $R_C > R_S$

Explanation: In Fig. 2, $\Delta A_1A_2A_3$ and $\Delta A_1A_4A_3$ are the two neighboring DTs. The red dotted circles denote the ECs and the green dotted circles denote the IECs. The line segment C_1C_2 joins the centers of the two ECs of the two DTs. Here C_1C_2 intersects the common side A_1A_3 of the two DTs. Again, from the figure, R_S being the sensing radius, length of $A_1A_3 > (2 * R_S)$. If we take any point on C_1C_2 then its distance from any vertices of $\Delta A_1A_2A_3$ and $\Delta A_1A_4A_3$ is greater than R_s. Further, it can be found that no point on C_1C_2 is covered by any sensor and C_1 and C_2 are from different ECs. Thus, the uncovered regions in the two ECs belong to the same coverage-hole.

Based on the above two theorems, first coverage-hole detection is carried out. When coverage-hole is identified, the IECs are formulated as described above.

3.3 *Comparison in Hole-Detection*

In this section, we have compared the coverage-hole detection using IEC property over the DT property. Even though ECs are capable of identifying the coverage-holes but the hole-area is not equal to the EC. Generally, the size of EC is much larger than the corresponding coverage-hole. From Fig. 2, it is obvious that the red dotted lines describing the ECs are much larger than the actual coverage-holes. That is why IEC concept is used here. The area estimation of the coverage-hole is done by combining the IECs as given in [13].

4 Simulation and Results Analysis

For simulation purpose we have used MATLAB R2015b. The simulations parameters are: number of nodes and dimension of sensing radius in a given RoI. The simulation in Fig. 4a, b shows two different situations in a given RoI of 10 m with number of sensor nodes 20 and 120 and the sensing radius 20 m and 10 m, respectively.

Based on the above simulations, it is clear that in case of Fig. 4a, N is less and R_S is higher, but number of EC and IEC is lower, so number of coverage-hole is also less. On the other hand, in Fig. 4b, N is higher and R_S is lesser, still number of EC and IECs are larger which increases the number of coverage-hole. We have run each simulation for several times to check for the average number of coverage-holes and hole-discovery time. Figure 5 shows the average number of coverage-holes detected in EC concept and in IEC concept. It can be seen that the average number of coverage-hole detected in IEC concept is much lesser than EC concept. This is because the IECs show the actual hole-area accurately. On the contrary, the ECs are much larger than the hole-area and the same hole-area can be denoted by multiple ECs. Figure 6

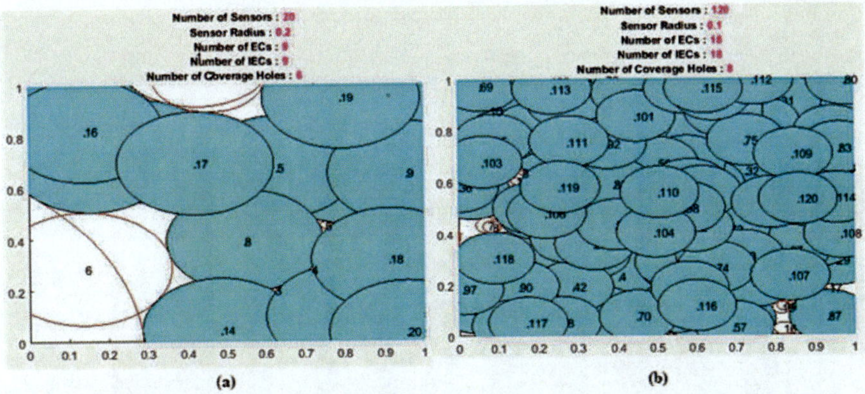

Fig. 4 Coverage-hole detection with **a** $R_S = 20\,\text{m}$ and $N = 20$ and **b** $R_S = 10\,\text{m}$ and $N = 120$

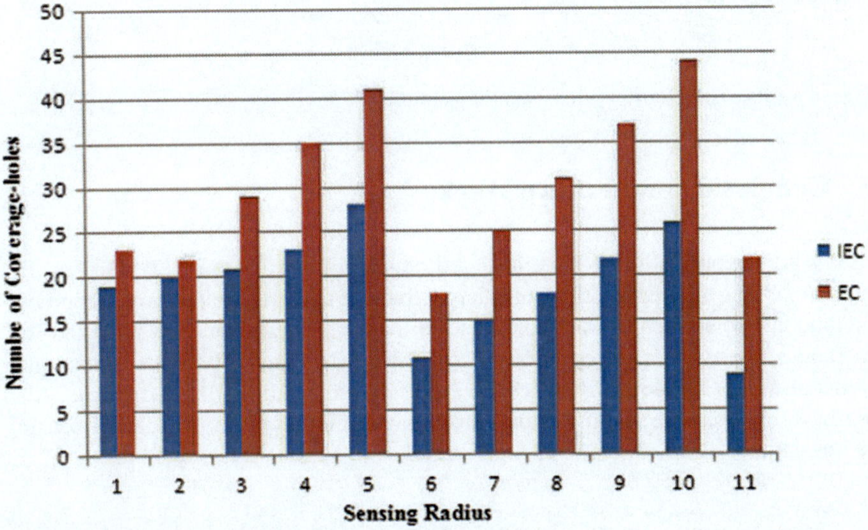

Fig. 5 Avg. no. of coverage-hole detection with IEC and EC

shows the average hole-discovery time in EC concept and in IEC concept. It can be found that the average hole-discovery time in IEC concept is much higher. This is because in this process first the ECs are determined and based on them the IECs are identified which increases the average hole-discovery time.

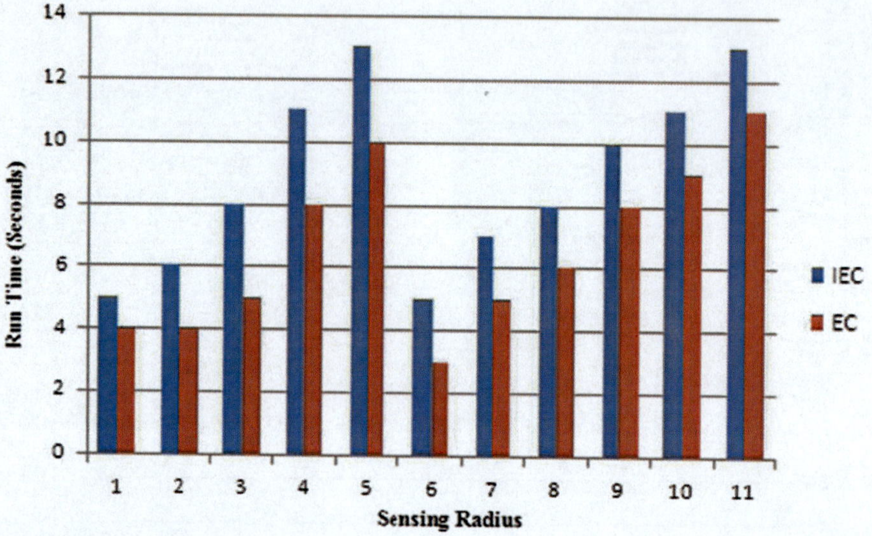

Fig. 6 Average hole-discovery time with EC and IEC

5 Conclusion and Future Work

In this paper, basically we have detected coverage-holes in the RoI based on EC property. Since the area of EC is much larger than the actual coverage-hole, hence IEC method is also applied to check the area estimation. Simulation shows that the IEC method is much accurate in comparison to DT and EC method in hole-identification in terms of number of coverage-holes, area estimation, and hole-discovery time. Also we have checked the identification of coverage-hole based on the size of sensing radius and no. of sensor nodes in a given RoI.

References

1. Kar K, Banerjee S (2003) Node placement for connected coverage in sensor networks
2. Ratnasamy S, Karp B, Shenker S, Estrin D, Govindan R, Yin L, Yu F (2003) Data-centric storage in sensornets with ght, a geographic hash table. Mobile Networks Appl 8(4):427–442
3. Aslam J, Butler Z, Constantin F, Crespi V, Cybenko G, Rus D (2003) Tracking a moving object with a binary sensor network. In: Proceedings of the 1st international conference on Embedded networked sensor systems. ACM, pp 150–161
4. Zhang H, Hou JC et al (2005) Maintaining sensing coverage and connectivity in large sensor networks. Ad Hoc Sens Wirel Networks 1(1–2):89–124
5. Lorincz K, Malan DJ, Fulford-Jones TR, Nawoj A, Clavel A, Shnayder V, Mainland G, Welsh M, Moulton S (2004) Sensor networks for emergency response: challenges and opportunities. IEEE Pervasive Comput (4):16–23

6. Roman R, Zhou J, Lopez J (2006) Applying intrusion detection systems to wireless sensor networks. In: IEEE consumer communications & networking conference (CCNC 2006)
7. Das S, Debbarma MK (2019) A survey on coverage problems in wireless sensor network based on monitored region. In: Advances in data and information sciences. Springer, pp 349–359
8. Ramazani S, Kanno J, Selmic RR, Brust MR (2016) Topological and combinatorial coverage hole detection in coordinate-free wireless sensor networks. IJSNet 21(1):40–52
9. Das S, DebBarma MK (2018) Hole detection in wireless sensor network: a review. In: Recent findings in intelligent computing techniques. Springer, pp 87–96
10. Kanno J, Buchart JG, Selmic RR, Phoha V (2009) Detecting coverage holes in wireless sensor networks. In: 2009 17th Mediterranean conference on control and automation. IEEE, pp 452–457
11. Ghosh P, Gao J, Gasparri A, Krishnamachari B (2014) Distributed hole detection algorithms for wireless sensor networks. In: 2014 IEEE 11th international conference on mobile ad hoc and sensor systems. IEEE, pp 257–261
12. Li W, Zhang W (2015) Coverage hole and boundary nodes detection in wireless sensor networks. J Network Comput Appl 48:35–43
13. Li W, Wu Y (2016) Tree-based coverage hole detection and healing method in wireless sensor networks. Comput Networks 103:33–43
14. Zhang G, Qi C, Zhang W, Ren J, Wang L (2017) Estimation and healing of coverage hole in hybrid sensor networks: a simulation approach. Sustainability 9(10):1733
15. Koriem S, Bayoumi M (2018) Detecting and measuring holes in wireless sensor network. J King Saud Univ Comput Inf Sci
16. Das S, Debbarma MK (2019) Node position estimation for efficient coverage hole-detection in wireless sensor network. Computación y Sistemas 23(1):185

An Energy-Efficient Resource Allocation System Using OFDM DAS Model for LTE Applications

B. Kiranmai, V. Lavanya, and S. S. Kiran

Abstract In wireless communications, the demand for wireless throughout and communication reliability as well as the user density will always increase. In this regard, one of the important technology enables for 5G network is massive MIMO technology, which is a special case of multiuser MIMO with an excess of Base Station (BS) antennas. In view of emerging networks, the energy consumption is a critical concern to meet enormous service expectations. The main limitation in this communication system is effectiveness of the spectrum utilization. To increase the effectiveness of the communication system, different techniques are to be used to increase the system throughput and decrease the power consumption. In this paper, a new less complexity suboptimal algorithm is used to separate subcarrier allocation and power distribution. An OFDM-Distributed Antenna System (DAS) is used to get more energy efficiency for Long-Term Evolution (LTE).

1 Introduction

1.1 Introduction to OFDM

A multi-carrier modulation technique that transmits data over variety of orthogonal subcarriers is termed Orthogonal Frequency Division Multiplexing (OFDM). OFDM breaks the information to be sent into tiny chunks, allocating every sub-data stream to a subcarrier and therefore the information is transmitted in parallel orthogonal

B. Kiranmai (✉) · S. S. Kiran
Department of Electronics and Communication Engineering, Lendi Institute of Engineering and Technology, Vizianagaram, India
e-mail: kbabburu@gmail.com; jahnavib2006@gmail.com

S. S. Kiran
e-mail: lendi.sskiran@gmail.com

V. Lavanya
Department of Electronics and Communication Engineering, M.V.G.R.C.E, Vizianagaram, India
e-mail: lavanyavadda@gmail.com

S. C. Satapathy et al. (eds.), *Communication Software and Networks*, Lecture Notes in Networks and Systems 134, https://doi.org/10.1007/978-981-15-5397-4_57

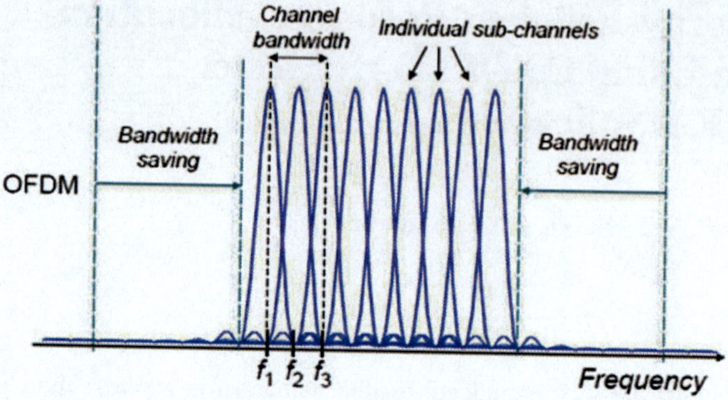

Fig. 1 Frequency spectrum of OFDM

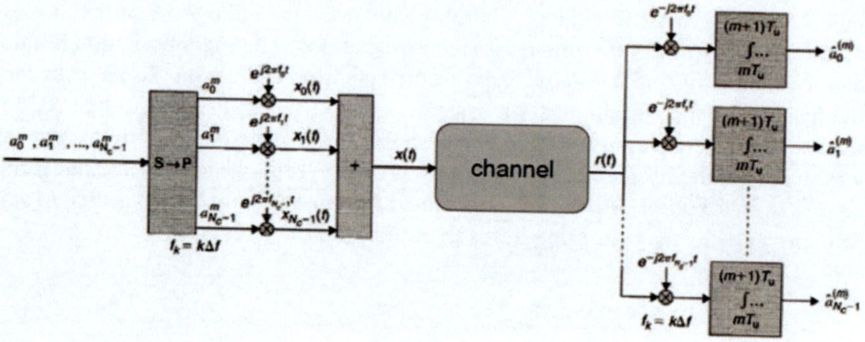

Fig. 2 Basic structure of OFDM communication system

subcarriers. These carriers are orthogonal to each other and may be packed tight rending the channel into narrowband channels allows important simplification of equalizer style in multipath environments. Versatile bandwidths are enabled through scalable variety of subcarriers. Frequency spectrum of OFDM is shown in Fig. 1.

The basic OFDM structure, which consists set of modulators and correlators is shown in Fig. 2. The subcarrier spacing selection being equal to the per carrier symbol rate $1/T_U$ with simple FFT processor that uses a simple and low complexity implementation.

1.1.1 Orthogonality Principle

The important property of the OFDM signal is the orthogonality between the subcarriers. Orthogonal means "perpendicular" or at "right angle". An exponential signal $\{e^{j2\pi f_k t}\}_{k=0}^{N-1}$ which have distinct subcarriers at $f_k = k/T_{\text{sym}}$ for $0 \leq t \leq T_{\text{sym}}$ and these signals are said to be orthogonal, if the integral of the products for their common

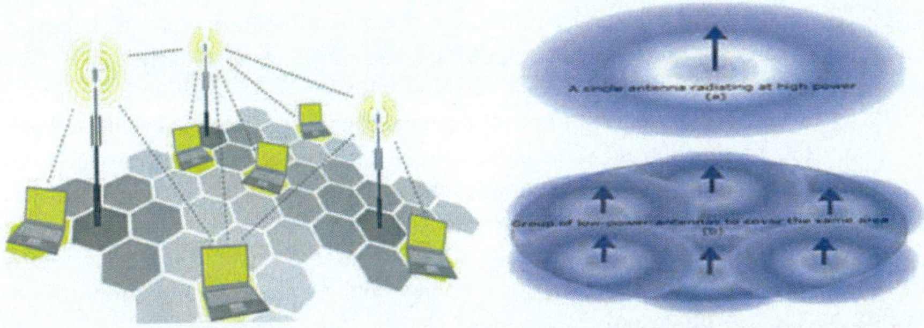

Fig. 3 The layout of the distributed antenna systems configuration

period is zero, that is

$$\frac{1}{T_{\text{sym}}}\int_{0}^{T_{\text{sym}}} e^{j2\pi f_k t} e^{-j2\pi f_i t}\,\mathrm{d}t = 1 \text{ for all } k = i$$
$$= 0 \text{ for all } k \neq i \qquad (1)$$

Equation (1) shows the orthogonality condition for the OFDM signal to be ICI-free.

1.2 Distributed Antenna Systems (DAS)

A distributed antenna system may be a network of distinct antenna nodes connected to a base station by using optical fiber and that has wireless service at intervals on a geographical area. In consideration of emerging technologies like 5G wireless communication systems has to handle huge data which requires energy consumption.

In DAS a number of the antennas are placed at the central base station (BS) and the radio remote units (RAUs) are the other antennas said to be distributed throughout the cell. These RAUs are controlled by the central base stations connected via high-bandwidth, less delay contributed connections like fiber optics [2, 3]. And The RAUs contain a low-complex processor, up/down converters, and LNAs and also having least intelligence of their own [1] (Fig. 3).

1.3 Energy Efficiency

Nowadays mobile traffic is growing drastically, therefore energy efficiency systems are incredibly necessary to satisfy the needs of the 5G wireless communications. Lot of analysis has been work out the literature for wireless networks, and for a general energy efficiency and spectral efficiency (SE) within the downlink OFDM networks, etc. [4, 5]. In spite of this, energy efficiency resource allocation equal distribution among the mobile stations exploitation OFDM with DAS wasn't studied.

In this context, an energy-efficient resource allocation for downlink multiuser OFDM DAS with equal distribution over composite fading channels has been used. The main aim of the optimization is to extend the energy efficiency of the total transmit power of every remote accessing unit with equal distribution of information rates and bit error rates (BER). However obtaining best results computationally advanced as a result of the optimization is a drawback. To extend the Energy efficiency (EE), less-complex suboptimal rule is developed by maintaining proportional equality for the downlink multiuser OFDM DAS.

In this paper, the remaining sections are arranged as follows, Sect. 2 describes the fundamental structure of multiuser OFDM DAS system and power utilization models. Section 3 depicts the implementation of less-complex suboptimal rule. Simulation results and conclusions are given in Sects. 4 and 5.

2 Proposed Structure

2.1 Energy Efficiency of an OFDM DAS System

The basic model of OFDM DAS is shown in Fig. 4.

Figure 4 shows the fundamental circular layout of downlink multiuser OFDM DAS with n subcarriers, k mobile stations, and m remote access units. All mobile stations and remote access units are equipped with one antenna. The CSI is obtainable at each transmitter and receiver. The remote accessing unit 1 as shown within Fig. 4

Fig. 4 Layout of the OFDM DAS design

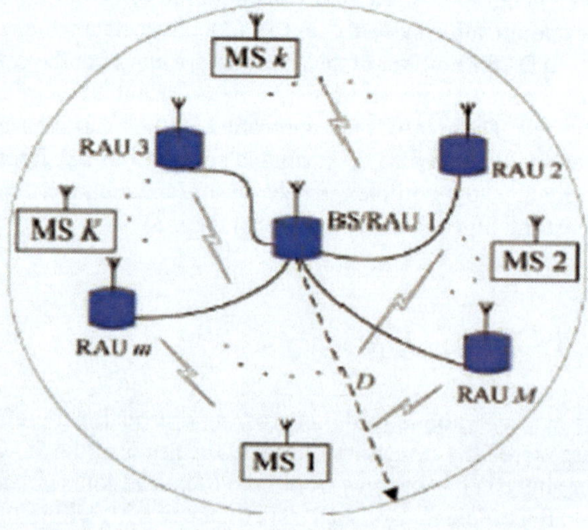

is a base station (BS). Every remote accessing unit is physically connected with base station/remote accessing unit 1 via an optical fiber. The SNR ration of k mobile stations on n subcarrier from m remote accessing units m may be expressed as

$$\mathrm{SNR} = \frac{P_{k,n,m}}{\sigma_z^2}[h_{k,n,m}^2] \tag{2}$$

Equation (2) shows the SNR of mobile stations k, subcarriers n and m RAUs Eq. (2) can also be expressed as

$$h_{k,n,m} = g_{k,n,m}\, w_{k,m} \tag{3}$$

Here $g_{k,n,m}$ indicates the small-scale fading of a wireless channel $w_{k,m}$ indicates the large-scale fading and independent of $g_{k,n,m}$. The large-scale fading can be expressed as

$$\omega_{k,m} = \sqrt{\frac{C S_{k,m}}{d_{k,m}^{\alpha}}} \tag{4}$$

In Eq. (4) the value of α lies between 3 and 5 and is called the path-loss exponent and d_k is the distance between mobile stations and the remote accessing units, where C is the mean path gain at reference distance $d_{k,m} = 1$ km, and $s_{k,m}$, is a lognormal shadow fading variable, a continuous data rate adaptation is used in mobile stations, then overall Spectrum Efficiency of k mobile stations be able to written as

$$SE_k = \frac{1}{N}\sum\nolimits_{n=1}^{N}\sum\nolimits_{m=1}^{M}\log_2(1 + \beta_{SNR,k,n,m}) \tag{5}$$

The Eq. (5) shows the spectral efficiency of mobile station sk, and $\beta = -1.5/(\ln(5P_{\mathrm{BER}}))$ is assumed to be constant for a particular P_{BER} condition.

2.2 *Power utilization in OFDM DAS Model*

While designing a system for energy efficiency one has to consider the total power utilization of the system. According to the OFDM DAS system, it contains three important parts one is power utilization of amplifiers, second power utilization of remote sensing units, and the third power utilization of the fiber optic transmission [7, 8]. So the total power utilization can be expressed as

$$P_{\mathrm{total}} = \frac{P_t}{\tau} + P_c + M P_o \tag{6}$$

The Eq. (6) is the total power utilization of amplifiers, RAUs, fiber optic transmission in this model. This equation can be rewritten by including the *k* number of mobile stations, *m* RAUs and *n* subcarriers as shown in the Eq. (7)

$$P_t = \sum_{k=1}^{k} \sum_{n=1}^{n} \sum_{m=1}^{m} p_{k,n,,m} \tag{7}$$

$$p_{k,n,m} = \min\left\{\left[\frac{1 - \sum_{k=1}^{K} \lambda_k}{\left(\frac{w}{\tau} + \lambda_1\right) N \ln_2} - \frac{1}{H_{k,n,m}}\right], p_m^{\max}\right\} \tag{8}$$

2.3 Energy Efficiency of an OFDM DAS Model

As per the definition of the energy efficiency in the literature is the ratio to the total data rate to the total power utilization (bits/J/Hz) [6], so the total power utilization for a DAS model is given in the Eq. (6) and the spectrum efficiency or the overall data rate is given in the Eq. (5).

Substituting these equations then the energy efficiency of an OFDM DAS model can be represented in the Eq. (8)

$$\eta_{\mathrm{EE}} = \frac{\mathrm{SE}}{\frac{P_t}{\tau} + P_{c+MP_o}} \tag{9}$$

The Eq. (9) shows the energy efficiency and SE is the spectrum efficiency as specified herein Eq. (5), and it can be written as

$$SE = \sum_{k=1}^{k} SE_k \tag{10}$$

for *k* mobile stations.

2.4 Energy Efficiency Optimization

The energy efficiency of OFDM DAS model be able to expressed in Eq. (9) as

$$\eta_{EE} = \frac{1}{N} T \tag{11}$$

where T is ratio of power utilization to the total power.

3 Resource Allocation for OFDM DAS Model

3.1 Subcarrier Allocation

Multiuser communications require multiple access techniques, an OFDM is used as a multiple access theme by permitting coinciding frequency-separated transmissions to/from multiple mobile terminals. The subcarrier allocation techniques are basically two different types, one is consecutive (or localized) frequency mapping and the other is distributed frequency mapping. The following figure shows the subcarrier allocation techniques (Fig. 5).

In the distributed frequency mapping a single subcarriers belonging to one link are spread across the whole OFDM bandwidth. Two factors determine the selection of the OFDM subcarrier (SC) spacing, Δf is, it is preferable to have as small SC

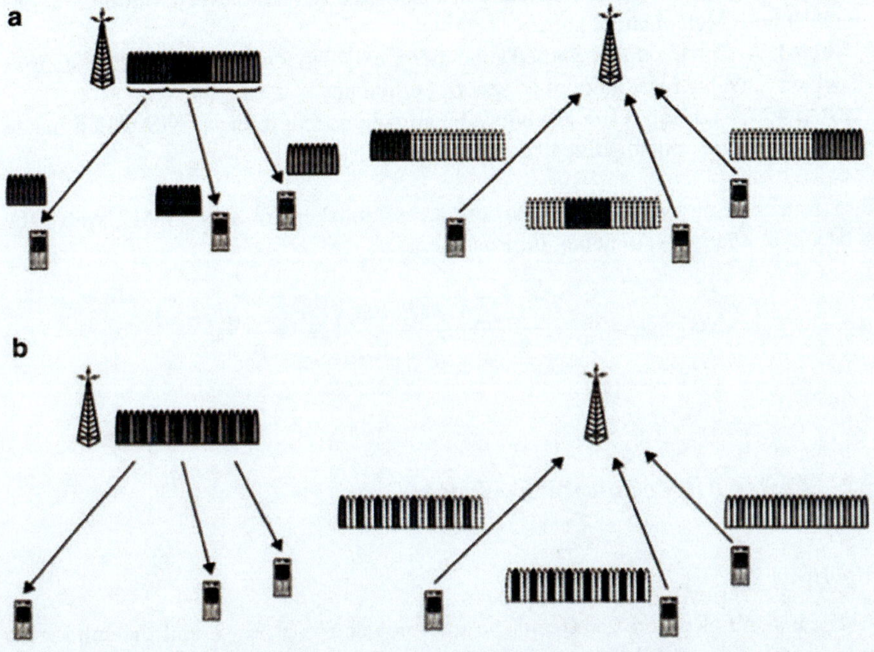

Fig. 5 **a** Constructive frequency mapping. **b** Distributed frequency mapping

spacing as possible, since this gives a long symbol period and consequently, the relative cyclic prefix overhead will be minimized and the other is too small SC spacing will increase the sensitivity to Doppler spread and completely different sorts of frequency inaccuracies.

Doppler spread is typical for mobile systems and the amount is directly limited by the relative velocity between the mobile and the base station. Frequency variations due to Doppler spread lead to losing the orthogonality in the receiver and ICI occurs. Dependent on the targeted mobility for the system and the allowed amount of ICI, the sub-carrier allocation can be selected.

The following steps show the optimization modeling:

- The distribution of subcarriers for RAUs: By using the Eq. (3), measure the accessing possibility between *k* MSs and *m* RAUs. Then the subcarrier assignment as follows:

  Subcarrier assignment:

Algorithm 1:

Step 1: Every RAU will be allocated with the mobile stations and with the subcarriers and with the purpose of improving the overall SE, the residual subcarriers are assigned while preserving rough relativity by assuming one and the same power distribution among the subcarriers.
Step 2: Assign the undistributed subcarriers and RAUs to every Mobile Stations (MS) with high channel gain.
Step 3: Assign the undistributed subcarriers and also RAUs to the Mobile stations which have the least amount of spectral efficiency.
Step 4: Assign the residual undistributed subcarriers to every MS which has to get at the most one unallocated subcarriers.

Subsequent to the MSs, the subcarriers are established for every RAU, we've the subsequent energy—efficiency improvement

$$\max_{\mathrm{p}} \eta_{\mathrm{EE}} = \frac{1}{N} \frac{\sum_{k=1}^{K} \sum n, \varepsilon \Omega_m \log_2\left(1 + p_{k,n,m} h_{k,n,m}\right)}{\frac{1}{N} \sum_{k=1}^{K} \sum m \varepsilon \Omega_m p_{k,n,m} + P_c + M P_o} \tag{12}$$

3.2 Power Allocation for Each RAU

Algorithm 2:

1: First initialization $i = 0$ and the total power $p_{k,n,m} = 0$, and the subcarrier spacing, $\lambda_1^{(i)} = 1$, $\lambda_k^{(i)} = 1/1000$, for all mobile stations $k = 2, \ldots, K$, $n = 1, 2, \ldots, N$;

2: Secondly initialize mobile stations = 1 and subcarrier = 1;
3: Compute $p_{k,n,m}$ according to Eqs. (8).
4: Then increment i, revise $\lambda_1^{(i+1)}$ and $\lambda_k^{(i+1)}$ according the equation $\lambda = \sqrt{\frac{p_{k,n,m(i)}}{2+p_{k,n,m(i)}}}$
5: If the multipliers λ_1 and λ_k are convergent, go back to the power p and stop the algorithm 1; Repeat the process from Step 2.

4 Simulation Results

The energy efficiency of this system was evaluated by using Monte Carlo simulations. The number of remote accessing units is 5, subcarriers are 52, the cell radius is 1 km, the path-loss component is 3.7, and standard deviation of large-scale shadow fading is 8 dB (Figs. 6, 7, 8, and 9).

5 Conclusions

The energy efficiency for 5G wireless applications can be increased with this new less-complex suboptimal algorithm. This technique converges to the optimum resolution at less number of iterations and incontestable the trade-off among the energy efficiency (EE) and spectrum efficiency (SE). With this new technique, the spectral efficiency of the wireless communication system has been improved from approximately 45 k bits/s/Hz to 162 k bits/s/Hz and the energy efficiency of the system is 16.2 bits/J/Hz for spectrum efficiency of 10 bits/s/Hz.

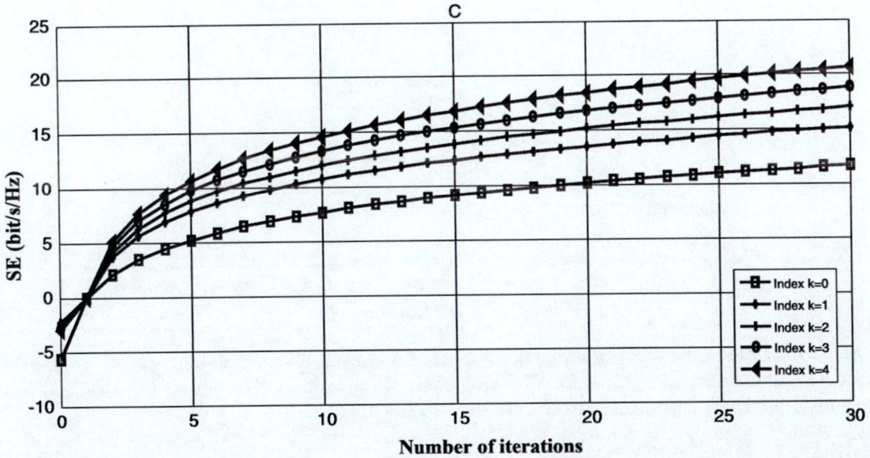

Fig. 6 This simulation result describes the spectrum efficiency improved with RAUs = 5 and subcarriers $n = 52$

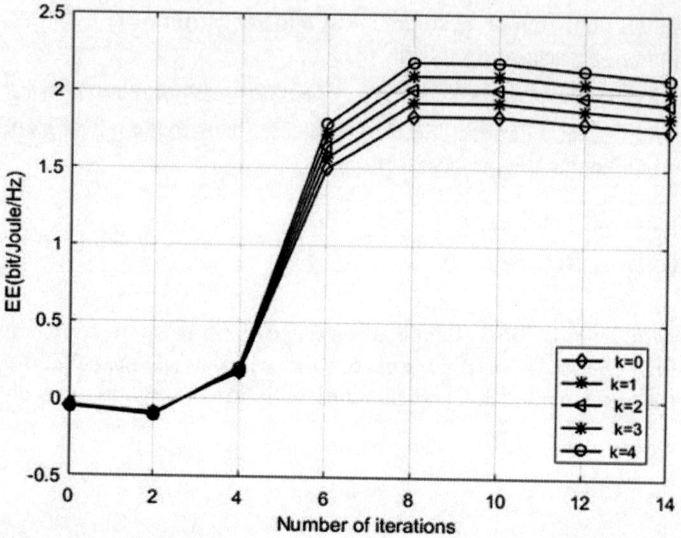

Fig. 7 Energy efficiency has equally distributed among the all used Mobile stations with RAUs = 5 and the subcarriers 52

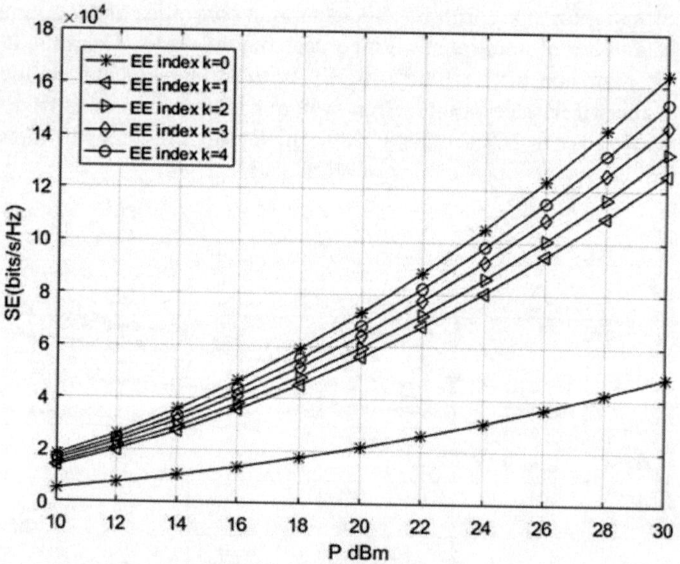

Fig. 8 Describes the spectrum efficiency has been improved with the new proposed technique, energy resource allocation equal for all fading channels consisting of multiuser OFDM DAS and spectrum efficiency is compared to be very less when RAUs $k = 0$

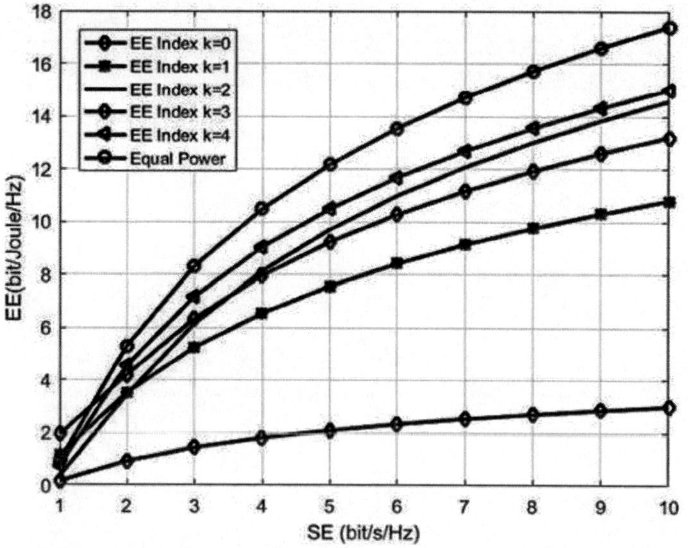

Fig. 9 Shows the increased energy efficiency is improved with the increased spectral efficiency with the new technique, it has been compared to be very less for RAUs $k = 0$

References

1. You X-H (2010) Cooperative distributed antenna systems for mobile communications. IEEE Wirel Commun 17(3):35–43
2. Zhu l (2011) Performance comparison between distributed antenna and microcellular systems. IEEE J Sel Areas Commun 29(6):1151–1163
3. Kim H (2012) Transmission schemes based on sum rate analysis in distributed antenna systems ieee trans. Wirel Commun 11(3):1201–1209
4. Shen Z-K (2005) Adaptive resource allocation in multiuser OFDM systems with proportional rate constraints. IEEE Trans Wirel Commun 4(6):2726–2737
5. Miao W (2011) Distributed interference-aware energy-efficient power optimization ieee trans. Wirel Commun 10(4):1323–1333
6. Xiong C (2011) Energy and spectral-efficiency tradeoff in downlink OFDMA networks ieee trans. Wirel Commun 10(11):3874–3886
7. Feng DQ (2012) A survey of energy-efficient wireless communications IEEE communication. Surveys Tuts 15(1):167–178
8. Deng L (2013) Unified energy efficiency and spectral efficiency trade off metric in wireless networks. IEEE Commun Lett 17(1):55–58

Blockchain Based Proof of Existence (PoE) Application for Educational Certificate Verification

K. Ravi Kishore, G. Jyostna, Mahesh U. Patil, P. R. Lakshmi Eswari, and E. Magesh

Abstract In this digital era, most of the documents such as Educational Certificates, Memorandum of Understanding (MoU)s/Agreements, Driving Licenses, Health Records, Employee Service Records, Sale Deeds and Property Registration Records, Birth and Death Certificates, Tax Returns, and so on are being generated and maintained as digital artifacts. But, people are misusing this facility by generating and submitting fake documents. The organization or person, whoever is receiving the documents, is unable to verify the authenticity and integrity of the submitted documents, so verification of these documents is a challenging task. To address this problem, Blockchain technology can be leveraged. Blockchain technology provides document authenticity, ownership, immutability, and tamper-proofness to the data recorded in it. Out of these digital documents, educational certificates are one of the most important entities to be secured as people keep hearing news about fake certificates. C-DAC has considered this as an important use case and implemented a Blockchain based solution for it. The organization, which is issuing the educational certificates, should issue certificates through Blockchain. Once a certificate is recorded on the Blockchain, it is persistent. Anyone can verify it by providing the unique identifier of the certificate such as student roll number, transaction ID, or cryptographic hash of the certificate. This will ease the verification process for the document receivers for verifying the authenticity and integrity of the document. And it also eliminates the role of third party verification agencies, which saves cost and increases the efficiency of verification.

K. Ravi Kishore (✉) · G. Jyostna · M. U. Patil · P. R. Lakshmi Eswari · E. Magesh
Centre for Development of Advanced Computing (C-DAC), Hyderabad, India
e-mail: ravikishorek@cdac.in

G. Jyostna
e-mail: gjyostna@cdac.in

M. U. Patil
e-mail: maheshp@cdac.in

P. R. Lakshmi Eswari
e-mail: prleswari@cdac.in

E. Magesh
e-mail: magesh@cdac.in

S. C. Satapathy et al. (eds.), *Communication Software and Networks*, Lecture Notes in Networks and Systems 134, https://doi.org/10.1007/978-981-15-5397-4_58

1 Background and Motivation

Nowadays, documents such as educational certificates and birth certificates are very important and mandatory for availing many of the services; starting from obtaining the admissions in the school to apply for a visa to travel abroad. Everywhere, people submit the required documents to avail the services but the document receivers are unable to verify the authenticity of submitted documents. Also, few service providers seek the attested copies of documents by gazetted officers/self-attested to confirm the authenticity and in a few cases, service providers are opting for third party verification.

Considering the importance of educational certificates, which are required to get admission for higher education, to apply for a job, and also to apply for a visa to travel abroad, proving the authenticity of educational certificates is of utmost importance. Providers may have their own way of verifying the authenticity of documents. Recruitment companies are opting for third party background verification of the documents which are provided by the candidates. Institutions offering admission for higher education are opting to provide transcripts from the respective universities in addition to the copy of certificates. To apply for a government job, it is required to get documents signed by a gazetted officer. Likewise, there are many places where these educational certificates are submitted and the authenticity and integrity of them are being verified by the document receivers.

Many of the job seekers are creating fake educational certificates and using them for job applications [1]. Verifying the authenticity of the produced certificates is a challenging task. This is creating havoc to the companies, if the document produced by the job seeker is identified as fake one during the background verification. So, it is very important to provide a solution for avoiding the difficulties in verifying the documents. This can be addressed by using one of the emerging technologies, Blockchain Technology, which makes it possible to verify the document authenticity and ownership details.

2 Introduction to Blockchain Technology

Blockchain is one of the emerging technologies, which plays a significant role in enabling trust, transparency, and security. It is the technology evolved by combining the best features of various existing technologies, namely Distributed Ledger, Cryptographic Hash (Integrity Measurement), Public Key Infrastructure, Peer to Peer Network, and Consensus.

The word Blockchain indicates a chain of blocks. A block is the fundamental unit of a Blockchain, which represents a set of transactions that have occurred in a particular time period. In addition, a block also contains metadata information such as total size, the version number of the block format, link to the previous block, root hash of all the transactions, and timestamp. A cryptographic hash of the block is treated as block ID, which is used to uniquely identify any block in the Blockchain. To maintain the chain of blocks, the hash of the previous block is placed in the current block and the hash of the current block will be placed in the next block. If there is a change of any transaction, automatically the hash of transaction changes and so the root hash (part of metadata) of the block gets changed. And so the Hash for the entire block also gets changed. Hash of the block was previously placed as part of the metadata in the next block. Because of the change in the Hash of the current block, the whole chain will break. On the other hand, to make the chain again, hashes of all the subsequent blocks have to be recreated, and which is practically not feasible. This makes Blockchain as an Immutable and Tamper-proof entity.

In case, if the details of any artifact (which is already recorded in Blockchain) is to be updated, rather than modifying the existing transaction, it can be made as another transaction on the same artifact. And from that point of time onwards, updated details will become the current state of the artifact. All the transactions performed on an artifact can be retrieved at any point of time. This provides traceability and audit ability for the operations held on an artifact.

Once an artifact is recorded in Blockchain, anyone can prove the existence of it at a particular instance of time with specific content [2]. While recording the details in Blockchain, cryptographic hash, timestamp, and other metadata information are stored. This is termed as Proof of Existence (PoE).

3 Related Works

Addressing the issue of document authenticity, the authors of [3] proposed not to store the digital asset on the Blockchain. Instead only store the proof that a digital asset has been certified (or signed) by an institution on the Blockchain.

The authors of [4] have proposed a novel approach for certificate verification using Ethereum Blockchain. Organizations should first register with the system and then it can create certificates. Once a certificate is created, it is stored in IPFS and gets a unique hash. Details of the certificate including unique hash are stored in the blockchain and resultant ID will be shared with the students. Students can verify the document using the ID at any time.

The authors of [5] have explained the advantages of Blockchain Technology in all functions like data storage, verifying the data, maintaining the privacy and accountability of all data and transactions performed within the blockchain, reducing the security threats and fraudulent activities, and moreover keeping the network decentralized.

Blocksign [2] application is offering online service for signing and verifying any document, contract, or agreement. It prompts the user to upload the document and then records it in their Blockchain and issues a receipt. Whenever the user wishes to verify the document, it prompts to upload the same document and verifies its existence in the Blockchain and responds accordingly.

Literature for verifying educational certificates is provided by the other researchers or authors [2, 4, 5]. In addition to the Literature, C-DAC has moved one step ahead and implemented an application for issuing the educational certificates and marksheets of it's Post Graduation(PG) Diploma courses. In addition to the available literature, this application provides more features such as i) Sharing certificates directly to the recruiters via email and they can verify them from Blockchain and ii) Even without sharing details from the application, anyone can verify the student certificate details by providing the student roll number.

4 Design and Implementation Details

Considering the importance of certificate verification, at C-DAC, Blockchain based Certificate Verification (BCV) application is developed. This application is intended for Issuing the Certificates and Marksheets of C-DAC's PG Diploma course students through Blockchain. Generally, Blockchain applications are intended to store the verification data but not the complete file/document. For storing the document(s), blockchain employs a traditional storage mechanism/solution. But, it suffers from a single point of failure, high latency in accessing the files and thus, application becomes inefficient. BCV application stores only integrity measurement of the uploaded files in the Blockchain, which will be used for verification purposes and stores the original file in a distributed storage in an encoded format. This Distributed Storage is termed as Proof of Storage (PoS).

The overall architecture of BCV application is shown in Fig. 1. C-DAC issues certificates and marksheets and records metadata of them in Blockchain and stores the original file in a Proof of Storage (PoS) and then shares a copy of recorded documents to students via emails. The detailed procedure of recording documents in Blockchain

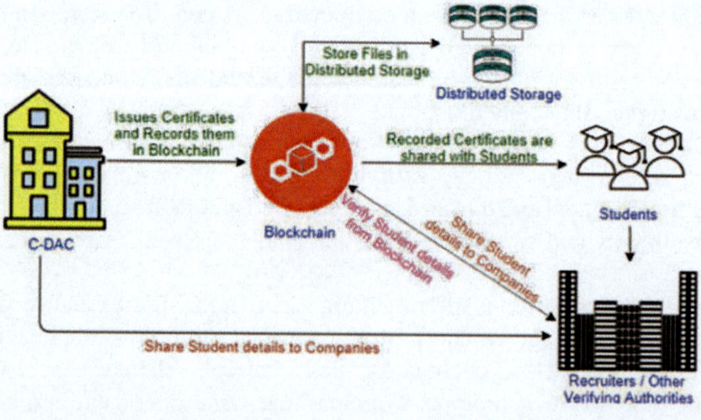

Fig. 1 Overall architecture of BCV app

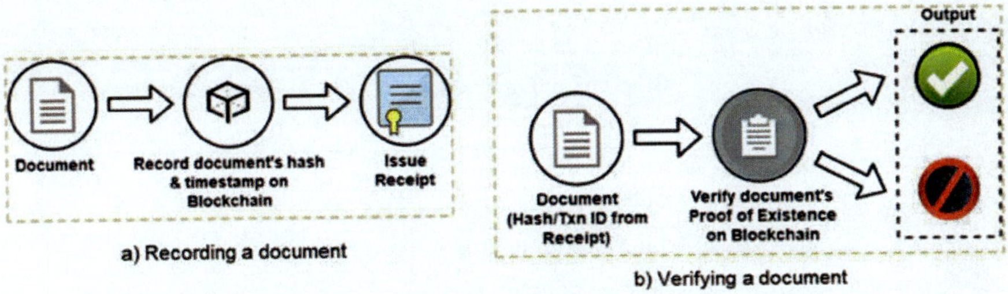

Fig. 2 Recording and verifying documents to/from blockchain

is provided below. Whenever a document is uploaded, the cryptographic hash of it is calculated and is recorded in Blockchain along with the current timestamp and other metadata as shown in Fig. 2a. Once the document is recorded successfully in the Blockchain, it issues a receipt with the corresponding details such as timestamp, transaction ID, and the hash of the document, issuer, and to whom it is issued.

Whenever a document is to be verified from Blockchain, either the corresponding document or hash of the document or transaction ID of the document could be provided to the application and it retrieves the corresponding details from the Blockchain and displays the result to the verifier as shown in Fig. 2b.

BCV Application is implemented in the architecture shown in Fig. 3. BCV Application interfaces with the REST APIs for their operations such as User Authentication, Recording a document on Blockchain, and verifying a document from Blockchain. These REST APIs communicate with the smart contract layer. A smart contract is the one that maintains the business procedures, rules, and penalties as programmed pieces of functions. Smart contract validates the request for

Fig. 3 High level architecture of BCV application

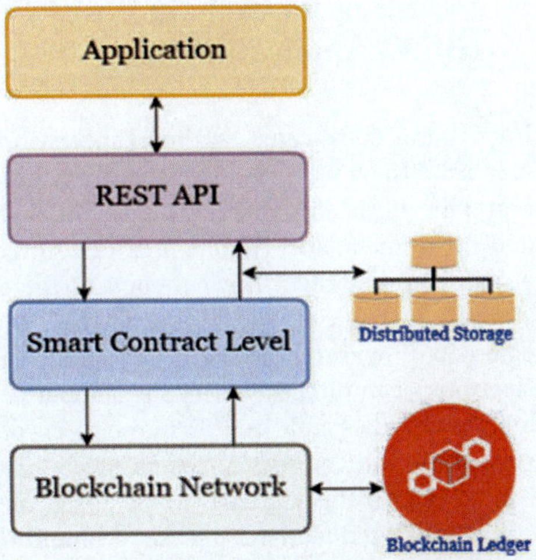

Table 1 Blockchain network configuration

Blockchain network configuration	
Blockchain framework	Hyperledger fabric
Type of blockchain	Permissioned Blockchain
Number of organizations	3
Number of peers per organizations	2
Total number of peer nodes	$3 \times 2 = 6$
Number of orderers	1
Number of channels	1

Record/Verify operations as per its rules and then takes a decision to either allow or deny the operation.

Smart contract communicates with the underlying Blockchain network. A Blockchain network can contain any number of participants (which are known as nodes or peers). Every network participant contains its own copy of Blockchain Ledger.

A transaction proposal can be initiated at this level and it is broadcasted to all the network participants (i.e., nodes or peers). Network participants will validate the transaction proposal against the data present in their own copy of ledgers and endorses the proposal with their concern (validation result). The peer which has initiated the transaction proposal will combine the proposal with all the endorsements received and sends it to the Orderer. Based on the endorsement policy, Orderer will take the action either to include this transaction in the Blockchain Ledger or not. Blockchain Network configuration for this application is provided in Table 1.

5 Details of Blockchain Based Certificate Verification (BCV) Application

For issuing certificates and marksheets through Blockchain, initially, these documents are to be uploaded into the application. BCV application calculates the cryptographic hash and is recorded in Blockchain along with the current timestamp and other metadata. Actual files are stored in Proof of Storage (PoS), which is a distributed storage. Later a copy of the documents is sent to the corresponding students through emails. Either C-DAC or students can share the student certificate details with recruiters or any other verification agencies through Blockchain. Recruiters can directly verify the student details from Blockchain by providing the unique details (such as roll number, cryptographic hash, and transaction ID) of students' certificates. Blockchain gives the complete list of operations held on that roll number if the search is made based on roll number; otherwise, it displays the corresponding document in case of inputting a hash or transaction ID. This helps in assuring the authenticity and integrity of the documents(certificates or marksheets in BCV application).

BCV application provides the option to Sign in for (i) various user roles of C-DAC ACTS (Advanced Computer Training School), (ii) Student Corner, and (iii) Verifier Corner. In addition, the dashboard displays the statistics of the application such as "Total number of transactions, Registered ACTS users, Number of Students, and the Total number of Users including Blockchain level users".

5.1 ACTS Sign

BCV application have five roles for ACTS Sign in as mentioned below

Administrator

Administrator is responsible for creating new users and disable/enable the user accounts. The dashboard of the Administrator is shown in Fig. 4a. Administrator can list the existing users and also can create new users as shown in Figs. 4a and 5.

a) Admin Dashboard b) Training Coordinator Dashboard

Fig. 4 Dashboards of admin and training coordinator

Fig. 5 New user creation by admin

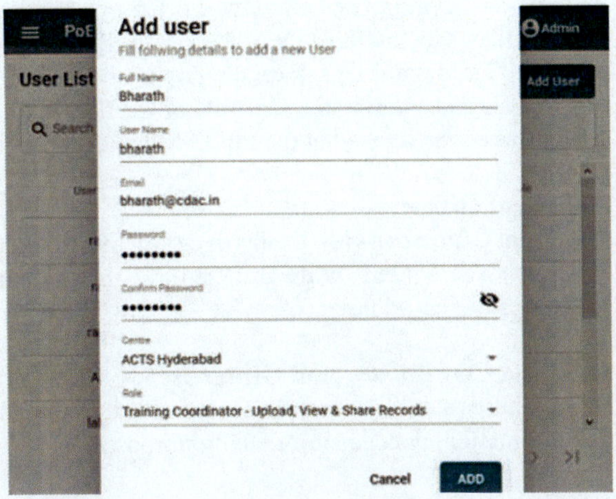

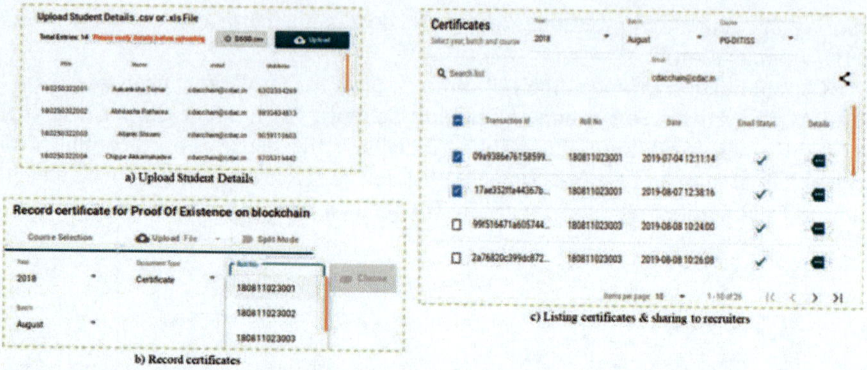

Fig. 6 Training coordinator operations

Training Coordinator

Training Coordinator is responsible for recording student details, their certificates, and marksheets in the Blockchain. The dashboard of Training coordinator is shown in Fig. 4b. By Clicking on Upload Student details, student details will be uploaded and stored in the application as shown in Fig. 6a.

To upload the student certificates/marksheets, the Upload Certificates icon could be used in the dashboard. It prompts for a few details namely year, batch, and course and then displays the list of student roll numbers and gives the browse option to choose the certificate/marksheet file from the system as shown in Fig. 6b. Once all the options are selected and the file is chosen from the file system, click on the upload button to record them in Blockchain. Only the cryptographic hash of the uploaded document will be recorded in Blockchain and the original file content will be stored in a Distributed storage in an encoded format.

Training Coordinator can view the list of uploaded certificates and even share the certificates to any third party for verification through email as shown in Fig. 6c. Training Coordinator can share the documents to verifiers by selecting them using checkboxes and entering the email ID of verifiers in the Email field and then clicking on the share icon located at the top right-hand side of the page.

Placement Officer

Placement Officer can view the list of certificates uploaded by the training coordinator and he can share them with recruiters through email. It is a similar operation shown in the training coordinator role.

Technical Coordinator and Center Head

Both of these roles can view the list of certificates uploaded by the training coordinator for information and decision-making purposes.

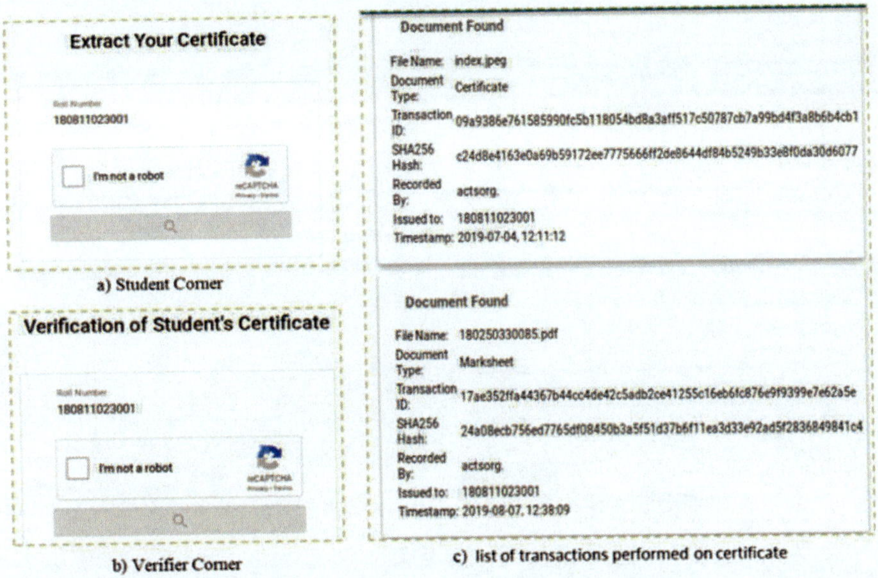

Fig. 7 Student and verifier corner

5.2 *Student Corner and Verifier Corner*

Students and Verifiers can get a copy of the certificate from the Application, which is stored in a distributed storage from this option by providing the Roll number as shown in Fig. 7a, b. Once the student roll number is provided, it will extract all the transactions (uploads and updates) performed on the corresponding certificate or marksheet as shown in Fig. 7c.

5.3 *Verification Through the Email Shared by Institute*

The documents shared by the roles Training coordinator and/or placement officer will send the corresponding document details to third party agencies through email as shown in Fig. 8a.

Email lists the shared roll numbers and displays Roll No, Document Type, and verification Link in the email content. By clicking on the link, it redirects to the application and displays the details as shown in Fig. 8b, c. From the verification result, the verifier can download the verification receipt as well as the copy of the uploaded document.

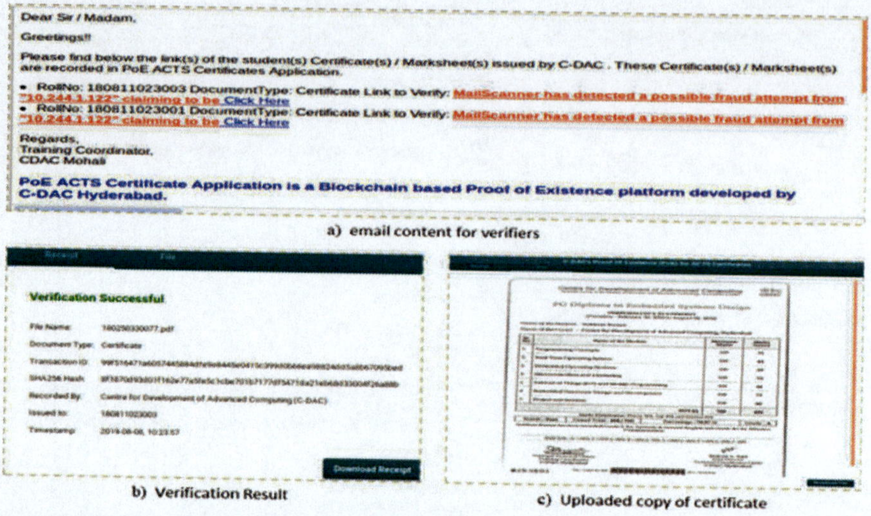

Fig. 8 Verification through the email shared by institute

6 Storage and Performance Metrics

As an experimentation, about 200+ students' certificates and marksheets are recorded in BCV application. While uploading the documents, below statistics are noted

6.1 Blockchain Ledger and Storage Statistics

Table 2 shows the statistics when various numbers of files are recorded on Blockchain.

It includes Blockchain Ledger size and Distributed Storage size when various files (namely 5, 10, and 20 files) are uploaded and recorded on Blockchain. The statistics provided in Table 2 are depicted as a comparison chart in Fig. 9.

Table 2 Blockchain ledger and storage statistics

No. of documents	Size of files (KB)	Blockchain ledger size (KB)	Distributed storage size (KB)
5	523	1.54	2362.373
10	481.2	2.68	3169.543
20	1016.8	8.05	4795.571

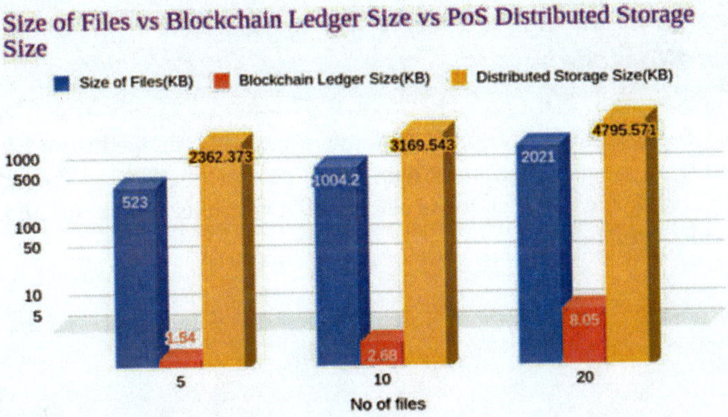

Fig. 9 Blockchain ledger and storage statistics

6.2 *Time Consumed for Recording Documents in Blockchain*

Experimentation is performed for recording various number of documents in Blockchain and the results are shown in the Table 3. And Fig. 10 compares Size of files and time consumed to record them in Blockchain.

Table 3 Time consumed for recording documents in the blockchain

No of files	Time consumed (s)	Size of files (KB)
1	3.64	75.1
5	19.9	505.5
10	40.82	1012.9
20	81.38	1997.9

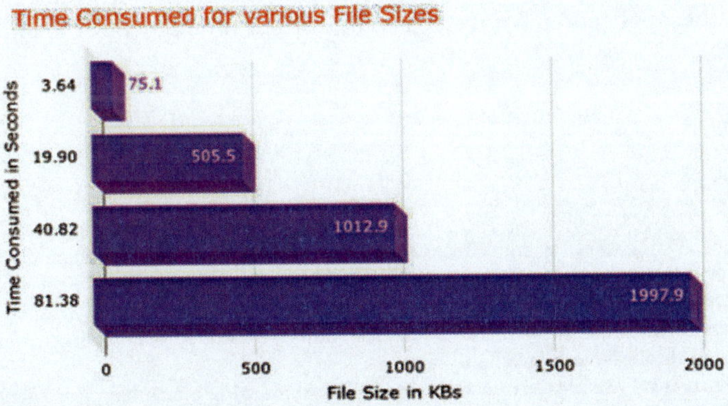

Fig. 10 Time consumed for recording in blockchain

6.3 Time Consumed for Document Verification from Blockchain

Whenever a document verification is required, it has to be fetched from Blockchain and verified from the provided input data. Time taken for retrieving the details of a single document during Verification is calculated for various documents and Average (Mean) is calculated below.

$$\mu = \frac{\sum_{i=1}^{N} X_i}{N} \tag{1}$$

Numerator in Eq. (1) represents the sum of all values (i.e., time consumed for retrieving document details). Denominator in Eq. (1) represents the number of individual times

$$\mu = \frac{\sum(116.63, 112.33, 103.28, 105.93, 105.69, 102.12, 100.61, 105.012, 103.47, 102.43)}{10},$$

$$\mu = 105.75\text{ms}$$

7 Conclusion

Blockchain based Certificate Verification (BCV) is the application developed for issuing and verifying the student certificates with the help of Blockchain technology. Tamper-Proofness, Authenticity, and Traceability are the highlighted features of Blockchain. This application helps in assuring the genuinity and authenticity of certificates to the verifiers. The underlying Blockchain technology stack is not only limited to this application, but it can also be utilized for many other applications in the future.

References

1. Document Verification Using Blockchain, https://medium.com/blockchainexpert-blog/document-verification-using-blockchain-a02c059ed97b
2. The best way to sign documents, https://blocksign.com/
3. Document Certification Through the Blockchain, https://www.martinstellnberger.co/document-certification-through-the-blockchain
4. Nitin Kumavat DJ (2019) Certificate verification system using blockchain. Int J Res Appl Sci Eng Technol (IJRASET) 7(IV):53–57
5. Blockchain Timestamp and Document Verification, https://www.blockchainexpert.uk/blog/blockchain-timestamp-and-document-verification

An Optimized Speed Adaptive Beacon Broadcast Approach for Information Dissemination in Vehicular Ad hoc Networks

Sandeep Kad and Vijay Kumar Banga

Abstract Vehicular Ad hoc Networks (VANETs) generally comprises vehicular nodes that communicate with one another mostly without utilizing any infrastructure. Multiple applications of VANETs needs to deliver information to the regions of interest and broadcasting is an efficacious communication paragon to rely upon. A simple broadcast suffers from redundancy, high resource utilization, and a number of collisions resulting in the broadcast storm problem. The current network situation depending upon the flow of traffic plays a decisive role where in dense scenarios the number of rebroadcasts should be minimum to avoid redundancy, contention and collisions and in a sparse environment, the number of rebroadcast should be high to improve the reachability of information. We propose an infrastructure-less Optimized Speed Adaptive Beacon Broadcast (OSABB) approach that calculates the information forwarding probability on the basis of the number of vehicular nodes in the vicinity. Simulations are done for OSABB and results are compared with two state-of-the-art existing approaches.

1 Introduction

Vehicular Ad hoc Networks, a prime component in an intelligent transport system aims to provide a pleasant experience to the travelers. It focusses to contribute towards the safety of human beings by supporting applications that assist drivers in the process of decision-making while they are behind the wheels. The vehicular nodes in VANETs communicate with each other directly or via fixed infrastructure. Due to the possibility of inter-vehicle communication, vehicular nodes are able to share important information like a warning to the drivers in case of any untoward event such as an accident is noticed, lane change alerts, emergency braking lights,

S. Kad (✉)
Inder Kumar Gujral Punjab Technical University, Kapurthala, India
e-mail: kadsandeep@yahoo.com

V. K. Banga
Amritsar College of Engineering and Technology, Amritsar, India
e-mail: vijaykumar.banga@gmail.com

S. C. Satapathy et al. (eds.), *Communication Software and Networks*, Lecture Notes in Networks and Systems 134, https://doi.org/10.1007/978-981-15-5397-4_59

out of sight obstacles and curve speed warnings on the road [1] etc. The affected vehicular nodes can disseminate a message to vehicles approaching the affected site. The information can be propagated well in time over multiple vehicular nodes, so that drivers can take alternative routes and can save both time and fuel. Academia and the automobile industry are taking keen interest to provide a secure platform to value human lives. Countries like Japan and Sweden are working aggressively to become a "zero traffic fatality" nation by 2020 [2]. In VANETs, vehicular nodes establish a pure ad hoc communication with other vehicular nodes or roadside infrastructure. Disseminating information in VANETs is a cumbersome task due to the limited availability of wireless radio communication range. The information to be disseminated is usually broadcasted which is further rebroadcasted so that it can be delivered to vehicles at far-off places. The rebroadcasting in VANET leads to the broadcast storm problem [3] and impacts scalability as well. This affects the delivery of information to the target vehicular nodes in terms of reachability and delay. Apart from this, many of the existing approaches suffers from network partition and fragmentation issues [4].

In this paper, we propose an optimized speed adaptive beacon broadcast (OSABB) approach in which a probabilistic forwarding mechanism is used which dynamically decides rebroadcasting of information on the basis current vehicular density in the vicinity. The remainder section of this paper is organized as follows. Section 2 reveals a summary of the related work. Section 3 discusses the assumptions and proposed approach followed by its implementation and evaluation through simulations in Sect. 4. At last Sect. 5 concludes the paper, highlighting the outcomes and future directions.

2 Related Work

Several information dissemination approaches have been proposed by researchers considering the distance between vehicles, the density of vehicles, and the lifetime between vehicles. We are discussing a few of them in this section. CAREFOR [5] is a distributed approach where a vehicle upon receiving a message forward it keeping in view the predefined probability which is calculated on the basis of current vehicle density in the vicinity, the distance between the sender and probable forwarding nodes and transmission range of the subsequent hop. NDDP [6] is a forwarding relay selection approach for multi-hop dissemination. The selection of potential forwarding vehicles is done considering the distance between the sender vehicle and potential relay vehicle, the relative velocity of the relay vehicle, and signal interference between the vehicles. AddP [7] is a beacon adaptive approach, where the beaconing process is controlled depending upon the density of vehicles in the vicinity. Using coding theory, warning messages are aggregated to reduce redundant broadcasts. In ADM [8] each vehicle node decides of its own, whether to forward the information or not. This decision is made keeping in consideration the priority level of the message to be disseminated and the network density. The approach can reduce latency but redundant messages are still high. In DPS [9], partition sizes are computed in such a way that every partition must contain one vehicle which leads to shorter 1-hop delays. To select the forwarding relay vehicle with lower transmission delay and packet collisions, this approach uses density of vehicles in the vicinity as well as inter-vehicle distance. Beacons are transmitted to share this information. The drawback of this approach is that no beacon congestion control mechanism is used. AMD [10] is an infrastructure-less multi-directional timeslot-based scheme that supports information dissemination for both urban as well as the highway environment. DRIVE [4] disseminates information in the region of concern, where the vehicle inside the sweet spot is preferred to disseminate the information to larger number vehicles in the vicinity. The approach aims to alleviate broadcast storm problem maintaining low overheads and delays.

3 Proposed Approach

In this section, we will discuss the proposed Optimized Speed Adaptive Beacon Broadcast (OSABB) approach, but before that, we put forward some of the assumptions made while implementing this work. We have considered a multilane road with intersections, where vehicles are moving in either direction. These vehicles are equipped with GPS and loaded with road maps providing information about intersections and layout of roads. Each vehicle is fitted with wireless communication devices that posses certain radio communication range within which the vehicle can communicate with other vehicles directly. The radio signals are impacted by the presence of skyscrapers, trees, and other constructions that are there on the road side which can obstruct the radio communication range of vehicles. Vehicles broadcast periodic beacons disseminating information about their vehicle id, position, speed an direction of movement and list of ID's of messages received. Vehicles upon receiving the beacons, stores this information with themselves. When a beacon is received from the same vehicle again, the receiving vehicles updates the already stored entry. As already presented in our previous work, i.e., speed adaptive beacon broadcast (SABB) [11], the beaconing process is controlled keeping in view the traffic flow on the road and is briefly discussed below. The relationship between the speed of the vehicle (S), the density of vehicles (V_{den}) and traffic flow (T_{flow}) on road in [12] is considered and speed ratio (S_r) is defined in Eq. 1 as

$$S_r = \frac{S_c}{S_f} \tag{1}$$

where S_c indicates current speed of vehicles and S_f is the maximum speed that is permitted on the road for vehicles as per norms. A linear relationship between speed of vehicles and vehicle density (V_{den} is current density of traffic and V_{jam} is a maximum limit of the number of vehicles) [12] on road is given in Eq. 2 as

$$\frac{S_\mathrm{c}}{S_\mathrm{f}} = 1 - \frac{V_\mathrm{den}}{V_\mathrm{jam}} \tag{2}$$

The periodic beaconing is controlled depending upon the speed of the vehicle. The period of the link between vehicles is dependent upon the relative speed between the vehicles and in case a vehicle moves at higher speed, it will observe disconnection with its neighbors within a short period. So the beaconing interval of such a vehicle should be lower as compared to vehicles that are moving at relatively stable speed along with its neighboring vehicles. The beacon interval (B_intr) in Eq. 3 is given as

$$B_\mathrm{intr} = \propto + \left(\left(1 - \frac{S_\mathrm{c}}{S_\mathrm{f}} \right) \times 10 \right) + \mathrm{Rand}\,[0, 0.002] \tag{3}$$

where α is the time taken between two successive beacons by a vehicle. A small random time interval between 0 and 2 ms is added to break synchronization. The beacon interval will be lesser for vehicles which are moving at high speed whereas it will be lower in case the movement is relatively the same as that of its peers in the vicinity. To disseminate the information, the source vehicle broadcast's the information and this information is received by vehicles that are in the radio transmission range of the source vehicle. It is to be noted that all 1-hop neighbors might not be able to get this broadcast due to the presence of the obstacles on the roadside as radio signals get obstructed. As there is a need to disseminate the information beyond 1-hop neighbors, so multi-hop communication is to be carried out, where an optimal forwarding relay vehicle is to be identified so that the information can be delivered to the vehicles beyond the 1-hop radio range. The SABB approach selects the forwarding relay vehicle by taking into consideration the parameters like the distance of the relay vehicle from the sender, the density of vehicles in the vicinity, and the current position of the vehicle. When the vehicles receive information to be broadcasted further, all the recipients calculate their waiting period (WP) as shown in Eq. 4 so that they can disseminate this information further. These vehicles initiate their waiting timer. The vehicle present at the intersection is preferred in comparison to other vehicles because this vehicle will be able to disseminate the message in multiple directions as the radio transmissions will be less affected by obstacles.

$$\mathrm{WP} = \begin{cases} \beta\left(1 - \frac{\mathrm{dist}_i}{\mathrm{TR}}\right) + \frac{\beta}{2}\left(1 - \frac{V_\mathrm{den}}{V_\mathrm{jam}}\right) + \mathrm{Rand}\,[0, 0.002]\ \text{if vehicle is at the intersection} \\ \left(\beta + (1-\beta)\left(1 - \frac{\mathrm{dist}_i}{\mathrm{TR}}\right)\right) + \frac{1}{2}\left((\beta) + (1-\beta)\left(1 - \frac{V_\mathrm{den}}{V_\mathrm{jam}}\right)\right) + \mathrm{Rand}\,[0, 0.002]\text{otherwise} \end{cases} \tag{4}$$

where β indicates the priority assigned to different components involved in evaluating WP, dist_i specifies the distance between the vehicle i and sender vehicle, TR indicates the transmission range. A vehicle before rebroadcasting a message checks its information table to have an idea about vehicle density in its periphery. This is done, when in sparse scenarios, the number of neighbors is less, the vehicle can initiate request beacon to gather information about the other vehicles availability in the vicinity. In OSABB, during the waiting period, if the vehicular nodes receives the same message again, they do not suppress the rebroadcast, instead the probability of rebroadcasting the information is evaluated. When the information is received by the vehicle for the first time, the counter (ctr) for this message is set to 1 by vehicle. When the same message is received again by the vehicle during its waiting period then the counter is further incremented by 1. This counting process continues till either the counter reaches the threshold or the waiting time has elapsed and the vehicle can transmit the message. If ctr crosses the threshold, the message is discarded by the vehicle and not rebroadcasted as shown in Eq. 5. An exponential function is proposed for evaluating the forwarding probability of the message. The evaluation of an optimal value of forwarding probability is a tedious task as the higher value of probability would lead redundant broadcast, whereas a low value will impact the reachability of the message. In a scenario that is dense, the resulting forwarding probability should be lower. In a sparse environment, the number of vehicle nodes in the vicinity is low, therefore to have better reachability, high forwarding probability is expected. It is observed in [3] that multiple rebroadcast (k-times) does not yield the desired results for ex., when $k = 2$, the additional information coverage is almost 61% and when $k = 4$, this coverage falls below 5%. Many of the counter-based approaches have fixed the threshold value of counter like in [13] the threshold value of 6 is considered optimal. In [14] threshold is varied depending upon the number of vehicles where in a sparse environment the threshold varies between 2 and 5, and at higher traffic densities the threshold is 2. In our approach we are proposing a dynamic threshold value that varies with current traffic density. The threshold is set to $6S_r$. Here S_r is evaluated as shown in Eq. 2. The value 6 is taken from the approach mentioned in [13]. The evaluation of forwarding probability (P_f) depending upon ctr is given in Eq. 5 below:

$$\mathrm{WP} = \begin{cases} 1 & \text{if ctr} = 1 \\ \left|1 - e^{\frac{2}{ctr^2}}\right| & \text{if } 2 \leq \text{ctr} \leq \text{threshold} \\ 0 & \text{if ctr} \geq \text{threshold} \end{cases} \tag{5}$$

The algorithm for evaluating the forwarding probability is shown in Fig. 1.

4 Performance and Evaluation

In this section, the performance of the OSABB approach is evaluated using OMNeT++ [15] and SUMO [16]. The VANET environment is modeled using Veins framework and OSABB is compared to DRIVE [4] and AMD [10] approaches. A scenario of 2 km × 2 km of Manhattan is selected using an open street map [17]. The vehicle density ranging from 20 and 100 vehicles per sq. km is considered. The vehicles speed is varied depending upon traffic density using the Kraus mobility model. The performance of the OSABB approach is evaluated on the basis of metrics like

Input: counter *ctr* keeps track of number of times m is received by vehicle *i* during its waiting period (*WP*), threshold is the limit which if ctr attains then message is not rebroadcasted
Output: forwarding probability for rebroadcasting a message
start
set *ctr = 0;*
while *(WP > 0)* do
increment the ctr by 1 if vehicle *i* receives message m
endwhile
if *ctr < threshold*
switch (*ctr*)
case *ctr=1*, set $P_f = 1$
case *2<=ctr<=threshold*, set $P_f = \left|1 - e^{\frac{2}{ctr^2}}\right|$
else
case *ctr > threshold*, set $P_f = 0$
endif
end

Fig. 1 Algorithm to calculate probability of message forwarding

message delivery ratio (MDR), delay, number of transmissions, and collisions. The simulation parameters are listed in Table 1.

As evident in Fig. 2 initially when the number of vehicles on the road is less, MDR is quite low. However, with the rise in the number of vehicles on the road, the delivery ratio is improving. OSABB is performing better in comparison to other approaches at all densities, the reason for this can be attributed to the selection of forwarding vehicle depending upon it position, i.e., in case a suitable relay node is available at the intersection, it is preferred for rebroadcasting. The performance of DRIVE might have got affected because of the absence of the relay node in the sweet spot when the number of vehicles is less. However, there is an improvement when the vehicle density rises. Figure 3 depicts the delay in delivery of information, where it can be noted that in all the three approaches, when the traffic density is low, the delay is higher. As DRIVE relies on the availability of vehicles in the sweet

Table 1 Simulation parameters

Parameter	Value
Frequency band	5.9 GHz
Transmission range	200 m
Transmission power	0.98 mW
Beacon size	32 bits
Data message size	2048 bytes
α	150 ms
β	0.5
Propagation model	Nakagami-m
Bit rate	6 Mbps
Simulation time	900 s
Number of runs	30

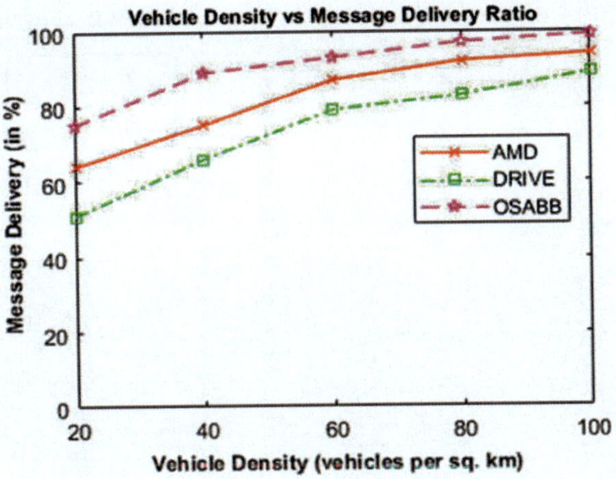

Fig. 2 Message delivery ratio

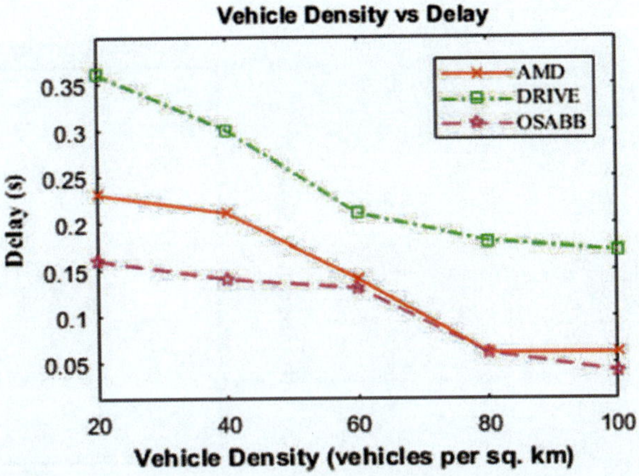

Fig. 3 Delay

spot, the absence of a vehicle impacts the information delivery, which is visible while evaluating message delivery as well. The performance of OSABB and AMD is comparable, but the delay is still on the higher side which might be because of the obstacles present in the urban scenarios. At some of the traffic densities, OSABB performs better than AMD, the credit for this can be attributed to the selection of vehicle present at the intersection as relay vehicle. It can be observed in Fig. 4 that the number of transmissions in DRIVE and AMD is quite high, whereas OSABB performs better significantly. The better selection of relay vehicles as well as the probabilistic forwarding mechanism has contributed to a great extent. At high traffic flows, the number of transmissions observed is almost half in comparison to DRIVE that itself validates the superiority of OSABB approach. Figure 5 depicts the number of collisions where all the approaches are comparable to one another. OSABB is impacted at certain densities because of the result of probability of forwarding. This,

Fig. 4 Number of transmissions

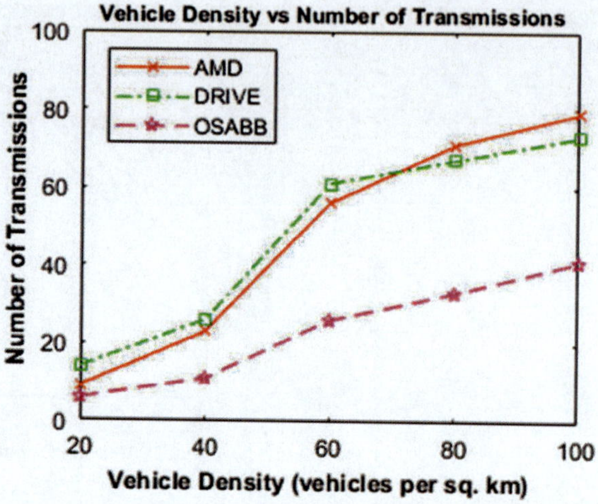

Fig. 5 Number of collisions

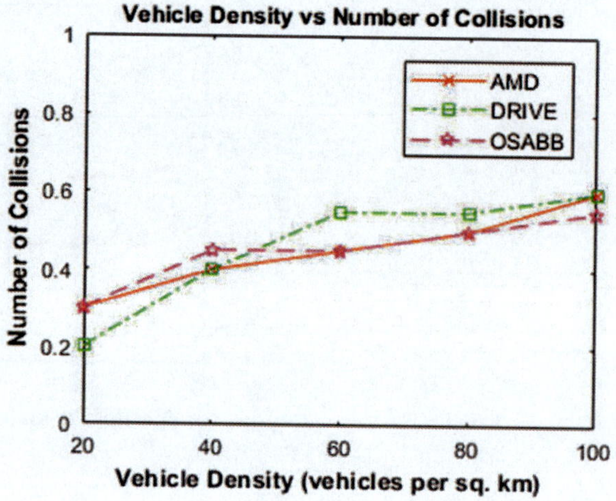

however, is balanced to some extent because of controlled beaconing resulting in reduced load, thus controlling congestion. At higher densities, OSABB performs better than the other two.

5 Conclusion

In this paper, we have presented the OSABB approach that mitigates the broadcast problem in an efficient manner. In this approach, the forwarding probability is evaluated using an exponential function while keeping in view the vehicle density in the vicinity. OSABB is evaluated with metrics like message delivery ratio, number of transmissions, number of collisions and delay. To validate the proposed approach, the outcomes are compared to DRIVE and AMD. It has been observed that the proposed approach performs better in the majority of the vehicle density scenarios for all the metrics. At higher density OSABB almost attains a 100% delivery ratio whereas at lower densities the delivery is at least 20% improved than AMD and DRIVE. The number of transmissions is 50% less in the proposed approach than DRIVE at high density. In terms of delay and number of collisions, the overall performance of OSABB is better than the other two. In the future, we aim to have a more complex road layout depicting a realistic environment to evaluate this approach at varying traffic speeds as well.

Acknowledgments The authors gratefully acknowledge IKG Punjab Technical University, Kapurthala, India and Amritsar College of Engineering and Technology, Amritsar, India, to carry out this research work.

References

1. Li F, Wang Y (2007) Routing in vehicular ad hoc networks: a survey. IEEE Veh Technol Mag 2:12–22
2. Zekri A, Jia W (2018) Heterogeneous vehicular communications: a comprehensive study. Ad Hoc Netw 75–76:52–79
3. Ni S-Y, Tseng Y-C, Chen Y-S, Sheu J-P (1999) The broadcast storm problem in a mobile ad hoc network. In: Proceedings of the 5th international conference on Mobile computing and networking. ACM Press, Washington, pp 151–162
4. Villas LA, Boukerche A, Maia G, Pazzi RW, Loureiro AAF (2014) DRIVE: an efficient and robust data dissemination protocol for highway and urban vehicular ad hoc networks. Comput Netw 75:381–394
5. Akamatsu R, Suzuki M, Okamoto T, Hara K, Shigeno H (2014) Adaptive delay-based geocast protocol for data dissemination in urban VANET. In: 7th international conference on mobile computing and ubiquitous networking. pp 141–146
6. Ghorai C, Banerjee I (2018) A robust forwarding node selection mechanism for efficient communication in urban VANETs. Veh Commun 14:109–121
7. Oliveira R, Montez C, Boukerche A, Wangham MS (2017) Reliable data dissemination protocol for VANET taffic safety applications. Ad Hoc Netw 63:30–44
8. Abdou W, Darties B, Mbarek N (2015) Priority levels based multi-hop broadcasting method for vehicular ad hoc networks. Ann Telecommun 70:359–368
9. Rayeni MS, Hafid A, Sahu PK (2015) Dynamic spatial partition density-based emergency message dissemination in VANETs. Veh Commun 2:208–222
10. Schwartz RS, Scholten H, Havinga P (2013) A scalable data dissemination protocol for both highway and urban vehicular environments. EURASIP J Wirel Commun Networking
11. Kad S, Banga VK (2019) Speed adaptive beacon broadcast for information dissemination in vehicular ad hoc networks. Int J Innovative Technol Explor Eng 8:11
12. Chowdhury MA, Sadek A (2010) Fundamentals of intelligent transportation systems planning. Artech House (2010)
13. Haas ZJ, Halpern JY (2006) Li Li: Gossip-based ad hoc routing. IEEE/ACM Trans Networking 14:479–491

14. Tseng Y-C (2003) Adaptive approaches to relieving broadcast storms in a wireless multihop mobile ad hoc network. IEEE Trans Comput 52:13
15. OMNe++ Discrete event simulator, https://omnetpp.org
16. Krajzewicz D, Erdmann J, Behrisch M, Bieker L (2012) Recent development and applications of SUMO—simulation of urban mobility, 128–138
17. OpenStreetMap, https://www.openstreetmap.org

Wavelet Packet Transform-Based Image Classification for Computer-Aided Glaucoma Diagnosis Using Naïve Bayes Classifier

Anisha Rebinth and S. Mohan Kumar

Abstract In this study, optic nerve head image classification for glaucoma diagnosis using Wavelet Packet Transform (WPT) and Naïve Bayes Classifier (NBC) is presented. Initially, the input optic nerve head images are decomposed by WPT which produces high and low-frequency sub-bands. From these sub-bands, entropy features like Shannon and sure entropy are extracted. These extracted features are classified by NBC. The performance is evaluated using classification accuracy, specificity and sensitivity. Results show that better classification accuracy of 94% and also, the sensitivity of 89% and specificity of 99% are obtained by using WPT-based entropy features and NBC for glaucomatous image classification system.

1 Introduction

Glaucoma is an eye disorder due to irregular high pressure in the eye. It causes vision loss if it is not treated at the earliest. Discrete Orthogonal Stockwell Transform (DOST) based glaucoma image classification is described in [1]. The input optic nerve head images are decomposed by DOST. The DOST coefficients are classified by random forest classifier. Glaucoma detection using optic nerve head image is described in [2]. Features are extracted by wavelet transform. The segmentation is made by hill-climbing algorithm and k-means clustering. Glaucoma detection based on texture feature extraction is described in [3]. At first, features are extracted by Markov random field and Gray Level Co-occurrence Matrix (GLCM). Support Vector Machine (SVM) is used for classification. Detection of optic cup and disk for glaucoma diagnosis is discussed in [4]. The features like cup to disk ratio, cup to disk area ratio, rim to disk area ratio and horizontal to vertical cup to disk ratio are extracted. The classification is made by K-Nearest Neighbor (KNN).

A. Rebinth (✉) · S. M. Kumar
New Horizon College of Engineering, Bangalore, India
e-mail: anisha.rebinth@gmail.com

S. M. Kumar
e-mail: drsmohankumar@gmail.com

S. C. Satapathy et al. (eds.), *Communication Software and Networks*, Lecture Notes in Networks and Systems 134, https://doi.org/10.1007/978-981-15-5397-4_60

The detection of glaucoma using optic nerve head images is described in [5]. The combination of distance transformation and thresholding is used to locate the optic disk center. Principal Component Analysis (PCA) is used to obtain the eigenvector. NBC is used for classification. Glaucoma diagnosis using SVM is presented in [6]. The input optic nerve head images are preprocessed by Gaussian filter. The PCA is used as a feature extractor. The classification is made by SVM classifier. Glaucoma diagnosis from optic nerve head images using Empirical Wavelet Transform (EWT) is discussed in [7]. Initially, the optic nerve head images are decomposed by EWT and correntropy features are extracted. The features are then selected by t-value feature selection algorithm. The classification is made by least square SVM. Glaucoma image classification based on wavelet energy features is discussed in [8]. The features are extracted by different wavelet tranform (daubechies, Symlet and biorthogonal) filters. The classification is made by NBC, SVM, random forest classifier.

The prediction of glaucoma and diabetic retinopathy through data mining techniques and retinal image analysis is presented in [9]. The input optic nerve head images are preprocessed by average filtering, then histogram equalization is applied. GLCM is used for feature extraction. Random tree and decision tree algorithm are used as classifiers. Glaucoma classification based on texture features and neural networks is described in [10]. At first, texture features are extracted and they are reduced by PCA algorithm. The neural network is used for the classification. Glaucoma diagnosis using optic nerve head images is described in [11]. The region of interest is extracted from input image. The features are extracted by texture and first-order statistical approach. The efficient features are selected and are classified by KNN, SVM, NBC and Linear Discriminant Analysis (LDA) algorithms. Ocular disorders based on invariant moment analysis and fractal in optic nerve head image is described in [12]. The features are extracted by Hu's invariant moments and fractal dimension. The LDA is used for classification.

In this study, glaucoma image classification using optic nerve head images based on WPT and NBC is described. The main contribution is the development of computer-aided gluacoma diagnosis system using multi-resolution analysis and probabilistics classification. It uses WPT-based entropy features and NBC classifier for early diagnosis of glaucoma. The organization of paper is as follows: The classification system for glaucoma diagnosis is discussed in Sect. 2. The performance analysis is made in Sect. 3 with Confusion Matrix (CM) and Receiver Operating Characteristics (ROC). The last section concludes the system for glaucoma diagnosis.

2 Glaucomatous Image Classification System

In this section, optic nerve head image classification system using WPT and NBC is presented for the diagnosis of glaucoma. Figure 1 shows the workflow of the classification system.

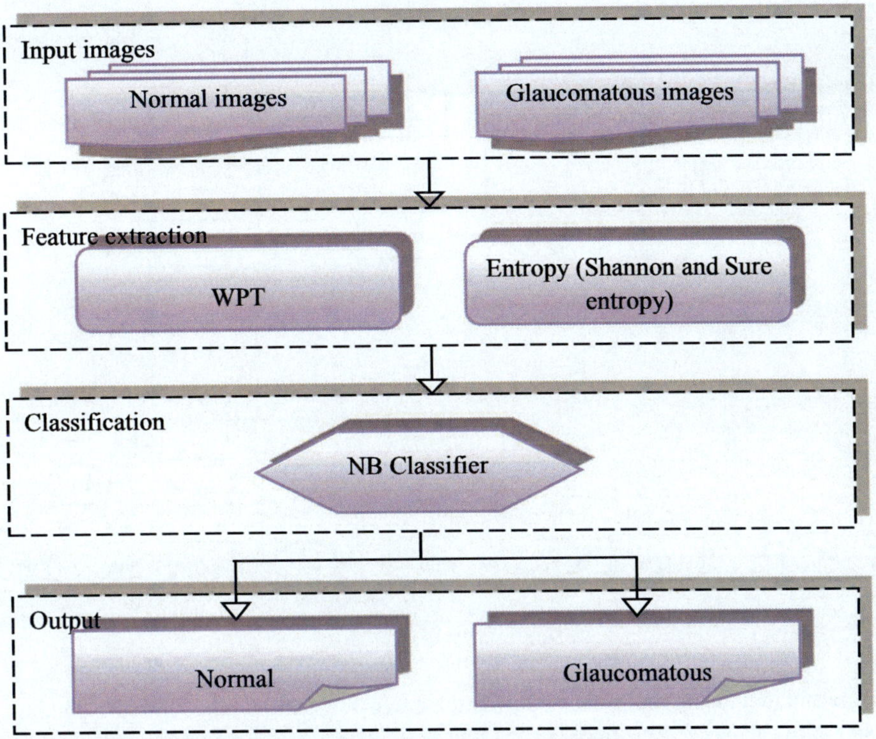

Fig. 1 Block diagram of glaucoma classification system

The first stage is the feature extraction stage where WPT is used to extract dominant features and in the second stage NBC is used for classification where the given optic nerve head image is classified as normal or glaucomatous.

2.1 WPT Decomposition

WPT is also known as optimal sub-band tree structuring. It passes more discrete-time signals comparing to wavelet transform. WPT is used in other fields like fingerprint recognition [13] and genetic watermarking [14]. WPT creates full binary tree by both approximation coefficients and detailed coefficients. Figure 2 shows the WPT decomposition for three levels.

In Fig. 2, $s[m]$ represents low-frequency components and $t[m]$ represents high-frequency components. For m levels of WPT decomposition produces 2 m different sets of coefficients opposed to wavelet transform. The total number of coefficients is

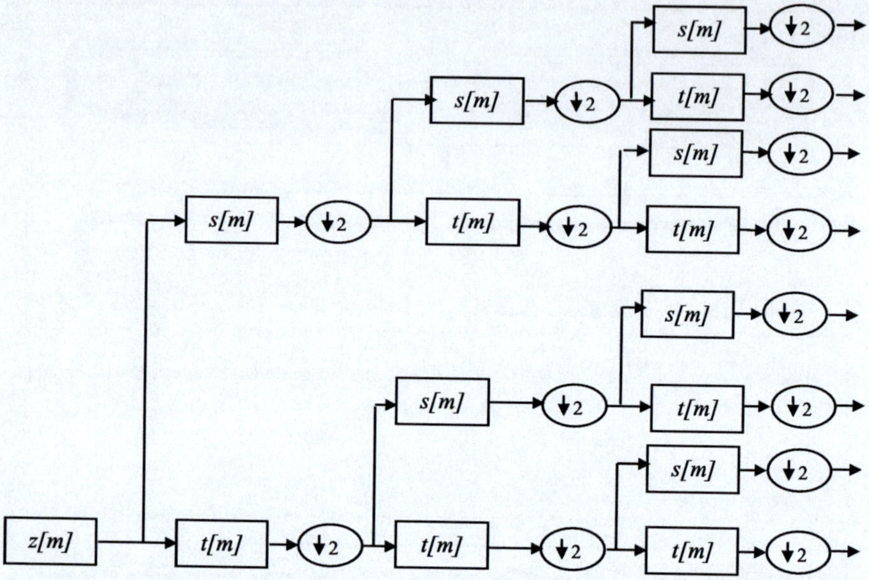

Fig. 2 WPT decomposition over three levels

same and there is no redundancy during the down-sampling process in WPT decomposition. The WPT decomposes the input optic nerve head images and produce the high- and low-frequency components. From the coefficients, features are extracted.

2.2 *Feature Extraction by WPT*

From WPT sub-bands, the entropy features like Shannon and sure entropy are extracted. Shannon entropy is also used in heart sound classification [15] which is given by

$$R(S) = -\sum_{k=0}^{J-1} R_k^2 \log\left(R_k^2\right) \tag{1}$$

where R_k is the probability of a given WPT coefficients of sub-band k with $\log(0) = 0$. The sure entropy is defined by

$$|R_k| \leq \omega \rightarrow \text{ent(sure)} = \sum_k \min\left(R_k^2, \omega^2\right) \tag{2}$$

where ω is a positive threshold value computed based on Steins unbiased risk estimate. The sure entropy is also used in image enhancement [16].

2.3 Classification Stage by NBC

NBCs are a group of probabilistic classifiers based on Bayes theorem and commonly used as a machine learning algorithm. It is used in many recognition algorithms such as biometric recognition [17] and facial expression analysis [18]. NBC needs a number of linear parameters and variables for learning stage. In NBC, the value of a particular feature is independent of any other feature in the given class. NBC performs better for high-dimensional feature space, i.e., which has large amounts of data points because searching the parameters for probability functions can be done quickly. The Bayes rule is defined by

$$g\left({}^{h_l}/_{y}\right) = \frac{g(h_l)g(y/h_l)}{g(y)} \tag{3}$$

3 Performance Analysis

In this section, the performance of WPT and NBC-based system for glaucoma diagnosis is analyzed with the help of 200 optic nerve head images (100 normal and 100 abnormal) [19]. The images have only the optic nerve head with a resolution of 256 × 256 pixels. Figure 3 shows the normal and glaucoma images from database.

To extract features from optic nerve head images, WPT is applied to input normal and abnormal images which produce different sub-bands. Then, entropy features like Shannon and sure entropy features are extracted from sub-band coefficients. The classification is made by NBC using the extracted entropy features. To evaluate the classification system, the performance measure such as accuracy, sensitivity and specificity are computed. Figure 4 shows the CMs of WPT and NBC-based system for glaucoma diagnosis. The accuracy, sensitivity and specificity are defined by

$$\text{Accuracy} = \frac{\text{True Positive} + \text{True Negative}}{\text{True Positive} + \text{True Negative} + \text{False Positive} + \text{False Negative}} \tag{4}$$

$$\text{Sensitivity} = \frac{\text{True Positive}}{\text{True Positive} + \text{False Negative}} \tag{5}$$

$$\text{Specificity} = \frac{\text{True Negative}}{\text{True Negative} + \text{False Positive}} \tag{6}$$

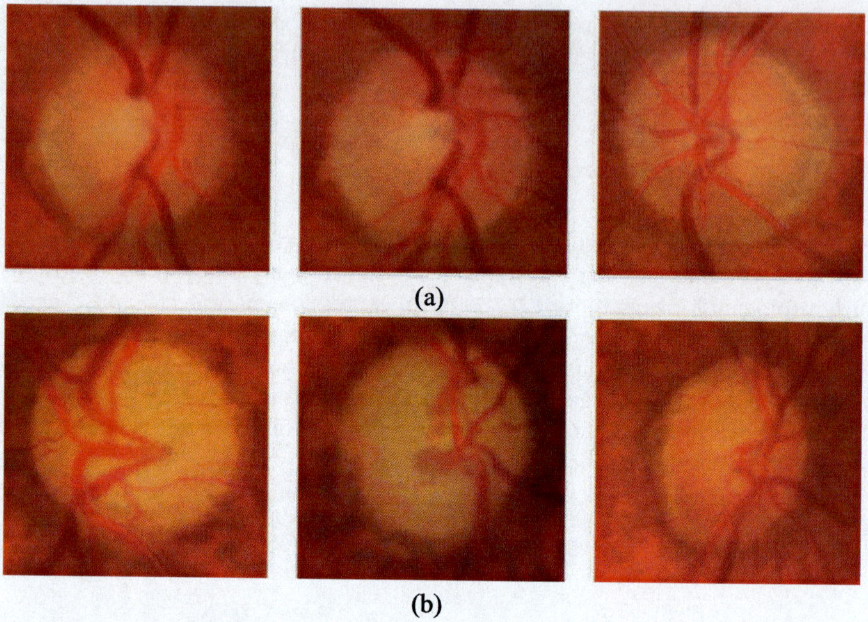

Fig. 3 Optic nerve head images **a** Normal **b** Abnormal

From the CMS in Fig. 4, it is observed that 94% of classification accuracy is obtained when Shannon and sure entropy of WPT are fused together. Also, their sensitivity and specificity are 89 and 99%. By using Shannon entropy, the system gives 82% classification accuracy and by sure entropy; it gives 90% classification accuracy. The classification accuracy is increased up to 8% by the fusion technique. Figure 5 shows the ROC curves of WPT and NBC-based system for glaucoma diagnosis.

From the above ROC curves, the performance of fused features occupies maximum area of 0.94 than their individual features performance (Shannon 0.82 and sure entropy 0.9). Figure 6 shows the performances of WPT + NBC-based system for glaucomatous image classification graphically.

It is observed from the experimental results that 25 abnormal images are wrongly classified while using Shannon entropy and it is reduced to 15 images by sure entropy. Further, the fusion of Shannon and sure entropy reduces the number of wrongly classified abnormal images into 11 only and one normal image is wrongly classified. It is also observed from Fig. 6 that, the better accuracy by using the fusion approach is 94. The results show that the fusion approach gives better performances for the diagnosis of glaucoma.

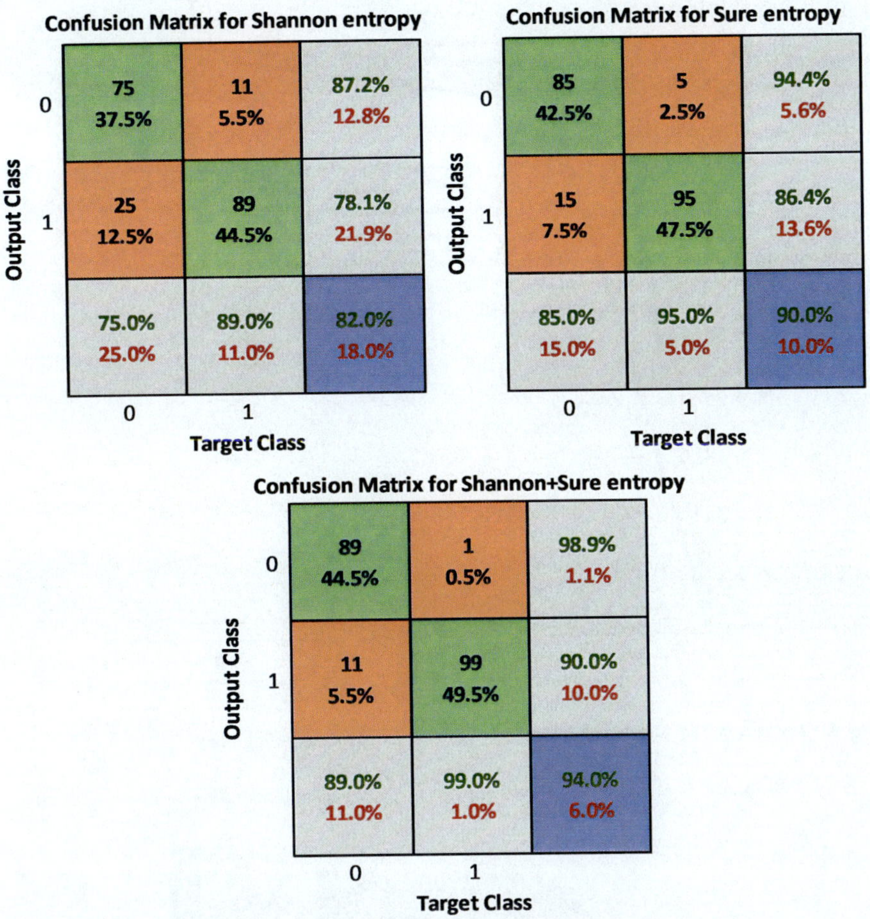

Fig. 4 CMs of WPT and NBC-based system for glaucoma diagnosis

4 Conclusion

In this paper, WPT and NBC-based optic nerve head image classification system for glaucoma diagnosis is presented. From WPT, different entropies such as Shannon and sure entropy are extracted as features. The performance of the system is analyzed using these features individually as well as fusion approach by NBC. Results show that the system classifies 188 images correctly from 200 images tested. The sensitivity of the system is 89% and specificity is 99%. Also it is noted that fusion of Shannon and sure entropy provides better performance than their individual counterpart.

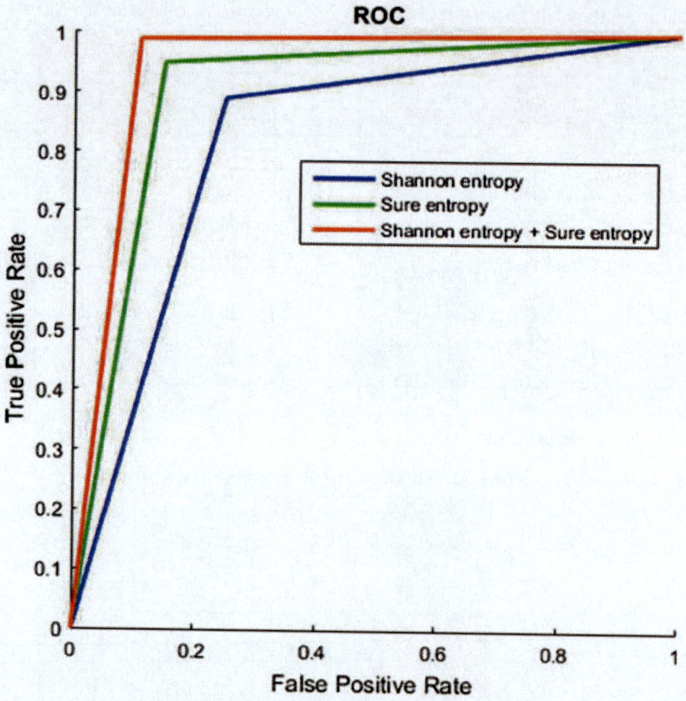

Fig. 5 ROC curves of WPT and NBC-based system for glaucoma diagnosis

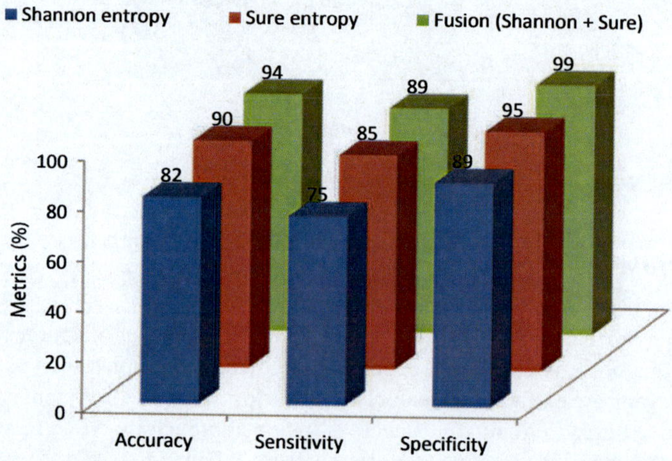

Fig. 6 Performancesof WPT + NBC-based system for glaucomatous image classification

References

1. Kannan KG, Ganeshbabu TR (2017) Glaucoma image classification using discrete orthogonal stockwell transform. Int J Adv Signal Image Sci 3(1):1–6
2. Ganeshbabu TR (2015) Computer aided diagnosis of glaucoma detection using digital fundus image. Int J Adv Signal Image Sci 1(1):1–11
3. Kavya N, Padmaja KV (2017) Glaucoma detection using texture features extraction. In: 51st asilomar conference on signals, systems, and computers. IEEE, pp 1471–1475
4. Lotankar M, Noronha K, Koti J (2015) Detection of optic disc and cup from color retinal images for automated diagnosis of glaucoma. In: IEEE up section conference on electrical computer and electronics (UPCON). IEEE, pp 1–6
5. Xiong L, Li H, Zheng Y (2014) Automatic detection of glaucoma in retinal images. In: 9th IEEE conference on industrial electronics and applications. IEEE, pp 1016–1019
6. Thangaraj V, Natarajan V (2017) Glaucoma diagnosis using support vector machine. In: International conference on intelligent computing and control systems (ICICCS). IEEE, pp 394–399
7. Maheshwari S, Pachori RB, Acharya UR (2016) Automated diagnosis of glaucoma using empirical wavelet transform and correntropy features extracted from fundus images. IEEE J Biomed Health Inf 21(3):803–813
8. Dua S, Acharya UR, Chowriappa P, Sree SV (2011) Wavelet-based energy features for glaucomatous image classification. In: IEEE transactions on information technology in biomedicine, vol 16(1). pp 80–87
9. Ramani RG, Balasubramanian L, Jacob SG (2012) Automatic prediction of diabetic retinopathy and glaucoma through retinal image analysis and data mining techniques. In: 2012 international conference on machine vision and image processing (MVIP). IEEE, pp 149–152
10. Yadav D, Sarathi MP, Dutta MK (2014) Classification of glaucoma based on texture features using neural networks. In: Seventh international conference on contemporary computing (IC3). IEEE, pp 109–112
11. Muthmainah MU, Nugroho HA, Winduratna B (2018) Analysis of retinal fundus images for classification of glaucoma. In: 1st international conference on bioinformatics, biotechnology, and biomedical engineering-bioinformatics and biomedical engineering, vol 1. IEEE, pp 1–6
12. Hutson N, Karan A, Adkinson JA, Sidiropoulos P, Vlachos I, Iasemidis L (2016) Classification of ocular disorders based on fractal and invariant moment analysis of retinal fundus images. In: 32nd Southern biomedical engineering conference (SBEC). IEEE, pp 57–58
13. Bharkad S, Kokare M (2013) Fingerprint matching using discreet wavelet packet transform. In: 2013 3rd IEEE international advance computing conference (IACC). IEEE, pp 1183–1188
14. Rebinth A, Mohan Kumar S (2018) Automated detection of retinal defects using image mining-a review. Eur J Biomed Pharm Sci 5:189–194
15. Rebinth A, Mohan Kumar S (2018) Automated segmentation of retinal images-a review. J Adv Res Dyn Control Syst 10(5):1881–1885
16. Chen YH, Huang HC (2009) Genetic watermarking based on wavelet packet transform. In: Ninth international conference on hybrid intelligent systems, vol. 1. IEEE, pp 262–265
17. Moukadem A, Dieterlen A, Brandt C (2013) Shannon entropy based on the s-transform spectrogram applied on the classification of heart sounds. In: IEEE international conference on acoustics, speech and signal processing. IEEE, pp 704–708
18. Rebinth A, Mohan Kumar S (2019) A deep learning approach to computer aided diagnosis of glaucoma. In: IEEE international conference on recent advances on energy efficient computing and communication
19. Li C, Zhou Y, Ouyang C, Tian L (2012) X-ray image enhancement based on fuzzy sure entropy in LabVIEW. In: 5th international conference on biomedical engineering and informatics. IEEE, pp 395–398

A Robust Method for Multi-algorithmic Palmprint Recognition Using Exponential Genetic Algorithm-Based Feature Selection

Aruna Kumari Palisetty and Jaya Suma Gogulamanda

Abstract Biometrics is widely used in authentication systems to improve their security. One of the leading traits is palmprint thanks to its extensive user's acceptability, accuracy, security as well as to its relatively inexpensive cost. Although palmprint authentication systems know certain maturity, still some challenging tasks need more researches. The integration of multiple representations of palmprint at the representation level provides highly accurate information about palm which led to robust palmprint recognition. This integration creates a problem of substantial dimensional feature space which consumes recognition time, space, and sometimes the accuracy. Feature selection addresses this problem effectively. Therefore, this paper mainly concentrates on the improvement of the Genetic Algorithm (GA)-based feature selection approach for a robust palmprint recognition by including exponential function in the searching process and altering the fitness function, which includes recognition rate, the impact of selected and non-selected features. Experiments on CASIA and IITD databases have shown significant improvement in recognition accuracy along with the reduction of space and time requirements with the proposed Exponential Genetic Algorithm(EGA) compared with GA and Principal Component Algorithm (PCA).

1 Introduction

Multibiometric system, a system based on either different modalities or various samples of the same trait or several instances of a single modality or different representations of single modality, seems to be the best way to overcome the problems caused by unimodal systems [1]. Indeed it restricts from forgery of the modalities to breach the security. A robust authentication is required against frauds with the

A. K. Palisetty (✉) · J. S. Gogulamanda
JNTUK University College of Engineering, Viziangaram, Vizianagaram, AP, India
e-mail: arunakumarip.cse@jntukucev.ac.in

J. S. Gogulamanda
e-mail: gjsuma.it@jntukucev.ac.in

S. C. Satapathy et al. (eds.), *Communication Software and Networks*, Lecture Notes in Networks and Systems 134, https://doi.org/10.1007/978-981-15-5397-4_61

improved performance of the system for most of the security applications. Among various modalities, palmprint gains more focus because of its high user acceptance, steady line shreds of evidence, low insensitivity, and low-resolution imaging [2]. A robust palmprint-based multibiometric authentication system can be developed by considering the different representations of palmprint; these representations are obtained by applying various feature extraction algorithms on palmprint. The pieces of evidence collected from palmprint should be integrated at various stages, namely, pixel level, representation level, score level, and decision level.

The representation or feature-level integration gives highly qualitative information of biometric modality against the rest of the levels, which is useful in constructing a robust system [1]. But in this integration stage, the different representations to be combined may contain heterogeneous feature space which is incompatible for integration. Normalizing the varied feature space produces new feature spaces which are compatible for integration [1]. In addition to this incompatibility, this stage provides ample dimensional feature space after integration which creates problems like increase in time and space requirements and sometimes reduces the system performance instead of improvement. This can be handled by applying dimensionality reduction methods which can be broadly classified as feature extraction and feature selection [3, 4]. It is important to distinguish between both notions. While feature selection methods output a subset of the original features without any further change, feature extraction algorithms transform the input features into a completely different space. Although the feature extraction is more general and the transformation mapping may provide features with better discriminatory ability than the best subset of the input features, new features may not have clear (physical) interpretation. The feature selection is beneficial especially for problems where some sensory inputs are likely to carry a little of useful information for the class discrimination, or if there are very strong correlations between sets of input observables so that very similar information is repeated in several variables, or if measurements on an examined object (or process) are costly and it is advisable to reduce their number. Furthermore, features keep their original physical meaning because no transformation of data is made. This may be important for a better problem understanding in some applications (e.g., in medicine) as only relevant information is analyzed.

The literature has studied various feature extraction, and feature selection approaches in different fields of applications. Feature extraction methods like PCA, Linear Discriminant Analysis (LDA), and Independent Component Analysis (ICA) are investigated in several applications including biometric recognition [5–7]. The feature subset selection approaches like sequential floating forward selection and backward selection methods, sequential forward and backward approaches, exhaustive search, branch and bound search method are suffering from problems like nesting effect, local optimum, and moreover not suitable for large datasets like palm biometric data [8–10]. A robust feature selection method for large scale data like biometric data is needed without any problems as mentioned earlier approaches. Evolutionary Computation (EC) approaches are able to address this feature selection problem in biometric data as it has robust global searching mechanisms [11, 12].

A multi-tree Genetic Programming (GP) method with classification has shown better performance than filter and wrapper methods [13]. GA with fuzzy objective function has investigated for feature selection [14]. For the selection of optimal features from Microarray gene expression data has been applied in [15], which concentrated on reducing computational complexity and fast convergence. A combination of filter and wrapper approaches based on GA has been proposed where three strategies called greedy, sequential, and improvement first strategies are examined and shown that GA produces the best results than remaining methods. The literature shows that GA is producing the best results as a feature selection in various applications [16].

This paper aimed to gain the advantages of representation-level fusion to build robust multi-algorithmic palmprint recognition system with EGA-based feature selection to overcome challenges in representation-level integration. Gayatri and Ramamoorthy [17] proposed a palmprint recognition using the feature-level fusion of four features, namely, energy, contrast, homogeneity, correlation extracted using Gabor Wavelet. But texture features of palmprint contains more discriminative data for better classification compared to the above four features. Zhang et al. [18] proposed palmprint recognition using feature-level integration of 2D-Gabor filter and 2D-Log Gabor filter, where the shreds of evidence extracted from filters are divided into sub-images and calculated the standard deviation (SD) of each sub-image. These SD's of all sub-images of texture image forms feature vector. Even though this approach has produced better results than the unimodal system, due to the calculation of SD, it losses advantage of qualitative and highly discriminative information about palmprint. Hence this work attempted to integrate texture features of palmprint at the representation level.

The proposed system architecture has shown in Fig. 1. The claimed palmprint image has been preprocessed to get fixed dimension palmprint Region of Interest (ROI) image. Then two different approaches based on 2D-Gabor filter and 2D-Log Gabor filter have been applied to obtain different feature representations from ROI. The extracted feature vectors have integrated at the representation level. Three feature reduction approaches called PCA, GA, and proposed EGA have applied. The reduced feature vector has been matched with the stored palmprint template database using

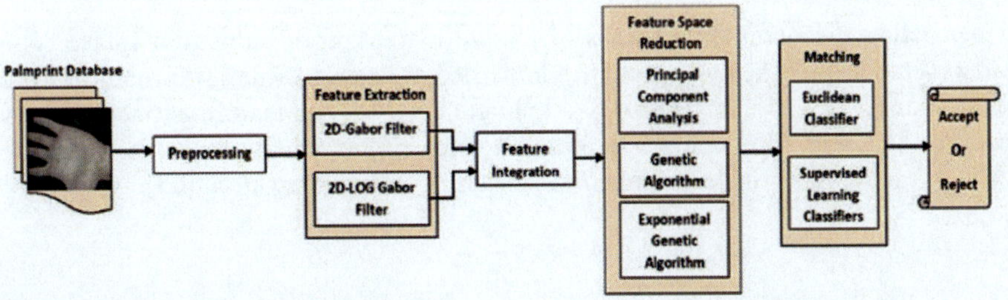

Fig. 1 Proposed system architecture

either Euclidean distance-based classifier or supervised learning classifiers to detect genuine or imposter claim.

This paper has planned as follows: Sect. 2 gives how preprocessing of palmprint image has been carried to get ROI image. Different feature extraction algorithms applied to the ROI image has presented in Sect. 3. The integration of different obtained feature spaces has presented in Sect. 4. The process of dimensionality reduction by PCA, GA, and proposed EGA is presented in Sect. 5. The experimental setup, results, and analysis have discussed in Sect. 6. The work has concluded in Sect. 7.

2 Preprocessing

It is the process of aligning various palmprint images and segmenting middle region (ROI) for further process. Major preprocessing approaches establishes a coordinate system based on key points between fingers. This preprocessing mainly consists of five steps Binarization, Contour extraction of hand/palm, Key points identification, Coordinate system establishment, and Extraction of ROI [19]. All preprocessing algorithms differ from third step onwards until they are all the same [19]. Key points can be identified using tangent-based approach [18], bisector-based approach (midpoint) approach [20], wavelet-based approach [21].

The tangent-based method considers two boundaries- one from point finger and middle finger and the other from ring finger and last finger—as two convex curves. The intersections are considered as two key points for establishing the coordinate system. This approach gives many advantages. They depend on a very short boundary around the bottom of fingers. Therefore it is robust to incomplete fingers and the presence of rings [19]. Because of these advantages, the tangent-based method has applied to extract ROI from palmprint image. To simplify the further processing and for uniform template size, ROI has cropped to 100×100 pixels.

3 Feature Extraction

This section discusses the process of feature extraction of palm ROI image. The Gabor filter is previously applied in biometrics to extract texture information from palmprint [7, 22–24], fingerprint [7, 15], and iris [7]. The texture extraction from palm using Gabor filter comprises principle lines, ridges, and wrinkles, etc. Because of this reason, it has widely applied in palmprint feature extraction [25].

3.1 2D—Gabor Filter

The literature shows Gabor-based feature extraction that has extensively used in various application of pattern recognition. Unstable contrast and brightness of images are better handled by the Gabor function and gives the location of time-frequency exactly [24, 26]. Because of these advantages, the following Gabor filter bank has applied to iris texture extraction [7, 26]:

$$g(a, b; \theta, \varphi, \sigma, \gamma, \lambda) = \exp\left(\frac{a^2 + \gamma^2 b^2}{2\sigma^2}\right) + \exp\left(i\left(2\pi\frac{a}{\lambda} + \varphi\right)\right) \quad (1)$$

where

$$a = a\cos\theta + b\sin\theta$$

$$b = -a\sin\theta + b\cos\theta$$

θ signifies the orientation of the normal to parallel stripes of a Gabor function, φ is the phase offset, λ specifies the sinusoidal factor wavelength, α is the standard deviation of the Gaussian envelope, and γ is the spatial aspect ratio [26].

3.2 2D-Log Gabor Filter

Because of time/space and frequency invariance, symmetry on the log frequency axis, Log Gabor filter has systematically investigated and applied for texture-based feature extraction [27]. The Log Gabor filter has applied by using the following formula [7, 24]:

$$G(\rho, \theta, a, b) = \exp\left(\frac{-1}{2}\left(\frac{\rho - \rho b}{\sigma a}\right)^2\right) + \exp\left(\frac{-1}{2}\left(\frac{\theta - \theta_{ab}}{\sigma\theta}\right)^2\right) \quad (2)$$

In which (ρ, θ) are the log-polar coordinates, a and b gives orientation and scale, the pair $(\rho k, \theta pk)$ corresponds to the frequency center of the filters, and $(\sigma\rho, \sigma\theta)$ is the angular and radial bandwidths.

4 Integration of Feature Spaces

This section explains how the different features collected from the palmprint image by applying two distinct feature extraction approaches have integrated at the feature level. The texture features extracted from palmprint by using 2D-Gabor filter and 2D-Log Gabor filter are compatible with each other. The texture analysis of 100 × 100 palmprint ROI image obtained by applying 2D-Gabor filter and 2D-Log Gabor filter produces 12 different images of size 100 × 100 each; this texture has brought to a single image of size 100 × 100 by using horizontal and vertical downsampling. Further, it converted into a row feature vector of size 10000. The weighted average of feature vectors produces integrated feature space, as shown below:

$$\text{featurevector}_{\text{integrated}} = \frac{(w_1 \times \text{featurevector}_{\text{gabor}}) + (w_2 \times \text{featurevector}_{\text{loggab}})}{2} \tag{3}$$

Here, $\text{featurevector}_{\text{integrated}}$ represents the feature vector after integration, $\text{featurevector}_{\text{gabor}}$ indicates the feature vector formed from the 2D-Gabor feature extraction, $\text{featurevector}_{\text{loggab}}$ indicates the feature vector created from the 2D-Log Gabor feature extraction. w_1 and w_2 are the weights assigned for Gabor features and log Gabor features. Experimentally these weights w_1 and w_2 are fixed to 0.6 and 0.4, respectively.

5 Reduction of High Dimensional Integrated Feature Space

The integration of two feature vectors at the representation level creates the dimensionality problem. This work solves this problem with existing PCA, GA, with new fitness function approaches and proposed EGA method.

5.1 *Principal Component Analysis*

PCA is a dimensionality reduction approach based on subspace projection and widely applied for image compression and recognition problems [28]. PCA has been used for extracting features from face [29–31] and enforced as a reduction strategy in various biometric recognition like face, signature, fingerprint, palm print before matching [7, 32, 33]. PCA is a linear data reduction technique and projects the original data into new dimensional space with maximum variability. The projected data is a collection of principal components which represents new dimensions of the data.

5.2 *Exponential Genetic Algorithm-Based Feature Space Reduction*

Genetic Algorithm is a population-based search algorithm applied in various applications like feature selection [34], routing protocol [35], parameter selection [36], etc. The main motto behind GA selection is that it is simple which requires less computation time, even though a large number of dimensions are present it identifies optimal features with limited time, mutation operation provides flexibility of small changes which cause good results, the number of parameters to be selected are less, and finally it applies probabilistic selection rather than deterministic.

GA mimics the evolution of man and which includes fitness, reproduction, mutation, and crossover [37]. The set of chromosomes known as population denotes solutions. For the selection of essential features from integrated palm feature space, chromosomes are represented in a binary form with a size equivalent to the number of dimensions in the integrated feature space. Here, 1 in chromosome denotes the selected feature, and 0 indicates a non-selected feature. Initially, chromosomes have randomly initialized, and then the evolution process continues as generations. In each generation, chromosomes are selected based on the selection method, and then the quality of the selected chromosomes are computed as fitness value. Based on the quality the selected chromosomes undergoes crossover and mutation operations to generate a new set of chromosomes and forms a new population. This process continues for either a fixed number of generations or some criteria. In the end, the chromosome with the best quality presents the optimum features for robust palmprint recognition.

This work has adopted GA for reducing fused feature space. Each palmprint template after feature-level fusion has encoded as chromosome, where each feature has encoded as genes. Randomly, the initial population of random size has initialized.

Fitness Function

The fitness of each chromosome has evaluated by using Eq. (4). Here, the Recognition Accuracy (RA) of the dataset formed by considering the selected features from the chromosome Chi has calculated by applying the C4.5 classification algorithm.

$$\text{fit}(Ch_i) = RA + n_{\text{selected}} * \left(\frac{N_{DB}}{n}\right) \tag{4}$$

where n_{selected} is the number of selected features in the given chromosome Ch_i. N_{DB} represents the total number of iris samples in dataset and n represents the number of dimensions or features in the dataset. By applying the Roulette Wheel selection method and then using a single-point crossover, mutation operations new population has generated.

New Solution Based on Exponential Order

The chromosomes produced after mutation are further undergone exponential order-based process to produce a new solution. The moto behind the inclusion of exponential function in GA is it provides a reliable solution, causes a quick way of finding optimality because of exponential weights, improves the diversity of solution.

$$\mathrm{Exp}(f) = n^{(p-1)} \tag{5}$$

where p indicates the chromosome rank based on its quality, n denotes the fitness value of each chromosome is how many times greater than in earlier chromosome.

The exponential function is multiplied with each chromosome. Before multiplication, each gene of the chromosome is converted to an integer by mapping "0" to a random number between (0, 0.5) and "1" to a random number between (0.51, 1). New solution after multiplication is converted to binary vector as in initialization process. The new solution is now replaced with the old one.

6 Experimental Results and Analysis

This section describes the experimental environment in which the proposed systems have tested evaluated. The experiments are performed on two different databases, namely, the CASIA palmprint database and Indian Institute of Technology Delhi (IITD) palmprint database. CASIA database consists of palmprint images collected from 312 individual persons. For each person, 8 images have captured for each hand. From this database, 6 samples have been selected from each hand of an individual person to evaluate the proposed system. The IITD database contains palmprint images of left and right hands collected from 230 different persons. From each user, 5 to 6 samples from the left and the right hand have captured. Since each hand of the same person contains utterly different pattern from another hand of the same person, here the samples from each hand has been considered as samples individual subjects; 460 subjects have selected, and from each subject 3 samples were chosen for the evaluation process. The experiments carried on a system with i7 processor CPU @1.8 GHz;16 GB RAM and implemented with Matlab 9.5.

Table 1 presents the recognition accuracy and the reduced number of features without and with different reduction strategies. These results have taken at FAR = 0.01% for the two databases—the results have shown that any reduction approach is giving an improvement in the recognition rate. PCA is producing better reduction compared to the other two. But, along with the amount of reduction recognition rate must be acceptable. Because of this reason, the proposed EGA-based reduction approach is given both adequate reductions in feature space and recognition rate.

Table 1 Recognition accuracy and computation time for processing dataset with euclidian classifier without and with different reduction strategies

Reduction approach	Recognition accuracy		Reduced number of features	
	IITD DB	CASIA DB	IITD DB	CASIA DB
Without reduction	80.2	81.5	10000	10000
PCA	83.4	85.6	1590	1982
GA	89.5	90.1	2458	2698
Proposed EGA	91.7	92.03	2190	2206

Figures 2 and 3 give the performance of the palmprint system with and without reduction approach on four different classification algorithms, namely SMO, C4.5, Naïve Bayes, and Random Forest. Figure 2 gives the classification accuracy on IITD database; Fig. 3 presents classification performance on CASIA database; on CASIA and IITD databases multi-algorithmic palmprint system using EGA as feature selection method has produced the best accuracies of 98.2%, 97.9% with SMO classifier and C4.5 classifier, respectively. In two databases SMO and C4.5 are giving best with very close classification accuracies. When compared to Euclidean classifier, supervised learning classifiers has produced the best accuracy results with and without reduction strategies. As already mentioned, the amount of reduction with reasonable

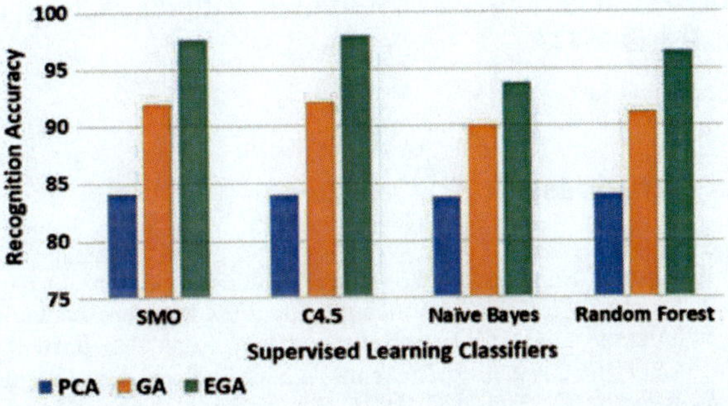

Fig. 2 Classification performance of multi-algorithmic palmprint system with and without applying reduction strategies for IITD DB

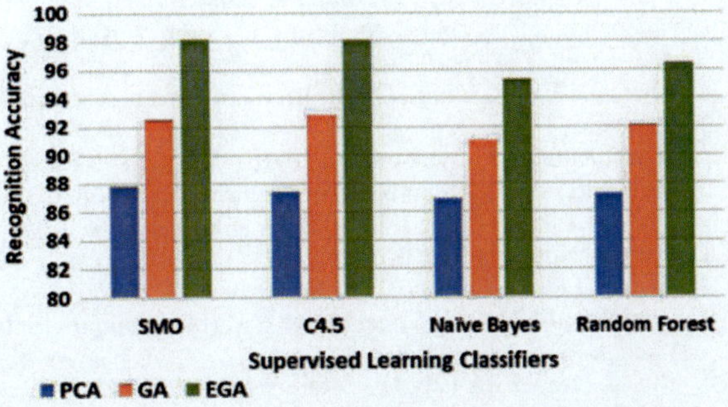

Fig. 3 Classification Performance of multi-algorithmic palmprint system with and without applying reduction strategies for CASIA DB

recognition rate is acceptable. So, EGA-based reduction strategy has given noticeable results in terms of both the amount of reduction and recognition accuracy.

7 Conclusion

Multibiometric systems produce robust recognition systems based on representation-level fusion with a solution to high dimensional feature space after fusion. This work has addressed this solution in terms of presenting different feature space reduction approaches after feature-level fusion in the multi-algorithmic palmprint recognition system. This work proposed EGA to find essential features for robust palmprint recognition. And a new fitness function was designed by including the recognition performance with selected features in chromosome and contribution of non-selected features along with database size. Basic GA with new fitness function has produced better performance than PCA, and the proposed EGA when effectively combined with designed fitness function generated best results compared to PCA and GA. The results showing improvement discrimination of genuine and an imposter in both Euclidean and supervised learning classifiers with EGA. Even though PCA producing more than 85% reduction EGA generated more than 75% reduction in feature space with 98.2% recognition accuracy. Among supervised learning classifiers SMO and C4.5 algorithms, discriminated palmprints efficiently compared to the other two. In the future, this can be improved with the inclusion of mathematical powers like calculus in GA.

References

1. Ross A, Nandakumar K, Jain AK (2006) Hand book of multibiometrics. Springer, Berlin
2. Zhang D (2004) Palmprint authentication. Kluwer Academic Publishers. ISBN:1-4020-8097-2
3. Fukunaga K (1990) Introduction to statistical pattern recognition. Computer science and scientific computing, 2nd edn. Academic Press, San Diego, CA, USA
4. Raudys SJ, Jain AK (1991) Small sample size effects in statistical pattern recognition: recommendations for practitioners. IEEE Trans Pattern Anal Mach Intell 13(3):252–264
5. Shi JX, Gu XF (2010) The comparison of iris recognition using Principal Component Analysis, Independent Component Analysis and Gabor Wavelets. In: IEEE, international conference on computer science and information technology, 2010
6. Zhang D, Jing X, Yang J (2006) Biometric image discrimination (BID) technologies. IGP/IRM Press
7. Aruna Kumari P, Jaya Suma G (2016) Experimental study of feature reduction using PCA in multi-biometric systems based on feature level fusion. In: International conference on advances in electrical, electronics and systems engineering. pp 109–114
8. Pudil P, Novovičová J, Kittler J (1994) Floating search methods in feature selection. Pattern Recogn Lett 15(11):1119–1125
9. Ferri F, Pudil P (1994) Comparative study of techniques for large-scale feature selection. Pattern Recogn Pract IV 1994:403–413
10. Bins J, Draper BA (2001) Feature selection from huge feature sets. In: Proceedings of IEEE International Conference on Computer Vision, vol 2. pp 159–165

11. Xue B, Zhang M, Browne WN (2013) Particle swarm optimization for feature selection in classification: a multi-objective approach. IEEE Trans Cybernet 43(6):1656–1671
12. Nguyen HB, Xue B, Liu I, Zhang M (2014) PSO and statistical clustering for feature selection: a new representation. In: Dick G et al. (eds) Simulated evolution and learning. SEAL 2014. Lecture notes in computer science, vol 8886. Springer, Cham
13. Chakraborty B (2002) Genetic algorithm with fuzzy fitness function for feature selection. In: Proceedings of IEEE ISIE, 2002, vol 1. pp 315–319
14. Benerjee M, Mitra S, Banka H (2007) Evolutionary rough feature selection in gene expression data. IEEE Trans Syst Man Cybernet 37(4):622–632
15. Chin YJ, Ong TS, Goh MKO, Hiew, BY (2009) Integrating palmprint and fingerprint for identity verification. In: Third international conference on network and system security. pp 437–442
16. Muni D, Pal N, Das J (2006) Genetic programming for simultaneous feature selection and classifier design. IEEE Trans Syst Man Cybernet 36(1):106–117
17. Gayatri R, Ramamoorthy P (2012) Palmprint recognition using Feature level fusion. J Comput Sci 8(7):1049–1061
18. Zhang D, Kong W-K, You J, Wong M (2003) Online palmprint recognition. IEEE Trans Pattern Anal Mach Intell 25(9):1041–1050
19. Kong A et al (2009) A survey of palmprint recognition. Pattern Recogn 42:1408–1418
20. Li W, Zhang D, Xu Z (2002) Palmprint identification by fourier transform. Int J Pattern Recognit Artif Intell 16(4):417–432
21. Han C, Cheng HL, Lin CL, Fan KC (2003) Personal authentication using palmprint features. Pattern Recogn 36:371–381
22. Zheng P, Sang N (2009) Using phase and directional line features for efficient palmprint authentication. In: 2nd international congress on image and signal processing. pp 1–4
23. Huang Y, Benesty J, Chen (2009) Using the pearson correlation coefficient to develop an optimally weighted cross relation based blind SIMO identification algorithm. In: IEEE international conference on acoustics, speech and signal processing. pp 3153–3156
24. Aruna Kumari P, Jaya Suma G (2016) Palmprint recognition using PCA and weighted feature level fusion of 2D–gabor and log-gabor features. Int J Control Theory Appl 9(17):8643–8650
25. Prasad MVNK, Kavati I, Adinarayana B (2014) Palmprint recognition using fusion of 2D-Gabor and 2D log-gabor features. In: Martínez Pérez G, Thampi SM, Ko R, Shu L (eds) Recent trends in computer networks and distributed systems security. SNDS 2014. Communications in computer and information science, vol 420. Springer, Berlin, Heidelberg
26. Lim S, Lee K, Byeon O, Kim T (2001) Efficient iris recognition through improvement of feature vector and classifier. ETRI J 23(2):61–70
27. Kong WK, Zhang D, Li W (2003) Palmprint feature extraction using 2-D gabor filters. Pattern Recogn 36:2339–2347
28. Hidayat E (2011) A comparative study of feature extraction using PCA and LDA for face recognition. In: 2011 7th international conference on information assurance and security (IAS), 12/2011
29. Turk M, Pentland A (1991) Eigenfaces for recognition. J Cogn Neurosci 3(1):71–86
30. Chen T, Jessie Hsu Y, Liu X, Zhang W (2002) Principle component analysis and its variants for biometrics. IEEE ICIP 2002, pp 61–64
31. Shah JH, Sharif M, Raza M, Azeem A (2013) A survey: linear and nonlinear PCA based face recognition techniques. Int Arab J Inf Technol 10(6)
32. Kerdprasop N, Chanklan R, Hirunyawanakul A, Kerdprasop K (2014) An empirical study of dimensionality reduction methods for biometric recognition. In: 7th international conference on security technology IEEE 2014. pp 26–29
33. Meng J, Yang Y (2012) symmetrical two-dimensional PCA with image measures in face recognition. Int J Adv Robotic Syst 9
34. Aruna Kumari P, Jaya Suma G (2019) FGANN: a hybrid approach for medical diagnosing. In: Computational intelligence and big data analytics. Springer briefs in applied sciences and technology

35. Rama Rao A, Valli Kumari V, Satyananda Reddy C (2016) Exponential genetic algorithm-based stable and load-aware qos routing protocol for MANET. J Intell Syst 26(4):669–682
36. Guo Fan, Peng Hui, Tang Jin (2016) Genetic algorithm-based parameter selection approach to single image defogging. Inf Process Lett 116(10):595–602
37. Melanie M (1999) An introduction to genetic algorithms. The MIT Press, A Bradford Book

Signature Extraction Using Connected Component Labeling

Venkata Sai Nitesh Virothi and Meghana Santoshi Janapareddy

Abstract Signature is a means of self-identification and authentication for humans. There are various algorithms used in order to extract the signature from several documents like bank cheques, certificates, and many more. Signatures have to be accepted only from the intended person. Different works have been carried out in the fields of offline recognition systems over the past few years. There are many algorithms and approaches used till date which have their own advantages and disadvantages. The algorithm used here is "Connected Component Analysis." It is used to detect the signature from the complete document provided.

1 Introduction

Because of the usage and importance of signature role, the people who want to manipulate the signatures are increasing everyday. As technology is evolving day by day, many illicit activities are also being developed, like fluent forgery, copied forgery, and many more. In order to make such activities, inefficacious modern algorithms in Machine Learning are used for Signature Recognition. There are basically two categories of signature recognition, namely online recognition and offline recognition. Online categories make use of dynamic information of a signature captured using a pen tablet or any stylus at the time of the signature. There are numerous techniques for online signature recognition; for instance, we have Artificial Neural Network and Dynamic Time Wrapping techniques. However, research on offline extraction or recognition has not shown any adequate outcomes. Offline extractions usually work on the signature that is basically of a scanned image. As mentioned, the input image is scanned from a piece or part of a written paper which can be a letter or any image with signature.

V. S. N. Virothi (✉) · M. S. Janapareddy
Anil Neerukonda Institute of Technology and Sciences, Visakhapatnam, Andhra Pradesh, India
e-mail: niteshvirothi19@gmail.com

M. S. Janapareddy
e-mail: meghanajs99@gmail.com

S. C. Satapathy et al. (eds.), *Communication Software and Networks*, Lecture Notes in Networks and Systems 134, https://doi.org/10.1007/978-981-15-5397-4_62

2 Related Work

In this field of signature recognition, there has been a lot of research carried out. It all started in 2004 where the University of Hong Kong hosted the first international signature recognition, but the event was solely for online detection of signature.

Chaudhari along with two others proposed a system to recognize signatures using neural networks and fuzzy minimum–maximum algorithm [1]. During that approach, the information was gathered by data acquisition; the first steps in the system are information gathering and data preprocessing, and moreover, Hu's method, i.e., seven moment invariant, reported an increase in the accuracy from 53 to 92% when one increases the number of signatures in a class. Ismail along with his team proposed a system with Principle Component Analysis (PCA) for offline systems. First K-Nearest Neighbor (KNN) classifier is used followed by Artificial Neural Network (ANN) for authentication. Before running both the tasks, the PCA applies image preprocessing and feature extraction whose result of using PCA vs not using PCA for False Recognition Rate (FRR) is 5% when PCA is used. Another verification system proposed by Piyush Shanker and Rajagopalan which uses Modified Dynamic Time Warping (DTW) stated that a threshold of 1.5 produces a system with acceptance rate of 0.33 for casual forgery and 19.6 for skilled and 25% rejection rate for original signature. An algorithm developed by Kennard has 26% ERR for 2D geometric warp. The proposed system was evaluated using many offline and online signatures of Chinese and Dutch by Liwicki; it provided a good performance.

An extraction based on splitting of signature was proposed by Wai and Aung [2]. The signature in the right corner at the bottom of image or a bank cheque is extracted and the image is transformed into binary form by Ostu's technique and is bound in the box of rectangular shape, then it checks for interesting pixel and finds the center of gravity of the signature image and divides the signature into four parts; each part is actually checked for an interesting pixel and subdivided into 64 subparts or blocks. From each sub-block, features are extracted which are robust like pixel density and angle.

Further, Ahmed focused on feature extraction methods for offline signature recognition [3]. Here, they used projection-based Discrete Random Transform (DRT); signature preprocessing is done making use of Otsu's method, and the data is trained using DTW with euclidean distance. By doing so, they obtained optimal performances in memory storage and time for processing, and outcomes when using the combination of horizontal and vertical projections gave a False Rejection Rate (FRR) of 8.4, False Acceptance Rate (FAR) of 5.6, Total Error Rate (TER) of 14.0, and Equal Error Rate (EER) of 7.6.

Tomar and Singh [4] used neural networks with inputs as directional feature and energy density of a signature to extract it for offline systems. These results are compared with the very basic and simple implementation version of these methods. It was found that this proposed work was very effective compared to the above two methods, particularly for a less number of training samples.

3 Implementation and Methodology

The first step that takes place is scanning a document or a certificate that contains signature. As mentioned earlier, the algorithm can extract the signatures from any scanned documents using “connected component analysis.” So, what is this connected component algorithm then? In image processing, this algorithm recognizes regions of connected pixels that usually got the same value, i.e., this connected components scans a given input image with a signature and then the pixels are grouped into components where they are based on pixel connectivity, i.e., all these picture elements share similar pixel intensity values which reveals that they are connected with one another. So, when all teams are established, every pixel will be marked with a color (color labeling) or with a gray scale depending on the component it has been assigned to. Now labeling of connected elements and assorted dissociates are central to many machine-driven image analysis applications. The image is scanned from top to bottom and left to right to identify the pixel regions that are actually connected,i.e. adjacent pixels sharing constant value V. It works on binary or gray-level images and different connectivity measures. Here, it assumes binary image input and 8-connectivity where there is a mask created and checks every pixel and its surrounding values using 8-connectivity technique. The operator scans the image by moving over a row until it arrives to a point p (where p at any stage is the pixel to be labeled in the scanning) for which $V = \{1\}$, then it examines the surrounding four neighbors of p that is left, above and couple of upper diagonal which have already been encountered in the scan.

According to the details, the “p” is classified when there is only one neighbor with “1,” then its label is allocated to p; suppose more than one neighbor has value equal to “1,” then any one of the labels is assigned and equivalencies are noted; if all are 0, a new label will be allocated.

After the scan, the identical label pair area units were sorted into equivalence category (classes and a distinct label), the final step is a second scan of the image after which each label is replaced by its equivalence categories. Where the labels might be completely different grey’s or can be colors. Thus, the connected components can be found and labeled by an exciting functionality that is provided by Scikit image library. We used this library to check the scanned input documents; we figured out that the biggest connected component is the signatures. So, if we are able to extract the biggest components, we can identify the signatures from the entire document. However, there is a possibility of getting some undesired lines or different shapes that have big connected components. So, in order to overcome this, a threshold value is used. Threshold value to detect the outliers are calculated by us, that is, any kind of lines, structure, and texts which does not belong to the signatures while many experiments are carried out. An equation that is obtained is calculated based on results during experimenting, which works so good for almost all the scanned documents that are of A4 size. Because all the bank cheques, letters, or certificates are usually of A4 size. A threshold value to remove attached pixels are comparatively smaller to variable that is initialized to A4 size scanned documents.

$$[a4\_constant = ((average/84.0) * 250.0) + 100]$$

where a4_constant is an initialized variable.

$$[ax\_constant = \left(\begin{array}{l} (average/constant\_parameter\_1) * (constant\_parameter\_2) \\ + (constant\_parameter\_3) \end{array} \right)]$$

x stands for scanned document size such as A4 or A0; it can be modified for A0 and so on by configuring the constants:

- Constant_parameter_1
- Constant_parameter_2
- Constant_parameter_3

Many experiments with different parameter values can be performed until the highest accuracy is achieved.

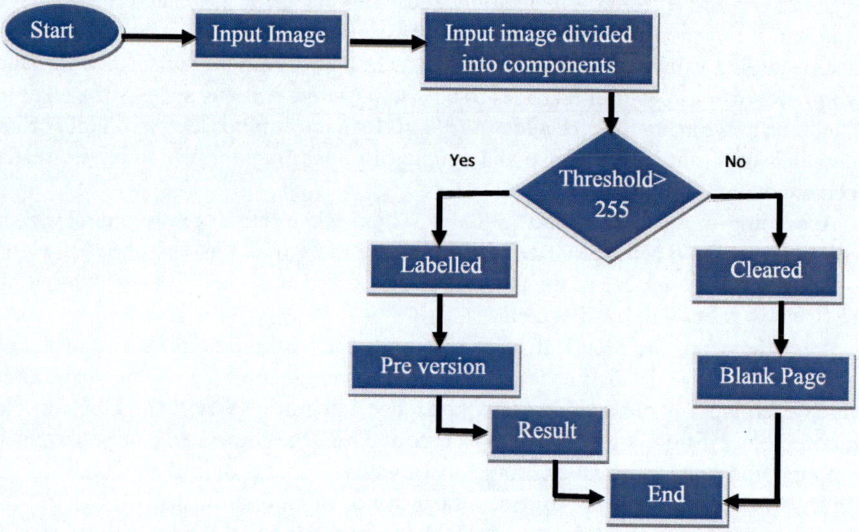

4 Result and Comparative Analysis

Any document of size A4 with a signature in it is taken as shown in the Fig. 1. Initially, all the required packages are imported and then the document is given as an input and then the whole document is scanned. While scanning the input, the algorithm scans the image and groups its pixels into components based on pixel connectivity (similar values) with the help of which it finds the largest characters. As usually signatures

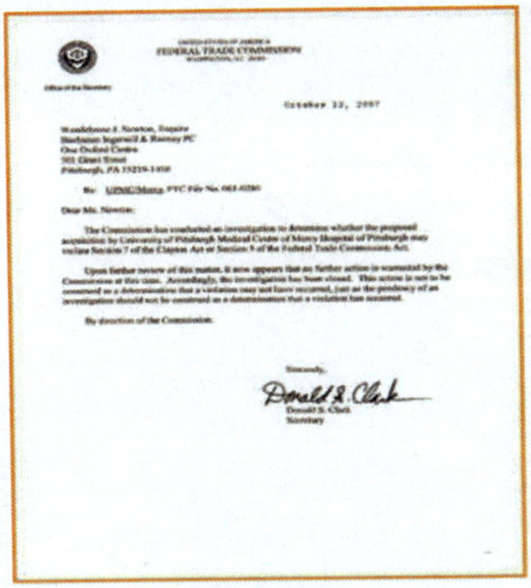

Input Image

(a):Output Image

(b):Enhanced output image

Fig. 1 Input image. **a** Output image, **b** Enhanced output image

are always represented or signed in large characters compared to actual text in the document. In Fig. 1a, as we can observe, the size of the text is comparatively less than the signature. The algorithm takes regions with large enough area and extracts the signature from the document. The Signatures are performed in a variety of conditions like sitting, standing, and fast conditions. The facts show that signatures signed at different conditions have an effect on the shape of the signature. After conducting several experiments and training data, we concluded that the threshold value is around 255, i.e., 127 to 255.

After scanning the above letter, i.e., Fig. 1, the signature is extracted and the output is as follows:

Dataset
Coming to the dataset, we trained the machine using a number of images with signature that are available online in Google. We took these images which are given as an input as shown below in Fig. 2. These images can be in any format like jpg, png, jpeg. etc. Even documents with signature can also be used as an input but we need to convert it to above-mentioned formats. The machine is trained with hundreds of different inputs to make sure the output is more accurate Figs. 2, 3, 4, and 5.

The outputs obtained indicates that the extraction is accurate and is approximately 90%.

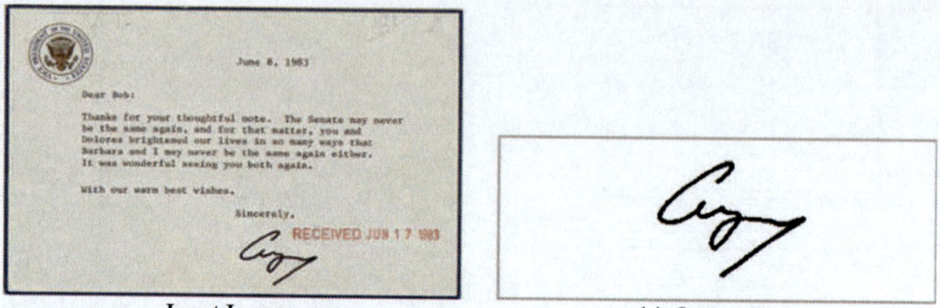

June 8, 1983

Dear Bob:

Thanks for your thoughtful note. The Senate may never be the same again, and for that matter, you and Dolores brightened our lives in so many ways that Barbara and I may never be the same again either. It was wonderful seeing you both again.

With our warm best wishes,

Sincerely,

RECEIVED JUN 17 1983

Input Image

(a):Output Image

Fig. 2 Input image. **a** Output image

4.1 Comparative Analysis

See Table 1.

5 Conclusion and Future Scope

As the method or the algorithm used is connected component labeling, we are able to extract the signature with significant accuracy; the extracted signature still has to be processed for more precise results. The method used here is simple and easy to implement. Unlike other methods and algorithms, this approach does not extract any unwanted lines or shapes as it contains a threshold value, so the result will be more specific. Yet if the threshold value changes or it may vary due to the size of signature, then the result might vary but the threshold provided here in this process is more reliable. Yet in the above-mentioned process, the results obtained is error free when the document is with a good resolution and the signature is clear enough. But when the document resolution is low or if the signature is not clear enough, then the output accuracy may vary, i.e., it may decrease. Maybe in the coming years and due to the advancement of technology, the output accuracy will definitely be high. In the future, this extraction technique will be used in many scenarios; for instance, while filling the details of a document, at the end you need to sign the document. So, extracting the signature using this technique helps us to directly upload the signature.

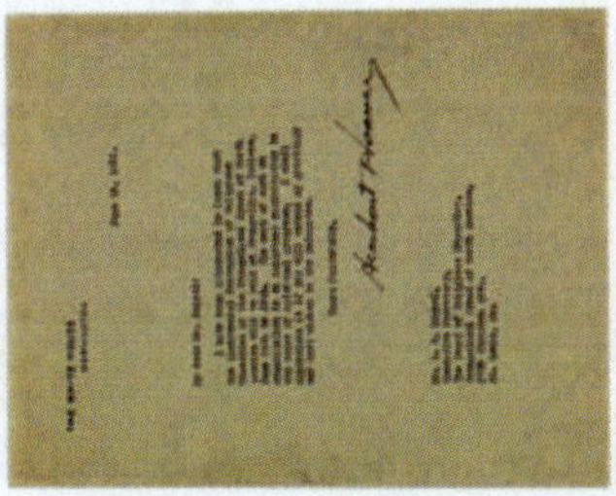

(a):Output Image

Fig. 3 Input image. **a** Output image

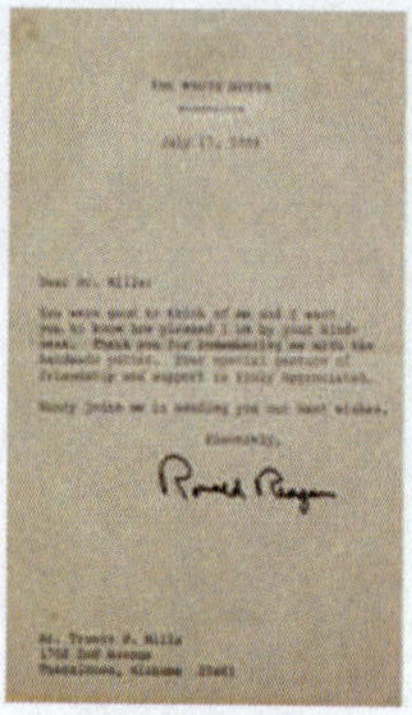

Input Image

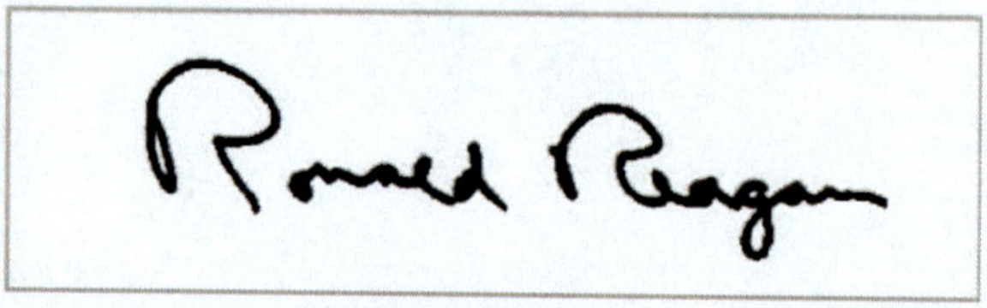

(a):Output Image

Fig. 4 Input image. **a** Output image

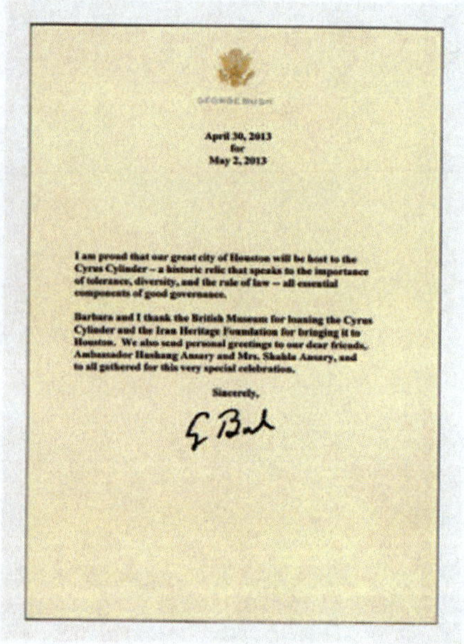

April 30, 2013
for
May 2, 2013

I am proud that our great city of Houston will be host to the Cyrus Cylinder – a historic relic that speaks to the importance of tolerance, diversity, and the rule of law – all essential components of good governance.

Barbara and I thank the British Museum for loaning the Cyrus Cylinder and the Iran Heritage Foundation for bringing it to Houston. We also send personal greetings to our dear friends, Ambassador Hushang Ansary and Mrs. Shahla Ansary, and to all gathered for this very special celebration.

Sincerely,

Input Image

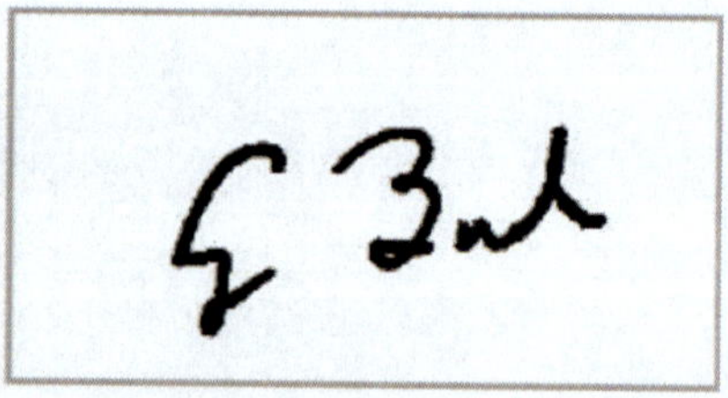

(a):Input Image

Fig. 5 Input image. **a** Input image

Table 1 Comparative Analysis of distinct algorithms for signature extraction

S. No	Algorithm	Advantage	Disadvantage
1.	Compression networks and positional cuttings. Gulzar et al. [5]	Gives 98% success rate.	Poor performance for signature that is not in the training phase. Usually, the failure to verify a signature was due to bad image quality and more similarity between pair of signatures
2.	Keypoint detection method proposed hyperspectral unmixing technique. Iqbal and Khurshid [6]	Detection-based technique gives results that are promising if the sign is not overlapping with other parts of the image or any input document, i.e., printed text, lines, stamps, etc.	The keypoint detection method decreases the performance in case the signature is overlapping with other parts
3.	Hashed break point chunking algorithm by Raphael et al. [7]	Signature extraction in order to detect document overlap is effective, and the execution is quite faster and accurate to initially check documents for plagiarism.	The documents that appear similar are further examined
4.	Dichotomy transformation, and boosting feature selection	In this paper, the lowest error rates among the SV systems and the proposed approach provide a very higher level of performance which is comparatively faster than the optimal overproduce and chosen approach.	Here, they used both ESC and DPDF; one performs better at lower resolution and other at higher resolution, respectively So, both are important and essential factors. If one among them does not provide satisfactory results, then the accuracy may be degraded
5.	Madasu et al. [8]	This approach has an extraction rate of almost 99.26% with training samples of 211 images	But it was not able to entirely extract a single signature from an image or bank cheque. And, if the signature is not signed at the specific or particular mentioned area, then this approach fails to extract the signature

(continued)

Table 1 (continued)

6.	A part-based feature extraction method Speeded Up Robust Features (SURF). Ahmed et al. [3]	This algo is based on the characteristics that are extracted from different sections of the scanned image and then it will disregard all the global aspects and finally it is durable to various variations of a picture or image.This method requires limited number of training samples	The precision of the method is quite low when performed on images with logos
7.	Neural network backpropagation. Babita [9]	Accuracy of recognition 89 to 98%	Increase in number of classes decreases the accuracy.Features considered is not suitable for database with large number of classes
8.	Support Vector Machine(SVM) Speed Up Robust Feature(SURF) Segmentation-based Fractal Texture Analysis (SFTA). Kaur et al. [10]	SURF is good at handling image with blurring or rotation. SURF is poor at handling image with viewpoint	The only disadvantage is if the threshold value is changed or if size of the signature is modified then the output may be incorrect
9.	kNN based on the features that were stored in the database. Umniati et al. [11]	28/40 signatures identified	Algorithm fails when there is inconsistency of person writing the signature
10.	Two-phase connected component labeling approach. Cüceloğlu1 and Oğul [12]	Gradient-based and LTP features are more useful in classifying signature segments	Missing true positives is usually caused by the intervening signatures with other texts or figures of the document, especially when the signature is much smaller than the opposite intervened portion or the faint fonts because of low scan quality or resolution

References

1. Chaudhari BM, Barhate AA, Bhole AA (2009) Signature recognition using fuzzy min-max neural network. In: International conference on control, automation, communication and energy conservation, vol 1. pp 414–420
2. Aung SL, Wai HH (2013) Feature extraction for offline signature verification system. Int J Comput Commun Eng Res (IJCCER) 1(3):84–87
3. Ahmed S, Malik MI, Liwicki M, Dengel A (2012) Signature segmentation from document images. In: Proceedings ICFHR-12, international conference on frontiers on handwriting. pp 423–427

4. Tomar M, Singh P (2011) An intelligent network for offline signature verification using chain code. In: Proceedings of advanced computing: first international conference on computer science and information technology part 3. CCSIT, Bangalore, India, pp 10–22
5. Khuwaja GA, Laghari MS (2011) Offline handwritten signature recognition. In: World academy of science: engineering and technology international journal of computer and information engineering, vol 5, no 11. pp 1304–1307
6. Iqbal K, Khurshid K (2017) Automatic signature extraction from document images using hyperspectral unmixing. In: Proceedings of the Pakistan academy of sciences a. physical and computational sciences, vol 54, no 3. pp 269–276
7. Finkel RA, Zaslavsky A, Monostori K, Schmidt H (2002) Signature extraction for overlap detection in documents. In: proceedings of 25th Australian computer science conference. Monash University, Melbourne, Australia, pp 59–64
8. Madasu VK, Mohd. Hafizuddin Yusof M, Hanmandlu M, Kubik K (2003) Automatic extraction of signatures from bank cheques and other documents. In: VIIth digital image computing: techniques and applications. pp 591–600
9. Babita P (2015) Online signature recognition using neuralnetwork. J Elect Electron Syste 4(3)
10. Kaur M, Jindal S (2016) Survey on offline signature recognition techniques. Int J Eng Trends Technol (IJETT) 36(6):309–313
11. Umniati N, Mutiara AB, Kusuma TM, Widodo S (2018) Component-connected feature for signature identification. Int J Adv Sci Eng Inf Technol 8(3):757–761
12. Cüceloğlul İ, Oğul H (2014) Detecting handwritten signatures in scanned documents. In: 19th computer vision winter workshop

Bitcoin Price Prediction and Analysis Using Deep Learning Models

Temesgen Awoke, Minakhi Rout, Lipika Mohanty, and Suresh Chandra Satapathy

Abstract Cryptocurrencies are a digital way of money in which all transactions are held electronically. It is a soft currency which doesn't exist in the form of hard notes physically. Here, we are emphasizing the difference of fiat currency which is decentralized that without any third-party intervention all virtual currency users can get the services. However, getting services of these cryptocurrencies impacts on international relations and trade, due to its high price volatility. There are several virtual currencies such as bitcoin, ripple, ethereum, ethereum classic, lite coin, etc. In our study, we especially focused on a popular cryptocurrency, i.e., bitcoin. From many types of virtual currencies, bitcoin has a great acceptance by different bodies such as investors, researchers, traders, and policy-makers. To the best of our knowledge, our target is to implement the efficient deep learning-based prediction models specifically long short-term memory (LSTM) and gated recurrent unit (GRU) to handle the price volatility of bitcoin and to obtain high accuracy. Our study involves comparing these two time series deep learning techniques and proved the efficacy in forecasting the price of bitcoin.

1 Introduction

Virtual currencies are a form of cryptocurrency which is an impressive technical achievement in digital marketing, nevertheless. Virtual currencies live on, and they couldn't fully replace fiat or conventional currencies. In the current study, we

T. Awoke · M. Rout (✉) · L. Mohanty · S. C. Satapathy
School of Computer Engineering, KIIT Deemed to be University, Bhubaneswar, Odisha, India
e-mail: minakhi.rout@gmail.com

T. Awoke
e-mail: temesgenawoke5@gmail.com

L. Mohanty
e-mail: mohanty.lipika93@gmail.com

S. C. Satapathy
e-mail: sureshsatapathy@gmail.com

S. C. Satapathy et al. (eds.), *Communication Software and Networks*, Lecture Notes in Networks and Systems 134, https://doi.org/10.1007/978-981-15-5397-4_63

are trying to show an interesting new perspective from which view of economics questions surrounding currency governance, the characteristics of money, political economy of financial intermediaries, and the nature of currency computation.

Virtual currencies become the most favorable and used for commercial enterprise transactions all over the world [1, 2]. The popularity is due to its innovative characteristics such as transparency, simplicity, and increasing acceptance through the world [3]. In the current time, bitcoin is the popular flourishing virtual currency. Reported to the website https://bitcoin.org, viewed on April 19, 2019, the virtual currency market value is close to 90 billions of dollars, but it varies from time to time. Bitcoin is a peer-to-peer cryptocurrency in which all transactions are not regulated or controlled by any third party. Third-party intervention between customers is impossible. It is highly volatile market price working 24/7. Market capitalization of bitcoin is increased through time to time. In the current time, more than 71 billions of dollars publicly traded. Due to its open-source nature, clear, transparent, simple, and time is saving which leads all virtual currencies in the world.

Bitcoin is a worldwide and most popular cryptocurrency, first introduced in 2008 and exploited as open source in 2009 by a person called Satoshi Nakamoto, but it became highly popular in 2017. Bitcoin functions as a decentralized moderate of electronic cash, with transactions proved and transcribed in a public distributed ledger (blockchain) without any third-party intervention. Transaction blocks consist of secure shell algorithm which is used to connect each other, and blocks are served as a non-editable data which is recorded when the transaction is being held. Then any virtual currency especially bitcoin has been adopted by the people, and the virtual currency market trend has been growing up.

The popularity of bitcoin is increased within a short period of time. Different technologies and business companies are joined with bitcoin. As different researchers assured that after 2015 around 100,000 technology and business companies have started the bitcoin market. Some of the popular companies which are joined with bitcoin are Amazon, Microsoft, Overstock, Dell, and others [1]. Many works have been done to predict time series, as well as BTC value. However, any deep learning models have not been much used yet to predict the BTC price value. Knowing the deep learning models become state-of-the-art neural network architecture that improves prediction accuracy in various domains including time series, we consider applications of deep learning to predict the BTC price value. In coming sections, we will explore previous works done on BTC price prediction, discuss deep learning models to predict the time series, and focus on three main articles which will serve as foundation of our work.

Primarily, the main challenge of bitcoin exchange rate is its high rate of price fluctuation. High price volatility implies a certain measure should be taken to predict the price of bitcoin accurately. Knowing the forecasting activity is necessary to tell about the future price of bitcoin and build trust as well as acceptance throughout the world. Influenced by a variety of factors, such as political system, public relations, and market policy of a country, can determine economical role of bitcoin and international relation of countries on different market strategies. Lastly, doesn't have an official road map: few key challenges and developments coming up for bitcoin prediction are in consistent, because there is no clear description of the exchange platform on which the transactions related to buying and selling are not regulated. The objective of our current study is to forecast the bitcoin price with improved efficiency using deep learning models and minimizing the risks for investors as well as policy-makers.

2 Related Works

Researches on the prediction of cryptocurrencies using machine learning are not much enough, especially on deep learning models. According to the research of 2016, more than 600 papers have been published on this topic. Our literature survey covers work done on bitcoin (BTC) price prediction using different techniques, the need, and evaluation of recurrent neural network (RNN) and its system architecture. Dennys et al. [4] used different attribute selection mechanisms to get the most important features and applied machine learning methods such as artificial neural network (ANN), support vector machine (SVM), and recurrent neural network (RNN) as well as k-means clustering in the bitcoin price prediction. However, one limitation of this study is only focused on the investors. Policy-makers should be considered as a major partner of the system because cryptocurrency can change the dynamics of world economy. Sean McNally et al. [5] used Bayesian optimized recurrent neural network and LSTM to predict the direction of Bitcoin price in USD. They also used ARIMA model to compare the deep learning methods. In Atsalakis et al. [6], this research focuses on computational intelligence method especially hybrid neuro-fuzzy controller in order to predict the exchange rate of bitcoin. This model used neuro-fuzzy approach and artificial neural networks. Goodfellow et al. [7] proposed a deep direct reinforcement learning framework for financial signal representation and trading. They combined the reinforcement learning (RL), deep learning (DL), and their current deep neural network (NN) to generate precise prediction results. They validate the proposed approach using commodity future markets as well as stock market data. Madan et al. [8] tried to predict the price of bitcoin using machine learning and investigate the trends of BTC surrounding. They used 25 attributes relating to bitcoin to forecast the daily price variation. In Lahmiri et al. [9], they implemented machine learning algorithms to predict the exchange rate of daily price of high data availability cryptocurrencies such as BTC, ripple, and digital cash. They applied RNN and GRNN (Generalized Regression Neural Network) to get the accurate prediction rate of high liquidity cryptocurrencies. Saxena et al. [10] investigated the minimum accuracy of bitcoin price using LSTM and ARIMA model. Paresh kumar et al. [11] suggested that bitcoin has a negative impact on market inflation. It is not predictable; therefore, bank of Indonesia should warn not to invest on bitcoin. Not only this but also different government authorities including police should prevent bitcoin marketing in Indonesia, and the objective of this study is to control the effects of cryptocurrency on the monetary system. Pant et al. [12] state that socially constructed ideas in a twitter about virtual currency have straight or sidelong impact over all the market analyses of virtual currencies. This study focuses on forecasting the fluctuated value of bitcoin by sentiment analysis and identifying the relationship between positive and negative sentiments. Nivethitha et al. [13] proposed the future stock price prediction using LSTM machine learning algorithm. They especially focused on time series prediction because it is a basic for share price prediction and other financial prediction models. And comparing with that of existing model ARIMA, LSTM algorithm provides efficient and accurate results. Roth et al. [14] assured that bitcoin is the new and most popular virtual currencies, while the security and its volatility rate are debatable. This study makes it functional for the peer-to-peer transaction of bitcoin through the network and the blockchain technology. Phaladisailoed et al. [15] used various machine learning algorithms to predict the bitcoin price more efficiently.

3 Proposed Methodology

The proposed methodology considers two different deep learning-based prediction models to forecast daily price of bitcoin by identifying and evaluating relevant features by the model itself. After applying both the models for bitcoin prediction, we can determine which model is much more accurate for the future fulfillment of our target and select appropriate parameters to obtain a better performance. In this work, we have proposed deep learning mechanisms such as LSTM and GRU which are the latest and efficient techniques for the forecasting of bitcoin price. As bitcoin is the most popular cryptocurrency, the price volatility issue should be handled within a short period of time. The process of prediction starting from collecting data till the forecasting of bitcoin price is depicted in Fig. 1.

3.1 RNN

RNN is a deep neural network characterized as a recurrent connection between the input and output of its neurons or layers and capable of learning sequences designed to capture temporal contextual information along time series data. They have recently gained popularity in deep learning due to their ability to overcome the limitation of existing neural network architecture where it comes to learn over long sequences. Two common RNN networks are LSTM and GRU and presented in the subsequent sections.

3.1.1 LSTM

LSTMs are explicitly designed to avoid the long-term dependency problem. Remembering information for long periods of time is practically their default behavior, not something they struggle to learn. All recurrent neural networks have the form of

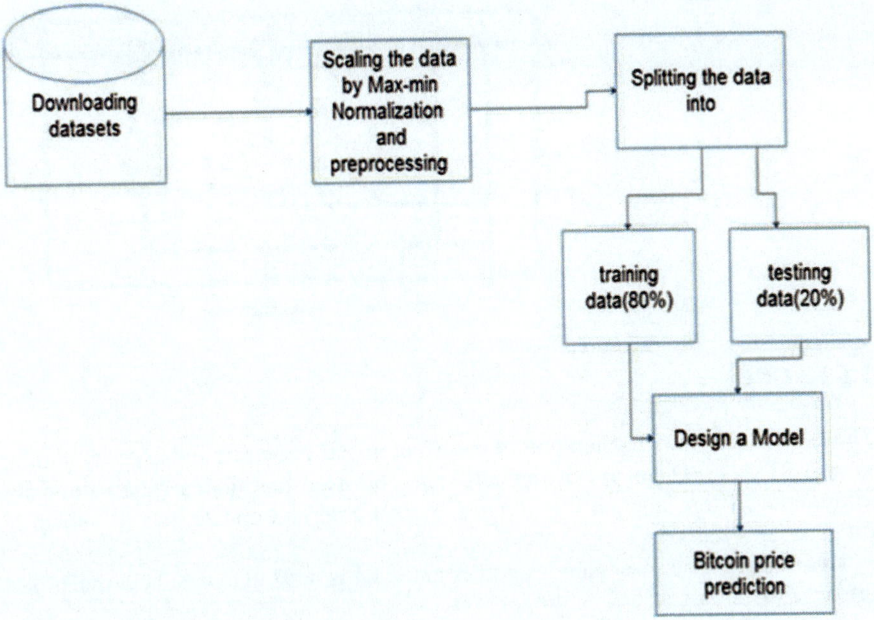

Fig. 1 Block diagram of proposed workflow

a chain of repeating modules of neural network. In standard RNNs, this repeating module will have a very simple structure, such as a single *tan h* layer.

The deep learning LSTM neural networks overcome the problems with RNN related to vanishing gradients, by replacing nodes in the RNN with memory cells and gating mechanism. In this regard, it is an attractive deep learning neural architecture mostly on the account of its efficacy in memorizing long- and short-term temporal information simultaneously, and it can be viewed the same in LSTM architecture depicted in Fig. 2.

Fig. 2 LSTM architecture

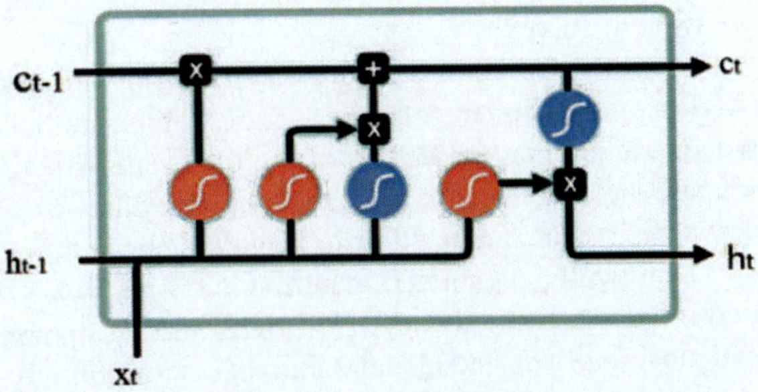

Fig. 3 GRU architecture

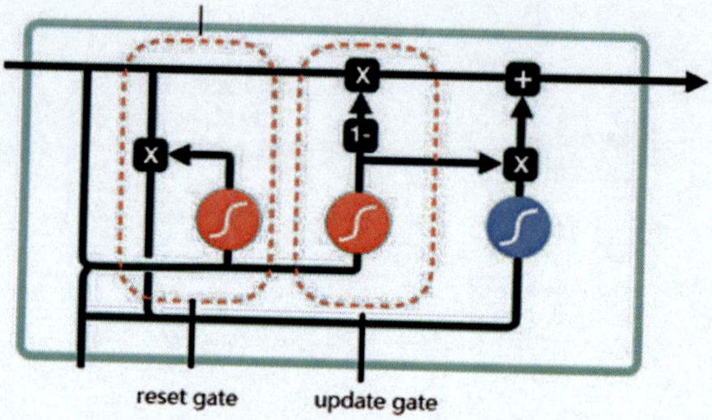

3.1.2 GRU

The GRU is the newer generation of recurrent neural networks and is pretty similar to an LSTM. GRU got rid of the cell state and used the hidden state to transfer information. It has also only two gates, a reset gate and update gate as shown in Fig. 3.

Reset Gate: The reset gate is another gate that is used to decide how much past information to forget.

Update Gate: The update gate acts similar to the forget and input gate of an LSTM. It decides what information to throw away and what new information to be added.

4 Simulation Results and Analysis

4.1 Data and Data Set Preparation Method

Data preparation is the process of collecting, combining, organizing, and structuring data, and then it can be considered as data visualization, analytics, and data mining with machine learning applications. It is critical to feed accurate data for the problem we want to solve.

Data set preparation is a crucial step in machine learning. As we mentioned before, the data preparation impacts the accuracy of the predictions. Therefore, in this section, we should explain the details of the data sets. We will expose the methods used to prepare the data in scope of our model. The dataset used for this research consists of daily price value collected from Kaggle website https://www.kaggle.com.

The overall data collection period is from January 1, 2014 to February 20, 2018. In this dataset, there are seven attributes such as opening price, high price, low price, and closing prices and also the market cap of publicly traded outstanding shares.

Table 1 Comparision of compilation time required by both the deep learning-based models

Model	Compilation time (ms)	Epoch
LSTM	53	100
GRU	5	100

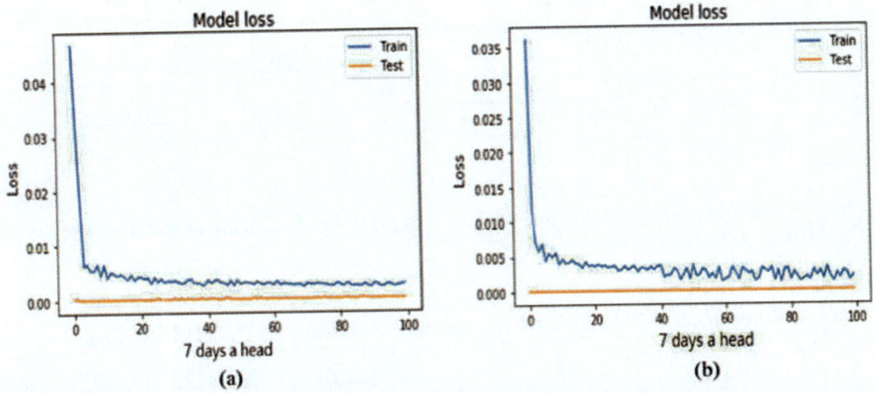

Fig. 4 **a** MSE graph obtained using LSTM model **b** MSE graph obtained using GRU model

4.2 *Results and Discussion*

The proposed model of LSTM and GRU price prediction of bitcoin was trained, and the predictions were carried out for popular cryptocurrency. The accuracy of the proposed LSTM as well as GRU model is investigated by finding the root mean square error (RMSE) and mean absolute percentage error (MAPE) to determine which model has better accuracy. We observed from the resultant Table 1 that LSTM takes greater compilation time than GRU model.

The MSE value obtained for 7 days ahead from both the models is plotted and shown in Fig. 4, and it is clearly observed that GRU is converging faster and steady than the LSTM model. From Fig. 5a, b, it is discovered that the variation of actual price and predicted price is more in LSTM than the GRU.

4.3 *Performance Measures*

One of the common ways to compare the time series models is to measure their performance for short- and long-term prediction. To validate the performance of these two models, we have used MAPE (Mean Absolute Percentage Error) and RMSE (Root Mean Square Error) as performance measure. These error values are obtained using LSTM and GRU and listed in Table 2.

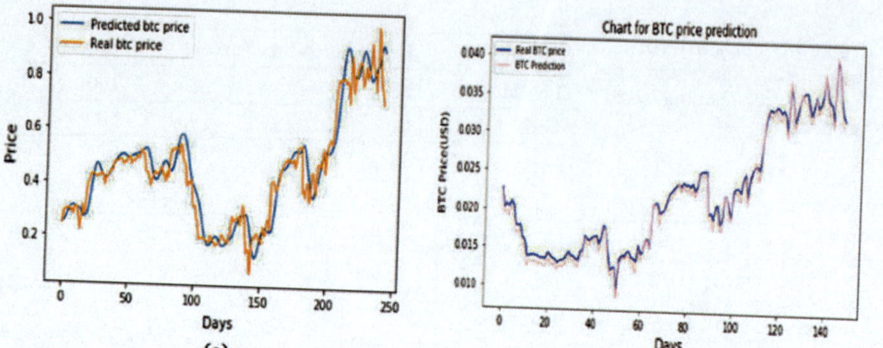

(a) (b)

Fig. 5 Comparison of actual and predicted bitcoin price during training phase of LSTM (**a**) and GRU (**b**)

Table 2 Comparison of RMSE and MAPE value obtained using LSTM and GRU models

Window size	Number of days ahead	LSTM		GRU	
		RMSE	MAPE	RMSE	MAPE
1	1	0.092	0.068	0.075	0.065
5	3	0.079	0.057	0.065	0.046
7	5	0.081	0.060	0.087	0.062
12	7	0.045	0.030	0.051	0.035
15	15	0.067	0.048	0.067	0.058

From this study, we found that the GRU-based forecasting model is more appropriate in order to forecast time series data of highest price volatility. As we have observed, from Table 2 and Fig. 6 the prediction accuracy of the LSTM is better at window size of 12 and days ahead of 7. However, in the rest of window sizes and days ahead, GRU model is more efficient than that of LSTM models and the comparison actual and predicted bitcoin price obtained.

5 Conclusion and Future Work

Bitcoin is the most popular decentralized way of virtual currency which has a great role in the free market economy and avoids the intermediary of another third party between customers. The main objective of our study is to forecast the bitcoin price with improved efficiency using deep learning models and minimizing the risks for the investors as well as policy-makers. We have implemented two deep learning techniques such as LSTM and GRU as prediction models. The study reveals that the GRU model is the better mechanism for time series cryptocurrency price prediction

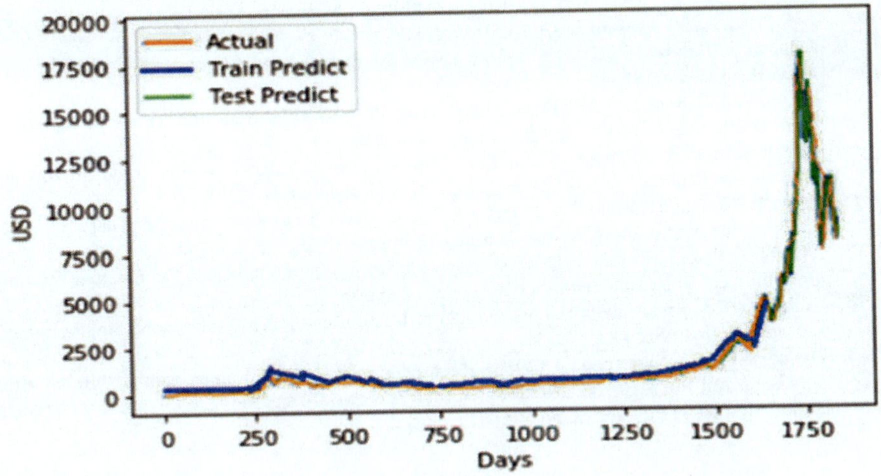

Fig. 6 Actual and predicted bitcoin price in terms of USD obtained using LSTM model

and takes lower compilation time. LSTM and GRU models are more capable of recognizing long-term dependencies. In this study, we have only compared to basic deep learning-based models, i.e., LSTM and GRU. However, it needs to investigate further to enhance the accuracy of the deep learning-based prediction models by considering different parameters in addition to the previous one. Features such as political system, public relations, and market policy of a country can affect and determine the price volatility of cryptocurrency. In our study, we have not considered other cryptocurrencies such as ripple, ethereum, lite coin, and others. We will enhance the model by applying on these cryptocurrencies so the model becomes a stable one. Fuzzification can also be incorporated at the input layer by considering the degree of participation of each of the features in the prediction.

References

1. Nakamoto S et al. (2008) Bitcoin: a peer-to-peer electronic cash system, 2008. Working Paper from www.bitcoin.org
2. Urquhart A (2016) The inefficiency of Bitcoin. Elsevier, pp 80–82
3. Jang H, Lee J (2017) An empirical study on modeling and prediction of bitcoin prices with bayesian neural networks based on blockchain information. 5427—5437
4. Dennys CA, Mallqui RAF (2018) Predicting the direction, maximum, minimum and closing prices of daily bitcoin exchange rate using machine learning techniques. Int J Soft Comput (IJSC) 596–606
5. McNally S, Roche J, Caton S (2018) Cryptocurrency forecasting with deep learning chaotic neural networks. IEEE, pp 339–343
6. Atsalakis GS, Atsalaki IG, Pasiouras, F, Zopounidis C (2019) Bitcoin price forecasting with neuro-fuzzy techniques. Elsevier, pp 770–780
7. Goodfellow I, Bengio Y, Courville A (2016) Deep learning. MIT press

8. Madan I, Saluja S, Zhao A(2015) Automated bitcoin trading via machine learning algorithms. 1–5. http://cs229.stanford.edu/proj2014/Isaac\%
9. Lahmiri S, Bekiros S (2019) Cryptocurrency forecasting with deep learning chaotic neural networks. Elsevier, 35–40
10. Saxena A, Sukumar TR (2018) Predicting bitcoin Price using lstm and compare its predictability with Arima model. Int J Pure Appl Math 2591–2600
11. Paresh Kumar N, Narayan S, Rahman RE, Setiawan I (2019) Bitcoin price growth and Indonesia's monetary system. Emerg Mark Rev 38:364–376
12. Pant DR, Neupane P, Poudel A, Pokhrel AK, Lama BK (2018) Recurrent neural network based bitcoin price prediction by twitter sentiment analysis. IEEE, pp 128–132
13. Nivethitha P, Raharitha P (2019) Future stock price prediction using LSTM machine learning algorithm. Int Res J Eng Technol (IRJET) 1182–1186
14. Roth N (2015) An architectural assessment of bitcoin: using the systems modeling language. Procedia Comput Sci 44:527–536
15. Phaladisailoed T, Numnonda T (2018) Machine learning models comparison for bitcoin price prediction. In: 2018 10th international conference on information technology and electrical engineering (ICITEE). IEEE, pp 506–511

Trajectory Planning in 2D Space Using Evolutionary Algorithms

Divyanshi Raghuvanshi, Anand Kumar Sahay, Junali Jasmine Jena, Lipika Mohanty, and Suresh Chandra Satapathy

Abstract Planning optimal trajectory in a 2D space with large number of nodes or coordinates is a challenging task. Path planning has various applications in the aviation industry, UAV, mobile robots, etc. So path planning algorithms should be efficient enough to manage the associated constraints such as time, travelling cost and energy in an optimal way. Evolutionary algorithm can be very effective in addressing such challenges. This paper reviews various approaches proposed for path planning of UAV using evolutionary algorithms. Formulating the appropriate fitness function according to the requirement of the problem is the major task, which affects the accuracy of the solution.

1 Introduction

The use of 'Automated traversing' object has increased in recent time. This automated traversing object could be a UAV or a mobile robot, which has to traverse a certain path for fulfillment of the purpose. It has wide applications in military missions, monitoring weather forecasting, traffic control, rescuing people and lot more. It may happen that UAV has to be used in a littered and obstacle-rich environment, for instance, in a metropolitan area and therefore, it is very much essential for a UAV

D. Raghuvanshi · A. K. Sahay · J. J. Jena (✉) · L. Mohanty · S. C. Satapathy
School of Computer Engineering, Kalinga Institute of Industrial Technology-(Deemed to be University), Bhubaneswar, Odisha, India
e-mail: junali.jenafcs@kiit.ac.in

D. Raghuvanshi
e-mail: divyanshiraghuvanshi02@gmail.com

A. K. Sahay
e-mail: anand.123sahay@gmail.com

L. Mohanty
e-mail: lipika.mohantyfcs@kiit.ac.in

S. C. Satapathy
e-mail: suresh.satapathyfcs@kiit.ac.in

S. C. Satapathy et al. (eds.), *Communication Software and Networks*, Lecture Notes in Networks and Systems 134, https://doi.org/10.1007/978-981-15-5397-4_64

to adopt a path planning mechanism to make sure that the path which is traversed by UAV should be free from collision and path length must be optimal [1]. One of the major problems in trajectory planning of these objects is that they have to traverse between the starting and the ending point, and there are chances that they may encounter a lot of obstacles and these obstacles can be fixed at one location or vary during the flight. To overcome the above problem, it is necessary to plan a path where automated objects could overcome the rough path, by optimizing and finding the feasible path.

This paper is based on planning the trajectory of automated traversing objects, in two-dimensional space, with the use of evolutionary algorithm. Over the last decade, evolutionary algorithms have emerged as very effective optimization and searching techniques. Evolutionary algorithms are a part of evolutionary computations and also belong to a group of modern heuristics-based search technique. As they have inherited the properties like flexibility and robustness from evolutionary computation, they can efficiently solve global optimization problems. In many applications, high complexity evolutionary algorithms have been used [2], such as Rathbun et al. [3] proposed an evolution based path planning algorithm for autonomous motion of a UAV through uncertain environments.

2 Evolutionary Algorithms

Path planning algorithms based on evolutionary algorithms use a population formed of possible solutions and gradually modify them to perform the stochastic search. The population is initialized randomly. At each step, various operations are performed upon the population. Then the cost function is used as a measure of fitness. Based on the fitness, population is either modified or selected for the next iteration. The same operations are repeated until stopping criteria are met. Figure 1 explains the basic

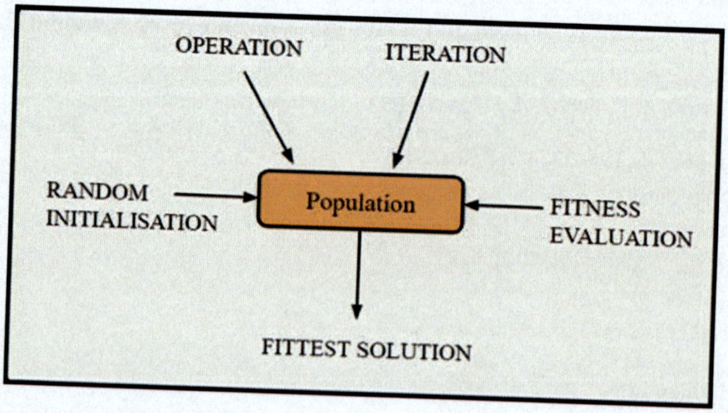

Fig. 1 Working of evolutionary algorithms

Table 1 Name of evolutionary algorithms and from where they have been inspired by

Sl no.	Year	Name of the paper	Name of the authors	Inspired by
1.	1988	Genetic algorithms [2]	Goldberg and Holland	Charles Darwin's theory of natural evolution
2.	1995	Particle swarm optimization [4]	James Kennedy and Russell Eberhart	Flocking of birds
3.	1997	Differential evolution [5]	Rainer Storn, Kenneth Price	Theory of evolution
4.	1999	Ant colony optimization [6]	Dorigo and Di Caro	The searching behaviour of ants, especially the pheromone communication between ants for finding the best path between the ants and the food source
5.	2009	Artificial bee colony [7]	Karaboga, Akay	The intelligent searching behaviour of honey bee
6.	2011	Teaching–learning-based optimization [8]	Rao, Savsani, Vakharia	The result of the guidance of a teacher on the performance of students/learner in class
7.	2016	Social group optimization [9]	Suresh Satapathy, Anima Naik	Basic concept of humans' social behaviour for solving a complex problem

working principle of evolutionary algorithms. It is not the best solution always, but an optimal solution, whose optimality can be increased, by modifying the control parameters, according to the complexity and scalability of the problem.

Table 1 depicts some of the evolutionary algorithms and inspirations from where they have been taken from.

Genetic algorithm (GA) was coined in 1970 by John H. Holland and his student [2]. It was inspired by Charles Darwin's theory of natural evolution. GA uses several operators such as crossover and mutation to get the solution to a problem. Particle swarm optimization (PSO) was coined in 1995 by James Kennedy [4]. The theory was inspired by the flocking of birds. It is a method where the problem gets the optimal solution by repeatedly trying to enhance the particle best solution with regard to a given measure of quality. Ant colony optimization (ACO) algorithm was developed by Marco Dorigo in 1992 in his Ph.D. [6]. He was inspired by the searching behaviour of ants, especially the pheromone communication between ants for finding the best path between the ants and the food source. Also, it helps us solve the problem which can be optimized for finding the best path. Differential evolution (DE) is a new heuristic approach for minimizing the nonlinear and non-differentiable continuous space functions [5]. DE was proposed by Rainer Storn in 1996 and was inspired by the theory of evolution. The extensive tests demonstrated that it converges faster with more accuracy than any other existing global optimization methods. Artificial bee colony (ACO) was developed by Dervis Karaboga in the year 2005, inspired by the searching behaviour of bees [7]. It commonly uses the parameters such as size of colony and cycle number should be maximum. It is an optimization technique that provides search on the basis of population procedure in which bees modify food's positions with time and they discover places with food resources with higher amount of nectar. Social group optimization is a population-based optimization technique [9] which is motivated by the concept of humans' social behaviour for solving a highly complex problem. It is observed that in the population-based category, social group optimization will enhance the class of effective and efficient evolutionary optimization techniques and will provide researchers a large scope to choose this in various problems and applications. Teaching–learning-based optimization was proposed by Rao et al. in 2011 [8]. Like other algorithms, it was also inspired by the effective guidance of the teacher on the performance by the students of the class. To get the global solution, TLBO uses a population (class of learners) of solutions so it also falls in the category of population-based method.

2.1 Application of Evolutionary Algorithm in Trajectory Planning

Evolutionary algorithms were inspired by the mechanism of how various species of nature interact and develop intelligence to solve their problem optimally. To maximize the function's quality, a set of elements (possible solutions) are generated randomly in function domain. The quality function is now the fitness function which is applied to the main problem domain. On the basis of the fitness function, the candidates with best fitness value are selected or modified for the next generation. This is realized by applying some biological operators inspired by the biological processes. This is repeatedly done until the best solution is found.

A suitable fitness function can be defined for trajectory planning that can be optimized using evolutionary algorithm. The designing of fitness function is the main challenge and the crucial part for optimizing the path planning. Figure 2 depicts the application of evolutionary algorithms to path planning. Though we have some specific algorithms like ACO, which is specifically designed for path planning, we can also apply other evolutionary algorithms if we can design an appropriate fitness function because most of the evolutionary algorithms focus on optimizing the fitness function. The more optimal the fitness value is, the more optimized the trajectory will be. Table 2 shows some of the evolutionary algorithms used for trajectory planning in UAV, mobile robots, etc.

Arantes et al. derived a heuristic approach for UAV path planning by using genetic algorithm under critical situation. The algorithm defines a new path to land UAV by considering all the constraints. Similarly, Atencia et al. solved complex multi-UAV problem by using multi-objective genetic algorithm [11]. It defined various hybrid fitness functions which satisfy the constraints and try to find the optimal solution by

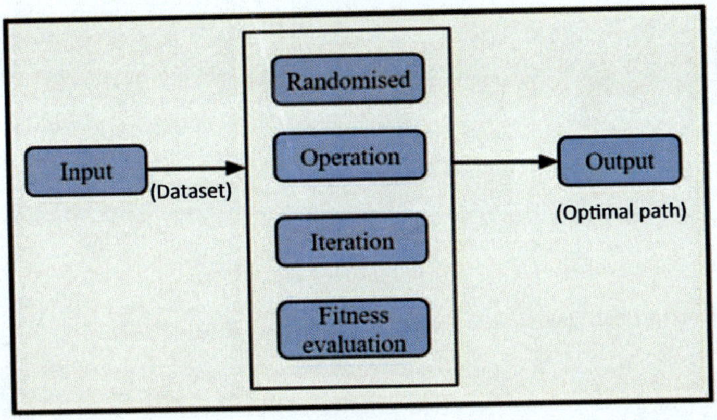

Fig. 2 Working of evolutionary algorithms in trajectory planning

Table 2 Evolutionary algorithms in trajectory planning

Sl no.	Year	Name of the authors	Evolutionary algorithms
1.	2017	Arantes et al. [10]	Genetic algorithm (GA)
2.	2016	Atencia et al. [11]	Genetic algorithm (GA)
3.	2017	Ghamry et al. [12]	Particle swarm optimization (PSO)
4.	2017	Phunga et al. [13]	Particle swarm optimization (PSO)
5.	2019	Mirjalili et al. [14]	Ant colony optimization (ACO)
6.	2017	Chen et al. [15]	Ant colony optimization (ACO)
7.	2017	Pan et al. [16]	Artificial bee colony optimization (ABCO)
8.	2018	Liu et al. [7]	Artificial bee colony optimization (ABCO)

checking whether the solution is valid or not. Ghamry et al. with the help of PSO coined a forest fire fighting application by using multi-UAV [12]. The major challenge in firefighting mission was to carry out tasks with higher efficiency and efficiently minimize amount of time. In the similar manner, Phunga et al. enhanced more by considering vision-based surface [13]. The algorithm assumes the fire spots are pre-detect and by considering input constraints while surpassing UAV collisions during motion. Mirjalili et al. with the help of ant colony optimization (ACO) designed the trajectory planning for AUV [14]. It also included that the algorithm may all achieve the optimal path for UAV when the obstacle number is increased gradually from one number to many numbers. Similarly, Chen et al. considered path planning by avoiding the obstacles in path planning [15]. Pan et al. optimized the route planning of UAV by using artificial bee colony (ABCO) and also focused on saving memory [16]. Liu et al. also enhanced his approach by using ABCO by making the application of evacuation of crowd in buildings [7].

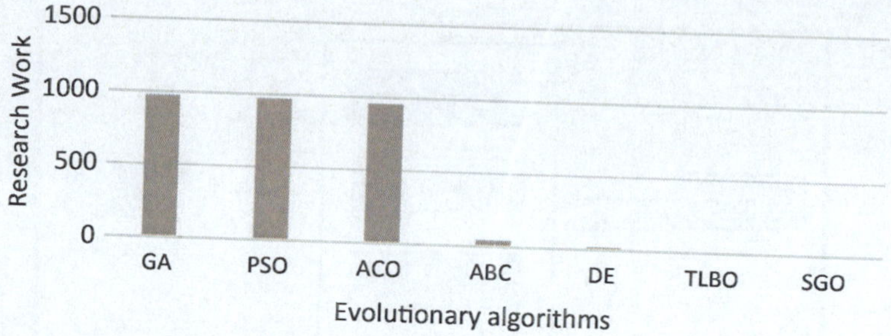

Fig. 3 Graph showing approximate number of research work based on some of the evolutionary algorithms

3 Conclusion and Future Work

There have been various approaches to find an optimal path while planning a trajectory, but solving it by using evolutionary algorithms makes the process simple and efficient. Evolutionary algorithms are easy to implement, simple, and time-efficient. Hence, they can be applied for real-time trajectory planning. There are several optimization algorithms which have not been used for trajectory planning till date. Figure 3 represents an approximate number of research works using some of the evolutionary algorithms taken from Google Scholar database (https://scholar.google.com/). Hence, application of other algorithms to this problem may serve as an emerging area of research.

References

1. Debnath SK, Omar R, Latip NBA (2019) A review on energy efficient path planning algorithms for unmanned air vehicles. In: Computational science and technology. Springer, Singapore, pp 523–532
2. Vikhar PA (2016) Evolutionary algorithms: a critical review and its future prospects. In: 2016 international conference on global trends in signal processing, information computing and communication (ICGTSPICC). IEEE, pp 261–265
3. Rathbun D, Kragelund S, Pongpunwattana A, Capozzi B (2002) An evolution based path planning algorithm for autonomous motion of a UAV through uncertain environments. In: Proceedings of the 21st digital avionics systems conference, vol 2. IEEE, pp 8D2–8D2
4. Eberhart R, Kennedy J (1995) Particle swarm optimization. In: Proceedings of the IEEE international conference on neural networks, vol 4. pp 1942–1948
5. Storn R, Price K (1997) Differential evolution–a simple and efficient heuristic for global optimization over continuous spaces. J Global Optim 11(4):341–359
6. Parsons S (2005) Ant colony optimization by Marco Dorigo and Thomas Stützle, MIT press, 305 pp., $40.00, ISBN 0-262-04219-3. Knowl Eng Rev 20(1):92–93
7. Liu H, Xu B, Lu D, Zhang G (2018) A path planning approach for crowd evacuation in buildings based on improved artificial bee colony algorithm. Appl Soft Comput 68:360–376

8. Rao RV, Savsani VJ, Vakharia DP (2011) Teaching–learning-based optimization: a novel method for constrained mechanical design optimization problems. Comput Aided Des 43(3):303–315
9. Satapathy S, Naik A (2016) Social group optimization (SGO): a new population evolutionary optimization technique. Comp Intell Syst 2(3):173–203
10. Silva Arantes JD, Silva Arantes MD, Motta Toledo CF, Júnior OT, Williams BC (2017) Heuristic and genetic algorithm approaches for UAV path planning under critical situation. Int J Artif Intell Tools 26(01):1760008
11. Ramirez-Atencia C, Bello-Orgaz G, R-Moreno MD, Camacho D (2017) Solving complex multi-UAV mission planning problems using multi-objective genetic algorithms. Soft Comput 21(17):4883–4900
12. Ghamry KA, Kamel MA, Zhang Y (2017) Multiple UAVs in forest fire fighting mission using particle swarm optimization. In: 2017 international conference on unmanned aircraft systems (ICUAS). IEEE, pp 1404–1409
13. Phung MD, Quach CH, Dinh TH, Ha Q (2017) Enhanced discrete particle swarm optimization path planning for UAV vision-based surface inspection. Auto Constr 81:25–33
14. Mirjalili S, Dong JS, Lewis A (2020) Ant colony optimizer: theory, literature review, and application in AUV path planning. In: Nature-inspired optimizers. Springer, Cham, pp 7–21
15. Chen J, Ye F, Jiang T (2017) Path planning under obstacle-avoidance constraints based on ant colony optimization algorithm. In: 2017 IEEE 17th international conference on communication technology (ICCT). IEEE, pp 1434–1438
16. Pan TS, Dao TK, Pan JS (2017) An unmanned aerial vehicle optimal route planning based on compact artificial bee colony. In: Advances in intelligent information hiding and multimedia signal processing. Springer, Cham, pp 361–369

Use of Evolutionary Algorithms for Detection of Fatal Diseases via DNA Micro-array Classification: A Review

Junali Jasmine Jena and Suresh Chandra Satapathy

Abstract Classification of DNA micro-array has been a popular approach of disease detection. Various fatal diseases such as cancer (colon, lungs, breast, and other parts) and tumor (brain and other parts) could be identified using this method. But choosing an appropriate set of gene expression for classification, out of huge gene set, is very much essential for enhancing identification accuracy and is an NP-Hard problem. Evolutionary algorithms have always been efficient in providing optimal solutions to NP-Hard problems. In this paper, various evolutionary algorithms based techniques for gene-expression analysis have been reviewed, explaining their basic working principle and objective, as well as some frequently used gene-expression datasets have been discussed.

1 Introduction

Human body has its own reaction mechanism to the malicious or malignant processes that happen inside the body. If these processes would be considered at the molecular level, then cells and tissues come into action. Gene-expression coding is a very popular technique of extracting molecular level information from DNA micro-arrays. These gene expressions are obtained by processing DNA micro-array and converting them into forms of table. Rows of the table indicate various genes from which data has been extracted and columns indicate some special features and characteristics [1]. Hence, it is an obvious fact that some gene expressions of malicious cells or tissues would be different from the normal ones.

Gene-expression analysis is based on identifying the differentiating features between normal and abnormal cells and tissues and basing on that, classifying a sample [2]. But this job of differentiation is not so easy. The number of genes is very

J. J. Jena (✉) · S. C. Satapathy
School of Computer Engineering, Kalinga Institute of Industrial Technology (Deemed to be University), Bhubaneswar, Odisha, India
e-mail: junali.jenafcs@kiit.ac.in

S. C. Satapathy
e-mail: suresh.satapathyfcs@kiit.ac.in

S. C. Satapathy et al. (eds.), *Communication Software and Networks*, Lecture Notes in Networks and Systems 134, https://doi.org/10.1007/978-981-15-5397-4_65

large, and each gene expression depicts various features and characteristics. Among those, there may be present such characteristics that are completely irrelevant, there may be a group of features that are redundant, or there may be some gene expressions wrongly coded (noisy data). Thus, finding those subsets of genes, which could effectively represent their respective type in the best possible way, is tough. Also, the minimal subset of genes is always preferred, as it would incur less computational cost and time. Hence, segregating effective and minimal subset of genes belongs to the group of NP-Hard problems. The rest of the paper has been organized as follows: Background study, Review of Literature, Benchmark Dataset, and Conclusion.

2 Background Study

2.1 *Gene Expression from DNA Micro-Array*

First, a very large number of gene sequences of a sample are arranged over a microscopic slide. Then, mRNA (messenger RNA) of that cell is collected, which is generated on the activation of genes. mRNA is then labeled using RT (Reverse Transcriptase Enzyme), which generates cDNA (Complementary DNA) followed by attaching FN (Fluorescent Nucleotides) to the cDNA. Normal and malignant cells are attached to different fluorescent nucleotides. cDNA is then placed in the DNA micro-array slide, which will form binding with their respective DNA and leave the fluorescent tag. The intensity of the color of fluorescence indicates the activeness of Gene, which is measured with a high-performance scanner. Gene may be more active toward malignant DNA or normal DNA or may exhibit equal activeness toward both [3]. Through the processing of those micro-array images, tables are formed. A brief idea of this process could be drawn from Fig. 1.

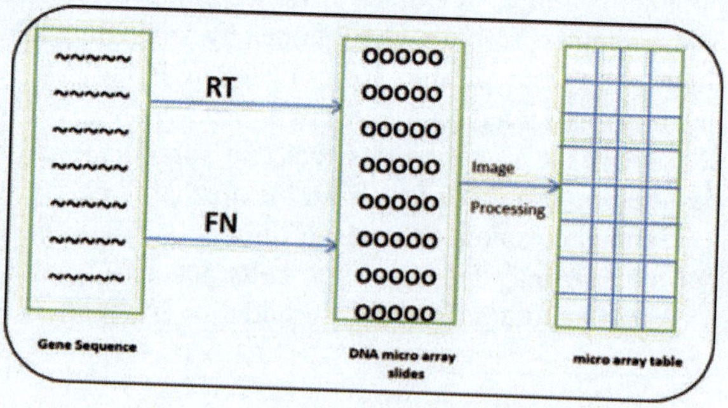

Fig. 1 Conversion of gene sequence to micro-array table

2.2 Gene Selection

It is basically done by three approaches: filter approach, wrapper approach, and hybrid approach. The filter approach uses statistical methods for relevance evaluation such as ranking methods and Information gain. Learning models are used by the Wrapper approach, such as greedy search algorithms and stochastic search algorithms, for evaluation of gene relevance. The hybrid approach is the combination of both the filter approach and the wrapper approach for gene relevance estimation.

2.3 Classification

To estimate the efficiency of the selected genes, classification models are used and evaluation is done by considering various performance metrics such as accuracy, specificity, and LOOCV.

Evolutionary algorithms could be used to enhance and optimize both the gene classification and selection approach. The next section reviews several gene-expression analysis techniques using evolutionary algorithms.

3 Literature Review

Evolutionary algorithms have been one of the most efficient approaches to find optimal solutions to NP-Hard problems. As optimal gene selection has been treated as one of the NP-Hard problems, hence evolutionary algorithms have got significant applications in this field. Wahde et al. [4] reviewed various techniques where evolutionary algorithms based gene-expression classification was done, basically GA-based techniques. Pal et al. [5] reviewed several techniques based on evolutionary algorithms, in the field of bio-informatics, where gene-expression analysis forms one of the parts.

Xu et al. [6] proposed a technique that consists of Binary-PSO(Particle Swarm Optimization) for gene selection and EAM-Ellipsoid AATMAP for classification using LOOCV (Leave-out-one cross-validation) as a performance metric. Akadi et al. [7] used mRMR(Minimum Redundancy–Maximum Relevance), which is a filter approach and GA (Genetic Algorithm), which is a wrapper approach, for gene selection and Naive Bayes and SVM classification on five benchmark gene dataset. They also used LOOCV as a measure of performance. Mohamad et al. [8] analyzed gene expression of ten benchmark gene datasets by using Improved Binary-PSO for gene selection and SVM for classification with LOOCV as a metric of performance measure. Saraswathi et al. [9] used ICGA (Integer-Coded Genetic Algorithm) for gene selection and PSO for estimating optimal weights for ELM (Extreme Learning Machine).

Similarly, Srivastava et al. [10] proposed a technique which selected genes based on weighted-ranking of genes and classified using DFA-SVM (Discreet Firefly Algorithm based Support Vector Machine) and DFA-RF (Discreet Firefly Algorithm based Random Forest). Mishra et al. [11] used BAT algorithm for finding optimal weights of FLANN (Functional Link Artificial Neural Network) for the classification of genes. Sahu et al. [12] clustered genes using the K-Means algorithm followed by ranking according to SNR (Signal-to-Noise Ratio) and afterward used SVM, k-NN (k-Nearest Neighbor), and PNN (Probabilistic Neural Network) for classification. Yang et al. [13] also clustered genes by K-means algorithm followed by selection using PSO and classification using ELM.

Chen et al. [14] in their paper used PSO for gene selection and Decision Tree for classification. Diaz et al. [15] proposed GA-based gene selection for lung cancer classification using SVM and ANN. Tabakhi et al. [16] used ACO (Ant Colony Optimization) for gene selection and considered variance as a relevance function. Elyasigomari et al. [17] selected genes using COA (Cuckoo Optimization Algorithm) based GA and classified using SVM and MLP (Multi-Layer Perceptron). Alshamlan et al. [18] used mRMR and ABC (Artificial Bee Colony) optimization for selection and SVM for classification. Similarly, they used hybridization of GA and ABC, i.e., GBC, for gene selection followed by SVM classification [19]. Kar et al. [20] used PSO for finding optimal k-values for k-NN classification which would help to search in a better neighborhood.

Das et al. [21] used PSO for selection, where the FCM index (Fuzzy C-Means) was used as a fitness function and Silhouette index as a performance metric. Chinnaswamy et al. [22] proposed a hybridized technique of correlation coefficient and PSO for gene selection. Vosooghifard et al. [23] selected genes using GWO (Grey-Wolf Optimization) followed by classification using a Decision Tree. Gurav et al. [24] proposed a hybrid approach. First, they selected the gene subset using GSO (Glowworm Swarm Optimization), then ranked the features based on information gain followed by classification via SVM. Garro et al. [25] used ABC for gene selection, and SVM, MLP, and RBF (Radial basis feed-forward network) for classification. Lu et al. [26] selected best expressing genes using MIM (Mutual Information Maximization) and AGA (Adaptive Genetic Algorithm) which automatically adjusts crossover and mutation probability for better adaptation to the search space. Wang et al. [27] used an efficient weight-based feature selection method for enhancing the accuracy of gene classification. Motieghadar et al. [28] proposed a hybridized approach based on GA and Learning Automata for gene selection. Similarly, Shahbeig et al. [29] used the hybridization of TLBO (Teaching–Learning based Optimization) and adaptive-fuzzy PSO for gene selection for breast cancer detection. Panda [30], in his paper, used ESO (Elephant Search optimization) for feature selection and k-NN for classification. Mohapatra et al. [31] used Modified Cat Swarm Optimization and Kernel ridge regression for gene analysis. The discussed technologies and methodologies have been briefed in Table 1.

Table 1 List of some research works involving evolutionary algorithms for gene-expression analysis

Sl. no.	Names of author	Year	Technology used
1	Xu et al. [6]	2007	Binary-PSO, EAM-Ellipsoid AATMAP
2	Akadi et al. [7]	2011	mRMR-GA, Naive Bayes, SVM
3	Mohamad et al. [8]	2011	Improved Binary-PSO, SVM
4	Saraswathi et al. [9]	2011	ICGA-PSO, ELM
5	Mishra et al. [11]	2012	BAT Algo, FLANN
6	Sahu et al. [12]	2012	K-means-SNR, SVM, k-NN, PNN
7	Srivastava et al. [10]	2013	DFA-SVM, DFA-RF
8	Yang et al. [13]	2013	K-means-PSO, ELM
9	Chen et al. [14]	2014	PSO, Decision tree
10	Diaz et al. [15]	2014	GA, SVM, ANN
11	Gurav et al. [24]	2014	GSO, SVM
12	Elyasigomari [17]	2015	COA-GA, SVM, MLP
13	Alshamlan et al. [18]	2015	mRMR-ABC, SVM
14	Kar et al. [20]	2015	PSO, k-NN
15	Alshamlan et al. [19]	2015	GBC, SVM
16	Das et al. [21]	2015	FCM-PSO
17	Tabakhi et al. [16]	2015	ACO
18	Vosooghifard et al. [23]	2015	GWO, Decision tree
19	Chinnaswamy et al. [22]	2016	Correlation-PSO
20	Garro et al. [25]	2016	ABC, SVM, MLP, RBF
21	Mohapatra et al. [31]	2016	Modified Cat SO, KRR
22	Wang et al. [27]	2017	Weight-based feature selection
23	Motieghader et al. [28]	2017	GA-learning automata
24	Shahbeig et al. [29]	2017	TLBO-adaptive-fuzzy PSO
25	Panda et al. [30]	2017	Elephant search optimization, k-NN
26	Lu et al. [26]	2017	MIM, AGA

4 Benchmark Dataset

There are some benchmark datasets that are followed by most of the researchers, for estimating the effectiveness of their proposed approach. Some of the datasets are listed in Table 2.

Table 2 List of some benchmark datasets

Sl. no.	Name of dataset	Features	Class	Instance
1	DLBCL	5470	2	77
2	Prostrate_Tumor	10509	2	102
3	Lung cancer I	12600	5	203
4	SRBCT	2309	4	83
5	Leukemia (ALLAML)	7129	2	72
6	Colon	2000	2	62
7	Central nervous system	7129	2	60
8	Breast cancer Wisconsin	10	2	699
9	Brain tumor I	5920	5	90
10	Lymphoma	62	3	62
11	Global cancer map	16063	14	190
12	NCI60	1416	9	60

5 Conclusion and Future Work

This paper focused on those techniques of gene-expression analysis that used the working principle of evolutionary algorithms. But several other algorithms such as SGO (Social Group Optimization) [32], GO (Grasshopper Optimization) [33], and SELO (Socio-Evolutionary Learning Optimization) [34] have been proposed by researchers which still have not been applied to this context, which makes this field an emerging application area. An approximate status of application of some evolutionary algorithms in gene analysis has been shown in Fig. 2. (Data collected from https://scholar.google.com/). Applying other algorithms or new hybrid techniques may pave way for more promising results.

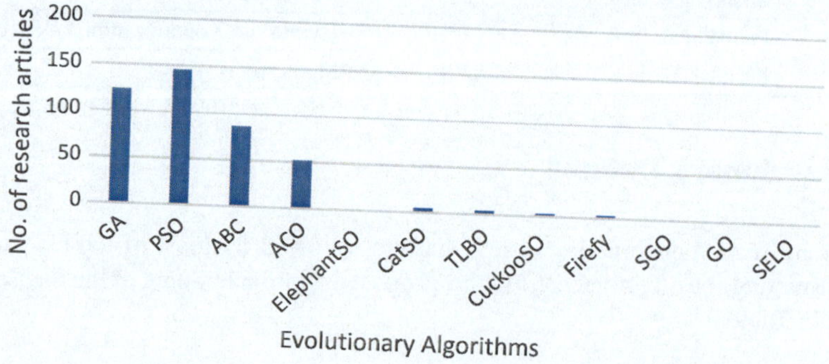

Fig. 2 Histogram of no. of research articles based on evolutionary algorithms

References

1. Jauhari S, Rizvi SAM (2014) Mining gene expression data focusing cancer therapeutics: a digest. IEEE/ACM Trans Comput Biol Bioinf (TCBB) 11(3):533–547
2. Tillinghast GW (2010) Microarrays in the clinic. Nat Biotechnol 28(8):810
3. Singh RK, Sivabalakrishnan M (2015) Feature selection of gene expression data for cancer classification: a review. Proc Comput Sci 50:52–57
4. Wahde M, Szallasi Z (2006) A survey of methods for classification of gene expression data using evolutionary algorithms. Exp Rev Mol Diagnost 6(1):101–110
5. Pal SK, Bandyopadhyay S, Ray SS (2006) Evolutionary computation in bioinformatics: a review. In: IEEE Transactions on Systems, Man, and Cybernetics, Part C (Applications and Reviews), vol 36(5). pp 601–615
6. Xu R, Anagnostopoulos GC, Wunsch DC (2007) Multiclass cancer classification using semisupervised ellipsoid ARTMAP and particle swarm optimization with gene expression data. IEEE/ACM Trans Comput Biol Bioinf 4(1):65–77
7. El Akadi A, Amine A, El Ouardighi A, Aboutajdine D (2011) A two-stage gene selection scheme utilizing MRMR filter and GA wrapper. Knowl Inf Syst 26(3):487–500
8. Mohamad MS, Omatu S, Deris S, Yoshioka M (2011) A modified binary particle swarm optimization for selecting the small subset of informative genes from gene expression data. IEEE Trans Inf Technol Biomed 15(6):813–822
9. Saraswathi S, Sundaram S, Sundararajan N, Zimmermann M, Nilsen-Hamilton M (2011) ICGA-PSO-ELM approach for accurate multiclass cancer classification resulting in reduced gene sets in which genes encoding secreted proteins are highly represented. IEEE/ACM Trans Comput Biol Bioinf (TCBB) 8(2):452–463
10. Srivastava A, Chakrabarti S, Das S, Ghosh S, Jayaraman VK (2013) Hybrid firefly based simultaneous gene selection and cancer classification using support vector machines and random forests. In: Proceedings of seventh international conference on bio-inspired computing: theories and applications (BIC-TA 2012). Springer, Berlin, pp 485–494
11. Mishra S, Shaw K, Mishra D (2012) A new meta-heuristic bat inspired classification approach for microarray data. Proc Technol 4:802–806
12. Sahu B, Mishra D (2012) A novel feature selection algorithm using particle swarm optimization for cancer microarray data. Proc Eng 38:27–31
13. Yang S, Han F, Guan J (2013) A hybrid gene selection and classification approach for microarray data based on clustering and PSO. In: International conference on intelligent computing. Springer, Berlin, Heidelberg, pp 88–93
14. Chen KH, Wang KJ, Wang KM, Angelia MA (2014) Applying particle swarm optimization-based decision tree classifier for cancer classification on gene expression data. Appl Soft Comput 24:773–780
15. Diaz JM, Pinon RC, Solano G (2014) Lung cancer classification using genetic algorithm to optimize prediction models. In: IISA 2014, The 5th international conference on information, intelligence, systems and applications. IEEE, pp 1–6
16. Tabakhi S, Najafi A, Ranjbar R, Moradi P (2015) Gene selection for microarray data classification using a novel ant colony optimization. Neurocomputing 168:1024–1036
17. Elyasigomari V, Mirjafari MS, Screen HR, Shaheed MH (2015) Cancer classification using a novel gene selection approach by means of shuffling based on data clustering with optimization. Appl Soft Comput 35:43–51
18. Alshamlan, H., Badr, G., & Alohali, Y. (2015). mRMR-ABC: a hybrid gene selection algorithm for cancer classification using microarray gene expression profiling. *Biomed research international, 2015*
19. Alshamlan HM, Badr GH, Alohali YA (2015) Genetic bee colony (GBC) algorithm: a new gene selection method for microarray cancer classification. Comput Biol Chem 56:49–60
20. Kar S, Sharma KD, Maitra M (2015) Gene selection from microarray gene expression data for classification of cancer subgroups employing PSO and adaptive K-nearest neighborhood technique. Expert Syst Appl 42(1):612–627

21. Das R, Saha S (2015) Gene expression classification using a fuzzy point symmetry based PSO clustering technique. In: 2015 second international conference on soft computing and machine intelligence (ISCMI). IEEE, pp 69–73
22. Chinnaswamy A, Srinivasan R (2016) Hybrid feature selection using correlation coefficient and particle swarm optimization on microarray gene expression data. In: Innovations in bio-inspired computing and applications. Springer, Cham, pp 229–239
23. Vosooghifard M, Ebrahimpour H (2015) Applying grey wolf optimizer-based decision tree classifer for cancer classification on gene expression data. In: 2015 5th International Conference on Computer and Knowledge Engineering (ICCKE). IEEE, pp 147–151
24. Gurav A, Nair V, Gupta U, Valadi J (2014) Glowworm swarm based informative attribute selection using support vector machines for simultaneous feature selection and classification. In: International conference on swarm, evolutionary, and memetic computing. Springer, Cham, pp 27–37
25. Garro BA, Rodríguez K, Vázquez RA (2016) Classification of DNA microarrays using artificial neural networks and ABC algorithm. Appl Soft Comput 38:548–560
26. Lu H, Chen J, Yan K, Jin Q, Xue Y, Gao Z (2017) A hybrid feature selection algorithm for gene expression data classification. Neurocomputing 256:56–62
27. Wang H, Jing X, Niu B (2017) A discrete bacterial algorithm for feature selection in classification of microarray gene expression cancer data. Knowl Based Syst 126:8–19
28. Motieghader H, Najafi A, Sadeghi B, Masoudi-Nejad A (2017) A hybrid gene selection algorithm for microarray cancer classification using genetic algorithm and learning automata. Inf Med Unlock 9:246–254
29. Shahbeig S, Helfroush MS, Rahideh A (2017) A fuzzy multi-objective hybrid TLBO–PSO approach to select the associated genes with breast cancer. Sig Process 131:58–65
30. Panda M (2017) Elephant search optimization combined with deep neural network for microarray data analysis. J King Saud Univ Comput Inf Sci
31. Mohapatra P, Chakravarty S, Dash PK (2016) Microarray medical data classification using kernel ridge regression and modified cat swarm optimization based gene selection system. Swarm Evol Comput 28:144–160
32. Satapathy S, Naik A (2016) Social group optimization (SGO): a new population evolutionary optimization technique. Comp Intell Syst 2(3):173–203
33. Mafarja M, Aljarah I, Heidari AA, Hammouri AI, Faris H, Ala'M AZ, Mirjalili S (2018) Evolutionary population dynamics and grasshopper optimization approaches for feature selection problems. Knowl Based Syst 145:25–45
34. Kumar M, Kulkarni AJ, Satapathy SC (2018) Socio evolution & learning optimization algorithm: a socio-inspired optimization methodology. Future Gener Comput Syst 81:252–272

An Efficient JAYA-Based Clustering Technique

Prabhu Prasad Dev, Priya Mishra, and Anasua Banerjee

Abstract This paper proposes a partitioned clustering technique using a relatively new evolutionary computational technique known as JAYA optimization algorithm. K-means clustering techniques happen to be widely researched partitional clustering algorithm in data mining literature. Despite its ease of implementation and widespread application, it has various shortcomings. The efficiency of k-means is heavily dependent on selection of initially cluster centroids. This apart, the time complexity of k-means is dependent on the size of the data sets to be clustered. In this work, we propose JAYA, a population-based approach which has scope to choose many candidate centroids and the initialization time and processed to evolve optimal cluster centroids and through simulation is carried out to benchmark data sets, and a comparison study is taken that JAYA-based algorithm is able to provide optimal performance, i.e., intra-cluster distance (ICD). The results are tabulated in the paper.

1 Introduction

Clustering is the process of separating the data sets into different types of groups, consisting of similar type data points. The grouping of similar object into a set is known as cluster. Here, an object in one cluster is to be different from another object of another cluster to be compared. The concept of cluster cannot be easily defined and that is the reason we use several algorithms which are available for clustering. These algorithms are different from each other in their properties. Suppose for example we can get a clear view of clustering when we go to a food store we see different types

P. P. Dev (✉) · P. Mishra · A. Banerjee
School of Computer Engineering, Kalinga Institute of Industrial Technology (KIIT), Deemed to be University, Bhubaneswar, India
e-mail: prasaddev97@gmail.com

P. Mishra
e-mail: mishrap367@gmail.com

A. Banerjee
e-mail: anasua123.banerjee@gmail.com

S. C. Satapathy et al. (eds.), *Communication Software and Networks*, Lecture Notes in Networks and Systems 134, https://doi.org/10.1007/978-981-15-5397-4_66

of vegetables and fruits displayed in the store. Clustering is to intrinsically group unlabeled data. Clustering is used in amazon and Netflix.

There are three types of clustering such as exclusive clustering, overlapping clustering, and hierarchical clustering. Exclusive clustering is also known as hard clustering where item belongs exclusively to one cluster. For example, if there are two circles with boundaries one with pink color and another with blue color, now the points with pink color will be inside the pink circle and the points with blue color will be inside the blue circle. This is called exclusive clustering (or hard clustering). The another type of clustering is overlapping clustering which is also known as soft clustering where the items belong to multiple clusters. Here, under this clustering, an object can belong to many clusters. For example, again if there are two circles of boundaries with pink color and another with blue color, there are some points inside that two circles the same as their boundary color. All the points are interlinked with other as well as both circles are overlapped where some points of blue color are interlinked with some points of the circle with boundaries having pink color. This is called overlapping clustering (or soft clustering). The another type of clustering is hierarchical clustering where individual clusters are formed from multiple data points and after some steps multiple individual clusters are combined into a single cluster, for example, if we will take six data points to let in first step, first and second data points are similar and fourth and fifth data points are similar, then combination of first and second is combined with third c, and then combination of fourth and fifth combination is combined with sixth cluster. Then in the final step, all clusters are combined into a single cluster.

Here, we have taken to techniques K-means clustering and JAYA which is briefly explained below.

2 K-means Clustering

K-means is exclusive clustering technique whose goal is to group similar data points into a cluster, where K represents the number of clusters in data sets. We will take the target as K which refers to the number of seeds (centroids) we need in the data sets. K-means clustering is an extensively used technique for data clustering analysis. We can take an example on cricket match. Here, the first task is to identify the batsman and bowler, and the data contains runs and wickets gained in the previous matches. So the bowler will have more wickets, and the batsman will have higher runs.

The working of k-means clustering is as follows [2]:

1. Determine number of clusters K.
2. Initialize centroids by doing random permutation of the iris dataset and then selecting K data points for the centroids.
3. Repeat the iteration until there is no change to the centroids.
4. Calculate the distance from each data point (x_i) to centroid (y_i) of one cluster.
5. Allocate each data point to the nearest cluster.

$$d(x, y) = \sqrt{\sum_{i=1}^{n} (x_i - y_i)^2}$$

6. Compute the intra-cluster distance (ICD) for the clusters by the following formula:

$$ICD = \frac{\sum_{k=1}^{n} d(k)}{k}$$

where k = number of clusters, $d(k)$ = distance from kth cluster centroid to all the data points within kth cluster.

Figure 1 depicts the flowchart of the k-means clustering technique.

The major **shortcoming** with k-means algorithm is that it contains clustering anomaly which means that before clustering we have to choose the random initial

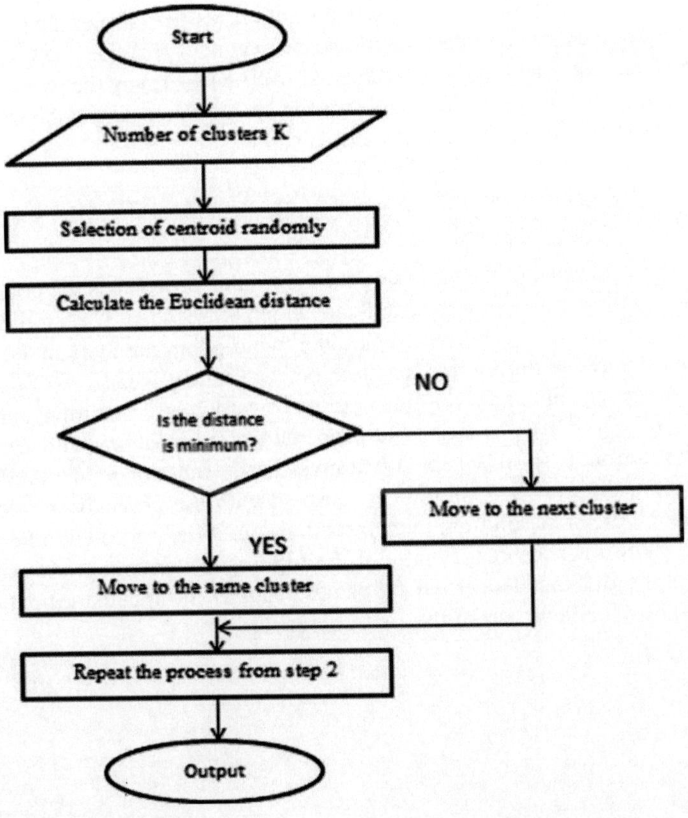

Fig. 1 K-means clustering flowchart

seeds from the data sets. The data sets depend upon the multiplier factor. If the data is of varying size and density, then clustering is difficult.

3 JAYA Algorithm

JAYA is a simple, powerful optimization algorithm proposed by R Venkata Rao in 2015 for solving the constrained and unconstrained optimization problems [1]. It is a specific parameter less algorithm as it requires only common parameters like population size, number of design variables, etc. It is a gradient-free optimization algorithm. It is a population-based method which repeatedly modifies a population of individual solutions. This technique tries to move closer to success (i.e., reaching the best solution) and to refrain from the failure (i.e., moving away from the worst solution).

Here, $f(c)$ is the objective function which is the intra-cluster distance which is to be reduced. Let us assume that in each iteration *i*, there exists number of design variables as "*m*", and number of candidate solutions as "*n*" (size of the population, *k* = 1, 2, ..., *n*). In entire candidate solutions, best value of $f(c)$ (i.e., $f(c)_b$) is obtained by the best candidate and the worst value of $f(c)$ (i.e., $f(c)_w$) by the worst candidate.

If $C_{i,j,k}$ is the value of the *i*th variable for the *j*th candidate during the *k*th iteration, then this value is to be updated as per Eq. (1):

$$C'_{i,j,k} = C_{i,j,k} + \text{rand}_{1,i,k}(C_{i,b,j} - |C_{i,j,k}|) - \text{rand}_{2,i,k}(C_{i,w,j} - |C_{i,j,k}|) \quad (1)$$

where $C_{i,b,j}$ is the value of the variable *i* for the candidate which is the best and $C_{i,w,j}$ is the value of the variable *i* for the candidate which is the worst. $C'_{i,j,k}$ is the modified value of $C_{i,j,k}$ and $\text{rand}_{1,j,i}$ and $\text{rand}_{2,j,i}$ are the two random numbers in the range [0, 1] for the *i*th variable during the *k*th iteration. The term "$+\text{rand}_{1,i,k}$ ($C_{i,b,j}-|C_{i,j,k}|$)" indicates the propensity of the solution to move toward the best solution, and the term "$-\text{rand}_{2,i,k}$ ($C_{i,w,j}-|C_{i,j,k}|$)" indicates the propensity of the solution to move away from the worst solution. $C'_{i,j,k}$ is accepted if it gives better function value. At the end of each iteration, the modified solution is checked with the previous solution if it is better than the previous solution; then, modified solution is accepted; otherwise, the previous solution is accepted. Finally, if the termination criteria are satisfied, then the optimal solution is stated; else these values become the input to the next iteration. Figure 2 shows the flowchart of the JAYA algorithm.

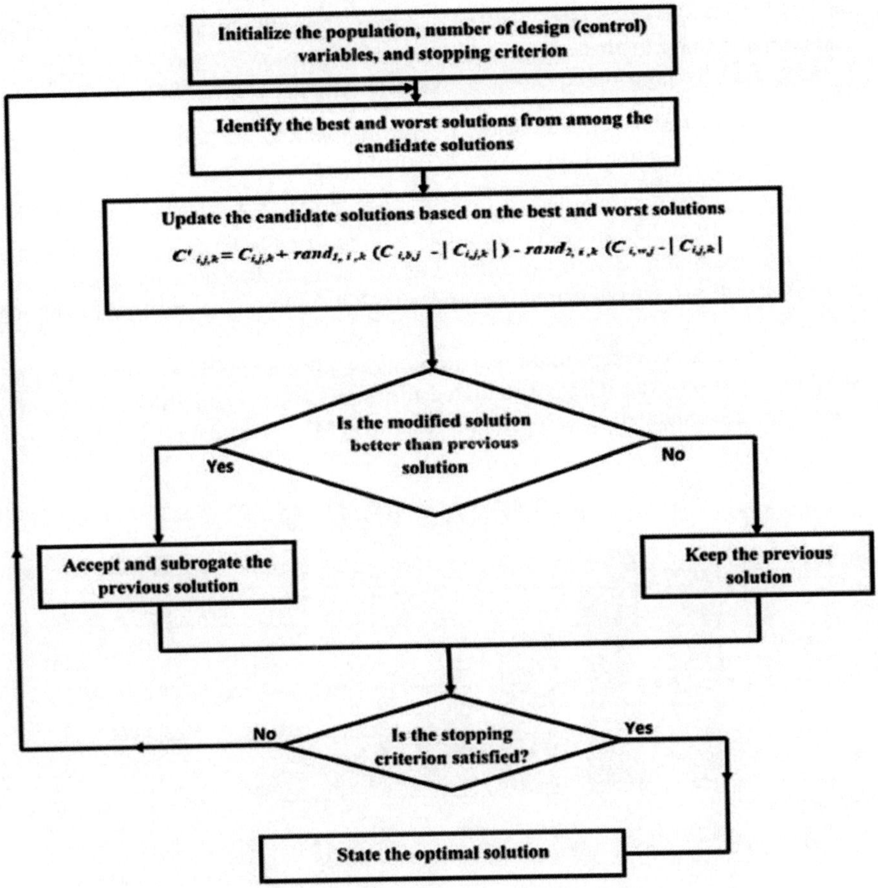

Fig. 2 JAYA algorithm flowchart

4 Experimental Results

Here we have considered iris dataset as an input data. We have initialized population size (n) as 5, number of design variable (m) as 4 and cluster size is taken as 3 in k-means algorithm. Number of design variables is 4 as in iris dataset it has four features as sepal length, petal length, sepal width, petal width). Then three centers (cluster centroid-1, cluster centroid-2, cluster centroid-3) from each of five candidates of iris dataset are randomly chosen. Each candidate (C1, C2, C3, C4, C5) is fed into k-means clustering algorithm which calculates the distance from each data point to centroid. Finally, intra-cluster distance (ICD) is calculated as per the following formula:

$$ICD = \frac{\sum_{k=1}^{n} d(k)}{k}$$

where k = number of clusters, $d(k)$ = distance from kth cluster centroid to all the data points within kth cluster.

Here, we have considered "k" as 3. So, ICD = $(d1 + d2 + d3)/3$.

$d1$ = sum of distance from cluster-1 centroid to each data point within cluster1.
$d2$ = sum of distance from cluster-2 centroid to each data point within cluster2
$d3$ = sum of distance from cluster-3 centroid to each data point within cluster3.

Among all the ICDs, the minimum ICD is considered as best candidate, i.e., best center and maximum ICD is considered as worst candidate, i.e., worst center. At every 20 iterations, we compared the minimum ICD of k-means and JAYA as shown in Fig. 3.

Here, we have taken termination or stopping condition criteria as 100th iteration. At 100th Iterations, the ICD is calculated for both k-means and JAYA. The results of k-means are tabulated in Table 1, and results of JAYA are tabulated in Table 2.

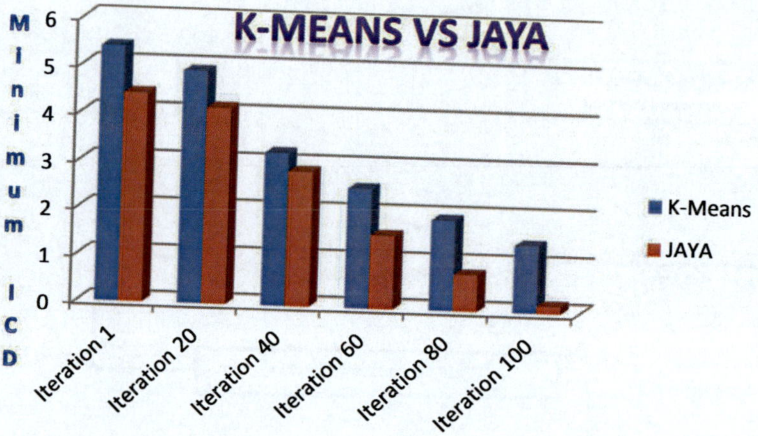

Fig. 3 Comparison of k-means and JAYA in terms of best ICD

Table 1 At 100th iteration, result of k-means

C	Cluster centroid-1	Cluster centroid-2	Cluster centroid-3	ICD	Status
C1	7.7 3.8 6.7 2.2	5.7 3.0 4.2 1.2	5.1 3.5 1.4 0.3	1.43	*Best*
C2	4.9 2.5 4.5 1.7	5.8 4.0 1.2 0.2	6.7 3.1 4.7 1.5	1.67	
C3	6.1 3.0 4.9 1.8	4.9 3.1 1.5 0.1	4.7 3.2 1.6 0.2	5.07	*Worst*
C4	5.4 3.9 1.7 0.4	6.3 2.7 4.9 1.8	5.0 3.6 1.4 0.2	1.62	
C5	5.7 2.8 4.5 1.3	5.0 3.4 1.6 0.4	6.4 2.9 4.3 1.3	2.75	

Table 2 At 100th iteration, result of JAYA algorithm

C	Cluster centroid-1	Cluster centroid-2	Cluster centroid-3	ICD	Status
C1	7.7 3.8 6.7 2.2	5.7 3.0 4.2 1.2	5.1 3.5 1.4 0.3	0.16	*Best*
C2	4.9 2.5 4.5 1.7	5.8 4.0 1.2 0.2	6.7 3.1 4.7 1.5	0.20	
C3	6.1 3.0 4.9 1.8	4.9 3.1 1.5 0.1	4.7 3.2 1.6 0.2	1.16	*Worst*
C4	5.4 3.9 1.7 0.4	6.3 2.7 4.9 1.8	5.0 3.6 1.4 0.2	1.62	
C5	5.7 2.8 4.5 1.3	5.0 3.4 1.6 0.4	6.4 2.9 4.3 1.3	0.04	

5 Conclusion

In this paper, the k-means clustering technique has experimented with an iris data set as an appropriate example. We found that thc JAYA algorithm is more efficient than k-means clustering technique. Besides this, JAYA algorithm proposed that technique shows that the ICD is less in comparison to a k-means algorithm. Moreover, in the future, the design technique can be implemented with different datasets to prove its significance.

References

1. Rao RV (2016) Jaya: a simple and new optimization algorithm for solving constrained and unconstrained optimization problems. Int J Ind Eng Comput 7(1):19–34
2. Singh SP, Yadav A Study of K-means and enhanced K-means clustering algorithm

Online Voting System Using Blockchain

Anjali Prajapati and Vandana Reddy

Abstract One of the major areas in technical development is blockchain and bitcoin. These technologies have enabled many simulations in in-hostile applications that have major issues with security and integrity of data. To provide more relevance to the available cyberphysical systems in the dimension of security, the blockchain technology offers a major help. If the present scenario is considered, we have multiple day-to-day applications that have been simulated and require more security enhancement. For example, the E-voting systems are a trend and their security features have to be upgraded to authenticate both systems and processes. The present research paper focuses on the same application and aims to provide security upgradation by proposing a working model of e-voting systems.

1 Introduction

In this new era of growing technologies, everyone is expecting a very much comfortable life which definitely needs security. Everything nowadays is online and can be easily by seating at home with just one click or tap with a mobile or a laptop. Things starting from buying vegetables to ordering an SUV can be done online. Since everyone is too busy with their work and responsibilities, they want a life where their work should be done within a moment without doing anything, or just with a single click. The authors have already seen many changes in our day-to-day life which is very much different from the old days, for example, if you want to do a payment from your bank account, there is no need of standing in a long queue outside the ATM and withdraw cash and pay the amount; you can just go to a shop, buy whatever you want, and pay the amount using your card or online (using Paytm, Freecharge, etc.). There are a lot of applications and websites which allow you to do all types of transactions not only easy but with full security and efficiency.

All this is possible because of blockchain and bitcoins. The basic idea of this block is to divide the database into different blocks and give the access to the authorized

A. Prajapati (✉) · V. Reddy
Faculty of Engineering, Christ (Deemed to be University), Bangalore, India
e-mail: Anjali.prajapati@btech.christuniversity.in

S. C. Satapathy et al. (eds.), *Communication Software and Networks*, Lecture Notes in Networks and Systems 134, https://doi.org/10.1007/978-981-15-5397-4_67

people and whenever someone is in the need of the data, that person and request for the data from the authorized person and if the authorized person feels that the need is genuine, he/she can do the needful. Since the database is divided into different blocks, in case of any cheating or threads, the design of the database will be different from the design of the original database. Since the blocks are divided sequentially and the data stored in the blocks is hashed, sometimes, it is possible that the owner of the block, himself/herself, does not know the meaning of the data or the information present in it, also any alternation of data or information is least likely to happen, and hence it increases the security of the database. However, if there is any alteration or thread present in the block, it can be found very easily since the information that was stored is hashed.

India is a democratic country and now since everything in India is becoming fast, safe, and secure, the authors also require a system that has similar abilities, which is fast, safe, and secure and can be used wherever needed. A system which allows the nation to choose their leader, not only with their willingness but also with the assurance that their vote, is not being altered or changed by any means or by anyone. A system is completely independent of a manual system and is completely dependent on the algorithm on which it is running. An E-VOTING system can be used by any literate or illiterate person who is 18 or above in age where the system assures the person in front of it that know your vote is safe with it.

The e-voting system which is being introduced below takes the issue, which is mentioned above, into consideration and make sure that the elections performed in a country like India to elect a leader democratically are done fairly and with full safety and security.

2 Literature Review

Progressively computerized innovation in the present helped numerous individuals' lives. In contrast to the constituent framework, there are numerous customary employments of paper [1] in its execution. The part of security and straightforwardness is a danger from still across the board race with the customary framework (disconnected). General decisions still utilize a unified framework; there is one association that oversees it. A portion of the issues that can happen in conventional appointive frameworks is with an association that has full power over the database and framework; it is conceivable to mess with the database of impressive chances. Blockchain innovation is one of the arrangements, since it grasps a decentralized framework and the whole database is possessed by numerous clients. Blockchain itself has been utilized in the bitcoin framework known as the decentralized bank framework. By receiving blockchain in the conveyance of databases on e-casting ballot frameworks can lessen one of the deceiving wellsprings of database control. This exploration examines the account of casting a ballot result utilizing blockchain calculation from each place of decision. By building this [2] proposed shrewd contract of our own, the authors have prevailed with regard to moving e-casting a ballot to the blockchain stage and we tended to a portion of the crucial issues that heritage e-casting ballot frameworks have, by utilizing the intensity of the Ethereum arrange and the blockchain structure. Because of our preliminaries, the idea of blockchain and the security technique which it utilizes, to be specific permanent hash chains, has turned out to be versatile to surveys and decisions. This accomplishment may even make ready for other blockchain applications that have an effect on each part of human life. Now, ethereum and the savvy contracts, which made a standout among the most progressive leaps forward since the blockchain itself, upset the restricted view of blockchain as a cryptographic money (coin), and transformed it into a more extensive arrangement base for some Internet-related issues of the cutting edge world and may empower the worldwide utilization of blockchain. E-casting a ballot is as yet a disputable point inside both political and logical circles. Notwithstanding the presence of a couple of genuine models, the vast majority of which are still being used; a lot more endeavors were either neglected to give the security and protection highlights of a customary decision or have genuine ease of use and adaptability issues. Despite what might be expected, blockchain-based e-casting a ballot arrangement includes the one the authors have executed utilizing the keen contracts and the ethereum system, address (or may address with applicable adjustments), the majority of the security concerns, similar to protection of voters, trustworthiness, check and non-revocation of votes, and straightforwardness of tallying. However, there are likewise a few properties that can't be tended to exclusively utilize the blockchain, for instance, validation of voters (on the individual dimension, not on the record level) requires extra systems to be coordinated, for example, utilization of biometric factors. E-VOTING is among the key open divisions that can be upset by blockchain innovation. The thought in blockchain empowered e-casting a ballot (BEV) is straightforward. To utilize an advanced cash similarity, BEV [3] issues every voter a "wallet" containing a client certification. Every voter gets a solitary "coin" speaking to one chance to cast a ballot. Making a choice exchanges the voter's coin to a competitor's wallet. A voter can spend his or her coin just once. Be that as it may, voters can change their vote before a preset due date. Here, the authors contend that blockchains may address two of the most predominant worries in casting a ballot today: voter access and voter extortion. The thought is as per the following. Qualified voters cast a vote namelessly utilizing a PC or cell phone. BEV utilizes an encoded key and carefully designed individual IDs. For instance, the versatile e-casting a ballot stage of the Boston-based startup Voatz utilizes brilliant biometrics and constant ID check. The general population record binds each cast ticket to an individual voter and sets up a perpetual, changeless record. No awful on-screen character can take part in detestable exercises in light of the fact that such exercises will be apparent on the record or remedied by a shared agreement organize. To trade off the system, programmers would need to effectively hack the vast majority of the squares (records with exchange re-ropes) before new squares were presented. The blockchain's review trail guarantees that no vote has been changed or evacuated and that no fake and ill-conceived casting a ballot have been included.

In paper [4], the creators present the total agreement procedure of POV. In the POV, part hubs have the privilege to cast a ballot, and the hinder with high votes is valid, making the legitimate square interesting. Some extraordinary hubs keep running for the privilege of creating hinders in request to execute straightforwardly with no outsider middle person in a consortium blockchain arrange. We break down the security, exchange conclusiveness, and probability of bifurcation. As shown in the paper, POV can get magnificent execution with ultra-low inactivity in exchange check. Hubs can be assaulted and progress toward becoming enemies. We expect that a few (not exactly half) of the association individuals' machines have been assaulted. The framework can endure under half key hubs being assaulted. Enemies may fashion exchanges or act like a typical one. Systems might be divided. In any case, the enemy can't break and produce the marks. The hubs in the union utilize elite and dependable machines and working frameworks, with the goal that the partnership hubs are less inclined to be assaulted than conventional hubs. System Time Protocol (NTP) server is utilized to synchronize the season of urgent hubs. At the point when a key hub is restarted, it will initially alter its opportunity to synchronize with the NTP server, at that point partake in accord process worked of blockchain.

Design

An architecture is given below to show the working of the model (Fig. 1).

Fig. 1 Flowchart of the model

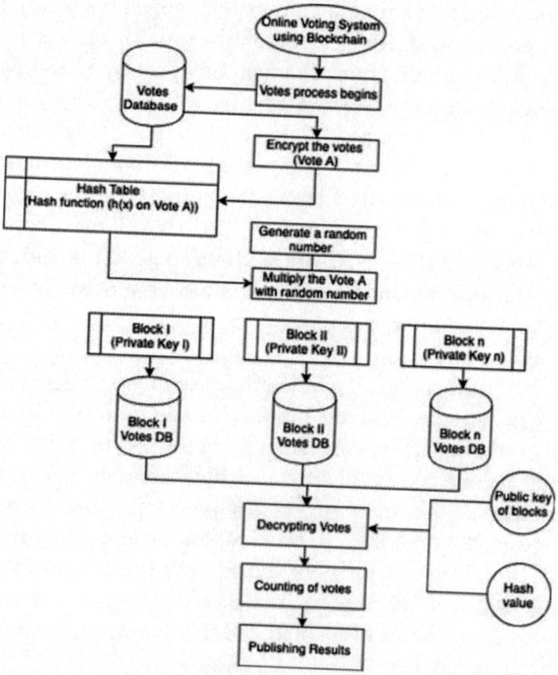

The flowchart design clearly shows that the working model is divided into different steps which are assigned with different works and functions. The description of each and every step is given below:

A. Person X: This is the first part of the flowchart which represents a random person who is here to give his/her vote. Each person whoever is giving vote will be provided with a unique ID (e.g., −10010101). This will make sure that a person cannot give vote more than once, and also identification of the vote corresponding to that person becomes easy which will be hashed later for storage.
B. Vote A: This part of the flowchart represents the vote given by the person X which will be hashed and stored in the database at the back end, and will be counted later for finding the total number of votes belonging to a particular party or candidate.
C. Hashing (Converting the vote to a hashed value): This is one of the most important parts of the model; this part of the model will take care of the conversion of the vote to a hashed value. There are a lot of ways through which hashing can be done but here, linear probing, which is a type of open addressing method of hashing through which we can avoid collision while creating the hash table, is used because it provides better cache performance and there is absolutely no requirement of links. Since everything (here, every vote) is stored in one table or database, the process of counting becomes easy and efficient.
D. Store it in database: This is the most crucial step of the design as it deals with the database and its design. After converting the votes to hashed value, it will be stored in the database at the back end and once it is stored at the backend or in the database, it will be freezed and no more changes will be allowed hereafter. Here, the most important thing which has to be taken care of is the design of the database; it is very much important as any alteration in the design of the database will indicate any sort of cracking or misplacement of the votes.
E. Blockchain creation and distribution: In this step, the database will be divided into 'n' number of blocks and after the division it will be distributed to the authorized and trusted people, but even they would not able to make any changes to the block as it is freezed. These blocks are created and distributed just to keep them safe and refrain from any kind of miss-happening.
F. Decryption: The vote can be extracted only by an authorized person with the help of the hash value and the public key of the block.
G. Calculation of Votes and Publishing the Results: The final step of the model deals with the counting and declaring number of votes belonging to each party or the candidate; once everything is done, the results can be published.

3 Implementation

The proposed methodology consists of a very simple process where a random person X who is 18 years or above in age will be assigned with a unique ID and that person will give his/her Vote A online which will be recorded. Later on, the vote will be converted to a hashed value using linear probing which is a type of open addressing method of hashing, and thereafter the votes will be stored in the database at the backend with a defined design of the database, after which the process of block creation and distribution will be performed and once that is done, the process of validation of the database will be performed to find out if there is any error or fault in the design of the database; once it is confirmed that there is no error or fault in the design of the database, once the validation process is done, the process of counting and declaring the number of the votes will start and at the end the result will be forwarded to the commission.

A. *The unique ID generation will be done by the function given below:*

The will take the serial value of the person giving vote and covert it to a unique ID by converting it to a binary number.

```
        static void id(int n)
    {
         int[] binaryNum = new int[1000];

         int i = 0;
         while (n > 0)
         {
              binaryNum[i] = n % 2;
              n = n / 2;
              i++;
         }

    }
```

B. *The hashing will be done using the following function:*

```
    private int hash(String key)
    {
        return key.hashCode() % maxSize;
    }
    public void insert(String key, String val)
    {
        int tmp = hash(key);
        int i = tmp;
        do
        {
            if (keys[i] == null)
            {
                keys[i] = key;
                vals[i] = val;
                currentSize++;
                return;
            }
            if (keys[i].equals(key))
            {
                vals[i] = val;
                return;
            }
            i = (i + 1) % maxSize;
        } while (i != tmp);
    }
```

4 Results and Comparative Study

The system accepts a vote and encrypts it with a private key and a hash, thereby stored in different blocks which can only be accessed by a person with public key and the hash value. The method proposed takes better compared to all other existing systems for voting system as it takes much less time to get executed; also the safety of the system is ensured as the votes are hashed and can only be accessed by the authorized person (Fig. 2).

5 Future Scope

The unique identity generation can be done using a random function, and it will be better if the unique ID is a combination of both numbers and alphabets, i.e., alphanumeric. This will help the system in covering more number of voters as the

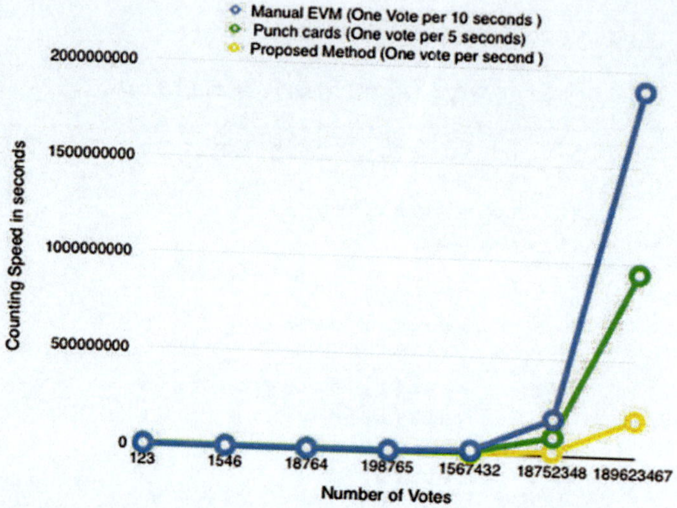

Fig. 2 Comparison of online voting system with EVM and punch cards system

binary ID generation may restrict us covering all the voters as it depends upon the size of the machine that is being used.

6 Conclusion

The system proposed above is an e-voting system using the idea and concept of blockchain, the e-voting system can be used for making the election fast, safe, as well as efficient. The system proposed above is easy to use and can be used whenever and wherever the voter wants to. Also, it makes the votes safe and secure.

References

1. Hanifatunnisa R, Rahardjo B (2017) Blockchain based E-voting recording system design. IEEE
2. Koç AK, Yavuz E, Çabuk UC, Dalklç G (2018) Towards secure E-voting using ethereum blockchain. IEEE
3. Kshetri N, Voas J (2018) Blockchain-enabled E-voting. IEEE
4. Li K, LiHui H, Hou H, Li K, Chen Y (2017) Proof of vote: a high-performance consensus protocol based on vote mechanism & consortium blockchain. In: 2017 IEEE 19th international conference on high performance computing and communications; IEEE 15th international conference on smart city; IEEE 3rd international conference on data science and systems

Data Communication Through Single-Board Computers

Prakruti Mallayya Bilagi, Vidyadhar S. Melkeri, and Gauri Kalnoor

Abstract Software-Defined Radio (SDR) is a reconfigurable radio developed by the combination of controller and an RF front end, which has a broad application in the communication system and broadcasting the information. SDR receiver consists of an analog solely to digital converter, connected to the antenna. In the latest, a high prototype of SDR is formed utilizing the Universal Software Radio Peripherals (USRP). Here the work is focused on the development of an SDR transmitter relating the single-board computer, Raspberry pi along‘ with other components such as a sound card and for SDR receiver, Realtek-SDR (RTL-SDR) for the full-duplex communication. In order to develop the SDR, C and C++ programming language are employed. The principle of the working process is on the modulation of frequency, which can also be further implemented for other modulation techniques.

1 Introduction

In the present day, communication is studied using analog circuitry by using integrated circuits like XR2206 or by some coding in MATLAB in the academic institutions. The experiments are performed and the results are viewed on the oscilloscope. But we fail to examine the output waveform in real time. The modulation techniques are currently being implemented and studied on the Universal Software Radio Peripheral (USRP), the research-oriented tool kits which are of high cost and sophisticated to deal with. Raspberry pi is a system on chip that has huge clock

P. M. Bilagi · V. S. Melkeri (✉)
Department of Electronics and Communication, Central University of Karnataka, Gulbarga, Karnataka, India
e-mail: vidyadharsm@gmail.com

P. M. Bilagi
e-mail: pmbilagi@gmail.com

G. Kalnoor
Department of Computer Science Engineering, Central University of Karnataka, Gulbarga, Karnataka, India
e-mail: kalnoor.gauri@gmail.com

S. C. Satapathy et al. (eds.), *Communication Software and Networks*, Lecture Notes in Networks and Systems 134, https://doi.org/10.1007/978-981-15-5397-4_69

generation capacity, making use of it as a carrier signal one can modulate the audio files in .wave format and transmit it using frequency modulation technique in real time. The approach is mentioned by Marcin Kondej in which using C++ coding and Linux commands he generates an FM signal and transmits the audio file in real time. This work includes the study and understands the FM generation and modifies it to obtain other modulation techniques.

1.1 Software-Defined Radio

Software-Defined Radio (SDR) is a collective term that refers to radio systems in which almost all of the functionality provided with the Physical Layer (PL) is executed in software using Digital Signal Processing (DSP) algorithms. An ideal SDR receiver will have a very small hardware front end; only an antenna and a high-speed GHz sampler are capable of capturing and digitizing a wide band of radio frequencies. Any demodulation, synchronization, decoding, or decryption required to recover information contained within a received signal would be performed in software that is executed on a superfast, dedicated processing device.

2 Literature Review

In [1] Mallari et at. presented how an RTL-SDR along with a GNU-radio can build a Wide Band Frequency Modulation receiver. An comparison between the low-cost RTL-SDR and a high cost is provided but effective USRP. In this paper GNU-radio is used as the software platform and built a project that has the blocks for RTL-SDR source, Frequency Translation Filter, WBFM Receive, Rational Resampler, Audio Sink, GUI FFT Sink. They successfully simulated an FM-receiver using a low-cost RTL-SDR.

In [2] Sruthi M B et al, proposed a windows-based platform to receive all the signals from DC to 1700 MHz. In their proposed model, an RealTek Software-Defined Radio is used. Which can receive all signals from Dc to 1700 MHz, for this to happen mixer is used. The transmitter is made of a personal computer connected to an antenna along with a mixer. The local oscillator provides the 106.250 MHz of input frequency. An inbuilt high-pass filter in the mixer will permit only the higher frequency of the signals to pass through. At the receiver end an RTL-SDR is used. The audio signal is converted to intermediate frequency and a signal is produced by mixing with a mixed, such that it can be tuned by an RTL-SDR. At the receiver end signals are collected and using RTL sharp and making some adjustment, waterfall spectrum diagram is displayed.

In [3] Saber et al. developed a module to generate a frequency in free bandwidth (ISM), then deploy it temporarily, without creating any interference to the primary user. An Raspberry pi as the FM transmitter using FHSS method and RTL-SDR along with MATLABSimulink are meant for the reception. The idea is to detect the presence of the PU signal by using local measurements and observations. The spectrum analyzer block in MATLAB to calculate the spectrum energy.

3 Prime Components

3.1 Realtek-SDR

It is a USB RF receiver, which is a low cost and very much handy device to use. Earlier these devices were used for receiving the Digital Video Broadcast-Terrestrial (DVB-T) signals only, later by configuring them into different modes it was discovered to be used as generic (receive only) SDR. Figure 1 shows the RTL-SDR used in this project for receiving the FM signals.

Not just the FM radio signal, but also UHF/DTV signals, Digital Audio Broadcast radio, 2G, 3 G, 4G, and Industrial Scientific and Medical (ISM) bands, etc. can be received based on the usage of the particular antenna and the signal present in the vicinity of the device. The functioning of the RTL-SDR depends on the principle of superheterodyne receivers, which is a popular design for receiving a wide range of frequency band signals. We need to demodulate the required signal and remove the noise before we start demodulating the received signals. For this purpose, a frequency selective filter is required. In the next stage, an intermediate frequency is generated by applying multiple filtering and amplification process to produce a fixed pre-specified frequency. Then the signal is passed to demodulation block which involves digital processing. The functionalities of the superheterodyne receiver are implemented by using two different integrated circuits:

- **Tuner**: It is an integrated circuit that is responsible for the down-conversion of the received baseband signal into intermediate frequency using analog signal processing techniques. The input signal is initially fed to a Low Noise Amplifier (LNA).

Fig. 1 RealTek-SDR

- **The RTL2832U**: The function of this IC is to sample the IF signal from the tuner by an 8-bit ADC and perform the signal processing tasks like decimation using an FIR low-pass filter and downsampling using a Digital Down Converter (DDC). It also contains the USB controller responsible for the handling of the USB 2.0 interface to the RTL-SDR. [4].

3.2 *Raspberry Pi*

It a miniaturized computer having all the essential components required to run the operating system on a single chip. It is designed in such a way that it is low cost and easy to use. It is designed using a Broadcom system on chip compatible with an ARM microcontroller and on chip GPU. The speed of the processor ranges from 700 MHz to 1.4 GHz and has RAM of up to have 1 GB. It supports external SD card for storing OS and other software and programs. The boards have got one to four USB ports.

3.2.1 Advanced Linux Sound Architecture

Advanced Linux Sound Architecture (ALSA) is a software framework and part of the Linux kernel, providing device drivers for sound card and an Application Programming Interface (API). The ALSA was developed because the Linux kernel sound drivers were causing lag in their functioning. The audio sound, which consists of varying air pressure, is converted into electric form by the transducer (mic). At regular intervals of time, samples are collected and converted into discrete values by analog-to-digital converters, known as sample rate. Further by using digital to analog converters, the original sound is produced back on the output transducers like speaker.

Sound Buffers and Data Transfer:

The recorded samples from the sound card are stored in the hardware buffer which is present in the sound card, when this buffer gets full it generates an interrupt. Then these buffers are transferred to the application buffers in the memory by the Direct Memory Access (DMA). Similarly, during playback, the buffers are transferred from memory to sound card's hardware buffer. The data in the buffer wraps to the start if it reaches the end of buffer. Hence a pointer is necessary to maintain track of current position, both in hardware and application buffers. The large size of the buffer causes the latency in transmission, so the ALSA has split the buffer into chunks of periods.

Transmission, therefore, takes place in the units of period. A period is made of frames which contain the data captured at a particular time interval. Figure 2 shows the construction of periods using frames and samples. There are two ways to store left and right channels data into the frames: one is interleaved mode where the information of left and right channels are stored alternatively within the frame while

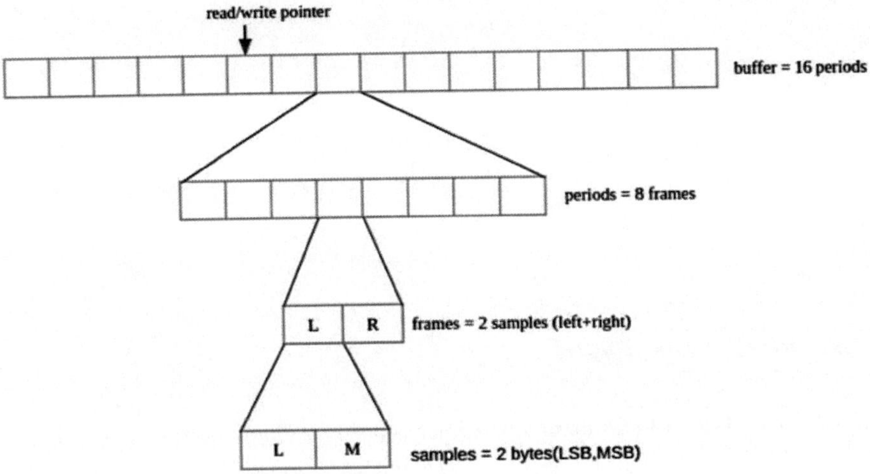

Fig. 2 Alsa buffer

in non-interleaved mode, first the information of one channel is stored completely followed by the next channel information. The transmission of data occurs between the hardware buffer and the application buffer when the sound card is activated.

4 Methodology

The transmitter has a package of codes to convert the RPi as an FM transmitter. It also has sample audio in .wave format, being message file that will be transmitted. It also has a MakeFile describing the information regarding a host machine that can run the code. The code has the following set of files:

PCM wave header.h, Audioformat.h, Error reporter.h, Error reporter.CPP, Main.CPP, Transmitter.h Transmitter.CPP, Wave reader.h, Wave reader.CPP, Make-file, Acoustic guiter.wav.

The entire working of the FM is illustrated using the block diagram in Fig. 3. From the block diagram, we can understand that the main function calls the transmitter block which in turn calls the wave reader block which will read the data from the audio file and feeds it to the transmitter block. This data is modulated using the transmitter block. Then the modulated signal is transmitted and the information regarding the operation can be examined/displayed on the command line. We will see how these particular blocks are declared and how they perform to produce the FM on the Raspberry pi.

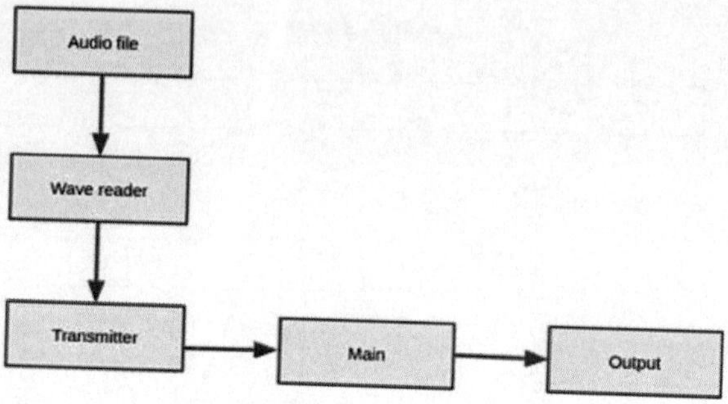

Fig. 3 Block diagram

4.1 PCM Wave Header

It is the structure of multimedia file format which is divided into a header and data. The header is followed by the data chunks in the time format. The canonical wave file format is shown in Table 1. The description of each field is provided in the respective column. These fields are declared in a structure named as PCM WaveHeader under the header file, PCM wave header.

Table 1 Canonical PCM file format description

Positions	Sample value	Field name	Descriptions
1–4	ChunkID	RIFF	The file is marked as the RIFF file. Each character are 1 byte long
5–8	Chunkize	File size	Overall size of the file
9–12	Format	Wave	It describes the header type of the file
13–16	Subchunk1 ID	fmt	Marker of the format chunk
17–20	Subchunk1 size	16	Specifies the length of the format which is declared above
21–22	AudioFormat	1	It specifies the type of format: 1 is assigned for PCM; It is a 2-byte integer
23–24	NumChannels	2	It specifies the number of channels: mono or stereo
25–28	SampleRate	44100	It specifies the sample rate, usually it is 44100 for CD and 48000 for DAT
29–32	Byte rate	176400	It is a product of sample rate, bits per sample and channel, divided over 8
33–34	Block align	4	It is a product of bits per sample and channel
35–36	BitsPerSample	16	It provides the bits per sample
37–40	Subchunk2ID	Data header	It marks the start of the data chunk header
41–44	Subchunk2Size	Integer	It gives the span of the data section
44–...	Data	Data	It has the raw sound data

4.2 Audio Format

It is a header file that has the basic information about the audio files like channel, bits per sample, and sample rate. It is declared as a structure under the title of audio format in the header file. We fill these fields with the actual values during the run time of the code. We are allowed only to use .wave files for the transmission. Any audio file which is not a .wave file will generate an error message, and the execution of the program is aborted.

4.3 Error Reporter

It is a header that declares the class that has functions to generate any error that occurs during the compilation or execution of the program. These errors are generated using a built-in function c str(), which will return the null-terminated stream of characters. The exception handling is provided using what() and throw() to avoid the unpredictable termination of the program.

4.4 Transmitter

The header transmitter has a class that contains the declaration of functions (few as public and few as private) whose definition is defined in the transmitter block. This block plays a major role in the generation of FM using a single-board computer. The working of this block can be explained using the block diagram in Fig. 4. It first declares the necessary headers.

Each block of the transmitter has been declared using particular functions in the program. Here goes the declaration of each block.

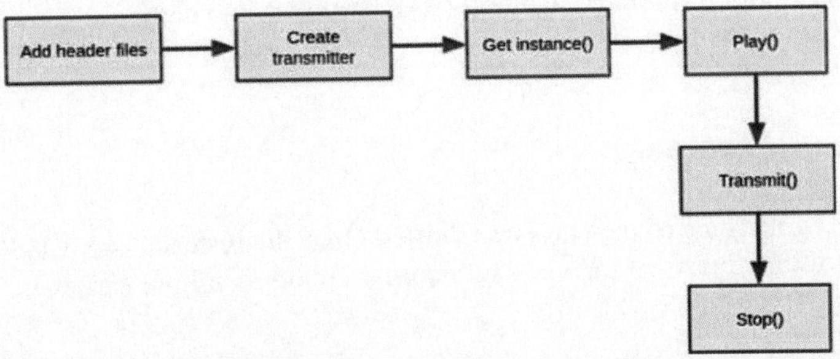

Fig. 4 Block diagram of the transmitter system

```
int memFd;
if ((memFd = open("/dev/mem", O_RDWR | O_SYNC)) < 0) {
    throw ErrorReporter("Cannot open /dev/mem (permission denied)");
}

peripherals = mmap(NULL, 0x002FFFFF, PROT_READ | PROT_WRITE,
                   | MAP_SHARED, memFd, isBcm2835 ? 0x20000000 : 0x3F000000);
close(memFd);
if (peripherals == MAP_FAILED) {
    throw ErrorReporter("Cannot obtain access to peripherals (mmap error)");
}
```

Fig. 5 Opening memory and allocating memory in Raspberry pi

4.4.1 Header Block

Here few headers are declared such as "sstream", "cmath", etc. which will help to use the functions that are declared under these headers. We will next define the base address for the GPIO base, clock base, clock divider base, and the counter buffer base. Then two more functions named ACCESS and ACCESS64 are defined which will generate the physical address when called. Both the functions take two input arguments, one being the base address and the other one is offset address. It returns the physical address that is generated by its definition. We will then declare the variables of the transmitter class named transmitting, restart, clockDivisor, frameOffset, peripherals, and a vector buffer with initial values of 0 and NULL depending on the type of variable to avoid the occurrence of any garbage value.

4.4.2 Create Transmitter

It is a function in the transmitter block that will make use of the Linux commands *uname −m −r* to detect the host system. Once the host machine is recognized, we open memory using */dev/mem*, a pseudo driver for accessing memory in the Linux file system and assign memory space of 0 ×002FFFFF and provide offset address of the GPIO pins of the Raspberry pi. Figure 5 shows how memory is accessed and assigned an address value. Here memFd is a pseudo driver name and peripherals are the name of the memory block.

4.4.3 Instance Creation

We create an instance of the class transmitter. Once the instance is created, it has all the fields of the objects of the class transmitter including all the variables.

4.4.4 Play

It is a block which reads an audio file in wave format and starts playing it, simultaneously it creates the thread to transmit the audio file. The function is bit complex. The functions begin with creation of an object and read the audio data from the .wave file and store it in the newly created object. Then the clock divisor is set, which is necessary to enable the clock divisor register.

- The clock divisor is set to 2^{11} kHz which will be divided by the input frequency. The clock divisor will divide the system clock to produce the desired output frequency.
- We create vector frames to store the values of the audio data.
- We then declare a variable called buffer frames and initialize it with a value of the SampleRate of the audio file. As we have seen in Table 1 the sample rate will be 44100/2 Hz for monotype.
- Create a vector and read the audio information frames by frames into it.
- Create a thread to start transmission. Keep on updating vector frames.

4.4.5 Transmit

To enable the transmission, we need to set the GPIO pin-4 on the Raspberry pi to its alternate function 0 and declare a general purpose clocks control register whose description is given in the datasheet of the Raspberry pi. [5] We need to enable the Clock Manager General Purpose Clocks Control to modulate the clock frequency of the Raspberry pi. This is done in the transmit function block of the transmitter section. Figure 6 shows how this declaration is done in the code.

We need to use frequency from the clock of Raspberry pi, which is modulated by the message signal that needs to be transmitted. In our case, the message is audio file in wave format. This is done by activating a 32 bit Clock Manager General Purpose Clock Divisor as shown in Fig. 7. This declaration is done by accessing the address of this register. We write the password and the integer part of this register by multiplying it with the value of the audio signal. The declaration of code is shown in Fig. 8. This declaration starts the transmission on the GPO pin-4. The transmission continues until the counter register gets empty. Hence, we need to fill the register

```
ACCESS(peripherals, GPIO_BASE) = (ACCESS(peripherals, GPIO_BASE) & 0xFFFF8FFF) | (0x01 << 14);
ACCESS(peripherals, CLK0_BASE) = (0x5A << 24) | (0x01 << 9) | (0x01 << 4) | 0x06;
```

Fig. 6 Declaration of clock manager general purpose clocks control in code

```
ACCESS(peripherals, CLK0DIV_BASE) = (0x5A << 24) | ((clockDivisor) - (int)(round(value * 16.0)));
while (temp >= offset) {
    asm("nop");
    current = ACCESS64(peripherals, TCNT_BASE);
    offset = (current - start) * (sampleRate) / 1000000;
```

Fig. 7 Clock manager general purpose clocks control register description [5]

Bit Number	Field Name	Description	Read/ Write	Reset
31-24	PASSWD	Clock Manager password "5a"	W	0
23-12	DIVI	Integer part of divisor This value has a minimum limit determined by the MASH setting. See text for details. To avoid lock-ups and glitches do not change this control while BUSY=1.	R/W	0
11-0	DIVF	Fractional part of divisor To avoid lock-ups and glitches do not change this control while BUSY=1.	R/W	0

Fig. 8 Declaration of clock manager general purpose clocks control in code

Fig. 9 Declaration of stop function in code

```
void Transmitter::stop()
{
    forceStop = true;
}
```

stack continuously with the data to be transmitted. Thus the Clock Manager and General Purpose Clock Divisor register are declared successively in the code. When the execution is stopped, all the data frames are deleted.

4.4.6 Stop

When the transmission is aborted by pressing Ctrl+X buttons from the keyboard, the stop() function executes. It just restores back the value of a variable which is key variable to start the transmission. The variable states the status of the transmission. If the variable is set false, then the transmission stops. This declaration is shown in Fig. 9. The function plays a major role in the program. The transmission does not stop unless we press Ctrl+X button, though we close the transmission tab.

4.5 *Wave Reader*

It is a class that has few functions and variables declared in its header file wave reader.h, defined in wave reader CPP file. Figure 10 shows the function present in the wave reader file.

- **Wave reader**: It is a function in which the data from the audio file in the PCM header/from the sound card will be written in the memory for fast access.
- **Read data**: This function reads the data which is present in the PCM wave header. It keeps track of the data by using variables like bytesToRead, headerOffSet. It takes the information from the ChunkId, ChunkSize from the PCM header and

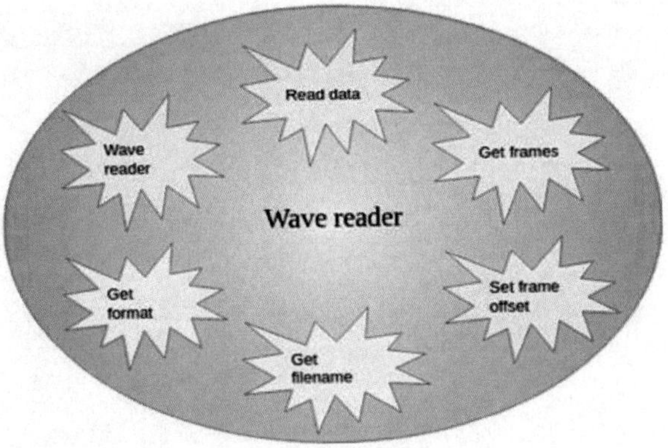

Fig. 10 Wave reader class components

keeps hopping the number of bytes to be read. We can directly use the ALSA card for the transmission as well.

- **Get frames**: It takes the input argument, frame count, and the status of transmission. Whenever the function is called, it pushes the data in the vector frame and returns it.
- **Set frame offset**: It is the function that sets the offset in the opened file using the information of the file descriptor. The frame offset is given as an input argument to this function. It used the *lseek* Linux system call and sets the required offset when called.
- **Get format**: It is a simple function that returns the format of the audio file. It returns a structure of the audio file that contains the information regarding the channels, sample rate, and bits per sample.

5 Receiver

An RTL-SDR is connected to Raspberry pi to receive the FM signals that are being sent from the receiver Raspberry pi. Then we need to tune RTL-SDR to receive the particular frequency. The antenna captures the signal and the received signals are down-converted from passband signals to intermediate frequency passbands by tuner. Then the signals are sampled and given for further digital processing. And finally, it plays the output in audio form.

6 Results

Transmitter: Figure 11 shows the FM transmitter setup, need to just connect a cable to the 7th pin on the board (GPIO-4). Then we need to run the FM code package. It will create an executable file called FM transmitter. Then run the following command

Fig. 11 Wave reader class components

to start the transmission . cd FM transmitter sudo ./fmtransmitter –f [frequency] –r [filename].

In the above code –f indicates the transmitting frequency and –r refers to the audio file in .wav format. We need to provide frequency and audio file name in the parenthesis. I have run the following command by specifying 89.9 as transmission frequency and acoustic.wave being my audio file. (Like "Sudo ./fm transmitter –f 89.9 –r acoustic music.wave") Once the program starts running without any errors then, we can see the status of the transmission as shown in Fig. 12.

Receiver: To make the Raspberry pi as a receiver, we need an RTL-SDR connected to it or receive it in simple phone as shown in Fig. 13, then we download the UHD drivers and few of the dependent packages. Then we run the code that is used to receive the FM signals from the RTL-SDR. The following line of code will receive the FM signal and plays the received track in Raspberry pi. sudo .rtlfm –f tx frequency –M wbfm –s 200000 –r 48000 –|aplay –r 48000 –f S16 LE.

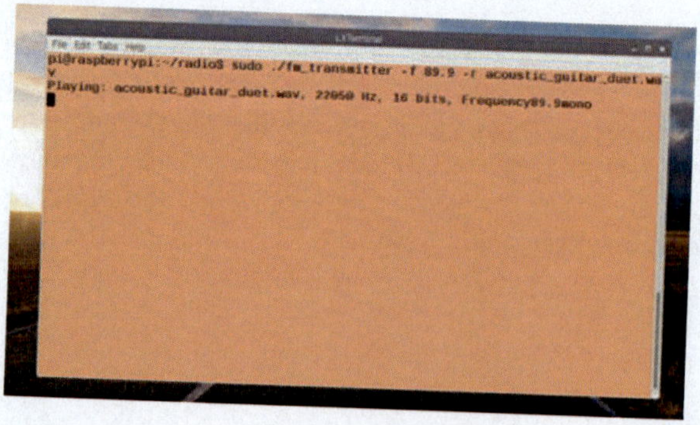

Fig. 12 Output on the screen when the transmission is taking place

Fig. 13 Output on the screen when the transmission is taking place

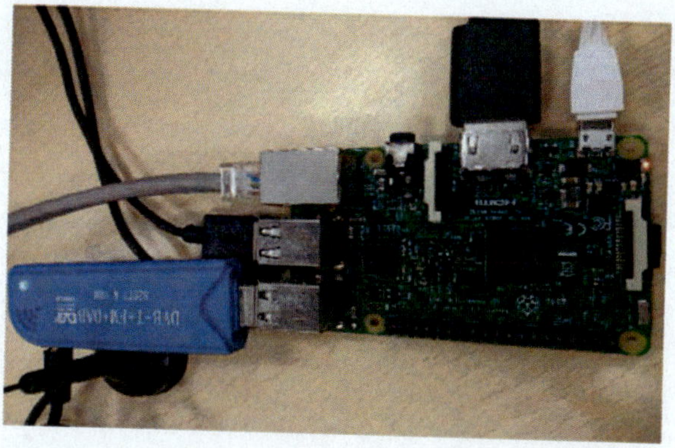

Acknowledgments I like to thank *Dr. Vidydhar S Melkeri* for his constant moral support during my project work. The project has taught a new way of learning and enhancing my knowledge.

I express my sincere gratitude to our coordinator *Dr. V Sandeep*, School of Engineering for providing necessary facilities and enjoyable working experience, and also for, *Dr. Layak Ali* for his motivation. I thank the staff, seniors, and friends at Central University of Karnataka.

I am highly indebted to my parents for their unquantified support and fortify me throughout life.

References

1. Vachhani K, Mallari RA Experimental study on wide band FM receiver using GNURadio and RTL-SDR. In: Presented by 2015 international conference on advances in computing, communications, and informatics (ICACCI)
2. Sruthi MB, Abirami M, Manikkoth A, Gandhiraj R, Soman KP Low-cost digital transceiver design for Software Defined Radio using RTL-SDR. Amrita Vishwa Vidyapeetham, Coimbatore, India
3. Saber M, Aroussi HK, El Rharras A, Saadan R Raspberry Pi and RTL-SDR for spectrum sensing based on FM real signals
4. Introduction to RTL-SDR part-I, principles and Hardware
5. Datasheet of BCM2835 ARM peripherals

Cloud Computing—Security, Issues, and Solutions

Ritesh Sharma, Mahendra Kumar Gourisaria, and S. S. Patra

Abstract Security is the most concerned aspect of cloud computing because data is located in different places around the globe and new threats are arising day by day. Data privacy and security protection are the most important concerns in cloud computing technology and are related to both hardware and software. This paper gives an overview of security issues, challenges, and proposed solutions. A very clear and classified overview is presented in this paper with respect to cloud security. Today's cloud computing provides the best and most efficient solutions to the Information and Communication Technology (ICT) industry, but the security problems are like nightmare for the cloud service providers as well as for the customers. We have also described various service and deployment models and identified major issues and challenges. This paper has also proposed some of the vital solutions with respect to privacy and security and also focus on various vulnerabilities and known security threats and attacks.

1 Introduction

Cloud computing is the means of accessing the application as utilities over the Internet as a service based on a service level agreement on pay-per-use basis. It allows us to create, configure, manipulate, customize, and access the application online [1]. The cloud technology is a buzzword in the IT industry due to the tremendous development in the cloud computing industry. Cloud shows the approach of a scattered system

R. Sharma (✉) · M. K. Gourisaria
School of Computer Engineering, KIIT Deemed to be University, Bhubaneswar, Odisha 751024, India
e-mail: sharmaritesh3312@gmail.com

M. K. Gourisaria
e-mail: mkgourisaria2010@gmail.com

S. S. Patra
School of Computer Application, KIIT Deemed to be University, Bhubaneswar, Odisha 751024, India
e-mail: sudhanshupatra@gmail.com

S. C. Satapathy et al. (eds.), *Communication Software and Networks*, Lecture Notes in Networks and Systems 134, https://doi.org/10.1007/978-981-15-5397-4_70

which consists of a set of virtual machines that can be easily provided to the customer in the runtime to meet his/her varying resource requirements. An abstraction layer is introduced by the cloud between the servers or physical storage and the user whose data is processed in the cloud [2]. Cloud computing is very scalable technology as it has everything for everyone from high-scale corporate companies to entrepreneurs and small start-ups.

Data is very significant so far as security is concerned regardless of the infrastructure used and cloud computing is not an exception; instead, it has some add-ons in security concerns as the cloud infrastructure is highly distributed and multi-tenant in nature [1]. It is said by someone that we can outsource responsibility but we cannot outsource accountability. Security is one of the most complex tasks to apply in cloud computing. Identity and Access Management (IAM) and privacy is another challenge of the cloud environment. Data protection and storage are other vital challenges in cloud computing. But, unfortunately, the efforts are limited from the side of operators when it comes to security, which is, in fact, the most concerned issue of cloud computing. The rest of the paper is organized as follows. Section 2 focuses on the related work done by other authors. The cloud computing architecture along with different services and deployments is described in Sect. 3. Section 4 describes the different cloud security issues. Section 5 details about the solutions of the various security issues followed by conclusion in Sect. 6.

2 Related Survey

In [3], the authors broadly focus on application- and virtualization-related security issues. In [4, 5], the authors present a comprehensive requirements such as transparency, confidentiality, accountability, integrity, and assurance. The authors in [6] have proposed a security framework for RFID technology. The different security issues related to the service delivery models of the clouds are taken care in [7]. Pandey et al. in [8] propose a framework for data security in cloud environment which is mainly based on cloud storage. Hashizume in [9] discussed the security issues on the basis of SPI (SaaS, PaaS, IaaS) service models. The issues of data management and governance with respect to control in the cloud computing are discussed in [10]. Zissi et al. in [11] propose the concept of cryptography-based third-party solution to maintain confidentiality in data communication. Younis and Kifayat in [12] focus on secure cloud computing for critical infrastructure. The contribution in [13] discusses the techniques of intrusion detection, whereas a security analysis of open-source cloud software platforms is discussed in [14]. Shahzad in [15] discusses a case study based on DOS attack while using AWS. Chen and Zhao [16] have given their analysis by focusing on data segregation and privacy protection.

3 Cloud Computing Architecture

The architecture of cloud computing refers to the components and sub-components involved in cloud computing. Basically, these components are a collection of front-end platforms (client-side devices such as mobile devices or desktops), and back-end platforms (server, storage), a network (Internet, intranet, inter-cloud) and a cloud-based delivery model.

3.1 *Front-End Platform*

Front-end platforms or cloud client platforms are the thin clients, fat clients, servers, mobile, and tablet devices which communicate with the cloud via a virtual session, a web browser, or an application (middleware).

3.2 *Cloud Storage*

The data is stored in an online data center based on virtual network which is accessed by multiple clients over the Internet. Generally, there are following configurations based on deployment of cloud, viz., public cloud, private cloud, community cloud, or combination of these called hybrid cloud.

3.3 *Service Models in Cloud Computing*

NIST (National Institute of Standard and Technology) identified three basic cloud service delivery models that provide services at different levels of a business model [17].

3.3.1 Software as a Service (SaaS)

SaaS is described as the service provided to the customer by the CSP (cloud service provider) to use its software applications which is deployed on the cloud infrastructure and numerous users can access the software application via a thin client-side setup like web browsers [18]. Ultimately, the customer will be provided with complete application software on demand. The cloud service provider is fully responsible for managing cloud infrastructure, operating system, networks, servers, storage, or even some of the application configurations. Some of the examples of SaaS are Google Docs, Gmail, Amazon EC2, Dropbox, etc.

3.3.2 Platform as a Service (PaaS)

PaaS is an application development environment which is provided to the cloud customer by the cloud service provider (CSP) to run, deploy, and manage applications without bothering about the actual infrastructure associated with the service. The customer will develop their own software applications, which will run on the cloud service provider's infrastructure [19]. In other words, the cloud service provider gives its customers the capability to develop their own applications using the tools and the programming languages and deploy his/her application on the cloud infrastructure [7]. Some of the examples of PaaS are Apache Stratus, Heroku, Microsoft Azure, Google App Engine, Force.com, etc.

3.3.3 Infrastructure as a Service (IaaS)

Iaas (often called hardware as a service) is defined as a service model that provides virtualized computing resources by means of hardware, software, and networking devices over the Internet. In this delivery model, the customer has the control over deployed services, operating system, and some selected components of the network, whereas the total responsibility to manage the whole cloud infrastructure is on the cloud service provider [4]. The customer in IaaS will pay to the CSP on a pay-per-use basis. In IaaS, the CSP provided a server, virtual server with a unique IP address, and an Application Programming Interface (API) to start, access, stop, and configure storage and virtual servers. Some examples of IaaS are Digital Ocean, Amazon AWS, Google Compute Engine, etc. (Table 1).

3.4 Deployment Models

Apart from the three service models (IaaS, PaaS, SaaS), there is another classification of cloud models which is based on deployment. Cloud integrators play a very significant role as it determines the right cloud computing path for a specific organization.

Table 1 Comparison study of service models of cloud

Model	Used by	Examples
Software as a service	End user	Google docs, Gmail, Dropbox
Platform as a service	Software developers	Azure, App Engine, Heroku
Infrastructure as a service	System administrators	DigitalOcean, AWS, compute Engine

3.4.1 Public Cloud

In this model, the computing services are provided to the customers by a cloud service provider and the services may be dedicated (single tenant) or shared (multi-tenant). Generally, the cloud service provider owns the infrastructure of the cloud and the infrastructure resides within the data centers of cloud provider [20]. Public cloud is advantageous when it comes to scalability, as it is bigger than the enterprise cloud and hence it provides facility to scale tremendously on-demand. Amazon EC2, Amazon S3, Google Apple Engine, and Microsoft Azure come under the preview of public cloud.

3.4.2 Private Cloud

In this model, the cloud services are provisioned to the customer over a secured and distinct channel to a single organization for the dedicated usage. The private cloud model provides security and offers greater control which was lacking in case of the public cloud [20]. There are two types of private clouds (i) internal private clouds and (ii) external private clouds. In the case of internal private clouds, the cloud infrastructure resides within the data center of the organization. This model provides great security and control but is somewhat limited when it comes to scalability and size [3, 20].

3.4.3 Community Cloud

The community cloud is a collaborative cloud model in which the cloud infrastructure is shared by a group of organizations of a specific community and the same interests [20]. The cloud is either managed internally or by a cloud service provider which is either internal or external. The costs are distributed over a few users than a public cloud but more than the private cloud. This model meets some of the security concerns that were lacking in the public cloud model [4, 20].

3.4.4 Hybrid Cloud

This cloud deployment model comprises both private and protected cloud. This model provides high scalability and protection as it has the ability to connect two or more cloud models [21]. The hybrid cloud model integrates the in-house IT infrastructure to meet unique and business-specific needs. It allows one to extend either the capabilities or capacity of a cloud model by integrating it with another cloud model [18, 21]. The main advantage of this model is that a company has to pay for extra resources only when they are needed.

4 Cloud Security Issues

The IT industry is rapidly migrating toward cloud computing platforms for computing and organizing data for accessibility and control over data. But, some of the cloud computing issues like security, resources, control of data, network-related issues, and other internal issues must not be underestimated [22]. This paper shows a big picture of the most concerned cloud computing issue worldwide, i.e., security (as shown in Fig. 1).

4.1 Privacy and Confidentiality

When users visit the sensitive data, the cloud services can prevent potential adversary from inferring the user's behavior. It takes care of when users are going to access the data, how the user is accessing the data, and to what extent the users should be allowed to manipulate it [23]. How guarantee data replication in a consistent way. As cloud computing architecture is highly distributed in nature, it also has a high risk of losing the privacy confidentiality of stored data [24]. The data stored inside the cloud resides in the remote servers and there is a high probability that the private data may be exposed to other users [24]. These privacy and confidentiality issues may be classified into two types.

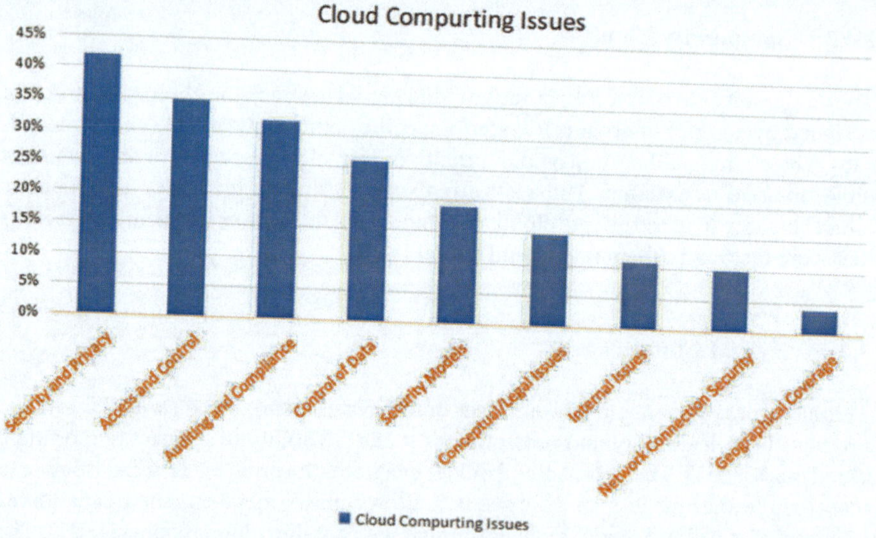

Fig. 1 Comparison of cloud computing issues

4.1.1 Insider Threats

This issue occurs due to the malignant cloud provider user, or the malignant cloud customer user, or may be due to the malignant third-party user who supports either the customer or the cloud service provider [25]. In this issue, the wicked users of either of the companies use their authenticated identity and expose the private data of the customer to the outsiders.

4.1.2 Outsider Threats

This issue occurs due to the malignant outsiders who directly attack the servers of the cloud service provider and exposes the data. They may be the hackers or the unauthorized personalities of any other organization [10, 25]. They use the techniques like SQL injections or some other hacking techniques to attack the servers of the cloud service provider [25].

4.1.3 Data Leakage

The issue of data leakage may be caused among the competitor organizations using the same cloud service due to some hardware fault or human errors and that will lead to a compromise in data privacy [10, 25]. It is an unauthorized transfer of information from one data center to the outside world. It can be accomplished by data hiding.

4.2 Multi-tenancy

Multi-tenancy is an obvious result of trying to achieve a cost-efficient and economic cloud infrastructure as it introduces the virtualization and resource sharing between different customers in the cloud [26]. Multi-tenancy is the most concerned issue in cloud computing security domain. Multi-tenancy is differently seen in various cloud computing models. In Software as a Service (SaaS), the customers are not having control over the underlying infrastructure [26].

Multiple customers are utilizing the same software or service which may lead to attacks in case of SaaS. In IaaS (Infrastructure as a Service), the customers have the provisions of controlling the cloud infrastructure but here also the multi-tenancy gets originated as multiple customers are using Virtual Machines (VM) and these VMs are derived from the same physical machine [26].

4.3 Data Segregation

As the cloud infrastructure is distributed in nature, all the resources are shared in cloud computing by multiple customers, the data segregation is a topic for concern [9]. The data of one customer must be kept separate from the other user in order to avoid intrusion of data. The data segregation issues are mostly originated by incorrectly defined security parameters and improper configurations of virtual machines and hypervisors. Vulnerabilities with data segregation can be detected by using some of the famous tests like insecure storage, SQL injections, and data validation [9, 10].

4.4 Data Integrity

Data integrity refers to the system which only allows the authorized user to manipulate the data. In cloud, the stored data can be easily accessed via Internet and the attackers will have an advantage, so this loophole should also be taken into consideration so far as data security is concerned [27]. The system should be built in such a way that the unauthorized user must not be allowed to manipulate the data stored in the cloud [27]. RAID and digital signature can be applied to maintain data integrity. Database which supports ACID properties can somehow maintain data integrity. Data segregation, user access, and data quality are also matter of concern.

4.5 Data Availability

Data availability aims to serve the user data on request of the user anytime and anywhere [28]. As cloud computing has a distributed and complicated architecture, there may be a possibility of system crashes like virtual machine crashes, networking crashes, storage crashes, hard disk damage, and IDC fire [28] which leads to data unavailable. Customer's data will not be served to the customer against his/her request and this will lead to the performance issues of the cloud. The data availability issues also get originated due to a denial-of-service problem. Due to some of the vulnerabilities in the cloud infrastructure such as network issues and application crashes, the denial-of-service issues may occur which will lead to data unavailability [28]. Weak recovery procedure is also one of the main reasons of data unavailability.

4.6 Authentication and Authorization

Authentication and authorization is one of the most important aspects of the data storage systems [29]. Authentication can be achieved by entering valid user name and password. Similarly, cloud system also requires highly secured authentication and authorization services in order to avoid malicious and unauthorized user data access [10, 29]. After logging into the system, a user must be authorized for doing certain task. The authentication and authorization issues harm the security of the cloud. Data breaching is another incident in which sensitive, confidential, or, otherwise, protected data is disclosed or accessed in an unauthorized manner. Data breaching leads to the leakage of sensitive data of an organization which will lead to privacy issues. Data breaches are mainly caused by any unauthorized personnel of the cloud service provider [29].

4.7 *Virtualization*

Virtualization is a very integral part of cloud computing. Virtualization refers to the separation of underlying resources into different virtual machines having their own distributed resources. By using the concept of virtualization, one can run multiple machines derived from a single provider machine (PM) [1]. Virtual-machine-related attacks highly affect the IaaS service of the cloud [11]. Some of the virtualization-related security issues are as follows.

4.7.1 VM Image Sharing

VM image is used to instantiate virtual machines. To share the VM images, a shared image repository is used or a user can own a dedicated VM image. As repository is used for the management of VM images [1, 30], a malignant user can misuse this feature with the intention to inject a faulty code inside the virtual machine, which can lead to severe security issues [1]. For example, a used VM which contains malware from the previous user can be shared to another user without cleaning the malware which will lead to a serious security issue.

4.7.2 VM Escape

Virtual machines are managed by a Virtual Machine Manager (VMM) but some VMs or malignant user's escapes from the supervision of VMM [30]. VMM monitors all the virtual machines and the users using the underlying resources such as hardware and any malicious code inside the VMM may lead to serious issues in the VM management and data leakage [30].

4.7.3 VM Migration

VM migration refers to the process of transferring a running VM between different physical machines without disconnecting the client [1]. In this process, the status of running VM is copied to the destination VMM from the source VMM and the process is again continued in the destination. The main reasons for VM migration are fault tolerance, load balancing, and maintenance. While transferring the VM, the data and code of VM can be disclosed by the attacker [31].

4.7.4 Cross VM Side-Channel Attacks

This attack is very common in the case of virtualization and leads a very adverse effect as far as security is concerned. This kind of attack is possible when two virtual machines are running on a single physical machine. The victim and the attacker are using the same processor and a cache memory which may be cost-efficient but it leads to a serious security issue of virtualization [31]. When the VM execution of the victim is alternated by the attacker, the information about the victim's behavior can be tracked by the attacker. This is enough to break the security barrier of any virtual machine.

4.8 Miscellaneous Security

4.8.1 Denial-of-Service (DoS) Attacks

In cloud computing, availability is always a concern for cloud customers. Denial-of-service attacks refer to the refusal of the cloud services by the cloud due to excessive traffic in the cloud because of network layer distributed attacks [32]. This threat is generated when the cloud is not in a position to respond to the cloud customer's action due to heavy traffic in the network. A specific case of this attack is when an anonymous user queries something on the cloud by creating thousands of ambiguous objects, and then the cloud is not in a position to respond [32].

4.8.2 Mobile Device Attacks

As cloud computing is emerging day by day, the connectivity is no more limited to desktop and laptops, it is also available in smartphones. The use of mobiles is also increasing and the mobile phones are now available with the traditional features of the desktop and laptops like rich APIs that support background activities and communication to social network, large data storage capabilities, and wireless Internet connectivity. These features help the attackers to attack the mobile devices via Internet spyware, worms, etc.

4.8.3 Social Networking Attacks

As the use of social media platforms is increasing day by day, the people are connecting to each other very rapidly. So, in this era of social networking, there is a high probability that the cloud service provider user and the attacker are connected to each other via a social media platform. So, the attacker can collect information of that user and can make use of it by establishing trust and use that online information to recognize the roles and relationships of users in the cloud, which is helpful in the attacking.

5 Security Issues and Solutions

As discussed earlier, there are many security issues in cloud computing which cannot be ignored as far as the performance of cloud and data privacy of customer are concerned in the big organizations [31]. These security issues must be taken into consideration when migrating toward cloud computing. A general overview of solutions to cloud computing security is presented below (Table 2).

6 Conclusion

The cloud computing technology is rapidly growing in the IT industry. Data security and privacy are the most important issues of cloud computing. Reduction in processing and data storage cost are mandatory for any organization for decision-making. Trust between CSP and user is the very important without which no one will store their valuable information on the cloud. In this paper, we have discussed a number of different techniques proposed by different authors for data security and protection. However, these techniques need lot of improvement. We have also discussed the solutions of various challenges. We have to provide more dynamic solutions to make it acceptable by the cloud service consumers.

As new solutions are arising day by day, the old and traditional solutions to cloud computing security issues are not capable of eradicating the new arising cloud security problems. So, in order to ensure secured cloud computing system, new researches should be done based on the new issues. In a world with decreasing resource per individual along with increasing demand, cloud solves the hardware problem pertaining to storage. In spite of having serious security concerns in the cloud computing environment, we cannot ignore the pros and immense potential of cloud computing technology. The cloud computing features like resource management, distributed nature, multi-tenant architecture, virtualization, storage management, etc. must not be ignored as they provide special computing facilities to the developer and the business organizations.

Table 2 Solutions to cloud security issues

Security issue	Proposed solutions
1. Privacy and confidentiality	A renowned and reputed cloud service provider must be chosen for cloud computing solutions in order to avoid the insider attacks that are performed by the cloud service provider users. Regular auditing must be done by the customer to check whether there is any malicious component present inside the organization or not [33]. Clear contract between the cloud service provider and the cloud customer must be made to ensure a good relationship between each other, which will avoid the problems of data privacy and confidentiality [33]
2. Multi-tenancy	The use of multi-tenant architecture must be made limited and must be used when necessary. The cloud service providers must impose proper mechanisms for abstraction between multiple tenants in the cloud. Proper scheduling and resource allocation algorithms must be implemented to avoid the problems of data security among the multiple tenants in the cloud [17]
3. Data segregation	Proper data segregation mechanisms must be imposed to ensure the separation of data of different users
4. Data integration	Proper authorization of users must be done in the cloud in order to avoid malicious users
5. Data availability	A regular auditing must be done by the cloud service provider to check the resources crash such as VM crash, storage device crash, and all other resource crashes. A regular backup facility must be present in the cloud to avoid serious problems like data loss [34]
6. Authentication and authorization	Proper authentication and authorization mechanism must be imposed by the customer and cloud service provider in the services in order to avoid the unauthenticated cloud customer user. Data encryption must be implemented to avoid data breaches [29, 34]
7. Virtualization-related issues	Efficient resource management mechanisms must be implemented in order to avoid problems like VM image sharing Optimized Virtual Machine Manager (VMM) should be used to avoid the VM escape problem. Optimized scheduling algorithms must be imposed to schedule resources properly and avoid cross VM side-channel attacks
8. Miscellaneous security issues	Mobile devices must be made secured to ensure the security to the cloud connectivity. Secured Application Programming Interfaces (APIs) must be developed to handle the severe attacks of the hackers. Penetration testing must be done by the developers of the cloud customer to test different kinds of attacks by the attackers

References

1. Basu S, Bardhan A, Gupta K, Saha P, Pal M, Bose M, Basu K, Chaudhury S, Sarkar P (2018) Cloud computing security challenges & solutions-A survey, pp 347–356. https://doi.org/10.1109/ccwc.2018.8301700
2. Gourisaria MK, Patra SS, Khilar PM (2016) Minimizing energy consumption by task consolidation in cloud centers with optimized resource utilization. Int J Electr Comput Eng 6(6):3283
3. Holloway I, Todres L (2003) The status of method: Flexibility, consistency and coherence. Qual Res 3(3):345–357
4. Gul I, ur Rehman A, Islam MH (2011) Cloud computing security auditing. In: Proceedings of the 2011 the 2nd international conference on next generation information technology (ICNIT). Gyeongju, Korea, pp 143–148
5. Chen Z, Yoon J (2010) IT auditing to assure a secure cloud computing. In: Proceedings of the 2010 6th world congress on services (SERVICES-1). Miami, FL, USA, pp 253–259
6. Kardas S, Celik S, Bingöl MA, Levi A (2013) A new security and privacy framework for RFID in cloud computing. In: Proceedings of the 5th IEEE international conference on cloud computing technology and science. Bristol, UK
7. Kandukuri BR, Paturi VR, Rakshit A (2009) Cloud security issues. In: Proceedings of the IEEE international conference on services computing, 2009 (SCC ′09). Bangalore, India, pp 517–520
8. Pandey A, Tugnayat RM, Tiwari AK (2013) Data security framework for cloud computing networks. Int J Comput Eng Technol 4(1):178–181
9. Hashizume K, Rosado DG, Fernández-Medina E, Fernandez EB (2013) An analysis of security issues for cloud computing. J Internet Serv Appl 4:5
10. Mahmood Z (2011) Data location and security issues in cloud computing. In: Proceedings of the 2nd international conference on emerging intelligent data and web technologies (EIDWT ′11). IEEE, pp 49–54
11. Zissis D, Lekkas D (2012) Addressing cloud computing security issues. Future Gener Comput Syst 28(3):583–592
12. Younis MYA, Kifayat K (2013) Secure cloud computing for critical infrastructure: a survey. Tech. Rep., Liverpool John Moores University, Liverpool, UK
13. Modi C, Patel D, Borisaniya B, Patel H, Patel A, Rajarajan M (2013) A survey of intrusion detection techniques in cloud. J Netw Comput Appl 36(1):42–57
14. Popovic O, Jovanovic Z, Jovanovic N, Popovic R (2011) A comparison and security analysis of the cloud computing software platforms. In: Telecommunication in modern satellite cable and broadcasting services (TELSIKS), vol 2, pp 632–634
15. Shahzad F (2014) State-of-the-art survey on cloud computing security challenges, approaches and solutions. In: International conference on emerging ubiquitous systems and pervasive networks (EUSPN - 2014), vol 37, pp 357–362
16. Chen D, Zhao H (2012) Data security and privacy protection issues in cloud computing. In: Proceeding of the international conference on computer science and electronics engineering (ICCSEE ′12), vol 1. Hangzhou, China, pp 647–651
17. Dumitrache M, Sandu IE, Barbu DC (2017) An integrated cloud computing solution for romanian public-sector entities: ICIPRO Project. Stud Inform Control
18. Zhang W, Chen Q (2010) From E-governmentto C-government via cloud computing. In: 2010 international conference on E-Business and EGovernment. Guangzhou, pp 679–682
19. Gourisaria MK, Patra SS, Khilar PM (2018). Energy saving task consolidation technique in cloud centers with resource utilization threshold. In: Progress in advanced computing and intelligent engineering. Springer, Singapore, pp 655–666
20. Sainadh PSV, Kumar US, Reddy SH (2017) Security issues in cloud computing. Int J Modern Trends Sci Technol 03(01):125–130
21. Kumar PR, Raj PH, Jelciana P (2017) Exploring security issues and solutions in cloud computing services—a survey. Cybern Inf Technol

22. Goswami V, Patra SS, Mund GB (2012) Performance analysis of cloud with queue-dependent virtual machines. In: 2012 1st international conference on recent advances in information technology (RAIT). IEEE, pp 357–362
23. Sun Y, Zhang J, Xiong Y, Zhu G (2014) Data security and privacy in cloud computing. Int J Distrib Sens Netw Article ID—190903
24. (2014) Recent trends in computer networks and distributed systems security. Springer Nature America, Inc.
25. Sen J (2014) Security and privacy issues in cloud computing. IGI Global
26. AlJahdali H, Abdulaziz A, Garraghan P, Townend P, Lau L, Xu J (2014) Multi-tenancy in cloud computing. In: 2014 IEEE 8th international symposium on service oriented system engineering, pp 344–351
27. Rao RV, Selvamani K (2015) Data security challenges and its solutions in cloud computing. In: International conference on intelligent computing, communication & convergence (ICCC-2015)
28. Zhou M, Zhang R, Xie W, Qian W, Zhou A (2010) Security and privacy in cloud computing: a survey. In: 2010 sixth international conference on semantics, knowledge and grids
29. Zhang N, Liu D, Zhang Y (2013) Research on cloud computing security. In: International conference on information technology and applications. Chengdu, pp 370–373
30. Bhadauria R, Chaki R, Chaki N, Sanyal S (2014) A survey on security issues in cloud computing. ACTA Technica Corviniensis—Bulletin of Engineering
31. Aldossary S, William A (2016) Data security, privacy, availability and integrity in cloud computing: issues and current solutions. Int J Adv Comput Sci Appl 7(4)
32. Birje MN, Challagidad PS, Goudar RH, Tapale MT (2017) Cloud computing review: concepts, technology, challenges and security 6(7)
33. Tiwari PK, Mishra B (2012) Cloud computing security issues, challenges and solutions. Int J Emerg Technol Adv Eng
34. Gupta G, Laxmi PR, Sharma S (2014) A survey on cloud security and techniques

Prediction of Sedimentation in a Watershed Using RNN and SVM

Abinash Sahoo, Ajit Barik, Sandeep Samantaray, and Dillip K. Ghose

Abstract Sediment transport in rivers generally occurs at time of severe proceedings linked with strong precipitation and high flow of rivers. Traditional ways to collect data in high-risk conditions are dangerous and expensive when contrasted to measurement of water discharge. Because of a variety of controlling aspects on river sediment transport, shaping a suitable input arrangement to develop suspended sediment load (SSL) model for forecasting sediment capacity is extremely significant for water resources management. Present work emphasizes on applicability of support vector machine (SVM) and recurrent neural network (RNN) for its appropriateness in modelling relation amid river stage, discharge and sediment load. Model efficiency was assessed utilizing root mean square error (RMSE), mean square error (MSE) and coefficient of determination. Outcomes of SVM were contrasted with those of ANN and it could be seen that SVM can be utilized as a competent tool to predict sediment yield.

1 Introduction

Requirement to simulate short- and long-term sediment yield is significant to manage catchment characteristics. Intricate nature of process like sediment yield depends variably on properties of watershed and rainfall layout, and reliance on a variety of additional features makes it complicated for predicting and estimating sediment

A. Sahoo (✉) · S. Samantaray · D. K. Ghose
Department of Civil Engineering, NIT Silchar, Assam, India
e-mail: bablusahoo1992@gmail.com

S. Samantaray
e-mail: samantaraysandeep963@gmail.com

D. K. Ghose
e-mail: dillipghose_2002@yahoo.co.in

A. Barik
Department of Civil Engineering, CET Bhubaneswar, Bhubaneswar, Odisha, India
e-mail: ajitbarik200@gmail.com

S. C. Satapathy et al. (eds.), *Communication Software and Networks*, Lecture Notes in Networks and Systems 134, https://doi.org/10.1007/978-981-15-5397-4_71

with pleasing accurateness. Therefore, need for developing alternative models for sediment yield simulation using available data is more important in present age.

Lafdani et al. [1] investigated capability of SVM and ANN models for predicting daily SSL in Doiraj River, located in western part of Iran. Cimen [2] introduced a novel method to estimate regression while coming across water resources management and for predicting the SSL for corresponding water discharge values. Kisi [3] developed least-square SVM model for accurately estimating daily SSL at Eel River in California. Misra et al. [4] analysed applicability of SVM for simulating monthly and weekly sediment yield from Vamsadhara River Basin, situated in south India. Jain [5] compared performance of SVM with ANN and data used is from two gauging stations on the Mississippi River. Samantaray and Sahoo [6–8] applied different neural networks to predict hydrological constraints of watershed. Sharma et al. [9] urbanized SVM and ANN for runoff and sediment yield prediction of a catchment in Nepal. Sadeghpour Haji et al. [10] investigated accuracy of WSVM and SVM models for forecasting SS in one day ahead by using data of Yadkin River. Wang and Traore [11] explored accurateness of time-lagged recurrent network (TLRN) to forecast SSL stirring at regular intervals throughout storm events in Kaoping River situated in Southern Taiwan. Boukhrissa et al. [12], Tfwala and Wang [13] investigated efficiency of statistical models for improving precision of streamflow-suspended sediment relation to estimate SSL. Saharia and Bhattacharjya [14] developed a competent vibrant ANN that accounted morphometric characteristics for Dikrong River Catchment. Liu et al. [15] built a wavelet ANN for predicting SSL of Kuye River, in Yellow River catchments of China. Evaluation of suspended sediment load through several neural networks [16, 17]. Sadeghifar et al. [18] studied RNN to predict depth of wave on basis of collected data and measured sea waves of Caspian Sea, Iran. Objective of this research is to predict sediment load of a watershed using various neural network techniques.

2 Study Area and Data

Angul District is located in centre of Odisha. The district covers 6232 sq.km area, having latitude and longitudes of 20.50 N 85.00 E. It is situated at an elevation of 195 m (640 ft) above sea level. Climate of this region is greatly diverse having average annual precipitation of 1421 mm. This city has a tropical climate. When compared with winter, the summers have much more rainfall. Temperature here averages 26.9 °C with record high of 50.9 and record low of 5.2 (Fig. 1).

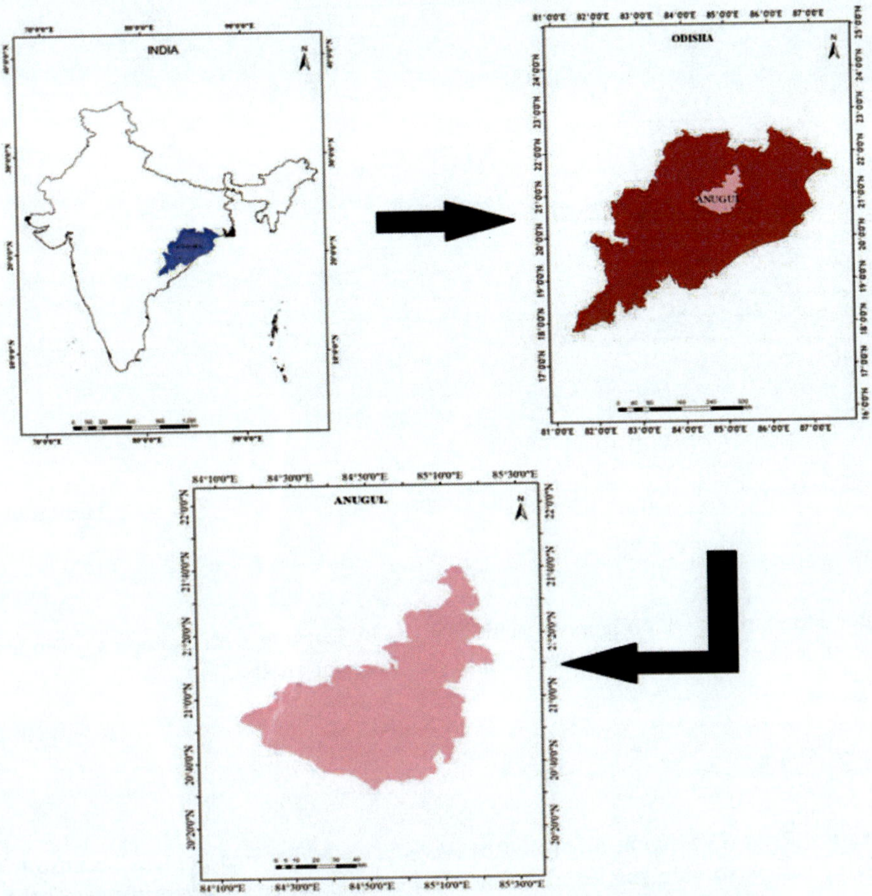

Fig. 1 Planned watershed

3 Methodology

3.1 RNN

RNNs (Fig. 2) comprises no less than single feedback relation where output is fed back to input so as to flow in a circle. This is different when compared with feedforward NNs, which consists of no loop and output is solitarily connected to input elements of succeeding layers. In accordance to Martens and Sutskever, RNNs could accumulate data and facts regarding time, making them appropriate to predict application on basis of time. These networks have previously been productively used in numerous time series experimentations. Similar to human brain, ANNs are

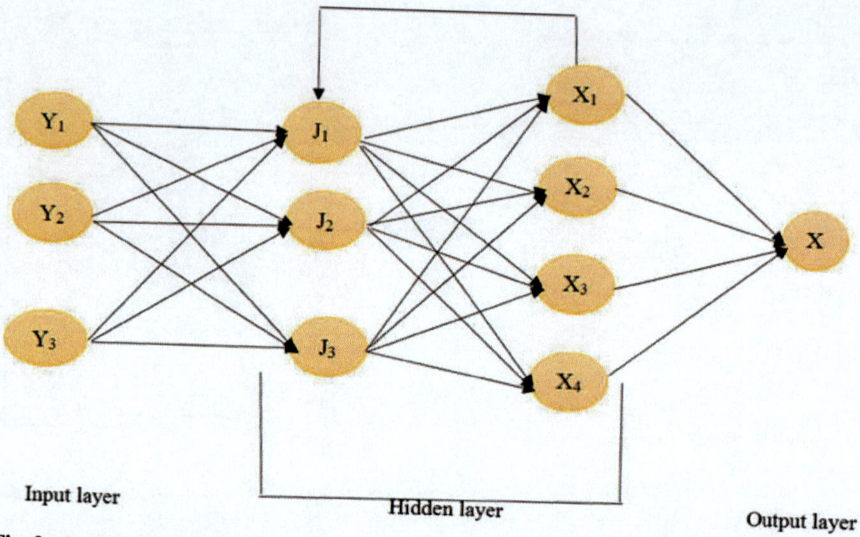

Fig. 2 Architecture of RNN

good in learning which helps to achieve goals by learning with examples (of input arrangement) adjusting weights on links amid network nodes.

3.2 SVM

Basics about SVM were urbanized by Vapnik. SVM became popular because of its numerous striking characteristics and potential experiential performance. These models consist of a set of connected supervised learning techniques utilized to classify and degeneration. SVM employs structural risk minimization (SRM) theory more keenly than empirical risk minimization (ERM) theory executed by the majority conventional models which also include ANNs. By screening input data as two vector sets in an n-dimensional space, SVM constructs an unravelling hyperplane in the space that increases margin amid two vector sets (Fig. 3). For calculating margin, two corresponding hyperplanes are built, one on either surface of unravelling hyperplane, that are "pushed up against" the vector sets.

3.3 Model Preparation

Sediment load data with up to five lag time on daily basis is retained from IMD Bhubaneswar, India for monsoon, from 1990 to 2017. Data from 1990 to 2009 and 2009 to 2017 are engaged to train and test model efficiency. MSE, RMSE and R^2 are assessed for performance of model.

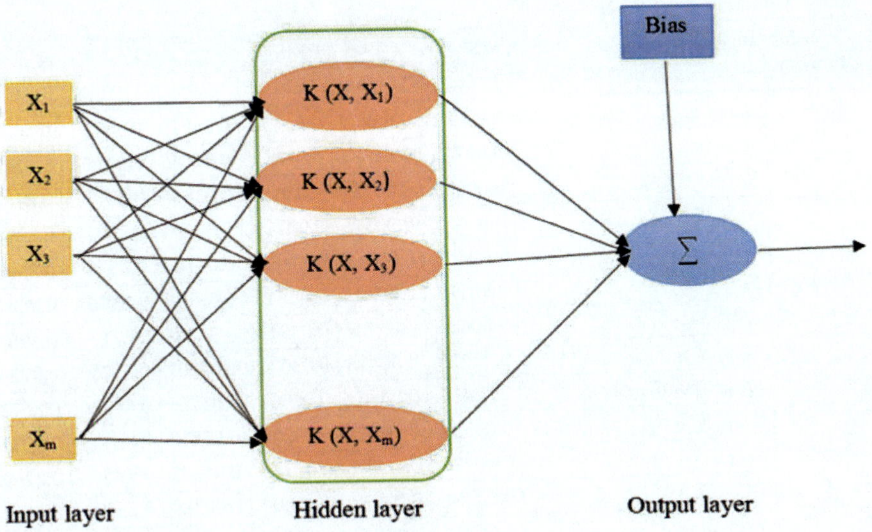

Fig. 3 Model of SVM

$$R^2 = \left[\frac{n\sum ab - (a)(\sum b)}{\sqrt{[n\sum a^2 - (\sum a)^2][n\sum b^2 - (\sum b)^2]}}\right]^2 \tag{1}$$

4 Outcomes of Model

For RNN, outcomes are discussed below for Angul gauge station. In case of Tansig transfer function, the best value of R^2 is 0.9198 and 0.8936 for testing and training phases. Similarly, for Logsig and purelin, the best value of R^2 is 0.9007 and 0.8873 for testing phases. Correspondingly, it gives 0.8805 and 0.8559 for training phases (Table 1).

The performances of SVM model with different input for five proposed watersheds are presented in Table 2. Three evaluating parameters MSE, RMSE and R^2 are estimated for testing and training phases which are explained below. While SVM5 model is considered, the paramount assessment of R^2 is 0.952 and 0.979 for training and testing phases.

Actual versus predicted sediment load of Angul watershed using RNN and SVM techniques for training and testing phases are presented in Fig. 4.

Table 1 Outcomes of RNN

Transfer function	Input	MSE		RMSE		R^2	
		Training	Testing	Training	Testing	Training	Testing
Tansig	S_{t-1}	0.00881	0.00927	0.03434	0.04876	0.8845	0.9009
	S_{t-1}, S_{t-2}	**0.00748**	**0.00861**	**0.01643**	**0.02645**	**0.8936**	**0.9198**
	S_{t-1}, S_{t-2},S_{t-3}	0.00389	0.00558	0.02037	0.03856	0.8713	0.8962
	S_{t-1}, S_{t-2},S_{t-3},S_{t-4}	0.00415	0.00575	0.02117	0.03094	0.8597	0.8794
	S_{t-1}, S_{t-2},S_{t-3},S_{t-4}, S_{t-5}	0.00409	0.00641	0.02895	0.03167	0.8336	0.8569
Logsig	S_{t-1}	**0.00172**	**0.00201**	**0.03845**	**0.05979**	**0.8805**	**0.9007**
	S_{t-1}, S_{t-2}	0.00886	0.00987	0.03913	0.04765	0.8716	0.8963
	S_{t-1}, S_{t-2},S_{t-3}	0.00225	0.00418	0.03072	0.05001	0.8525	0.8726
	S_{t-1}, S_{t-2},S_{t-3},S_{t-4}	0.00417	0.00567	0.02967	0.03765	0.8488	0.8673
	S_{t-1}, S_{t-2},S_{t-3},S_{t-4}, S_{t-5}	0.00359	0.00402	0.03645	0.05812	0.8336	0.8557
Purelin	S_{t-1}	0.00773	0.00847	0.02785	0.05802	0.7904	0.8269
	S_{t-1}, S_{t-2}	0.00891	0.00932	0.03004	0.06380	0.8317	0.8579
	S_{t-1}, S_{t-2},S_{t-3}	**0.00779**	**0.00834**	**0.03215**	**0.04448**	**0.8559**	**0.8873**
	S_{t-1}, S_{t-2},S_{t-3},S_{t-4}	0.00546	0.00774	0.03586	0.06138	0.7997	0.8114
	S_{t-1}, S_{t-2},S_{t-3},S_{t-4}, S_{t-5}	0.00226	0.00375	0.03916	0.05723	0.8226	0.8459

Table 2 Outcomes of SVM

Model	Input	Training			Testing		
		MSE	RMSE	R^2	MSE	RMSE	R^2
SVMI	S_{t-1}	0.508	90.52	0.851	0.432	87.13	0.889
SVM2	S_{t-1}, S_{t-2}	0.516	88.02	0.877	0.558	85.86	0.901
SVM3	S_{t-1}, S_{t-2},S_{t-3}	0.572	85.69	0.903	0.589	82.47	0.936
SVM4	S_{t-1}, S_{t-2},S_{t-3},S_{t-4}	0.601	81.75	0.937	0.647	78.09	0.953
SVM5	S_{t-1}, S_{t-2},S_{t-3},S_{t-4}, S_{t-5}	0.627	77.88	0.952	**0.698**	**75.68**	**0.979**

5 Conclusion

Present research demonstrated comparison of SVM and RNN models for predicting SSL. Outcomes reveal that SVM performs better and acts as an appropriate sediment estimation tool. Generally, potential of identifying importance of input parameters helps SVM as a useful for pre-processing of input also to forecast SSL. Even though ANN methods demonstrate very fine effectiveness in spite of shortage of data from a basin, these models are extremely dependent on used model and outline for training phase and any occurrence or incident exterior to range training, and these models perform very weak in predicting preferred event.

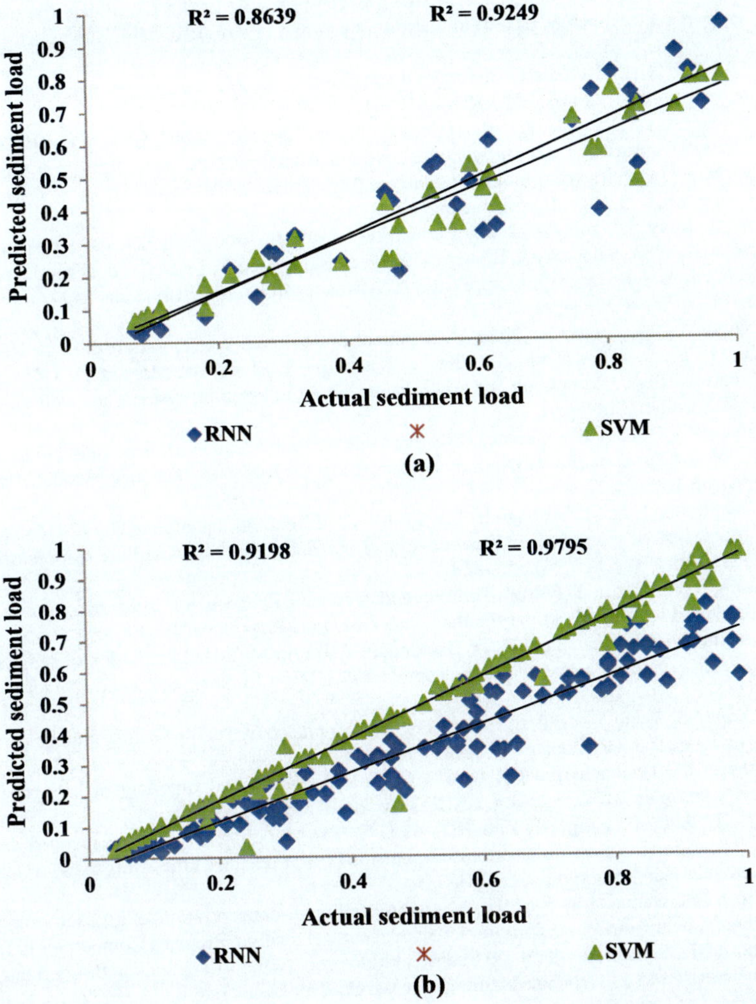

Fig. 4 Actual versus predicted runoff via RNN and SVM for **a** training phase and **b** testing phase

References

1. Lafdani EK, Nia AM, AhmadI A (2013) Daily suspended sediment load prediction using artificial neural networks and support vector machines. J Hydrol 478:50–62
2. Cimen M (2008) Estimation of daily suspended sediments using support vector machines. Hydrological Sciences–Journal–des Sciences Hydrologiques 53(3):655–666
3. Kisi O (2012) Modeling discharge-suspended sediment relationship using least square support vector machine. J Hydrol 456–457:110–120

4. Misra D, Oommen T, Agarwal A, Mishra SK, Thompson AM (2009) Application and analysis of support vector machine based simulation for runoff and sediment yield. Biosyst Eng 103:527–535
5. Jain SK (2012) Modeling river stage-discharge-sediment rating relation using support vector regression. Hydrol Res 43(6):851–861
6. Samantaray S, Sahoo A (2020) Appraisal of runoff through BPNN, RNN, and RBFN in tentulikhunti watershed: a case study. In: Satapathy S, Bhateja V, Nguyen B, Nguyen N, Le DN (eds) Frontiers in intelligent computing: theory and applications. Advances in intelligent systems and computing, vol 1014. Springer, Singapore
7. Samantaray S, Sahoo A (2020) Estimation of runoff through BPNN and SVM in Agalpur Watershed. In: Satapathy S, Bhateja V, Nguyen B, Nguyen N, Le DN. (eds) Frontiers in intelligent computing: theory and applications. Advances in intelligent systems and computing, vol 1014. Springer, Singapore
8. Samantaray S, Sahoo A (2020) Assessment of sediment concentration through RBNN and SVM-FFA in arid watershed, India. In: Satapathy S, Bhateja V, Mohanty J, Udgata S (eds) Smart intelligent computing and applications. Smart innovation, systems and technologies, vol 159. Springer, Singapore
9. Sharma N, Zakaullah M, Tiwari H Kumar D (2015) Runoff and sediment yield modeling using ANN and supportvector machines: a case study from Nepal watershed, Model. Earth Syst Environ 1:23
10. Sadeghpour Haji M, Mirbagheri SA, Javid AH, Khezri M, Najafpour GD (2014) A wavelet support vector machine combination model for daily suspended sediment forecasting. IJE TRANS C: Aspects 27(6):855–864
11. Wang YM, Traore S (2009) Time-lagged recurrent network for forecasting episodicevent suspended sediment load in typhoon prone area. Int J Phys Sci 4(9):519–528
12. Boukhrissa ZA, Khanchoul K, Le Bissonnais Y, Tourki M (2013) Prediction of sediment load by sediment rating curve and neural network (ANN) in El Kebir catchment, Algeria. J Earth Syst Sci 122(5):1303–1312
13. Tfwala SS, Wang YM (2016) Estimating sediment discharge using sedimentrating curves and artificial neural networks in the Shiwen River, Taiwan. Water 8:53
14. Saharia M, Bhattacharjya RK (2012) Geomorphology-based time-lagged recurrent neural networks for runoff forecasting. KSCE J Civ Eng 16(5):862–869
15. Liu QJ, Shi ZH, Fang NF, Zhu HD, Ai L (2013) Modeling the daily suspended sediment concentration in a hyperconcentrated river on the Loess Plateau, China, using the Wavelet–ANN approach. Geomorphology 186:181–190
16. Ghose DK, Samantaray S (2018) Modelling sediment concentration using back propagation neural network and regression coupled with genetic algorithm. Procedia Comput Sci 125:85–92
17. Ghose DK, Samantaray S (2019) Sedimentation process and its assessment through integrated sensor networks and machine learning process. Computational Intelligence in Sensor Networks, Studies in Computational Intelligence 776:473–488
18. Sadeghifar T, Nouri Motlagh M, Torabi Azad M, Mohammad Mahdizadeh M (2017) Coastal wave height prediction using Recurrent Neural Networks (RNNs) in the south Caspian Sea. Marine Geodesy 40(6):454–465

Prophecy of Runoff in a River Basin Using Various Neural Networks

Sneha Raichel Jimmy, Abinash Sahoo, Sandeep Samantaray, and Dillip K. Ghose

Abstract Rainfall–runoff is an extremely complex procedure because of nonlinear and multidimensional dynamics, and therefore not easy for modeling. In recent years, artificial intelligence methods are utilized to model hydrogeological time series. Present study considers nonlinear multiple regressions (NMR), feedforward back-propagation neural network (FFBPN), and adaptive neuro-fuzzy inference system (ANFIS) to predict runoff of Thiruvananthapuram watershed, Kerala. Infiltration and evapotranspiration loss, precipitation, and average temperature are taken into input scenario while runoff acts as output for model. The estimation results obtained by using neuro-fuzzy technique are tested and contrasted to those of artificial neural networks (ANNs). Root mean squared errors (RMSE) and coefficient of determination (R^2) are utilized as assessing criterion to evaluate the model performances. Based on research finding ANFIS provides superlative value for RMSE and R^2 0.9625 and 0.9814, but for NMR and FFBPNN it delivers 0.9376 and 0.9592, respectively. Assessment outcomes show that ANFIS is better suited to apply for runoff estimation.

1 Introduction

Rainfall–Runoff models play an important part in operational flood management events such as to warn and forecast flood-related problems and design of hydraulic systems. The major intent to forecast runoff is providing most precise prediction regarding potential unidentified flood situation and giving caution to all concerned

S. R. Jimmy (✉) · A. Sahoo · S. Samantaray · D. K. Ghose
Department of Civil Engineering, NIT Silchar, Silchar, Assam, India
e-mail: sjrachel7@gmail.com

A. Sahoo
e-mail: bablusahoo1992@gmail.com

S. Samantaray
e-mail: samantaraysandeep963@gmail.com

D. K. Ghose
e-mail: dillipghose_2002@yahoo.co.in

S. C. Satapathy et al. (eds.), *Communication Software and Networks*, Lecture Notes in Networks and Systems 134, https://doi.org/10.1007/978-981-15-5397-4_72

population with sufficient lead time. Although many theoretical and physical models are accepted, ANN models are regarded to be most functional and practical tool which helps operational hydrologists predominantly in hydrological data insufficiency circumstances.

Remesan et al. [1] suggested an integrated hybrid model to predict runoff of Breu watershed, Southwest England. Moghaddamnia et al. [2] explored techniques of estimation of evaporation based on ANN and ANFIS in arid Sistan Province situated in southeast of Iran. Samantaray and Sahoo [3–5] applied various neural networks for hydrological modeling. Kuok et al. [6] proposed an algorithm for training FFNN, namely, particle swarm optimization which was used for modeling runoff and rainfall relation at Sungai Bedup Basin. He et al. [7] applied various data-driven techniques such as ANN, ANFIS, and SVM to forecast stream flow in semiarid region of northwestern China. Güldal and Tongal [8] investigated best fit prediction model from recurrent neural network (RNN) and ANFIS along with classical stochastic models to forecast the lake levels during anthropogenic activities and hydrometeorological changes in the Lake Egirdir, Turkey. Gautam and Holz [9] explored forecasting and simulation of Sieve Basin in Italy using ANFIS-based rainfall–runoff model. Firat and Gungor [10] developed a river flow forecasting system using ANFIS on River Great Menderes, situated in Turkey. Elshorbagy et al. [11] predicted the runoff in Red River Valley located in southern Manitoba, Canada using artificial neural network and compared the result with the traditional methods. Assessment of sediment load is done via various ANN techniques [12, 13]. Chiang et al. [14] compared two neural networks such as static feedforward and dynamic feedforward to model rainfall–runoff relation of Lan-Yang River in Taiwan. Chandwani et al. [15] presented how NNs, genetic algorithms (GA), and fuzzy logic could be used in modeling rainfall–runoff relation. Asadi et al. [16] predicted the runoff of Aghchai catchment situated in northwest Iran by proposing a hybrid intelligent model which includes genetic algorithms and Levenberg–Marquardt algorithm (LM) for studying FFNN. Rajurkar et al. [17] did a study on Narmada River in Madhya Pradesh using a linear multi-input single-output model (MISO) along with artificial neural network to model its rainfall–runoff relation during monsoon floods. To predict runoff of a watershed using various neural network techniques is the main constraint of this research.

2 Study Area and Data

Thiruvananthapuram watershed, Kerala is considered for research area. Thiruvananthapuram division is located between at 8.17° N–8.54° N and 76.41° E–77.17° E. It spans over an area of 2192 km^2. The district comprises three geographical regions such as lowlands, midlands, and highlands. The climate is generally hot tropical. The mean maximum temperature of the district goes up to 35 °C and the mean minimum temperature comes down to 20 °C. The total annual average rainfall is 150 cm per year (Fig. 1).

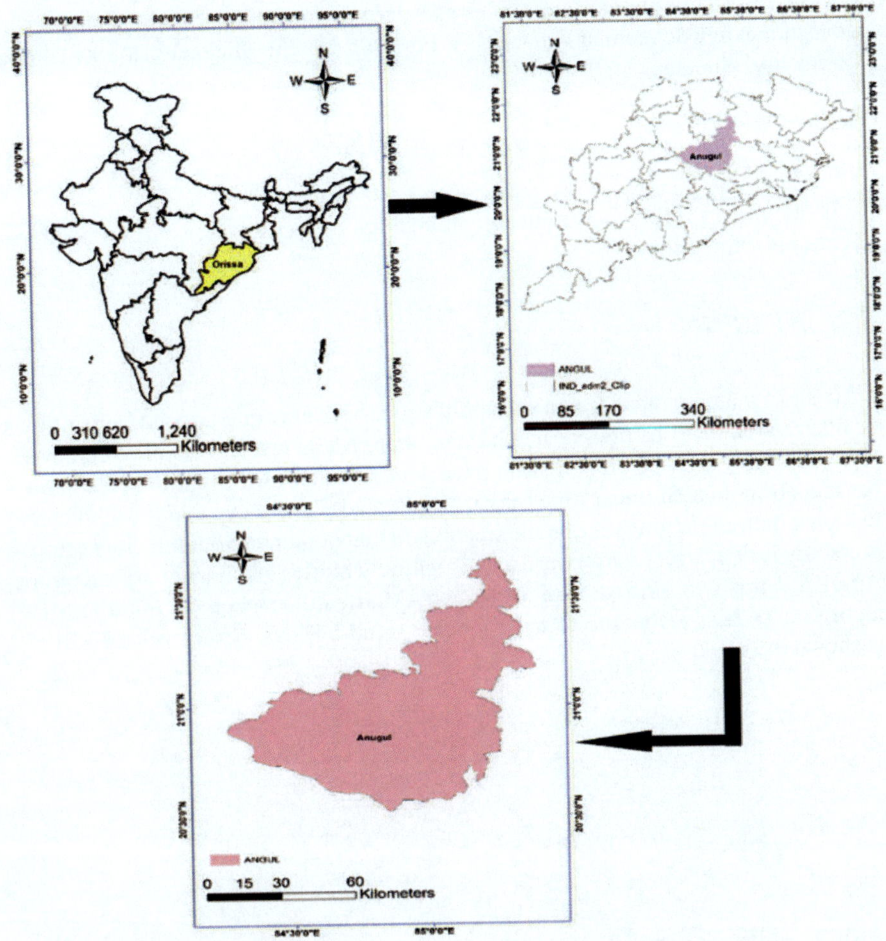

Fig. 1 Proposed watershed

3 Methodology

3.1 NMR

It is a form of analysis where observational data are modeled through function that is a combination of nonlinear constraints. This consists of one or more independent variables. Data are close-fitting through successive approximations. A general form of regression is $Y = f(x, \beta)$.

Function f is nonlinear in modules of vector of constraints β, which narrates an independent and dependent variable x, y. For example, the Michaelis–Menten model for enzyme kinetics:

$$f(x, \beta) = \frac{\beta_1 x}{\beta_2 + x} \tag{1}$$

It cannot be expressed as a linear combination of the two βs, so it is a nonlinear regression expression.

3.2 FFBPNN

On basis of mathematical design and application, ANNs are encouraged from billions of interrelated neurons present in a human brain. Many researchers have confirmed that ANN may well be an enhanced proficient substitute for traditional techniques for modeling then on linear time series. NNs are very much capable in significantly mapping datasets having set of numeric input to that of numeric output. Used setup in this study is two-layer feedforward having sigmoid hidden and linear output neurons. Applied setup was skilled with Levenberg–Marquardt backpropagation (LMBP) algorithm as best technique to train modest sized FFNNs. Structure of FFBPNN is shown in Fig. 2.

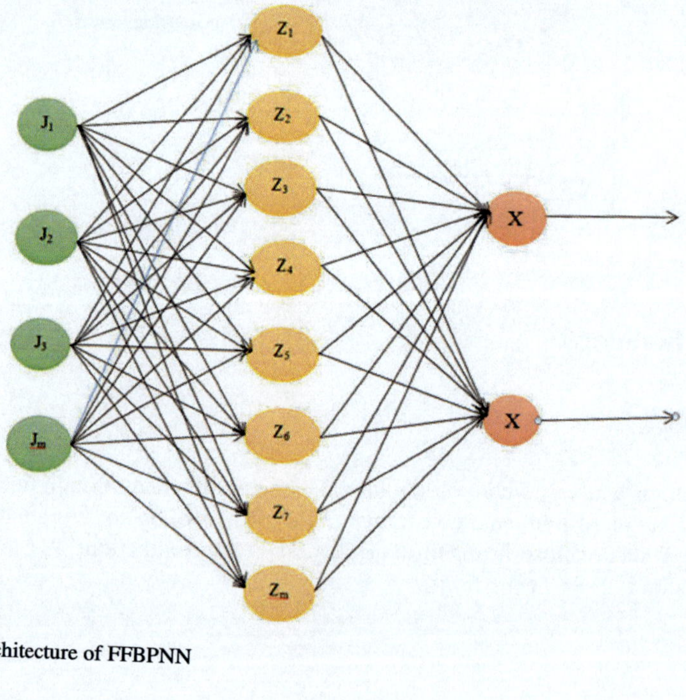

Fig. 2 Architecture of FFBPNN

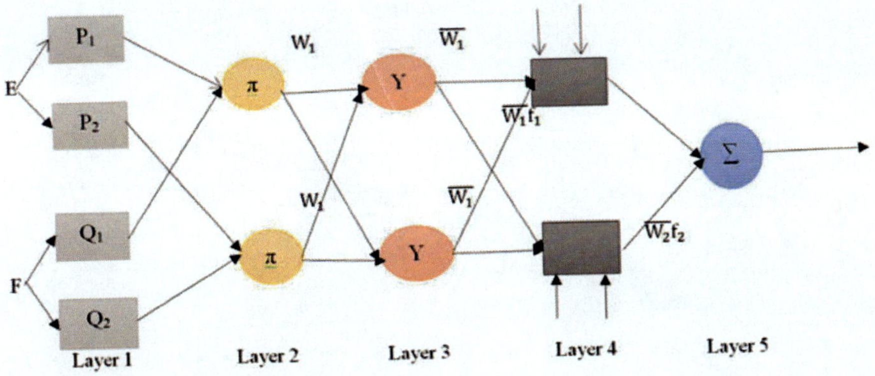

Fig. 3 Architecture of ANFIS

3.3 *ANFIS*

Though ANN is relatively authoritative to model a variety of real-world tribulations, it too has its inadequacies. If input data are not as much precise or uncertain, ANN would be stressed in handling the data and hence ANFIS might be found to be an improved choice. Jang first projected ANFIS and used its theory and ethics effectively to a lot of tribulations [13, 48]. ANFIS is a soft computing technique where a certain input–output dataset is articulated in a FIS. FIS employs a nonlinear map starting from input to output space. Mapping is achieved by an amount of fuzzy if-then regulations, where every rule illustrates local behavior of map. Fuzzy membership constraints are optimized either by using a BP algorithm or by combination of both backpropagation and least square method. Effectiveness of FIS is dependent on projected constraints (Fig. 3).

3.4 *Data Preparation and Model*

Precipitation, evapotranspiration, infiltration loss, and average temperature are taken as input constraint and water table depth for output.

Three numerical constraints, i.e., coefficient of determination (R^2), root mean square error (RMSE), and mean square error (MSE), have been used for model performance. MSE can be assessed by equation,

$$\text{MSE} = \left[\frac{1}{M}\sum_{i=1}^{m}(Z_a - Z_p)^2\right] \quad (2)$$

where Z_a: actual output and
Z_p: predicted output.

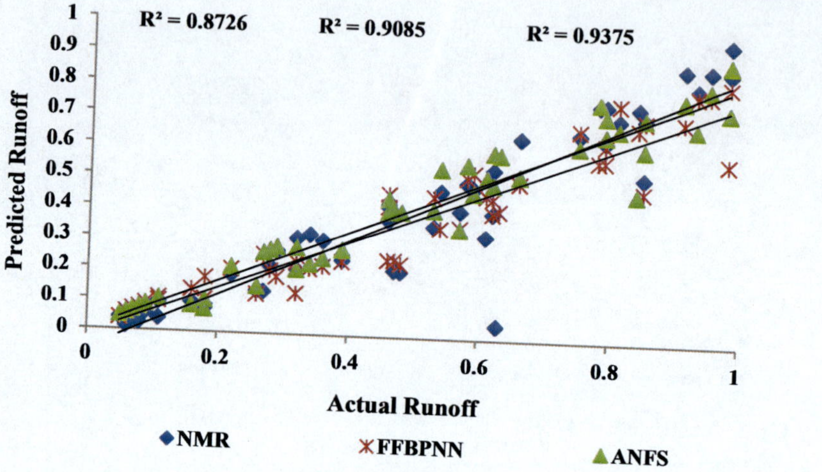

(a)

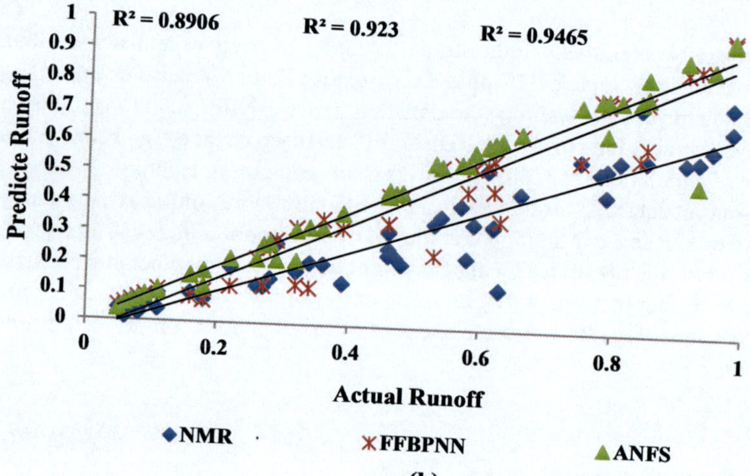

(b)

Fig. 4 Actual versus predicted runoff using NMR, CFBPNN, and ANFIS for **a** training phase and **b** testing phase

4 Results and Discussion

All transfer function performances measured applying NMR, ANFIS, and FFBPNN are shown in Tables 1, 2, and 3, respectively. Three diverse transfer functions for FFBPNN and six diverse membership functions for ANFIS are taken into consideration for finding the superlative model which helps in predicting runoff efficiently. It can be noticed from table that all models presented suitable outcomes for predicting

Table 1 Performance value of NMR

Best model input	Model parameters	MSE		RMSE		R^2	
		Training	Testing	Training	Testing	Training	Testing
Precipitation Avg. temperature Evapotranspiration Infiltration loss	$a =$ 920.9310 $b = -2.9391$ $c = 0.7255$ $d = -$ 191858.8636	0.000536	0.0004	0.013654	0.037653	0.8726	0.8906

runoff. In case of FFBPNN, Tan-sig transfer function with 4-8-1 architecture provides paramount value of coefficients 0.9452 and 0.9618 in training and testing phases. But in case of ANFIS, Gbell transfer function gives prominent values of R^2 as 0.9557 and 0.9802 for training and testing phases, respectively. Performance values of all transfer functions are given below. Actual versus predicted value for R^2 is presented in Fig. 4.

5 Conclusion

In this study, NMR, ANN, and ANFIS models have been projected and assessed for daily rainfall–runoff modeling. Three statistical parameters were deemed for performance assessment of chosen models. It can be seen that projected ANFIS model is competent enough to produce finest efficiency in opposition to NMR and ANN model. Outcomes of present research have a greater insinuation in area of hydrological modeling. Projected method is reasonably common and may well be useful for different modeling setups for different conditions. Present findings will help in stimulating more investigation and additional tests for improving proposed amalgam modeling method in other watersheds utilizing various model engines.

Table 2 Performance value of FFBPNN

Model input	Sigmoid function	Architecture (L-M-N)	MSE		RMSE		R^2	
			Training	Testing	Training	Testing	Training	Testing
Precipitation Avg. temperature Evapotranspiration Infiltration loss	Tan-sig	4-2-1	0.000815	0.000299	0.027743	0.052745	0.9248	0.9427
		4-4-1	0.000791	0.000718	0.026814	0.050128	0.8922	0.9169
		4-6-1	0.000729	0.000113	0.042163	0.077134	0.8836	0.9011
		4-8-1	**0.000617**	**0.000878**	**0.032258**	**0.066345**	**0.9452**	**0.9618**
	Log-sig	4-2-1	0.000501	0.000125	0.018949	0.044865	0.8508	0.8778
		4-4-1	0.000536	0.000265	0.033812	0.061379	0.8635	0.8856
		4-6-1	**0.000677**	**0.000902**	**0.025158**	**0.051836**	**0.9264**	**0.9435**
		4-8-1	0.000561	0.000801	0.033754	0.060017	0.8992	0.9137
	Purelin	4-2-1	0.000294	0.000315	0.038216	0.061082	0.8609	0.8969
		4-4-1	0.000746	0.000368	0.026388	0.050017	0.8441	0.8712
		4-6-1	**0.000477**	**0.000259**	**0.022613**	**0.044765**	**0.8994**	**0.9114**
		4-8-1	0.000387	0.000156	0.018861	0.040023	0.9008	0.9355

Table 3 Performance value of ANFIS

Function	MSE		RMSE		R^2	
	Training	Testing	Training	Testing	Training	Testing
Tri	0.000595	0.000878	0.073546	0.057158	0.8775	0.9051
Trap	0.000998	0.000945	0.078521	0.082761	0.8991	0.9217
Gbell	**0.000134**	**0.000834**	**0.069132**	**0.071387**	**0.9557**	**0.9802**
Gauss	0.000648	0.000176	0.058151	0.037864	0.9358	0.9663
Gauss2	0.000859	0.000654	0.077822	0.051637	0.9244	0.9558
Pi	0.000435	0.000834	0.069124	0.020016	0.9051	0.9335

References

1. Remesan R, Shamim MA, Han D, Mathew J (2009) Runoff prediction using an integrated hybrid modelling scheme. J Hydrol 372:48–60
2. Moghaddamnia A, Gousheh G, Piri J, Amin S, Han D (2009) Evaporation estimation using artificial neural network and adaptive neuro-fuzzy inference system techniques. Adv Water Resour 32:88–97
3. Samantaray S, Sahoo A (2020a) Appraisal of runoff through BPNN, RNN, and RBFN in Tentulikhunti Watershed: a case study. In: Satapathy S, Bhateja V, Nguyen B, Nguyen N, Le DN (eds) Frontiers in intelligent computing: theory and applications. Advances in intelligent systems and computing, vol 1014. Springer, Singapore
4. Samantaray S, Sahoo A (2020b) Estimation of runoff through BPNN and SVM in Agalpur Watershed. In: Satapathy S, Bhateja V, Nguyen B, Nguyen N, Le DN (eds) Frontiers in intelligent computing: theory and applications. Advances in intelligent systems and computing, vol 1014. Springer, Singapore
5. Samantaray S, Sahoo A (2020c) Assessment of sediment concentration through RBNN and SVM-FFA in Arid Watershed, India. In: Satapathy S, Bhateja V, Mohanty J, Udgata S (eds) Smart intelligent computing and applications. Smart innovation, systems and technologies, vol 159. Springer, Singapore
6. Kuok KK, Harun S, Shamsuddin SM (2010) Particle swarm optimization feedforward neural network for modeling runoff. Int J Environ Sci Technol 7(1):67–78
7. He Z, Wen X, Liu H, Du J (2014) A comparative study of artificial neural network, adaptive neuro-fuzzy inference system and support vector machine for forecasting river flow in the semiarid mountain region. J Hydrol 509:379–386
8. Güldal V, Tongal H (2010) Comparison of recurrent neural network, adaptive neuro-fuzzy inference system and stochastic models in Eğirdir Lake level forecasting. Water Resour Manage 24:105–128
9. Gautam DK, Holz KP (2001) Rainfall-runoff modelling using adaptive neuro-fuzzy systems. J Hydroinformatics 03(1)
10. Firat M, Gungor M (2007) River flow estimation using adaptive neuro fuzzy inference system. Math Comput Simul 75:87–96
11. Elshorbagy A, Simonovic SP, Panu US (2000) Performance evaluation of artificial neural networks for runoff prediction. J Hydrol, Eng 5:424–427
12. Ghose DK, Samantaray S (2018) Modelling sediment concentration using back propagation neural network and regression coupled with genetic algorithm. Procedia Comput Sci 125:85–92
13. Ghose DK, Samantaray S (2019) Sedimentation process and its assessment through integrated sensor networks and machine learning process. In: Computational intelligence in sensor networks, studies in computational intelligence, vol 776, pp 473–488

14. Chiang Y-M, Chang L-C, Chang F-J (2004) Comparison of static-feedforward and dynamic-feedback neural networks for rainfall–runoff modeling. J Hydrol 290:297–311
15. Chandwani V, Vyas SK, Agrawal V, Sharma G (2015) Soft computing approach for rainfall-runoff modelling: a review. Aquatic Procedia 4:1054–1061
16. Asadi S, Shahrabi J, Abbaszadeh P, Tabanmehr S (2013) A new hybrid artificial neural network for rainfall-runoff process modeling. Neurocomputing 05.023
17. Rajurkar MP, Kothyari UC, Chaube UC (2009) Artificial neural networks for daily rainfall-runoff modelling. Hydrol Sci J 47(6):865–877

Application of Artificial Neural Networks and Multiple Linear Regression for Rainfall–Runoff Modeling

Wajahat Annayat, Ajay Gupta, Kondepogu Ravi Prakash, and Briti Sundar Sil

Abstract Present study makes an effort to estimate the runoff as a function of rainfall, temperature, and humidity of the Barak River. Two approaches of artificial neural network (ANN) (i) feedforward backpropagation neural network (FFBPNN) and (ii) layer recurrent neural network (LRNN) are used to predict runoff during the monsoon period. The evaluation criteria considered for both of the models are the mean square error (MSE), root mean square error (RMSE), and coefficient of determination (R^2). FFBPNN performs best with architecture 4-6-1 following Tan-sig transfer function, which possesses MSE training and testing values 0.00012 and 0.00116, RMSE training and testing values 0.01072 and 0.03401, and R^2 training and testing values 0.9714 and 0.9645. LRNN performs best with architecture 4-7-1 following Tan-sig transfer function, which possesses MSE training and testing values 0.00021 and 0.00069, RMSE training and testing values 0.01449 and 0.02617, and R^2 training and testing values 0.9467 and 0.9380. Further, the comparison of FFBPNN, LRNN, and multiple linear regression techniques (MLR) is done and the result shows that all the approaches perform best and gives good R^2 values. Findings of this study could be used for planning and management of hydraulic structures in neighborhood of watershed.

W. Annayat (✉) · A. Gupta · K. R. Prakash · B. S. Sil
Department of Civil Engineering, NIT Silchar, Silchar, Assam, India
e-mail: wajahatannayat123@gmail.com

A. Gupta
e-mail: 19ajay93@gmail.com

K. R. Prakash
e-mail: kondepoguravi@gmail.com

B. S. Sil
e-mail: britisil@gmail.com

S. C. Satapathy et al. (eds.), *Communication Software and Networks*, Lecture Notes in Networks and Systems 134, https://doi.org/10.1007/978-981-15-5397-4_73

1 Introduction

The increasing demand for water in most parts of the world is due to high population growth, industrialization, and urbanization [1]. Due to this influence, river systems and watersheds have changed resulting in high loss of property damages and life due to flooding [2]. Water availability estimation plays a great role in the planning of water resources project. Information about runoff from a hydrologic catchment is required for the design of hydrological engineering and management purposes [3]. Hydrologists are trying to forecast the stream flow for various resolves like flood control, irrigation, water supply, water quality, drainage, recreation, power generation, and wildlife propagation [4].

ANNs work on principle of biological nervous systems. It consists of massively parallel systems which are self-possessed of processing elements that are connected by links of varying weights [5]. The ability of ANN to learn from sufficient data pairs makes it possible for solving complex problems such as nonlinear modeling, pattern recognition, association, classification, and control. ANN since the past years is used successfully in hydrology-related areas such as rainfall–runoff modeling [6–10]. ANN is capable of modeling rainfall–runoff relationship since it has the ability to generalize patterns in ambiguous and noisy input data that synthesize a complex model without probability distributions or prior knowledge [11]. ANN is the diffused application for forecasting of the flood as it is used as a black-box hydrological model, at time scale which ranges from 1 day to 1 year [12]. ANN models are strong predicting tools for rainfall–runoff relationship parameters. Both linear and nonlinear systems are modeled by ANN and are used in various aspects of science and engineering [13]. In areas of planning and management of water resource, the results will support as decision-making.

The two factors in which ANN depends are the network topology and the algorithm that is used to specify the value weights value connecting the nodes. The research on ANN focuses on brain modeling for many computational tasks which did not perform well by traditional serial computers [14, 15]. Out of the various artificial neural networks, the most promising networks that are used in the present work are FFBPNN and LRNN for training and testing of data. In FFBPNN, the connection of the weight feeds the activation in a direction forward from input to output layer. The criteria used for evaluation in the model are MSE, RMSE, and R^2. Both FFBPNN and LRNN gave good MSE, RMSE, and R^2 values for prediction of runoff at the Barak Basin watershed.

Therefore, objective of this study is to estimate the catchment runoff by FFBPNN and LRNN using hydrological parameters such as precipitation, minimum temperature, maximum temperature, and relative humidity. Furthermore, the second goal is to associate results of FFBPNN and LRNN model with MLR techniques to examine the effect of such inputs on the model performance.

2 Study Area and Data

Barak River starts off from Japvo Mountain of Manipur Hills at an elevation of 3,015 m and follows the course south all the way through hilly terrain up to Tipaimukh close to the tri-junction of the three states: Assam, Manipur, and Mizoram. At this point, the river takes a hairpin curve and enters into the plains of Cachar District of Assam and figures the border of states of Assam and Manipur up to Jirimat, slight upstream of Lakhimpur. The river then flows westward of the Barak Valley of Assam. Finally, it enters into Bangladesh where it is known as the Surma River and Kushiyara River which is later called as Magna River. Major parts covered by Barak River are northeastern India, Bangladesh, and Myanmar. The Barak Basin lies between 89°50′ E to 94°0′ E and 22°44′ N to 25°58′ N. The data used in the present study such as precipitation, maximum temperature, minimum temperature, relative humidity, and runoff are collected from 'Central Water Commission' (CWC), Shillong from the year 2001 to 2011. The data from 2001 to 2007 is used for training and the data from 2008 to 2011 is used for testing. Location map of the study area is shown in Fig. 1.

3 Methodology

3.1 *ANN*

It is an alternate computational approach which is based on study of the brain and its emerging properties. There are three layers of ANN, viz., (i) an input, (ii) hidden, and (iii) output layer. From external sources, input layer receives information and passes for processing the network. Information from input level is sent to hidden level for processing of work, and then the output level receives the processed information and sends result out to an external receptor. The two factors in which ANN depends are the network topology and the algorithm that is used to specify the value of the weight connecting the nodes.

3.2 *FFBPNN*

For proper functioning, process of ANN an organized technique is used called FFBPNN. To analyze error gradient of network with respect to network's modifiable weights, backpropagation is used. Backpropagation, for the most part, permits quick conjunction on reasonable error for local minima. Backpropagation networks are usually of multiple layers with one input, hidden, and output layer. Architecture of FFBPNN is shown in Fig. 2.

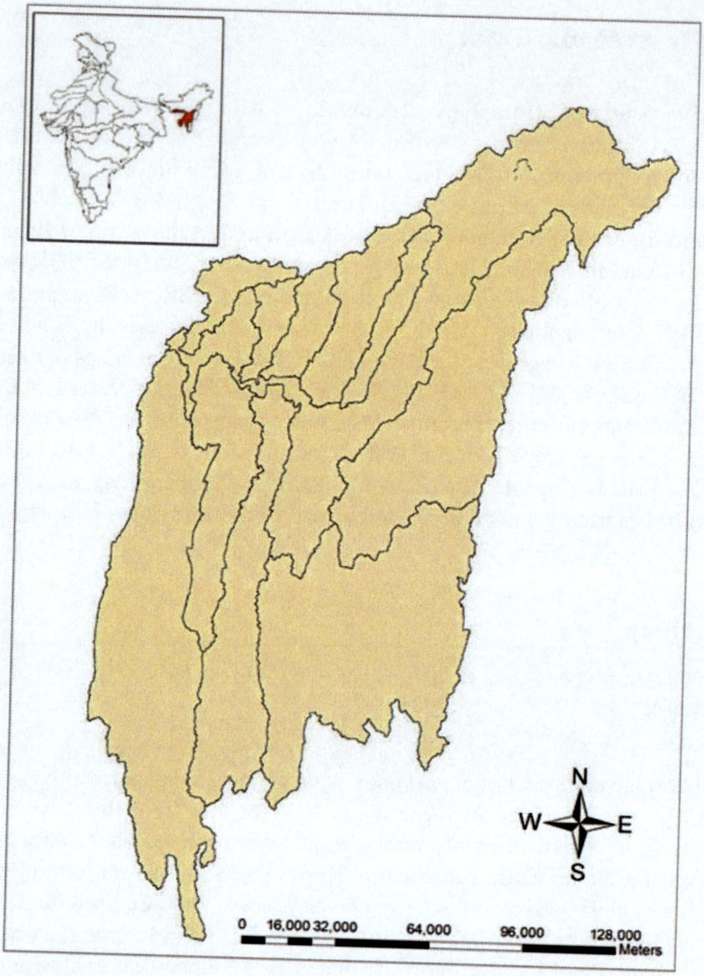

Fig. 1 Location map of study area

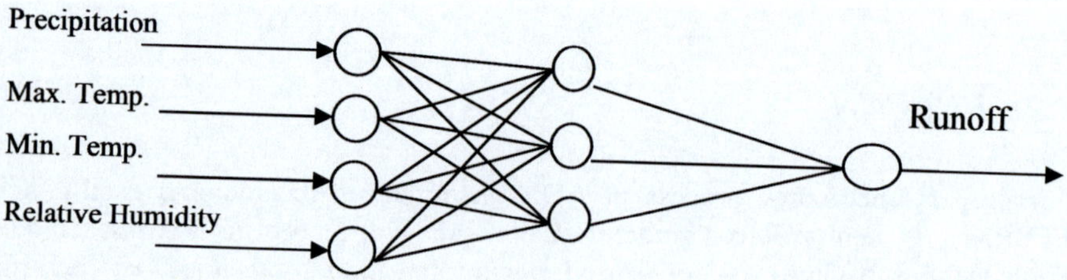

Fig. 2 BPNN model with four inputs

3.3 LRNN

It involves four layers such as input, hidden, context, and an output layer. Every input unit is interconnected to every hidden unit as to each context unit. There is an equal number of hidden layer and context layer which are connected to each other. In fact, the connection downward from the hidden layer allows storing the outputs of hidden layer by the context layer at every time step and then they are fed back by the upward links as extra inputs. Thus, the connections of layer recurrent allow the hidden layers to regenerate the data at various time steps. The architecture of FFBPNN is shown in Fig. 3.

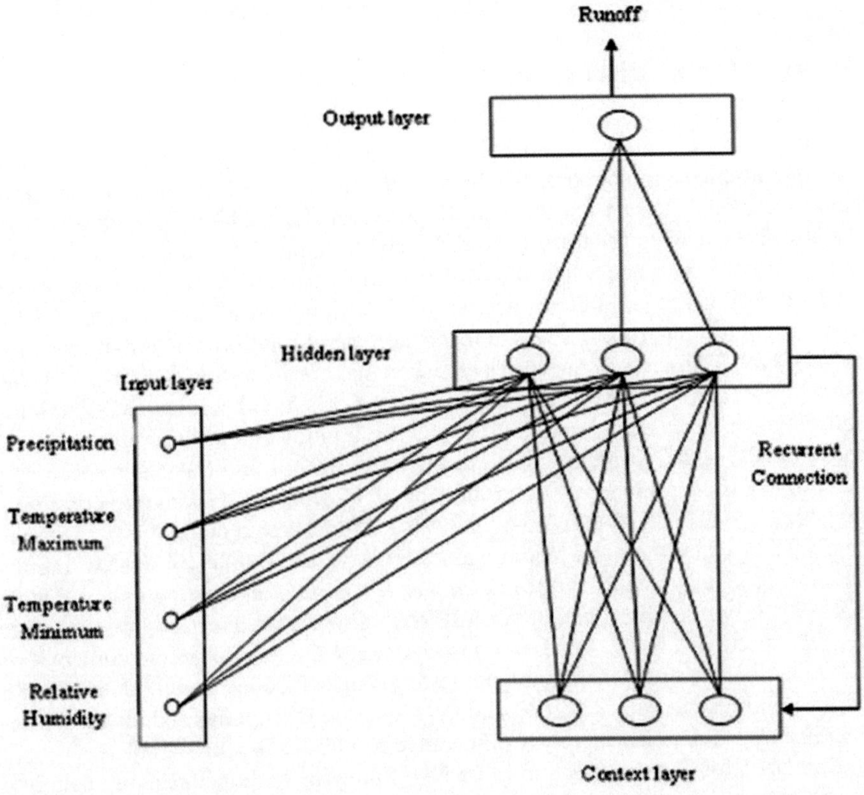

Fig. 3 Architecture of LRNN model with four inputs

3.4 MLR

It is a technique in which a linear equation is fitted to observed data which is a relationship between two or more independent variables and a dependent variable. Each independent variable x is related to dependent variable y. The relation between dependent and independent variables is of the following form:

$$Y = a_0 + a_1x_1 + a_2x_2, \ldots, a_px_p \pm e \tag{1}$$

where Y = Dependent variable; $a_0 = y$ intercept; a_1, a_2, and a_p = slopes associated with $x_1, x_2, \ldots, x_p$; x_1, x_2 and x_p = values of independent variables; e = the error.

4 Results and Discussion

A combination of various architectures is made with tangential sigmoidal, purelin, and logarithmic sigmoidal transfer function in order to check performance and efficiency of model. Data from 2001 to 2007 is used for training the data and from 2008 to 2011 is used for testing the data. The training and testing phases of both models were terminated when the MSE was minimal and R^2 was highest. Based on some evaluation conditions such as MSE, RMSE, and R^2 the final goal of the training and testing processes was to reach the desired solution. For Tan-sig function in FFBPNN, following architectures are taken into consideration, viz., 4-2-1, 4-3-1, 4-4-1, 4-5-1, and 4-6-1. Model architecture of 4-6-1 is found best with MSE training and testing values 0.00012 and 0.00116, RMSE training and testing values 0.01072 and 0.03401, and R^2 training and testing values 0.9714 and 0.9645, respectively. For Log-sig function in FFBPNN, following architectures are taken into consideration, viz., 4-2-1, 4-3-1, 4-4-1, 4-5-1, and 4-6-1. Model architecture of 4-5-1 is found best with MSE training and testing values 0.00016 and 0.000910, RMSE training and testing values 0.0126 and 0.11225, and R^2 training and testing values 0.9645 and 0.9413. For Purelin function in FFBPNN, following architectures are taken into consideration, viz., 4-2-1, 4-3-1, 4-4-1, 4-5-1, and 4-6-1. Model architecture of 4-4-1 is found best with MSE training and testing values 0.00069 and 0.00066, RMSE training and testing values 0.2643 and 0.02569, and R^2 training and testing values 0.9589 and 0.9345, respectively. Comparative results are shown in Table 1.

Similarly, for Tan-sig function in LRNN, following architectures are taken into consideration, viz., 4-2-1, 4-5-1, 4-6-1, 4-7-1, and 4-8-1. Model architecture of 4-7-1 is found best with MSE training and testing values 0.00021 and 0.00069, RMSE training and testing values 0.01449 and 0.02617, and R^2 training and testing values 0.9467 and 0.9389, respectively. For Log-sig function in LRNN, following architectures are taken into consideration, viz., 4-2-1, 4-5-1, 4-6-1, 4-7-1, and 4-8-1. Model architecture of 4-6-1 is found best with MSE training and testing values 0.00048 and

Table 1 MSE, RMSE, and R^2 with various combinations of architectures using FFBPNN at Barak River

Model input	Transfer function	Architecture	MSE		RMSE		R^2	
			Training	Testing	Training	Testing	Training	Testing
Precipitation Max. temperature Min. temperature Relative humidity	Tan-sig	4-6-1	0.00012	0.00116	0.01072	0.03401	0.9714	0.9645
	Log-sig	4-5-1	0.00016	0.00091	0.0126	0.11225	0.9645	0.9413
	Purelin	4-4-1	0.00069	0.00066	0.02643	0.02569	0.9589	0.9345

Table 2 MSE, RMSE, and R^2 with various combinations of architectures using LRNN at Barak River

Model input	Transfer Function	Architecture	MSE		RMSE		R^2	
			Training	Testing	Training	Testing	Training	Testing
Precipitation Maximum temperature Minimum temperature Relative humidity	Tan-sig	4-7-1	0.00021	0.00069	0.01449	0.02617	0.9467	0.9389
	Log-sig	4-6-1	0.00048	0.00099	0.02186	0.3154	0.9478	0.9178
	Purelin	4-8-1	0.0007	0.00056	0.02649	0.02359	0.9378	0.9278

0.00099, RMSE training and testing values 0.02186 and 0.03154, and R^2 training and testing values 0.9478 and 0.9178, respectively.

For Purelin function in LRNN, following architectures are taken into consideration, viz., 4-2-1, 4-5-1, 4-6-1, 4-7-1, and 4-8-1. Model architecture of 4-8-1 is found best with MSE training and testing values 0.0007 and 0.00056, RMSE training and testing values 0.02649 and 0.02359, and R^2 training and testing values 0.9378 and 0.9287, respectively. Comparative results are shown in Table 2.

Further, the comparison of FFBPNN, LRNN, and multiple linear regression techniques (MLR) is done and the result shows that all the approaches perform best and give good R^2 values. Multiple linear regression (MLR) technique is used between the datasets at the Barak Basin with the R^2 in training and testing as 0.8452 and 0.8036, respectively, as shown in Fig. 4.

4.1 Observed Versus Simulated Runoff Assessment for the Barak Watershed

The graphical results in terms of observed and predicted runoff values are shown in Fig. 5 from FFBPNN model and LRNN model. Estimated peak runoff is 784.53 and

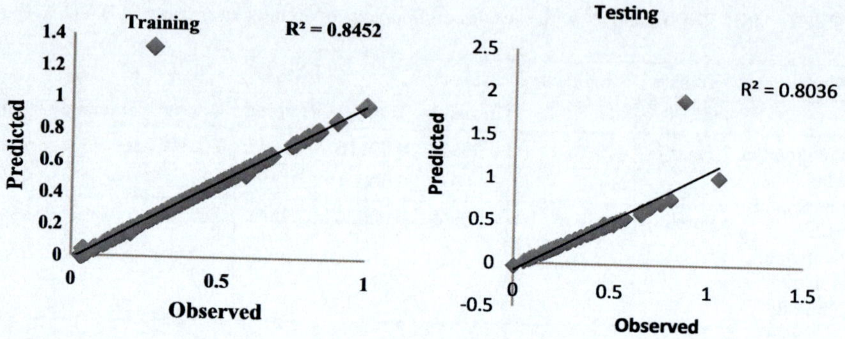

Fig. 4 Observed runoff versus predicted runoff using MLR technique

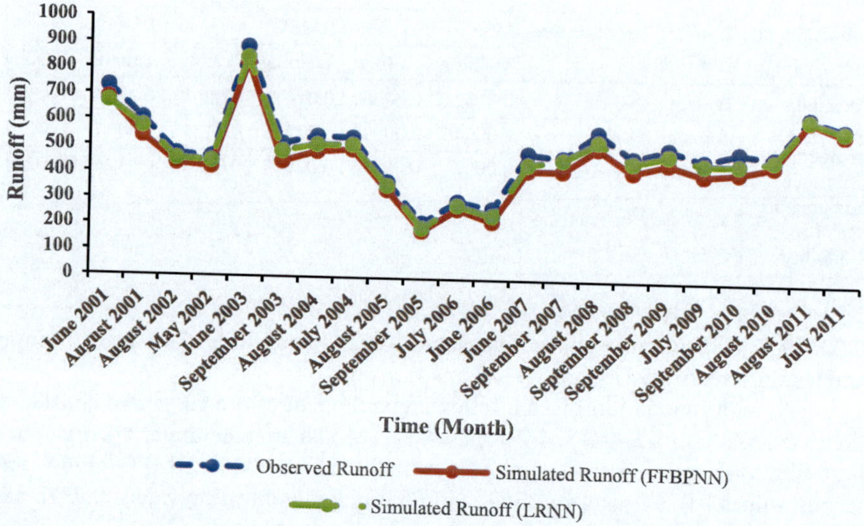

Fig. 5 Observed runoff versus simulated runoff using FFBPNN and LRNN at Barak River

847.53 mm for FFBPNN and LRNN, respectively, against the actual peak runoff of 884.45 mm.

5 Conclusion

Various parameters like rainfall, temperature, and humidity are considered for estimating runoff at the Barak Basin. The study indicates that both FFBPN and LRNN were quite effective in predicting the runoff with evaluation criteria of MSE, RMSE, and R^2. FFBPNN performs best with architecture 4-6-1 following Tan-sig transfer function. Similarly, LRNN performs best with architecture 4-7-1 following Tansig transfer function. Further, multiple linear regressions show a better correlation between datasets at the Barak Basin with the R^2 in training and testing as 0.8452 and 0.8036, respectively. The authors believe that the combination of FFBPNN, LRNN, and MLR model can be used by the planners and engineers for different purposes like preparedness for a possible flood event, flood mitigation, construction of dykes, embankments, disaster management, flood insurance, etc.

References

1. Wu F, Shi Q, Hasan SS, Shi C, Gibson J (2019) Urbanization and industrial transformation for improved water management. River Basin Manage 61–89
2. Tokar AS, Johnson PA (1999) Rainfall-runoff modelling using artificial neural networks. J Hydrol Eng 4(3):232–239
3. Wajahat A, Sundar SB (2018) Estimation and analysis of possible flood for the Silchar city-a case study. 11(1):25–32
4. Annayat W, Sil BS, Ajay G (2018) Development of rainfall-runoff model using FFBPNN and LRNN for Silchar city-a case study. Dis Adv 11:19–23
5. Smith J, Eli RN (1995) Neural-network models of the rainfall-runoff process. J Water Resour Plann Manage 121(6):499–508
6. Govindaraju RS. (2000) Artificial neural networks in hydrology. II: hydrologic applications. J Hydrol Eng 5(2):124–137
7. Samantaray S, Sahoo A (2020) Appraisal of runoff through BPNN, RNN, and RBFN in Tentulikhunti Watershed: a case study. In: Satapathy S, Bhateja V, Nguyen B, Nguyen N, Le DN (eds) Frontiers in intelligent computing: theory and applications. Advances in intelligent systems and computing, vol 1014. Springer, Singapore
8. Samantaray S, Sahoo A (2020) Estimation of runoff through BPNN and SVM in Agalpur Watershed. In: Satapathy S, Bhateja V, Nguyen B, Nguyen N, Le DN (eds) Frontiers in intelligent computing: theory and applications. Advances in intelligent systems and computing, vol 1014. Springer, Singapore
9. Samantaray S, Sahoo A (2020) Assessment of sediment concentration through RBNN and SVM-FFA in Arid Watershed, India. In: Satapathy S, Bhateja V, Mohanty J, Udgata S (eds) Smart intelligent computing and applications. Smart innovation, systems and technologies, vol 159. Springer, Singapore
10. Ghose DK, Samantaray S (2019) Sedimentation process and its assessment through integrated sensor networks and machine learning process. In: Mishra B, Dehuri S, Panigrahi B, Nayak A, Mishra B, Das H (eds) Computational intelligence in sensor networks. Studies in computational intelligence, vol 776. Springer, Berlin, Heidelberg
11. Tokar AS, Markus M (2000) Precipitation-runoff modelling using artificial neural networks and conceptual models. J Hydrol Eng 5(2):156–161
12. Ghose Dillip K, Samantaray Sandeep (2018) Modelling sediment concentration using back propagation neural network and regression coupled with genetic algorithm. Procedia Comput Sci 125(2018):85–92
13. Riad S, Mania J, Bouchaou L, Najjar Y (2004) Rainfall-runoff model using an artificial neural network approach. Math Comput Model 40(7–8):839–846
14. Chan NW (2015) Impacts of disasters and disaster risk management in Malaysia: the case of floods. In: Resilience and recovery in asian disasters. Springer, pp 239–265
15. Srinivasulu S, Jain A (2006) A comparative analysis of training methods for artificial neural network rainfall-runoff models. Appl Soft Comput 6(3):295–306

Runoff Prediction Using Hybrid Neural Networks in Semi-Arid Watershed, India: A Case Study

Sandeep Samantaray, Abinash Sahoo, Nihar Ranjan Mohanta, Paresh Biswal, and Umesh K. Das

Abstract Predicting runoff is a nonlinear and intricate procedure that is very much important to design canals, managing and arrangement of water usability, controlling flood, and prediction of soil erosion. Many methods to predict runoff on basis of hydro-meteorological and geomorphologic condition is available and being readily used. But recently, a number of soft computing methods have evolved for predicting runoff. The proposed study uses a crossbreed smart model which is an amalgamation of data pre-processing techniques, Genetic Algorithm (GA), and Support Vector Machine (SVM) algorithm for Neural Networks (NN). This manuscript explores the efficiency of SVM evolved NN to forecast rainfall-runoff and usability of this for predicting runoff in Balangir watershed. For an affluent organization of water resources in semi-arid province, predicting one-day lead runoff is very much essential. At a specified instant for predicting runoff, input variables taken into consideration are rainfall and runoff that are witnessed on earlier time period. Genetic operator is cautiously considered for optimizing the NN to avoid untimely meeting and variation problems.

S. Samantaray (✉) · A. Sahoo · U. K. Das
Department of Civil Engineering, NIT Silchar, Silchar, Assam, India
e-mail: samantaraysandeep963@gmail.com

A. Sahoo
e-mail: bablusahoo1992@gmail.com

U. K. Das
e-mail: umeshdas96@gmail.com

N. R. Mohanta · P. Biswal
Department of Civil Engineering, GIET Gunpur, Gunpur, Odisha, India
e-mail: niharthenew@gmail.com

P. Biswal
e-mail: biswalparesh3@gmail.com

S. C. Satapathy et al. (eds.), *Communication Software and Networks*, Lecture Notes in Networks and Systems 134, https://doi.org/10.1007/978-981-15-5397-4_74

1 Introduction

Assessment of water availability has a significant role in water resource planning and management. Understanding watershed hydrology and determining runoff yield is very much necessary to plan any watershed development. Runoff calculation that results from rainfall in a watershed is the foremost step for estimating accessibility of water. Runoff length calculated in a river might be of small phase or extended phase which depends on watershed characteristics. Rainfall-runoff modeling is one among many complex and incomprehensible hydrologic phenomena because of unpredictability in time and space intricacy in catchment characteristics with numerous rainfall stages. Precision in prediction of hydrology and water resources has an enormous impact on city development, land use, and civil project plan. Rainfall–runoff model also impacts significantly in managing flood events, for example, hydraulic structure design and predicting flood.

Sedki et al. [1] examined efficiency of GA-based Neural Network (NN) to forecast rainfall-runoff and use fullness of the model for predicting runoff in semi-arid watershed, Morocco. Samantaray and Sahoo [2–4] used several neural networks to forecast various constraints. Asadi et al. [5] projected a smart model which is a mixture of GAs and Levenberg–Marquardt (LM) for learning feedforward NNs to predict runoff. Behzad et al. [6] presented a novel model known as SVM on basis of structural risk minimization standard, which reduces generalized risk error as contradictory to empirical risk minimization standard utilized by customary ANNs. Sivapragasam et al. [7] projected a simple and efficient method to predict runoff value of Tryggevælde watershed, Denmark and rainfall value of Singapore on basis of Singular Spectrum Analysis (SSA) coupled with SVM. Bray and Han [8] explored the use of SVM to forecast flood, which focused on identifying an appropriate model formation and its applicable constraints to develop rainfall-runoff model. Duong et al. [9] investigated enactment of SWAT, Recurrent Fuzzy Neural Network (RFNN), and a development of RFNN-GA that is an amalgam of RFNN and GA for predicting runoff of Srepok River, Vietnam. Ding and Dong [10] presented a hybrid approach indicated by WT-PSR-GA-NN that is a composition of Wavelet Transform (WT), Phase Space Reconstruction (PSR), NN, and GA for forecasting rainfall-runoff. Ghose and Samantaray [11, 16] used two NN methods: Feed-Forward Back Propagation Network (FFBPN) and Radial Basis Function Network (RBFN), for forecasting runoff as a function of precipitation, temperature, and evapotranspiration loss. Okkan and Serbes [12] examined usability and potential of LS-SVM to predict magnitude of runoff of Tahtali and Gordes catchments in Izmir city, Turkey. Samantaray and Ghose [17] used Recurrent Neural Network, and Radial Basis Neural Network to predict runoff. Savic et al. [13] introduced genetic programming that is a development of calculating techniques which gives apparent and prearranged structure recognition, to model rainfall-runoff for predicting the flow of Kirkton catchment, Scotland (U.K.). Sharifi et al. investigated and found efficient solutions to determine the best input combination for runoff model using Gamma Test (GT), onward assortment, and aspect study. Ghose and Samantaray [11, 14] employed several neural networks to forecast sediment load in a watershed. Sharma et al. [15] used SVMs for simulating runoff and sediment yield on a daily basis to predict sediment concentration of Kankaimai watershed, Nepal. The objective of this research is to predict runoff using SVM and GA techniques.

2 Study Area

Balangir is the third prevalent city situated in Western region of Odisha, India as shown in Fig. 1. The district covers 6,575 km^2 area. Coordinates of the proposed area is 20.72° N 83.48° E. Climate is categorized to be tropical monsoon. In summer, this area experiences more rainfall in comparison to winter. The mean temperature

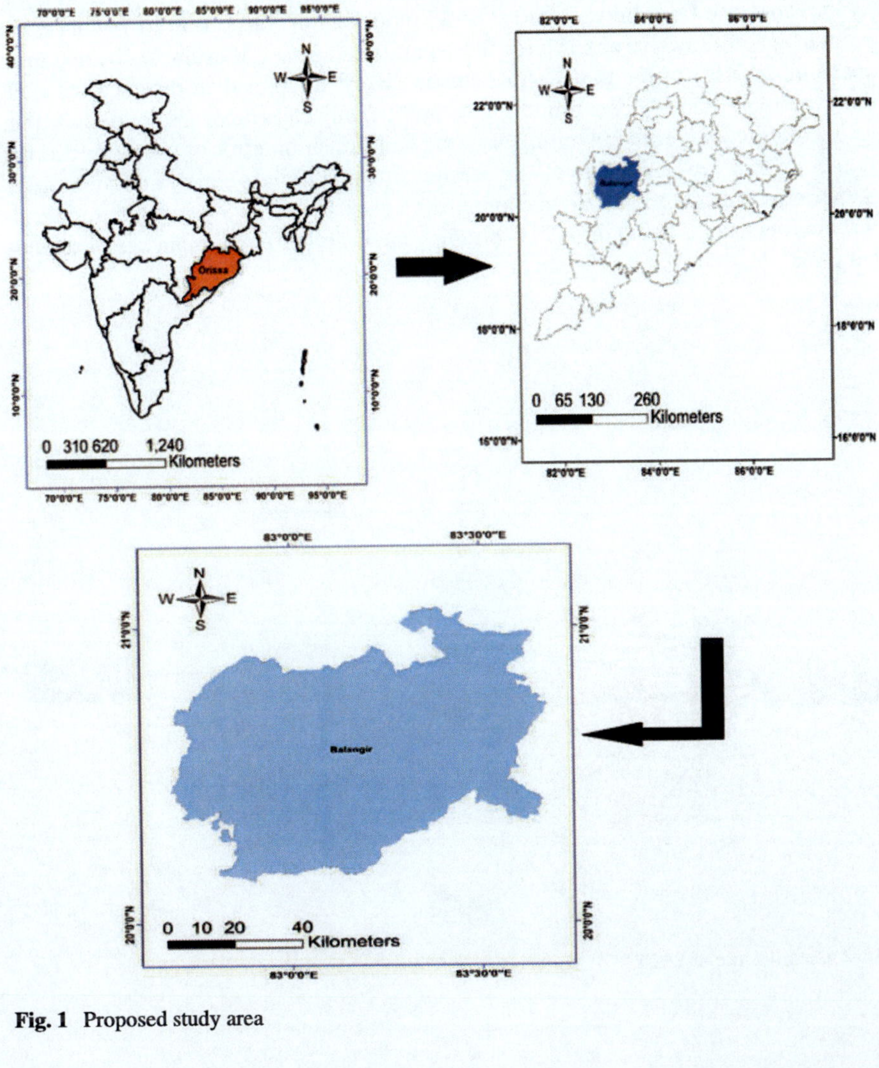

Fig. 1 Proposed study area

in summer is 34.8 °C and the warmest month is May, whereas in winter mean temperature is 16.6 °C and the coolest month is December.

3 Methodology

3.1 SVM

SVMs are a connected set of supervised learning techniques utilized for categorization and regression. Inn-dimensional space, SVM constructs an extrication hyperplane in space taking in view two sets of vector input data that exploit border amid the data sets (Fig. 1). For calculating border amid two data sets, two corresponding hyperplanes are built on each side. SVM model is an illustration of instances as point in space, drawn so as to divide the separate categories of examples by apparent space making it as broad as feasible. Instinctively, the prevalent distance between the hyperplane and adjacent training-data point of any class provides better partition, as in common bigger the margin, lesser the simplification error of classifier. Kernel functions that are applied in SVM are linear, polynomial, radial, and sigmoidal. SVM helps in categorizing text and hypertext, classifying imagery, recognizing handwritten fonts and is being broadly useful in genetic and different branches of science (Fig. 2).

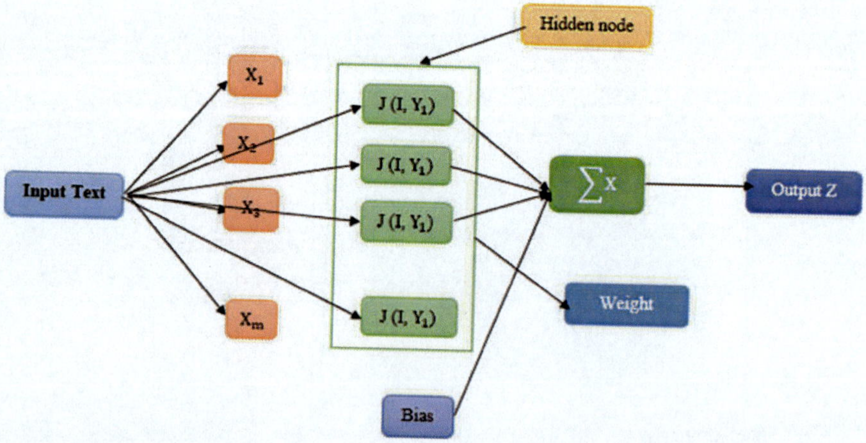

Fig. 2 Architecture of SVM

3.2 GA

GA is a search method on basis of normal assortment principles and inheritance. GA encodes decision variables of an exploration problem to restricted-span alphabet of firm cardinality. Strings that are applicant answers for exploration problem are known as chromosomes, alphabets as genes, and gene values are named as alleles. Taking an instance, such as a traveling salesman problem, chromosomes symbolize routes, and gene might signify a city. Contrary to conventional optimizing methods, GAs toil by means of parameter coding, as compared to the parameter itself. For finding better results and implementing usual assortment, there is a necessity for computing to distinguish among better and poor results. The computation done can be an objective function which may be an arithmetical model or computer imitation, or may be a biased function wherever individual persons select enhanced answers compared to the bad ones. In reality, computation of fitness must conclude an applicant answer, that can be consequently utilized by GAs for guiding the development of better answers.

3.3 Model Formulation and Performance

The monthly rainfall, average monthly temperature, evapotranspiration are collected from India meteorological department, Bhubaneswar for monsoon months (May–October), from 1997–2016. To train and to test the network, data from 1997–2010 and from 2011–2016 are utilized, respectively. Daily data are transformed to monthly data that is utilized for developing model. Input and output data are scaled in a way which allows each data to fall in a specific array prior to training. The procedure concerned is known as normalization so as to bind normalized values in a range between 0 and 1. Normalization equation applied to scale data is

$$X_t = 0.1 + \frac{X - X_{\min}}{X_{\max} - X_{\min}} \tag{1}$$

where X_t = altered data series, X = original input data series, $X_{\min}$ = minimum of original input data series, $X_{\max}$ = maximum of original input data series.

3.4 Evaluating Criteria

Evaluating criteria for ascertaining top model are coefficient of determination (R^2), Mean Square Error (MSE), and root mean square error RMSE. For choosing the perfect model for the proposed study, the clause is MSE, RMSE should be smallest, and R^2 should be maximum.

$$R^2 = \left[\frac{n\sum xy - (\sum x)(\sum y)}{\sqrt{\left[n\sum x^2 - (\sum x)^2\right]\left[n\sum y^2 - (\sum y)^2\right]}}\right]^2 \tag{2}$$

$$\mathrm{MSE} = \frac{1}{N}\sum_{i=1}^{N}(\hat{y}_l - y_i)^2$$

$$\mathrm{RMSE} = \left[\frac{1}{N}\sum_{i=1}^{N}(\hat{x}_l - x_i)^2\right]^{1/2} \tag{3}$$

4 Results and Discussions

Here for evaluating model effectiveness, different architectures are utilized for establishing model performance. For each architecture, MSE, RMSE, and R^2 are used to evaluate performance for both training and testing. In Table 1, results for GA of Balangir watershed are presented. Various combinations of P_t, T_t, H_t, P_{t-1}, T_{t-1}, and H_{t-1} parameters are considered to compute performance. Among them, when P_t, T_t, H_t, P_{t-1}, T_{t-1}, and H_{t-1} are applied as model input that gives MSE training and testing value 0.000487 and 0.002489, RMSE training and testing value 0.03291 and 0.04185, and R^2 training and testing value 0.9781 and 0.9381, respectively. The complete results are shown in Table 1.

Similarly, for SVM the results are discussed below for the station. Different architectures are considered to compute performance. For P_t, T_t, H_t, P_{t-1}, T_{t-1}, and H_{t-1} input, model gives the best value of performance which possesses MSE training and testing value 0.000589 and 0.005025, RMSE training and testing value 0.02453

Table 1 Results of GA at Balangir

Model input	MSE		RMSE		R^2	
	Training	Testing	Training	Testing	Training	Testing
P_t	0.000613	0.005321	0.037142	0.055276	0.8834	0.8995
P_t, T_t	0.000842	0.005794	0.034563	0.062871	0.9012	0.9234
P_t, T_t, H_t	**0.000483**	**0.002375**	**0.033167**	**0.045372**	**0.9254**	**0.9433**
P_t, T_t, H_t, P_{t-1}	0.000591	0.004729	0.028256	0.055839	0.9375	0.9514
$P_t, T_t, H_t, P_{t-1}, T_{t-1}$	**0.000737**	**0.006483**	**0.041585**	**0.068284**	0.9473	0.9501
$P_t, T_t, H_t, P_{t-1}, T_{t-1}, H_{t-1}$	0.000178	0.004088	0.037891	0.065261	**0.9512**	**0.9601**

Table 2 Results of SVM at Balangir

Model input	MSE		RMSE		R^2	
	Training	Testing	Training	Testing	Training	Testing
P_t	**0.000765**	**0.006365**	**0.040321**	**0.069052**	0.9217	0.9468
P_t, T_t	0.001326	0.004065	0.036538	0.063578	0.9386	0.9564
P_t, T_t, H_t	0.000936	0.008276	0.062765	0.081538	**0.9416**	**0.9687**
$P_t, T_t, H_t, P_{t-1},$	0.000727	0.007126	0.077378	0.092754	0.9564	0.9745
$P_t, T_t, H_t, P_{t-1}, T_{t-1}$	0.000665	0.001876	0.028267	0.073814	0.9604	0.9794
$P_t, T_t, H_t, P_{t-1}, T_{t-1}, H_{t-1}$	**0.000576**	**0.005015**	**0.021648**	**0.072648**	**0.9638**	**0.9801**

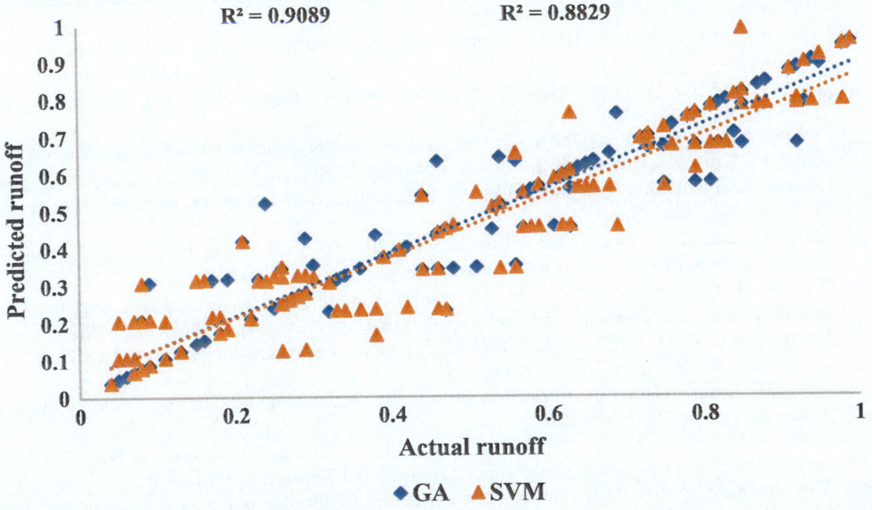

Fig. 3 Actual versus predicted runoff using GA and SVM

and 0.07049, and R^2 training and testing value 0.9941 and 0.9345, respectively. Performance values are considered for computation accessible in Table 2 (Fig. 3).

5 Conclusion

SVM is a new machine learning methodology that has substantial features counting on the fact that requisite on kernel and nature of optimization problem resulting in exclusively universal optimal, top simplification performance, and avoidance from congregating to local most favorable result. Application of SVM is contrasted with GA to predict daily runoff at Balangir watershed. Even though both techniques are based on data-driven models, running time is significantly quicker with identical or superior precision for SVM as compared to GA. Among the two methods, SVM provides the finest value of R^2 as 0.9638 and 0.9801 for training and testing phase when P_t, T_t, H_t, P_{t-1}, T_{t-1}, H_{t-1} are considered as input. But in case of GA, it gives R^2 value 0.9781 and 0.9381 for training and testing phase.

References

1. Sedki A, Ouazar D, El Mazoudi E (2009) Evolving neural network using real coded genetic algorithm for daily rainfall–runoff forecasting. Expert Syst Appl 36(3):4523–4527
2. Samantaray S, Sahoo A (2020) Appraisal of runoff through BPNN, RNN, and RBFN in Tentulikhunti Watershed: a case study. In: Satapathy S, Bhateja V, Nguyen B, Nguyen N, Le DN (eds) Frontiers in intelligent computing: theory and applications. Advances in intelligent systems and computing, vol 1014. Springer, Singapore
3. Samantaray S, Sahoo A (2020) Estimation of runoff through BPNN and SVM in Agalpur Watershed. In: Satapathy S, Bhateja V, Nguyen B, Nguyen N, Le DN (eds) Frontiers in intelligent computing: theory and applications. Advances in intelligent systems and computing, vol 1014. Springer, Singapore
4. Samantaray S, Sahoo A (2020) Assessment of sediment concentration through RBNN and SVM-FFA in Arid Watershed, India. In: Satapathy S, Bhateja V, Mohanty J, Udgata S (eds) Smart intelligent computing and applications. Smart innovation, systems and technologies, vol 159. Springer, Singapore
5. Asadi S, Shahrabi J, Abbaszadeh P, Tabanmehr S (2013) A new hybrid artificial neural networks for rainfall–runoff process modeling. Neurocomputing 121:470–480
6. Behzad M, Asghari K, Eazi M, Palhang M (2009) Generalization performance of support vector machines and neural networks in runoff modeling. Expert Syst Appl 36(4):7624–7629
7. Sivapragasam C, Liong SY, Pasha MFK (2001) Rainfall and runoff forecasting with SSA–SVM approach. J Hydroinformatics 3(3):141–152
8. Bray M, Han D (2004) Identification of support vector machines for runoff modelling. J Hydroinformatics 6(4):265–280
9. Duong HN, Nguyen HT, Vaclav S, Sanghyuk L (2016) A comparative study of SWAT, RFNN and RFNN-GA for predicting river runoff. Indian J Sci Technol 9(17):10–12
10. Ding H, Dong W (2016) Chaotic feature analysis and forecasting of Liujiang River runoff. Soft Comput 20(7):2595–2600
11. Ghose DK, Samantaray S (2018) Modelling sediment concentration using back propagation neural network and regression coupled with genetic algorithm. Procedia Comput Sci 125:85–92
12. Okkan U, Serbes ZA (2012) Rainfall–runoff modeling using least squares support vector machines. Environmetrics 23(6):549–564
13. Savic DA, Walters GA, Davidson JW (1999) A genetic programming approach to rainfall-runoff modelling. Water Resour Manage 13(3):219–231
14. Ghose DK, Samantaray S (2019a) Sedimentation process and its assessment through integrated sensor networks and machine learning process. In: Computational intelligence in sensor networks, Studies in computational intelligence 776:473–488
15. Sharma N, Zakaullah M, Tiwari H, Kumar D (2015) Runoff and sediment yield modeling using ANN and support vector machines: a case study from Nepal watershed. Model Earth Syst Environ 1(3):23
16. Ghose DK, Samantaray S (2019) Estimating runoff using feed-forward neural networks in scarce rainfall region. In: Smart intelligent computing and applications. Springer, Singapore, pp 53–64
17. Samantaray S, Ghose DK (2019) Dynamic modelling of runoff in a watershed using artificial neural network. In: Smart intelligent computing and applications. Springer, Singapore, pp 561–568

Runoff is a Key Constraint Toward Water Table Fluctuation Using Neural Networks: A Case Study

Bijay Narayan Sarkar, Sandeep Samantaray, U. Kumar, and Dillip K. Ghose

Abstract Monitoring water table fluctuations is essential, and it is even more important to predict the groundwater level to plan for the future needs. Though it is challenging task to model groundwater fluctuations due to its nonlinear nature, Artificial Neural Network (ANN) is the most robust tool to monitor and forecast groundwater level, where other empirical model fails. In this study, a casecade forward neural network and co-adaptive neuro-fuzzy inference system architecture has been designed and trained to learn the past water table fluctuations, to predict the future groundwater level. The model efficiency is analyzed and validated through the field observed value. After proper validation, the model is applied to predict the future water levels in the wells.

1 Introduction

One of the major sources of water in many regions is groundwater, in arid regions it is generally the only source of water. In most of the countries, more than 50 % of the drinking water comes from the subsurface. The groundwater table fluctuates as a result of external and internal sources like rainfall, pumping, etc. In many regions water tables are declining due to excessive pumping. It is crucial to know how long the groundwater resources will last and also to evaluate the actual amount of recharge in present-day and groundwater mining. Groundwater is a very important resource to provide drinking and agriculture water mainly in arid and semi-arid regions by using

B. N. Sarkar (✉) · S. Samantaray · U. Kumar · D. K. Ghose
Department of Civil Engineering, NIT Silchar, Silchar, Assam, India
e-mail: bijaysarkar242@gmail.com

S. Samantaray
e-mail: samantaraysandeep963@gmail.com

U. Kumar
e-mail: upendra_kumar72@rediffmail.com

D. K. Ghose
e-mail: dillipghose_2002@yahoo.co.in

S. C. Satapathy et al. (eds.), *Communication Software and Networks*, Lecture Notes in Networks and Systems 134, https://doi.org/10.1007/978-981-15-5397-4_75

tube wells, deep wells, pumps, etc. These resources of water are called aquifer and commonly having a high quality of water, which usually do not need any chemical treatment, and commonly are free of bacterial diseases.

Daliakopoulos et al. [1] studied efficiency of seven different ANN configurations for forecasting groundwater level in the Messara Valley. Nayak et al. [2] investigated the water table depth prediction using ANN by modeling the groundwater levels of Central Godavari Delta System. Samantaray and Sahoo [3–5] employed several neural networks to predict climatic indices. Kisi and Shiri [6] studied groundwater depth forecasting using wavelet-neuro-fuzzy technique and ANFIS conjunction technique by obtaining data from the US Illinois State Water Survey of two wells, Bondville and Perry. Shirmohammadi et al. [7] studied prediction of groundwater level for different data operating techniques like system identification, time series, and adaptive ANFIS models in Mashhad plain, Khorasan Razavi province, Iran. Ghose and Samantaray [8] researched on development hydrological models for flow and prediction of each month of monsoon period using ANN, tested on Mahanadi river. Ghose and Samantaray [9] aimed on prediction of variation of groundwater table comparing with the change of humidity, temperature, and precipitation as a data to the ANN models for Mahanadi river basin Odisha, India. Taormina et al. [10] studied simulations of hourly groundwater levels using application of FFNs at Lagoon of Venice, Italy. Sahoo and Jha [11] evaluated the success of two data-operated approaches for forecasting of spatio-temporal water-level distribution, using the data of Konan groundwater basin located in Shikoku Island, Japan. Shiri et al. [12] explored different types of techniques for forecasting of groundwater level like Gene Expression Programming (GEP), ANN, ANFIS, and Support Vector Machine (SVM), using data of Hoengchon well located in Republic of Korea. Maiti and Tiwari [13] predicted modeling of groundwater level fluctuations by comparing the advantages and disadvantages of three modern soft computing techniques such as ANN, Bayesian ANN, and ANFIS. Hosseini and Nakhaei [14] designed Genetic Algorithm-Back Propagation (GA-BP) and ANN-Back Propagation (ANN-BP) models to determine the groundwater levels in the Shabestar Plain located in the East Azerbaijan. Tanty and Desmukh [15] studied different hydrological problems like prediction of rainfall-runoff, stream-flow prediction, groundwater hydrology, etc. using artificial neural networks. Khaki et al. [16] studied application of the ANN and ANFIS for simulation groundwater level in the Langat Basin, Malaysia. Khaki et al. [17] examined the accuracy of monthly groundwater level forecasts by researching the effectiveness of the steepness coefficient in the sigmoid function of a developed *ANN* model at Langat Basin, Malaysia. Samantaray and Ghose [18] studied to estimate the sediment concentration and to calculate the suspended sediment load by ANN approach. Das et al. [19] determined the aquifer performance by developing an advanced approach using BPNN, RBFN, RNN, and ANFIS using wells of Mahulpali village located in Sambalpur district in Odisha, India. Ghose and Samantaray [20] studied on the prediction of the hydrological modeling using regression and backpropagation neural network at Suktel River basin. Nourani and Mousavi [21] introduced a hybrid Artificial Intelligence (AI) meshless model for spatiotemporal groundwater level modeling for Miandoab plain, northwest of Iran. Prediction of water table fluctuation in Sonitpur watershed, Assam through various neural networks is the objective of research content.

2 Study Area and Data

Sonitpur watershed is considered for study area in the research with coordinates 26.63° N and 92.8° E. It occupies over an area of 5324 km^2 (2,056 sq mi), located on the northern banks of Brahmaputra river.

Brahmaputra, Jiabharali, Gabharu, Borgang, and Buroi are the major rivers in the district. The study area falls under the Tropical Rainforest climate region, which has an average temperature of 27 °C. The precipitation in this area is heavy about 3,000 mm (9ft) and above, mostly in months of January–June which is both a blessing and a misfortune. Average precipitation of all months is at least 60 mm and with an average temperature of above 18 °C in the cold months of a year (Fig. 1).

3 Methodology

3.1 CFBPNN

A perceptron connection is that which is formed by having direct relation between input and output but in FFNN connection shaped by indirect relationship between output and input. Shape of connection is nonlinear; however, activation functions in hidden layer. If we joined connection formed by perceptron and multilayer network, then a network is formed with direct connection between output and input layer, aside from connection incidentally. Network which is built from this association pattern is called Cascade Forward Neural Network (CFNN).

From Fig. 2, it shows that input and output have a direct relationship. The outcome of this kind of relationship is that network weight to be appraised increases is same as the neurons in input level. As with FFNN, backpropagation algorithm on CFNN also includes three stages: feed forward of input pattern, weight adjustment, and error counting. The process continues after the feed forward stage with error calculation. After this step, according to specified stop criteria, update weights and recalculate. Continue this step until no error occurs or iteration stops. In this section, we briefly study the CFNN model for weighting adjustments by conjugate gradient optimization method.

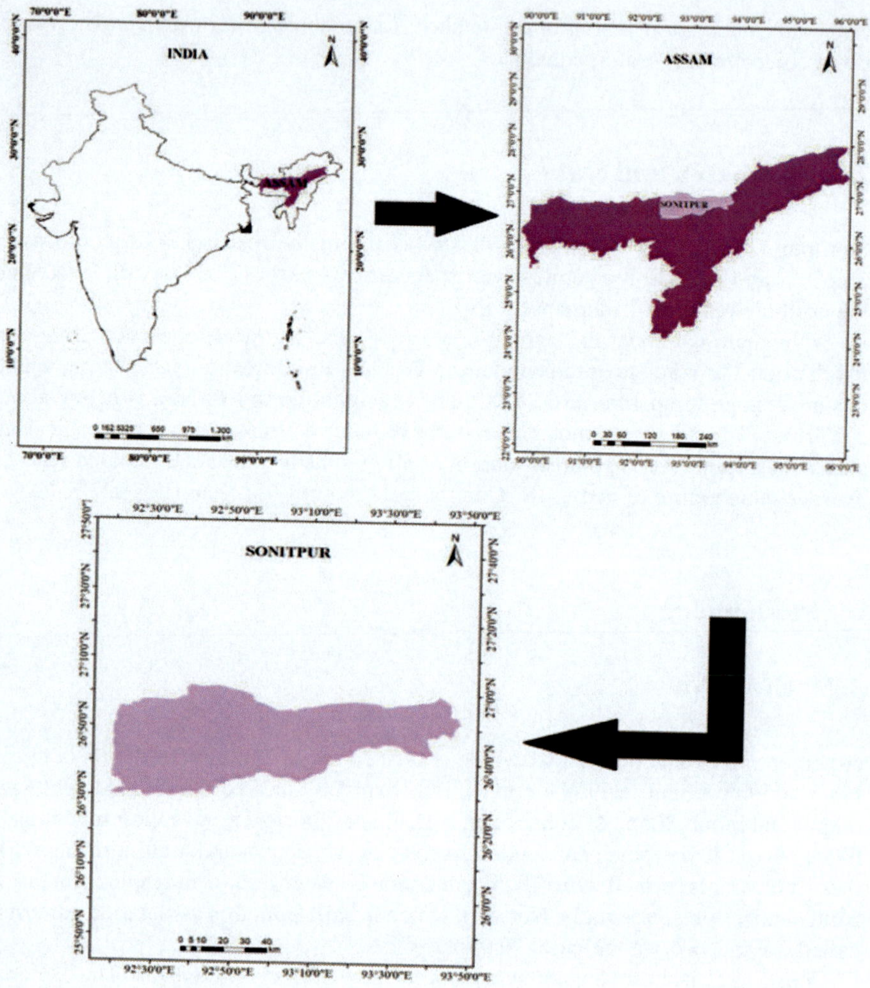

Fig. 1 Recommended research area

3.2 CANFIS

In CANFIS, the basic ideas of ANFIS, its predecessor has been expanded. In this concept of ANFIS, it has been expended to any number of input/output pairs [1–5]. In addition, CANFIS yields advantages from nonlinear fuzzy rules. This CANFIS realizes the sugeno type or TSK) fuzzy inference accomplishing fuzzy if-then rules (Fig. 3).

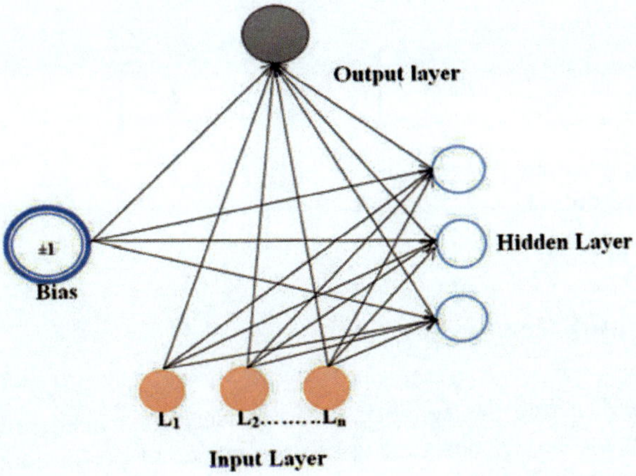

Fig. 2 Architecture of CFBPNN

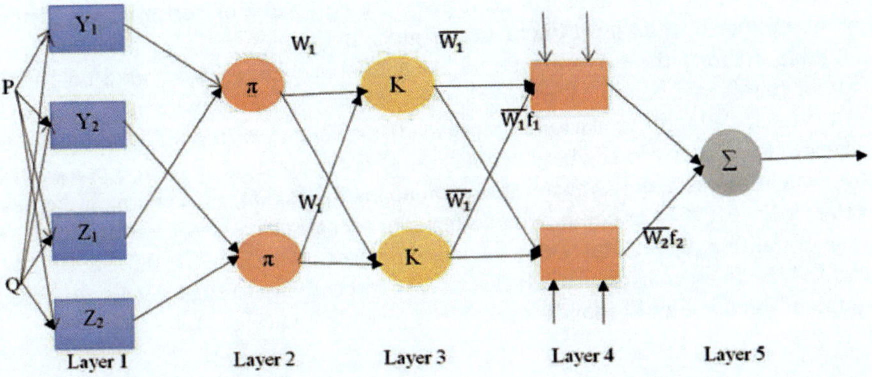

Fig. 3 Architecture of CANFIS

3.3 Model Formulation

Various metrological data like precipitation, average temperature, evapotranspiration, and humidity data are possessed from IMD, India for the monsoon seasons, from 1990 to 2017. Data from 1990 to 2009 are applied to train, and data from 2009 to 2017 are used to test network. Root mean square error, mean square error, and coefficient of determination are employed to evaluate model performance

$$\text{MSE} = \left[\frac{1}{A}\sum\nolimits_{i=1}^{a}(I_a - I_p)^2\right] \quad (1)$$

where I_a: actual output
I_p: predicted output.

4 Results and Discussion

For training and testing the network, 180 data sets are considered spanning over 1988–2017 for the model design. Superlative values of performance with various scenarios are given in Table 1.

From the above result, the best values R^2 are 0.8098, 0.8668, 0.8921, 0.9046, and 0.9117 for scenario I to V in case of testing phases. For scenario V when runoff is added with the previous scenario it gives the best value of performance. Actual versus predicted water level fluctuation is given in Fig. 4.

Table 2 shows the performance of five model scenarios. It is found that addition of runoff in scenario IV progresses proficiency of model, whereas addition of evapotranspiration in scenario III improves the efficiency of model as compared to scenario I and scenario II where neither runoff nor evapotranspiration is considered for evaluation of model. Above estimation appearances that for scenario V the best value of R^2 is 0.9614 as compared to other four scenarios in case of testing phase. In case of training phase, superlative values of R^2 are 0.8587, 0.8901, 0.9198, 0.9287, and 0.9326 for scenarios I, II, III, IV, and V, respectively. Actual versus predicted graph of fluctuation of groundwater table is given below (Fig. 5).

Table 1 Results of CFBPNN

Scenario	Input	Training period			Testing period		
		MSE	RMSE	R^2	MSE	RMSE	R^2
I	P_t	0.00099	0.02115	0.7535	0.00106	0.02918	0.8098
II	P_t, H_t	0.00102	0.02312	0.8067	0.00141	0.03442	0.8668
III	P_t, H_t, T_{avg}	0.00111	0.02557	0.8258	0.00158	0.03678	0.8921
IV	$P_t, H_t, T_{\text{avg}}, E_t$	0.00126	0.02634	0.8412	0.00163	0.03687	0.9046
V	$P_t, H_t, T_{\text{avg}}, E_t, Q_1$	0.00139	0.02712	0.8578	0.00169	0.03701	0.9117

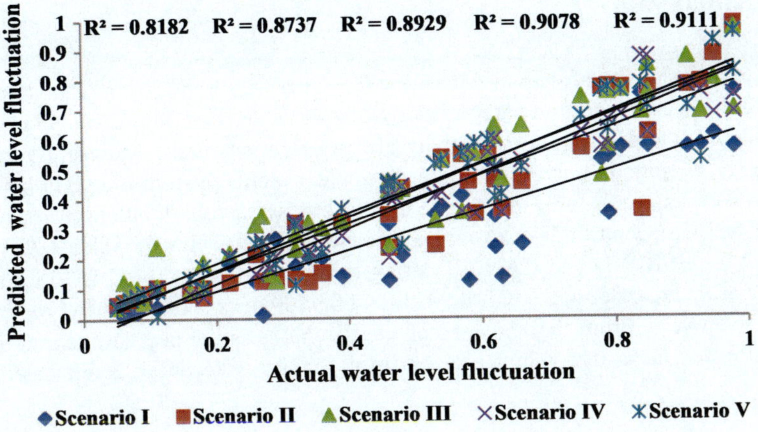

Fig. 4 Actual versus predicted water level fluctuation of five scenarios using CFBPNN

Table 2 Results of CANFIS

Scenario	Input	Training period			Testing period		
		MSE	RMSE	R^2	MSE	RMSE	R^2
I	P_t	0.00141	0.02765	0.8587	0.00295	0.03668	0.8856
II	P_t, H_t	0.00156	0.03665	0.8901	0.00325	0.03712	0.9137
III	P_t, H_t, T_{avg}	0.00167	0.03741	0.9198	0.00342	0.04019	0.9411
IV	P_t, H_t, T_{avg}, E_t	0.00179	0.03889	0.9287	0.00366	0.04527	0.9509
V	P_t, H_t, T_{avg}, E_t, Q_1	0.00199	0.03991	0.9326	0.00387	0.04768	0.9614

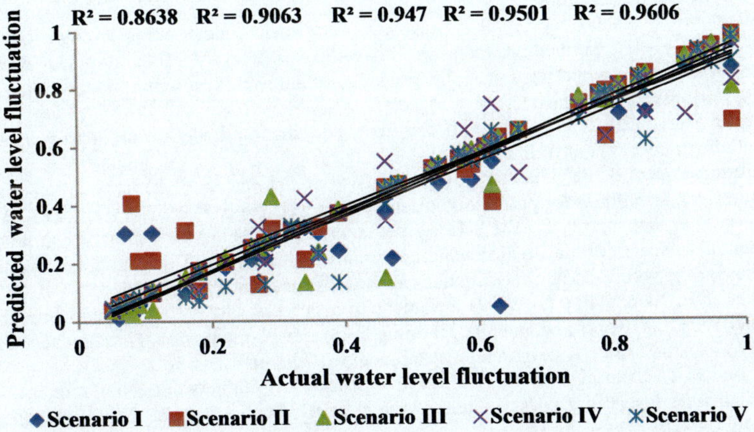

Fig. 5 Actual versus predicted water level fluctuation of five scenarios using CANFIS

5 Conclusion

The goal of this study was to determine the CFBPNN and CANFIS as a possible tool for forecasting groundwater level in Sonitpur watershed, India. Many factors are considered like precipitation, discharge from irrigation canal, evaporation, relative humidity, average temperature, and groundwater recharge from the plain boundary were taken as inputs, and the output is future water levels fluctuation of Sonitpur watershed. In scenario V, it provides improved results with R^2 value 0.9117 and 0.9614 for CFBPNN and CANFIS during testing phase. But for CANFIS, the best values of R^2 are 0.8856, 0.9137, 0.9411, and 0.9509 in case of scenario I to scenario IV, respectively. From the above research, discharge is the key component toward groundwater fluctuation, which is also helpful for other relevant research watersheds.

References

1. Daliakopoulos IN, Coulibaly P, Tsanis IK (2004) Groundwater level forecasting using artificial neural networks. J Hydrol 309:229–240
2. Nayak PC, Rao YRS, Sudheer KP (2005) Groundwater level forecasting in a shallow aquifer using artificial neural network approach. Water Resour Manage 20:77–90
3. Samantaray S, Sahoo A (2020a) Appraisal of runoff through BPNN, RNN, and RBFN in Tentulikhunti Watershed: a case study. In: Satapathy S, Bhateja V, Nguyen B, Nguyen N, Le DN (eds) Frontiers in intelligent computing: theory and applications. Advances in intelligent systems and computing, vol 1014. Springer, Singapore
4. Samantaray S, Sahoo A (2020b) Estimation of runoff through BPNN and SVM in Agalpur Watershed. In: Satapathy S, Bhateja V, Nguyen B, Nguyen N, Le DN (eds) Frontiers in intelligent computing: theory and applications. Advances in intelligent systems and computing, vol 1014. Springer, Singapore
5. Samantaray S, Sahoo A (2020c) Assessment of sediment concentration through RBNN and SVM-FFA in Arid Watershed, India. In: Satapathy S, Bhateja V, Mohanty J, Udgata S (eds) Smart intelligent computing and applications. Smart innovation, systems and technologies, vol 159. Springer, Singapore
6. Kisi O, Shiri J (2012) Wavelet and neuro-fuzzy conjunction model for predicting water table depth fluctuations. Hydrol Res
7. Shirmohammadi B, Vafakhah M, Moosavi V, Moghaddamnia A (2012) Application of several data-driven techniques for predicting groundwater level. Water Resour Manage 27:419–432
8. Ghose DK, Samantaray S (2019) Integrated sensor networking for estimating groundwater potential in scanty rainfall region: challenges and evaluation. In: Computational intelligence in sensor networks, studies in computational intelligence, vol 776, pp 335–352
9. Ghose DK, Samantaray S (2019) Sedimentation process and its assessment through integrated sensor networks and machine learning process. In: Computational intelligence in sensor networks, studies in computational intelligence, vol 776, pp 473–488
10. Taormina R, Chau KW, Sethi R (2012) Artificial neural network simulation of hourly groundwater levels in a coastal aquifer system of the Venice lagoon. Eng Appl Artif Intell 25:1670–1676
11. Sahoo S, Jha MK (2013) Groundwater-level prediction using multiple linear regression and artificial neural network techniques: a comparative assessment. Hydrogeol J 21:1865–1887
12. Shiri J, Kisi O, Yoon H, Lee KK, Nazemi AH (2013) Predicting groundwater level fluctuations with meteorological effect implications—a comparative study among soft computing techniques. Comput Geosci 56:32–44

13. Maiti S, Tiwari RK (2013) A comparative study of artificial neural networks, Bayesian neural networks and adaptive neuro-fuzzy inference system in groundwater level prediction. Environ Earth Sci 71:3147–3160
14. Hosseini Z, Nakhaei M (2014) Estimation of groundwater level using a hybrid genetic algorithm neural network. Pollution 1(1):9–21
15. Tanty R, Desmukh TS (2015) Application of artificial neural network in hydrology—a review. Int J Eng Res Technol (IJERT) 2278-0181
16. Khaki M, Yusoff I, Islami N (2015) Simulation of groundwater level through artificial intelligence system. Environ Earth Sci 73:8357–8367
17. Khaki M, Yusoff I, Islami N, Hussin NH (2016) Artificial neural network technique for modeling of groundwater level in Langat Basin, Malaysia. Sains Malaysiana 45(1):19–28
18. Samantaray S, Ghose DK (2018) Evaluation of suspended sediment concentration using descent neural networks. Procedia Comput Sci 132:1824–1831
19. Das UK, Samantaray S, Ghose DK, Roy P (2019) Estimation of aquifer potential using BPNN, RBFN, RNN, and ANFIS, In: Smart intelligent computing and applications, smart innovation, systems and technologies, vol 105, pp 569–576
20. Ghose DK, Samantaray S (2018) Modelling sediment concentration using back propagation neural network and regression coupled with genetic algorithm. Procedia Comput Sci 125:85–92
21. Nourani V, Mousavi S (2016) Spatiotemporal groundwater level modeling using hybrid artificial intelligence-meshless method. J Hydrol

Estimation of Water Table Depth Using Wavelet-ANFIS: A Case Study

Sriharsha Sridharam, Abinash Sahoo, Sandeep Samantaray, and Dillip K. Ghose

Abstract Predicting fluctuation in groundwater level (GWL) is extremely significant in managing water resources problems and hydrological phenomenon. The objective of present study is comparison of ANFIS and Wavelet-ANFIS models to forecast GWL for altered period of prediction and finding model effectiveness based on their performances. mean absolute error (MAE), root mean square error (RMSE) and correlation coefficient (R^2) statistical tools were utilized to evaluate accurateness of the aforesaid models. Results indicate that Wavelet-ANFIS model with R^2 of 0.9628 performed better than ANFIS model with R^2 0.9247. It can be seen that wavelet transform helps in improving the efficiency of ANFIS model to forecast GWL. It is observed that the forecast done by Wavelet-ANFIS is more precise than those by ANFIS.

1 Introduction

Groundwater in India is one of the key sources of human and other living beings water consumption in the world, mainly in arid and semi-arid regions where the rainfall is low and land surface water is scarce. The study of groundwater table fluctuations is very prominent in agricultural sector, as it could be a major role-playing element in sensitive crops. To assess the condition of aquifers and water level in terms of space and time and taking necessary actions for the improvement of aquifer, studies

S. Sridharam (✉) · A. Sahoo · S. Samantaray · D. K. Ghose
Department of Civil Engineering, NIT Silchar, Silchar, Assam, India
e-mail: sriharshasridharam@gmail.com

A. Sahoo
e-mail: bablusahoo1992@gmail.com

S. Samantaray
e-mail: samantaraysandeep963@gmail.com

D. K. Ghose
e-mail: dillipghose_2002@yahoo.co.in

S. C. Satapathy et al. (eds.), *Communication Software and Networks*, Lecture Notes in Networks and Systems 134, https://doi.org/10.1007/978-981-15-5397-4_76

regarding groundwater levels are for sure required. Forecasting groundwater levels by explicit model is essential for supervision of aquifers.

Kisi and Shiri [1] improved wavelet-neuro-fuzzy model with the combination of discrete wavelet transform and neuro-fuzzy model and conjuncted by giving various input combinations of groundwater depth data of Bondville and Perry wells. Banerjee et al. [2] estimated groundwater for maintaining salinity by using Feed Forward-ANN model which was verified by the field data of real time. They trained the model for 2 years and predicted for 5 years which was compared with both Saturated–Unsaturated Transport and real-time field data resulting in accurate prediction. Ghose and Samantaray [3, 4] applied neural networks to forecast sediment load. Barzegar et al [5] forecasted the groundwater level for one, two and three months step-ahead by using 367 monthly GWL datasets in the Maragheh–Bonab plain Iran. By evaluating hybrid wavelet group models, they found that ELM models are superior to GMDH models. Chang et al. [6] predicted GWL in a basin using soft computing techniques by combining the self organized map (SOM) and Nonlinear Autoregressive with Exogenous Inputs network which extracted the complex input–output patterns of ground aquifer systems. The results concluded that the SOM-NARX model is much reliable in predicting groundwater levels. Samantaray and Sahoo [7–9] used various soft computing techniques to forecast climatic parameters. Ebrahimi et al. used ANN, support vector regression and multi linear regression (MLR) of GWL data obtained from two wells of Qom plain in Iran. Results stated that the Meyer mother wavelet and Db5 wavelets gave better results. Emamgholizdeh et al. [10] forecasted groundwater level by using ANN and ANFIS in Bastam Plain in Iran. For predicting groundwater level hydrogeological and hydrological parameters of nine years data sets are taken and the results stated that ANFIS model gives better results than ANN. Fallah-Mehdipour et al. [11] predicted groundwater levels of Karaj plain, Iran in three observation wells by using artificial intelligence tools ANFIS and genetic programming. Results show that GP performs better than ANFIS for the given wells. Moosavi et al. [12] predicted groundwater level in a watershed with the help of ANN, ANFIS, Wavelet-ANN and Wavelet-ANFIS. Results showed that Wavelet-ANFIS gives more reliable than the other three models in terms of accuracy. Gorgij et al. [13] projected water table in East Azerbaijan, Iran and the simulation was done with 13 piezometers by a WANN for the next twelve months using genetic programming. Nourani et al [14] used a Self Organizing Map technique to identify clusters of GWL which are spatially homogeneous, and FFNN is applied to model one and multi-step-ahead GWLs. It is observed that FFNN model attached with SOM method is much useful in decreasing input variable's dimensionality and FFNN models' complexity. Nourani and Mousavi [15] illustrated the comparison of groundwater level pattern at different time scales by using ANN and ANFIS in the first step and piezometer, rainfall and runoff parameters in second step and finally multiquadric radial basis function to differentiate groundwater flow equations. GWL prediction for different forecasting periods is done by applying ANFIS models and several data-driven methods which include time series and system identification models [16].

In recent years, wavelet analysis is fetching as a usual tool to investigate confined disparities of influence of time series within. Obtaining time–frequency space through the disintegration of time series, determining the leading modes of unpredictability and so as the disparity of these modes in terms of time is empowered. The objective of the research is to predict groundwater level in an arid watershed, India using hybrid neural network.

2 Study Area

The study region is located in between 17°23′ N/17°55′ N latitude and 82°30′ E/82°53′ E longitude in Visakhapatnam of Andhra Pradesh, India. The proposed watershed is surrounded with an area of 681.96 km^2 and populated with more than 2 million people. It is a semi-arid climate region having 20 °C minimum temperature during in winter and 36 °C maximum temperatures during March–May in summer, respectively (Fig. 1).

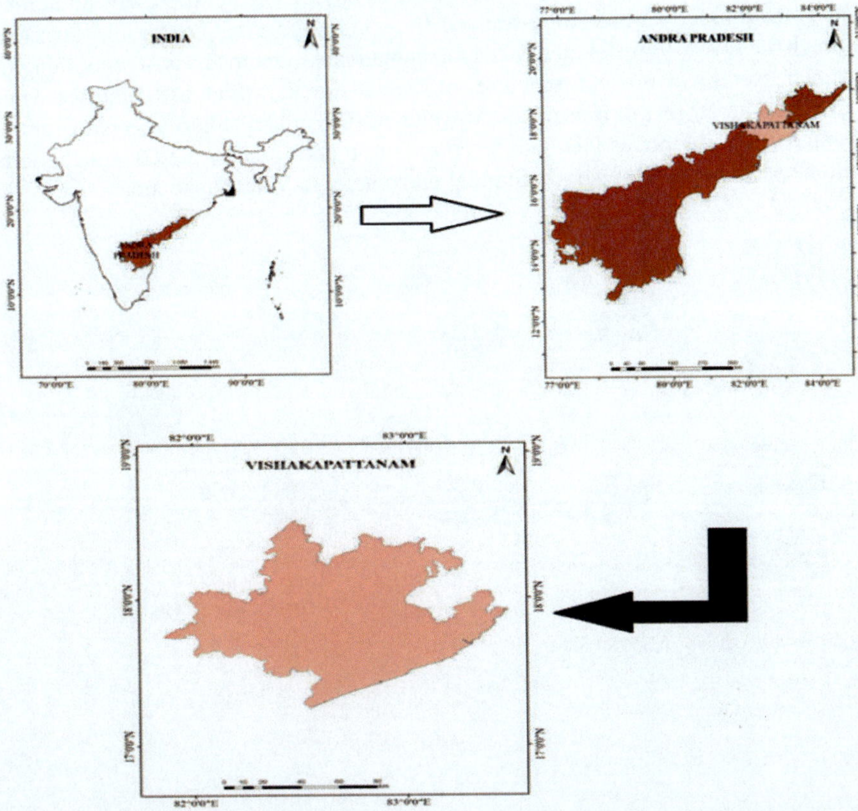

Fig. 1 Proposed research work

3 Methodology

3.1 *ANFIS*

By the use of the data sets similar to that of ANN models, the following model is accomplished and verified. In the present effort, testing is done for membership functions (MFs) of six different types are verified for these models of ANFIS, i.e. Tri, Trap, Gbell, Gauss, Gauss2, Pi. A detail of the schematic diagram of ANFIS is depicted in Fig. 2.

3.2 *Wavelet-ANFIS*

After determining all the major parameters, the ANFIS model by the FCM rule generator in addition to the DWTMRA disintegrated time series of the normal GL and P are combined to have the Wavelet-ANFIS model. Further precisely, numerous amalgamations of various time-wadded GL- and P-information are disintegrated through the DWT. Wavelet theory is a mathematical model in terms of smoothing of a signal, system, or process with a set of special signals, which are small waves or wavelets. Therefore, this type of best Wavelet-ANFIS model of hybrid version for the average GL-variations is then later used to put on and forecast the GL-fluctuations with respect to space at every individual piezometer in Visakhapatnam.

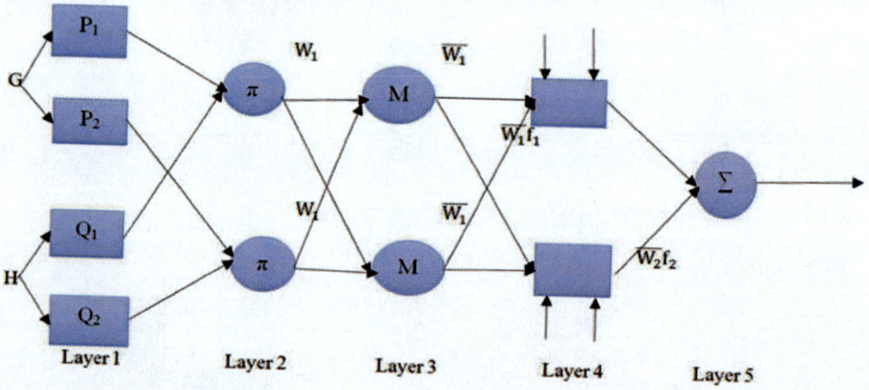

Fig. 2 Architecture of ANFIS

3.3 Model Formulation

Various meteorological data on daily basis are retained from meteorological department of India for monsoon seasons (1990–2017). Data from 1990–2009 are employed to train and data from 2009–2017 are engaged to test model efficiency. MSE, RMSE, R^2 are assessed for performance of model.

$$R^2 = \left[\frac{n\sum cd - (c)(\sum d)}{\left[n\sum c^2 - (\sum c)^2\right]\left[n\sum d^2 - (\sum d)^2\right]} \right]^2 \tag{1}$$

4 Results and Discussions

Comparing results of Wavelet-ANFIS and ANFIS concludes that Wavelet-ANFIS improves model efficacy by a sensible quantity that can be observed in Tables 1 and 2. By examining through performance evaluation constraints, Wavelet-ANFIS gives paramount value for all the classification than ANFIS.

Table 1 Comparative performance of ANFIS

Scenario	Membership function	Training			Testing		
		MAE	RMSE	R^2	MAE	RMSE	R^2
Infiltration loss, precipitation, temperature, water table depth, evapotranspiration	Pi	0.000613	0.058254	0.7993	0.000402	0.030217	0.8416
	Gauss	0.000713	0.057216	0.7815	0.000514	0.029678	0.8358
	Gbell	0.000576	0.068432	0.8851	0.000389	0.046821	0.9247
	Gauss2	0.000821	0.067832	0.8317	0.000674	0.048762	0.9028
	Tri	0.000635	0.060247	0.8046	0.000426	0.052768	0.8661
	Trap	0.000703	0.072456	0.8327	0.000587	0.058762	0.8806

Table 2 Comparative performance of Wavelet-ANFIS

Scenario	Membership function	Training			Testing		
		MAE	RMSE	R^2	MAE	RMSE	R^2
Precipitation, Infiltration loss, temperature, Evapotranspiration, water table depth	Pi	0.000647	0.053492	0.8365	0.000475	0.039434	0.8862
	Gauss	0.000662	0.055234	0.8229	0.000551	0.040289	0.8795
	Gbell	0.000528	0.071161	0.9201	0.000409	0.054941	0.9628
	Gauss2	0.000601	0.074846	0.8857	0.000705	0.057506	0.9437
	Tri	0.000499	0.061481	0.8476	0.000509	0.040655	0.9085
	Trap	0.000606	0.078436	0.8758	0.000751	0.056259	0.9251

From the above table, we found that Gbell function gives the best value of performance with R^2 value 0.8851 and 0.9247 for training and testing phases. Consequentially for other transfer functions, the value of R^2 lies within a range from 0.8416 to 0.9028 in testing phases.

Table 2 represents the overall performance of all six membership functions for hybrid ANFIS network. Here Gbell function represents the best values of R^2 as 0.9201 and 0.9628 for training and testing phases, which is more than the other membership function. Similarly for other membership function best value of R^2 are lies between ranges from 0.8229 to 0.8857 for training phases. Actual versus predicted water table depth for both training and testing phases is displayed in Fig. 3.

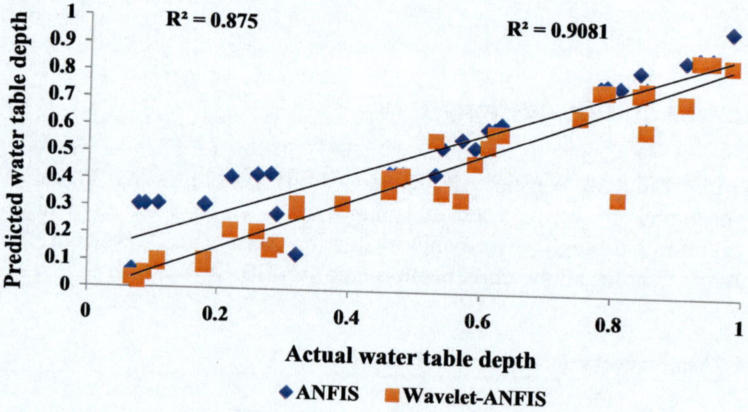

(a)

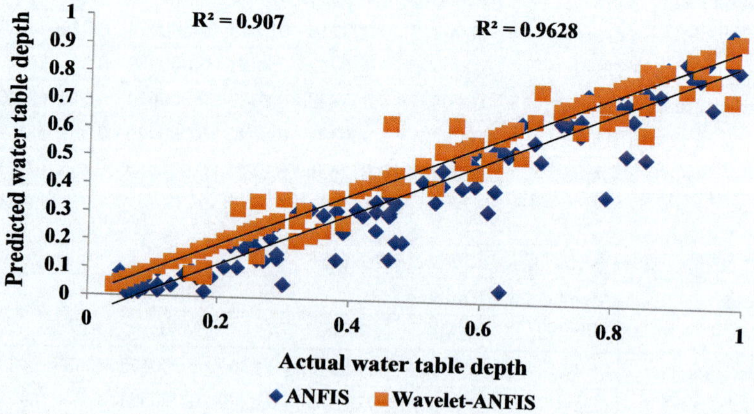

(b)

Fig. 3 Actual versus predicted water table depth for **a** testing and **b** training phases

5 Conclusion

The accurateness of Wavelet-ANFIS method to forecast short-term (one-, two- and three-day-ahead) GWL was explored in this research. Also, it exhibited that data pre-processing could result in finding better GWL prediction model. Outcomes obtained from this study revealed that Wavelet-ANFIS model performed best in predicting GWL as compared to ANFIS. Utmost benefit of an ANN is its skill in modelling multifaceted nonlinear relation without any previous assumptions regarding the nature of connection. It can thus be said that the Wavelet-ANFIS presents an advanced substitute to ANFIS for improving on how to simulate input–output relation and forecast groundwater level.

References

1. Kisi O, Shiri J (2012) Wavelet and neuro-fuzzy conjunction model for predicting water table depth fluctuations. Hydrol Res 43(3):286–300
2. Banerjee P, Singh VS, Chatttopadhyay K, Chandra PC, Singh B (2011) Artificial neural network model as a potential alternative for groundwater salinity forecasting. J Hydrol 398(3–4):212–220
3. Ghose DK, Samantaray S (2018) Modelling sediment concentration using back propagation neural network and regression coupled with genetic algorithm. Procedia Comput Sci 125:85–92
4. Ghose DK, Samantaray S (2019) Sedimentation process and its assessment through integrated sensor networks and machine learning process. In: Computational intelligence in sensor networks, studies in computational intelligence, vol 776, pp 473–488
5. Barzegar R, Fijani E, Moghaddam AA, Tziritis E (2017) Forecasting of groundwater level fluctuations using ensemble hybrid multi-wavelet neural network-based models. Sci Total Environ 599:20–31
6. Chang FJ, Chang LC, Huang CW, Kao IF (2016) Prediction of monthly regional groundwater levels through hybrid soft-computing techniques. J Hydrol 541:965–976
7. Samantaray S, Sahoo A, (2020a) Appraisal of runoff through BPNN, RNN, and RBFN in Tentulikhunti Watershed: a case study. In: Satapathy S, Bhateja V, Nguyen B, Nguyen N, Le DN (eds) Frontiers in intelligent computing: theory and applications. Advances in intelligent systems and computing, vol 1014. Springer, Singapore
8. Samantaray S, Sahoo A (2020b) Estimation of runoff through BPNN and SVM in Agalpur Watershed. In: Satapathy S, Bhateja V, Nguyen B, Nguyen N, Le DN (eds) Frontiers in intelligent computing: theory and applications. Advances in intelligent systems and computing, vol 1014. Springer, Singapore
9. Samantaray S, Sahoo A (2020c) Assessment of sediment concentration through RBNN and SVM-FFA in Arid Watershed, India. In: Satapathy S, Bhateja V, Mohanty J, Udgata S (eds) Smart intelligent computing and applications. Smart innovation, systems and technologies, vol 159. Springer, Singapore
10. Emamgholizadeh S, Moslemi K, Karami G (2014) Prediction the groundwater level of bastam plain (Iran) by artificial neural network (ANN) and adaptive neuro-fuzzy inference system (ANFIS). Water Resour Manage 28(15):5433–5446
11. Fallah-Mehdipour E, Haddad OB, Mariño MA (2013) Prediction and simulation of monthly groundwater levels by genetic programming. J Hydro-Environ Res 7(4):253–260
12. Moosavi V, Vafakhah M, Shirmohammadi B, Behnia N (2013) A wavelet-ANFIS hybrid model for groundwater level forecasting for different prediction periods. Water Resour Manage 27(5):1301–1321

13. Gorgij AD, Kisi O, Moghaddam AA (2016) Groundwater budget forecasting, using hybrid wavelet-ANN-GP modelling: a case study of Azarshahr Plain, East Azerbaijan, Iran. Hydrol Res 48(2):455–467
14. Nourani V, Alami MT, Vousoughi FD (2015) Wavelet-entropy data pre-processing approach for ANN-based groundwater level modeling. J Hydrol 524:255–269
15. Nourani V, Mousavi S (2016) Spatiotemporal groundwater level modeling using hybrid artificial intelligence-meshless method. J Hydrol 536:10–25
16. Shirmohammadi B, Vafakhah M, Moosavi V, Moghaddamnia A (2013) Application of several data-driven techniques for predicting groundwater level. Water Resour Manage 27(2):419–432

Estimation of Flood in a River Basin Through Neural Networks: A Case Study

Abinash Sahoo, Ujjawal K. Singh, Mavoori Hitesh Kumar, and Sandeep Samantaray

Abstract Climate change has had worst and extreme impacts all over the world. Due to rise in global temperature some region faces drought and then a sudden bout of excessive rainfall. Rainfall in excess causes one of the most destructive and dangerous natural hazard called flooding that causes serious damage to life and property on earth every year. There are several complexities in nature and pattern of floods which makes flooding an important and challenging task for the researcher. To solve this problem, there are three Artificial Neural Networks (ANNs) techniques namely Support Vector Machine (SVM), Back Propagation Neural Network (BPNN) and Radial Basis Function Network (RBFN). These techniques are capable of modelling nonlinear and complex systems. The capability of these techniques is presented in this paper. In this research, to measure the performance of models, three performance criteria, including a coefficient of determination (R^2), mean square error and root mean square error are utilized. The result shows that the SVM model performs best among the three models and can be accepted as a suitable and appropriate method to predict flood.

A. Sahoo (✉) · U. K. Singh · S. Samantaray
Department of Civil Engineering, NIT Silchar, Silchar, Assam, India
e-mail: bablusahoo1992@gmail.com

U. K. Singh
e-mail: ujjawalsinghjarula@gmail.com

S. Samantaray
e-mail: samantaraysandeep963@gmail.com

M. H. Kumar
Department of Civil Engineering, GIET Gunpur, Gunpur, Odisha, India
e-mail: kumar.hitesh646@gmail.com

S. C. Satapathy et al. (eds.), *Communication Software and Networks*, Lecture Notes in Networks and Systems 134, https://doi.org/10.1007/978-981-15-5397-4_77

1 Introduction

Globally, floods are natural disasters that cause serious damage to both lives and property frequently. It is estimated by the global rainfall data that 40 % of the total economic losses by all kinds of disasters are due to flooding. Keeping this thing in mind and for important decision-making a study on flood water level forecasting is significant. Therefore, there is an importance of methods that forecast flood accurately and timely. Conventional techniques to establish relation amid input and output data requires understanding the behaviour of system, yet there exist complexities and highly nonlinear properties in relationship.

Adnan et al. [1] proposed a model to predict flood water level at downstream station utilizing BPNN at Sungai Batu Pahat, for improving prediction values. Chen et al. [5] proposed Decision Group BPNN to improve prediction abilities in accordance to deterministic BPNN model using data of Wu-Shi catchment positioned in mid of Taiwan. Feng and Lu [6] projected a novel arrangement for forecasting flood level with connected application, on basis of ANN method which gives better outcomes in terms of its performance and effectiveness increasing the ability to forecast flood at an early stage. Ghose and Samantaray [9, 10] applied various techniques to predict climatic constraints in a watershed. Ghorbani et al. [8] investigated discharge time series utilizing SVM and ANN and contrasted their efficiency with conventional models taking water level and release data from Big Cypress River, Texas, USA. Kia et al. [11] urbanized a flood model taking different flood causing aspects with use if ANN methods and geographic information system to model and simulate flood-affected regions located in the south of Peninsular Malaysia. Kisi and Cimen [12] examined accurateness of wavelet and SVM hybrid model to forecast monthly streamflow using data from Goksudere River, Turkey. Suliman et al. [18] focused on flood forecasting models utilizing ANN and models utilizing SVM and reviewed both techniques and compared their outcomes with those of investigational outcomes. Singh et al. [17] investigated prospective of SPNN and M5 model tree on basis of the regression way for estimating annual flood. Zakaria and Shabri [20] investigated the prospective of SVM to forecast streamflow at ungauged locations and compared its efficacy with Multiple Linear Regression (MLR) based on the results obtained. Samantaray and Sahoo [14–16] employed different ANN techniques to predict hydrological parameters in a basin. Fernando and Shamseldin [7] studied exigent difficulty in interpreting hydrological formulations of inner functioning of ANNs by digging out information from their resolutions using data from River Brosna, Ireland. Ruslan et al. [13] provided a substitute way which is an ANN tool competent enough to model nonlinear and multifaceted arrangement to predict correct flood level using Kelang river data.

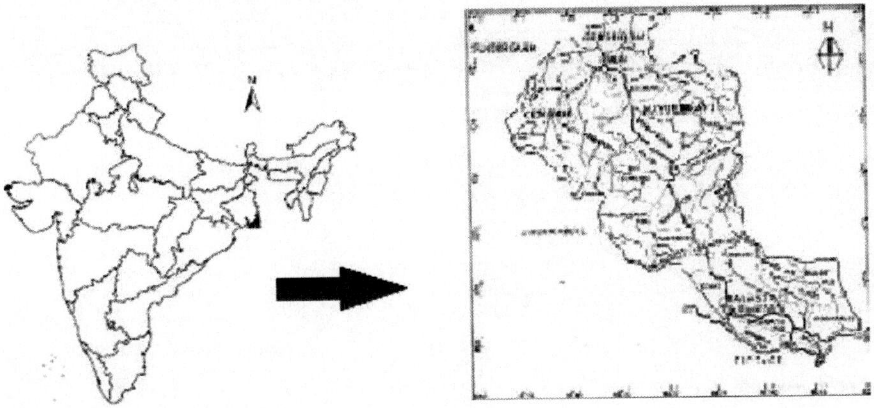

Fig. 1 Planned watershed

2 Research Area

Brahmani is a key wandering waterway in Odisha, India. Brahmani is formed by coming together of Sankh and Koel streams near the town of Rourkela at 22°15′ N and 84°47′ E, and surges in the course of Sundargarh, Deogarh, Dhenkanal, Cuttack, Jajpur and Kendrapara. Jointly with Baitarani, it appears as a big delta prior to draining to the Bay of Bengal at Dhamra (Fig. 1).

3 Methodology

3.1 RBFN

It was first commenced to the field of ANN by Broomhead and Lowe [2]. RBFN is encouraged by local refrained response seen in biological neurons. RBFN is one class of feed-forward neural networks. This network is now being extensively utilized worldwide for function estimation or classification for solving nonlinear exertions since it has a simple topological structure and because of its capability in performing learning procedure in unambiguous approach. In addition, because of its benefits like higher reliability, faster convergence and small errors, RBFN has been significantly applied to forecast hydrological phenomenon. RBFNs are good at learning in identifying network formation arrangement and subsequently look for variables of the model (Fig. 2).

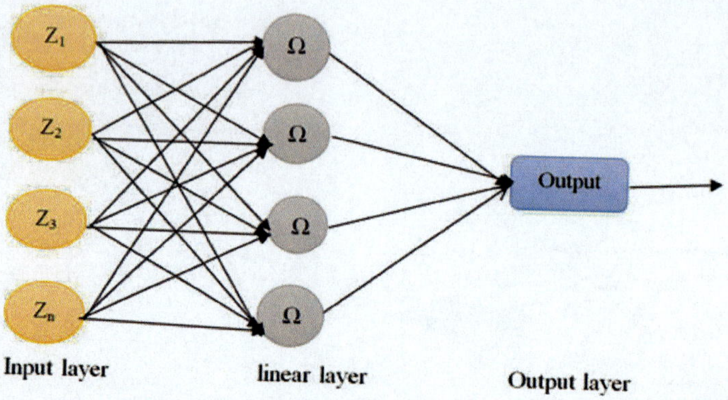

Fig. 2 Architecture of RBFN

3.2 *BPNN*

Primary and basic feature of ANN is its skill to learn, allocate memory and corresponds process, finally leading to fault forbearance. BPNN is learning model having all the properties associated with ANN. Process of BPNN is to find error at output level that transmits backward to input level via hidden level in network to acquire final anticipated outputs. To determine the weight of network and adjustment in weight of interconnections and to minimize output error gradient descent method is mostly utilized. Most references associated with flood forecasting concentrated on obtaining a deterministic BPN model effective for flood forecasting (Fig. 3).

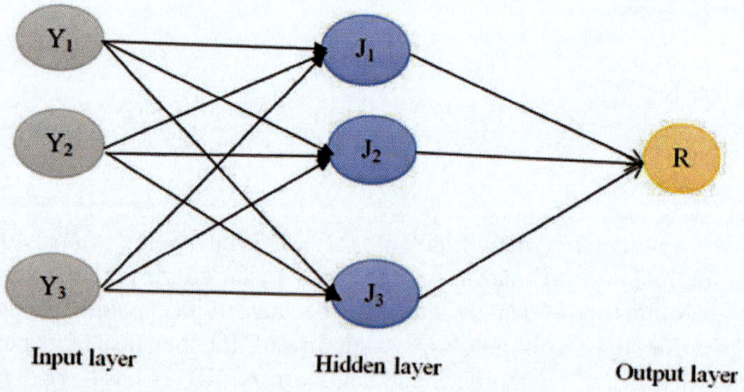

Fig. 3 Model of BPNN

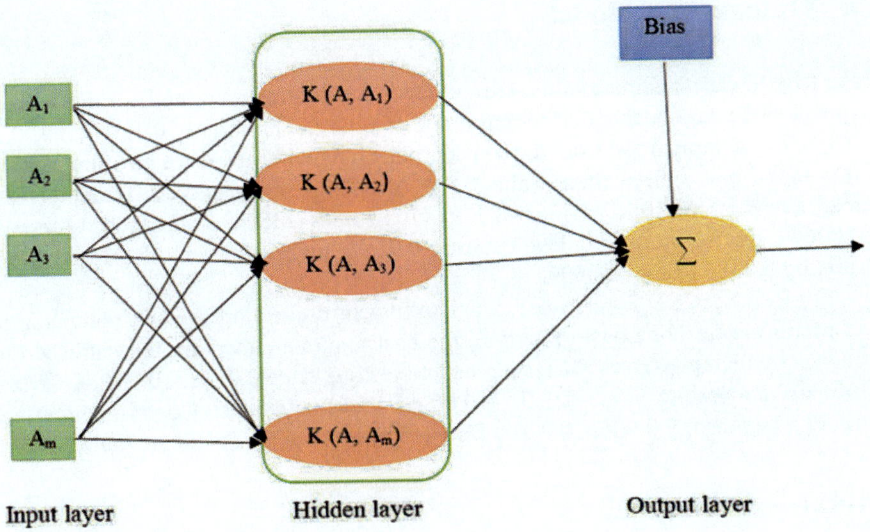

Fig. 4 Architecture of SVM

3.3 SVM

SVM technique works on the basis of statistical learning theory [19]. SVM is one among different ANNs that is being increasingly accepted and paid attention in classifying pattern arrangement and estimating nonlinear regression [3] because it performs in a very simplification manner. SVM comprises a support vector set and kernel function. Support vectors jointly with kernel generate function estimation. Numerous investigations on application of SVM to model different hydrological phenomena exhibited an improved performance (Fig. 4).

3.4 Model Preparation

Flood data with up to five lag time on monthly basis is retained from IMD, Bhubaneswar for monsoon seasons (1990–2017). Data from 1990–2009 are engaged to train and data from 2009–2017 are involved to test model efficiency. MSE, RMSE, R^2 are assessed for the performance of model.

$$R^2 = \left[\frac{n\sum kl - (k)(\sum l)}{\sqrt{[n\sum k^2 - (\sum k)^2][n\sum l^2 - (\sum l)^2]}} \right]^2 \quad (1)$$

4 Outcomes of Model

For RBFN, various spread values are applied for simulation presented in Table 1. Here spread values are considered within range of 0–1, i.e. preferably 0.1, 0.2, 0.4, 0.8, 0.9 for predicating flood from the considerable input parameters for mapping output. It is found that with a spread value 0.2 the RBFN appearances superlative performance with architecture having 5-0.2-1 which possess MSE training 0.00511 testing 0.00699, RMSE training 0.05997 testing 0.03573 and coefficient of determination training 0.8385 testing 0.8509.

For BPNN, 5-2-1, 5-4-1, 5-6-1, 5-8-1 architectures are employed for computation of performance. For Logsig function, the best model architecture is established to be 4-2-1 which possesses MSE training and testing value 0.00283, 0.00427, RMSE training and testing value 0.06998, 0.05901 and coefficient of determination for training and testing 0.9066, 0.9268. Detailed results are accessible in Table 2.

Table 1 Results of RBFN

Input	Architecture	MSE		RMSE		R^2	
		Training	Testing	Training	Testing	Training	Testing
$R_t, R_{t-1}, R_{t-2}, R_{t-3}, R_{t-4}, R_{t-5}$	5-0.1-1	0.00545	0.00681	0.05732	0.03368	0.8234	0.8443
	5-0.2-1	**0.00511**	**0.00699**	**0.05997**	**0.03573**	**0.8385**	**0.8509**
	5-0.4-1	0.00476	0.00537	0.05662	0.03254	0.8066	0.8276
	5-0.8-1	0.00441	0.00508	0.05178	0.03006	0.7729	0.7996
	5-0.9-1	0.00453	0.00513	0.06852	0.03711	0.7971	0.8103

Table 2 Results of BPNN

Input	Transfer function	Architecture	MSE		RMSE		R^2	
			Training	Testing	Training	Testing	Training	Testing
$R_t, R_{t-1}, R_{t-2}, R_{t-3}, R_{t-4}, R_{t-5}$	Tansig	4-2-1	0.00302	0.00487	0.06265	0.05822	0.8748	0.8956
		4-4-1	0.00476	0.00565	0.06913	0.05991	0.8817	0.9021
		4-6-1	**0.00252**	**0.00335**	**0.05528**	**0.04657**	**0.8986**	**0.9119**
		4-8-1	0.00297	0.00387	0.06127	0.05375	0.8804	0.9009
	Logsig	4-2-1	**0.00283**	**0.00427**	**0.06998**	**0.05901**	**0.9066**	**0.9268**
		4-4-1	0.00227	0.00368	0.05167	0.04123	0.8807	0.9027
		4-6-1	0.00379	0.00532	0.06315	0.05897	0.8976	0.9111
		4-8-1	0.00258	0.00381	0.04879	0.03357	0.8723	0.8905
	Purelin	4-2-1	0.00287	0.00378	0.04712	0.03116	0.8895	0.9096
		4-4-1	0.00218	0.00324	0.04004	0.02998	0.8527	0.8834
		4-6-1	0.00236	0.00356	0.04223	0.03212	0.8634	0.8929
		4-8-1	**0.00302**	**0.00427**	**0.05335**	**0.04356**	**0.8372**	**0.8687**

Table 3 Results of SVM model

Input	MSE		RMSE		R^2	
	Training	Testing	Training	Testing	Training	Testing
R_t	0.00356	0.00424	0.02546	0.03654	0.8508	0.8667
R_t, R_{t-1}	0.00487	0.00547	0.02998	0.03981	0.8743	0.8943
R_t, R_{t-1}, R_{t-2}	**0.00597**	**0.00623**	**0.03546**	**0.04823**	**0.8897**	**0.9005**
$R_t, R_{t-1}, R_{t-2}, R_{t-3}$	**0.00612**	**0.00782**	**0.04786**	**0.05734**	**0.9065**	**0.9221**
$R_t, R_{t-1}, R_{t-2}, R_{t-3}, R_{t-4}$	**0.00745**	**0.00799**	**0.05512**	**0.06623**	**0.9231**	**0.9395**
$R_t, R_{t-1}, R_{t-2}, R_{t-3}, R_{t-4}, R_{t-5}$	**0.00882**	**0.00901**	**0.06732**	**0.07834**	**0.9482**	**0.9593**

Similarly, for SVM, all the performance indicators are described in Table 3. Paramount value of R^2 for input R_t, R_{t-1}, R_{t-2}, R_{t-3}, R_{t-4}, R_{t-5} is 0.9482, and 0.9593 in training and testing phase. Correspondingly for other input scenario paramount value of R^2 for testing and training phase are described below.

From the above research, we found that among five scenarios while five days lag is added with rainfall, all rainfall-flood correlation models perform best as compared to four-day lag rainfall. Actual versus predicted flood graph for all three technics is presented in Fig. 5.

5 Conclusion

Performance of RBFN, BPNN, SVM methods is presented in this research in an effort for providing substantiation in predicting flood level. The study reveals that SVM model performs better over RBFN and BPNN models in terms of flood prediction. Outcomes of the present study revealed that ANN models achieved superior results during training, and plunged down in terms of its superiority than that of SVM at time of testing; though, this concept cannot be universal for all other hydrological factors. SVM was observed to have achieved superior consequences in terms of its performance and efficacy. However, there still lies certain difficulty in research to forecast flood warnings. ANN techniques also permit to simulate several parameters in input layer as well as output layer. This is extremely vital in calculating flood magnitude and frequency as stage, discharge and other hydrological parameters often have many influential roles.

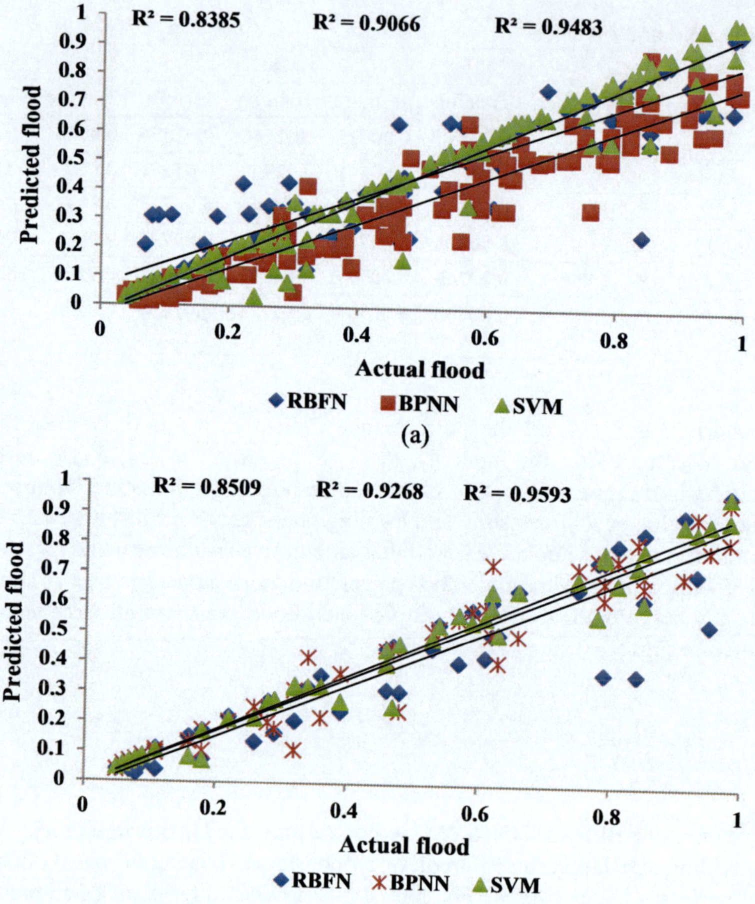

(b)

Fig. 5 Actual versus predicted flood for **a** training **b** testing phase

References

1. Adnan R, Ruslan FA, Samad AM, Zain ZM (2012) Flood water level modelling and prediction using artificial neural network: case study of Sungai Batu Pahat in Johor. In: IEEE control and system graduate research colloquium. IEEE, pp 22–25
2. Broomhead DS, Lowe D (1988) Radial basis functions, multi-variable functional interpolation and adaptive networks (No. RSRE-MEMO-4148). Royal Signals and Radar Establishment Malvern (United Kingdom)
3. Cao LJ, Tay Francis EH (2003) Support vector machine with adaptive parameters in financial time series forecasting. IEEE Trans Neural Networks 14:1506–1518
4. Chang LC, Chang FJ, Wang YP (2009) Auto-configuring radial basis function networks for chaotic time series and flood forecasting. Hydrol Process: Int J 23(17):2450–2459
5. Chen CS, Chen BPT, Chou FNF, Yang CC (2010) Development and application of a decision group back-propagation neural network for flood forecasting. J Hydrol 385(1–4):173–182

6. Feng LH, Lu J (2010) The practical research on flood forecasting based on artificial neural networks. Expert Syst Appl 37(4):2974–2977
7. Fernando DA, Shamseldin AY (2009) Investigation of internal functioning of the radial-basis-function neural network river flow forecasting models. J Hydrol Eng 14(3):286–292
8. Ghorbani MA, Khatibi R, Goel A, FazeliFard MH, Azani A (2016) Modeling river discharge time series using support vector machine and artificial neural networks. Environ Earth Sci 75(8):685
9. Ghose DK, Samantaray S (2019) Sedimentation process and its assessment through integrated sensor networks and machine learning process. In: Computational intelligence in sensor networks, studies in computational intelligence, vol 776, pp 473–488
10. Ghose DK, Samantaray S (2019) Integrated sensor networking for estimating groundwater potential in scanty rainfall region: challenges and evaluation. In: Computational intelligence in sensor networks, studies in computational intelligence, vol 776, pp 335–352
11. Kia MB, Pirasteh S, Pradhan B, Mahmud AR, Sulaiman WNA, Moradi A (2012) An artificial neural network model for flood simulation using GIS: Johor River Basin, Malaysia. Environ Earth Sci 67(1):251–264
12. Kisi O, Cimen M (2011) A wavelet-support vector machine conjunction model for monthly streamflow forecasting. J Hydrol 399(1–2):132–140
13. Ruslan FA, Samad AM, Zain ZM, Adnan R (2013) Modelling flood prediction using Radial Basis Function Neural Network (RBFNN) and inverse model: a comparative study. In: IEEE international conference on control system, computing and engineering, pp 577–581
14. Samantaray S, Sahoo A (2020) Appraisal of runoff through BPNN, RNN, and RBFN in Tentulikhunti Watershed: a case study. In: Satapathy S, Bhateja V, Nguyen B, Nguyen N, Le DN (eds) Frontiers in intelligent computing: theory and applications. Advances in intelligent systems and computing, vol 1014. Springer, Singapore
15. Samantaray S, Sahoo A (2020) Estimation of runoff through BPNN and SVM in Agalpur Watershed. In: Satapathy S, Bhateja V, Nguyen B, Nguyen N, Le DN (eds) Frontiers in intelligent computing: theory and applications. Advances in intelligent systems and computing, vol 1014. Springer, Singapore
16. Samantaray S, Sahoo A (2020) Assessment of sediment concentration through RBNN and SVM-FFA in Arid Watershed, India. In: Satapathy S, Bhateja V, Mohanty J, Udgata S (eds) Smart intelligent computing and applications. Smart innovation, systems and technologies, vol 159. Springer, Singapore
17. Singh KK, Pal M, Singh VP (2010) Estimation of mean annual flood in Indian catchments using backpropagation neural network and M5 model tree. Water Resour Manage 24(10):2007–2019
18. Suliman A, Nazri N, Othman M, Abdul M, Ku-Mahamud KR (2013) Artificial neural network and support vector machine in flood forecasting: a review. In: Proceedings of the 4th international conference on computing and informatics, ICOCI, pp 28–30
19. Vapnik VN (1998) Statistical learning theory. Wiley, New York
20. Zakaria ZA, Shabri A (2012) Streamflow forecasting at ungaged sites using support vector machines. Appl Math Sci 6(60):3003–3014

Assessment of Flow Discharge in a River Basin Through CFBPNN, LRNN and CANFIS

Sriharsha Sridharam, Abinash Sahoo, Sandeep Samantaray, and Dillip K. Ghose

Abstract Forecasting of river flow is an important process which provides primary and basic information about problems regarding design and operations of river System. A progressive utilization of extended records of rainfall and other climatic data is possible because of its availability. Artificial Neural Network (ANN) methods have been applied comprehensively for past few decades in forecasting of streamflow, and it has been proven that ANN techniques are far better than other forecasting methods. An appropriate length selection of training datasets is complicated, and there is no certainty in predictions of trained ANNs with new sets of data which makes this method more complicated. There are three different methods of ANN namely, Layered Recurrent Neural Network (LRNN), Coactive Neuro-Fuzzy Inference System (CANFIS) and Cascade Forward Back Propagation Neural Network (CFBPNN) are used for streamflow forecasting and results are evaluated. Based on the results, CANFIS was found to be better than other ANN techniques in monthly flow forecasting.

1 Introduction

In various water resources planning, development of design and activities of operation and maintenance there is always needed of accurate forecasting of river flow. The process of river flow is highly nonlinear and complex in nature. Several models have proposed by researchers to forecast river flow with reasonable performance.

S. Sridharam (✉) · A. Sahoo · S. Samantaray · D. K. Ghose
Department of Civil Engineering, NIT Silchar, Silchar, Assam, India
e-mail: sriharshasridharam@gmail.com

A. Sahoo
e-mail: bablusahoo1992@gmail.com

S. Samantaray
e-mail: samantaraysandeep963@gmail.com

D. K. Ghose
e-mail: dillipghose_2002@yahoo.co.in

S. C. Satapathy et al. (eds.), *Communication Software and Networks*, Lecture Notes in Networks and Systems 134, https://doi.org/10.1007/978-981-15-5397-4_78

Hydrodynamic models give a sound physical basis for this motive and have capability to simulate a wide range of flow situations. However, for developing countries such as India, there is scarcity of accurate stream geometry data which is required by these proposed models.

Rani and Parekh [6] presented an ANN approach to forecast reservoir water of Sukhi Reservoir project. Malik and Kumar [5] compared potential of soft computing and statistical techniques for simulating daily discharge on Pranhita basin, India. Samantaray and Sahoo [9–11] employed various algorithms for predicting hydrological indices. Tfwala et al. [12] explored accurateness of ANN in estimating missing flow records. Akrami et al. [1] employed ANFIS and modified ANFIS to increase rainfall forecasting effectiveness. Fereydooni et al. [3] applied ANN and Stochastic models for envisaging monthly flow discharge of Ghara–Aghaj River, Iran. Chang et al. [2] investigated reliability and accuracy of floodwater storage of a sewer-pumping system in Taipei City, Taiwan. Samantaray and Ghose [8] used ANN techniques to forecast climatic constraints in a basin. Tran et al. [13] investigated two basic ANN models, namely, FFNN and LRNN to forecast streamflow attempting to comprehend why ANN models were used successfully in some streamflow forecasting studies but not always. Ghose and Samantaray [4] applied soft computing techniques to predict groundwater potential. Rezaeian-Zadeh et al. [7] associated four ANN processes for predicting monthly discharge in Kohkiloye and Boyer-Ahmad Province, Iran. Objective of this research is to predict sediment load of a watershed using various neural network techniques.

2 Study Area and Data

The Penna also known as Pennar is a river in southern part of India. The starting point of Penna is Nandi Hills, Karnataka state, and runs through states Andhra Pradesh and then finally reaches to the Bay of Bengal. Penna River rises at 13.55° N 77.60° E at 11 km southwest of Chikkaballapur, Karnataka. The length of Penna River is 597 km (371 mi) long, with covering a drainage basin area of 55,213 km^2. Average annual rainfall is 500 mm (Fig. 1).

3 Methodology

3.1 LRNN

LRNN is a recurrent neural network related to FFNN, excluding that hidden or output layer has a recurrent connection with a tap delay connected with it. It permits network to have an infinite dynamic response time series. This network is similar to time and distributed delay neural network. Moreover, time delay is like to FFNN, except that

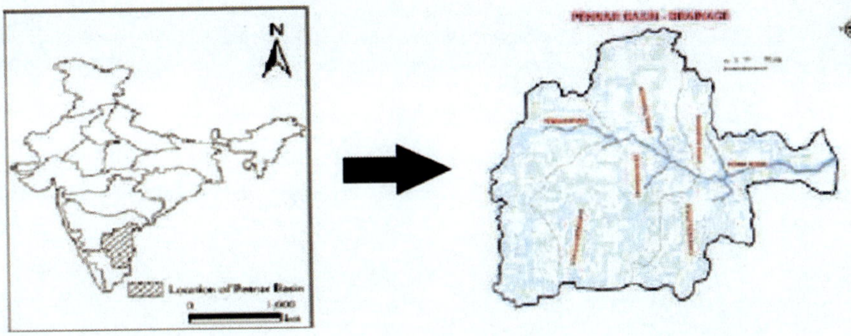

Fig. 1 Planned watershed

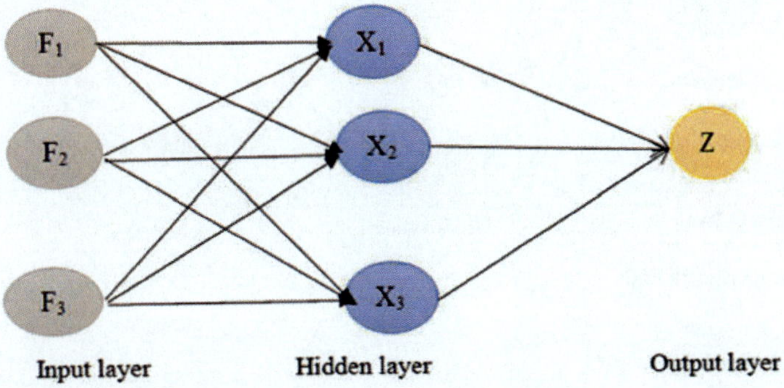

Fig. 2 Architecture of LRNN

input weight has an associated tapped delay line at input which consents network to have finite dynamic responses to time series input samples (Fig. 2).

3.2 CFBPNN

First layer weights initiate from input. Subsequently, each layer has biases and the last level is called output. Using MATLAB NN toolbox, weights and biases of each of layers are initialized. Alteration is a procedure which appraises weights with stipulated learning function. If additional connection is added, the speed of trained model is increased (Fig. 3).

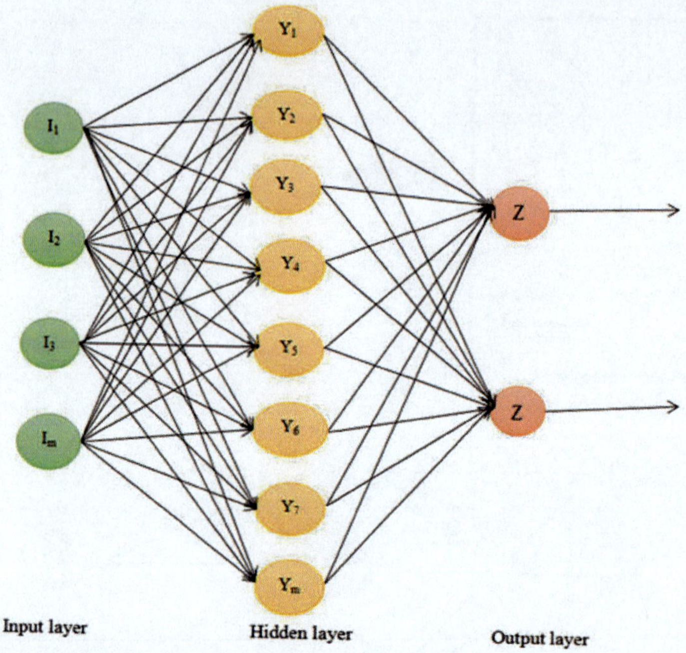

Fig. 3 Model of CFBPNN

3.3 *CANFIS*

It is a generalized form of ANFIS, and it may be observed as universal constraints of several nonlinear functions. CANFIS is highlighted by assimilating neural networks with fuzzy inference systems in a single topology. Architecture of CANFIS involving of fuzzification, rule, normalization, defuzzification and summation layer is shown in Fig. 4.

Various indices like MSE, RMSE, R^2 are assessed for the performance of model.

$$R^2 = \left[\frac{n \sum ab - (a)\left(\sum b\right)}{\sqrt{\left[n \sum a^2 - \left(\sum a\right)^2\right]\left[n \sum b^2 - (b)^2\right]}} \right]^2 \quad (1)$$

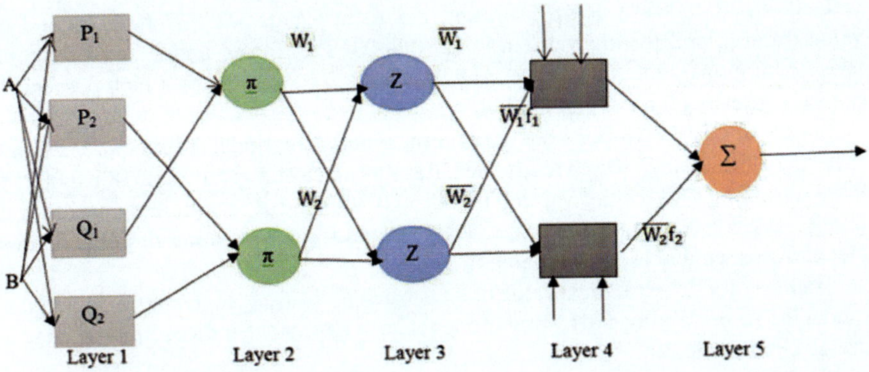

Fig. 4 Model of CANFIS

4 Outcomes of Model

For LRNN, Tansig, Logsig, Purelin with five scenarios are considered for performance evaluation revealed in Table 1. For Logsig function paramount model architecture is found to be R_t, I_t, $T_{\max}$, E_t, Q_t which possess MSE training and testing

Table 1 Outcomes of LRNN

Transfer function	Scenario	MSE		RMSE		R^2	
		Training	Testing	Training	Testing	Training	Testing
Tansig	R_t	0.00096	0.00115	0.02654	0.04365	0.7943	0.8136
	R_t, I_t	0.00108	0.00123	0.02998	0.04769	0.8009	0.8334
	R_t, I_t, $T_{\max}$	0.00112	0.00135	0.03154	0.05004	0.8265	0.8553
	R_t, I_t, $T_{\max}$, E_t	0.00120	0.00141	0.03986	0.05832	0.8445	0.8702
	R_t, I_t, $T_{\max}$, E_t, Q_l	0.00125	0.00157	0.04521	0.06013	0.8667	0.8856
Logsig	R_t	0.00109	0.00125	0.03013	0.04997	0.8189	0.8472
	R_t, I_t	0.00113	0.00137	0.03228	0.05113	0.8344	0.8678
	R_t, I_t, $T_{\max}$	0.00124	0.00154	0.04231	0.05991	0.8557	0.8856
	R_t, I_t, $T_{\max}$, E_t	0.00132	0.00166	0.04998	0.06537	0.8721	0.8923
	R_t, I_t, $T_{\max}$, E_t, Q_l	0.00138	0.00178	0.05679	0.07034	0.8912	0.9003
Purelin	R_t	0.00092	0.00111	0.02456	0.04236	0.7865	0.8115
	R_t, I_t	0.00105	0.00119	0.02893	0.04627	0.8093	0.8224
	R_t, I_t, $T_{\max}$	0.00111	0.00129	0.03221	0.05098	0.8212	0.8496
	R_t, I_t, $T_{\max}$, E_t	0.00116	0.00142	0.04002	0.05867	0.8498	0.8611
	R_t, I_t, $T_{\max}$, E_t, Q_l	0.00121	0.00149	0.04623	0.06144	0.8578	0.8726

value 0.00138, 0.00178, RMSE training and testing value 0.05679, 0.07034 and R^2 value training and testing value 0.8912, 0.9003. For Tansig and Purelin function paramount model architecture is establish to be R_t, I_t, T_{max}, E_t, Q_t which possess R^2 for testing phases 0.8856 and 0.8724.

In case of CFBPNN, For Logsig function superlative model architecture is found to be R_t, I_t, T_{max}, E_t, Q_l which retain MSE training and testing value 0.00185, 0.00254, RMSE training and testing value 0.06348, 0.07993 and R^2 for training and testing 0.9097, 0.9392. Various scenarios are deliberated for computation of performance. Detailed consequences are available in Table 2.

Similarly for CANFIS best value of R^2 is when R_t, I_t, T_{max}, E_t, Q_l are considered as model input for training and testing phases. The performance indices for all scenarios are given in Table 3.

Actual versus predicted flow discharges of proposed gauge station are presented in Fig. 5.

Table 2 Outcomes of BPNN

Transfer function	Architecture	MSE		RMSE		R^2	
		Training	Testing	Training	Testing	Training	Testing
Tansig	R_t	0.00114	0.00166	0.03254	0.05167	0.8501	0.8639
	R_t, I_t	0.00127	0.00186	0.03985	0.05846	0.8672	0.8874
	R_t, I_t, T_{max}	0.00141	0.00197	0.04421	0.06127	0.8865	0.9025
	R_t, I_t, T_{max}, E_t	0.00153	0.00202	0.05013	0.06889	0.8923	0.9108
	R_t, I_t, T_{max}, E_t, Q_l	**0.00165**	**0.00211**	**0.05167**	**0.07009**	**0.9065**	**0.9226**
Logsig	R_t	0.00128	0.00156	0.03989	0.05634	0.8516	0.8875
	R_t, I_t	0.00143	0.00199	0.04401	0.05889	0.8661	0.9008
	R_t, I_t, T_{max}	0.00156	0.00205	0.04987	0.06288	0.8753	0.9126
	R_t, I_t, T_{max}, E_t	0.00171	0.00223	0.05768	0.07009	0.8912	0.9298
	R_t, I_t, T_{max}, E_t, Q_l	**0.00185**	**0.00254**	**0.06348**	**0.07993**	**0.9097**	**0.9392**
Purelin	R_t	0.00115	0.00126	0.03367	0.05247	0.8391	0.8657
	R_t, I_t	0.00124	0.00145	0.03776	0.05774	0.8513	0.8721
	R_t, I_t, T_{max}	0.00137	0.00168	0.04345	0.06004	0.8609	0.8996
	R_t, I_t, T_{max}, E_t	0.00159	0.00216	0.05043	0.06934	0.8802	0.9113
	R_t, I_t, T_{max}, E_t, Q_l	**0.00161**	**0.00237**	**0.05254**	**0.07112**	**0.8993**	**0.9178**

Table 3 Outcomes of CANFIS

Scenario	Input	Training			Testing		
		MSE	RMSE	R^2	MSE	RMSE	R^2
I	R_t	0.00137	0.01743	0.8448	0.00284	0.02687	0.8775
II	R_t, I_t	0.00163	0.02173	0.8885	0.00305	0.03042	0.9095
III	$R_t, I_t, T_{\max}$	0.00177	0.02796	0.9056	0.00332	0.03756	0.9342
IV	$R_t, I_t, T_{\max}, E_t$	0.00189	0.03251	0.9301	0.00387	0.04021	0.9497
V	$R_t, I_t, T_{\max}, E_t, Q_l$	0.00201	0.03678	0.9516	0.00401	0.04687	0.9652

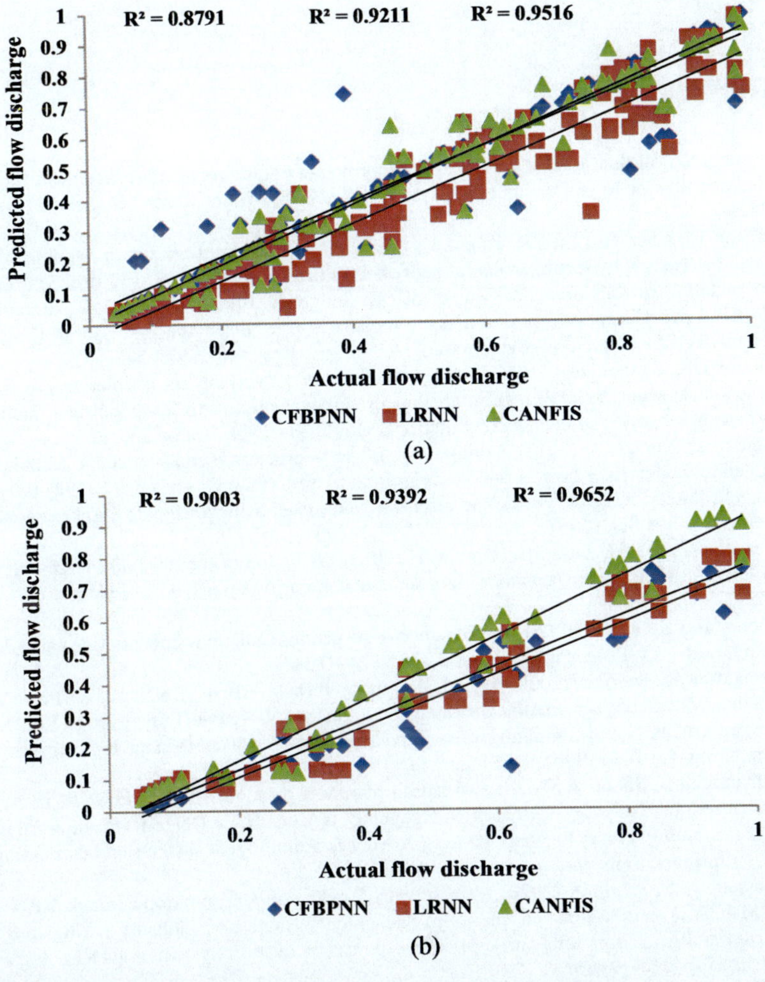

Fig. 5 Actual versus predicted flow discharge for **a** training **b** testing phases

5 Conclusion

Improvement in efficiency of model in forecasting of river flow is the most important in water resources planning and management. NN models have emerged as powerful tools in hydrologic forecasting. Accuracy of three ANN models applied in this study has been explored for anticipating river flows. Results proved that CANFIS model provided more accurate forecasting results than regular CFBPNN and LRNN models and also accomplished better in terms of its effectiveness. Supplementary exploration is recommended to predict daily flows at numerous gauge stations and use other network arrangements, and transfer functions.

References

1. Akrami SA, El-Shafie A, Jaafar O (2013) Improving rainfall forecasting efficiency using modified adaptive neuro-fuzzy inference system (MANFIS). Water Resour Manage 27(9):3507–3523
2. Chang FJ, Chen PA, Lu YR, Huang E, Chang KY (2014) Real-time multi-step-ahead water level forecasting by recurrent neural networks for urban flood control. J Hydrol 517:836–846
3. Fereydooni M, Rahnemaei M, Babazadeh H, Sedghi H, Elhami MR (2012) Comparison of artificial neural networks and stochastic models in river discharge forecasting,(Case study: Ghara-Aghaj River, Fars Province, Iran). Afr J Agric Res 7(40):5446–5458
4. Ghose DK, Samantaray S (2019) Integrated sensor networking for estimating groundwater potential in scanty rainfall region: challenges and evaluation. In: Computational intelligence in sensor networks, studies in computational intelligence, vol 776, pp 335–352
5. Malik A, Kumar A (2018) Comparison of soft-computing and statistical techniques in simulating daily river flow: a case study in India. J Soil Water Conserv 17(2):192–199
6. Rani S, Parekh F (2014) Predicting reservoir water level using artificial neural network. Int J Innovative Res Sci, Eng Technol 3(7):14489–14496
7. Rezaeian-Zadeh M, Tabari H, Abghari H (2013) Prediction of monthly discharge volume by different artificial neural network algorithms in semi-arid regions. Arab J Geosci 6(7):2529–2537
8. Samantaray S, Ghose DK (2018) Evaluation of suspended sediment concentration using descent neural networks. Procedia Comput Sci 132:1824–1831
9. Samantaray S, Sahoo A (2020) Appraisal of runoff through BPNN, RNN, and RBFN in Tentulikhunti Watershed: a case study. In: Satapathy S, Bhateja V, Nguyen B, Nguyen N, Le DN (eds) Frontiers in intelligent computing: theory and applications. Advances in intelligent systems and computing, vol 1014. Springer, Singapore
10. Samantaray S, Sahoo A (2020) Estimation of runoff through BPNN and SVM in Agalpur Watershed. In: Satapathy S, Bhateja V, Nguyen B, Nguyen N, Le DN (eds) Frontiers in intelligent computing: theory and applications. Advances in intelligent systems and computing, vol 1014. Springer, Singapore
11. Samantaray S, Sahoo A (2020) Assessment of sediment concentration through RBNN and SVM-FFA in Arid Watershed, India. In: Satapathy S, Bhateja V, Mohanty J, Udgata S (eds) Smart intelligent computing and applications. Smart innovation, systems and technologies, vol 159. Springer, Singapore
12. Tfwala SS, Wang YM, Lin YC (2013) Prediction of missing flow records using multilayer perceptron and coactive neurofuzzy inference system. Sci World J

13. Tran HD, Muttil N, Perera BJC (2011) Investigation of artificial neural network models for streamflow forecasting. In: 19th International congress on modelling and simulation (MODSIM2011). Modelling and simulation society of Australia and New Zealand Inc. (MSSANZ), pp 1099–1105